Hypothesis Tests for One or Two Sample Means

The one-sample z-test:

$$z = \frac{\bar{X} - \mu}{\sigma/\sqrt{N}}$$

Formula 7.2

The one-sample t-test:

$$t = \frac{\bar{X} - \mu}{s/\sqrt{N}}$$

Formula 8.3

The matched t-test:

$$t = \frac{\bar{D}}{s_D/\sqrt{N}}$$

Formula 13.1

The t-test for two equal-sized samples:

$$t = \frac{(\bar{X}_1 - \bar{X}_2) - (\mu_1 - \mu_2)}{\sqrt{(s_1^2 + s_2^2)/N}}$$

Formula 9.8

The separate-variances t-test:

$$t = \frac{(\bar{X}_1 - \bar{X}_2) - (\mu_1 - \mu_2)}{\sqrt{\dfrac{s_1^2}{N_1} + \dfrac{s_2^2}{N_2}}}$$

Formula 9.4

The pooled-variances t-test:

$$t = \frac{(\bar{X}_1 - \bar{X}_2) - (\mu_1 - \mu_2)}{\sqrt{\dfrac{(N_1 - 1)s_1^2 + (N_2 - 1)s_2^2}{N_1 + N_2 - 2} \left(\dfrac{1}{N_1} + \dfrac{1}{N_2}\right)}}$$

Formula 9.7A

Confidence interval for the population mean (using t distribution):

$$\mu_{\text{lower}} = \bar{X} - t_{\text{crit}}s_{\bar{X}}$$
$$\mu_{\text{upper}} = \bar{X} + t_{\text{crit}}s_{\bar{X}}$$

Formula 8.5

Confidence interval for the difference of two population means:

$$\mu_1 - \mu_2 = (\bar{X}_1 - \bar{X}_2) \pm t_{\text{crit}}s_{\bar{X}_1 - \bar{X}_2}$$

Formula 9.9

EXPLAINING
PSYCHOLOGICAL
STATISTICS

EXPLAINING PSYCHOLOGICAL STATISTICS

Barry Cohen
New York University

Brooks/Cole Publishing Company

I(T)P™ An International Thomson Publishing Company

Pacific Grove • Albany • Bonn • Boston • Cincinnati • Detroit • London • Madrid • Melbourne
Mexico City • New York • Paris • San Francisco • Singapore • Tokyo • Toronto • Washington

Sponsoring Editor: *Jim Brace-Thompson*
Marketing Team: *Gay Meixel, Romi Fineroff*
Marketing Representative: *Connie Jirovsky*
Editorial Assistant: *Jodi Hermans, Terry Thomas*
Production Coordinator: *Kirk Bomont*
Project Management: *Lifland et al., Bookmakers*
Manuscript Editor: *Quica Ostrander*
Interior Design: *Quica Ostrander*

Interior Illustration: *Gail Magin*
Cover Design: *Roy R. Neuhaus*
Cover Photo: *Anton Want/Tony Stone Images*
Art Coordinator: *Quica Ostrander*
Indexer: *Denise Throckmorton*
Typesetting: *Weimer Graphics*
Cover Printing: *Color Dot*
Printing and Binding: *Quebecor/Fairfield*

For more information, contact:

BROOKS/COLE PUBLISHING COMPANY
511 Forest Lodge Road
Pacific Grove, CA 93950
USA

International Thomson Publishing Europe
Berkshire House 168-173
High Holborn
London WC1V 7AA
England

Thomas Nelson Australia
102 Dodds Street
South Melbourne, 3205
Victoria, Australia

Nelson Canada
1120 Birchmount Road
Scarborough, Ontario
Canada M1K 5G4

International Thomson Editores
Campos Eliseos 385, Piso 7
Col. Polanco
11560 México D. F. México

International Thomson Publishing GmbH
Königswinterer Strasse 418
53227 Bonn
Germany

International Thomson Publishing Asia
221 Henderson Road
#05-10 Henderson Building
Singapore 0315

International Thomson Publishing Japan
Hirakawacho Kyowa Building, 3F
2-2-1 Hirakawacho
Chiyoda-ku, Tokyo 102
Japan

Printed in the United States of America

10 9 8 7 6 5 4 3 2 1

Library of Congress Cataloging-in-Publication Data

Cohen, Barry H., [date]
  Explaining psychological statistics / Barry H. Cohen.
    p.   cm.
  Includes bibliographical references and index.
  ISBN 0-534-20076-1
    1. Psychometrics.   2. Psychology—Mathematical models.
  3. Statistics—Study and teaching (Higher)   I. Title.
  BF39.C56   1995
  150′.1′5195—dc20   95-35752
                                                  CIP

DEDICATION

*To the memory of
Rose and Charles
and
Fannie and Louis,
without whose journey to America
this book would not have been possible,
and
to Leona*

Contents

3

Measures of Central Tendency 68

4

Measures of Variability 100

7

Introduction to Hypothesis Testing: The One-Sample z-Test
191

8

The One-Sample t-Test
227

9

The *t*-Test for Two Independent Sample Means
258

10

Statistical Power 294

**PART IV   HYPOTHESIS TESTS INVOLVING TWO
MEASURES ON EACH SUBJECT   323**

11

Linear Correlation 325

14
One-Way Independent ANOVA
447

15
Multiple Comparisons
490

18

**Two-Way
Mixed Design
ANOVA
608**

PART VI   NONPARAMETRIC STATISTICS   653

21

Statistical Tests for Ordinal Data
717

PREFACE TO THE INSTRUCTOR

This text evolved from my lecture notes for a one-semester statistics course (Intermediate Psychological Statistics) required of students in the M.A. psychology program at New York University. Many of my students were former English, theater, music, or arts majors headed for doctoral programs in clinical psychology. Most of these students were very bright but had very weak mathematical backgrounds. This text reflects my efforts to teach sophisticated statistical concepts to highly intelligent students who either had had no mathematical training since high school or could scarcely remember anything of their undergraduate statistics course, even if they had taken it recently. In preparing the text, I decided to include all of the introductory material that would be required if this text were used for an undergraduate course. At the same time, enough advanced material has been included to make this text suitable for a course on the master's level.

Order of Topics

The topics follow the usual order for an introductory course, with a few exceptions. First, correlation (and regression) is not included among the descriptive statistics. Correlation coefficients are so often used to test hypotheses in psychological research, especially in the "softer" areas, that it seemed counterproductive to introduce this topic as a descriptive statistic only to return to it later under the category of significance testing. Also, the topic of bivariate statistics can interrupt the smooth flow from explanations of the z score and the normal distribution to the use of those concepts for one-sample hypothesis testing. Similarly, the topic of probability as applied to discrete events can also interrupt the flow from descriptions of the normal distribution to explanations of hypothesis tests based on that distribution. Therefore, in this text, discrete probability is not discussed until Chapter 19, when it is needed for an explanation of nonparametric statistics. On the other hand, the rules of probability as applied to smooth distributions are relevant to parametric hypothesis testing, so these rules are presented at the end of the part of the text that deals with descriptive statistics.

The most unusual change in the normal order of topics is the placement of the information on the matched (or paired) t-test after the chapters on correlation and regression, rather than immediately after either the chapter on the one-sample t-test or the chapter on the t-test for two independent samples. The rationale for this placement is that the power of the matched t-test, and the degree to which it is better than the independent-samples t-test, can best be explained in terms of linear correlation. (And the placement of the material on correlation was constrained by the considerations mentioned above.) For my purposes, the conceptual link between the matched t-test and correlation is more important than the computational link between the matched t-test and the one-sample t-test. However, as students generally enter the course with a good intuitive understanding of correlation, it would not create any problems to cover the material on the matched t-test immediately after the material on the one-sample or the two-sample t-test.

ABCD Format

You will notice that each chapter is divided into four sections, labeled A, B, C, and D. Section A focuses on definitional formulas—their structure and their relationship to other definitional formulas—so that students can gain some insight into why and how statistical formulas work. Students seem to have better retention of statistical formulas that do not appear arbitrary. Section B presents computational formulas and step-by-step procedures, adding the details students need to analyze data and interpret results. Section B also includes information on how to report statistical results in the latest APA format, and most of the B sections contain an excerpt from a journal article illustrating the use and reporting of the test discussed in that chapter. Section C presents optional material that is usually more advanced and that can be omitted without any loss in continuity. Section D provides a comprehensive summary of the previous three sections, along with definitions of key terms and a list of all the major formulas in that chapter.

The Content of the Text

The text is organized into six parts. Part I contains only one chapter, which introduces all the fundamental concepts of the text. Section A describes the different scales of measurement, which will be referred to chiefly in Chapters 3 and 4 and again in Part VI. Chapter 1 also introduces some concepts of experimental design. (This text contains a good deal of information on experimental design woven into the discussions of statistical tests.) Section B introduces the rules of summation, which will be used many times in the text to illustrate the derivation of one formula from another. Section C introduces the notion of double and triple summations. These are used only in Chapter 14, Section C, but are presented in order to prepare the student for more advanced statistical texts.

Part II covers descriptive statistics. Chapter 2, on frequency distributions, can easily be omitted in its entirety for more advanced courses with no loss in continuity. Section A describes the construction of frequency distributions and the finding of percentiles for ungrouped distributions, and Section B goes over the same material for the more complex case of grouped distributions. Because beginning students often find the concept of linear interpolation confusing and intimidating, this topic has been placed in Section C, which also includes a discussion of Tukey's stem-and-leaf display—not because it is more advanced, but because it is newer and less often encountered than the frequency distribution. In Chapter 3, Sections A and B contain the traditional material on the measurement of central tendency. Section C discusses Tukey's box-and-whisker plot as a means of identifying outliers and suggests what can be done to lessen the impact of outliers. Sections A and B of Chapter 4 contain material on the measurement of variability. Section C includes a description of the standard measures of skewness and kurtosis—not because these measures are frequently used in psychological research, but because they are frequently included in statistical software packages and because they serve to deepen the students' understanding of how distributions can differ. Section A of Chapter 5 combines a description of standardized scores with a description of the normal distribution, and Section B demonstrates how these topics can be used together to solve practical problems. Finally, Section C presents the basic rules of probability (including conditional probability) as applied to smooth distributions, in preparation for the discussions of hypothesis tests in Part III.

Part III covers hypothesis tests involving one or two sample means. An entire chapter, Chapter 6, is devoted to the critical concepts of the sampling distribution and the Central Limit Theorem. The logic of null hypothesis testing is introduced in Section A of Chapter 7, and a step-by-step procedure to carry out hypothesis tests is introduced in Section B. Section C attempts to give the student a deeper understanding of p values as conditional probabilities and the fallacies that can arise in dealing with the topic of null hypothesis testing. In Chapter 8, Section A describes the one-sample t-test, and Section B focuses on interval estimation and confidence intervals for a single population mean. Section C contains some more advanced information concerning sampling distributions and the properties of estimators, but this material is meant only to prepare students for more advanced statistical treatments and is not needed in later chapters. Chapter 9, in Sections A and B, explains both the separate-variances and the pooled-variances t-tests, and Section C contains a discussion of when to use which test. Finally, Chapter 10 is devoted to the central concept of power. Section A explains power conceptually, and Section B delves into the calculations of power analysis. Note that the value of the power calculations goes beyond their actual use in experimental design; these calculations foster a deeper understanding of null hypothesis testing (e.g., knowing when a failure to reject the null hypothesis should be taken seriously). Section C deals with the serious concerns that have been raised about the overreliance on null hypothesis testing, and refers to classic and recent journal articles that have contributed to this important debate. Some defense of null hypothesis testing is offered.

The theme uniting the chapters of Part IV is that each chapter deals with a situation in which there are two dependent measures for each subject. Chapter 11 deals with linear correlation, including its limitations and its uses as a research tool. In addition to its discussion of power in relation to testing correlation coefficients for significance, Section C introduces Fisher's Z transformation and its applications for finding confidence intervals and comparing correlations from two independent samples. Sections A and B of Chapter 12 present the basic concepts and mechanics of linear regression, including confidence intervals for predictions. Section C covers two major topics: the point-biserial r and multiple regression. Point-biserial r is described as an important complement to the t value from a two-sample test and a useful alternative measure of effect size. Multiple regression is covered in some detail for the case of two predictors, and then the complications that arise with three or more predictors are briefly outlined. Instructors who choose to cover an introduction to multiple regression may want to postpone this topic until the end of the course because of its complexity. Finally,

Chapter 13 describes the matched, or repeated-measures, *t*-test and includes a good deal of material on experimental design. The calculation of the matched *t*-test is shown in terms of both Pearson's correlation coefficient and the direct-difference method. Section C demonstrates the connection between the power of the matched *t*-test and the degree of linear correlation between the two sets of scores.

Part V is devoted to the analysis of variance—both one-way and two-way designs. Section A of Chapter 14 develops the one-way ANOVA as a generalization of the *t*-test, and Section B presents computational formulas and other details. Section C deals with the concepts of power and effect size as applied to ANOVA and the use of the *F* ratio for testing homogeneity of variance. Chapter 15 discusses various methods used to control Type I errors when following an ANOVA with multiple comparisons but focuses on the computational details of two tests in particular: Fisher's protected *t*-tests and Tukey's "honestly significant difference." Section C tackles the more difficult concepts behind complex comparisons, planned comparisons, and orthogonal contrasts. Section A of Chapter 16 develops the two-way ANOVA by adding a grouping factor to a one-way ANOVA. Section B concentrates on computational formulas and on the varieties of experimental designs that call for a two-way ANOVA. Two important topics are presented in Section C: follow-up tests with and without a significant interaction, and the unweighted means solution for an unbalanced design. Section A of Chapter 17 shows how the one-way repeated measures ANOVA can be analyzed as though it were a two-way ANOVA with one observation per cell and draws connections between the one-way RM ANOVA and both the one-way independent-samples ANOVA and the matched *t*-test. As usual, Section B deals with computational formulas, experimental design, and publishing practices. Section C covers several important topics, including the Geisser-Greenhouse correction for a lack of sphericity, post hoc comparisons, and techniques for counterbalancing. Finally, Chapter 18 deals with the commonly encountered two-way ANOVA with one between-group factor and one within-subjects factor. Section C of this chapter, in addition to discussing post hoc comparisons and violations of assumptions, introduces two important advanced topics: analysis of covariance and multivariate analysis of variance. Both are discussed in terms of their simplest cases, and the section gives an indication of the complexity these designs can attain.

Part VI introduces the most often used of the nonparametric statistical procedures. Section A of Chapter 19 describes the binomial distribution, and Section B applies this distribution to a nonparametric alternative to the matched *t*-test: the sign test. Section C reviews the probability rules first presented in Chapter 5 and applies them to discrete events. Permutations and combinations are also introduced as ways of counting to determine the probabilities of discrete events. Chapter 20 describes the chi-square test, with Section A covering the one-way (goodness-of-fit) test and Section B detailing the two-way (independence) test. Section C deals with several topics, the most important of which is the measurement of the strength of association between the variables in a chi-square test. Finally, Chapter 21 presents two of the most common statistical tests for ordinal data: the Wilcoxon (Mann-Whitney) rank-sum test and the Spearman correlation coefficient. Section C describes two additional tests for multigroup designs: the Kruskal-Wallis test and the Friedman test.

The Content of Introductory Versus Intermediate Courses

For an introductory, one-semester course, I would suggest skipping Section C in all chapters, with the possible exception of chapters 2, 5, 6, 7, and 10. Chapter 10 can be skipped entirely (though I recommend including at least Section A from that chapter), as can Chapters 18 and 19 and possibly Chapters 17 and 21, depending on the instructor's priorities. For an intermediate (i.e., master's level) course, I would suggest including Section C for all chapters but eliminating Chapter 2 entirely. In fact, Chapters 1 through 6 could be reviewed briefly in one lecture.

Appendixes

Appendix A contains all the statistical tables that are needed to cover the material in this text. Appendix B consists of a two-part review of the basic arithmetic and algebraic operations that are required to work the exercises in each chapter. You may want to urge your students to take the diagnostic quiz at the beginning of Appendix B during the first week of your course, so students will know as soon as possible if they need to review basic math. Appendix C presents a statistical decision tree that helps to answer the frequently asked question, Which statistical test should I use? If used near the end of the course, this appendix can aid students in summarizing and organizing what they have learned. Appendix C also contains exercises that students can use to test their own ability to choose the appropriate test for a particular experimental design. Finally, Appendix D contains answers to selected exercises (those marked with asterisks in the text) or, in some cases, selected parts of selected exercises. Note that answers are always given for exercises from earlier chapters that are referred to in later chapters. If an exercise you would like to assign refers to a previous exercise, it is preferable to assign the previous exercise as well, but it is not necessary; students can use Appendix D to obtain the answer to the previous exercise for comparison purposes.

Supplements

Exercises for which answers are not given in Appendix D are solved completely in the Instructor's Manual. Also included in the manual is a rationale for each exercise (whether it is answered in Appendix D or not), explaining the principles that are being illustrated by that exercise and detailing how that exercise relates to the other exercises and the concepts being taught in the course. Finally, the Instructor's Manual contains teaching tips for new instructors.

If you plan to use computer data analysis as an adjunct to your course, students can purchase a Computer Guide separately. The Computer Guide contains sample programs and detailed explanations of how to use SPSS and SAS for data analysis. The Computer Guide covers both mainframe and PC/DOS versions of both SPSS and SAS, as well as SPSS for Windows. The chapters of the guide match the chapters of the text in terms of the statistical procedures described. However, the guide contains its own exercises based on real data sets. In the context of these exercises, students learn to use the data-handling features of SPSS and SAS as well as the basic statistical procedures.

Barry Cohen

PREFACE TO THE STUDENT

About the Size of This Text

One of the first things you will have noticed about this text is how thick it is. This book is larger than most introductory or intermediate statistics texts, and some students may find the sheer number of pages intimidating. The book does contain a great deal of information about statistics, but there are several reasons why the number of pages can give you a misleading impression. First, note that concrete analogies and detailed verbal explanations take up more room than mathematical language and statistical jargon. This text is not designed to be mathematically elegant; it is designed to communicate with students whose mathematical background and experience with psychological research are minimal. (If your mathematics background is weak, be sure to take the quiz in Appendix B.) Second, a good deal of redundancy is built into the structure of the text. You may have noticed in skimming through the pages that each chapter is divided into four sections. Material in one section may be covered from a different angle in another section of the same chapter. Third, the text is comprehensive enough to be used either as an undergraduate text for a first course in statistics or as an intermediate text for a course on the master's level. Your instructor may not assign all of the chapters or all of the sections of all of the chapters. But the early chapters may be useful for master's students (as a refresher course), and the later chapters may be useful to undergraduates (for future reference).

The Goals of This Text

Most psychology students take a course in statistics because it is required—and it is required because statistics plays an important role in most areas of psychological research. Any research you conduct after your statistics course will strengthen and deepen your understanding of statistical concepts. And even if you do not have the opportunity or inclination to conduct research after graduation, you will want to read the results of research published in psychological journals. Therefore, this text not only teaches you how to use statistics to summarize and draw conclusions from the psychological data you collect in the future, it also emphasizes how to read and interpret the statistical results published in psychological journals. This emphasis should prove especially useful to the many students who will eventually enter careers in which they will need to be critical consumers of research results presented in a statistical form. On the other hand, those who enter careers in which they will be collecting psychological data will undoubtedly perform most of their statistical analyses using sophisticated computer software. These students will appreciate this text's emphasis on understanding the meaning of statistical results rather than on the details of calculating those results.

Although I wrote this text for the student with little mathematical background and perhaps some anxiety about math, I have not given up on the goal of teaching the basic mathematical structure of statistical formulas. Rather, I have undertaken to explain some fairly complex mathematical relationships verbally. In my experience, the material in a statistics course is more memorable if the rationale for each formula and the connections between apparently dissimilar formulas are made clear. To attain the goals described above, I have used a unique ABCD structure in each chapter.

The ABCD Format

Section A provides the *conceptual foundation* for each chapter. I believe that in order to teach the concepts completely, it is necessary to explain the structure of the basic statistical formulas. However, the emphasis in this section is not on computational shortcuts or the details of performing a statistical test. Rather, the focus is on definitional formulas: what the different pieces of a formula represent and how one formula is related to those covered in previous chapters. The prose in Section A can be a bit "long-winded," because I am using simple language, trying to stay concrete, and expressing the same concept in several ways in order to get my point across. The exercises at the end of Section A help to solidify the truly fundamental concepts in your mind before you encounter the additional details presented in Section B.

Section B presents the *basic statistical procedures* for each chapter. This section covers essentially the same material as Section A but includes important details and qualifications, and the emphasis is on step-by-step procedures for performing statistical tests. Alternative formulas are introduced because they are either easy to use with a calculator or frequently encountered in other texts or manuals. The exercises at the end of Section B ensure that you can perform the statistical tests covered in the chapter and also remind you of the important concepts related to those tests.

Section C contains *optional material* that should be tackled only after you have successfully completed the exercises of Section B. Often the material in Section C is more advanced conceptually and potentially confusing. In some cases, concepts or procedures are in Section C because they are not often used or because there is a good chance that your instructor will ignore them in order to devote time to more central issues. Your instructor may want to cover Section C for some chapters but not others. However, the material in Section C is always useful, and you may decide to try some of the C sections not assigned by your instructor. The exercises at the end of this section will help you test your mastery of this additional material.

Finally, Section D contains a *summary* of each of the three preceding sections. This text does not come with a separate study guide; the D section of each chapter is designed to function as a built-in study guide. The summary of Section A is designed to be quite thorough, and if you absorbed the concepts of that section on the first reading you may not have to read Section A over again for study purposes—the summary should be sufficient. It is recommended that you read the summary of Section A immediately after reading Section A for the first time and before proceeding with Section B. The summary of Section B streamlines the procedures of that section and presents an additional example to broaden your understanding. The summary will often be sufficient as a guide to solving the exercises for Section B. The summary of Section C can serve not only as a review of that material but also as a basis for deciding whether you want to tackle that section on your own. If you decide not to read an unassigned Section C, the summary can give you some slight familiarity with material you may need to learn in the future. Section D concludes with definitions of the most important terms in the chapter and a list of the important formulas. Any term that is defined at the end of a chapter appears in **boldface** in the text and in the index, and the page on which it is defined appears in boldface in the index, as well.

Additional Features

The chapters of this text are grouped into six parts, each based on a theme. Each part begins with its own introduction that will orient you to the theme and give you some idea of the content of each section in each chapter. In addition, each chapter begins by listing the symbols, concepts, and formulas from previous chapters that you will need to use in that chapter. You will notice the cumulative nature of the material in this text; most of what you learn in the early chapters will be used again in subsequent chapters.

Appendixes

Appendix A contains all of the statistical tables you will need to solve the exercises in each chapter. Appendix B contains a review of the arithmetic and algebraic tools you will need to solve the exercises. If you think that your math background is weak, or you don't remember much of it, take the diagnostic quiz at the beginning of Appendix B as soon as possible. If you need to refresh your math, it is best to do this before you become immersed in your statistics course. The math review and exercises contained in Appendix B will be all you need to handle the mathematics of the statistical tests in this text. Appendix C is a statistical decision tree that can help you decide which statistical summary or test to use for a particular situation. This appendix will be most useful near the end of your statistics course, to summarize and organize what you have learned and help you to apply your knowledge to the analysis of actual research designs. Finally, Appendix D contains the answers to selected exercises—those marked by an asterisk in the text. (When exercises that are starred have multiple parts, Appendix D may include answers for only some of the parts.) Many of the exercises in later chapters of the text refer to exercises in earlier chapters and ask you to compare answers or use earlier results as a stepping stone to solving the later exercise. Later exercises refer only to earlier exercises whose answers are in Appendix D. Therefore, when an exercise refers to a previous one that was not assigned, or that you could not solve, you can look up its answer in Appendix D and use that answer to help you solve the later exercise.

As a preview of the content of this text, you may want to read the Preface to the Instructor.

Acknowledgments

This text began in the summer of 1988 when Dr. R. Brooke Lea, then my teaching assistant, both encouraged and assisted me in transferring my lectures to written form for class distribution. Dr. Lea deserves much credit as well as gratitude for the existence of this book.

Over the past seven years the text has evolved with the help of many interested students and teaching assistants, including Ann Demarais, Andrew Hilford, Naomi Kubo, Jeannette Levine, Kathryn Perko, Katya Rascovsky, Kay Rothman, Rosalind Sackoff, Jeremy Stone, Inna Subotin, Jennifer Thomas, Kay Wenzel, and Cheryl Williams. Thanks to all those I mentioned, and both thanks and apologies to those I may have accidentally left out.

My lecture notes would never have become a published text without the support and encouragement of the editors at Brooks/Cole. Philip Curson was the first to make me think that publishing my text was a possibility, and Marianne Taflinger kept me going through the early stages of review and text development. James Brace-Thompson continued the regimen of patience and encouragement, and to all of them I owe my heartfelt thanks. Another debt of gratitude is owed the entire production staff at Brooks/Cole and Lifland et al., Bookmakers, especially Kirk Bomont, who oversaw the production, Quica Ostrander, who contributed doubly by copyediting the manuscript and designing the text, Denise Throckmorton, who prepared the index, and Roy Neuhaus, who designed the striking cover. Several instructors of statistics were enormously helpful in reviewing the text at various stages, offering useful suggestions and corrections: Linda Allred, East Carolina University; Karen Caplovitz Barrett, Colorado State University; Michael Biderman, University of Tennessee at Chattanooga; Dennis Bonge, University of Arkansas; Dennis Cogan, Texas Tech University; Lawrence T. DeCarlo, Fordham University; George Domino, University of Arizona; Rick Froman, John Brown University; Tom Gilovich, Cornell University; Debra K. Harless, Bethel College; Kelli Klebe, University of Colorado–Colorado Springs; John B. Murray, St. John's University; and Peter Taylor, Cornell University.

I am grateful to the Longman Group UK Ltd., on behalf of the Literary Executor of the late Sir Ronald A. Fisher, F.R.S., and Dr. Frank Yates, F.R.S., for permission to reproduce Table IV from *Statistical Tables for Biological, Agricultural and Medical Research,* 6th Ed. (1974).

Finally, the Department of Psychology at New York University was very supportive while I was writing this book. I especially want to thank Bettie Brewer, the Graduate Secretary, for helping me in countless ways. All of my family and friends deserve much thanks, as well, for their patience and understanding—most of all my wife, Leona, who deserves thanks also for tirelessly typing the entire manuscript.

Barry Cohen

EXPLAINING PSYCHOLOGICAL STATISTICS

Psychological Statistics: An Introduction

The purpose of Part I is to introduce important concepts that will be used throughout the remainder of this text. This part consists of only Chapter 1, because Chapter 1 contains material relevant to all of the other chapters and thus should stand alone. At first I was tempted to skip Part I entirely and dive into the material directly. The problem is that by its nature Part I must introduce difficult concepts briefly without giving you all of the detail necessary for complete understanding. For some of the concepts, you must wait for five or six more chapters to get the full explanation.

On the other hand, I believe that an overview of where the text is going provides you with a rudimentary framework, which can be strengthened and added to in subsequent chapters. With this approach, many of the sophisticated concepts encountered later in the text will seem at least a little bit familiar. With this in mind, you should not worry if a particular concept is not completely clear after studying Chapter 1. I expect that the concept will become clearer in a later chapter, when you have dealt with more of its details and implications.

Because of the unique nature of Chapter 1, the breakdown of the chapter into A, B, C, and D sections will deviate somewhat from the general plan described in the Preface to the Student. Although Section A will be more conceptual and Section B more procedural, they will not cover the same domain. Section A will introduce a wide range of concepts that are important throughout the text, whereas Section B will introduce the mechanics of summation notation, which form the basis for nearly all of the statistical formulas presented in this text. Section C presents optional advanced material that, in this case, will help you prepare to deal with more mathematical treatments of statistics that may be encountered elsewhere. Section D contains summaries of Sections A, B, and C as well as definitions of key terms.

CHAPTER

CONCEPTUAL FOUNDATION

If you have not already read the Preface to the Student, please do so now. Many readers have developed the habit of skipping the Preface, because it is often used by the author as a soapbox or as an opportunity to give his or her autobiography and to thank many people the reader has never heard of. The preface of this text is different and plays a particularly important role. You may have noticed that this book uses a unique form of organization (each chapter is broken into A, B, C, and D sections). The preface gives the rationale for this unique format and explains how you can derive the most benefit from it.

What Is (Are) Statistics?

An obvious way to begin a text about statistics is to pose the rhetorical question "What *is* statistics?" However, it is also proper to pose the question "What *are* statistics?"—because the term *statistics* can be used in at least two different ways. In one sense *statistics* refers to a collection of numerical facts, such as a set of performance measures for a baseball team (e.g., batting averages of the players) or the results of the latest U.S. census (e.g., the average size of households in the United States). So the answer to "What are statistics?" is that they are observations organized into numerical form.

In a second sense, *statistics* refers to a branch of mathematics that is concerned with methods for understanding and summarizing collections of numbers. So the answer to "What is statistics?" is that it is a set of methods for dealing with numerical facts. Psychologists, like other scientists, refer to numerical facts as *data*. The word "data" is a plural noun and always takes a plural verb, as in "the data *were* analyzed." (The singular form, "datum," is rarely used.) Actually, there is a third meaning for the term *statistics,* which distinguishes a statistic from a parameter. To explain this distinction, I have to contrast samples with populations, which I will do at the end of this section.

As a part of mathematics, statistics has a theoretical side that can get very abstract. This text, however, deals only with **applied statistics.** It describes methods for data analysis that have been worked out by statisticians but does not show how these methods were derived from more fundamental mathematical principles. For that part of the story, you would need to read a text on **theoretical statistics** (e.g., Hogg & Craig, 1965).

The title of this text uses the phrase "psychological statistics." This could mean a collection of numerical facts about psychology (e.g., how large a percentage of the population claims to be happy), but as you have probably guessed, it actually refers to those statistical methods that are commonly applied to the analysis of psychological data. Indeed, just about every kind of statistical method has been used at one time or another to analyze some set of psychological data. The methods presented in this text are the ones usually taught in an introductory or intermediate statistics course for psychology students, and they have been chosen because they are not only commonly used but also simple to explain. Unfortunately, some methods that are very commonly used in psychological research (e.g., factor analysis) are so complex that they can be dealt with only in advanced courses in statistics.

One part of applied statistics is concerned only with summarizing the set of data that a researcher has collected; this is called **descriptive statistics.** If all sixth graders in the United States take the same standardized exam, and you want a system for describing each student's standing with respect to the others, you need descriptive statistics. However, most psychological research involves relatively small groups of people from which inferences are drawn about the larger population; this application of statistics is called **inferential statistics.** If you have a random sample of 100 patients who have been taking a new antidepressant drug, and you want to make a general statement about the drug's possible effectiveness in the entire population, you need inferential statistics. This text begins with a presentation of several procedures that are commonly used to create descriptive statistics. Although such methods can be used just to describe data, it is quite common to use these descriptive statistics as the basis for inferential procedures. The bulk of the text is devoted to some of the most common procedures of inferential statistics.

Statistics and Research

The reason a course in statistics is nearly universally required for psychology students is that statistical methods play a critical role in most types of psychological research. However, not all forms of research rely on statistics. For instance, it was once believed that only humans make and use tools. Then chimpanzees were observed stripping leaves from branches before inserting the branches into holes in logs to "fish" for termites to eat (van Lawick-Goodall, 1971). Certainly such an observation has to be replicated by different scientists in different settings before becoming widely accepted as evidence of tool-making among chimpanzees, but statistical analysis is not necessary.

On the other hand, suppose you wanted to know whether a glass of warm milk at bedtime will help insomniacs get to sleep faster. In this case, the results are not likely to be obvious. You don't expect the warm milk to "knock out" any of the subjects, or even to help every one of them. The effect of the milk is likely to be small and noticeable only after averaging the time it takes a number of subjects to fall asleep (the sleep latency) and comparing that to the average for a (control) group that does not get the milk. Descriptive statistics is required to demonstrate that there is a difference between the two groups, and inferential statistics is needed to show that if the experiment were repeated, it would be highly likely that the difference would be in the same direction. (If warm milk really has *no* effect on sleep latency, the next experiment would be just as likely to show that warm milk slightly increases sleep latency as to show that it slightly decreases it.)

Variables and Constants

A key concept in the above example is that the time it takes to fall asleep varies from one insomniac to another and also varies after a person drinks warm milk. Because sleep latency varies, it is called a *variable.* If sleep latency were the same for everyone, it would be a *constant,* and you really wouldn't need statistics to evaluate your research. It would be obvious after testing a few subjects whether the milk was having an effect. But, because sleep latency varies from person to person and from night to night, it would not be obvious whether a particular case of shortened sleep latency was due to warm milk or just to the usual variability. Rather than focusing on any one instance of sleep latency, you would probably use statistics to compare a whole set of sleep latencies of people who drank warm milk with another whole set of people who did not.

   In the field of physics there are many important constants (e.g., the speed of light, the mass of a proton), but most human characteristics vary a great deal from person to person. The number of chambers in the heart is a constant for humans (4), but resting heart rate is a variable. Many human variables (e.g., beauty, charisma) are easy to observe but hard to measure precisely or reliably. Because the types of statistical procedures that can be used to analyze the data from a research study depend in part on the the way the variables involved were measured, we turn to this topic next.

Scales of Measurement

Measurement is a system for assigning values to observations in a consistent and reproducible way. When most people think of measurement, they think first of physical measurement, in which numbers and measurement units (e.g., minutes and seconds for sleep latency) are used in a precise way. However, in a broad sense, measurement need not involve numbers at all.

Nominal Scales

For instance, facial expressions can be classified by the emotions they express (e.g., anger, happiness, surprise). The different emotions can be considered values on a **nominal scale;** the term *nominal* refers to the fact that the values are simply named, rather than assigned numbers. (Some emotions can be identified quite reliably, even across diverse cultures and geographical locations; see Ekman, 1982.) If numbers are assigned to the values of a nominal scale, they are assigned arbitrarily and therefore cannot be used for mathematical operations. For example, the *Diagnostic and Statistical Manual* of the American Psychiatric Association (the latest version is DSM-IV) assigns a number as well as a name to each psychiatric diagnosis (e.g., the number 300.3 designates obsessive-compulsive disorder). However, it makes no sense to use these numbers mathematically; for instance, you cannot average the numerical diagnoses of all the members in a family to find out the "average" mental illness of the family. Even the order of the assigned numbers is arbitrary; the higher DSM-IV numbers do not indicate more severe diagnoses.

Many variables that are important to psychology (e.g., gender, type of psychotherapy) can be measured only on a nominal scale, so we will be dealing with this level of measurement throughout the text. Nominal scales are often referred to as *categorical scales* because the different levels of the scale represent distinct categories; each object measured is assigned to one and only one category. A nominal scale is also referred to as a *qualitative* level of measurement, as each level has a different quality and therefore cannot be compared with other levels with respect to quantity.

Ordinal Scales

A quantitative level of measurement is being used when the different values of a scale can be placed in order. For instance, an elementary school teacher may rate the handwriting of each student in a class as excellent, good, fair, or poor. Unlike the categories of a nominal scale, these designations have a meaningful order, and therefore constitute an **ordinal scale.** One can add the percentage of students rated excellent to the percentage of students rated good, for instance, and then make the statement that a certain percentage of the students have handwriting that is "better than fair."

Often the levels of an ordinal scale are given numbers, as when a coach rank-orders the gymnasts on a team based on ability. These numbers are not arbitrary like the numbers that may be assigned to the categories of a nominal scale; the gymnast ranked number 2 *is* better than the gymnast ranked number 4, and gymnast number 3 is somewhere between. However, the rankings cannot be treated as real numbers; that is, it cannot be assumed that the third-ranked gymnast is midway between the second and the fourth. In fact, it could be the case that the number 2 gymnast is much better than either number 3 or 4, and that number 3 is only slightly better than number 4 (as shown in Figure 1.1 on page 6). Although the average of the numbers 2 and 4 is 3, the average of the abilities of the number 2 and 4 gymnasts is not equivalent to the abilities of gymnast number 3.

A typical example of the use of an ordinal scale in psychology is when photographs of human faces are rank-ordered for attractiveness. A less obvious example is the measurement of anxiety by means of a self-rated questionnaire (on

Figure 1.1

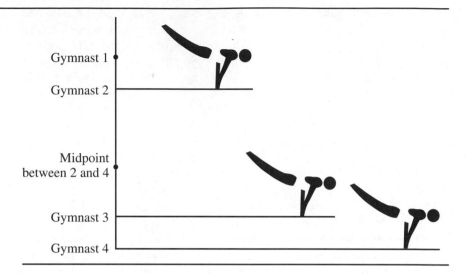

which subjects indicate the frequency of various anxiety symptoms in their lives using numbers corresponding to never, sometimes, often, etc.). Higher scores can generally be thought of as indicating greater amounts of anxiety, but it is not likely that the anxiety difference between subjects scoring 20 and 30 is going to be exactly the same as the anxiety difference between subjects scoring 40 and 50. Nonetheless, scores from anxiety questionnaires and similar psychological measures are usually dealt with mathematically by researchers, as though they were certain the scores were equally spaced throughout the scale.

The topic of measurement is a complex one to which entire textbooks have been devoted, so we will not delve further into measurement controversies here. For our purposes, you should be aware that when dealing with an ordinal scale (when you are sure of the order of the levels but not sure that the levels are equally spaced), you should use statistical procedures that have been devised specifically for use with ordinal data. The descriptive statistics that apply to ordinal data will be discussed in the next three chapters. The use of inferential statistics with ordinal data will not be presented until Chapter 21. (Though it can be argued that inferential ordinal statistics should be used more frequently, such procedures are not used very often in psychological research.)

Interval and Ratio Scales

In general, physical measurements have a level of precision that goes beyond the ordinal property described above. We are confident that the inch marks on a ruler are equally spaced; we know that considerable effort goes into making sure of this. Because we know that the space, or interval, between 2 and 3 inches is the same as that between 3 and 4 inches, we can say that this measurement scale possesses the *interval property* (see part a of Figure 1.2). Such scales are based

Figure 1.2

Interval and Ratio Scales

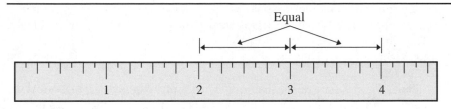

a. Interval scale

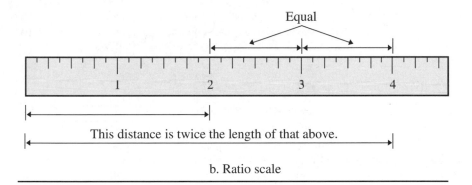

b. Ratio scale

on *units* of measurement (e.g., the inch); a unit at one part of the scale is always the same size as a unit at any other part of the scale. It is therefore permissible to treat the numbers on this kind of scale as actual numbers and to assume that a measurement of three units is exactly halfway between two and four units.

In addition, most physical measurements possess what is called the *ratio property*. This means that when your measurement scale tells you that you now have twice as many units of the variable as before, you really *do* have twice as much of the variable. Measurements of sleep latency in minutes and seconds have this property. When a subject's sleep latency is 20 minutes, it has taken that person twice as long to fall asleep as a subject with a sleep latency of 10 minutes. Measuring the lengths of objects with a ruler also involves the ratio property. Scales that have the ratio property in addition to the interval property are called **ratio scales** (see part b of Figure 1.2).

Whereas all ratio scales have the interval property, there are some scales that have the interval property but not the ratio property. These scales are called **interval scales.** Such scales are relatively rare in the realm of physical measurement; perhaps the best known examples are the Celsius (also known as centigrade) and Fahrenheit temperature scales. The degrees are equally spaced, according to the interval property, but one cannot say that something that has a temperature of 40 degrees is twice as hot as something that has a temperature of 20 degrees. The reason these two temperature scales lack the ratio property is that the zero point for each is arbitrary. Both scales have different zero points (0° C = 32° F), but in neither case does zero indicate a total lack of heat. (Heat comes from the motion of particles within a substance, and as long as there is some motion, there is some heat.) In contrast, the Kelvin scale of temperature is a true ratio scale, because its zero point represents *absolute* zero temperature—a total lack of heat. (Theoretically, the motion of internal particles has stopped completely.)

Though interval scales may be rare when dealing with physical measurement, they are not uncommon in psychological research. If we grant that IQ scores have the interval property (which is open to debate), we still would not consider IQ a ratio scale. It doesn't make sense to say that someone who scores a zero on a particular IQ test has no intelligence at all, unless intelligence is defined very narrowly. And does it make sense to say that someone with an IQ of 150 is exactly twice as intelligent as someone who scores 75?

Continuous Versus Discrete Variables

One distinction among variables that affects the way they are measured is that some variables vary continuously, whereas others have only a finite (or countable) number of levels with no intermediate values possible. The latter variables are said to be discrete (see Figure 1.3). A simple example of a **continuous variable** is height; no matter how close two people are in height, it is possible to find someone whose height is midway between those two people. (Quantum physics has shown that there are limitations to the precision of measurement, and it may be meaningless to talk of continuous variables at the quantum level, but these concepts have no practical implications for psychological research.)

An example of a **discrete variable** is the size of a family. This variable can be measured on a ratio scale by simply counting the family members, but it does not

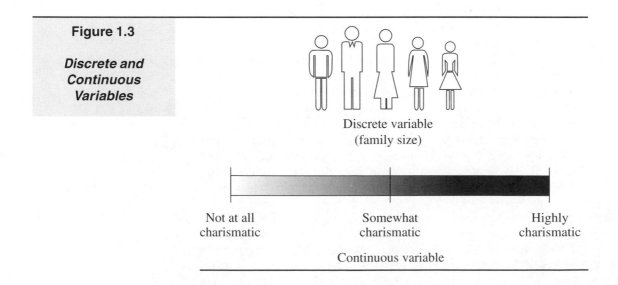

Figure 1.3

Discrete and Continuous Variables

Discrete variable
(family size)

Not at all
charismatic

Somewhat
charismatic

Highly
charismatic

Continuous variable

vary continuously—a family can have two children or three children, but there is no meaningful value in between. The size of a family will always be a whole number and never involve a fraction. The distinction between discrete and continuous variables affects some of the procedures for displaying and describing data, as you will see in the next chapter. Fortunately, however, the inferential statistics discussed in Parts III, IV, and V of this text are *not* affected by whether the variable measured is discrete or continuous. There are implications of this distinction for ordinal statistics, and these will be discussed in Chapter 21.

Scales Versus Variables

It is important not to confuse variables with the scales with which they are measured. For instance, the temperature of the air outside can be measured on an ordinal scale (e.g., the hottest day of the year, the third hottest day), an interval scale (degrees Celsius or Fahrenheit), or a ratio scale (degrees Kelvin); these three scales are measuring the same physical quantity but yield very different measurements. In many cases, a variable that varies continuously, such as charisma, can only be measured crudely, with relatively few levels (e.g., highly charismatic, somewhat charismatic, not at all charismatic). On the other hand, a continuous variable such as generosity can be measured rather precisely by the exact amount of money donated to charity in a year (which is at least one aspect of generosity). Although in an ultimate sense all scales are discrete, scales with very many levels relative to the quantities measured are treated as continuous for display purposes, whereas scales with relatively few levels are usually treated as discrete (see Chapter 2). Of course, the scale used to measure a discrete variable is always treated as discrete.

Parametric Versus Nonparametric Statistics

Because the kinds of statistical procedures described in Parts III, IV, and V of this text are just as valid for interval scales as they are for ratio scales, it is customary to lump these two types of scales together and refer to *interval/ratio scales* or interval/ratio data when some statement applies equally to both types of scales. The data from interval/ratio scales can be described in terms of distributions, which will be explained in detail in Chapter 2. These data distributions sometimes resemble well-known mathematical distributions, which can be described by just a few values called parameters. (I will expand on this point in Section C.) Statistical procedures based on distributions and their parameters are called **parametric statistics.** With interval/ratio data it is often (but not always) appropriate to use parametric statistics. Conversely, parametric statistics are truly valid only when you are dealing with interval/ratio data. The bulk of this text (i.e., Parts III, IV, and V) is devoted to parametric statistics. If your data have been measured on a nominal or ordinal scale, or your interval/ratio data do not meet the distributional assumptions of parametric statistics (which will be explained at the appropriate time), you should be using **nonparametric statistics** (described in Part VI).

Independent Versus Dependent Variables

Returning to the experiment in which one group of insomniacs gets warm milk before bedtime and the other does not, I need to point out that there are actually *two* variables involved in this experiment. One of these, sleep latency, has already been discussed; it is being measured on a ratio scale. The other variable is less obvious; it is group membership. That is, subjects "vary" as to which experimental condition they are in—some receive milk, and some do not. This variable, which in this case has only two levels, is called the **independent variable.** A subject's "level" on this variable—that is, which group a subject is placed in—is determined at random by the experimenter and is independent of anything that happens during the experiment. The other variable, sleep latency, is called the **dependent variable** because its value depends, at least partially (it is hoped), on the value of the independent variable. That is, sleep latency is expected to depend in part on whether the subject drinks milk before bedtime. Notice that the independent variable is measured on a nominal scale (the two categories are "milk" and "no milk"). However, because the dependent variable is being measured on a ratio scale, parametric statistical analysis can be performed. If neither of the variables were measured on an interval or ratio scale (for example, if sleep latency were categorized as simply less than or greater than 10 minutes), a nonparametric statistical procedure would be needed (see Part VI). If the independent variable were also being measured on an interval/ratio scale (e.g., amount of milk given), then you would still use parametric statistics, but of a different type (see Chapter 12). I will discuss different experimental designs as they become relevant to the statistical procedures I am describing (the relationship between experimental design and statistical procedures is summarized in Appendix C). For now, I will simply point out that parametric statistics can be used to analyze the data from an experiment, even if one of the variables is measured on a nominal scale.

Experimental Versus Correlational Research

It is important to realize that not all research involves experiments; much of the research in some areas of psychology involves measuring differences between groups that were not created by the researcher. For instance, insomniacs can be compared to normal sleepers on variables such as anxiety. If inferential statistics shows that insomniacs, in general, differ from normal people in daily anxiety, it is interesting, but we still do not know whether the greater anxiety causes the insomnia, the insomnia causes the greater anxiety, or some third variable (e.g., increased muscle tension) causes both. We cannot make causal conclusions, because we are not in control of who is an insomniac and who is not. Nonetheless, such *correlational* studies can produce useful insights and sometimes suggest confirming experiments.

To continue this example: If a comparison of insomniacs and normals reveals a statistically reliable difference in the amount of sugar consumed daily, these results suggest that sugar consumption may be interfering with sleep. In this case, correlational research has led to an interesting hypothesis that can be tested more conclusively by means of an experiment. A researcher randomly selects two groups of sugar-eating insomniacs; one group is restricted from eating sugar and the other is not. If the sugar-restricted insomniacs sleep better, that evidence supports the notion that sugar consumption interferes with sleep. If there is no sleep difference between the groups, the causal connection may be in the opposite direction (i.e., lack of sleep may produce an increased craving for sugar), or the insomnia may be due to some as yet unidentified third variable (e.g., maybe anxiety produces both insomnia *and* a craving for sugar). The statistical analysis is generally the same for both experimental and correlational research; it is the causal conclusions that differ.

Populations Versus Samples

In psychological research measurements are often performed on some aspect of a person. The psychologist may want to know about people's ability to remember faces, or solve anagrams, or experience happiness. The collection of all people who could be measured, or in whom the psychologist is interested, is called the **population.** However, it is not always people who are the subjects of measurement in psychological research; a population can consist of laboratory rats, mental hospitals, married couples, small towns, etc. Indeed, as far as theoretical statisticians are concerned, a population is just a set (often infinitely large) of numbers. The statistical procedures used to analyze data are the same regardless of where the numbers come from (as long as certain assumptions are met, as subsequent chapters will make clear). In fact, the statistical methods you will be studying in this text were originally devised to solve problems in agriculture, beer manufacture, human genetics, and other diverse areas.

If you had measurements for an entire population, you would have so many numbers that you would surely want to use descriptive statistics to summarize your results. This would also enable you to compare any individual to the rest of the population, compare two different variables measured on the same population, or even to compare two different populations measured on the same variable. More often, practical limitations will prevent you from gathering all of the measurements that you might want. In such cases you would obtain measurements for some subset of the population; this subset is called a **sample** (see Figure 1.4 on page 12).

Sampling is something we all do in daily life. If you have tried two or three items from the menu of a nearby restaurant and have not liked any of them, you do not have to try everything on the menu before deciding not to dine at that restaurant anymore. When you are conducting research, you follow a more formal

Figure 1.4

A Population and a Sample

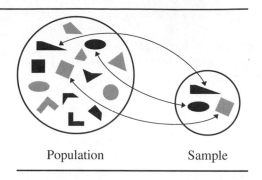

Population   Sample

sampling procedure. If you have obtained measurements on a sample, you would probably begin by using descriptive statistics to summarize the data in your sample. But it is not likely that you would stop there; usually you would then use the procedures of inferential statistics to draw some conclusions about the entire population from which you obtained your sample. Strictly speaking, these conclusions would be valid only if your sample was a *random sample.* In reality, truly random samples are virtually impossible to obtain, and most research is conducted on *samples of convenience* (e.g., students in an introductory psychology class who must either "volunteer" for some experiments or complete some alternative assignment). To the extent that one's sample is not truly random, it may be difficult to generalize one's results to the larger population. The role of sampling in inferential statistics will be discussed at greater length in Chapter 6.

Now we come to the third definition for the term *statistic.* A **statistic** is a value derived from the data in a sample, rather than a population. It could be a value derived from all of the data in the sample, such as the mean, or it could be just one measurement in the sample, such as the maximum value. If the same mathematical operation used to derive a statistic from a sample is performed on the entire population from which you selected the sample, the result is called a population **parameter,** rather than a sample statistic. As you will see, sample statistics are often used to make estimates of, or draw inferences about, corresponding population parameters.

Statistical Formulas

Many descriptive statistics, as well as sample statistics that are used for inference, are found by means of statistical formulas. Often these formulas are applied to all of the measurements that have been collected, so a notational system is needed for referring to many data points at once. It is also frequently necessary to add many measurements together, so a symbol is needed to represent this operation. Throughout the text, Section B will be reserved for a presentation of the nuts and bolts of statistical analysis. The first Section B will present the building blocks of all statistical formulas: subscripted variables and summation signs.

Exercises

1. Give two examples of each of the following. a) Nominal scale; b) Ordinal scale; c) Interval scale; d) Ratio scale; e) Continuous variable; f) Discrete variable.

*2. What type of scale is being used for each of the following measurements?

    a) Number of arithmetic problems correctly solved

    b) Class standing (i.e., one's rank in the graduating class)

    c) Type of phobia

    d) Body temperature (in °F)

    e) Self-esteem, as measured by self-report questionnaire

    f) Annual income in dollars

    g) Theoretical orientation toward psychotherapy

    h) Place in a dog show

    i) Heart rate in beats per minute

*3. Which of the following variables are discrete and which are continuous? a) The number of people in one's social network; b) Intelligence; c) Size of vocabulary; d) Blood pressure; e) Need for achievement

4. a) Give two examples of a population that does not consist of individual people.

    b) For each population described in part a, indicate how you might obtain a sample.

*5. A psychologist records how many words subjects recall from a list under three different conditions: large reward for each word recalled, small reward for each word recalled, and no reward.

    a) What is the independent variable?

    b) What is the dependent variable?

    c) What kind of scale is being used to measure the dependent variable?

6. Subjects are randomly assigned to one of four types of psychotherapy. The progress of each subject is rated at the end of 6 months.

    a) What is the independent variable?

    b) What is the dependent variable?

    c) What kind of scale is formed by the levels of the independent variable?

    d) Describe one type of scale that might be used to measure the dependent variable.

*7. Which of the following studies are experimental and which are correlational?

    a) Comparing pet owners with those who don't own pets on an empathy measure

    b) Comparing men and women with respect to performance on a video game that simulates landing a space shuttle

    c) Comparing subjects run by a male experimenter with subjects run by a female experimenter with respect to the number of tasks completed in one hour

    d) Comparing the solution times of subjects given a hint with those not given a hint

8. Which of the following would be called a statistic, and which a parameter?

    a) The average income for 100 U.S. citizens selected at random from various telephone books

    b) The average income of citizens in the United States

    c) The highest age among respondents to a sex survey in a popular magazine

Throughout the text, asterisks will be used in each section of exercises to indicate exercises whose answers appear in Appendix D.

Variables with Subscripts

Recognizing that statistics is a branch of mathematics, you should not be surprised that its procedures are usually expressed in terms of mathematical notation. For instance, you probably recall from high school math that a variable whose value is unknown is most commonly represented by the letter X. This is also the way a statistician would represent a random variable. However, to describe

**BASIC
STATISTICAL
PROCEDURES**

statistical manipulations with samples, we need to refer to collections of random variables. Because this concept is rather abstract, I will use a very concrete example.

When describing the characteristics of a city to people who are considering living there, a realtor typically gives a number of facts such as the average income and the size of the population. Another common statistic is the average temperature in July. The usual way to find the average temperature for the entire month of July is to take the average temperature for each day in July and then average these averages. The problem is, how do we represent the average temperature for a particular day in July? If we simply use X for each day and X for the whole month, it becomes too confusing to write a formula for finding the average temperature for the month. And since there are 31 days in July, we cannot use a different letter of the alphabet for each day. The solution is to use subscripts. The average temperature for July 1 can be written X_1, for July 2, X_2, and so on up to X_{31}. We now have a compact way of referring to 31 different variables. If we wanted to indicate a different type of variable, such as high or low temperature for each day, we would need to use a different letter (e.g., Y_1, Y_2, up to Y_{31}). If we want to make some general statement about the average temperature for any day in July without specifying which particular day, we can write X_i. The letter i used as a subscript stands for the word *index,* and can take the place of any numerical subscript.

The Summation Sign

To get the average temperature for the month of July, we must add up the average temperatures for each day in July and then divide by 31. Using the subscripts introduced above, the average temperature for July can be expressed as $(X_1 + X_2 + X_3 + \cdots + X_{31})/31$. (Note that because it would take up a lot of room to write out all of the 31 variables, dots are used to indicate that variables are left out.) Fortunately, there is a neater way of indicating that all the variables from X_1 to X_{31} should be added. The mathematical symbol that indicates that a string of variables is to be added is called the summation sign, and it is symbolized by the upper case Greek letter sigma (Σ). The summation sign works in conjunction with the subscripts on the variables in the following manner. First, you write $i = 1$ under the summation sign to indicate that the summing should start with the variable that has the subscript 1. (You could write $i = 2$ to indicate that you want to begin with the second variable, but it is rare to start with any subscript other than 1.) On top of the summation sign you indicate the subscript of the last variable to be added. Finally, next to the summation sign you write the letter that stands for the collection of variables to be added, using the subscript i. So the sum of the average temperatures for each day in July can be symbolized as follows:

$$\sum_{i=1}^{31} X_i$$

The above expression is a neat, compact way of telling you to perform the following steps:

1. Take X_i and replace i with the number indicated under the summation sign (in this case, X_1).
2. Put a plus sign to the right of the previous expression ($X_1 +$).
3. Write X_i again, this time replacing i with the next integer, and add another plus sign ($X_1 + X_2 +$).
4. Continue the above process until i has been replaced by the number on top of the summation sign ($X_1 + X_2 + X_3 + \cdots + X_{31}$).

If you wanted to write a general expression for the sum of the average temperatures on all the days of any month, you could not use the number 31 on top of the summation sign (e.g., June has only 30 days). To be more general, you could use the letter N to stand for the number of days in any month, which leads to the following expression:

$$\sum_{i=1}^{N} X_i$$

To find the average temperature for the month in question, we would divide the above sum by N (the number of days in that month). The whole topic of finding averages will be dealt with in detail in Chapter 3. For now we will concentrate on the mathematics of finding sums.

Summation notation can easily be applied to samples from a population, where N represents the sample size. For instance, if N is the number of people that are allowed by law on a particular elevator, and X_i is the weight of any one particular person, then the expression above represents the total weight of the people on some elevator that is full to its legal capacity. When statisticians use summation signs in statistical formulas, almost always $i = 1$ appears under the summation sign and N appears above it. Therefore, most introductory statistics texts leave out these indexes and simply write the summation sign by itself, expecting the reader to assume that the summation goes from $i = 1$ to N. Although mathematical statisticians dislike this lack of precision, I will, for the sake of simplicity, go along with the practice of leaving off the indexes from summation signs. In fact, in later chapters I will usually write X_i as X, unless it is important to remind you that there is really a whole series of Xs. In any case in which more than one subscript is being used (e.g., a double summation), I will include the appropriate indexes above and below the summation sign.

The summation sign plays a role in most of the statistical formulas in this text. To understand those formulas fully it is helpful to know several interesting mathematical properties of the summation sign. The most important of those properties will be presented in the remainder of this section.

Properties of the Summation Sign

The first property we will discuss concerns the addition of two collections of variables. Returning to our example about the temperature in July, suppose that you are interested in a temperature-humidity index (THI), which is a better indicator of comfort than temperature alone. Assume that the average THI for any day is just equal to the average temperature of that day (X_i) plus the average humidity of that day (Y_i) (although this is not the index that is usually used). Thus we can express the THI for any day as $X_i + Y_i$. If you wanted to add the THI for all the days in the month you could use the following general expression: $\sum(X_i + Y_i)$. It turns out that this expression produces the same result as adding the Xs and Ys separately. This leads to a first rule for dealing with summation signs.

Summation Rule 1A

$$\sum(X_i + Y_i) = \sum X_i + \sum Y_i$$

The rule works in exactly the same way for subtraction.

Summation Rule 1B

$$\sum(X_i - Y_i) = \sum X_i - \sum Y_i$$

Rule 1A works because if all you're doing is adding, it doesn't matter what order you use. Note that $\sum(X_i + Y_i)$ can be written as

$$(X_1 + Y_1) + (X_2 + Y_2) + (X_3 + Y_3) + \cdots + (X_N + Y_N)$$

If you remove the parentheses and change the order, as follows,

$$X_1 + X_2 + X_3 + \cdots + X_N + Y_1 + Y_2 + Y_3 + \cdots + Y_N$$

you can see that the above expression is equal to $\sum X_i + \sum Y_i$. The proof for Rule 1B is exactly parallel.

Sometimes the summation sign is applied to a constant: $\sum C_i$. In this case, we could write $C_1 + C_2 + C_3 + \cdots + C_N$, but all of these terms are just equal to C, the value of the constant. The fact that the number of Cs being added is equal to N leads to the following rule.

Summation Rule 2

$$\sum C = NC$$

In the equation above the subscript on the letter C was left off because it is unnecessary and is not normally used.

Quite often a variable is multiplied or divided by a constant before the summation sign is applied: $\sum CX_i$. This expression can be simplified without changing its value by placing the constant in front of the summation sign. This leads to the next summation rule.

Summation Rule 3

$$\sum CX_i = C\sum X_i$$

The advantage of this rule is that it reduces computational effort. Instead of multiplying every value of the variable by the constant before adding, we can first add up all the values, and then multiply the sum by the constant. You can see why Rule 3 works by writing out the expression and rearranging the terms:

$$\sum CX_i = CX_1 + CX_2 + CX_3 + \cdots + CX_N$$

The constant C can be factored out of each term and the rest can be placed in parentheses, as follows: $C(X_1 + X_2 + X_3 + \cdots + X_N)$. The part in parentheses is equal to $\sum X_i$, so the entire expression equals $C\sum X_i$.

The last rule presents a simplification that is *not* allowed. Because $\sum(X_i + Y_i) = \sum X_i + \sum Y_i$, it is tempting to assume that $\sum X_i Y_i$ equals $(\sum X_i)(\sum Y_i)$ but unfortunately this is *not* true. In the case of Rule 1A, only addition is involved, so the order of operations does not matter (the same is true with a mixture pf subtraction and addition). But when multiplication and addition are mixed together, the order of operations cannot be changed without affecting the value of the expression. This leads to the fourth rule.

Summation Rule 4

$$\sum(X_i Y_i) \neq \left(\sum X_i\right)\left(\sum Y_i\right)$$

This inequality can be demonstrated with a simple numerical example. Assume that $X_1 = 1$, $X_2 = 2$, $X_3 = 3$, $Y_1 = 4$, $Y_2 = 5$, $Y_3 = 6$.

$$\sum(X_i Y_i) = 1 \cdot 4 + 2 \cdot 5 + 3 \cdot 6 = 4 + 10 + 18 = 32$$
$$(\sum X_i)(\sum Y_i) = (1 + 2 + 3)(4 + 5 + 6) = (6)(15) = 90$$

As you can see, the two sides of the above inequality do not yield the same numerical value.

An important application of Rule 4 involves the case in which X and Y are equal, so that we have $\sum(X_i X_i) \neq (\sum X_i)(\sum X_i)$. Because $X_i X_i$ equals X_i^2 and $(\sum X_i)(\sum X_i) = (\sum X_i)^2$, a consequence of Rule 4 is that

$$\sum X_i^2 \neq (\sum X_i)^2$$

This is an important property to remember, because both terms play an important role in statistical formulas, and in some cases both terms appear in the same formula. The term on the left, $\sum X_i^2$, says that each X value should be squared *before the values are added.* If $X_1 = 1$, $X_2 = 2$, and $X_3 = 3$, then $\sum X_i^2 = 1^2 + 2^2 + 3^2 = 1 + 4 + 9 = 14$. On the other hand, the term on the right, $(\sum X_i)^2$, says that all of the X values should be added *before the total is squared.* Using the same X values as above, $(\sum X_i)^2 = (1 + 2 + 3)^2 = 6^2 = 36$. Notice that 36 is larger than 14. When all the values are positive $(\sum X_i)^2$ will always be larger than $\sum X_i^2$.

In this text, I will use only one summation sign at a time in the main formulas. Summation signs can be doubled or tripled to create more complex formulas (see Section C). There's no limit to the number of summation signs that can be combined, but matters soon become difficult to keep track of, so I will use other notational tricks to avoid such complications. Any mention of double or triple summations later in this text will be confined to Section C of the appropriate chapter.

Rounding Off Numbers

Whereas discrete variables can be measured exactly, the measurement of continuous variables always involves some rounding off. If you are using an interval or ratio scale, the precision of your measurement will depend on the unit you are using. If you are measuring height with a ruler in which the inches are divided into tenths, you must round off to the nearest tenth of an inch. When you report someone's height as 65.3 inches, it really means that the person's height was somewhere between 65.25 inches (half a unit below the reported measurement) and 65.35 inches (half a unit above). You can choose to round off to the nearest inch, of course, but you cannot be more precise than the nearest tenth of an inch.

Rounding off also occurs when calculating statistics, even if the data come from a discrete variable. If three families contain a total of eight people, the average family size is 8/3. To express this fraction in terms of decimals requires rounding off, because this is a number with repeating digits past the decimal point (i.e., 2.666 . . . and so on infinitely). When the original data come in the form of whole numbers, it is common to express calculations based on those numbers to two decimal places (i.e., two digits to the right of the decimal point). In the case of 8/3, 2.666 . . . is rounded off to 2.67. The rule is simple: When rounding to two decimal places, look at the digit in the third decimal place (e.g., 2.66<u>6</u>). If this digit is 5 or more, the digit to its left is raised by one and the rest of the digits are dropped (e.g., 2.666 becomes 2.67 and 4.5251 is rounded off to 4.53). If the digit in the third decimal place is less than 5, it is just dropped, along with any digits to its right (e.g., 7/3 = 2.333 . . . , which is rounded to 2.33; 4.5209 is rounded to 4.52).

The only exception to this simple rule occurs when the digit in the third decimal place is 5 and the remaining digits are all zero (e.g., 3/8 = .375). In this case, add one to the digit in the second decimal place if it is odd, and drop the remaining digits (.375 is rounded to .38); if the digit in the second decimal place is even, simply drop the digits to its right (.425 is rounded to .42). This convention is arbitrary, but it is useful in that about half the numbers will have an odd digit in the second decimal place and will be rounded up and the other half will be rounded down. Of course, these rules can be applied no matter how many digits to the right of the decimal point you want to keep. For instance, if you want to keep five such digits, you look at the sixth one to make your decision.

Extra care must be taken when rounding off numbers that will be used in further calculations (e.g., the mean family size may be used to calculate other statistics, such as a measure of variability). If you are using a calculator, you may want to jot down all the digits that are displayed. When this is not convenient, a good strategy is to hold on to two more decimal places than you want to have in your final answer. If you are using whole numbers and want to express your final answer to two decimal places, then your intermediate calculations should be rounded off to not less than four decimal places (e.g., 2.66666 would be rounded to 2.6667).

The amount of *roundoff error* that is tolerable depends on what your results are being used for. When it comes to homework exercises or exam questions, your instructor should give you some idea of what he or she considers a tolerable degree of error due to rounding. Fortunately, with the use of computers for statistical analysis, rounding error is rapidly disappearing as a problem in psychological research.

Exercises

The exercises below are based on the following values for two variables: $X_1 = 2$, $X_2 = 4$, $X_3 = 6$, $X_4 = 8$, $X_5 = 10$; $Y_1 = 3$, $Y_2 = 5$, $Y_3 = 7$, $Y_4 = 9$, $Y_5 = 11$.

* 1. Find the value of each of the following expressions.

a) $\sum_{i=2}^{5} X_i$ b) $\sum_{i=1}^{4} Y_i$ c) $\sum 5X_i$ d) $\sum 3Y_i$

e) $\sum X_i^2$ f) $\left(\sum 5X_i\right)^2$ g) $\sum Y_i^2$

h) $\left(\sum Y_i\right)^2$

* 2. Find the value of each of the following expressions.

a) $\sum(X + Y)$ b) $\sum XY$

c) $\left(\sum X\right)\left(\sum Y\right)$ d) $\sum (X^2 + Y^2)$

e) $\sum (X - Y)$ f) $\sum (X + Y)^2$

g) $\sum (X + 7)$ h) $\sum (Y - 2)$

3. Make up your own set of at least five numbers and demonstrate that $\sum X_i^2 \neq (\sum X_i)^2$.

* 4. Use the appropriate summation rule(s) to simplify each of the following expressions (assume all letters represent variables rather than constants).

a) $\sum(9)$ b) $\sum(A - B)$ c) $\sum(3D)$
d) $\sum(5G + 8H)$ e) $\sum(Z^2 + 4)$

5. Using the appropriate summation rules, show that, as a general rule, $\sum(X_i + C) = \sum X_i + NC$.

* 6. Round off the following numbers to two decimal places (assume digits to the right of those shown are zero).

a) 144.0135   b) 67.245   c) 99.707
d) 13.345   e) 7.3451   f) 5.9817
g) 5.997

7. Round off the following numbers to four decimal places (assume digits to the right of those shown are zero).

    a) .76995   b) 3.141627   c) 2.7182818

    d) 6.89996   e) 1.000819   f) 22.55555

*8. Round off the following numbers to one decimal place (assume digits to the right of those shown are zero).

    a) 55.555   b) 267.1919   c) 98.951

    d) 99.95   e) 1.444   f) 22.14999

OPTIONAL MATERIAL

Double Summations

In the example in Section B, we summed the average temperatures of each day in the month of July. It is easy to extend this example to sum the average temperatures for each of the twelve months of a particular year. However, if we continue using the notation of Section B, how do we indicate at any point which month we are dealing with? Were we to use different letters it would look like we were referring to different types of variables. The solution is to add a second subscript, so that the first subscript can represent the month (1 to 12), and the second subscript can represent the day. For instance, $X_{1,21}$ would stand for the average temperature for January 21, whereas $X_{12,1}$ would stand for December 1. (You can see the need for the comma separating the two subscripts.)

To represent any possible day, it is customary to write X_{ij}. (If a third subscript were needed, the letter k would be used, and so on.) The ij notation is called a double subscript, and it indicates a collection of variables that can be categorized along two different dimensions (e.g., days in a month and months in a year). According to this notation, the sum of temperatures for July would be written as

$$\sum_{j=1}^{31} X_{7j}$$

To sum temperatures for the entire year requires a double summation sign, as shown below,

$$\sum_{i=1}^{M} \sum_{j=1}^{N} X_{ij}$$

where M equals the number of months in the year and N equals the number of days in the given month. (I could have written 12 on top of the first summation sign, but I wanted to show what the double summation looks like in the general case. Also, strictly speaking I should have written N_i, because the number of days is different for different months, but that would make the expression look too complicated.)

The double summation sign indicates that the first step is to set i and j to 1. Then keep increasing j by 1 until you get to N. (For this example, $N = 31$; when N has been reached, January has been summed.) Next, i is increased to 2, and the second summation sign is once again incremented from 1 to N (now $N = 28$

or 29, and when *N* has been reached, February has been summed, and that sum can be added to the sum for January.) This process continues until the first summation sign (using index *i*) has been incremented from 1 to *M*.

The use of double summation signs in statistical formulas will be illustrated in Chapter 14, Section C. Although I will not otherwise use this complex notation in this text, it is worthwhile to expose you to it, because this is the notation you will encounter in more mathematically oriented statistics texts.

If you wanted to represent the above summation for a series of several years, a triple summation could be used:

$$\sum_{k=1}^{O} \sum_{i=1}^{M} \sum_{j=1}^{N} X_{kij}$$

In this case subscript *k* changes for each year, and *O* represents the total number of years to be summed.

Triple summations can be used to express the statistical formulas for a two-way analysis of variance (see Chapter 16), but this book will present formulas with a simpler (though less generalizable) notation.

Random Variables and Mathematical Distributions

The relatively simple rules of statistical inference that I will be explaining in this text were not derived for sampling finite populations of actual people and then measuring them for some characteristic of interest. Mathematical statisticians think in terms of **random variables** that may take on an infinite range of values, but are more likely to yield some values than others. If the random variable is continuous, and the population is infinitely large, the curve that indicates the relative likelihood of the values is called a *probability density function*. When the probability density function of some random variable corresponds to some well-understood mathematical distribution, it becomes relatively easy to predict what will happen when random samples of some fixed size are repeatedly drawn. In mathematical statistics, a parameter is a value that helps describe the form of the mathematical distribution that describes a population. (The parameter is usually a part of the mathematical equation that generates the distribution.) The methods for estimating or drawing inferences about population parameters from sample statistics are therefore called inferential parametric statistics; the methods of Parts III, IV, and V fall into this category. If you are dealing with discrete random variables that have only a few possible values or if you have no basis for guessing what kind of distribution exists for your variable in your population, it may be appropriate to use one of the procedures known as distribution-free, or nonparametric, statistics; the methods of Part VI fall in this category.

Much of the controversy surrounding the use of parametric statistics to evaluate psychological research arises because the distributions of many psychological variables, measured on actual people, do not match the theoretical mathematical distributions on which the common methods are based. Often the researcher has collected so few data points that the empirical distribution (i.e., the distribution of the data collected) gives no clear basis for determining which theoretical distribution would best represent the population. Moreover, using any theoretical distribution to represent a finite population of psychological measurements involves some degree of approximation. Fortunately, the procedures described in this text are applicable to a wide range of psychological variables (sometimes a data transformation may be helpful, as described in Chapter 3), and computer simulation studies have shown that the approximations involved usually do not produce errors large enough to be of practical significance. You can rest assured that I will not have much more to say about the theoretical basis for the applied statistics presented in this text, except to explain, where appropriate, the assumptions underlying the use of mathematical statistics to analyze the data from psychological research.

Exercises

1. a) Describe an example for which it would be useful to have a variable with a double subscript.

   b) Describe an example for which a triple subscript would be appropriate.

* 2. Imagine that there are only nine baseball players on a team and 12 teams in a particular league. If X_{ij} represents the number of home runs hit by the ith player on the jth team, write an expression using a double summation (including the indexes) to indicate the sum of all the home runs in the entire league.

3. Referring to exercise 2, assume Y_{ij} represents the number of hits other than home runs for each player.

   a) Write an expression to indicate the sum of all hits (including home runs) in the league.

   b) Assume Z_{ij} represents the number of times at bat for each player. Write an expression to indicate the *average* of all the batting averages in the league. (Batting average = number of hits/number of times at bat.)

* 4. A die is a cube that has a different number of spots (1 to 6) on each of its sides. Imagine that you have N dice in all and that X_{ij} represents the number of spots on the ith side of the jth die. Write an expression using a double summation (including the indexes) to indicate the sum of all the spots on all of the dice.

5. Perform the summation indicated by the expression you wrote in exercise 4 for $N = 10$.

SUMMARY

The Important Points of Section A

1. Descriptive statistics is concerned with summarizing a given set of measurements, whereas inferential statistics is concerned with generalizing beyond the given data to some larger potential set of measurements.

2. The type of descriptive or inferential statistics that can be applied to a set of data depends, in part, on the type of measurement scale that was used to obtain the data.

3. If the different levels of a variable can be named, but not placed in any specific order, then a *nominal scale* is being used. The categories in a nominal scale can be numbered, but the numbers cannot be used in any mathematical way—even the ordering of the numbers would be arbitrary.

4. If the levels of a scale can be ordered, but the intervals between adjacent levels are not guaranteed to be the same size, you are dealing with an *ordinal scale*. The levels can be assigned numbers, as when subjects or items are rank-ordered along some dimension, but these numbers cannot be used for arithmetical operations (e.g., we cannot be sure that the average of ranks 1 and 3, for instance, equals rank 2).

5. If the intervals corresponding to the units of measurement on a scale are always equal (e.g., the difference between 2 and 3 units is the same as between 4 and 5 units), the scale has the interval property. Scales that have equal intervals but do not have a true zero point are called *interval scales*.

6. If an interval scale has a true zero point (i.e., zero on the scale indicates a total absence of the variable being measured), then the ratio between two measurements will be meaningful (a fish that is 30 inches long is twice as long as one that is 15 inches long). A scale that has both the interval and the ratio properties is called a *ratio scale*.

7. A variable that has countable levels with no values possible between any two adjacent levels is called a *discrete variable*. A variable that can be measured with infinite precision (i.e., intermediate measurements are always possible) is called a *continuous variable*. In practice, most physical measurements are treated as continuous even though they are not infinitely precise.

8. The entire set of measurements about which one is concerned is referred to as a *population*. The measurements that comprise a population can be from individual people, families, animals, hospitals, cities, etc. A subset of a population is called a *sample,* especially if the subset is considerably smaller than the population and is chosen at random.

9. Values that are derived from and in some way summarize samples are called *statistics,* whereas values that describe a population are called *parameters.*

10. If at least one of your variables has been measured on an interval or ratio scale, and certain additional assumptions have been met, it may be appropriate to use *parametric statistics* to draw inferences about population parameters from sample statistics (Parts III, IV, and V of this text). If all of your variables have been measured on ordinal or nominal scales, or the assumptions of parametric statistics have not been met, it may be necessary to use *nonparametric statistics* (Part VI).

The Important Points of Section B

Summation Rule 1A

$$\sum(Y_i + Y_i) = \sum X_i + \sum Y_i$$

This rule says that when summing the sums of two variables (e.g., summing the combined weights of male and female members of tennis teams), you can get the same answer by summing each variable separately (sum the weights of all of the men and then the weights of all of the women) and then adding these two sums together.

Summation Rule 1B

$$\sum(X_i - Y_i) = \sum X_i - \sum Y_i$$

This rule says that when summing the differences of two variables (e.g., summing the height differences of male-female couples) you can get the same answer by summing each variable separately (sum the heights of all of the men and then the heights of all of the women) and then subtracting the two sums at the end.

Summation Rule 2

$$\sum C = NC$$

For instance, if everyone working for some company earns the same annual salary, C, and there are N of these workers, the total wages paid in a given year, $\sum C$, is equal to NC.

Summation Rule 3

$$\sum CX_i = C\sum X_i$$

For instance, in a company where the workers earn different annual salaries (X_i), if each worker's salary were multiplied by some constant, C, then the total wages paid during a given year, $\sum X_i$, would be multiplied by the same constant. Because the constant can be some fraction, there is no need to have a separate rule for dividing by a constant.

Summation Rule 4

$$\sum(X_iY_i) \neq \left(\sum X_i\right)\left(\sum Y_i\right)$$

An important corollary of this rule is that $\sum X_i^2 \neq (\sum X_i)^2$

Rules for Rounding Numbers

If you want to round to N decimal places, look at the digit in the $N + 1$ place.

a) If it is less than 5, do not change the digit in the Nth place.

b) If it is 5 or more, increase the digit in the *N*th place by one.

c) If it is 5 and there are no more digits to the right (or they are all zero), raise the digit in the *N*th place by one only if it is an odd number; leave the *N*th digit as is if it is an even number.

In all cases, the last step is to drop the digit in the $N + 1$ place and any other digits to its right.

The Important Points of Section C

1. Double subscripts can be used to indicate that a collection of variables differs along two dimensions. For instance, a double subscript can be used to represent test scores of school pupils, with one subscript indicating a student's seat number within a class and the other subscript indicating the number of the classroom (e.g., $X_{12,3}$ can represent a score for the twelfth student in room 3).

2. When a double summation sign appears in front of a variable with a double subscript, both subscripts are set to the first value, and then the second summation is completed. Next, the first subscript is incremented by one, and the second summation is completed again, and so on.

3. The procedures of mathematical statistics are derived from the study of random variables and theoretical mathematical distributions. Applying these procedures to psychological variables measured on finite populations of people involves some approximation, but if some reasonable assumptions can be made (to be discussed in later chapters), the error involved is not a practical problem.

Definitions of Key Terms

Applied statistics A branch of mathematics that takes methods derived from theoretical statistics and applies them to practical problems in the analysis of data.

Theoretical statistics A branch of mathematics in which methods are derived for summarizing and drawing inferences about collections of numbers.

Descriptive statistics A branch of statistics in which the goal is to summarize a collection of numbers in a useful way.

Inferential statistics A branch of statistics in which the goal is to draw inferences about a larger collection of numbers based on the information from a sample.

Nominal scale A scale of measurement in which the levels can be named or numbered, but the levels have no intrinsic order, and the numbers have no mathematical properties.

Ordinal scale A scale of measurement in which the levels are ordered but not necessarily equally spaced.

Ratio scale An interval scale that has a true zero point—i.e., zero on the scale represents a total lack of the variable being measured.

Interval scale A measurement scale in which the levels are equally spaced—that is, a unit at one part of the scale is equal to a unit at any other part of the scale. However, zero on the scale is not necessarily a true zero point.

Continuous variable Theoretically, a variable that can be measured with infinite precision; that is, no matter how close two measurements are, another one can be found between them.

Discrete variable A variable whose levels are finite, or countably infinite, and for which there are no intermediate values.

Parametric statistics A branch of statistics that deals with interval or ratio data and is often directed at drawing inferences about (or estimating) the parameters of a population distribution.

Nonparametric statistics A branch of statistics developed for use with ordinal or nominal data. Because of the less restrictive assumptions required, nonparametric statistics are sometimes preferred for dealing with interval/ratio data.

Independent variable A variable manipulated by an experimenter. A variable that has not been manipulated can be treated as an independent variable in statistical tests, but the behavior of the variable cannot lead to causal conclusions.

Dependent variable A variable that is measured, but not directly manipulated. Generally, a researcher hopes to find that the dependent variable has been affected by one or more independent variables.

Population A usually large, and sometimes infinite collection of numbers in which one is interested.

Sample A subset of a population (ideally, chosen at random) that may be used for drawing inferences about the population.

Statistic A value derived from the data in a sample. Sample statistics are often used to estimate corresponding population parameters.

Parameter A value that summarizes or describes a population.

Random variable A variable whose values depend on the outcome of random events; usually these values are not equally likely, but in the case of continuous variables, follow some *probability density function*.

Descriptive Statistics

Part II of this text is devoted to descriptive statistics, which is a set of techniques for summarizing or describing a group of numbers. There are many possible uses for descriptive statistics, but two distinct categories. First, descriptive statistics applies when you have measurements for everyone in the population of interest. For instance, you may have IQ scores for every student in an elementary school, and you may want to make some general statement about the level of IQ for that school. Or you may want to compare a particular student in that school to the group as a whole. Looking at a list of hundreds of IQ scores would not be very helpful. Descriptive statistics can help you make sense out of such a large set of data.

The other reason to use descriptive statistics is as a stepping stone to inferential statistics. Before inferences or estimations can be made it is usually necessary to summarize the measurements in a sample using the methods of descriptive statistics. This second use of descriptive statistics arises much more frequently in the work of psychological researchers. Government scientists sometimes work with data on an entire population (e.g., the U.S. census), but psychologists most often deal with the data from relatively small samples. And yet, these psychologists would like to make statements that apply, in general, to all people in the population being sampled. In most cases, the first step is to use the methods presented in Part II to summarize the data in the sample.

It is often very convenient to represent a set of thousands of numbers by one or two summary statistics. However, the potential danger in this type of summarizing is that, depending on the distribution of the numbers, some summary statistics can be quite misleading. Chapter 2 presents several methods for inspecting the distribution of your numbers, which is an important first step before using the summary statistics described in Chapters 3 through 5. Section A of Chapter 2 shows how to construct a frequency distribution when the variable of interest does not take on many values and how to locate any one score in the distribution. Graphing techniques for displaying the data are also outlined. Section B extends these techniques to the case in which the variable takes on so many different values that it becomes necessary to form groups (i.e., intervals) of values. Section C introduces the important topic of exploratory data analysis and one of its methods for inspecting data: the stem-and-leaf plot. Also, the mechanics of linear interpolation are described in terms of more accurately locating an individual score in a grouped distribution. The primary examples in Chapter 2 involve exam scores, but the numbers used in these examples could represent a wide range of psychological variables. Moreover, it cannot be overemphasized that the techniques in Chapter 2 are useful not only for describing a population of scores but for gaining a greater understanding of the data in a sample before computing a few summary statistics or engaging in inferential statistics.

Chapter 3 is about how to represent a group of numbers by a single number that expresses the overall location (i.e., the central tendency) of the set of numbers. Section A of Chapter 3 compares three well-known measures of central tendency: mode, median, and mean. In addition, Section A discusses factors affecting the shape of distributions and shows that distribution shape is, in turn, a factor that can affect one's choice of a measure of central tendency. Section B is chiefly concerned with exploring the mathematical properties of one of the measures of central tendency: the mean. Section C expands on the topic of exploratory data analysis introduced in Chapter 2, with a description of box-and-whisker plots. The emphasis is on the identification of extreme data points (outliers) in a sample. Also, Section C offers suggestions for dealing with outliers once they are found. An additional topic is the use of the percentile formula, presented in Chapter 2, for finding the median of a distribution.

Chapter 4 is concerned with measuring the spread, or variability, of the numbers in a distribution. The chapter shows that describing a set of numbers only by central tendency can be misleading; adding a second summary statistic (i.e., a measure of variability) can be very informative. Section A compares several measures of variability (e.g., range, semi-interquartile range, variance, standard deviation), and Section B goes into greater detail with respect to the mathematics of these measures, especially the variance and standard deviation. Section C is concerned chiefly with methods for measuring both the skewness (i.e., asymmetry) and the kurtosis (i.e., flatness or peakedness) of distributions. These methods are most often used to compare a given distribution with the theoretical normal distribution, which forms the basis of the parametric statistics presented in this text.

Chapter 5 presents some powerful uses for measures and concepts introduced in previous chapters. Section A introduces the pivotal concept of standardized scores, especially z scores. The z score locates any score in a distribution in terms of the mean and the standard deviation of that distribution. Then Section A describes the normal distribution and demonstrates how informative z scores can be when used in conjunction with the normal distribution. Section B covers the mechanics of using z scores to find various proportions of the normal distribution, and vice versa. Section C goes into greater detail about the mathematics of the normal distribution and then presents the rules of probability, culminating in an introduction to conditional probability. These probability rules underlie the inferential statistics described in the remainder of the text. These rules will be referred to and reviewed in Part VI, when they are needed explicitly to derive the procedures of nonparametric statistics.

FREQUENCY TABLES, GRAPHS, AND DISTRIBUTIONS

CHAPTER

CONCEPTUAL FOUNDATION

You will need to use the following from the previous chapter:

Symbols
Σ: Summation sign

Concepts
Continuous versus discrete variables

I used to give a diagnostic quiz during the first session of my course in statistics. The quiz consisted of ten multiple-choice questions requiring simple algebraic manipulations, designed to show whether students had the basic mathematical tools to handle a course in statistics. (I have since stopped giving the quiz, because the results were frequently misleading, especially when very bright students would panic and produce deceptively low scores.) Most students were curious to see the results of the quiz and to know how their performance compared to those of their classmates. To show how the class did on the quiz, about the cruelest thing I could have done would have been to put all of the raw scores, in no particular order, on the blackboard. This is the way the data would first appear after I graded the quizzes. For a class of 25 students, the scores would typically look like those in Table 2.1.

Table 2.1

8	6	10	9	6
6	8	7	4	9
6	2	8	6	10
4	5	6	8	4
7	8	4	7	6

You can see that there are a lot of 6s and 8s and not a lot of 10s or scores below 4, but this is not the best way to get a sense of how the class performed. A very simple and logical step makes it easier to understand the scores: put them in order. A string of scores arranged in numerical order (customarily starting with the highest value) is often called an **array**. Putting the scores from Table 2.1 into an array produces the array shown in Table 2.2.

Table 2.2

10	10	9	9	8
8	8	8	8	7
7	7	6	6	6
6	6	6	6	5
4	4	4	4	2

Frequency Distributions

The array in Table 2.2 is certainly an improvement, but the table could be made more compact. If the class contained 100 students, an array would be quite difficult to look at. A more informative way to display these data is in a **simple frequency distribution,** which is a table consisting of two columns. The first column lists all of the possible scores, beginning with the highest score in the array and going down to the lowest score. The second column lists the *frequency* of each score—that is, how many times that score is repeated in the array. You don't have to actually write out the array before constructing a simple frequency distribution, but doing so makes the task easier. Table 2.3 is a simple frequency distribution of the data in Table 2.2. *X* stands for any score, and f stands for the frequency of that score. Notice that the score of 3 is included in the table even though it has a frequency of zero (i.e., there are no 3s in the data array). The rule is to list all the possible scores from the highest to the lowest, whether a particular score actually appears in the data or not. To check whether you have included all your scores in the frequency distribution, add up all of the frequencies (i.e., Σf), and make sure that the total is equal to the number of scores in the array (i.e., check that $\Sigma f = N$).

X	f	Table 2.3
10	2	
9	2	
8	5	
7	3	
6	7	
5	1	
4	4	
3	0	
2	1	
$\Sigma f = 25$		

A simple frequency distribution is very helpful when the number of different values listed is not very high (nine in the case of Table 2.3), but imagine 25 scores on a midterm exam graded from 0 to 100. The scores might range from 64 to 98, requiring 35 different scores to be listed, at least ten of which would have zero frequencies. In that case a simple frequency distribution would not be much more informative than a data array. A better solution would be to group the scores into equal-sized intervals (e.g., 80–84, 85–89, etc.) and construct a *grouped frequency distribution.* Because the mechanics of dealing with such distributions can be complicated, I will save this topic for Section B.

The Cumulative Frequency Distribution

In order to evaluate his or her own performance in a class, a student will frequently ask, "How many students in the class had lower scores than mine?" To answer this question for any particular student you need only sum the frequencies for scores below that student's own score. However, you can perform a procedure that will answer that question for any student in the class: You can construct a **cumulative frequency distribution.** The *X* and f columns of such a distribution are the same as in the simple frequency distribution, but each entry in the cumulative frequencies (cf) column contains a sum of the frequencies for the corresponding score and all scores below. Table 2.4 shows the cumulative frequencies for the data in Table 2.3.

Table 2.4

X	f	cf
10	2	25
9	2	23
8	5	21
7	3	16
6	7	13
5	1	6
4	4	5
3	0	1
2	1	1

If a student attained a score of 7 on the quiz, we can look at the entry in the cf column for a score of 6 to see that this student performed better than 13 other students. The cf entry corresponding to a score of 7 (i.e., 16) answers the question, "How many scores are either lower than or tied with a score of 7?"

The mechanics of creating the cf column are easy enough. The cf entry for the lowest score is just the same as the frequency of that score. The cf for the next highest score is the frequency of that score plus the frequency of the score below. Each cf entry equals the frequency of that score plus the cf for the score below. For example, the cf for a score of 7 is the frequency of 7, which is 3, plus the cf for 6, which is 13: cf for $7 = 3 + 13 = 16$. The entry at the top of the cf column should equal the total number of scores, *N,* which also equals Σf.

The Relative Frequency and Cumulative Relative Frequency Distributions

Although it may be satisfying to know that you scored better than many other students, what usually matters in terms of getting good grades is what *fraction* of the class scored below you. Outscoring 15 students in a class of 25 is pretty good, because you beat 3/5 of the class. Having 15 students below you in a class of 100 is not very good, because in that case you have outperformed only 3/20, or .15,

Table 2.5

X	f	cf	rf	crf
10	2	25	.08	1.00
9	2	23	.08	.92
8	5	21	.20	.84
7	3	16	.12	.64
6	7	13	.28	.52
5	1	6	.04	.24
4	4	5	.16	.20
3	0	1	0	.04
2	1	1	.04	.04

of the class. The kind of table that can tell you what fraction of the scores are lower than yours is called a **cumulative relative frequency distribution.** There are two different ways to arrive at this distribution.

As a first step, you can create a **relative frequency distribution** by dividing each entry in the f column of a simple frequency distribution by *N*. The resulting fraction is the relative frequency (rf), and it tells you what proportion of the group attained each score. Notice that in Table 2.5, each rf entry was created by dividing the corresponding f by 25. The cumulative relative frequencies (crf) are then found by accumulating the rf's starting from the bottom, just as we did with the f column to obtain the cf entries. Alternatively, you can convert each entry in the cf column into a proportion by dividing it by *N*. (For example, the crf of .64 for a score of 7 can be found either by dividing 16 by 25 or by adding .12 to the crf of .52 for the score below.). Either way you get the crf column, as shown in Table 2.5. Note that the crf for the top score in the table must be 1.0—if it isn't, you made some kind of mistake (perhaps too much rounding off in lower entries).

The Cumulative Percentage Distribution

Let us again focus on a quiz score of 7. I pointed out earlier that by looking at the cf entry for 6 you can see that 13 students scored below 7. Now we can look at the crf entry for 6 to see that a score of 7 beats .52, or a little more than half, of the class (13/25 = .52). A score of 6, however, beats only .24 (the crf entry for 5), or about one-fourth, of the class. Sometimes people find it more convenient to work with percentages. If you want a **cumulative percentage frequency** (cpf) column, you need only multiply each crf entry by 100. A score of 7 is better than the scores of 52% of the class; a 6 beats only 24% of the scores. Because the cpf column is especially useful for describing scores in a group, let's look at Table 2.6 and focus only on that column. The entries in the cpf column have a special name: They are called **percentile ranks (PR).** By convention, a percentile rank (PR) is defined as the percentage of the group that is at or below a given score. To find the PR of a particular score we look straight across at the cpf entry, rather

	X	f	cpf
Table 2.6	10	2	100%
	9	2	92
	8	5	84
	7	3	64
	6	7	52
	5	1	24
	4	4	20
	3	0	4
	2	1	4

than looking at the score below. Thus, the PR of a score of 7 is 64; 64% of the group scored 7 or below. Similarly, the PR for 6 is 52. The way percentile ranks are found changes a bit when dealing with a continuous variable or when dealing with grouped frequency distributions, but the concept is the same, as you will see in Section B.

Percentiles

Instead of being concerned with the percentage at or below a particular score, sometimes you may want to focus on a particular percentage and find the score that has that percentile rank. For instance, before seeing the results of the diagnostic quiz, a professor might decide that the bottom 20% of the class must receive some remedial training on algebra, regardless of their actual scores on the quiz. That is, whether the whole class does well or poorly, whoever is in the bottom 20% will have to get help. In this case, we want to find the score in the distribution that has a PR of 20. You can see from Table 2.6 that a score of 4 has a PR of 20, so that is the score we are interested in. This score is called the twentieth *percentile*. Anyone with this score or a lower score will have to get algebra help.

A **percentile** can be defined as a score that has a given PR—the 25th percentile is a score whose PR is 25. In other words, a percentile is the score at or below which a given percentage of the group falls. The most interesting percentiles are either *quartiles* (i.e., 25%, 50%, or 75%) or *deciles* (i.e., 10%, 20%, etc.). Unfortunately, these convenient percentiles rarely appear as entries in a cpf column. In Table 2.6, the only convenient percentile is the twentieth. The score of 6 comes close to the 50th percentile (52%), and the score of 5 is a good approximation for the 25th percentile. On the other hand, the 75th percentile is almost exactly midway between 7 and 8. Section B describes how to use a graph to approximate percentiles (and also PRs) that do not appear in the cumulative percent frequency table. Then, Section C presents a mathematical procedure that uses *linear interpolation* to more precisely estimate percentiles (and PRs) that do not appear as entries in a table.

Graphs

The information in a frequency distribution table can usually be presented more clearly and dramatically in the form of a graph. A typical graph is made with two perpendicular lines, one horizontal and the other vertical. The values of some variable (X) are marked off along the horizontal line, which is also called the *horizontal axis* (or X axis). A second variable, labeled Y, is marked off along the *vertical axis* (or Y axis). When graphing a frequency distribution, the variable of interest (e.g., quiz scores) is placed along the X axis, and distance (i.e., height) along the Y axis represents the frequency count for each variable.

The Bar Graph

Probably the simplest type of graph is the **bar graph,** in which a rectangle, or bar, is erected above each value of X. The higher the frequency of that value, the greater the height of the bar. The bar graph is appropriate when the values of X represent a discrete rather than a continuous variable (as defined in Chapter 1). A good example of a variable that has discrete values is family size. Whereas quiz scores can sometimes be measured with fractions, family size is *always* a whole number. The appropriate way to display a frequency distribution of family size is with a bar graph. Imagine that the X values in Table 2.3 are not quiz scores but the sizes (number of parents plus number of children) of 25 randomly selected families in which the parents have been taking fertility drugs. The bar graph for these data is shown in Figure 2.1. Notice that the bars do not touch; we wouldn't want to give the impression that the values come from a continuous variable—that a family can be, for instance, between 3 and 4 in size.

The advantage of a bar graph as compared to a table should be clear from Figure 2.1; the bar graph shows at a glance how the family sizes are distributed among the 25 families. Bar graphs are also appropriate when the variable in question has been measured on a nominal or ordinal scale. For instance, if the 25 members of a statistics class were sorted according to eye color, the values along

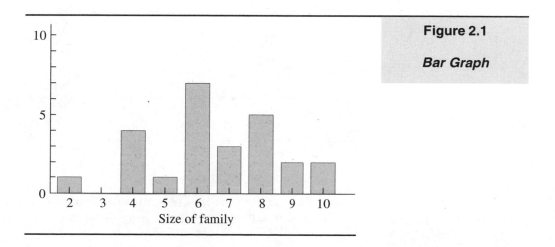

Figure 2.1

Bar Graph

the X axis would be blue, brown, green, etc. and the heights of the bars would indicate how many students had each eye color.

The Histogram

A slightly different type of graph is more appropriate if the variable is continuous. A good example of a continuous variable is height. Unlike family size, height varies continuously, and it is often represented in terms of fractional values. By convention, however, in the United States height is most commonly reported to the nearest inch. If you ask someone how tall she is, she might say, for example, 5 feet 5 inches, but you know she is rounding off a bit. It is not likely that she is *exactly* 5 feet 5 inches tall. You know that her height could be anywhere between 5 feet 4½ inches and 5 feet 5½ inches. Because height is continuous, a value like 5 feet 5 inches generally stands for an interval that goes from 5 feet 4½ inches (the lower **real limit**) to 5 feet 5½ inches (the upper real limit). When constructing a bar graph that involves a continuous variable, the bar for each value is drawn wide enough so that it goes from the lower real limit to the upper real limit. Therefore, adjacent bars touch each other. A bar graph based on a continuous variable, in which the bars touch, is called a **frequency histogram.** The data from Table 2.3 can be displayed in a histogram, if we assume that the X values represent inches above 5 feet for a group of 25 women whose heights have been measured. (That is, a value of 2 represents 5 feet 2 inches, or 62 inches; 3 is 5 feet 3 inches, or 63 inches; etc.). The histogram is shown in Figure 2.2. As with the bar graph, the heights of the bars represent the frequency count for each value. A glance at this figure shows you how the women are distributed in terms of height.

Figure 2.2

Frequency Histogram

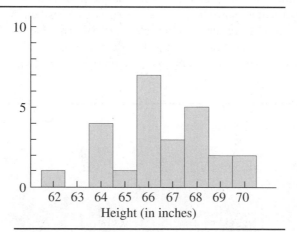

The Frequency Polygon

Some researchers find the bars of a histogram hard to look at, and prefer an alternative format, the **frequency polygon.** An easy way to think of a frequency

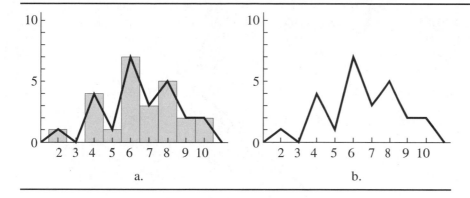

Figure 2.3

Frequency Polygon

polygon is to imagine placing a dot in the middle of the top of each bar in a histogram and connecting the dots with straight lines, as shown in part a of Figure 2.3. Of course, normally a frequency polygon is drawn without first constructing the histogram, as shown in part b of Figure 2.3. Notice that the frequency polygon is connected to the horizontal axis at the high end by a straight line from the bar representing the frequency count of the highest value, and is similarly connected at the low end. Thus, the area enclosed by the polygon is clearly defined and can be used in ways to be described later. A frequency polygon is particularly useful when comparing two overlapping distributions on the same graph. The bars of a histogram would only get in the way in that case.

Just as a simple frequency distribution can be displayed as a histogram or as a polygon, so too can the other distributions we have discussed: the relative frequency distribution, the cumulative frequency distribution, etc. It should be obvious, however, that a histogram or polygon based on a relative frequency distribution will have exactly the same shape as the corresponding graph of a simple frequency distribution—only the scale of the Y axis will change (since all of the frequency counts are divided by the same number, $N = \Sigma f$). Whether it is more informative to display actual frequencies or relative frequencies depends on the situation. If the group from which the data have been taken is very large, relative frequencies will probably make more sense—especially if the frequencies are being estimated from partial data.

The Cumulative Frequency Polygon

A **cumulative frequency polygon** (also called an *ogive*) has a very different shape than a simple frequency polygon. It can be quite useful in conjunction with a simple frequency distribution. (One of its most important uses, as a quick way of finding percentile ranks, will be discussed in Section B.) For one thing, the cf polygon never slopes downward as you move to the right in the graph, as you can see in Figure 2.4 (which was drawn using the same data as in all the examples above). That is because the cumulative frequency can never decrease; it can stay the same, if the next value has a zero frequency, but there are no negative frequency counts, so a cumulative frequency can never go down as the number of

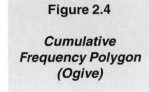

Figure 2.4

Cumulative Frequency Polygon (Ogive)

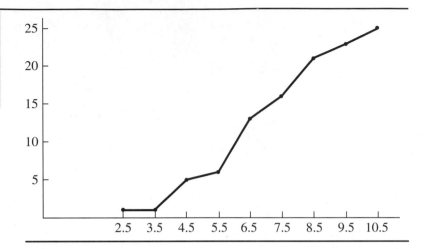

values increases. This is why it is awkward to use a histogram to display cumulative frequencies; in this case, the polygon is definitely easier to look at and interpret. Notice that in the cumulative frequency polygon the dots of the graph are not centered above the values being counted, but rather above the *upper real limit* of each value (e.g., 5 feet 4½ inches, instead of 5 feet 4inches). The rationale is that to make sure you have accumulated, for instance, all of the heights labeled 5 feet 4 inches, you have to include all measurements up to 5 feet 4½ inches. I will discuss the various graphs mentioned above again, when we apply these graphing techniques to grouped frequency distributions in Section B. For now, I will compare the relative frequency polygon with the concept of a **theoretical frequency distribution.**

Theoretical Distributions

The frequency polygon in part b of Figure 2.3 is not very smooth; it consists of straight lines and sharp angles, which at no point resemble a curve. However, if height were measured to the nearest tenth of an inch, and many more people were included in the distribution, the polygon would consist of many more lines, which would be shorter, and many more angles, which would be less sharp (see part a of Figure 2.5). If height were measured to the nearest hundredth of an inch on a large population, the frequency polygon would consist of very many tiny lines, and it would begin to look quite smooth (see part b of Figure 2.5). If we kept increasing the precision of the height measurements, eventually the frequency polygon could not be distinguished from a smooth curve (see part c of Figure 2.5). If only one gender were included, the distribution of heights would begin to resemble one of the well-known mathematical distributions. (The distribution gets more complicated when the heights of both genders are included, as you will see in Chapter 3.)

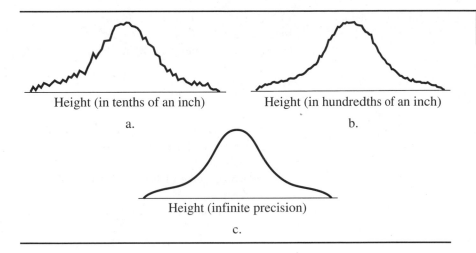

Height (in tenths of an inch)

a.

Height (in hundredths of an inch)

b.

Height (infinite precision)

c.

Figure 2.5

A mathematical distribution is determined by an equation and usually appears as a perfectly smooth curve. The best-known distribution is the *normal distribution,* which looks something like a bell viewed from the side (as in part c of Figure 2.5). With enough people (of one gender) in the distribution, the frequency polygon for height will look a lot like the normal curve (except that the true normal curve actually never ends, extending to infinity in both directions before touching the horizontal axis). This resemblance is important, because many advanced statistical procedures become fairly easy if you assume that the variable of interest follows a normal distribution. I will have much more to say about the normal distribution in Chapter 5, and about other mathematical distributions in later chapters.

Exercises

*1. A psychotherapist has rated all 20 of her patients in terms of their progress in therapy, using a 7-point scale. The results are shown in the table below.

Progress	f
Greatly improved	5
Moderately improved	4
Slightly improved	6
Unchanged	2
Slightly worse	2
Moderately worse	1
Greatly worse	0

a) Draw a bar graph to represent the above results. To answer the following questions, create relative frequency (rf), cumulative frequency (cf), cumulative relative frequency (crf), and cumulative percentage frequency (cpf) columns for the table.

b) What proportion of the patients was greatly improved?

c) How many patients did not improve (i.e., were unchanged or became worse)? What proportion of the patients did not improve?

d) What is the percentile rank of a patient who improved slightly? Of a patient who became slightly worse?

e) Which category of improvement corresponds to

the third quartile (i.e., 75th percentile)? To the first quartile?

* 2. A cognitive psychologist is training subjects to use efficient strategies for memorizing lists of words. After the training period, 25 subjects are each tested on the same list of 30 words. The numbers of words correctly recalled by the subjects are as follows: 25, 23, 26, 24, 19, 25, 24, 28, 26, 21, 24, 24, 29, 23, 19, 24, 23, 24, 25, 23, 24, 25, 26, 28, 25. Create a simple frequency distribution to display these data, and then add columns for rf, cf, crf, and cpf.

　　a) What proportion of the subjects recalled exactly 24 words?

　　b) How many subjects recalled no more than 23 words? What proportion of the total does this represent?

　　c) What is the percentile rank of a subject who scored 25? Of a subject who scored 27?

　　d) Which score is close to being at the first quartile? The third quartile?

3. A boot camp sergeant recorded the number of attempts each of 20 soldiers required to complete an obstacle course. The results were: 2, 5, 3, 1, 2, 7, 1, 4, 2, 4, 8, 1, 3, 2, 6, 5, 2, 4, 3, 1. Create a simple frequency table to display these data. Add columns for cf, rf, crf, and cpf. (Note: Because lower numbers reflect better scores, you may want to put the lowest number on top of the table.)

　　a) What proportion of the soldiers could complete the course on their first attempt?

　　b) What proportion of them needed four or more attempts?

　　c) What is the percentile rank of someone who needed five attempts?

　　d) What score is closest to being the third quartile?

4. An ethnographer surveyed 25 homes to determine the number of people per household. She found the following household sizes: 2, 1, 3, 5, 1, 4, 3, 2, 2, 6, 3, 4, 5, 1, 2, 4, 2, 7, 4, 6, 5, 5, 6, 6, 5. Construct a simple frequency table to display these results. Add columns for cf, rf, crf, and cpf.

　　a) What percentage of households have three or fewer members?

　　b) What household size corresponds to the 80th percentile?

　　c) How many households have only one member? What proportion does that correspond to?

　　d) What proportion of households have five or more members?

　　e) Draw a bar graph to represent the data.

* 5. A physics professor gave a quiz with 10 questions to class of 20 students. The scores were: 10, 3, 8, 7, 1, 6, 5, 9, 8, 4, 2, 7, 7, 10, 9, 6, 8, 3, 8, 5. Create a simple frequency table to display these results. Add columns for cf, rf, crf, and cpf.

　　a) How many students obtained a perfect score? What proportion does that represent?

　　b) What score is closest to the 50th percentile?

　　c) What is the percentile rank of a student who scored a 5? Of a student who scored a 9?

　　d) What proportion of the students scored 9 or more?

　　e) Draw a frequency histogram to represent the data.

BASIC STATISTICAL PROCEDURES

Grouped Frequency Distributions

Constructing a simple frequency distribution is, as the name implies, simple. Unfortunately, measurements on an interval/ratio scale usually result in too many different values for a simple frequency distribution to be helpful. The example of quiz scores was particularly convenient, because there were only eight different values. However, suppose the example involved 25 scores on a midterm exam graded from 0 to 100. Hypothetical scores are listed in the form of an array in Table 2.8.

　　In order to put these scores in a simple frequency distribution, we have to include all of the values from 98 down to 64, as shown in Table 2.9. The simple

					Table 2.8
98	96	93	92	92	
89	89	88	86	86	
86	85	85	84	83	
81	81	81	81	79	
75	75	72	68	64	

X	f	X	f	X	f	X	f	X	f	Table 2.9
98	1	91	0	84	1	77	0	70	0	
97	0	90	0	83	1	76	0	69	0	
96	1	89	2	82	0	75	2	68	1	
95	0	88	1	81	4	74	0	67	0	
94	0	87	0	80	0	73	0	66	0	
93	1	86	3	79	1	72	1	65	0	
92	2	95	2	78	0	71	0	64	1	

frequency distribution is obviously not very helpful in this case; in fact, it seems little better than merely placing the scores in order in an array. The problem, of course, is that the simple frequency distribution has too many different values. The solution is to group the possible score values into equal-sized ranges, called **class intervals.** A table that shows the frequency for each class interval is called a **grouped frequency distribution.** The data from Table 2.9 were used to form the grouped frequency distribution in Table 2.10. Notice how much more informative the frequency distribution becomes when scores are grouped in this way.

Class Interval	f	Class Interval	f	Table 2.10
95–99	2	75–79	3	
90–94	3	70–74	1	
85–89	8	65–69	1	
80–84	6	60–64	1	

Apparent Versus Real Limits

In order to describe the construction of a grouped frequency distribution, I will begin by focusing on just one class interval from Table 2.10—for example, 80–84. The interval is defined by its **apparent limits.** A score of 80 is the *lower apparent limit* of this class interval, and 84 is the *upper apparent limit.* If the variable is thought of as continuous, however, the apparent limits are not the *real* limits of the class interval. For instance, if the score values from 64 to 98 represented the heights of one-year-old infants in centimeters, any fractional value would be possible. In particular, any height above 79.5 cm would be rounded to 80 cm and included in the interval 80–84. Similarly, any height below 84.5 cm would be rounded to 84 and also included in the interval 80–84. Therefore, the *real limits* of the class interval are 79.5 (lower real limit) and 84.5 (upper real limit).

In general, the real limits are just half a unit above or below the apparent limits—whatever the unit of measurement happens to be. In the example of infant heights, the unit is centimeters. If, however, you were measuring the lengths of people's index fingers to the nearest tenth of an inch, you might have an interval (in inches) from 2.0 to 2.4, in which case the real limits would be 1.95 to 2.45. In this case, half a unit of measurement is half of a tenth of an inch, which is one twentieth of an inch, or .05. To find the width of a class interval (usually symbolized by i), we use the real limits, rather than the apparent limits. The width of the interval from 2.0 to 2.4 inches would be $2.45 - 1.95 = .5$ inch. In the case of the 80–84 interval we have been discussing, the width is $84.5 - 79.5 = 5$ cm (if the values are thought of as the heights of infants), not the 4 cm that the apparent limits would suggest. If the values are thought of as midterm grades, they will not include any fractional values (exams graded from 0 to 100 rarely involve fractions). Nevertheless, the ability being measured by the midterm is viewed as a continuous variable.

Constructing Class Intervals

Notice that the different class intervals in Table 2.10 do not overlap. Consider, for example, the interval 80–84, and the next highest one, 85–89. It is impossible for a score to be in both intervals simultaneously. This is important, because it would become very confusing if a single score contributed to the frequency count in more than one interval. It is also important that there is no gap between the two intervals; otherwise, a score could fall "between the cracks" and not get counted at all. Bear in mind that even though there appears to be a gap when you look at the apparent limits (80–84, 85–89), the gap disappears when you look at the real limits (79.5–84.5, 84.5–89.5). Perhaps you are wondering what happens if a score turns out to be exactly 84.5. First, when dealing with a continuous variable, the probability of any particular *exact* value (e.g., 84.500) arising is considered too small to worry about. In reality, however, measurements are not so precise, and such values do arise. In that case, a simple rule can be adopted, such as, any value ending in exactly .5 is placed in the higher interval if the number before the .5 is even.

Choosing the Class Interval Width

Before you can create a grouped frequency distribution, you must first decide how wide to make the class intervals. This is an important decision. If you make the class interval too large, there will be too few intervals to give you much detail about the distribution. For instance, suppose we chose to put the data from Table

Table 2.11

Class Interval	f
90–99	5
80–89	14
70–79	4
60–69	2

2.8 into a grouped frequency distribution in which i (the interval width) equals 10. The result would be as shown in Table 2.11. Such a grouping could be useful if these class intervals actually corresponded with some external criterion; for instance, the class intervals could correspond to the letter grades A, B, C, and D. However, in the absence of some external criterion for grouping, it is preferable to have at least ten class intervals in order to get a detailed picture of the distribution. On the other hand, if you make the class intervals too narrow, you may have so many intervals that you are not much better off than with the simple frequency distribution. In general, more than 20 intervals is considered too many to get a good picture of the distribution.

You may have noticed that Table 2.10, with only eight intervals, does not follow the recommendation of 10 to 20 intervals. There is, however, at least one other guideline to consider in selecting a class interval width: multiples of five are particularly easy to work with. In order to have a number of class intervals between 10 and 20, the data from Table 2.8 would have to be grouped into intervals with $i = 3$ or $i = 2$. The distribution with $i = 2$ is too similar to the simple frequency distribution (i.e., $i = 1$) to be of much value, but the distribution with $i = 3$ is informative, as shown in Table 2.12.

Finally, note that it is a good idea to make all of the intervals the same size. Although there can be reasons to vary the size of the intervals within the same distribution, it is rarely done, and this text will not discuss such cases.

Table 2.12

Class Interval	f	Class Interval	f
96–98	2	78–80	1
93–95	1	75–77	2
90–92	2	72–74	1
87–89	3	69–71	0
84–86	6	66–68	1
81–83	5	63–65	1

Finding the Number of Intervals for a Particular Class Width

Whether Table 2.12 is really an improvement over Table 2.10 depends on your purposes and preferences. In trying to decide which size class interval to use, you can use a quick way to determine how many intervals you will wind up with for a particular interval width. First, find the *range* of your scores, by taking the highest score in the array and subtracting the lowest score. (Actually, you have to start with the *upper real limit* of the highest score and subtract the *lower real limit* of the lowest score. If you prefer, instead of dealing with real limits, you can usually just subtract the lowest from the highest score and add 1.) For the midterm scores, the range is 98.5–63.5 = 35. Second, divide the range by a convenient interval width, and round up if there is any fraction at all. This gives you the number of intervals. For example, using $i = 3$ with the midterm scores, we get 35/3 = 11.67, which rounds up to 12, which is the number of intervals in Table 2.12. Note that if the range of your values is less than 20 to start with it is reasonable to stick with the simple frequency distribution, though you may want to use $i = 2$, if the number of scores in your array is small (which would result in many zero frequencies).

Choosing the Limits of the Lowest Interval

Having chosen the width of your class interval, you must decide on the apparent limits of the lowest interval; the rest of the intervals will then be determined. Naturally, the lowest class interval must contain the lowest score in the array, but that still leaves room for some choice. A useful guideline is to make sure that either the lower apparent limit or the upper apparent limit of the lowest interval is a multiple of i. (If the lower limit of one interval is a multiple of i, then all the lower limits will be multiples of i.) This is true in Table 2.12: The lower limit of the lowest interval (63) is a multiple of i, which is 3. It also would have been reasonable to start with 64–66 as the lowest interval, because then the upper limit (66) would have been a multiple of i. This approach leads to an entirely different set of intervals, as shown in Table 2.13.

Notice that in Table 2.13, two intervals are tied for having the highest frequency (both 85–87 and 79–81 have a frequency of 5). The interval with the highest frequency is called the *mode*; Table 2.13 therefore has two modes, so the distribution is called *bimodal*. (Modes and shapes of distributions will be discussed in much more detail in Chapter 3.) The bimodal structure of the distribution is evident in Table 2.13 but not in Table 2.12, which differs only in that it starts with a lower limit of 63 rather than 64.

Table 2.13	Class Interval	f	Class Interval	f
	97–99	1	79–81	5
	94–96	1	76–78	0
	91–93	3	73–75	2
	88–90	3	70–72	1
	85–87	5	67–69	1
	82–84	2	64–66	1

Choosing the limits of the lowest interval is a matter of judgment, and the decision can be made after seeing the alternatives. The third possibility for the lowest interval, 62–64, is not recommended, because neither the lower nor the upper limit is divisible by 3. Making one of the limits divisible by i will make some of the calculations involving the distribution easier.

Relative and Cumulative Frequency Distributions

Once a grouped frequency distribution has been constructed, it is easy to add columns for cumulative, relative, and cumulative relative frequencies, as described in Section A. These columns have been added to the grouped frequency distribution in Table 2.10 to create Table 2.14.

Interval	f	cf	rf	crf		Table 2.14
95–99	2	25	.08	1.00		
90–94	3	23	.12	.92		
85–89	8	20	.32	.80		
80–84	6	12	.24	.48		
75–79	3	6	.12	.24		
70–74	1	3	.04	.12		
65–69	1	2	.04	.08		
60–64	1	1	.04	.04		

Cumulative Percentage Distribution

Perhaps the most useful table of all is one that shows cumulative percent frequencies, because (as noted in Section A) such a table allows you to find percentile ranks (PR) and percentiles. The cumulative percent frequencies for the midterm scores are shown in the Table 2.15. It is important to note that the cumulative

Interval	f	pf	url	cf	cpf		Table 2.15
95–99	2	8%	99.5	25	100%		
90–94	3	12%	94.5	23	92%		
85–89	8	32%	89.5	20	80%		
80–84	6	24%	84.5	12	48%		
75–79	3	12%	79.5	6	24%		
70–74	1	4%	74.5	3	12%		
65–69	1	4%	69.5	2	8%		
60–64	1	4%	64.5	1	4%		

percentage entry for a particular interval corresponds to the *upper real limit* of that interval. For example, across from the interval 85–89 is the entry 80%. This means that a score of 89.5 is the 80th percentile (that is why the table includes a separate column for the upper real limit, labeled url). In order to score better than 80% of these in the class, a student must have a score that beats not only all the scores below the 85–89 interval, but all the scores *in* the 85–89 interval. And the only way a student can be sure of beating all the scores in the 85–89 interval is to score at the top of that interval: 89.5.

On the other hand, if you wanted to know what your percentile rank would be if you scored 79.5 on the midterm, you would look at the cumulative percent frequency entry for the 75–79 interval, which tells you that your PR is 24 (i.e., you beat 24% of the group). If you wanted to know the PR for a score of 67 or 81, or you wanted to know what score was at the 40th percentile, you could not find that information in Table 2.15. However, you could use a formula to estimate these answers. (This formula is based on *linear interpolation* and is described in Section C.) Or, you could use a graph, as shown below.

Graphing a Grouped Frequency Distribution

A grouped frequency distribution can be displayed as a histogram, like the one used to represent the simple frequency distribution in Section A (see Figure 2.2). In a graph of a grouped distribution, however, the width of each bar extends from the lower real limit to the upper real limit of the class interval that the bar represents. As before, the height of the bar indicates the frequency of the interval. (This is only true when all the class intervals have the same width, but as this is the simplest and most common arrangement, we will consider only this case.) A histogram for a grouped frequency distribution is shown in Figure 2.6, which is a graph of the data in Table 2.13.

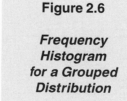

Figure 2.6

Frequency Histogram for a Grouped Distribution

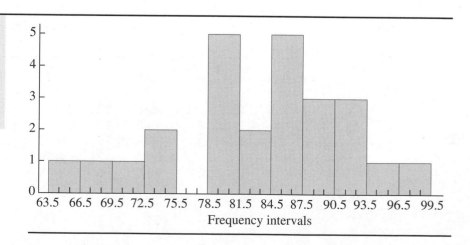

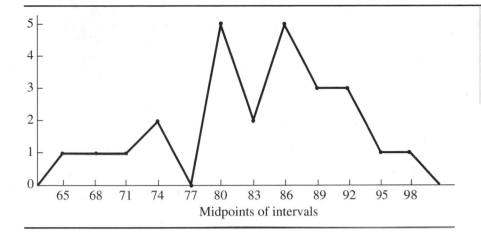

Figure 2.7

*Frequency Polygon
for a Grouped
Distribution*

If you prefer to use a frequency polygon, place a dot at the top of each bar of the histogram at the midpoint of the class interval. (A quick way to calculate the midpoint is to add the upper and lower apparent limits and divide by two—this also works with the real limits.) Place dots on the horizontal axis (to represent zero frequency) on either side of the distribution—that is, at the midpoint of the next interval below the lowest and above the highest, as shown in Figure 2.7. Connecting the polygon to these additional dots on either side closes the polygon, with the horizontal axis serving as one of the sides. Thus, the frequency polygon encloses a particular amount of area, which represents the total number of scores in the distribution. A third of that area, for example, would represent a third of the scores. I will have a lot more to say about the areas enclosed by frequency polygons and smooth distributions in Chapter 5.

Cumulative Percentage Polygons

To illustrate the difference between a grouped frequency distribution and a cumulative percentage distribution, the same data and the same class intervals that were graphed in Figure 2.7 are graphed as a cumulative percentage polygon in Figure 2.8 (page 48). The shape of the polygon would have been the same had it been based on cumulative frequencies, or cumulative relative frequencies, but using percentages allows you to find percentiles and percentile ranks with ease, as will be shown next.

Finding Percentiles and Percentile Ranks from a Graph

As indicated in Section A, the percentile, or PR, in which you are interested often falls between two of the entries in a table. In these common cases, a graph—in particular, a cumulative percentage polygon like the one in Figure 2.8—can help

Figure 2.8

Using an Ogive to Find Percentiles and Percentile Ranks

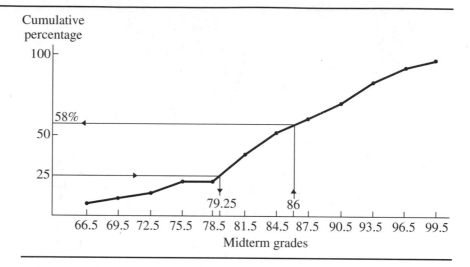

you to estimate the intermediate value that you are seeking. To find the percentile rank of any score, you first find the score on the *X* axis of the polygon, and then draw a vertical line from that score up to intersect the cumulative polygon. Then draw a horizontal line from the point of intersection to the *Y* axis. The percentage at the point where the horizontal line intersects the *Y* axis is the PR of the score in question.

Figure 2.8 illustrates this procedure for a score of 86. Follow the line up from 86 on the *X* axis until it hits the polygon, and then follow the line to the left. You see that the horizontal line intersects the *Y* axis at 58%, which is therefore the PR for a score of 86. The procedure for finding percentiles is exactly the opposite. For instance, to find the 25th percentile (i.e., the first quartile), start at the 25% point on the *Y* axis and move to the right on a horizontal line until you hit the polygon. From the point of intersection, go straight down to the horizontal axis. You should hit a score of about 79.25 on the *X* axis, as shown in Figure 2.8. This score is called the 25th percentile.

The accuracy of the above procedures depends on how carefully the lines are drawn. Drawing the graph to a larger scale tends to increase accuracy, but accuracy is not very important when you consider that the graph will change considerably based on the somewhat arbitrary choices of class width and the lower limit of the first class interval. Also, note that because the cumulative polygon consists of straight *lines,* the graphic technique just described results in linear interpolation (see Section C). The procedure can be made more accurate by fitting a curve to the points of the cumulative polygon (as shown in Figure 2.9), rather than connecting the points by straight lines. The curve in Figure 2.9 is more realistic, but how it is drawn depends inevitably on assumptions about how the distribution would look if it were smooth (i.e., if you had infinitely precise measurements of the variable on an infinitely large population).

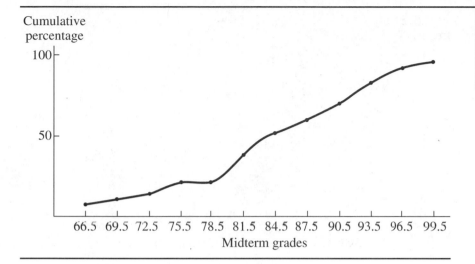

Figure 2.9

Smoothing the Ogive

Guidelines for Drawing Graphs of Frequency Distributions

Graphs of frequency distributions are not often published in psychological journals, but when they are several guidelines are generally followed to make the graphs easier to interpret. (These guidelines were derived from the style manual of the American Psychological Association and apply to any graph to be published.) The first guideline is that you should make the X axis longer than the Y axis by about 50%, so that the height of the graph is only about two-thirds of the width. (Some researchers suggest that the height be closer to three-quarters of the width. The exact ratio is not critical, but a proportion in this ballpark is considered easiest to interpret visually.) The second guideline is that the scores or measurement values should be placed along the horizontal axis and the frequency for each value indicated on the vertical axis. This creates a profile, like the skyline of a big city in the distance, that is easy to grasp. A third guideline is obvious: The units should be equally spaced on both axes (e.g., a single frequency count should be represented by the same distance anywhere along the Y axis). The fourth guideline is that the intersection of the X and Y axes should be the zero point for both axes, with numbers getting larger (i.e., more positive) as you move up or to the right. The fifth guideline is that you should choose a measurement unit and a scale (i.e., how much distance on the graph equals one unit) so that the histogram or polygon fills up nearly all of the graph, without at any point going beyond the axes of the graph.

Sometimes it is difficult to satisfy the last two guidelines simultaneously. Suppose you want to graph a distribution of normal body temperatures (measured to the nearest tenth of a degree Fahrenheit) for a large group of people. You

Figure 2.10

Frequency Histograms: Continuous Scale and Broken Scale

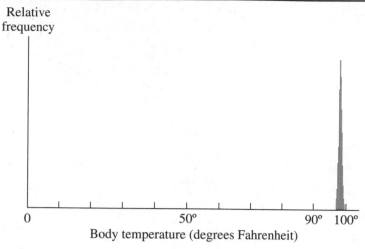

a.

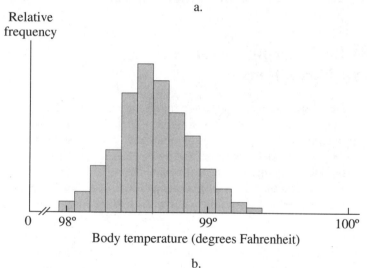

b.

would like to mark off the *X* axis in units of .1 degree, but if you start with zero on the left and mark off equal intervals, each representing .1 degree, you will have to mark off 1000 intervals to get to 100 degrees! Assuming that your distribution extends from about 97 degrees F to about 100 degrees F (98.6 degrees being average body temperature), you will be using only a tiny portion of the *X* axis, as indicated in part a of Figure 2.10. The solution to this dilemma is to increase the scale so that .1 degree takes more distance along the *X* axis and not to mark off units continuously from 0 to 97 degrees. Instead, you can indicate a break along the *X* axis, as shown in part b of Figure 2.10, so that the zero point can still be included but the distribution can fill the graph. Similarly, a break can be used on the *Y* axis if all the frequency counts are high, but do not differ greatly.

The sixth and last guideline is that both axes should be clearly labeled. In the case of a frequency distribution the *X* axis should be labeled with the name of the variable and the unit in which it was measured.

Exercises

*1. Below are the IQ scores for the 50 sixth-grade students in Happy Valley Elementary school. Construct the appropriate grouped frequency distribution, and add rf, cf, crf, and cpf columns (treat IQ as a continuous variable). 104, 111, 98, 132, 128, 106, 126, 99, 111, 120, 125, 106, 99, 112, 145, 136, 124, 130, 129, 114, 103, 121, 109, 101, 117, 119, 122, 115, 103, 130, 120, 115, 108, 113, 116, 109, 135, 121, 114, 118, 110, 136, 112, 105, 119, 111, 123, 115, 113, 117

a) Draw a frequency histogram to represent the above data.

b) Use approximation to find the first and third quartiles.

c) Use approximation to find the 40th and 60th percentiles.

d) Use approximation to find the percentile rank of a student whose IQ is 125; of a student whose IQ is 117; of a student whose IQ is 108.

*2. An industrial psychologist has devised an aptitude test for selecting employees to work as cashiers using a new computerized cash register. The aptitude test, on which scores can range from 0 to 100, has been given to 60 new applicants, whose scores appear below: 83, 76, 80, 81, 74, 68, 92, 64, 95, 96, 55, 70, 78, 86, 85, 94, 76, 77, 82, 85, 81, 71, 72, 99, 63, 75, 76, 83, 92, 79, 82, 69, 91, 84, 87, 90, 80, 65, 84, 87, 97, 61, 73, 75, 77, 86, 89, 92, 79, 80, 85, 87, 82, 94, 90, 89, 85, 84, 86, 56. Construct a grouped frequency distribution table for the above data.

a) Draw a frequency polygon to display the distribution of these applicants.

b) Draw a cumulative percentage polygon (i.e., an ogive) for the same data.

c) Suppose the psychologist is willing to hire only those applicants who scored at the 80th percentile or higher (i.e., the top 20%). Use the graph you drew in part b to find the appropriate cutoff score (i.e, find the 80th percentile).

d) Again use the ogive you drew to find the 75th and 60th percentiles.

e) If the psychologist wants to use a score of 88 as the cutoff for hiring, what percentile rank does an applicant need to be in to qualify? (Use your ogive to find the answer.)

f) Using your ogive again, find the percentile rank for a score of 81.

*3. A telephone company is interested in the number of long-distance calls its customers make. Company statisticians randomly selected 40 customers and recorded the number of long-distance calls they made the previous month. They found the following results: 17, 0, 52, 35, 2, 8, 12, 28, 9, 43, 53, 39, 4, 21, 17, 47, 19, 13, 7, 32, 6, 2, 0, 45, 4, 29, 5, 10, 8, 57, 9, 41, 22, 1, 31, 6, 30, 12, 11, 20. Construct a grouped frequency distribution for the above data.

a) Find the 50th percentile.

b) What percentage of customers made fewer than ten long-distance calls?

c) What is the percentile rank of a customer who made 50 calls?

d) What percentage of customers made 30 or more calls?

*4. A state trooper, interested in finding out the proportion of drivers exceeding the posted speed limit of 55 mph, measured the speed of 25 cars in an hour. Their speeds in mph were: 65, 57, 49, 75, 82, 60, 52, 63, 49, 75, 58, 66, 54, 59, 72, 63, 85, 69, 74, 48, 79, 55, 45, 58, 51. Create a grouped frequency distribution table for these data. Add columns for cf, rf, crf, and cpf.

a) Approximately what percentage of the drivers were exceeding the speed limit?

b) Suppose the state trooper only gave tickets to those exceeding the speed limit by 10 mph or more. Approximately what proportion of these drivers would have received a ticket?

c) Estimate the 50th percentile.

d) Estimate the first and third quartiles.

e) What is the approximate percentile rank of a driver going 62 mph?

*5. A psychologist is interested in the number of dreams people remember. She asked 40 subjects to write down the number of dreams they remember over the course of a month and found the following

results: 21, 15, 36, 24, 18, 4, 13, 31, 26, 28, 16, 12, 38, 26, 0, 13, 8, 37, 22, 32, 23, 0, 11, 33, 19, 11, 1, 24, 38, 27, 7, 14, 0, 13, 23, 20, 25, 3, 23, 26. Create a grouped frequency distribution for these data. Add columns for cf, rf, crf, and cpf. (Note: Treat the number of dreams as a continuous variable.)

a) Draw a frequency histogram to display the distribution of number of dreams remembered.

b) Suppose that the psychologist would like to select subjects who remembered 30 or more dreams for further study. How many subjects would she select? What proportion does this represent? Approximately what percentile rank does that correspond to?

c) Approximately what number of dreams corresponds to the 50th percentile?

d) Estimate the percentile rank of a subject who recalled 10 dreams; a subject who recalled 25 dreams.

**OPTIONAL
MATERIAL**

Stem-and-Leaf Displays

In this chapter I have been describing how a set of quiz scores or exam scores can be displayed so that it is easy to see how a group of people performed and how one particular individual performed in comparison to the group. Sometimes a group of scores represents the performance of a sample of the population, as described in the previous chapter (e.g., a group of allergy sufferers that has been given a new drug, and the scores are number of sneezes per hour). In this case, you would be less interested in individual scores than in some summary statistic you can use to extrapolate to the population of all allergy sufferers. However, the use of summary statistics depends on certain assumptions, including an assumption about the shape of the distribution for the entire population. Exploring the shape of the distribution for the sample can give you a clue about the shape of the population distribution and which types of summary statistics would be most valid. Thus, the graphs and tables described in this chapter can be useful not only for describing a particular group but also for exploring a sample, and the degree to which that sample may be representative of a population.

When dealing with a sample, however, it is tempting to jump quickly to the summary statistics and then to conclusions about populations. J. W. Tukey (1977), a well-known statistician, recommended a more extensive inspection of sample data before moving on to inferential statistics. He called this preliminary inspection **exploratory data analysis,** and he proposed a number of simple but very useful alternatives to the procedures described in this chapter. One of these, the *stem-and-leaf display,* will be described in this section. Another technique will be presented in Section C of Chapter 3.

One disadvantage of using a grouped frequency distribution to find percentile ranks is that you don't know where the individual scores are within each interval.

Of course, if you still have the raw data from which you constructed the grouped distribution, this is no problem. But if you do not have access to the raw data (perhaps because you are working from someone else's grouped distribution, or because you lost your own raw data), the grouped distribution represents a considerable loss of information. An alternative that has the basic advantage of a grouped frequency distribution without the loss of the raw data is a **stem-and-leaf display** (also called a *stem plot*). The simplest stem-and-leaf display resembles a grouped distribution with $i = 10$. I will use the data from Table 2.8 to construct this kind of display.

First, construct the *stem*. To do this you must notice that all of the scores in Table 2.8 are two-digit numbers. The first (or leftmost) of the two digits in each score is referred to as the "leading" digit, and the second (or rightmost) digit is the "trailing" digit. The stem is made from all of the leading digits by writing these digits in a vertical column starting with the lowest at the top, as shown below:

6
7
8
9

The 6 stands for scores in the sixties, the 7 for scores in the seventies, etc. Note that as with the simple frequency distribution, the 7, for examble, is included even if there were no scores in the seventies. However, 5 is not included because there are no scores in the fifties or below 50.

Now you are ready to add the leaves to the stem. The leaves are the trailing digits, and they are written to the right of the corresponding leading digit, in order with higher numbers to the right, even if there are duplicates, as shown in Table 2.16.

The 2559 next to the 7 means that the following scores have 7 as their first digit: 72, 75, 75, 79. You can check this in Table 2.8. Compare the display in Table 2.16 with the grouped distribution in Table 2.11. The grouped distribution tells you only how many scores are in each interval. The stem-and-leaf display gives you the same information, plus you can see just where the scores are within each interval. Moreover, the stem-and-leaf display functions like a histogram on its side, with the horizontal rows of numbers (the leaves) serving as the bars. As long as the leaf numbers are all equally spaced, the length of any row will be proportional to the frequency count corresponding to the adjacent leading digit.

Stem	Leaf													**Table 2.16**
6	4	8												
7	2	5	5	9										
8	1	1	1	1	3	4	5	5	6	6	6	8	9	9
9	2	2	3	6	8									

Table 2.17	Stem	Leaf
	6*	4
	6.	8
	7*	2
	7.	5 5 9
	8*	1 1 1 1 3 4
	8.	5 5 6 6 6 8 9 9
	9*	2 2 3
	9.	6 8

(For example, there are twice as many scores in the seventies as in the sixties, so the row next to the 7 should be twice as long as the row next to the 6.)

The stem-and-leaf display in Table 2.16 has a serious drawback, as you have probably noticed. It has the same problem pointed out with respect to Table 2.11: too few categories, and therefore too little detail about the shape of the distribution. Obviously, a class width of 10 will not always be appropriate. The solution suggested by Tukey (1977) for this situation is to divide each stem in half, essentially using $i = 5$. The stem is constructed by listing each leading digit twice, once followed by an asterisk (trailing digits 0 to 4 are included in this category) and once folllowed by a period (for digits 5 to 9). The leaves are then added to the appropriate parts of the stem, as shown in Table 2.17.

Compare Table 2.17 to Table 2.10. Table 2.10 gives you the same frequency count for each interval, but does not tell you where the raw scores are, and does not automatically show you the shape of the distribution.

Of course, your data will not always range conveniently between 10 and 100. Suppose you have recorded the weight (in pounds) of every young man who tried out for your college's football team last year. The numbers might range from about 140 to about 220. Using just the leading digits for your stem would give only the numbers 1 and 2 written vertically. (Even if you used the asterisk and period to double the length of your stem, as described above, your stem would still be too short.) Also, the leaves would consist of two digits each, which is not desirable. The obvious solution is to use the first two digits for the stem (14, 15, etc.) and the third digit for the leaf. On the other hand, if you are dealing with SAT scores, which usually range from 200 to 800, you can use the first digit as the stem, the second digit as the leaf, and simply ignore the third digit, which is always a zero anyway. Table 2.18 shows a stem-and-leaf display of hypothetical verbal SAT scores for 30 students in an introductory psychology class. The SAT scores range from a low of 370 (stem = 3, leaf = 7) to a high of 740 (stem = 7, leaf = 4).

Yet another advantage of the stem-and-leaf display is that a second distribution can easily be compared to the first as part of the same display. Suppose you wanted to compare the math SAT scores to the verbal SAT scores for the students in the hypothetical introductory psychology class. You can use the same stem,

Table 2.18

Stem	Leaf
3	7 8
4	0 2 2 5 7 8 9 9
5	0 0 1 3 4 4 5 5 5 8 8
6	1 1 2 5 6 9
7	0 2 4

but place the leaves for the second distribution to the left, as shown in Table 2.19. You can see from Table 2.19 that the math scores are somewhat lower than the verbal scores. If the data were real, one might speculate that the subject matter of psychology attracts students whose verbal skills are relatively better than their math skills. In subsequent chapters you will learn how to test such hypotheses. Before you are ready for hypothesis testing, however, you need to learn more about descriptive statistics in the next few chapters. The topic of exploratory data analysis, in particular, will be pursued further in Section C of Chapter 3.

Table 2.19

Math Scores	Stem	Verbal Scores
9 9 6 4	3	7 8
9 8 8 7 5 5 3 2 2 1	4	0 2 2 5 7 8 9 9
8 6 5 3 3 2 1 1 0	5	0 0 1 3 4 4 5 5 5 8
6 3 3 1 0	6	1 1 2 5 6 9
2 1	7	0 2 4

Linear Interpolation

Linear interpolation has many applications in statistics. Whenever you have a table that gives a Y value for each X value (e.g., a cumulative percentage for each class interval), and you want to know the Y value for some X that is *not* in the table—that is, the X value of interest is between two X values in the table—you will need to *interpolate* to find the corresponding Y value. Linear interpolation is the simplest way to do this, but you must assume that the relation between X and Y is linear. I will have a lot to say about linear relationships later in this text, but for now I will describe linear interpolation in terms of a frequency distribution so that the concept of a linear relation begins to make some sense.

   A grouped distribution is a good example of data for which you might need to use linear interpolation: You know how many scores are in each interval, but you don't know exactly where in the interval those scores are. (I am assuming that you don't have access to the raw data from which the frequency table was constructed, or that it is not convenient to refer back to the raw data.) The key

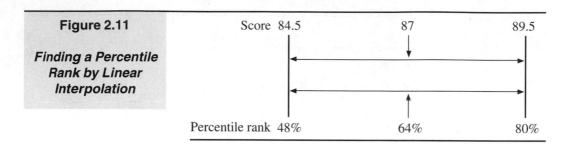

Figure 2.11

Finding a Percentile Rank by Linear Interpolation

assumption behind linear interpolation (in the absence of information about where the scores are in an interval, or just to simplify matters) is that the scores are spread evenly (i.e., linearly) throughout the interval.

Consider the interval 85–89 in Table 2.15, for which the frequency is 8. We assume that the 8 scores are spread evenly from 84.5 to 89.5, so that, for instance, four of the scores are between 84.5 and 87 (the *midpoint* of the interval), and the other four are between 87 and 89.5. This reasoning also applies to the percentages. The cumulative percentage at 84.5 is 48 (the cpf entry for 80–84), and at 89.5 it is 80 (the cpf entry for 85–89), as shown in Figure 2.11. On the basis of our assumption of linearity, we can say that the midpoint of the interval, 87, should correspond to a cumulative percentage midway between 48 and 80, which is 64% [(48 + 80)/2 = 128/2 = 64)]. Thus, the PR for a score of 87 is 64.

Finding Percentile Ranks by a Formula

A more complicated question to ask is: what is the PR for a score of 86 in Table 2.15? Since 86 is not right in the middle of an interval, we need to know how far across the interval it is. Then we can use linear interpolation to find the cumulative percentage that corresponds to that score. To go from the lower real limit, 84.5, to 86 we have to go 1.5 score points. To go across the entire interval requires 5 points (the width of the interval). So 86 is 1.5 out of 5 points across the interval; 1.5 out of 5 = 1.5/5 = .3, or 30%. A score of 86 is 30% of the way across the interval. That means that to find the cumulative percentage for 86, we must go 30% of the way from 48 to 80, as shown in Figure 2.12. From 48 to 80 there are 32 percentage points, and 30% of 32 is (.3)(32) = 9.6. So we have to add 9.6 percentage points to 48 to get 57.6, which is the PR for a score of 86. In sum, 86 is 30% of the way from 84.5 to 89.5, so we go 30% of the way from 48 to 80, which is 57.6. This procedure is expressed in Formula 2.1,

$$PR = cpf_{LRL} + \left(\frac{X - LRL}{i}\right) pf_X \qquad \text{Formula 2.1}$$

where X is the score whose PR you are trying to find, LRL is the lower real limit of the interval that X falls in, cpf_{LRL} is the cpf entry corresponding to the lower

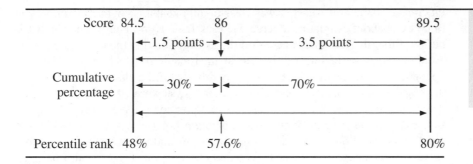

Figure 2.12

Finding a Percentile Rank by Linear Interpolation

real limit (i.e., the next lower interval), and pf_X is the percent contained within the target interval.

I will illustrate the use of the formula by finding the PR for a score of 75 in Table 2.15. Before using the formula, however, I recommend that you make a rough estimate of the answer. First, notice that 75 falls in the interval 75–79, and therefore its PR must be between 12 and 24. Because 75 is very close to the bottom of the interval (i.e., 74.5) its PR should be much closer to 12 than to 24. Thus, a rough estimate for 75 would be a PR of about 13 or 14.

Now we can use the formula, noting that X is 75, cpf_{LRL} is 12, LRL is 74.5, i is 5, and pf_X is 12.

$$ PR = 12 + \left(\frac{75 - 74.5}{5}\right)12 = 12 + \left(\frac{.5}{5}\right)12 = 12 + (.1)12 = 12 + 1.2 = 13.2 $$

Had the answer of 13.2 been very far from our rough estimate, we would have suspected a calculation error (or a wrong entry from the table) and rechecked the work.

Some people prefer to work out PR problems in terms of frequencies, converting to a percentage only at the end. This can be accomplished with Formula 2.2, which is very similar to Formula 2.1.

$$ PR = \left[cf_{LRL} + \left(\frac{X - LRL}{i}\right)f_X \right]\frac{100}{N} \qquad \text{Formula 2.2} $$

In Formula 2.2, cf_{LRL} is the cumulative frequency entry for the next lowest interval, f_X is the frequency within the target interval, and N is Σf. The part of the formula within the brackets finds the estimated cf corresponding to X; multiplying by 100/N converts this to a cumulative percentage. You can find the PR of 75 with this formula as follows:

$$ PR = \left[3 + \left(\frac{75 - 74.5}{5}\right)3\right]\frac{100}{25} = [3 + (.1)3]4 = [3.3]4 = 13.2 $$

Of course, you get the same answer. Bear in mind that it is not terribly important to be exact about estimating a percentile rank from a grouped frequency distribution. First, the estimate is based on the assumption that the scores are spread evenly throughout the interval (i.e., linear interpolation), which may not be true. Second, the estimate may be considerably different if the class interval width or the starting score of the lowest interval changes. However, I have spent a good deal of space on this topic because linear interpolation is used in several ways in statistics, and because beginning students might otherwise get confused by the topic. Now that I have described in detail how to find a PR corresponding to a score, it will be easy to describe how to find the score that corresponds to a given percentile rank.

Finding Percentiles by a Formula

Suppose you want to find the sixtieth percentile (i.e., the score for which the PR is 60) for the midterm exam scores. First, you can see from Table 2.15 that 60% lands between the entries 48% (corresponding to 84.5) and 80% (corresponding to 89.5). Because 60 is a bit closer to 48 than it is to 80, a rough estimate of the 60th percentile should be a bit closer to 84.5 than to 89.5—perhaps 86. To find exactly what proportion of the way from 48 to 80 you have to go to get to 60 (so that you can then go the same proportion of the distance between 84.5 and 89.5), you can use Formula 2.3, which is just a transformation of Formula 2.1:

$$X = \text{LRL} + \left(\frac{\text{PR} - \text{cpf}_{\text{LRL}}}{\text{pf}_X} \right) i \qquad \text{Formula 2.3}$$

In this formula, PR is the percentile rank of the score (X) you are trying to find, and X is the percentile. (The other symbols are defined as they were for Formula 2.1.) Plugging the appropriate values into formula 2.3, we get

$$X = 84.5 + \left(\frac{60 - 48}{32} \right) 5 = 84.5 + \frac{12}{32}(5) = 84.5 + 1.875 = 86.375$$

We have found that a score of 86.375 is at the 60th percentile; as this is close to our rough estimate of 86, there is no reason to suspect an error. However, you can check your work by using Formula 2.1 to find the PR of 86.375. If the PR turns out to be 60, you probably did not make a mistake (unless somehow you made compensating errors). Of course, you can also check any of your PR calculations by using Formula 2.3 to see if you get back to the original score.

If you prefer to work with frequencies and cumulative frequencies, you can use Formula 2.4 (which is the reverse of Formula 2.2) to find percentiles. (Although this is the procedure suggested by most other texts, I think it is simpler to work entirely with percentages.)

$$X = \text{LRL} + \left(\frac{cf_p - cf_{\text{LRL}}}{f_X} \right) i \qquad \text{Formula 2.4}$$

To use Formula 2.4, you must convert the percentile you wish to find to a cumulative frequency. First, divide by 100 to transform the percentile to a proportion, then multiply by N to find the corresponding cf (i.e., $cf_p = PN/100$). The remaining terms in Formula 2.4 should be familiar by now.

Exercises

Note: Some chapters will refer back to exercises from previous sections of the chapter or from earlier chapters for purposes of comparison. A shorthand notation, consisting of the chapter number and section letter followed by the problem number, will be used to refer to exercises. For example, exercise 3B2a refers to Chapter 3, Section B, exercise number 2, part a.

X	f	X	f
18	1	9	1
17	0	8	3
16	2	7	5
15	0	6	5
14	1	5	7
13	0	4	5
12	0	3	4
11	1	2	2
10	2	1	1

* 1. Construct a stem-and-leaf display for the data in Table 2.8.

2. Construct a stem-and-leaf display for the IQ data in exercise 2B1. Compare it to the frequency histogram you drew for the same data.

* 3. Construct a stem-and-leaf display for the aptitude test scores in exercise 2B2.

4. Use the appropriate formula to calculate all three quartiles for the data in the table below: (Hint: Each value for X can be assumed to represent a class that ranges from a half unit below to a half unit above the value shown; for example, $X = 16$ represents the range from 15.5 to 16.5.)

* 5. Calculate the answers to exercise 2B1, parts b, c, and d, using the appropriate formulas.

* 6. Calculate the answers to exercise 2B2, parts c through f, using the appropriate formulas.

7. Calculate the answers to exercise 2B4, parts c through e, using the appropriate formulas.

8. Calculate the answers to exercise 2B5, parts c and d, using the appropriate formulas.

SUMMARY

The Important Points of Section A

1. Often the first step in understanding a group of scores is to put them in order, thus forming an *array*.

2. If the number of different values in the group is not too large, a *simple frequency distribution* may make it easy to see where the various scores lie. To create a simple frequency distribution, write down all the possible scores in a column with the highest score in the group on top and the lowest on the bottom, even if some of the intermediate possible scores do not appear in the group. Add a second column in which you record the frequency of occurrence in the group for each value in the first column.

3. It is easy to create a cumulative frequency (cf) distribution from a simple frequency distribution: The cf entry for each score is equal to the frequency for that score plus the frequencies for all lower scores. (This is the same as saying that the cf for a given score is the frequency for that score plus the cf for the next lower score.) The cf entry for the highest score must equal $\Sigma f = N$ (the total number of scores in the group).

4 To convert a simple or cumulative frequency distribution to a relative or cumulative relative distribution, just divide each entry by N. The relative distribution tells you the proportion of scores at each value, and the cumulative relative distribution tells you what proportion of the scores is at or below each value.

5. Multiplying each entry of a cumulative relative frequency distribution by 100 gives a cumulative percentage distribution. The entries of the latter distribution are *percentile ranks* (PR); each entry tells you the percentage of the distribution that is at or below the corresponding score. A percentile, on the other hand, is the score corresponding to a particular cumulative percentage. For example, the 40th percentile is the score that has a PR of 40. If the percentile of interest does not appear in the table it can be estimated with the procedures outlined in Section B or Section C.

6. If the scores in a distribution represent a discrete variable (e.g., number of children in a family) and you want to display the frequency distribution as a graph, a *bar graph* should be used. In a bar graph, the heights of the bars represent the frequency counts, and adjacent bars do not touch. A bar graph is also appropriate for distributions involving nominal or ordinal scales (e.g., the frequency of different eye colors in the population).

7. When dealing with a continuous variable (e.g., height), the distribution can be graphed as a *histogram,* which is a bar graph in which adjacent bars *do* touch. In a histogram, the width of the bar that represents a particular value goes from the *lower* to the *upper real limit* of that value (e.g., the real limits of 60 inches would be 59.5 and 60.5 inches).

8. An alternative to the histogram is the frequency polygon, in which a point is drawn above each value. The height of the point above the value on the X axis represents the frequency of that value. These points are then connected by straight lines, and the polygon is connected to the X axis at either end to form a closed figure. It is usually easier to compare two polygons on the same graph (e.g., separate distributions for males and females) than two histograms. Cumulative frequency polygons are especially useful, as described in Section B.

9. A frequency polygon with many intermediate values can begin to look smooth, and possibly resemble a perfect mathematical distribution, such as the normal curve.

The Important Points of Section B

The procedure for constructing a grouped frequency distribution can be summarized in the following steps and guidelines.

Step 1. Choose the width (i) *of the class intervals.* First, find the range of scores by subtracting the lowest score in your distribution from the highest and adding one (i.e., range = highest score − lowest score + 1). Then divide the range by a convenient class width (a multiple of 5, or a number less than 5, if appropriate), and round up if there is any fraction, to find the number of intervals that would result. If the number of intervals is between 10 and 20, you have found an appropriate width; otherwise, try another convenient value for *i*. Note that an external criterion might dictate the size of the class intervals (e.g., letter grades based on predetermined scores).

Step 2. Choose the apparent limits of the lowest interval. The lowest interval must contain the lowest score in the distribution. In addition, it is highly desirable for the lower limit or the upper limit to be a multiple of the chosen interval width.

Step 3. All the class intervals should be the same size. Work upward from the lowest interval to find the remaining intervals; stop when you reach an interval that contains the highest score. Taking into consideration the *real* limits of the intervals, make sure that no two intervals overlap, and that there are no gaps between intervals (the *real* limits are half a unit of measurement above or below the apparent limits).

Step 4. Count the number of scores that fall between the real limits of each interval. Write the frequency count for each interval next to the limits of that interval.

 If the number of possible scores between the highest and lowest scores in the distribution is less than 20, a simple (i.e., ungrouped) frequency distribution may suffice. Furthermore, in some cases it may be desirable to have fewer than 10 or more than 20 intervals. You must use your judgment to determine the intervals that result in the most informative description of the data. Once a grouped frequency distribution has been constructed, it is easy to derive the following related distributions.

1. *Relative frequency distribution.* Divide each entry of the grouped frequency distribution by *N* (the total number of scores in the distribution = Σf). The sum of the entries should equal 1.0.

Table 2.20					
	171	166	157	153	151
	149	147	145	143	142
	140	137	137	135	135
	134	134	132	131	131
	128	128	127	126	126
	126	124	124	123	122
	122	121	120	120	119
	119	118	117	117	116
	115	114	114	113	112
	110	108	105	102	97

2. *Cumulative frequency distribution.* The cf entry for the lowest interval is just the frequency of that interval. The cumulative frequency for any higher interval is the frequency of that interval plus the cumulative frequency for the next lowest interval. The highest entry should equal N.

3. *Cumulative relative frequency distribution.* Divide each entry of the cumulative frequency distribution by N. The highest entry should equal 1.0.

4. *Cumulative percent frequency distribution.* Multiply each entry in the cumulative relative frequency distribution by 100. The highest entry should equal 100%.

An example will illustrate the above procedures. Suppose you are studying eating disorders in women, and want to know, as a basis for comparison, the weights for a random group of college-age women of average height. The weights, in pounds, for 50 hypothetical women appear in an array in Table 2.20.

Step 1. The range of scores equals $171 - 97 + 1 = 74 + 1 = 75$. For $i = 10$, the number of intervals would be $75/10 = 7.5$, which rounded up equals only 8. For $i = 5$, $75/5 = 15$. Because 15 is between 10 and 20 we will use $i = 5$.

Step 2. The lowest interval could be 95–99, as 95 is a multiple of i; the width of the interval (subtracting the *real* limits) would then be $99.5 - 94.5 = 5$. ($96 - 100$ could also be the lowest interval, as 100 is a multiple of i.

Step 3. The next interval above 95–99 is 100–104. Form intervals in this way until you arrive at the highest interval, which is 170–174, because this interval contains the highest score (174).

Step 4. Next to each interval, make tally marks to indicate how many of the scores in Table 2.20 fall in that interval. The sum of the tally marks is the frequency for a given interval.

Table 2.21

Interval	f	rf	cf	crf	pf	cpf
170–174	1	.02	50	1.00	2	100%
165–169	1	.02	49	.98	2	98
160–164	0	0	48	.96	0	96
155–159	1	.02	48	.96	2	96
150–154	2	.04	47	.94	4	94
145–149	3	.06	45	.90	6	90
140–144	3	.06	42	.84	6	84
135–139	4	.08	39	.78	8	78
130–134	5	.10	35	.70	10	70
125–129	6	.12	30	.60	12	60
120–124	8	.16	24	.48	16	48
115–119	7	.14	16	.32	14	32
110–114	5	.10	9	.18	10	18
105–109	2	.04	4	.08	4	8
100–104	1	.02	2	.04	2	4
95– 99	1	.02	1	.02	2	2

Table 2.21 shows the grouped frequency distribution for the 50 weights, along with relative (rf), cumulative (cf), cumulative relative (crf), percent frequency (pf), and cumulative percent frequency (cpf) distributions.

Guidelines for Graphs

1. The *Y* axis should be only about two-thirds as long as the *X* axis.

2. For frequency distributions, the variable of interest is placed along the *X* axis and the frequency counts (or relative frequency) are represented along the *Y* axis.

3. The measurement units are equally spaced along the entire length of both axes.

4. The intersection of the *X* and *Y* axes is the zero point for both dimensions.

5. Choose a scale to represent the measurement units on the graph so that the histogram or polygon fills the space of the graph as much as possible. Indicating a break in the scale on one or both axes may be necessary to achieve this goal.

6. Both axes should be clearly labeled and the *X* axis should include the name of the variable and the unit of measurement.

The relative frequency polygon and cumulative percentage polygon for the data in Table 2.21 have been graphed in parts a and b of Figure 2.13, according to the guidelines listed above.

Figure 2.13

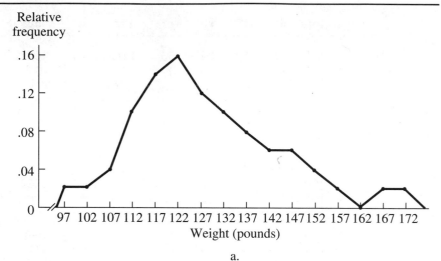

Relative frequency

a.

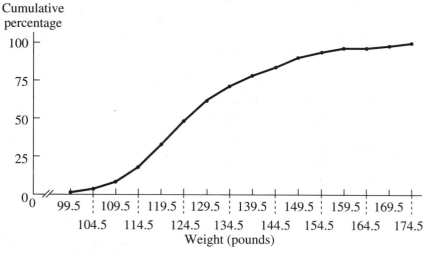

Cumulative percentage

b.

The Important Points of Section C

1. *Exploratory data analysis* can be an important step before computing sample statistics and drawing inferences about populations. One of the exploratory techniques discussed by Tukey (1977) is called a *stem-and-leaf display,* which is similar to a simple frequency distribution.

2. When dealing with double-digit numbers, the possible first (i.e., leading) digits are written in a column, with the lowest on top, to form the stem. The leaves are the second (i.e., trailing) digits that actually appear in the group of scores. All of the trailing digits (even repetitions) are written in a row to the right of the leading digit to which they correspond.

3. A stem-and-leaf display has the advantages of preserving all of the raw score information, as well as serving as a histogram on its side (the length of the row of trailing digits shows how often the leading digit has occurred).

4. If the stem is too short, each leading digit can be used twice, once for low trailing digits (i.e., 0 to 4) and once for high trailing digits (i.e., 5 to 9). If the scores are in the three-digit range, it is possible to use the first digits for the stem, the second digits for the leaves, and to ignore the third digits entirely.

5. Formulas based on linear interpolation were presented for finding exact percentile ranks and percentiles.

Definitions of Key Terms

Array A list of scores in numerical order (usually from highest to lowest).

Simple frequency distribution A list of all possible scores from the highest to the lowest in a group, together with a frequency count for each possible score.

Cumulative frequency distribution A distribution in which the entry for each score (or interval) is the total of the frequencies *at or below* that score (or interval).

Cumulative relative frequency distribution A cumulative distribution in which each cumulative frequency is expressed as a proportion of the total N.

Relative frequency distribution A distribution in which the frequency of a score (or interval) is expressed as a proportion of the total number of scores in the distribution (i.e., each f is divided by $N = \Sigma f$).

Cumulative percentage frequency distribution A cumulative distribution in which each cumulative frequency is expressed as a percentage of the total N (i.e., the cumulative relative frequency is multiplied by 100).

Percentile rank (PR) The PR of a score is the percentage of the scores in the distribution that are at or below that score (i.e., the PR is the cumulative percentage associated with that score).

Percentile A score that has a specified PR (e.g., the 25th percentile is the score whose PR is 25). Of most interest are deciles (e.g., 10%, 20%, etc.) or quartiles (25%, 50%, or 75%).

Bar graph A graph in which vertical bars represent values corresponding to a discrete variable and adjacent bars are kept separate. In a graph representing a frequency distribution, the heights of the bars are proportional to the frequencies of the corresponding values.

Real limits If the values in a distribution represent values of a continuous variable, the *lower real limit* is one-half of a measurement unit below the lower apparent limit, and the *upper real limit* is one-half of a measurement unit above the upper apparent limit.

Frequency histogram A bar graph of a frequency distribution in which the variable of interest is considered to be continuous, and therefore adjacent bars touch (each bar extends horizontally from the lower to the upper real limit of the score or interval represented).

Frequency polygon A graph of a frequency distribution for a continuous variable, in which a point is drawn above the *midpoint* of each interval (or score) at a height proportional to the corresponding frequency. These points are connected by straight lines, which are connected to the horizontal axis at both ends of the distribution.

Cumulative frequency polygon Also called an *ogive*; a frequency polygon in which the points represent cumulative frequencies and are drawn above the upper real limit for each interval (or score).

Theoretical distribution A distribution based on a mathematical formula, which a particular frequency polygon may or may not resemble.

Class interval A range of scores that represents a subset of all the possible scores in a group.

Grouped frequency distribution A list of (usually equal-sized) class intervals that do not overlap, and together include all of the scores in a distribution, coupled with the frequency count associated with each class interval.

Apparent limits The lowest score value included in a class interval is called the *lower apparent limit,* and the highest score value is the *upper apparent limit.*

Exploratory data analysis (EDA) A set of techniques devised by J. W. Tukey (1977) to use when inspecting data before attempting to summarize or draw inferences.

Stem-and-leaf display One of Tukey's EDA methods; a stem-and-leaf display resembles a frequency distribution but retains the raw scores. Leading digits form the vertical stem and trailing digits form the horizontal rows (or leaves).

Linear interpolation A method for finding an intermediate value for a variable, based on the assumption that the variable changes in a linear fashion from one point to another (e.g., when trying to find the PR of a score within an interval, it is assumed that the frequency of that interval is spread linearly across the entire interval).

Key Formulas

Percentile ranks by linear interpolation (after a percent frequency and a cumulative percent frequency distribution have been constructed):

$$PR = cpf_{LRL} + \left(\frac{X - LRL}{i} \right) pf_X \qquad \text{Formula 2.1}$$

Percentile ranks by linear interpolation (after a cumulative frequency distribution has been constructed):

$$PR = \left[cf_{LRL} + \left(\frac{X - LRL}{i} \right) f_X \right] \frac{100}{N} \qquad \text{Formula 2.2}$$

Percentiles by linear interpolation (after a percent frequency and a cumulative percent frequency distribution have been constructed):

$$X = LRL + \left(\frac{PR - cpf_{LRL}}{pf_X} \right) i \qquad \text{Formula 2.3}$$

Percentiles by linear interpolation (after a cumulative frequency distribution has been constructed and the desired percentile has been converted to a cumulative frequency):

$$X = LRL + \left(\frac{cf_p - cf_{LRL}}{f_X} \right) i \qquad \text{Formula 2.4}$$

3

C H A P T E R

CONCEPTUAL FOUNDATION

You will need to use the following from previous chapters:

Symbols
Σ: Summation sign

Concepts
Frequency histograms and polygons

Procedures
Rules for using the summation sign

In Chapter 2 I began with an example in which I wanted to tell a class of 25 students how well the class had performed on a diagnostic quiz and make it possible for each student to evaluate how his or her score compared to the rest of the class. As I demonstrated, a simple frequency distribution, especially when graphed as a histogram or a polygon, displays the scores at a glance, and a cumulative percentage distribution makes it easy to find the percentile rank for each possible quiz score. However, the first question a student is likely to ask about the class performance is, "What is the average for the class?" And it is certainly a question worth asking. Although the techniques described in Chapter 2 provide much more information than does a simple average for a set of scores, the average is usually worth reporting. In trying to find the average for a group of scores, we are looking for one spot that seems to be the center of the distribution. Thus, we say that we are seeking the **central tendency** of the distribution. But, as the expression "central tendency" implies, there may not be a single spot that is clearly and precisely at the center of the distribution. In fact, there are several procedures for finding the central tendency of a group of scores, and which procedure is optimal can depend on the shape of the distribution involved. In this chapter I will describe the common ways that central tendency can be measured and discuss the manner in which the shape of a distribution can influence the choice of a measure for central tendency.

The Arithmetic Mean

When most students ask about the average on an exam, they have in mind the value that is obtained when all of the scores are added and then divided by the total number of scores. Statisticians call this value the **arithmetic mean,** and it is symbolized by the Greek letter μ (mu, pronounced "myoo") when it refers to the mean of a population. Beginning with Chapter 6 we will also be interested in the mean for a sample, in which case the mean is symbolized either by a bar over the letter representing the variable (e.g., $\overline{X}$, called "X bar"), or by the capital letter M, for mean. There are other types of means, such as the harmonic mean (which will be introduced in Chapter 10) and the geometric mean, but the arithmetic mean is by far the most commonly used. Therefore, when I use the term "mean" without further specification, it is the arithmetic mean to which I am referring. The arithmetic mean, when applicable, is undoubtedly the most useful measure of central tendency. However, before we consider the many statistical properties of the mean, we need to consider two lesser known, but nonetheless useful, measures of central tendency.

The Mode

Often the main purpose in trying to find a measure of central tendency is to characterize a large group of scores by one value that could be considered the most typical of the group. If you want to know how smart a class of students is (perhaps because you have to prepare to teach them), you would like to know how smart the "typical" student in that class is. If you want to know how rich a country is, you might want to know the annual income of a "typical" family. The simplest and crudest way to define the most typical score in a group is in terms of which score occurs with the highest frequency. That score is called the **mode** of the distribution.

The mode is easy to find once you have constructed a frequency distribution. Take another look at Table 2.3, the simple frequency distribution of quiz scores in Chapter 2, Section A. To find the mode, we need only look down the frequencies (f) column and find the highest number. In Table 2.3, the highest frequency is 7; the score that corresponds to the highest frequency is the mode. In this case, a quiz score of 6 is the mode, because it is the most frequently occurring score in the group. It is perhaps even easier to identify the mode when a frequency distribution has been displayed as a histogram or a graph. Simply look for the highest bar in the histogram, or the highest point in the polygon—the score that is directly below that highest bar or point is the mode.

The Mode of a Grouped Distribution

The mode is defined in the same way for a grouped distribution as for a simple distribution, except with a grouped distribution the mode is the most frequently occurring *interval* (or the midpoint of that interval) rather than a single score. For instance, in Table 2.12 (Chapter 2, Section B), the highest frequency (6) corresponds to the interval 84–86, the midpoint of which (85) is therefore considered the mode. However, when the same data are regrouped starting with a different lowest interval (but with the same interval width), as in Table 2.13, there is no longer a single mode. The highest frequency (5) occurs twice in Table 2.13— once for the interval 85–87 and once for 79–81. This is why I referred to that distribution as *bimodal*. Clearly, a drawback to using the mode with grouped distributions is that the mode can depend on the way the scores are grouped.

Advantages and Disadvantages of the Mode

In general, the mode is not considered a very reliable measure of central tendency. Consider the simple frequency distribution in the two lefthand columns of Table 3.1. The mode of this distribution is 7, because that score has the highest frequency (6). However, if just one of the students who scored a 7 was later found really to have scored a 4, the mode would move from a score of 7 to a score of 4 (as shown in the two righthand columns of Table 3.1). Naturally, we would like to see more stability in our measures of central tendency.

When dealing with interval/ratio scales, it seems that the only advantage of the mode as a measure of central tendency is the ease with which it can be found. In the age of computers, this is not a significant advantage. However, the mode has the unique advantage that it can be found for any kind of measurement scale. In fact, when dealing with nominal scales, other measures of central tendency (such as the mean) cannot be calculated; the mode is the *only* measure of central tendency in this case. For instance, suppose you are in charge of a psychiatric emergency room, and you want to know the most typical diagnosis of a patient coming for emergency treatment. You cannot take the "average" of 20 schizophrenics, 15 depressives, etc. All you can do to assess central tendency is to find the most frequent diagnosis (e.g., schizophrenia may be the *modal* diagnosis in the psychiatric emergency room).

One more use of the mode that is worth noting is in describing the shape of a distribution. It is very common for a smooth, or nearly smooth, distribution to have only one mode; such distributions are described as **unimodal** (see Figure 3.1). On the other hand, a distribution that contains two distinct subgroups (e.g., men and women measured on strength of grip) may have two different modes (one for each subgroup), as shown in Figure 3.2. Such a distribution is described as **bimodal.** It is possible for a smooth distribution to have three or more modes, but this is very unlikely to occur with a large set of real data.

Table 3.1	*X*	f	*X*	f
	10	1	10	1
	9	0	9	0
	8	3	8	3
	7	*6*	7	5
	6	5	6	5
	5	5	5	5
	4	5	4	*6*
	3	2	3	2
	2	1	2	1
	1	1	1	1
	0	1	0	1

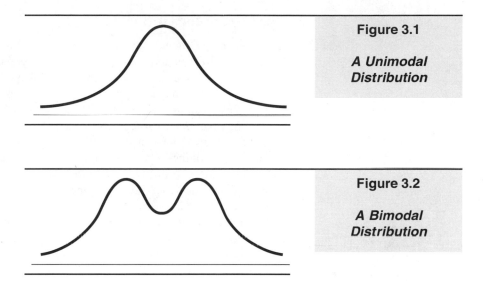

Figure 3.1

A Unimodal Distribution

Figure 3.2

A Bimodal Distribution

The number of modes is only one aspect of a distribution's shape. Other ways to describe the shape of a distribution will be explained below.

The Median

If you are looking for one score that is in the middle of a distribution, a logical score to focus on is the score that is at the 50th percentile (i.e., a score whose PR is 50). Half of all the scores in the distribution will be less than this value (by definition), and half will be greater. The score that is the 50th percentile has a special name; it is called the **median.** The median is a very useful measure of central tendency, as you will see, and it is very easy to find. If the scores in a distribution are arranged in an array (i.e., in numerical order), and there are an *odd* number of scores, the median is literally the score in the middle. If there are an *even* number of scores, the median is the average of the two middle scores (assuming that the scores are on an interval/ratio scale).

For instance, it is easy to find the median for the quiz scores used as an example in Chapter 2, Section A. Look at Table 2.2, in which the scores are arranged in order, and find the middle score. Because there are a total of 25 scores, the middle score is the 13th—with 12 scores above it and 12 scores below it. Thus, the middle score in Table 2.2 is 6, which is therefore the median (as well as the mode, in this case). If the lowest score in Table 2.2 were dropped, there would be an even number of scores ($N = 24$), and therefore *two* scores in the middle (the 12th and the 13th). The median would then be the average of the two middle scores, which are 6 and 7—so the median in this case would equal $(6 + 7)/2 = 6.5$. To find the median for a grouped distribution, you can use the procedures for finding any percentile, either by formula or by graph.

The Median for Ordinal Data

Unlike the mode, the median cannot be found for a nominal scale, because the values (e.g., different psychiatric diagnoses) do not have any inherent order (e.g., we cannot say which diagnoses are "above" manic-depressive and which "below"). However, if the values can be placed in a meaningful order, you are then dealing with an ordinal scale, and the median *can* be found for ordinal scales. For example, suppose that the coach of a debating team has rated the effectiveness of members of the team on a scale from 1 to 10. The data in Table 2.3 could represent those ratings. Once the scores have been placed in order, the median is the middle score. (Unfortunately, if there are two middle scores and you are dealing with ordinal data, it is not proper to average the two scores, though this is sometimes done anyway as an approximation.) Even though the ratings from 1 to 10 cannot be considered equally spaced, we can assume, for example, that the debaters rated between 1 and 5 are all considered less effective than one who is rated 6. Thus, we can find a ranking or rating such that half the group is below it and half above, except for those who are tied with the middle score (or one of the two middle scores). The median is more informative if there are not many ties. In general, having many tied scores diminishes the usefulness of an ordinal scale, as will be made clear in Part VI of this text.

The Median and the Area of a Distribution

As I mentioned above, the mode is particularly easy to find from a frequency polygon—it is the score that corresponds to the highest point. The median also bears a simple relationship to the frequency polygon. If a vertical line is drawn at the median on a frequency polygon, so that it extends from the horizontal axis until it meets the top of the frequency polygon, the area of the polygon will be divided in half. This is because the median divides the total number of scores in half, and the area of the polygon is proportional to the number of scores.

To better understand the relation between the frequency of scores and the area of a frequency polygon, take another look at a frequency histogram (see Figure 3.3). The height of each bar in the histogram is proportional to the frequency of

Figure 3.3

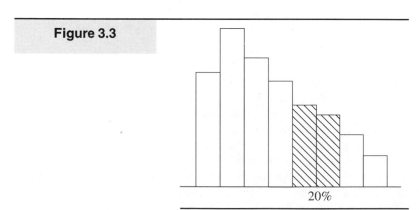

20%

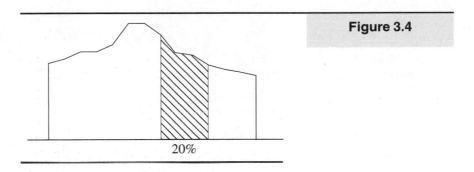

Figure 3.4

20%

the score or interval that the bar represents. (This is true for the simplest type of histogram, which is the only type we will consider.) Because the bars all have the same width, the area of each bar is also proportional to the frequency. You can imagine that each bar is a building and that the taller the building, the more people live in it. The entire histogram can be thought of as the skyline of a city; you can see at a glance where (in terms of scores on the X axis) the bulk of the people live. All the bars together contain all the scores in the distribution. If two of the bars, for instance, take up an area that is 20% of the total, then you know that 20% of the scores fall in the intervals represented by those two bars.

A relationship similar to the one between scores and areas of the histogram bars can be observed in a frequency polygon. The polygon encloses an area that represents the total number of scores. If you draw two vertical lines within the polygon, at two different values on the X axis, you enclose a smaller area, as shown in Figure 3.4. Whatever proportion of the total area is enclosed between the two values (.20 in Figure 3.4) is the proportion of the scores in the distribution that fall between those two values. We will use this principle to solve problems in Chapter 5. At this point I just wanted to give you a feeling for why a vertical line drawn at the median divides the distribution into two equal areas.

Skewed Distributions

One advantage of the median over the mean is that the median can be used with ordinal scales. The calculation of the mean, on the other hand, rests on the assumption of equal intervals, which may not be true for an ordinal scale. Both the mean and the median, however, can be found when working with an interval/ ratio scale; which is the better measure of central tendency depends in part on the shape of the distribution. I have already discussed one aspect of a distribution's shape: the number of modes. Another aspect of shape that is particularly relevant to choosing the median or the mean as a measure of central tendency is *skewness.* A distribution is **skewed** if the bulk of the scores are concentrated on one end of the scale, with relatively few scores on the other end. When graphed as a frequency polygon, a skewed distribution will look something like those in

Figure 3.5

Skewed Distributions

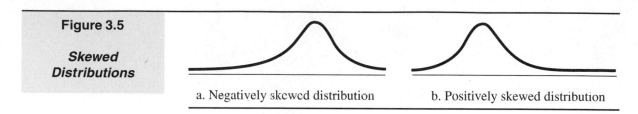

a. Negatively skewed distribution　　　　b. Positively skewed distribution

Figure 3.5. The distribution in part a of Figure 3.5 is said to be **negatively skewed,** whereas the one in part b is called **positively skewed.**

To remember which shape involves a negative skew and which a positive skew, think of the **tail of the distribution** as a long, thin skewer. If the skewer points to the left (in the direction in which the numbers eventually become negative), the distribution is negatively "skewered" (i.e., negatively skewed); if the skewer points to the right (the direction in which the numbers become positive), the distribution is positively skewed.

The description above of the relation between the area of a polygon and the proportion of scores can help you understand the skewed distribution. A section of the tail with a particular width will have a relatively small area, and therefore relatively few scores. A section with the same width in the thick part of the distribution (the "hump") will have a lot more area, and thus a lot more scores.

The Median and Mean of a Skewed Distribution

An important advantage of the median is in representing the central tendency of a skewed distribution. First, consider a symmetrical, unimodal distribution, as depicted in part a of Figure 3.6. The median is exactly in the center, right at the mode. Because the distribution is symmetrical, there is the same amount of area on each side of the mode. Now let's see what happens when we turn this distribution into a positively skewed distribution by adding a few high scores, as shown in part b of Figure 3.6. Adding a small number of scores on the right increases the area on the right slightly. In order to have the same area on both sides, the median must move to the right a bit. Notice, however, that the median does not have to move very far along the X axis. Because the median is in the thick part of the distribution, moving only slightly to the right shifts enough area to compensate for the few high scores that were added. (See how the shaded area on the right end of the graph in part b of Figure 3.6 equals the shaded area between the median and the mode.) Thus, the median is not strongly affected by the skewing of a distribution, and that can be an advantage in describing the central tendency of a distribution.

In fact, once you have found the median of a distribution, you can take a score on one side of the distribution and move it much further away from the median. As long as the score stays on the same side of the median, you can move it out as

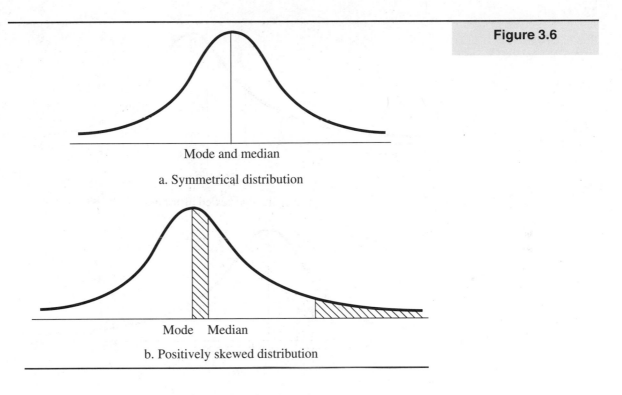

Figure 3.6

Mode and median

a. Symmetrical distribution

Mode   Median

b. Positively skewed distribution

far as you want — the median will not change its location. This is *not* true for the mean. The mean is affected by the numerical value of every score in the distribution! Consequently the mean will be pulled in the direction of the skew, as illustrated in parts a and b of Figure 3.7. The median does not move at all unless a score crosses the median (or a new score is added to one side of the distribution). When the distribution is negatively skewed (part a of the figure), the mean will be to the left of (i.e., more negative than) the median, whereas the reverse will be true for a positively skewed distribution (part b of the figure). Conversely, if you find both the mean and the median for a distribution, and the median is higher (i.e., more positive) the distribution has a negative skew; if the mean is higher, the skew is positive. If the mean and median are the same, the distribution is symmetric around its center.

Examples of Skewed Distributions

Let us consider an example of a skewed distribution for which choosing a measure of central tendency has practical consequences. There has been much publicity in recent years about the astronomical salaries of a few superstar athletes. However, the bulk of the professional athletes in any particular sport (we'll use baseball as our example) are paid a more reasonable salary. Therefore, the

Figure 3.7

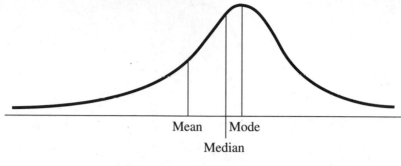

a. Negatively skewed distribution

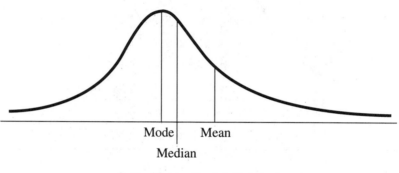

b. Positively skewed distribution

distribution of salaries for major league baseball players is positively skewed, as shown in Figure 3.8. When the Players' Association is negotiating with management, guess which measure of central tendency each side prefers to use? Of course, management points out that the average (i.e., mean) salary is already quite high. The players can point out, however, that the high mean is caused by the salaries of relatively few superstars, and that the mean salary is not very

Figure 3.8

Annual Salaries of Major League Baseball Players

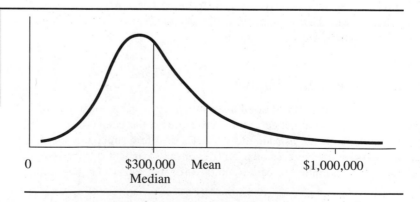

representative of the majority of players. The argument of the Players' Association would be that the median provides a better representation of the salaries of the majority of players. In this case it seems that the players are correct. However, the mean has some very useful properties, which will be explored in detail in Section B.

Quite a few psychological variables often produce skewed distributions. Among those with a positive skew, reaction time (RT) is one of the most common. In a typical RT experiment, the subject waits for a signal before hitting a response button; the time between the onset of the signal and the depression of the button is recorded as the reaction time. There is a physiological limit to how quickly a subject can respond in such a situation. (The limit is longer if the subject must respond differently to different signals.) After the subjects have had some practice, most of their responses will cluster just above this limit; a few responses will take considerably longer. The occasional long RTs reflect momentary fatigue or inattention and result in the positive skew.

Floor and Ceiling Effects

Positively skewed distributions are likely whenever there is a limit on values of the variable at the low end but not the high end, or when the bulk of the values are clustered near the lower limit rather than the upper limit. This kind of one-sided limitation is called a **floor effect.** The RT measurements mentioned above represent one example of a floor effect; the RT distribution would have a shape similar to the distribution shown in part b of Figure 3.5. Another example is measurements of clinical depression in a large random group of college students. A third example is scores on a test that is too difficult for the group being tested; many scores would be near zero and there would be only a few high scores.

The opposite of a floor effect is, not surprisingly, a **ceiling effect,** which occurs when the scores in a distribution approach an upper limit but are not near any lower limit. A ceiling effect will result in a negatively skewed distribution. A rather easy exam will show a ceiling effect, with most scores near the maximum and relatively few scores (e.g., only those of students who didn't study or didn't do their homework) near the low end (as depicted in Figure 3.9). For

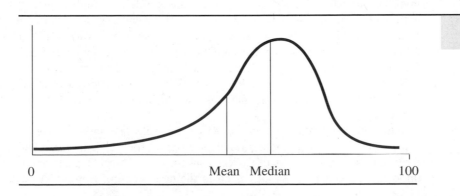

Figure 3.9

example, certain tests are given to patients with brain damage or chronic schizo-phrenia to assess their orientation to the environment, knowledge of current events, etc. Giving such a test to a random group of adults will produce a nega-tively skewed distribution (such as the one shown in part a of Figure 3.5). For descriptive purposes, the median is often preferred to the mean whenever either a floor or a ceiling effect is exerting a strong influence on the distribution.

Undeterminable Scores

Although theoretically a variable may have a limit on one end but no limit on the other end, practical considerations will invariably impose some limit on both ends. For instance, whereas RT is theoretically limited only on the low end, an experimenter will not wait forever for a subject to respond. Usually some arbi-trary limit is imposed on the high end—for example, "If the subject does not respond after 10 seconds, record 10 s as the RT and go on to the next trial." Calculating the mean would be misleading, however, if any of these 10-second responses were included. First, these 10-second responses are really **undeter-minable scores**—the researcher doesn't know how long it would have taken for the subject to respond. Second, averaging in a few 10-second responses with the rest of the responses, which are less than 1 second, can produce a mean that misrepresents the results. On the other hand, the median will not change if the response is recorded as 10 s or 100 s (assuming that the median is less than 10 s to begin with). Thus, when some of the scores are undeterminable, the median has a strong advantage over the mean as a descriptive statistic.

Open-Ended Categories

Sometimes when data are collected for a study, some of the categories are delib-erately left **open-ended.** For instance, in a study of AIDS awareness, subjects might be asked how many sexual partners they have had in the last 6 months, with the highest category being "ten or more." Once a subject has had at least ten different partners in the given period, it may be considered relatively unim-portant to the study to determine exactly how many more than ten partners were involved. (Perhaps the researchers fear that the accuracy of numbers greater than ten could be questioned.) However, this presents a problem for calculating the mean. It would be misleading to average in the number ˙10 when the subject reported having ten *or more* partners. Of course, this is not a problem for finding the median; all of the subjects reporting ten or more partners would simply be tied for the highest position in the distribution. (A problem in determining the median would arise only if as many as half the subjects reported ten or more partners.)

   Although the median provides a very reasonable description of central ten-dency when the distribution is strongly skewed or open-ended, researchers pre-fer to use the mean because of its very convenient statistical properties (the most important of which will be discussed in the next section). However, to use the mean in the situations described above, the researcher would need to make certain assumptions or actually transform the data in some systematic way (see Section C).

Exercises

* 1. It is not possible for more than half of the distribution to be above the a) mode; b) median; c) mean; d) mean, median, or mode.

2. Which of the following is not appropriate for describing the central tendency of ordinal data? a) Mode b) Median c) Mean d) All of the above are appropriate.

* 3. A negatively skewed distribution can arise when you are dealing with a) a floor effect; b) a ceiling effect; c) a reaction time experiment; d) a very difficult exam.

* 4. Select the measure of central tendency (mean, median, or mode) that would be most appropriate for describing each of the following hypothetical sets of data.
　a) Heart rates for a group of women before they start their first aerobics class
　b) Religious preferences of delegates to the United Nations
　c) Annual incomes of baseball players, with the highest category being "$1 million or more"
　d) Types of phobias exhibited by patients attending a phobia clinic

　e) Amounts of time subjects spend solving a classic cognitive problem; some subjects are unable to solve it
　f) Height in inches for a group of boys in the first grade

5. Describe a realistic situation in which you would expect to obtain each of the following. a) A negatively skewed distribution; b) A positively skewed distribution; c) A bimodal distribution

* 6. Referring to the data from exercise 2A1, which rating category is the mode of the distribution? Which category comes closest to being the median?

7. Referring to the data from exercise 2A2, which score is the mode of the distribution? Without grouping the frequencies, can you determine which score is at the median?

* 8. A midterm exam was given in a large introductory psychology class. The median score was 85, the mean was 81, and the mode was 87. What kind of distribution would you expect from these exam scores?

Calculating the Mean

In Section A the arithmetic mean was defined informally as the sum of all of the scores divided by the number of scores added. It is more useful to express the mean as a formula in terms of the summation notation that was presented in the first chapter. The formula for the *population mean* is

$$\mu = \frac{\sum_{i=1}^{N} X_i}{N} = \frac{1}{N} \sum_{i=1}^{N} X_i$$

which tells you to sum all the Xs from X_1 to X_N before dividing by N. (You would get the same answer if you divided each X_i by N and then added the fractions,

BASIC STATISTICAL PROCEDURES

but it would be a lot more work.) If you simplify the summation notation by leaving off the indexes (as I promised I would in Chapter 1), you end up with Formula 3.1:

$$\mu - \frac{\Sigma X}{N}$$ Formula 3.1

The procedure for finding the mean of a sample is exactly the same as the procedure for finding the mean of a population, as shown by Formula 3.2 for the sample mean:

$$\overline{X} = \frac{\Sigma X}{N}$$ Formula 3.2

Suppose that the following set of scores represents measurements of clinical depression in seven normal college students: 0, 3, 5, 6, 8, 8, 9. We will use Formula 3.1 to find the mean: $\mu = 39/7 = 5.57$. (If we had considered this set of scores a sample, we would have used Formula 3.2 and of course obtained the same answer, which would have been referred to by $\overline{X}$, the symbol for the sample mean.) Note that the mean depression score is a little less than the median (6), suggesting that the distribution has a slight negative skew. To appreciate the sensitivity of the mean to extreme scores, imagine that all of the students have been measured again and all have attained the same rating as before, except for the student who had scored 9. This student has become clinically depressed, and therefore receives a new rating of 40. Thus, the new set of scores is 0, 3, 5, 6, 8, 8, 40. The new mean is $\mu = 70/7 = 10$. Note that the median is still 6, but now the skew is positive and quite evident.

The Mean of a Frequency Distribution

As another example, let us find the mean for the data in Table 2.1. The sum of the 25 scores is 164, so $\mu = 164/25 = 6.56$. The mean is very close to the median, because the distribution is nearly symmetrical. We obtained the sum of the scores by simply adding all the numbers in Table 2.1. However, if a simple frequency distribution has already been constructed (as in Table 2.3), there is a shortcut procedure for finding ΣX, and therefore for finding μ. First, a column labeled fX, which consists of each score multiplied by its frequency, is added to Table 2.3, which then becomes Table 3.2. Notice that ΣfX is the sum of all the Xs. The mean of a frequency distribution is given by Formula 3.3:

$$\mu = \frac{\Sigma fX}{\Sigma f}$$ Formula 3.3

Formula 3.3 represents a shortcut because instead of adding the number 6, for instance, a total of seven times, you multiply 6 by 7. This procedure can also be used to find the mean of a grouped frequency distribution; in that case, the

X	f	fX		
10	2	20		**Table 3.2**
9	2	18		
8	5	40		
7	3	21		
6	7	42		
5	1	5		
4	4	16		
3	0	0		
2	1	2		
	25	$\Sigma fX = 164$		

midpoint of each interval would be used for the value of X, in finding fX for that interval. Unfortunately, when you apply this procedure to a grouped distribution, the mean that you find will probably be a little off, because the midpoint of an interval is only an approximation of the scores in that interval, and the approximation will change with different choices of class intervals. If the raw data are available, the mean should be calculated directly from the raw scores. With the availability of computers for data analysis, you probably won't need the computational shortcut of using a grouped frequency distribution to find the mean, and therefore you won't have to sacrifice the accuracy of finding the mean directly.

The Weighted Mean

The statistical procedure for finding the **weighted mean,** better known as the *weighted average,* has many applications in statistics as well as in real life. I will begin this explanation with the simplest possible example. Suppose a professor who is teaching two sections of statistics has given a diagnostic quiz at the first meeting of each section. One class has 30 students who score an average of 7 on the quiz, whereas the other class has only 20 students who average an 8. The professor wants to know the average quiz score for all of the students taking statistics (i.e., both sections combined). The simplest approach would be to take the average of the two section means (i.e., 7.5), but as you have probably guessed, this would give you the wrong answer. The correct thing to do is to take the *weighted* average of the two section means. It's not fair to count the class of 30 equally with the class of 20 (imagine giving equal weights to a class of 10 and a class of 100!). Instead, the larger class should be given more *weight* in finding the average of the two classes. The amount of weight should depend on the class size, as it does in Formula 3.4 below. Note that Formula 3.4 could be used to average together any number of class sections or other groups, where N_i is the number of scores in one of the groups and $\bar{X}_i$ is the mean of that group. The formula uses the symbol for the sample mean, because weighted averages are often applied to samples in order to make better guesses about populations.

$$\overline{X}_w = \frac{\sum N_i \overline{X}_i}{\sum N_i} = \frac{N_1\overline{X}_1 + N_2\overline{X}_2 + \cdots}{N_1 + N_2 + \cdots} \qquad \text{Formula 3.4}$$

We can apply Formula 3.4 to the means of the two statistics sections.

$$\overline{X}_w = \frac{(30)(7) + (20)(8)}{30 + 20} = \frac{210 + 160}{50} = \frac{370}{50} = 7.4$$

Notice that the weighted mean (7.4) is a little closer to the mean of the larger class (7) than to the mean of the smaller class. The weighted average of two groups will always be between the two group means and closer to the mean of the larger group. For more than two groups, the weighted average will be somewhere between the smallest and the largest of the group means.

Formula 3.4 for the weighted mean may remind you of Formula 3.3 for finding the mean of a grouped distribution. It should, because the same principle is involved in both cases. The mean of a simple frequency distribution is really just a weighted average of the possible scores, where the frequency of a score is its "weight."

The weighted average procedure is a kind of shortcut; to understand the procedure fully you should take a look at the long version. In the case of the two sections of the statistics course, the weighted average indicates what the mean would be if the two sections were combined into one large class of 50 students. To find the mean of the combined class directly, you would need to know the sum of scores for the combined class and then to divide it by 50. To find $\sum X$ for the combined class, you would need to know the sum for each section. You already know the mean and N for each section, so it is easy to find the sum for each section. First, take another look at Formula 3.2 for the sample mean.

$$\overline{X} = \frac{\sum X}{N}$$

If you multiply both sides of the equation by N, you get $\sum X = N\overline{X}$. (Note that it is also true that $\sum X = N\mu$; we will use this equation below.) You can use this new equation to find the sum for each statistics section. For the first class, $\sum X_1 = (30)(7) = 210$, and for the second class, $\sum X_2 = (20)(8) = 160$. Thus, the total for the combined class is $210 + 160 = 370$, which divided by 50 is 7.4. Of course, this is the same answer we obtained with the weighted average formula. What the weighted average formula is actually doing is finding the sum for each group, adding all the group sums to find the total sum, and then dividing by the total number of scores from all the groups.

One of the most common applications of the weighted average, at least for college students, is for finding a grade point average (GPA). Suppose that a student earned five As, three Bs, and two Cs in his or her first year in college. It's clear that this student has better than a B average—perhaps a B+—but to find the GPA more precisely we need to convert the letter grades to numerical values. The most common scheme is A = 4, B = 3, C = 2, D = 1, and F = 0. The weights are the number of credits earned in each grade category. Consider the simplest case, in which all courses carry the same number of credits, so the weights are just the number of courses in each grade category. You can use Formula 3.4 for a weighted average to find the GPA:

$$\overline{X}_w = \frac{(5)(4.0) + (3)(3.0) + (2)(2.0)}{5 + 3 + 2} = \frac{20 + 9 + 4}{10} = \frac{33}{10} = 3.3$$

If grades such as A− or B+ are included, they must also be converted to numerical equivalents—for example, A− = 3.7, B+ = 3.3, etc. If all courses are not worth the same number of credits, then the actual number of credits must be used as the weight for each grade category, instead of the number of courses.

Properties of the Mean

1. *If a constant is added (or subtracted) to every score in a distribution, the mean is increased (or decreased) by that constant.* For instance, if the mean of a midterm exam is only 70, and the professor decides to add 10 points to every student's score, the new mean will be 70 + 10 = 80 (i.e., $\overline{X}_{new} = \overline{X}_{old} + C$). The rules of summation presented in Chapter 1 prove that if you find the mean after adding a constant to every score (i.e., $\Sigma(X + C)/N$), the new mean will equal $\overline{X} + C$. First, note that $\Sigma(X + C) = \Sigma X + \Sigma C$ (according to Summation Rule 1A). Next, note that $\Sigma C = NC$ (according to Summation Rule 3). So,

$$\frac{\Sigma(X + C)}{N} = \frac{\Sigma X + \Sigma C}{N} = \frac{\Sigma X + NC}{N} = \frac{\Sigma X}{N} + \frac{NC}{N} = \frac{\Sigma X}{N} + C = \overline{X} + C$$

(A separate proof for subtracting a constant is not necessary; the constant being added could be negative without changing the proof.)

2. *If every score is multiplied (or divided) by a constant, the mean will be multiplied (or divided) by that constant.* For instance, suppose that the average for a statistics quiz is 7.4. Now imagine that the professor wants the quiz to count as one of the exams in the course. (Also imagine that the class has voted to go along with this.) The professor multiplies each student's quiz score by 10 (so that scores are on a scale of from 1 to 100 instead of from 1 to 10). The mean is also multiplied by 10, so the mean of the new "exam" scores is 7.4 · 10 = 74.

We can prove that this property also holds in general. The mean of the scores after multiplication by a constant is $(\Sigma CX)/N$. By Summation Rule 2A, you know that $\Sigma CX = C\Sigma X$, so

$$\frac{\Sigma CX}{N} = \frac{C\Sigma X}{N} = C\frac{\Sigma X}{N} = C\overline{X}$$

There is no need to prove that this property also holds for dividing by a constant, because the constant in the above proof could be less than 1.0 without changing the proof.

   3. *The sum of the deviations from the mean will always equal zero.* To make this idea concrete, imagine that a group of waiters has agreed to share all of their tips. At the end of the evening, each waiter puts his or her tips in a big bowl; the money is counted and then divided equally among the waiters. Because the sum of all the tips is being divided by the number of waiters, each waiter is actually getting the mean amount of tip. Any waiter who had pulled in more than the average tip would lose something in this deal, whereas any waiter whose tips for the evening were initially below the average would gain. These gains or losses can be expressed symbolically as deviations from the mean, $X_i - \mu$, where X_i is the amount of tips collected by the ith waiter and μ is the mean of the tips for all the waiters. The property above can be stated in symbols as $\sum(X_i - \mu) = 0$.

   In terms of the waiters, this property says that the sum of the gains must equal the sum of the losses. This makes sense—the gains of the waiters who come out ahead in this system come entirely from the losses of the waiters who come out behind. Note, however, that the *number* of gains does not have to equal the *number* of losses. For instance, suppose that ten waiters decided to share tips and that nine waiters receive $10 each and a tenth waiter gets $100. The sum will be $(9 \cdot 10) + 100 = \$90 + \$100 = \$190$, so the mean is $\$190/10 = \$19$. The nine waiters who pulled in $10 each will each gain $9, and one waiter will lose $81 ($100 − $19). But although there are nine gains and one loss, the total amount of gain $(9 \cdot \$9 = \$81)$ equals the total amount of loss $(1 \cdot \$81 = \$81)$. This demonstrates that in a positively skewed distribution, the majority of scores can be below the mean (the reverse is true for a negatively skewed distribution).

   The property above can also be proven to be generally true. First, note that $\sum(X_i - \mu) = \sum X_i - \sum \mu$, according to Summation Rule 1B. Because μ is a constant, $\sum \mu = N\mu$ (Summation Rule 3), so $\sum(X_i - \mu) = \sum X_i - N\mu$. As demonstrated above in the discussion of the weighted average, $\sum X_i = N\mu$, so $\sum(X_i - \mu) = N\mu - N\mu = 0$.

   4. *The sum of the squared deviations from the mean will be less than the sum of squared deviations around any other point in the distribution.* As a simple example, consider three numbers: 1, 3, 8. The mean of these numbers is $12/3 = 4$. The deviations from the mean (i.e., $X_i - 4$) are -3, -1, and $+4$. (Note that these sum to zero, as required by property 3 above.) The squared deviations are 9, 1, and 16, the sum of which is 26. If you take any other number besides the mean (4), the sum of the squared deviations from that number will be more than 26. For example, the deviations from 3 (which happens to be the median) are -2, 0, $+5$; note that these do *not* sum to zero. The squared deviations are 4, 0, and 25, which sum to 29. Proving that this property is always true is a bit tricky, but the interested reader can find such a proof in a more advanced text (e.g., Hays, 1994). This property, often called the *least-squares property,* is a very important one and will be mentioned in the context of several statistical procedures later in this text.

Choosing a Measure of Central Tendency

In deciding which measure of central tendency to use, the first consideration is the type of measurement scale that is being employed. If the scale has the interval or ratio property, either the mode, median, or mean can be found. The mode has the advantages of being easiest to find (although this is a minor advantage in the age of computers) and of corresponding to an actual score in the distribution (the most frequent score). As you have seen, the median for a grouped frequency distribution, or the mean, is likely to be a fractional value that does not correspond to an actual score (unless you are dealing with a very large population). If the distribution is smooth and has a simple structure (e.g., there is only one "hump"), the mode can be a reasonable indicator of central tendency, but if the distribution is rather bumpy (which is especially likely when the sample is not large), the mode can be very misleading and unreliable. Therefore, when dealing with interval/ratio data, either the mean or the median is preferred.

For descriptive purposes, the median is preferred when the distribution is strongly skewed. Moreover, the mean cannot be calculated if the distribution contains an open-ended category or undeterminable scores at one end, so the median must be used in such cases. In the absence of strong skewing or undeterminable scores, the mean is considered preferable to the median, largely because of the least-squares property described earlier. Furthermore, the mean has the advantage of being affected by the numerical value of every score in the distribution (this can turn into a disadvantage if there are extreme scores). Finally, the mean is usually the most reliable measure of central tendency when extrapolating from a sample to a population; for symmetric distributions, the means of many samples will tend to vary less than the medians of those samples.

Note, however, that whereas the mean minimizes the sum of the squared deviations, it is the median that minimizes the sum of the absolute deviations. In the example involving three numbers (1, 3, 8), the deviations from the mean (i.e., 4) were -3, -1, and $+4$, whereas the deviations from the median (i.e., 3) were -2, 0, and $+5$. I have already shown that the first set of deviations leads to a smaller sum when squared and then added. However, if we take the absolute values of the deviations and then add them, the sum of the deviations from the mean is 8 (3 + 1 + 4), but the sum of the deviations from the median is only 7 (2 + 0 + 5). It is the median that minimizes the sum of the *absolute* deviations. This is a useful property, especially for descriptive purposes, but the least-squares property is even more important, particularly for drawing statistical inferences. Therefore, the mean is the only measure of central tendency that will be discussed in the context of hypothesis testing, later in this text.

If you are dealing with an ordinal scale, then strictly speaking it is not valid to calculate a mean. The calculation of the mean assumes that the numerical values being averaged possess the interval property—that the units are equally spaced. Since the mean is not a valid measure of central tendency, the choice for measuring central tendency when dealing with an ordinal scale is between the mode

Figure 3.10

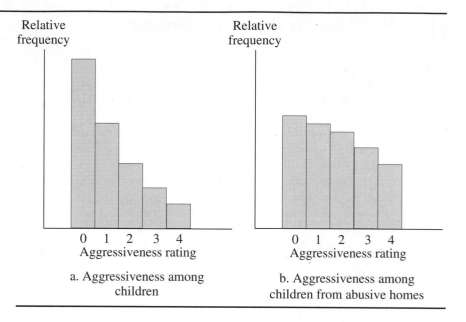

a. Aggressiveness among
children

b. Aggressiveness among
children from abusive homes

and the median. The median is clearly the better choice for nearly any distribution of ordinal data. For example, suppose the aggressiveness of school children being observed in the playground is rated on an ordinal scale (from zero, indicating that a child does not start fights at all, to 4, indicating that the child starts fights often and without provocation). The modal rating is likely to be the lowest rating, as illustrated in part a of Figure 3.10. This is not very informative with respect to the aggressiveness that *does* appear. If a subset of children, known to come from abusive homes, were rated and collected into a separate distribution, the mode might still be the lowest rating, even if there were an obvious increase in the amount of aggressiveness displayed by the group (see part b of Figure 3.10). Whereas the mode is the same and cannot distinguish between the distributions in parts a and b of Figure 3.10, a shift in the median reveals the hypothetical increase in aggressiveness of the abused children.

Finally, if you are dealing with a nominal scale, you have no choice about measures of central tendency—only the mode can be used. The median would require that the categories be put in order, but if the scale is truly nominal, there is no basis for ordering the categories.

Exercises

*1. Find the mean, median, and mode for the following set of numbers: 11, 17, 14, 10, 13, 8, 7, 14.

2. a) Calculate the mean for the following set of numbers: 17, 23, 20, 16, 19, 14, 13, 20. How does this

mean compare with the mean you found for exercise 1 above? Can you see how the data from exercise 1 were transformed to create the data for this exercise?

b) Calculate the mean for the following set of numbers: 51, 69, 60, 48, 57, 42, 39, 60. How does this mean compare with the mean you found for part a? Can you see how the data from part a were transformed to create the data for part b?

* 3. There are three fourth-grade classes at Happy Valley Elementary School. The mean IQ for the ten pupils in the gifted class is 119. For the 20 pupils in the regular class, the mean IQ is 106. Finally, the five pupils in the special class have a mean IQ of 88. Calculate the mean IQ for all 35 fourth-grade pupils.

4. Find the mean, median, and mode for the grouped frequency distribution that you constructed for the data in exercise 2B1. (Use the raw scores to find the mean.)

* 5. Find the mean, median, and mode for the grouped frequency distribution that you constructed for the data in exercise 2B2. (Use the raw scores to find the mean.)

6. A veterinarian is interested in the life span of golden retrievers. She recorded the age at death (in years) of the retrievers treated in her clinic. The ages were 12, 9, 11, 10, 8, 14, 12, 1, 9, 12.
a) Calculate the mean, median, and mode for age at death.
b) After examining her records, the veterinarian determined that the dog that had died at 1 year was killed by a car. Recalculate the mean, median, and mode without that dog's data.

c) Which measure of central tendency in part b changed the most, compared to the values originally calculated in part a?

* 7. A fifth-grade teacher calculated the mean of the spelling tests for his 12 students; it was 8. Unfortunately, now that he is ready to record the grades, one test seems to be missing. The 11 scores he has are 10, 7, 10, 10, 6, 5, 9, 10, 8, 6, 9. Find the missing score. (Hint: You can use property 3 of the mean.)

8. A psychology teacher has given an exam on which the highest possible score is 200 points. The mean score for the 30 students who took the exam was 156. Because there was one question that every student answered incorrectly, the teacher decides to give each student 10 extra points and then divide each score by 2, so the total possible score is 100. What will the mean of the scores be after this transformation?

* 9. A student has earned 64 credits so far, of which 12 credits are As, 36 credits are Bs, and 16 credits are Cs. If A = 4, B = 3, and C = 2, what is this student's grade point average?

10. The mean for a set of scores is 55. If 7 points are subtracted from each score, and then each score is multiplied by 3, what will be the value of the mean for the transformed set of scores?

Box-and-Whisker Plots

In this section I will be discussing the problem of extreme scores in a data set: specifically, how to define a score as extreme and what to do about such a score. You have already seen that extreme scores at one end of a distribution cause skewing and can render the mean misleading as a measure of central tendency. If you would like to use the mean because of its various desirable statistical properties, it can be helpful to identify, and possibly reduce the effect of, extreme scores. One way to identify skewing is by constructing a frequency polygon. However, this is not the best way to notice and identify particular extreme scores.

OPTIONAL MATERIAL

Figure 3.11

A Simple Box-and-Whisker Plot

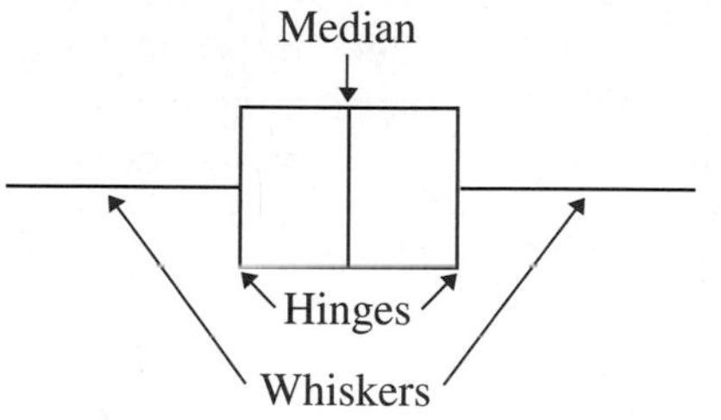

Median

Hinges

Whiskers

You could also use a stem-and-leaf display as described in Chapter 2, Section C, but the statistician who devised that procedure (Tukey, 1977) created another technique that is even more appropriate for identifying extreme scores. This procedure is called a **box-and-whisker plot,** or simply a *boxplot,* and it is a powerful technique for exploratory data analysis.

The boxplot is based on the median, because this measure is not affected by how extreme a score is. The sides of the box, which are called the **hinges,** are the 25th and 75th percentiles; the median, being the 50th percentile, is somewhere in between, but it need not be exactly midway. (Technically, the sides of the box are defined as the median of the scores above the 50th percentile and the median of the scores below the 50th percentile, but with a fairly large sample size, and not a lot of tied scores, there is very little error involved in just using the 25th and 75th percentiles.)

The lengths of the "whiskers" depend, to some extent, on the length of the box; the length of the box is called the **H-spread,** as it is the separation of the two hinges. Figure 3.11 illustrates the basic parts of a box-and-whisker plot. In order to draw the whiskers, it is first necessary to mark the outermost possible limits of the whiskers on either side of the box. These outermost limits are called the **inner fences.** To find each inner fence, you start at one of the hinges and then move away from the box a distance equal to 1.5 times the H-spread, as shown in Figure 3.12. However, the whiskers generally do *not* extend all the way to the inner fences. The end of the upper whisker is drawn at the highest value in the distribution that is not higher than the upper inner fence; this value, called an **adjacent value,** could turn out to be not even close to the inner fence. Similarly,

Figure 3.12

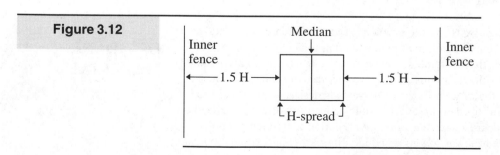

Inner fence

Median

Inner fence

← 1.5 H →   ← 1.5 H →

H-spread

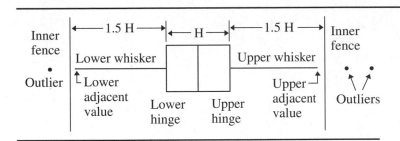

Figure 3.13

the lower whisker ends at the lowest (i.e., leftmost) value that is not past the lower inner fence. Finally, any scores in the distribution that fall beyond the ends of the whiskers are indicated as points in the plot; these points are called **outliers.** Figure 3.13 depicts all of the parts and dimensions of a box-and-whisker plot.

One of the main purposes of a boxplot is to provide a systematic procedure for identifying outliers. Because outliers can so strongly affect the location of the mean in a distribution, it is important to investigate these points. Sometimes an outlier is found to be the result of a gross error in recording or transcribing the data or some procedural mishap, such as a subject's misreading the instructions or becoming ill. Once such outliers are detected, perhaps with the help of a boxplot, they can be removed from the data at once, and any statistics that had been already calculated can be recalculated. Other outliers may represent legitimate data that just happen to be quite unusual (e.g., the score of a subject with a "photographic memory" in a picture recall experiment). If a data point is very unusual, but no error or mishap can be found to account for it, the value should *not* be removed unless you have chosen a procedure for "trimming" your data (see below). The unusual data point is part of the phenomenon being studied. However, you should make an effort to find the underlying causes for extreme values, so that you can study such data separately, or simply avoid collecting such data in future experiments.

The boxplot also lets you see the spread of a distribution, and the symmetry of that spread, at a glance. I will illustrate the construction of a boxplot for data from a hypothetical problem-solving experiment, in which the number of minutes taken to solve the problem is recorded for each subject. The data for 40 subjects are presented in the form of a simple frequency distribution in Table 3.3.

The first step is to find the median. As the 50th percentile was already found in exercise 2C4, we know that the median is about 5.7. In addition, the two hinges have already been calculated: the 25th percentile was found to be 4.1 and the 75th was found to be 7.83. Now that we have the two hinges, we can find the H-spread (or H, for short): H = 7.83 − 4.1 = 3.73. Next, we multiply H by 1.5 to get 5.6; this value is added to the upper hinge (7.83) to give the upper inner fence, which is therefore equal to 7.83 + 5.6 = 13.43. Normally, we would also subtract 5.6 from the lower hinge (4.1), but as this would give a negative number, and time cannot be negative, we will take 0 to be the lower inner fence.

Table 3.3	X	f	X	f
	18	1	9	1
	17	0	8	3
	16	2	7	5
	15	0	6	5
	14	1	5	7
	13	0	4	5
	12	0	3	4
	11	1	2	2
	10	2	1	1

Finally, we have to find the upper and lower adjacent values. The upper adjacent value is the highest score in the distribution that is not higher than 13.43 (i.e., the upper inner fence). From Table 3.3 we can see that 11 is the upper adjacent value. The lower adjacent value, in this case, is simply the lowest score in the distribution, which is 1. The outliers are the points that are higher than 11: 14, 16, 16, and 18; there is no room for outliers on the low side. The boxplot is shown in Figure 3.14; the skew of the distribution can be seen easily from this figure.

Figure 3.14

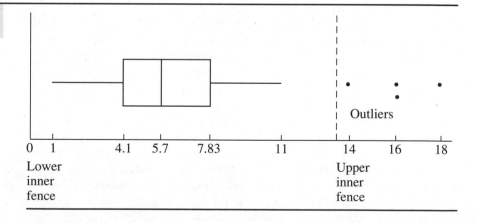

Dealing with Outliers

Now that we have identified four outliers, what do we do with them? First, we investigate the original data to see if an error has been made at any stage between measuring the response and constructing the boxplot. If we find that a particular data point results from an error, we correct the error if possible and then check to see if that data point is still an outlier. If we know that a data point results from an error but we have no way of fixing the error, we eliminate that data point. If the outlier is not a mistake, we check the records of our experimental procedure to see if we can account for the extreme score. Did the subject arrive late and rush through the initial experimental instructions? Was this the first time a new clock was used (perhaps the old one broke during the course of the experiment and the new one wasn't set properly)? If we can find no basis for eliminating the outlier (other than that it is an outlier), we must include it with the rest of our data.

If an outlier represents a value that is legitimate but very rare, it is not likely to arise in the next experiment. However, if a certain percentage of outliers can be expected, there are methods that can be used to reduce their impact. One method is *trimming*. Suppose previous research has determined that the most extreme 10% of the data points from a certain type of experiment are likely to be misleading—that is, not representative of the phenomenon being studied but attributable to extraneous factors, such as poor motivation, fatigue, cheating, etc. A 10% *trimmed* sample is created by automatically throwing out the top 5% and the bottom 5% of the data points. The mean of the trimmed sample is called a *trimmed mean.*

A related method for dealing with outliers is called *Winsorizing*. A percentage is chosen, as with trimming, but instead of being eliminated, the extreme values are replaced by the most extreme value that is not being thrown away; this is done separately for each tail unless the distribution is so skewed that there is no tail on one side. For instance, if in Table 3.3 the 18, 16, 16, and 14 were in the percentage to be trimmed, they would all be replaced by the value 11—the highest value not being trimmed. The *Winsorized sample* then has five 11s as its highest values, followed by two 10s, one 9, etc., as in Table 3.3. The trimmed mean is one of several measures of central tendency that are known as "robust estimators of location," because they are virtually unaffected by outliers. However, trimming or Winsorizing your samples will have some effects on the inferential statistics that may subsequently be performed on the data, so you will need to learn some advanced techniques in order to apply the procedures described above to the testing of hypotheses about populations. (A good starting place is Yuen & Dixon, 1973.)

A more sophisticated method for lessening the impact of extreme scores or marked skewing involves performing a data transformation. For instance, if your data contain a few very large scores, taking the square root of each of the values will reduce the influence of those large scores. Suppose you have recorded the number of minutes each subject takes to solve a problem, and the data for five subjects are 4, 9, 9, 16, 100 (you are a very patient researcher). The mean is 27.6, because of the influence of the one extremely high score. If you take the square root of each score, you obtain 2, 3, 3, 4, 10. Now, the mean is 4.4, and the skewing is much less extreme.

Other functions can be used to transform data, such as taking logarithms, arcsines, and reciprocals. As you might imagine, selecting the most appropriate data transformation is a rather advanced skill, and although I will mention the topic again when discussing inferential statistics, I will refer the interested reader to a more advanced text to learn which transformations are appropriate for different distributions and when to use them. The topics of skewness and outliers will be dealt with again in Section C of Chapter 4.

Finding the Median by a Formula

The median is exactly the same as the 50th percentile, so the procedures for finding the median are identical to the procedures for finding a percentile, as described in Chapter 2. Formula 2.3 would be appropriate for finding the median, but when used for this purpose it can be simplified somewhat to create the more specific Formula 3.5:

$$\text{Median} = \text{LRL} + \left(\frac{50 - \text{cpf}_{\text{LRL}}}{\text{pf}_X} \right) i \qquad \text{Formula 3.5}$$

Note that PR from Formula 2.3 has been replaced by 50, because the median is the 50th percentile. If you are working with only frequencies and cumulative frequencies, Formula 2.4 can be modified appropriately to yield Formula 3.6:

$$\text{Median} = \text{LRL} + \left(\frac{N/2 - \text{cf}_{\text{LRL}}}{f_X} \right) i \qquad \text{Formula 3.6}$$

In this case, cf_p has been replaced by $N/2$ because the cf for the median is always half of the total number of scores.

You can apply Formula 3.6 to find the median of the midterm exams discussed in Section B of Chapter 2. Because there are a total of 25 scores, the median is the score for which cf equals 12.5 (i.e., $N/2$). Table 2.15 reveals that 12.5 is between a cf of 12, which corresponds to 84.5, and a cf of 20, which corresponds to 89.5. Because 12.5 is only slightly higher than 12 (which is cf_{LRL} in this case), an estimate for the median should be just above 84.5—so about 85. The median will be in the interval from 84.5 to 89.5, so LRL = 84.5, and $f_X = 8$. Now you can use Formula 3.6:

$$\text{Median} = 84.5 + \left(\frac{12.5 - 12}{8} \right) 5 = 84.5 + .0625(5) = 84.5 + .3125 = 84.8$$

The calculated median, 84.8, is close enough to the estimate (85) that you can have confidence in the answer. Of course, the median can also be found by use of a cumulative percentage polygon; just follow the procedure discussed in Section B of Chapter 2 for finding any other percentile by graph.

Exercises

* 1. Use the formula to calculate the median for the data in a) exercise 2B1, b) exercise 2B3, and c) exercise 2B4.

* 2. Create a box-and-whisker plot for the data in exercise 2B1. Be sure to identify any outliers.

3. Create a box-and-whisker plot for the data in exercise 2B3. Be sure to identify any outliers.

4. Create a box-and-whisker plot for the data in exercise 2B4. Be sure to identify any outliers.

SUMMARY

The Important Points of Sections A and B

To illustrate the use of the three measures of central tendency described in this chapter, I will present a simple example involving data measured on a ratio scale. Suppose that you are a therapist running a series of support groups to help people cope with the loss of a loved one. You want to know how many close friends each of these people has, so you take a survey of those in the groups. The data are summarized in the simple frequency distribution given in Table 3.4.

What is the "typical" or "average" number of close friends for each person? In other words, how can you summarize the central tendency of this distribution with a single number? The simplest measure of central tendency is the *mode*. The mode is the score that occurs most frequently; the mode for the distribution in Table 3.4 is 2 because there are more 2s (f = 12) than any other score in the distribution. The mode of 2 is a reasonable way to characterize this distribution; you can say that it is quite common for people in this group to have two close friends. However, if no one in the group had more than three close friends, the mode would still be 2; the mode does not provide a good summary of the entire distribution.

X	f	cf		Table 3.4
6	3	40		
5	5	37		
4	6	32		
3	8	26		
2	12	18		
1	5	6		
0	1	1		
$\Sigma f = 40$				

The median is located in the middle of the distribution; half of the scores are higher in value than the median and half are lower. To find the median for the distribution in Table 3.4, you can use the methods of Chapter 2 for finding a percentile. (Finding the middle two scores would be somewhat misleading because of the many duplicate scores.) The median is the 50th percentile, so for this distribution you arc looking for a score for which cf = 20 ($N/2 = 40/2 - 20$). A cf of 20 is between cf = 18, corresponding to 2.5 (i.e., the upper real limit of a score of 2), and cf = 26, corresponding to 3.5. Using either a graph or a formula, the median comes out to 2.75. The median is higher than the mode because it is affected by the number of scores that are 3 and above. If no one in the group had more than three close friends, the median would be considerably lower than 2.75.

The arithmetic mean is the sum of all the scores divided by the number of scores being summed. For a frequency distribution, the sum can be obtained by multiplying each score by its frequency (i.e., fX) and then adding all these products. This sum is then divided, of course, by N, which also equals Σf. This procedure is made easier if you add an fX column to the distribution table, as shown in Table 3.5. Dividing ΣfX by Σf gives 120/40 = 3.0, which is the mean (μ) for this distribution. Notice that the mean is somewhat higher than the median (2.75) because the distribution has a positive skew. There is a floor effect in that no one can have fewer than zero friends, but there is no upper limit to the number of friends one can have.

The above procedure for finding the mean of a simple frequency distribution is equivalent to finding the weighted mean of the X values, where the weights are the frequencies corresponding to the X values. To find the mean of a grouped distribution, you follow the same procedure except that the X values are considered to be the midpoints of the intervals, which are each multiplied by the frequency of the interval.

Table 3.5	X	f	fX
	6	3	18
	5	5	25
	4	6	24
	3	8	24
	2	12	24
	1	5	5
	0	1	0
			$\Sigma fX = 120$

Properties of the Mean

1. If a constant is added (or subtracted) to every score in a distribution, the mean of the distribution will be increased (or decreased) by that constant (i.e., $\overline{X}_{new} = \overline{X}_{old} \pm C$).

2. If every score in a distribution is multiplied (or divided) by a constant, the mean of the distribution will be multiplied (or divided) by that constant (i.e., $\overline{X}_{new} = C\overline{X}_{old}$).

3. The sum of the deviations from the mean will always equal zero [i.e., $\Sigma(X_i - \mu) = 0$].

4. The sum of the squared deviations from the mean will be less than the sum of squared deviations from any other point in the distribution [i.e., $\Sigma(X_i - \mu)^2 < \Sigma(X_i - C)^2$, where C represents some location in the distribution other than the mean].

Advantages and Disadvantages of the Measures of Central Tendency

Advantages of the mode:

1. Easy to find
2. Can be used with any scale of measurement
3. The only measure that can be used with a nominal scale
4. Corresponds to an actual score in the distribution

Disadvantages of the mode (the following apply when the mode is used with ordinal or interval/ratio data):

1. Generally unreliable, especially when representing a relatively small sample (can change radically with only a minor change in the distribution)
2. Can be misleading; the mode tells you which score is most frequent, but tells you nothing about the other scores in the distribution. Radical changes can be made to the distribution without changing the mode.
3. Cannot be used easily with inferential statistics

Advantages of the median:

1. Can be used with either ordinal or interval/ratio data
2. Can be used even if there are open-ended categories or undeterminable scores on either side of the distribution
3. Provides a good representation of a typical score in a skewed distribution; is not unduly affected by extreme scores
4. Minimizes the sum of the absolute deviations (i.e., the sum of score distances from the median—ignoring sign—is less than the sum of score distances from any other location in the distribution)

Disadvantages of the median:

1. May not correspond to an actual score in the distribution (e.g., if there are an even number of scores, or tied scores in the middle of the distribution)
2. Does not reflect the values of all the scores in the distribution (e.g., an extreme score can be moved even further out without affecting the median)
3. Compared to the mean, it is less reliable for drawing inferences about a population from a sample, and harder to use with advanced statistics.

Advantages of the mean:

1. Reflects the values of all the scores in the distribution
2. Has many desirable statistical properties (see the discussion of Properties of the Mean in Section B)
3. Is the most reliable for drawing inferences and the easiest to use in advanced statistical techniques

Disadvantages of the mean:

1. Usually not an actual score in the distribution
2. Not appropriate for use with ordinal data
3. Can be misleading when used to describe a skewed distribution
4. Can be strongly affected by just one very extreme score (i.e., an outlier)

The Important Points of Section C

1. One technique of exploratory data analysis that is useful in displaying the spread and symmetry of a distribution and especially in identifying outliers is to construct a *box-and-whisker plot,* sometimes called just a *boxplot.*

2. The left and right sides of the box, called the *hinges,* are closely approximated by the 25th and 75th percentiles, respectively. The median is drawn wherever it falls *inside* the box. The distance between the two hinges is called the *H-spread.*

3. The furthest the whiskers can possibly reach on either side is determined by the *inner fences.* Each inner fence is 1.5 times the H-spread away from its hinge, so the separation of the inner fences is four times the H-spread (1.5 H on either side of the box equals 3 H, plus 1 H for the box itself).

4. The *upper adjacent value* is defined as the highest score in the distribution that is not higher than the upper inner fence. The *lower adjacent value* is defined analogously. The *upper whisker* is drawn from the upper hinge to the upper adjacent value; the *lower whisker* is drawn in a corresponding fashion.

5. Any scores in the distribution that are higher than the upper inner fence or lower than the lower inner fence are indicated as points in the boxplot. These values are referred to as *outliers.*

6. If an outlier is found to be the result of an error in measurement or data handling or is clearly caused by some factor extraneous to the experiment, then it can be legitimately removed from the data set. If the outlier is merely unusual it should not be automatically thrown out; investigating the cause of the extreme response may uncover some previously unnoticed variable that is relevant to the response being measured.

7. The influence of outliers or extreme skewing can be reduced by *trimming* the sample (deleting a fixed percentage of the most extreme scores) or *Winsorizing* the sample (replacing a fixed percentage of extreme scores with the most extreme value, in the same direction, that is not deleted). Finally, a *data transformation,* such as taking the square root or logarithm of each data value, can be used to reduce the influence of outliers.

8. The formulas for finding percentiles by linear interpolation presented in Chapter 2, Section C can be simplified for finding the median.

Definitions of Key Terms

Central tendency The location in a distribution that is most typical or best represents the entire distribution.

Arithmetic mean The value obtained by summing all the scores in a group (or distribution) and then dividing by the number of scores summed. It is the most familiar measure of central tendency and is usually referred to simply as "the mean" or "the average."

Mode The most frequent category, ordinal position, or score in a population or sample.

Unimodal Describing a distribution with only one major peak (e.g., the normal distribution is unimodal).

Bimodal Describing a distribution that has two roughly equal peaks (this shape usually indicates the presence of two distinct subgroups).

Median The location in a distribution that is at the 50th percentile; half the scores are higher in value than the median and the other half are lower.

Skewed distribution A distribution with the bulk of the scores close to one end and relatively few scores in the other direction.

Positively skewed Describing a distribution with a relatively small number of scores much higher than the majority of scores, but no scores much lower than the majority.

Negatively skewed The opposite of positively skewed (a distribution with a few scores much lower than the majority, but none much higher).

Tail of a distribution The part of a distribution that approaches the X axis, because there are relatively few scores with values in that vicinity. A positively skewed distribution has one long tail that points to the right (with a relatively short tail in the other direction), whereas a negatively skewed distribution has a long tail pointing to the left.

Floor effect A situation in which a lower limit on a measurement scale prevents some scores from becoming as low as they might otherwise be. This can produce a bunching of scores at the low end, and therefore a positively skewed distribution.

Ceiling effect The opposite of a floor effect—scores tend to bunch up at the high end of the scale, producing a negatively skewed distribution.

Undeterminable score A score whose value has not been measured precisely but is known to be above or below some limit (e.g., a subject who has 3 minutes in which to answer a question does not answer at all).

Open-ended category A measurement category that has no limit on one end (e.g., "10 or more").

Weighted mean The mean of two or more group means, also called the weighted average. This value is found by "weighting" (i.e., multiplying) each group mean by the size of its group, adding all the "weighted" means, and then dividing by the total weight (the sum of all the group sizes).

Box-and-whisker plot Also called a *boxplot*; a technique for exploratory data analysis devised by Tukey (1977). A boxplot allows you to see the spread and symmetry of a distribution and the position of any extreme scores at a glance.

Hinges The sides of the box in a boxplot, corresponding approximately to the 25th and 75th percentiles of the distribution.

H-spread The distance between the two hinges (i.e., the width of the box) in a boxplot.

Inner fences Locations on either side of the box in a boxplot that are 1.5 times the H-spread from each hinge. The distance between the upper and lower inner fence is four times the H-spread.

Adjacent values The *upper* adjacent value is the highest score in the distribution that is not higher than the upper inner fence of a boxplot; the *lower* adjacent value is similarly defined in terms of the lower inner fence. The upper whisker is drawn from the upper hinge to the upper adjacent value; the lower whisker is drawn from the lower hinge to the lower adjacent value.

Outlier In general, an extreme score standing by itself in a distribution. An outlier is more specifically defined in the context of a boxplot as any score that is beyond the reach of the whiskers on either side. The outliers are indicated as points in the boxplot.

Key Formulas

The arithmetic mean of a population:

$$\mu = \frac{\Sigma X}{N}$$

Formula 3.1

The arithmetic mean of a sample:

$$\overline{X} = \frac{\Sigma X}{N}$$

Formula 3.2

The mean of a frequency distribution (if the distribution is grouped, X is the midpoint of each interval):

$$\mu = \frac{\Sigma fX}{\Sigma f}$$

Formula 3.3

The weighted mean of two or more samples:

$$\overline{X}_w = \frac{\Sigma N_i \overline{X}_i}{\Sigma N_i} = \frac{N_1 \overline{X}_1 + N_2 \overline{X}_2 + \cdots}{N_1 + N_2 + \cdots}$$

Formula 3.4

The median of a distribution (if cumulative percentages have already been found):

$$\text{Median} = \text{LRL} + \left(\frac{50 - \text{cpf}_{\text{LRL}}}{\text{pf}_X} \right) i$$

Formula 3.5

The median of a distribution (based on the cumulative frequencies):

$$\text{Median} = \text{LRL} + \left(\frac{N/2 - \text{cf}_{\text{LRL}}}{\text{f}_X} \right) i$$

Formula 3.6

MEASURES OF VARIABILITY

You will need to use the following from previous chapters:

Symbols
Σ: Summation sign
μ: Population mean
$\overline{X}$: Sample mean

Concepts
Percentiles
Properties of the mean

Procedures
Rules for using the summation sign

4

CHAPTER

CONCEPTUAL FOUNDATION

Thus far I have discussed both the central tendency and the overall shape of a distribution, but I have left out a very fundamental dimension by which distributions differ. The following hypothetical situation will highlight the importance of this dimension.

You're an eighth-grade English teacher entering a new school, and the principal is giving you a choice of teaching either class A or class B. Having read Chapter 3 of this text, you inquire about the mean reading level of each class. (To simplify matters you can assume that the distributions of both classes are symmetrical.) The principal tells you that class A has a mean reading level of 8.0, whereas class B has a mean of 8.2. All else being equal, you are inclined to take the slightly more advanced class. But all is not equal. Look at the two distributions in Figure 4.1.

What the principal neglected to mention is that reading levels in class B are much more spread out. It should be obvious that class A would be easier to teach. If you geared your lessons toward the 8.0 reader, no one in class A is so much below that level that he or she would be lost, nor is anyone so far above that level

Figure 4.1

Mean Reading Levels in Two Eighth-Grade Classes

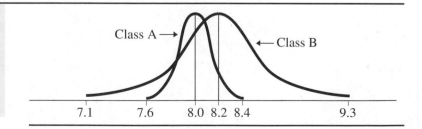

that he or she would be completely bored. On the other hand, teaching class B at the 8.2 level could leave many students either lost or bored.

The fact is that even in a symmetrical distribution the mean is not very representative of the scores, if the distribution contains a great deal of variability. The principal could have shown you both distributions to help you make your decision; the difference in variability (also called the *dispersion*) is so obvious that if you had seen the distributions you could have made your decision instantly. For less obvious cases, and for the purposes of advanced statistical techniques, it would be useful to measure the width of each distribution. However, there is more than one way to measure the spread of a distribution. This chapter is about the different ways of measuring variability.

The Range

The simplest and most obvious way to measure the width of a distribution is to subtract the lowest score from the highest score. The resulting number is called the **range** of the distribution. For instance, judging Figure 4.1 by eye, in class A the lowest reading score appears to be about 7.6 and the highest about 8.4. Subtracting these two scores we obtain $8.4 - 7.6 = .8$. However, if these scores are considered to be measured on a continuous scale, we should subtract the lower real limit of 7.6 (i.e., 7.55) from the upper real limit of 8.4 (i.e., 8.45) to obtain $8.45 - 7.55 = .9$. For class B, the lowest and highest scores appear to be 7.1 and 9.3, respectively, so the range would be $9.35 - 7.05 = 2.3$—considerably larger than the range for class A.

The major drawback to the range as a measure of variability is that, like the mode, it can be quite unreliable. The range can be changed drastically by moving only one score in the distribution, if that score happens to be either the highest or the lowest. For instance, adding just one excellent reader to class A can make the range of class A as large as the range of class B. But the range of class A would then be very misleading as a descriptor of the variability of the bulk of the distribution (see Figure 4.2). In general, the range will tend to be misleading

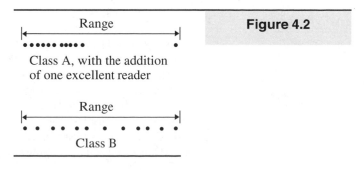

Figure 4.2

Range

Class A, with the addition of one excellent reader

Range

Class B

whenever a distribution includes a few extreme scores (i.e., outliers). Another drawback to the range is that it cannot be determined for an open-ended distribution or a distribution that contains undeterminable scores at one end or the other.

On the positive side, the range not only is the easiest measure of variability to find, it also has the advantage of capturing the entire distribution without exception. For instance, in designing handcuffs for use by police departments, a manufacturer would want to know the entire range of wrist sizes in the adult population, so that the handcuffs could be made to adjust over this range. It would be important to make the handcuffs large enough so that no wrist would be too large to fit but able to become small enough so that no adult could wriggle out and get free.

The Semi-Interquartile Range

There is one measure of variability that can be used with open-ended distributions and is virtually unaffected by extreme scores, because, like the median, it is based on percentiles. It is called the **interquartile (IQ) range,** and it is found by subtracting the 25th percentile from the 75th percentile. The 25th percentile is often called the first quartile and symbolized as Q1; similarly, the 75th percentile is known as the third quartile (Q3). Thus the interquartile range (IQ) can be symbolized as Q3 − Q1. The IQ range gives the width of the middle half of the distribution, therefore avoiding any problems caused by outliers or undeterminable scores at either end of the distribution. (You may recognize that the definition of the IQ range is the same as the definition of the H-spread of a box-and-whisker plot—see Chapter 3, Section C.) A more popular variation of the IQ range is the **semi-interquartile (SIQ) range,** which is simply half of the interquartile range, as shown in Formula 4.1:

$$\text{SIQ range} = \frac{Q3 - Q1}{2} \qquad \text{Formula 4.1}$$

The SIQ range is preferred because it gives the distance of a "typical" score from the median; that is, roughly half the scores in the distribution will be closer to the median than the length of the SIQ range, and about half will be further away. The SIQ range is often used in the same situations for which the median is preferred to the mean as a measure of central tendency, and it can be very useful for descriptive purposes. However, the SIQ range's chief advantage—its unresponsiveness to extreme scores—can also be its chief disadvantage. Quite a few scores on both ends of a distribution can be moved much further from the center without affecting the SIQ range. Thus the SIQ range does not always give an accurate indication of the width of the entire distribution (see Figure 4.3). Moreover, the SIQ range shares with the median the disadvantage of not fitting easily into advanced statistical procedures.

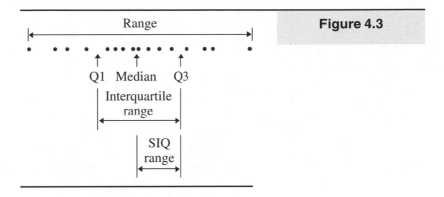

Figure 4.3

The Mean Deviation

The SIQ range can be said to indicate the "typical" distance of a score from the median. This is a very useful way to describe the variability of a distribution. For instance, if you were teaching an English class and were aiming your lessons at the middle of the distribution, it would be helpful to know how far off your teaching level would be, on the average. However, the SIQ range does not take into account the distances of *all* the scores from the center. A more straight-forward approach would be to find the distance of every score from the middle of the distribution and then average those distances. Let us look at the mathematics involved in creating such a measure of variability.

First, we have to decide on a measure of central tendency from which to calculate the distance of each score. The median would be a reasonable choice, but as we are developing a measure to use in advanced statistical procedures the mean is preferable. The distance of any score from the mean $(X_i - \mu)$ is called a **deviation score.** (A deviation score is sometimes symbolized by a lower case x; but in my opinion that notation is too confusing, so it will not be used in this text.) The average of these deviation scores would be given by $\Sigma(X_i - \mu)/N$. Perhaps you have noticed a problem with this expression. According to one of the properties of the mean, as presented in Section B of Chapter 3, $\Sigma(X_i - \mu)$ will always equal zero, which means that the average of the deviation scores will also always equal zero. This problem disappears when you realize that it is the distances we want to average, regardless of their direction (i.e., sign). What we really want to do is take the **absolute values** of the deviation scores before averaging, to find the "typical" amount by which scores deviate from the mean. (Taking the absolute values turns the minus signs into plus signs and leaves the plus signs alone; in symbols, $|X|$ means take the absolute value of X.) This measure is called the **mean deviation,** or average deviation, and it is found using Formula 4.2:

$$\text{Mean deviation} = \frac{\Sigma|X_i - \mu|}{N} \qquad \text{Formula 4.2}$$

To clarify the use of Formula 4.2, I will find the mean deviation of the following three numbers: 1, 3, 8. The mean of these numbers is 4. Applying Formula 4.2 yields:

$$\frac{|1 - 4| + |3 - 4| + |8 - 4|}{3} = \frac{|-3| + |-1| + |+4|}{3} = \frac{3 + 1 + 4}{3} = \frac{8}{3} = 2.67$$

The mean deviation makes a lot of sense and it should be easy to understand; it is literally the average amount by which scores deviate from the mean. Unfortunately, the mean deviation does not fit in well with more advanced statistical procedures. For one thing, you might recall from Chapter 3 that the sum of the absolute deviations (and therefore the average of the absolute deviations, as well) is smallest when the deviations are taken from the *median,* not the mean (If you calculate the mean deviation from the median of the three numbers above, which is 3, you get $(|1 - 3| + |3 - 3| + |8 - 3|)/3 = (2 + 0 + 5)/3 = 7/3 = 2.33$, which is less than the mean deviation from the mean.) Instead, the mean minimizes the sum of the *squared* deviations. This suggests yet another measure of dispersion, which will be described next.

The Variance

If you square all the deviations from the mean, instead of taking the absolute values, and sum all of these squared deviations together, you get a quantity called the **sum of squares (SS)**, which is less for deviations around the mean than for deviations around any other point in the distribution. (Note that the squaring eliminates all the minus signs, just as taking the absolute values did.) Formula 4.3 for SS is

$$SS = \Sigma(X_i - \mu)^2 \qquad\qquad \text{Formula 4.3}$$

If you divide SS by the total number of scores (*N*), you are finding the mean of the squared deviations, which can be used as a measure of variability. The mean of the squared deviations is most often called the **population variance,** and it is symbolized by the lower-case Greek letter sigma squared (σ^2; the upper-case sigma, Σ, is used as the summation sign). Formula 4.4 for the variance is as follows:

$$\sigma^2 = \frac{\Sigma(X_i - \mu)^2}{N} \qquad\qquad \text{Formula 4.4}$$

Because the variance is literally the mean of the squared deviations from the mean, it is sometimes referred to as a mean square, or MS for short. This notation is commonly used in the context of the analysis of variance procedure, as you will see in Part V of this text. Recall that the numerator of the variance formula is often referred to as SS; the relationship between MS and SS is expressed in Formula 4.5:

$$\sigma^2 = MS = \frac{SS}{N} \qquad \text{Formula 4.5}$$

It is certainly worth the effort to understand the variance, because this measure plays an important role in advanced statistical procedures, especially those included in this text. However, it is easy to see that the variance does not provide a very descriptive measure of the spread of a distribution. As an example, consider the variance of the numbers 1, 3, and 8.

$$\sigma^2 = \frac{(1-4)^2 + (3-4)^2 + (8-4)^2}{3} = \frac{3^2 + 1^2 + 4^2}{3}$$
$$= \frac{9 + 1 + 16}{3} = \frac{26}{3} = 8.67$$

The variance (8.67) is larger than the range of the numbers! This is because the variance is based on *squared* deviations. The obvious remedy to this problem is to take the square root of the variance, which leads to our final measure of dispersion.

The Standard Deviation

Taking the square root of the variance produces a measure that provides a good description of the variability of a distribution, and one that plays a role in advanced statistical procedures, as well. The square root of the population variance is called the **population standard deviation** (SD) and is symbolized by the lowercase Greek letter sigma (σ). (Notice that the symbol is *not* squared—squaring the standard deviation gives the variance.) Formula 4.6A for the standard deviation is

$$\sigma = \sqrt{\frac{\Sigma(X_i - \mu)^2}{N}} \qquad \text{Formula 4.6A}$$

An alternative way to express this relationship is

$$\sigma = \sqrt{\sigma^2} = \sqrt{\frac{SS}{N}} = \sqrt{MS} \qquad \text{Formula 4.6B}$$

Because σ is the square root of MS, it is sometimes referred to as the *root-mean-square* (RMS) of the deviations from the mean. At this point, you may be wondering why anyone would bother squaring all the deviations, if after averaging they plan to take the square root? First, we need to make it clear that squaring, averaging, and then taking the square root of the deviations is not the same as just averaging the absolute values of the deviations. If the two procedures were equivalent, the standard deviation would always equal the mean deviation. An example will show that this is not the case. The standard deviation of the numbers 1, 3, and 8 is equal to the square root of their variance, which was found above to be 8.67. So, $\sigma = \sqrt{8.67} = 2.94$, which is clearly larger than the mean deviation (2.67) for the same set of numbers.

The process of squaring and averaging gives extra weight to large scores, which is not removed by taking the square root. Thus the standard deviation is never smaller than the mean deviation, although the two measures can be equal. In fact, the standard deviation will be equal to the mean deviation whenever there are only two numbers in the set. In this case, both measures of variability will equal half the distance between the two numbers. (For example, for the numbers 10 and 20, $\sigma = 5$; you should check this with Formula 4.6A, and compare the result to the results of using Formula 4.2.)

I mentioned above that the standard deviation gives more weight to large scores than does the mean deviation. This is true because squaring a large deviation has a great effect on the average. This sensitivity to large scores can be a problem if there are a few very extreme scores in a distribution, which result in a misleadingly large standard deviation. If you are dealing with a distribution that contains a few extreme scores (whether low, high, or some of each), you should consider an alternative to the standard deviation, such as the mean deviation, which is less affected by extreme scores, or the semi-interquartile range, which is hardly affected at all. On the other hand, you could consider a method for eliminating outliers or transforming the data as suggested in Chapter 3, Section C.

The Variance of a Sample

Thus far the discussion of the variance and standard deviation has confined itself to the situation in which you are describing the variability of an entire population of scores (i.e., your interests do not extend beyond describing the set of scores at hand). Later chapters, however, will consider the case in which you have only a sample of scores from a larger population, and you want to use your description of the sample to extrapolate to that population. Anticipating that need, I will now consider the case in which you want to describe the variability of a sample.

To find the variance of a sample, you can use the procedure expressed in Formula 4.4, but it will be appropriate to change some of the notation. First, I will use s^2 to symbolize the sample variance, according to the custom of using English letters for sample statistics. Along these lines, the mean subtracted from each score will be symbolized as $\overline{X}$ instead of μ, because it is the mean of a sample. Thus Formula 4.4 becomes:

$$s^2 = \frac{\sum(X_i - \overline{X})^2}{N}$$

The Biased and Unbiased Sample Variances

The formula above represents a perfectly reasonable way to describe the variability in a sample, but a problem arises when the variance thus calculated is used to estimate the variance of the larger population. The problem is that the variance of the sample tends to underestimate the variance of the population. Of course, the variance of every sample will be a little different, even if all of the samples are the same size, and they are from the same population. Some sample variances will be a little larger than the population variance and some a little smaller, but unfortunately the average of infinitely many sample variances (when calculated by the formula above) will be *less* than the population variance. This tendency of a sample statistic consistently to underestimate (or overestimate) a population parameter is called *bias*. The sample variance as defined by the (unnumbered) formula above is therefore called a **biased estimator.**

Fortunately, the underestimation just described is so well understood that it can be corrected easily by making a slight change in the formula for calculating the sample variances. To calculate an **unbiased sample variance,** you can use Formula 4.7:

$$s^2 = \frac{\sum (X_i - \bar{X})^2}{N - 1} \qquad \text{Formula 4.7}$$

If infinitely many sample variances are calculated with Formula 4.7, the average of these sample variances *will* equal the population variance (σ^2).

Notation for the Variance and the Standard Deviation

You've seen that there are two different versions of the variance of a sample: biased and unbiased. Some texts use different symbols to indicate the two types of sample variances, such as an uppercase S for biased and a lowercase s for unbiased, or a plain s for biased and $\hat{s}$ (pronounced "s hat") for unbiased. I will adopt the simplest notation by assuming that the variance of a sample will always be calculated using Formula 4.7 (or its algebraic equivalent). Therefore, the symbol s^2 for the sample variance will always refer to the *unbiased* sample variance. Whenever the biased formula is used (i.e., the formula with N rather than $N - 1$ in the denominator), you can assume that the set of numbers at hand is being treated like a population, and therefore the variance will be identified by σ^2. On the other hand, when you are finding the variance of a population you are never interested in extrapolating to a larger group, so there would be no reason to calculate an unbiased variance. Thus when you see σ^2 you know that it was obtained by Formula 4.4 (or its equivalent), and when you see s^2 you know that Formula 4.7 (or its equivalent) was used.

As you might guess from the discussion above, using Formula 4.6A (or 4.6B) to find the standard deviation of a sample produces a biased estimate of the population standard deviation. The solution to this problem would seem to be to use the square root of the unbiased sample variance (as given in Formula 4.7) whenever you are finding the standard deviation of a sample. This produces a new formula for the standard deviation, as shown below.

$$s = \sqrt{\frac{\Sigma(X_i - \overline{X})^2}{N - 1}}$$
　　　　　　　　　　　　　　　　　　　　　　Formula 4.8

Surprisingly, this formula does not entirely correct the bias in the standard deviation, but fortunately the bias that remains is small enough to be ignored. Therefore, I will refer to s (defined by Formula 4.8) as the **unbiased sample standard deviation** and I will use σ (defined by Formula 4.6A) as the symbol for the standard deviation of a population.

Degrees of Freedom

The adjustment in the variance formula that made the sample variance an unbiased estimator of the population variance was quite simple: $N - 1$ was substituted for N in the denominator. Explaining why this simple adjustment corrects the bias described above is not so simple, but I can give you some feeling for why $N - 1$ makes sense in the formula. Return to the example of finding the variance of the numbers 1, 3, and 8. As you saw before, the three deviations from the mean are -3, -1, and 4, which add up to zero (as will always be the case). The fact that these three deviations must add up to zero implies that knowing only two of the deviations automatically tells you what the third deviation will be. That is, if you know that two of the deviations are -1 and -3, then you know that the third deviation must be $+4$ so that the deviations will sum to zero. (Put another way, if you are told that the mean of three numbers is 4 and that two of the numbers are 1 and 3, you should be able to figure out that the third number must be 8.) Thus only two of the three deviations are free to vary (i.e., $N - 1$); once two deviations have been fixed, the third is determined. The number of deviations that are free to vary is called the number of **degrees of freedom (df).** Generally, when there are N scores in a sample, df $= N - 1$. Once the deviation scores have been squared and summed (i.e., SS) for a sample, dividing by the number of degrees of freedom is necessary to produce an unbiased estimate of the population variance. This new notation can be used to create shorthand formulas for the sample variance and standard deviation, as shown below.

$$s^2 = \frac{\text{SS}}{N - 1} = \frac{\text{SS}}{\text{df}}$$
　　　　　　　　　　　　　　　　　　　　　　Formula 4.9

$$s = \sqrt{\frac{\text{SS}}{N - 1}} = \sqrt{\frac{\text{SS}}{\text{df}}}$$
　　　　　　　　　　　　　　　　　　　　　　Formula 4.10

In applying these formulas to the sample of three numbers (1, 3, 8), you do not have to recalculate SS, which was the numerator when we found σ^2 by Formula

4.5. Given that SS = 26, Formula 4.9 tells you that $s^2 = SS/(N - 1) = 26/2 = 13$, which is considerably larger than σ^2 (8.67). The increase from σ^2 to s^2 is necessary to correct the underestimation created by Formula 4.5 when estimating the true variance of the larger population. Formula 4.10 shows that $s = \sqrt{13} = 3.61$, which, of course, is considerably larger than σ (2.94).

The large differences between the biased and unbiased versions of the variance and standard deviation are caused by our unusually tiny sample ($N = 3$). As N becomes larger, the difference between N and $N - 1$ diminishes, as does the difference between σ^2 and s^2 (or σ and s). When N is very large (e.g., over 100), the distinction between the biased and unbiased formulas is so small that it is often ignored.

Exercises

*1. Which of the three most popular measures of variability (range, SIQ range, standard deviation), would you choose in each of the following situations?

a) The distribution is badly skewed with a few extreme outliers in one direction.

b) You are planning to perform advanced statistical procedures (e.g., draw inferences about population parameters).

c) You need to know the maximum width taken up by the distribution.

d) You need a statistic that takes into account every score in the population.

e) The highest score in the distribution is "more than 10."

2. Which measure of variability is not affected if the largest score in the distribution is made even larger? a) Range; b) SIQ range; c) Mean deviation; d) Variance; e) Standard deviation

*3. Which measure of variability is the least useful for descriptive purposes? a) Range; b) SIQ range; c) Mean deviation; d) Variance; e) Standard deviation

*4. a) Calculate the SS and variance (i.e., σ^2) for the following set of scores: 11, 17, 14, 10, 13, 8, 7, 14. (Note: These are the same scores used in exercise 3B1.)

b) Calculate the mean deviation and the standard deviation (i.e., σ) for the set of scores in part a.

5. How many degrees of freedom are contained in the set of scores in exercise 4? Calculate the unbiased

sample variance (i.e., s^2) and standard deviation (i.e., s) for that set of scores. Compare your answers to σ^2 and σ, which you found in exercise 4.

*6. The mean of a sample of five scores is 13. Four of the scores are 2, 17, 12, and 19.

a) What is the fifth score?

b) How many degrees of freedom are there in this sample?

c) Calculate the mean deviation and the sample standard deviation (s).

7. Eliminate the score of 2 from the data in exercise 6, and recalculate both MD and s. Compared to the values calculated in exercise 6, which of these two statistics changed more?

*8. The IQ scores for ten sixth-graders are 111, 103, 100, 107, 114, 101, 107, 102, 112, 109.

a) Calculate the range of these IQ scores.

b) Calculate the population standard deviation (σ) for these IQ scores.

*9. Calculate the range, SIQ range, mean deviation, and standard deviation (σ) for the following set of scores: 17, 19, 22, 23, 25, 26, 26, 27, 28, 28, 29, 30, 32, 35, 35, 36.

10. a) Calculate the range, SIQ range, mean deviation, and standard deviation (σ) for the following set of scores: 3, 8, 13, 23, 25, 26, 26, 27, 28, 28, 29, 30, 32, 41, 49, 56.

b) How would you describe the relationship between the set of data above and the set of data in exercise 9?

c) Compared to the values calculated in exercise 9, which measures of variability have changed the most, which the least, and which not at all?

BASIC STATISTICAL PROCEDURES

To illustrate the computational procedures commonly used to find each of the variability measures discussed in Section A, I will use an example that by now must be familiar to you: the 25 quiz scores that were discussed in the previous two chapters. Table 4.1 is a new version of Table 2.4, to which have been added several columns that are useful in finding percentiles.

The Range

To find the range of the distribution in Table 4.1, you simply take the upper real limit of the highest score in the Table (10.5) and subtract the lower real limit of the lowest score (1.5), which gives the range: $10.5 - 1.5 = 9$.

The Semi-Interquartile Range

To compute the SIQ range for the same data, begin by finding both the 25th (Q1) and 75th (Q3) percentiles. You can approximate these values from Table 4.1. Looking at the cpf column, you see that the 24th percentile corresponds to 5.5, so Q1 must be only slightly above 5.5. The 75th percentile, being almost exactly midway between the 64th and 84th percentiles, is therefore about halfway between 7.5 and 8.5, or about 8. You can use the methods of Chapter 2 to arrive at more precise values (Q1 = 5.54 and Q3 = 8.05).

The interquartile range is Q3 − Q1, which, in this case, equals $8.05 - 5.54 = 2.51$. Half of the distribution is contained within the $2\frac{1}{2}$-point range between Q1 and Q3. The semi-interquartile range is equal to $(Q3 - Q1)/2 = 2.51/2 = 1.255$. If the distribution were symmetric, the SIQ range would tell us that half the scores

Table 4.1				
X	f	cf	pf	cpf
10	2	25	8	100
9	2	23	8	92
8	5	21	20	84
7	3	16	12	64
6	7	13	28	52
5	1	6	4	24
4	4	5	16	20
3	0	1	0	4
2	1	1	4	4

were within $1\frac{1}{4}$ points either way from the median. Because the distribution in this example is skewed, that is not the case—but the SIQ range we calculated can be used as a very crude approximation. The fact that the IQ range (2.51) is so much less than half of the range (9) indicates that the scores are concentrated in the middle, rather than being spread more evenly across the full range. (With an even spread the IQ range should be about half as large as the range.) The degree to which scores in a distribution are clustered near the center, concentrated in the tails, or more evenly spread is known as **kurtosis,** and it can be measured by a single statistic. This topic is fairly advanced, and therefore will be discussed in Section C.

The Mean Deviation

The mean deviation (MD) is found by subtracting the mean from each score, taking the absolute value of the difference, adding all of these absolute values together, and then dividing the sum by the number of scores. If a simple frequency distribution has already been constructed, the mean deviation can be found using the same kind of shortcut that we used to find the mean in Chapter 3, Section B. The first step is to create a table that includes columns for the absolute value of the deviation from the mean ($|X_i - \mu|$) and for f times that value ($f|X_i - \mu|$). The mean (μ) of the 25 quiz scores was already found to be 6.56, so this value was subtracted from each score to arrive at the values in the columns in Table 4.2.

   (Note that $f|X - \mu|$ offers a shortcut in the calculations. Instead of adding the amount by which 8 deviates from the mean (that is, 1.44) five times, you can multiply 1.44 by 5 and then add the result.) By adding up the rightmost column (i.e., $f|X_i - \mu|$), you are simply adding up the absolute values of all the deviation scores. To get the mean deviation you divide this sum by N: MD = 40.56/25 = 1.62. The mean deviation is larger than the SIQ range, because the former is

X	f	$|X - \mu|$	$f|X - \mu|$	
				Table 4.2
10	2	3.44	6.88	
9	2	2.44	4.88	
8	5	1.44	7.20	
7	3	0.44	1.32	
6	7	0.56	3.92	
5	1	1.56	1.56	
4	4	2.56	10.24	
3	0	3.56	0	
2	1	4.56	4.56	
	25		40.56	

affected by the extreme scores in the distribution (e.g., one 2 and two 10s). If the extreme scores were made even more extreme, the mean deviation would be increased but the SIQ range would remain the same.

The Variance and Standard Deviation

The procedure for finding the variance (σ^2) of a simple frequency distribution is the same as the procedure for finding the mean deviation, except that the deviations from the mean are squared instead of being expressed as absolute values, as shown in Table 4.3. The sum of the rightmost column in Table 4.3 is the sum of the squared deviations from the mean (SS), which when divided by N, equals the variance. For this distribution, $\sigma^2 = 98.166/25 = 3.93$.

   As the previous section explained, the variance is not a useful descriptor of a distribution's variability. A much better descriptor is the standard deviation (σ), which is simply the square root of the variance. For this example, $\sigma = \sqrt{3.93} = 1.98$. Note that the standard deviation is somewhat larger than the mean deviation. This is because extreme scores are given extra weight by the squaring process. Making the extreme scores even more extreme would increase both the MD and σ, but it would increase σ more so, thus increasing the discrepancy between σ and MD.

Table 4.3	X	f	$(X - \mu)^2$	$f(X - \mu)^2$
	10	2	11.834	23.667
	9	2	5.954	11.907
	8	5	2.074	10.370
	7	3	0.194	0.582
	6	7	0.314	2.198
	5	1	2.434	2.434
	4	4	6.554	26.214
	3	0	12.674	0
	2	1	20.794	20.794
		25		98.166

Computational Formulas

The procedures for finding SS, which in turn forms the basis for calculating the variance and standard deviation, can be tedious. To arrive at the values in Table 4.3, we had to subtract the mean from each score (being careful not to round off too much), square the difference, and then sum these up. In the days when computers and electronic calculators were not available, it was important to devise formulas to make the computations easier and more efficient. Now that inexpensive electronic calculators find σ^2 and σ with the touch of one button after all the data have been entered, computational formulas are no longer important. However, for historical interest (and just in case you are caught in a situation where you must calculate these statistics by a formula—on an exam, perhaps), I will demonstrate computational formulas for σ^2 and σ.

Consider again Formula 4.3, which is known as the **definitional formula** for SS:

$$SS = \Sigma(X_i - \mu)^2$$

Formula 4.3 is also called the *deviational formula,* because it is based directly on deviation scores. Compare this formula to the easier **computational formula** for SS:

$$SS = \Sigma X^2 - N\mu^2 \qquad \text{Formula 4.11A}$$

Note that according to this formula only the term X^2 must be summed. The term $N\mu^2$ is subtracted only once, after ΣX^2 has been found. By contrast, subtracting the mean from each score and then squaring the difference requires a good deal more effort when calculating by hand. (Even if all of the scores are whole numbers, the mean probably isn't, so subtracting the mean from each score will require you to retain several places past the decimal point to maintain accuracy.) If it seems unlikely to you that Formula 4.3 is exactly equivalent to the much simpler Formula 4.11A, turn to Section C, which shows with a few steps of algebra how one formula can be transformed into the other.

Some statisticians might protest that Formula 4.11A is not as simple as possible, because it requires that the mean be computed first. I think that anyone would want to find the mean before assessing variability—but if you want to find SS more directly from the data, consider Formula 4.11B.

$$SS = \Sigma X^2 - \frac{(\Sigma X)^2}{N} \qquad \text{Formula 4.11B}$$

As I pointed out in Chapter 1, ΣX^2 and $(\Sigma X)^2$ are very different values; the parentheses in the latter term instruct you to add up all the X values *before* squaring (i.e., you square only once at the end), whereas in the former term you square each X before adding.

All you need to do to create a computational formula for the population variance (σ^2) is to divide Formula 4.11A by N; the result is shown in Formula 4.12A:

$$\sigma^2 = \frac{\sum X^2}{N} - \mu^2 \qquad\qquad \text{Formula 4.12A}$$

There is an easy way to remember this formula. The term $\sum X^2/N$ is the mean of the squared scores, whereas the term μ^2 is the square of the mean score. So the variance, which is the mean of the squared deviation scores, is equal to the mean of the squared scores minus the square of the mean score. An alternative formula that does not require you to compute μ first, is found by dividing Formula 4.11B by N, as shown below:

$$\sigma^2 = \frac{1}{N}\left[\sum X^2 - \frac{(\sum X)^2}{N}\right] \qquad\qquad \text{Formula 4.12B}$$

I will leave Formula 4.12B in the form above, not dividing each term by N, for easier comparison with the formula for the unbiased sample variance, which I will present shortly.

The formulas for the population standard deviation (σ) are found simply by taking the square root of each variance formula, as shown below:

$$\sigma = \sqrt{\frac{\sum X^2}{N} - \mu^2} \qquad\qquad \text{Formula 4.13A}$$

$$\sigma = \sqrt{\frac{1}{N}\left[\sum X^2 - \frac{(\sum X)^2}{N}\right]} \qquad\qquad \text{Formula 4.13B}$$

To illustrate the use of computational formulas, we will find σ again for the 25 quiz scores, using Formula 4.13A. First, we will need to square all the scores and add them; the values are listed in Table 4.4. The sum of the fX^2 column is 1174, which gives $\sum X^2$ for the formula. Since we know that μ for this distribution equals 6.56, we have all we need to use Formula 4.13A.

Table 4.4	X	f	X^2	fX^2
	10	2	100	200
	9	2	81	162
	8	5	64	320
	7	3	49	147
	6	7	36	252
	5	1	25	25
	4	4	16	64
	3	0	9	0
	2	1	4	4
		25		1174

$$\sigma = \sqrt{\frac{\Sigma X^2}{N} - \mu^2} = \sqrt{\frac{1174}{25} - 6.56^2} = \sqrt{46.96 - 43.03} = \sqrt{3.93} = 1.98$$

Of course, this is the same value we obtained with the definitional formula.

Unbiased Computational Formulas

When calculating the variance of a set of numbers that is considered a sample of a larger population, it is usually desirable to use a variance formula that yields an unbiased estimate of the population variance. An unbiased sample variance (s^2) can be calculated by dividing SS by $N - 1$, instead of by N. A computational formula for s^2 can therefore be derived by taking a computational formula for SS and dividing by $N - 1$. For instance, dividing Formula 4.11B by $N - 1$ produces Formula 4.14.

$$s^2 = \frac{1}{N-1}\left[\Sigma X^2 - \frac{(\Sigma X)^2}{N}\right] \qquad \text{Formula 4.14}$$

Note the similarity between Formula 4.14 and Formula 4.12B.

The square root of the unbiased sample variance is used as an unbiased estimate of the population standard deviation, even though, as I pointed out before, it is not strictly unbiased (the remaining slight bias is routinely ignored). Taking the square root of Formula 4.14 yields Formula 4.15 for the standard deviation of a sample (s).

$$s = \sqrt{\frac{1}{N-1}\left[\Sigma X^2 - \frac{(\Sigma X)^2}{N}\right]} \qquad \text{Formula 4.15}$$

If you have already calculated σ^2 for a set of numbers and you wish to find s^2, not much additional calculation is needed. First, multiply σ^2 by N to find SS (note that by multiplying both sides of Formula 4.5 by N you get $N\sigma^2 =$ SS). Then, to obtain s^2 you divide SS (i.e., $N\sigma^2$) by $N - 1$. This procedure is summarized below as Formula 4.16.

$$s^2 = \frac{N\sigma^2}{N-1} = \frac{N}{N-1}\sigma^2 \qquad \text{Formula 4.16}$$

Similarly, it is easy to find σ^2 if you already have s^2; reversing the procedure above with respect to N and $N - 1$ yields Formula 4.17.

$$\sigma^2 = \frac{(N-1)s^2}{N} = \frac{N-1}{N}s^2 \qquad \text{Formula 4.17}$$

Of course, the same tricks can be performed to find the standard deviation; just take the square root of the appropriate formula.

Properties of the Standard Deviation

Note: These properties apply equally to the biased and unbiased formulas.

1. *If a constant is added (or subtracted) from every score in a distribution, the standard deviation will not be affected.* In the previous chapter, I used the example of an exam on which the mean score was 70. The professor decided to add 10 points to each student's score, which caused the mean to rise from 70 to 80. Had the standard deviation been 15 points for the original exam scores, the standard deviation would still be 15 points after 10 points were added to each student's exam score. Because the mean "moves" with the scores, and the scores stay in the same relative positions with respect to each other, shifting the location of the distribution (by adding or subtracting a constant) does not alter its spread (see Figure 4.4). This can be shown to be true in general by using simple algebra. The standard deviation of a set of scores after a constant has been added to each one is

$$\sigma_{new} = \sqrt{\frac{\Sigma(X + C - \mu_{new})^2}{N}}$$

As I showed in the previous chapter, $\mu_{new} = \mu_{old} + C$. Therefore,

$$\sigma_{new} = \sqrt{\frac{\Sigma[(X + C) - (\mu_{old} + C)]^2}{N}}$$

Rearranging the order of terms gives the following expression:

$$\sigma_{new} = \sqrt{\frac{\Sigma(X - \mu_{old} + C - C)^2}{N}} = \sqrt{\frac{\Sigma(X - \mu_{old})^2}{N}} = \sigma_{old}$$

The above proof works the same way if you are subtracting, rather than adding, a constant.

2. *If every score is multiplied (or divided) by a constant, the standard devia-*

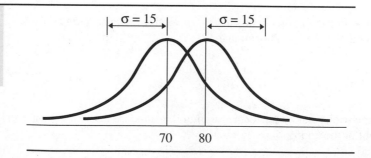

Figure 4.4

Adding a Constant to a Distribution

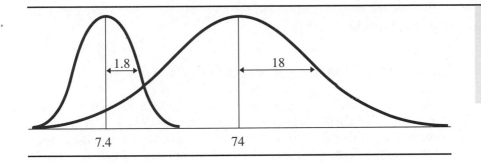

Figure 4.5

Multiplying a Distribution by a Constant

tion will be multiplied (or divided) by that constant. In describing a similar property of the mean in the previous chapter, I used an example of a quiz with a mean of 7.4; each student's score was multiplied by 10, resulting in an exam with a mean of 74. Had the standard deviation of the quiz been 1.8, the standard deviation after scores were multiplied by 10 would have been 18. Whereas adding a constant does not increase the spread of the distribution, multiplying by a constant does (see Figure 4.5). For example, quiz scores of 4 and 7 are spread by only three points, but after they are multiplied by 10 the scores are 40 and 70, which are 30 points apart. Once again we can show that this property is true in general by using some algebra and the rules of summation. The standard deviation of a set of scores after multiplication by a constant is

$$\sigma_{new} = \sqrt{\frac{\Sigma(CX_i - \mu_{new})^2}{N}}$$

As I showed in the previous chapter, $\mu_{new} = C\mu_{old}$. Therefore,

$$\sigma_{new} = \sqrt{\frac{\Sigma(CX_i - C\mu_{old})^2}{N}} = \sqrt{\frac{\Sigma[C(X_i - \mu_{old})]^2}{N}} = \sqrt{\frac{\Sigma C^2(X_i - \mu_{old})^2}{N}}$$

The term C^2 is a constant, so according to Summation Rule 2, we can move this term in front of the summation sign. Then a little bit of algebraic manipulation proves the above property.

$$\sigma_{new} = \sqrt{\frac{C^2\Sigma(X_i - \mu_{old})^2}{N}} = \sqrt{C^2}\sqrt{\frac{\Sigma(X_i - \mu_{old})^2}{N}} = C\sigma_{old}$$

3. *The standard deviation from the mean will be smaller than the standard deviation from any other point in the distribution.* This property follows from property 4 of the mean, as presented in Chapter 3. If SS is minimized by taking deviations from the mean rather than from any other location, it makes sense that σ, which is $\sqrt{SS/N}$, will also be minimized. Proving this requires some

algebra and the rules of summation; the proof can be found in some advanced texts (e.g., Hays, 1994, p.188).

Choosing a Measure of Variability

If you are dealing with interval/ratio data, you can calculate any of the variability measures described in this chapter. It is likely that you would start by finding the range, as a way of checking to see if there are any extreme scores in the distribution that might indicate a data recording error or a procedural mishap. Also, for practical purposes it might be useful to know the total width of the distribution so that all values could be accounted for in your analysis. However, because the range tends to be unreliable, it is not a good measure to use when drawing inferences about the population or for comparing one group to another. Moreover, the range cannot be found at all if there are open-ended or undeterminable scores at either end of the distribution. So even after finding the range (if that is possible), you would be very likely to calculate another measure of dispersion.

The semi-interquartile range is the only variability measure that can be calculated when there are undeterminable or open-ended scores, and it is the only one that is unaffected by the most extreme scores in a distribution. This measure is preferred when it is important to exclude the most extreme scores on either side of the distribution in order to obtain a measure of variability that is not misleading. (The above comments also apply to the interquartile range, but the SIQ range is generally preferred.)

The mean deviation is a good descriptive measure, but it is rarely used. Although the mean deviation is less sensitive to extreme scores than the standard deviation, the SIQ range is preferable to the mean deviation when a few very extreme scores are likely and could be misleading. And although the variance plays an important role in various statistical procedures, you would not use this measure to describe the spread of a distribution.

The standard deviation clearly represents the best combination of a good descriptor of variability and a useful measure for drawing inferences about populations. The standard deviation is also preferred because of its role in advanced statistical procedures. Its only drawback is its sensitivity to extreme scores. If there are a few very extreme scores, you should consider eliminating outliers, transforming the data, or using the mean deviation or SIQ range instead of the standard deviation.

Another decision to make when using either the variance or standard deviation is whether to use the biased formula (i.e., divide SS by N) or the unbiased formula (i.e., divide SS by $N - 1$). If you are using these measures only to describe a set of data, to compare values within the group, or to compare one set of data to a related set, the biased formulas are generally acceptable and are commonly used. It is when inferences about a larger population are to be made that the unbiased formulas become preferable.

Finally, if you plan to calculate the variance or standard deviation "by hand," whether in biased or unbiased form, there is still one decision left to make: You can use the definitional or the computational formula. It might seem that the computational formula would be the obvious choice, but consider an example for which the definitional formula is easier to use. Suppose you are looking at the body temperatures of children two days after the beginning of flu symptoms. The temperatures of five children (to the nearest degree Fahrenheit) were 103, 101, 105, 104, and 102. Using the computational formula would require squaring each of these numbers (e.g., $103^2 = 10,609$) as well as squaring their sum. With the definitional formula, the mean (in this case, 103) is subtracted from each data point, resulting in a set of manageably small deviation scores (i.e., 0, -2, $+2$, $+1$, -1). The definitional formula can be easier to work with when the numbers are large but clustered fairly close together. There is another way to make the numbers manageable in such a situation. We can make use of property 1 of the standard deviation, as stated earlier in this section. According to this property, you can subtract any constant you want from each score in the entire set without changing the standard deviation. In the above example, you could subtract 100 degrees from each data point to obtain the following set of numbers: 3, 1, 5, 4, 2. After the subtraction, the computational formula would be easier to use. Of course, the decision about which formula to use is only relevant when you are calculating by hand. If you are using an electronic calculator that finds the variance automatically or submitting the data to a computer program that calculates descriptive statistics, the formula has already been chosen for you. (But you still must decide whether you want to use the biased or the unbiased version.)

Exercises

* 1. For the data in exercise 2B1, a) calculate the range; b) calculate the IQ and SIQ ranges.

2. For the data in exercise 2B3, a) calculate the range; b) calculate the IQ and SIQ ranges.

* 3. For the data in exercise 2A5, a) calculate the IQ and SIQ ranges; b) calculate the mean deviation and the population standard deviation (σ); c) recalculate σ using the computational formula.

4. Recalculate σ for the IQ scores in Section A, exercise 8, using the computational formula (Hint: You can transform the scores to make your calculation easier.)

* 5. Use the computational formula to calculate the unbiased sample standard deviation for the following set of numbers: 21, 21, 24, 24, 27, 30, 33, 39.

6. a) Use the computational formula to calculate s for the following set of numbers: 7, 7, 10, 10, 13, 16, 19,

25. (Note: This set of numbers was created by subtracting 14 from each of the numbers in the previous exercise.)

b) Compare your answer to this exercise with your answer to the previous exercise. What general principle is being illustrated?

* 7. a) Use the computational formula to calculate s for the following set of numbers: 7, 7, 8, 8, 9, 10, 11, 13. (Note: This set of numbers was created by dividing each of the numbers in exercise 5 above by 3.)

b) Compare your answer to this exercise with your answer to exercise 5. What general principle is being illustrated?

8. a) For the data in exercise 5 use the definitional formula to calculate s around the *median* instead of the mean.

b) What happens to s? What general principle is being illustrated?

*9. If σ for a set of data equals 4.5, what is the corresponding value for s a) when $N = 5$? b) When $N = 20$? c) When $N = 100$?

10. If s for a set of data equals 12.2, what is the corresponding value for σ: a) when $N = 10$? b) When $N = 200$?

OPTIONAL MATERIAL

Derivation of a Computational Formula

As I mentioned in Section B, it is very convenient that the definitional formula for SS, which equals $\sum(X_i - \mu)^2$, also equals $\sum X^2 - N\mu^2$. In fact, it is so convenient that some students find it hard to believe. Fortunately, only a few steps of algebra and the application of some of the summation rules from Chapter 1 are needed to show that these two formulas are really equivalent. I will start with Formula 4.3 and show how to transform it into Formula 4.11A.

First, notice that $(X_i - \mu)^2$ stands for $(X_i - \mu)(X_i - \mu)$. If we do the multiplication indicated, our task will be simplified. (If you do not recall how to multiply two binomial expressions—and why would you?—you can check the algebra review in Appendix B to see that $(a - b)^2 = (a - b)(a - b) = a^2 - 2ab + b^2$.) In this case, $(X_i - \mu)^2 = X_i^2 - 2\mu X_i + \mu^2$. Now we put back the summation sign: $\sum(X_i^2 - 2\mu X_i + \mu^2)$. The first summation rule tells us that we can break this expression into three separate summations, as follows: $\sum X_i^2 - \sum 2\mu X_i + \sum \mu^2$. Of the three terms that result, the first can be left as is; it is already the same as the first term in the computational formula. The second term contains two constants (2 and μ), and according to Summation Rule 3 the constants can be placed in front of the summation sign: $2\mu\sum X_i$. Next, recall that $\sum X_i = N\mu$. (Just take the formula for the mean, $\mu = \sum X_i/N$, and multiply both sides by N.) So the second term $(\sum 2\mu X_i)$ actually equals $2\mu(N\mu) = 2N\mu^2$. The third term contains only a constant (a constant squared is still a constant), and following Summation Rule 2, $\sum \mu^2 = N\mu^2$.

Now we can rewrite the three-term expression we have been dealing with as $\sum X_i^2 - 2N\mu^2 + N\mu^2$. The second and third terms can be added together to yield $- N\mu^2$, so our final expression is $\sum X_i^2 - N\mu^2$. Dropping the index from X, we have $\sum X^2 - N\mu^2$, which is exactly the same as computational Formula 4.11A.

Measuring Skewness

This chapter has described several ways to measure the spread of a distribution. However, although I have discussed the concept of a skewed distribution, I have not mentioned that the degree of skewness can be measured. Quantifying skewness can be useful in deciding when the skewing is so extreme that you ought to transform the distribution or use different types of statistics. For this reason most statistical packages provide a measure of skewness when a full set of descriptive statistics is requested. Whereas the variance is based on the average of squared deviations from the mean, skewness is based on the average of *cubed* deviations from the mean:

$$\text{Skewness} = \frac{\Sigma(X_i - \mu)^3}{N}$$

Recall that when you square a number, the result will be positive whether the original number was negative or positive (e.g., $-2^2 = +4$ and $+2^2 = +4$). However, the cube, or third power, of a number has the same sign as the original number; if the number is negative, the cube will be negative ($-2^3 = -2 \cdot -2 \cdot -2 = -8$), and if the number is positive, the cube will be positive (e.g., $+2^3 = +2 \cdot +2 \cdot +2 = +8$). Deviations below the mean will still be negative after being cubed, and positive deviations will remain positive after being cubed. Thus skewness will be the average of a mixture of positive and negative numbers, which will balance out to zero *only* if the distribution is symmetric. (Note that the deviations will always average to zero before being cubed, but after being cubed they need not.) Any negative skew will cause the skewness measure to be negative, and any positive skew will produce a positive skewness measure. Unfortunately, like the variance, the measure of skewness does not provide a good description of a distribution, because it is in cubed units. Rather than taking the cube root of the above formula, you can derive a more useful measure of the skewness of a population distribution by dividing the above formula by σ^3 (the cube of the population standard deviation) to produce Formula 4.18:

$$\text{Population skewness} = \frac{\Sigma(X_i - \mu)^3}{N\sigma^3} \qquad \text{Formula 4.18}$$

Formula 4.18 has the very useful property of being dimensionless (cubed units are being divided, and thus canceled out, by cubed units); it is a "pure" measure of the shape of the distribution. Not only is this measure of skewness unaffected by adding or subtracting constants (as is the variance), it is also unaffected by multiplying or dividing by constants. For instance, if you take a large group of people and measure each person's weight in pounds, the distribution is likely to have a positive skew that will be reflected in the measure obtained from Formula 4.18. Then, if you convert each person's weight to kilograms, the *shape* of the distribution will remain the same (although the variance will be multiplied by a constant), and fortunately the skewness measure will also remain the same. The only drawback to Formula 4.18 is that if you use it to measure the skewness of a sample, the result will be a biased estimate of the population skewness. You would think the bias could be corrected by the following formula (where *s* is the unbiased standard deviation):

$$\text{Skewness} = \frac{\Sigma(X_i - \overline{X})^3}{(N - 1)s^3}$$

However, Formula 4.19 is needed to correct the bias completely:

$$\text{Sample skewness} = \frac{N}{N - 2} \frac{\Sigma(X_i - \overline{X})^3}{(N - 1)s^3} \qquad \text{Formula 4.19}$$

To illustrate the use of Formula 4.19, I will calculate the sample skewness of four numbers: 2, 3, 5, 10. First, using Formula 4.8 you can verify that $s = 3.56$, so $s^3 = 45.12$. Next, $\sum(X - \overline{X})^3 = (2 - 5)^3 + (3 - 5)^3 + (5 - 5)^3 + (10 - 5)^3 = -3^3 + -2^3 + 0^3 + 5^3 = -27 + (-8) + 0 + 125 = 90$. (Note how important it is to keep track of the sign of each number.) Now we can plug these values into Formula 4.19:

$$\text{Sample skewness} = \frac{4}{4 - 2} \frac{90}{(4 - 1)45.12} = 2\frac{90}{135.36} = 2(.665) = 1.33$$

As you can see, the skewness is positive. Although the total amount of deviation below the mean is the same as the amount of deviation above, one large deviation (i.e., $+5$) counts more than two small ones (i.e., -2 and -3).

Measuring Kurtosis

Two distributions can both be symmetric (i.e., with no skewness), unimodal, and bell-shaped, and yet not be identical in shape. (For one thing, "bell-shaped" is a crude designation—many variations are possible.) Moreover, these two distributions can even have the same mean and variance and still differ fundamentally in shape. The major way that such distributions can differ is in the degree of flatness that characterizes the curve. If a distribution tends to have relatively thick or heavy tails, and then bends sharply so as to have a relatively greater concentration of scores in the middle (more "peakedness"), that distribution is called a **leptokurtic distribution.** On the other hand, a distribution that tends to be flat, with relatively thin tails and less "peakedness," is called a **platykurtic distribution.** These two different shapes are illustrated in Figure 4.6, along with a distribution that is midway between in its degree of *kurtosis*—a **mesokurtic distribution.** (The normal distribution is mesokurtic.) Just as the measure of skewness is based on cubed deviations from the mean, the measure of kurtosis is based on deviations from the mean raised to the fourth power. This measure of kurtosis must then be divided by the standard deviation raised to the fourth power, in order to

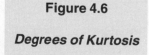

Figure 4.6

Degrees of Kurtosis

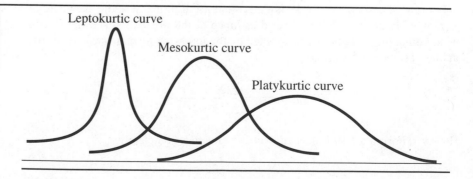

create a dimensionless measure of distribution shape that will not change if you add, subtract, multiply, or divide all the data by a constant. Formula 4.20 for the kurtosis of a population is shown below:

$$\text{Population kurtosis} = \frac{\sum(X_i - \mu)^4}{N\sigma^4} - 3 \qquad \text{Formula 4.20}$$

Besides the change from the third to the fourth power, you will notice another difference between the above formula and Formula 4.19 for skewness: the subtraction of 3 in the kurtosis formula. The subtraction appears in most kurtosis formulas in order to facilitate comparison with the normal distribution. Subtracting 3 ensures that the kurtosis of the normal distribution will come out to zero. Thus a distribution that has relatively fatter tails than the normal distribution will have a positive kurtosis (i.e., it will be leptokurtic), whereas a relatively thin-tailed distribution will have a negative kurtosis (it will be platykurtic). In fact, unless you subtract 3, the population kurtosis will never be less than +1. Therefore, after you subtract 3, population kurtosis can range from −2 to positive infinity.

As you might have learned to expect, applying Formula 4.20 to a sample in order to estimate the kurtosis of a population produces a biased estimate. The correction looks rather complicated, as shown in Formula 4.21 below:

$$\text{Sample kurtosis} = \frac{N(N + 1)}{(N - 2)(N - 3)} \frac{\sum(X_i - \overline{X})^4}{(N - 1)s^4} - 3\frac{(N - 1)(N - 1)}{(N - 2)(N - 3)}$$
$$\text{Formula 4.21}$$

Note that sample kurtosis is not defined if you have fewer than four numbers (you would wind up dividing by zero), and sample skewness is not defined for fewer than three numbers. Also, be aware that with small sample sizes, the estimates of population skewness or kurtosis are quite unreliable. (The reliability of population estimates will be discussed in Part III.)

I will illustrate the calculation of the sample kurtosis for the following four numbers: 1, 5, 7, 11. First, we find that the unbiased variance (s^2) of these numbers is 17.33. To find s^4, we need only square the variance: $17.33^2 = 300.44$. [Note that in general, $(x^2)^2 = x^4$.] Next, we find $\sum(X - \overline{X})^4 = (1 - 6)^4 + (5 - 6)^4 + (7 - 6)^4 + (11 - 6)^4 = -5^4 + (-1)^4 + 1^4 + 5^4 = 625 + 1 + 1 + 625 = 1252$. Now we are ready to use Formula 4.21:

$$\frac{4(4 + 1)}{(4 - 2)(4 - 3)} \frac{1252}{(4 - 1)300.44} - 3\frac{(4 - 1)(4 - 1)}{(4 - 2)(4 - 3)} = 10(1.389) - 3(4.5)$$
$$= 13.89 - 13.5 = .39$$

The calculated value of .39 suggests that the population from which the four numbers were drawn has slightly heavier tails than the normal distribution.

The most common reason for calculating the skewness and kurtosis of a set of data is to decide whether your sample comes from a population that is normally distributed. All of the statistical procedures in Parts III, IV, and V of this text are based on the assumption that the variable being measured has a normal distribution in the population. If your skewness and/or kurtosis measures demonstrate that there is little chance of the population distribution being close to normal, you may have to use different statistical procedures (such as those in Part VI of this text). The use of skewness and kurtosis measures in making decisions about the population distribution will be discussed further in Chapter 6, Section C.

Exercises

* 1. Calculate the population standard deviation and skewness for the following set of data: 2, 4, 4, 10, 10, 12, 14, 16, 36.

2. a) Calculate the population standard deviation and skewness for the following set of data: 1, 2, 2, 5, 5, 6, 7, 8, 18. (Note that this set was formed by halving each number in the previous exercise.)

b) How does each value calculated in part a compare to its counterpart calculated in the previous exercise? What general principle is being illustrated?

* 3. a) Calculate the population standard deviation and skewness for the following set of data: 1, 2, 2, 5, 5, 6, 7, 8. (Note that this set was formed by dropping the highest number from the set in exercise 2.)

b) What can you say about the effect of one extreme score on dispersion and skewness?

4. a) Calculate the sample skewness for the data in problem 3.

b) When the population skewness formula is applied to a sample, is the true population skewness being underestimated or overestimated?

* 5. a) Calculate the population kurtosis for the following set of data: 3, 9, 10, 11, 12, 13, 19.

b) Calculate the sample kurtosis for the data in part a. When the population kurtosis formula is applied to a sample, is the true population kurtosis being underestimated or overestimated?

6. a) Calculate the population kurtosis for the following set of data: 9, 10, 11, 12.

b) Compare your answer to your answer for part a of problem 5. What is the effect on kurtosis when you have extreme scores on both sides of the distribution?

SUMMARY

The Important Points of Section A

1. The simplest measure of the variability (or *dispersion*) of a distribution is the *range,* which is found by subtracting the highest score in the group (or its upper real limit) from the lowest score (or its lower real limit). Unfortunately, the range tends to be unreliable, and it cannot be found if there are open-ended or undeterminable scores on either side of the distribution.

2. The *semi-interquartile range* is half the distance between the first and third quartiles. The SIQ range has the advantage of being unaffected by extreme or undeterminable scores at either end of the distribution, but it does not play a role in advanced statistical procedures.

3. The *mean deviation* (MD) gives the average distance of scores from the mean, which is a useful way of describing the variability of a distribution. However, the average of the absolute deviations is smaller when the deviations are taken from the median than when they are taken from the mean, which is one of the reasons the MD is rarely used in advanced statistical procedures.

4. The *variance* is the average of the *squared* deviations from the mean, and although it plays an important role in advanced statistics, it does not provide a good description of the spread of a distribution. A more descriptive measure can be found by taking the square root of the variance, which is called the *standard deviation*.

5. The *standard deviation* is generally a bit larger than the mean deviation, and considerably larger if there are extreme scores in the distribution. Although the standard deviation lends itself to use in advanced statistics, its major drawback is its sensitivity to extreme scores.

6. The population variance formula, when applied to data from a sample, tends to underestimate the variance of the population. To correct this *bias,* the sample variance (s^2) is calculated by dividing the sum of squared deviations (SS) by $N - 1$, instead of by N. The symbol σ^2 will be reserved for any calculation of variance in which N, rather than $N - 1$, is used in the denominator.

7. The denominator of the formula for the unbiased sample variance, $N - 1$, is known as the *degrees of freedom* (df) associated with the variance, because once you know the mean, df is the number of deviations from the mean that are free to vary.

8. Although the sample standard deviation ($\sqrt{s^2} = s$) is not a perfectly unbiased estimation of the standard deviation of the population, the bias is so small that s is referred to as the unbiased sample standard deviation.

The Important Points of Section B

The following hypothetical situation will illustrate the calculation of the various measures of variability. You are a ninth-grade English teacher. During the first class of the fall term, you ask each of your 12 pupils how many books he or she has read over the summer vacation, to get some idea of their interest in reading.

The responses are as follows: 3, 1, 3, 3, 6, 2, 1, 7, 3, 4, 9, 2. Putting the scores in order will help in finding some of the measures of variability: 1, 1, 2, 2, 3, 3, 3, 3, 4, 6, 7, 9.

The Range. The range is the highest score minus the lowest. The number of books read ranged from 1 to 9, so the range is $9 - 1 = 8$. If the scale is considered continuous (e.g., 9 books is really anywhere between $8\frac{1}{2}$ and $9\frac{1}{2}$ books), then the range is the upper real limit of the highest score minus the lower real limit of the lowest score: $9.5 - 0.5 = 9$.

Interquartile Range. With such a small set of scores, grouping does not seem necessary to find the quartiles. Because $N = 12$, the 25th percentile is between the third and fourth scores. In this case, the third and fourth scores are both 2, so Q1 = 2. Similarly, the 75th percentile is between the ninth and tenth scores, which are 4 and 6, so Q3 = 5. The IQ range is $Q3 - Q1 = 5 - 2 = 3$.

Semi-Interquartile Range. The SIQ range is $(Q3 - Q1)/2$, which is half of the IQ range. For this example, the SIQ range is $3/2 = 1.5$.

Mean Deviation. This is the average of the *absolute* deviations from the mean. The mean for this example is $\mu = \Sigma X / N = 3.67$. The mean deviation is found by applying Formula 4.2 to the data:

$$\text{MD} = \frac{\Sigma |X_i - \mu|}{N}$$

$$= \frac{1}{N}[|-2.67| + |-2.67| + |-1.67| + |-1.67| + |-.67| + |-.67|$$

$$+ |-.67| + |-.67| + |.33| + |2.33| + |3.33| + |5.33|]$$

$$= \frac{1}{12}[23.66] = 1.97$$

Sum of Squares. This is the sum of the *squared* deviations from the mean, as found by Formula 4.3:

$$\text{SS} = \Sigma(X_i - \mu)^2$$
$$= (-2.67)^2 + (-2.67)^2 + (-1.67)^2 + \cdots$$
$$+ (2.33)^2 + (3.33)^2 + (5.33)^2$$
$$= 7.13 + 7.13 + 2.79 + 2.79 + .45 + .45 + .45 + .45 + .11$$
$$+ 5.44 + 11.11 + 28.44$$
$$= 66.75$$

Variance. This is the average of the *squared* deviations from the mean, as found by Formula 4.4 or 4.5:

$$\sigma^2 = \frac{\Sigma(X_i - \mu)^2}{N} = \frac{\text{SS}}{N} = \frac{66.75}{12} = 5.56$$

Standard Deviation. This equals the square root of the variance, as given by Formula 4.6A or 4.6B. For this example, $\sigma = \sqrt{5.56} = 2.36$.

Unbiased Sample Variance. If the 12 students in the English class were considered a sample of all ninth-graders, and you wished to extrapolate from the sample to the population, you would use Formula 4.7 to calculate the unbiased variance, s^2:

$$s^2 = \frac{\Sigma(X_i - \bar{X})^2}{N - 1} = \frac{66.75}{11} = 6.068$$

Unbiased Sample Standard Deviation. This is found by taking the square root of the unbiased sample variance: $s = \sqrt{s^2} = \sqrt{6.068} = 2.46$.

Computational Formula. We can demonstrate that the computational formula for SS (Formula 4.11A) gives the same value as the definitional formula (Formula 4.3). First, we find $\Sigma X^2 = 1^2 + 1^2 + 2^2 + 2^2 + 3^2 + 3^2 + 3^2 + 3^2 + 4^2 + 6^2 + 7^2 + 9^2 = 1 + 1 + 4 + 4 + 9 + 9 + 9 + 9 + 16 + 36 + 49 + 81 = 228$. Then, $(\Sigma X)^2 = 44^2 = 1936$.

$$SS = \Sigma X^2 - \frac{(\Sigma X)^2}{N} = 228 - \frac{1936}{12} = 228 - 161.33 = 66.67$$

Within rounding error, this is the same value for SS found above. (With the definitional formula we rounded off after each subtraction to make the calculation simpler. If you must calculate step-by-step, keep more digits after the decimal point for more accuracy.) Once SS has been calculated with the computational formula, the population variance or standard deviation (or the unbiased estimates of either) can be found with Formula 4.5, 4.6B, 4.9, or 4.10.

Properties of the Standard Deviation

1. If a constant is added (or subtracted) from every score in a distribution, the standard deviation will remain the same (i.e., $\sigma_{new} = \sigma_{old}$).

2. If every score is multiplied (or divided) by a constant, the standard deviation will be multiplied (or divided) by that constant (i.e., $\sigma_{new} = \sigma_{old}$).

3. The standard deviation around the mean will be smaller than it would be around any other point in the distribution.

Advantages and Disadvantages of the Major Measures of Variability

Advantages of the range:

1. Easy to calculate
2. Can be used with ordinal as well as interval/ratio data
3. Encompasses an entire distribution

Disadvantages of the range:

1. Depends on only two scores in the distribution and is therefore not reliable
2. Cannot be found if there are undeterminable or open-ended scores at either end of the distribution
3. Plays no role in advanced statistics

Advantages of the SIQ range:

1. Can be used with ordinal as well as interval/ratio data
2. Can be found even if there are undeterminable or open-ended scores at either end of the distribution
3. Is not affected by extreme scores or outliers

Disadvantages of the SIQ range:

1. Does not take into account all the scores in the distribution
2. Does not play a role in advanced statistical procedures

Advantages of the mean deviation:

1. Easy to understand (it is just the average distance from the mean)
2. Provides a good description of variability
3. Takes into account all scores in the distribution
4. Is less sensitive to extreme scores than the standard deviation

Disadvantages of the mean deviation:

1. Is smaller when taken around the median rather than the mean
2. Is not easily used in advanced statistics
3. Cannot be calculated with undeterminable or open-ended scores

Advantages of the standard deviation:

1. Takes into account all scores in the distribution
2. Provides a good description of variability
3. Tends to be the most reliable measure
4. Plays an important role in advanced statistical procedures

Disadvantages of the standard deviation:

1. Is very sensitive to extreme scores or outliers
2. Cannot be calculated with undeterminable or open-ended scores

The Important Points of Section C

1. By using the summation rules and some basic algebra, I demonstrated that $\sum(X_i - \mu)^2 = \sum X^2 - N\mu^2$. The latter expression often reduces the computations involved in finding the variance or standard deviation, but the former expression may be easier to use when the numbers are large and close together.

2. *Skewness* can be measured by cubing (i.e., raising to the third power) the deviations of scores from the mean of a distribution. The measure of skewness will be a negative number for a negatively skewed distribution, a positive number for a positively skewed distribution, and zero if the distribution is perfectly symmetric around its mean.

3. *Kurtosis* can be measured by raising deviations from the mean to the fourth power. If the kurtosis measure is set to zero for the normal distribution, positive kurtosis indicates relatively fat tails and more "peakedness" in the middle of the distribution (a leptokurtic distribution), whereas negative kurtosis indicates relatively thin tails and a lesser "peakedness" in the middle (a platykurtic distribution).

Definitions of Key Terms

Range The total width of a distribution, as measured by subtracting the lowest score (or its lower real limit) from the highest score (or its upper real limit).

Interquartile (IQ) range The width of the middle half of a distribution, as measured by subtracting the 25th percentile (Q1) from the 75th percentile (Q3).

Semi-interquartile (SIQ) range Half of the IQ range—roughly half of the scores in the distribution are within this distance from the median.

Deviation score The difference between a score and a particular point in the distribution. When I refer to deviation scores, I will always mean the difference between a score and the mean (i.e., $X_i - \mu$).

Absolute value The magnitude of a number, ignoring its sign; that is, negative signs are dropped, and plus signs are left as is. *Absolute deviation scores* are the absolute values of deviation scores.

Mean deviation The mean of the absolute deviations from the mean of a distribution.

Sum of squares (SS) The sum of the squared deviations from the mean of a distribution.

Population variance (σ^2) Also called the mean square (MS), this is the mean of the squared deviations from the mean of a distribution.

Population standard deviation (σ) Sometimes referred to as the root-mean-square (RMS) of the deviation scores, it is the square root of the population variance.

Biased estimator A formula for finding a sample statistic that tends to under- or over-estimate a population parameter (e.g., calculating SS/N for a sample will underestimate σ^2 for the larger population).

Unbiased sample variance (s^2) This formula (i.e., SS/$N - 1$), applied to a sample, provides an unbiased estimate of σ^2. When I refer in this text to the sample variance (or use the symbol s^2), I am referring to this formula.

Unbiased sample standard deviation (s) The square root of the unbiased sample variance. Although it is not a perfectly unbiased estimate of σ, I will refer to s (i.e., $\sqrt{s^2}$) as the unbiased standard deviation of a sample, or just the sample standard deviation.

Degrees of freedom (df) The number of scores that are free to vary after one or more parameters have been found for a distribution. When finding the variance or standard deviation of one sample, df = $N - 1$, because the deviations are taken from the mean, which has already been found.

Kurtosis The degree to which a distribution bends sharply from the central peak to the tails, or slopes more gently, as compared to the normal distribution.

Definitional formula A formula that provides a clear definition of a sample statistic or population parameter but may not be convenient for computational purposes. In the case of the variance or standard deviation, the definitional formula is also called the *deviational formula,* because it is based on finding deviation scores from the mean.

Computational formula An algebraic transformation of a definitional formula, which yields exactly the same value (except for any error due to intermediate rounding off) but reduces the amount or difficulty of the calculations involved.

Leptokurtic distribution A distribution whose tails are relatively fatter than in the normal distribution, and which therefore has a positive value for kurtosis (given that the kurtosis of the normal distribution equals zero).

Platykurtic distribution A distribution whose tails are relatively thinner than in the normal distribution, and which therefore has a negative value for kurtosis.

Mesokurtic distribution A distribution in which the thickness of the tails and the degree of flatness, or "peakedness," are comparable to the normal distribution, and which therefore has a value for kurtosis that is near zero.

Key Formulas

The semi-interquartile range after the 25th (Q1) and 75th (Q3) percentiles have been determined:

$$\text{SIQ range} = \frac{Q3 - Q1}{2} \qquad \text{Formula 4.1}$$

The mean deviation (after the mean of the distribution has been found):

$$\text{Mean deviation} = \frac{\Sigma |X_i - \mu|}{N} \qquad \text{Formula 4.2}$$

The sum of squares, definitional formula (requires that the mean of the distribution be found first):

$$SS = \Sigma (X_i - \mu)^2 \qquad \text{Formula 4.3}$$

The population variance, definitional formula (requires that the mean of the distribution be found first):

$$\sigma^2 = \frac{\Sigma (X_i - \mu)^2}{N} \qquad \text{Formula 4.4}$$

The population variance (after SS has already been calculated):

$$\sigma^2 = MS = \frac{SS}{N} \qquad \text{Formula 4.5}$$

The population standard deviation, definitional formula (requires that the mean of the distribution be found first):

$$\sigma = \sqrt{\frac{\Sigma (X_i - \mu)^2}{N}} \qquad \text{Formula 4.6A}$$

The population standard deviation (after SS has been calculated):

$$\sigma = \sqrt{\sigma^2} = \sqrt{\frac{SS}{N}} = \sqrt{MS} \qquad \text{Formula 4.6B}$$

The unbiased sample variance, definitional formula:

$$s^2 = \frac{\Sigma(X_i - \bar{X})^2}{N - 1} \qquad\qquad \text{Formula 4.7}$$

The unbiased sample standard deviation, definitional formula:

$$s = \sqrt{\frac{\Sigma(X_i - \bar{X})^2}{N - 1}} \qquad\qquad \text{Formula 4.8}$$

The unbiased sample variance (after SS has been calculated):

$$s^2 = \frac{SS}{N - 1} = \frac{SS}{df} \qquad\qquad \text{Formula 4.9}$$

The unbiased sample standard deviation (after SS has been calculated):

$$s = \sqrt{\frac{SS}{N - 1}} = \sqrt{\frac{SS}{df}} \qquad\qquad \text{Formula 4.10}$$

The sum of squares, computational formula (requires that the mean has been calculated):

$$SS = \Sigma X^2 - N\mu^2 \qquad\qquad \text{Formula 4.11A}$$

The sum of squares, computational formula (direct from raw data):

$$SS = \Sigma X^2 - \frac{(\Sigma X)^2}{N} \qquad\qquad \text{Formula 4.11B}$$

The population variance, computational formula (requires that the mean has been calculated):

$$\sigma^2 = \frac{\Sigma X^2}{N} - \mu^2 \qquad\qquad \text{Formula 4.12A}$$

The population variance, computational formula (direct from raw data):

$$\sigma^2 = \frac{1}{N}\left[\Sigma X^2 - \frac{(\Sigma X)^2}{N} \right] \qquad\qquad \text{Formula 4.12B}$$

The population standard deviation, computational formula (requires that the mean has been calculated):

$$\sigma = \sqrt{\frac{\Sigma X^2}{N} - \mu^2}$$ Formula 4.13A

The population standard deviation, computational formula (direct from raw data):

$$\sigma = \sqrt{\frac{1}{N}\left[\Sigma X^2 - \frac{(\Sigma X)^2}{N}\right]}$$ Formula 4.13B

The unbiased sample variance, computational formula (direct from raw data):

$$s^2 = \frac{1}{N-1}\left[\Sigma X^2 - \frac{(\Sigma X)^2}{N}\right]$$ Formula 4.14

The unbiased sample standard deviation, computational formula (direct from raw data):

$$s = \sqrt{\frac{1}{N-1}\left[\Sigma X^2 - \frac{(\Sigma X)^2}{N}\right]}$$ Formula 4.15

The unbiased sample variance (if the biased formula has already been used to find the variance):

$$s^2 = \frac{N\sigma^2}{N-1} = \frac{N}{N-1}\sigma^2$$ Formula 4.16

The biased sample variance (if the unbiased formula has already been used to find the variance):

$$\sigma^2 = \frac{(N-1)s^2}{N} = \frac{N-1}{N}s^2$$ Formula 4.17

Skewness of a population in dimensionless units:

$$\text{Population skewness} = \frac{\Sigma(X_i - \mu)^3}{N\sigma^3}$$ Formula 4.18

Sample skewness (i.e., estimate of population skewness based on sample data):

$$\text{Sample skewness} = \frac{N}{N-2}\frac{\Sigma(X_i - \bar{X})^3}{(N-1)s^3}$$ Formula 4.19

Kurtosis of a population in dimensionless units:

$$\text{Population kurtosis} = \frac{\sum(X_i - \mu)^4}{N\sigma^4} - 3 \qquad \text{Formula 4.20}$$

Sample kurtosis (i.e., estimate of population kurtosis based on sample data):

$$\text{Sample kurtosis} = \frac{N(N + 1)}{(N - 2)(N - 3)} \frac{\sum(X_i - \overline{X})^4}{(N - 1)s^4} - 3\frac{(N - 1)(N - 1)}{(N - 2)(N - 3)}$$

$$\text{Formula 4.21}$$

STANDARDIZED SCORES AND THE NORMAL DISTRIBUTION

You will need to use the following from previous chapters:

Symbols
Σ: Summation sign
μ: Population mean
σ: Population standard deviation
σ^2: Population variance

Concepts
Percentile ranks
Mathematical distributions
Properties of the mean and standard deviation

A

CONCEPTUAL FOUNDATION

A friend meets you on campus and says, "Congratulate me! I just got a 70 on my physics test." At first, it may be hard to generate much enthusiasm about this grade. You ask, "That's out of 100, right?" and your friend proudly says "Yes." You may recall that a 70 was not a very good grade in high school, even in physics. But if you know how low exam grades often are in college physics, you might be a bit more impressed. The next question you would probably want to ask your friend is, "What was the average for the class?" Let's suppose your friend says 60. If your friend has long been afraid to take this physics class and expected to do poorly, you should offer congratulations. Scoring 10 points above the mean isn't bad!

On the other hand, if your friend expected to do well in physics and is doing a bit of bragging, you would need more information to know if your friend has something to brag about. Was 70 the highest grade in the class? If not, you need to locate your friend more precisely within the class distribution to know just how impressed you should be. Of course, it is not important in this case to be precise about your level of enthusiasm, but if you were the teacher trying to decide whether your friend should get a B+ or an A−, more precision would be helpful.

z Scores

To see just how different a score of 70 can be in two different classes, even if both of the classes have a mean of 60, take a look at the two class distributions in Figure 5.1. (To simplify the comparison, I am assuming that the classes are large enough to produce smooth distributions.) As you can see, a score of 70 in class A is excellent, being near the top of the class, whereas the same score is not so impressive in class B, being near the middle of the distribution. The difference

Figure 5.1

Distributions of Scores on a Physics Test

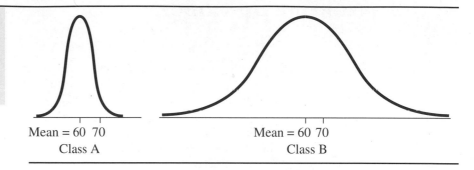

Mean = 60 70
Class A

Mean = 60 70
Class B

between the two class distributions is visually obvious—class B is much more spread out than class A. Having read the previous chapter, you should have an idea of how to quantify this difference in variability. The most useful way to quantify the variability is to calculate the standard deviation (σ). The way the distributions are drawn in Figure 5.1, σ would be about 5 points for class A, and about 20 for class B.

An added bonus from calculating σ is that, in conjunction with the mean (μ), σ provides us with an easy and precise way of locating scores in a distribution. In both classes a score of 70 is 10 points above the mean, but in class A those 10 points represent two standard deviations, whereas 10 points in class B is only half of a standard deviation. Telling someone how many standard deviations your score is above or below the mean is more informative than telling your actual (raw) score. This is the concept behind the **z score.** In any distribution for which the mean (μ) and standard deviation (σ) can be found, any raw score can be expressed as a *z* score by using Formula 5.1:

$$z = \frac{X - \mu}{\sigma}$$ Formula 5.1

Let us apply this formula to the score of 70 in class A:

$$z = \frac{70 - 60}{5} = \frac{10}{5} = +2$$

and in class B:

$$z = \frac{70 - 60}{20} = \frac{10}{20} = +.5$$

In a compact way, the *z* scores tell us that your friend's exam score is more impressive if your friend is in class A ($z = +2$) than if she is in class B ($z = +.5$). Note that the plus sign in these *z* scores is very important because it tells you that the scores are above rather than below the mean. If your friend had scored a 45 in class B, her *z* score would have been

$$z = \frac{45 - 60}{20} = \frac{-15}{20} = -.75$$

The minus sign in the above *z* score informs us that in this case your friend was three-quarters of a standard deviation *below* the mean. The *sign* of the *z* score tells you whether the raw score is above or below the mean; the *magnitude* of the *z* score tells you the raw score's distance from the mean in terms of standard deviations.

   z scores are called **standardized scores,** because they can be used to compare raw scores from different distributions. To continue the example above, suppose that your friend scores a 70 on the physics exam in class B and a 60 on a math exam in a class where $\mu = 50$ and $\sigma = 10$. In which class was your friend further above the mean? We have already found that your friend's *z* score for the physics exam in class B was $+.5$. The *z* score for the math exam would be

$$z = \frac{60 - 50}{10} = \frac{10}{10} = +1$$

Thus your friend's *z* score on the math exam is higher than her *z* score on the physics exam, so she seems to be performing better (in terms of class standing on the last exam) in math than in physics.

Finding a Raw Score from a *z* Score

As you will see in the next section, sometimes you want to find the **raw score** that corresponds to a particular *z* score. As long as you know μ and σ for the distribution, this is easy. You can use Formula 5.1 by filling in the given *z* score and solving for the value of *X*. For example, if you are dealing with class A (as shown in Figure 5.1) and you want to know the raw score for which the *z* score would be -3, you can use Formula 5.1 as follows:

$$-3 = \frac{X - 60}{5} \quad \text{so} \quad -15 = X - 60 \quad \text{so} \quad X = -15 + 60 = 45$$

   To make the calculation of such problems easier, Formula 5.1 can be rearranged in a new form that I will designate Formula 5.2:

$$X = z\,\sigma + \mu \qquad\qquad \text{Formula 5.2}$$

Now if you want to know, for instance, the raw score of someone in class A who obtained a *z* score of -2, you can use Formula 5.2 as follows:

$$X = z\sigma + \mu = -2(5) + 60 = -10 + 60 = 50$$

Note that you must be careful to retain the minus sign on a negative z score when working with a formula, or you will come up with the wrong raw score. (In the example above, $z = +2$ would correspond to a raw score of 70, as compared to a raw score of 50 for $z = -2$.) Some people find negative z scores a bit confusing, probably because most measurements in real life (e.g., height, IQ) cannot be negative. It may also be hard to remember that a z score of zero is not bad, it is just average (i.e., if $z = 0$, then the raw score $= \mu$). Formula 5.2 will come in handy for some of the procedures outlined in Section B. The structure of this formula also bears a strong resemblance to the formula for a confidence interval, for reasons that will be made clear when confidence intervals are defined in Chapter 8.

Sets of z Scores

It is interesting to see what happens when you take a group of raw scores (e.g., exam scores for a class) and convert all of them to z scores. To keep matters simple, we will work with a set of only four raw scores: 30, 40, 60, and 70. First, we need to find the mean and standard deviation for these numbers. The mean (μ) equals $(30 + 40 + 60 + 70)/4 = 200/4 = 50$. The standard deviation (σ) can be found by Formula 4.13A, after first calculating ΣX^2: $30^2 + 40^2 + 60^2 + 70^2 = 900 + 1600 + 3600 + 4900 = 11,000$. The standard deviation is found as follows:

$$\sigma = \sqrt{\frac{\Sigma X^2}{N} - \mu^2} = \sqrt{\frac{11,000}{4} - 50^2} = \sqrt{2750 - 2500} = \sqrt{250} = 15.81$$

Each raw score can now be transformed into a z score using Formula 5.1:

$$z = \frac{30 - 50}{15.81} = \frac{-20}{15.81} = -1.265$$

$$z = \frac{40 - 50}{15.81} = \frac{-10}{15.81} = -.6325$$

$$z = \frac{60 - 50}{15.81} = \frac{10}{15.81} = +.6325$$

$$z = \frac{70 - 50}{15.81} = \frac{20}{15.81} = +1.265$$

By looking at these four z scores, it is easy to see that they add up to zero, which tells us that the mean of the z scores will also be zero (zero divided by 4— or any other number—is always zero). This is not a coincidence. The mean for any complete set of z scores will be zero. This follows from Property 1 of the mean, as discussed in Chapter 3: If you subtract a constant from every score, the mean is decreased by the same constant. To form z scores, you subtract a constant (namely, μ) from all the scores before dividing. Therefore, this constant must also be subtracted from the mean. But since the constant being subtracted *is* the mean, the new mean is μ (the old mean) minus μ (the constant), or zero.

It is not obvious what the standard deviation of the four z scores will be, but it will be instructive to find out. We will use Formula 4.13A again, substituting z for X:

$$\sigma_z = \sqrt{\frac{\sum z^2}{N} - \mu_z^2}$$

The term that is subtracted is the mean of the z scores squared. But as you have just seen, the mean of the z scores is always zero, so this term drops out. Therefore, $\sigma_z = \sqrt{(\sum z^2/N)}$. The term $\sum z^2$ equals $(-1.265)^2 + (-.6325)^2 + (.6325)^2 + (1.265)^2 = 1.60 + .40 + .40 + 1.60 = 4.0$. Therefore, σ equals $\sqrt{z^2/N} = \sqrt{(4/4)} = \sqrt{1} = 1$. As you have probably guessed, this also is no coincidence. The standard deviation (σ) for a complete set of z scores will always be 1. This follows from two of the properties of the standard deviation described in Chapter 4. Property 1 implies that subtracting the mean (or any constant) from all the raw scores will not change the standard deviation. Then, according to Property 2, dividing all the scores by a constant will result in the standard deviation being divided by the same constant. The constant used for division when creating z scores *is* the standard deviation, so the new standard deviation is σ (the old standard deviation) divided by σ (the constant divisor), which always equals 1.

Properties of z Scores

I have just derived two important properties that apply to any set of z scores: (1) the mean will be zero, and (2) the standard deviation will be 1. Now we must consider an important limitation of z scores. I mentioned that z scores can be useful in comparing scores from two different distributions (in the example discussed earlier in the chapter, your friend performed relatively better in math than in physics). However, the comparison is reasonable only if the two distributions are similar in shape. Consider the distributions for classes D and E, shown in Figure 5.2. In the negatively skewed distribution of class D, a z score of $+2$

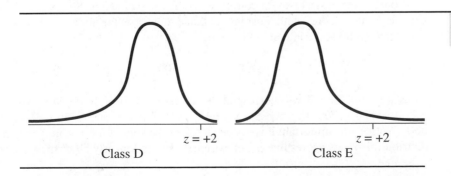

Figure 5.2

$z = +2$ $z = +2$

Class D   Class E

would put you very near the top of the distribution. In class E, however, the positive skewing implies that although there may not be a large percentage of scores above $z = +2$, there are some scores that are much higher.

Another property of z scores is relevant to the above discussion. Converting a set of raw scores into z scores will not change the shape of the original distribution. For instance, if all the scores in class E were transformed into z scores, the distribution of z scores would have a mean of zero, a standard deviation of 1, and exactly the same positive skew as the original distribution. We can illustrate this property with a simple example, again involving only four scores: 3, 4, 5, 100. The mean of these scores is $(3 + 4 + 5 + 100)/4 = 112/4 = 28$. The standard deviation is 41.58. (You should calculate this yourself for practice.) Therefore, the corresponding z scores (using Formula 5.1) are $-.6$, $-.58$, $-.55$, and $+1.73$. First, note the resemblance between the distribution of these four z scores and the distribution of the four raw scores: In both cases there are three numbers close together with a fourth number much higher. You can also see that the z scores add up to zero, which implies that their mean is also zero. Finally, you can calculate $\sigma = \sqrt{(\Sigma z^2/N)}$ to see that $\sigma = 1$.

SAT, *T*, and IQ Scores

For descriptive purposes, standardized scores that have a mean of zero and a standard deviation of 1.0 may not be optimal. For one thing, about half the z scores will be negative (even more than half if the distribution has a positive skew), and minus signs can be cumbersome to deal with; leaving off a minus sign by accident can lead to a gross error. For another thing, most of the scores will be between zero and 2, requiring two places to the right of the decimal point in order to have a reasonable amount of accuracy. Like minus signs, decimals can be cumbersome to deal with. For these reasons, it can be more desirable to standardize scores so that the mean is 500 and the standard deviation is 100. Because this scale is used by the Educational Testing Service (Princeton, New Jersey) to report the results of the Scholastic Assessment Test, standardized scores with $\mu = 500$ and $\sigma = 100$ are often called **SAT scores.** (The same scale is used to report the results of the Graduate Record Examination and several other standardized tests.)

Probably the easiest way to convert a set of raw scores into SAT scores is to first find the z scores with Formula 5.1 and then use Formula 5.3 below to transform each z score into an SAT score:

$$\text{SAT} = 100z + 500 \qquad\qquad \text{Formula 5.3}$$

Thus a z score of -3 will correspond to an SAT score of $100(-3) + 500 = -300 + 500 = 200$. If $z = +3$, then $\text{SAT} = 100(+3) + 500 = 300 + 500 = 800$. (Notice how important it is to keep track of the sign of the z score.) For any distribution of raw scores that is not extremely skewed, nearly all of the z scores will fall between -3 and $+3$; this means (as shown above) that nearly all the SAT scores will be between 200 and 800. There are so few scores that would lead to an SAT score below 200 or above 800 that generally these are the most extreme scores given. Because z scores are rarely expressed to more than two places beyond the decimal point, multiplying by 100 ensures that the SAT scores will not require decimal points at all. (From a psychological point of view, it feels better to score 500 or 400 on the SAT than to be presented with a zero or negative z score.)

Less familiar to students, but commonly employed for reporting the results of psychological tests, is the *T score*. The *T* score is very similar to the SAT score, as you can see from Formula 5.4:

$$T = 10z + 50 \hspace{3cm} \text{Formula 5.4}$$

A full set of *T* scores will have a mean of 50 and a standard deviation of 10. If z scores are expressed to only one place past the decimal point, the corresponding *T* scores will not require decimal points at all.

The choice of which standardized score to use is usually a matter of convenience and tradition. The current convention regarding intelligence quotient (IQ) scores is to use a formula that creates a mean of 100. The Stanford-Binet test uses the formula $16z + 100$, resulting in a standard deviation of 16, whereas the Wechsler test uses $15z + 100$, resulting in a standard deviation of 15.

The Normal Distribution

It would be nice if all variables measured by psychologists had identically shaped distributions, because then the z scores would always fall in the same relative locations, regardless of the variable under study. Although this is unfortunately not the case, it is useful that the distributions for many variables somewhat resemble one or another of the well-known mathematical distributions. Perhaps the best understood distribution with the most convenient mathematical properties is the normal distribution (mentioned in Chapter 2). Actually, you can think of the normal distribution as a family of distributions. There are two ways that members of this family can differ. Two normal distributions can differ either by having different means (e.g., heights of men and heights of women) and/or by having different standard deviations (e.g., heights of adults and IQs of adults). What all normal distributions have in common is the same shape—and not just any bell-like shape, but rather a very precise shape that follows an exact mathematical equation (see Section C).

Because all normal distributions have the same shape, a particular z score will fall in the same relative location on any normal distribution. Probably the most useful way to define relative location is to state what proportion of the distribution is above (i.e., to the right of) the z score and what proportion is below (to the left of) the z score. For instance, if $z = 0$, then .5 of the distribution (i.e., 50%) will be above that z score and .5 will be below it. (Because of the symmetry of the normal distribution, the mean and the median fall at the same location, which is also the mode.) A statistician can find the proportions above and below any z score. In fact, these proportions have been found for all z scores expressed to two decimal places (e.g., .63 or 0.63, 2.17, etc.) up to some limit, beyond which the proportion on one side is too small to care about. These proportions have been put into tables of the normal distribution, such as Table A.1 in Appendix A of this text.

The Standard Normal Distribution

Tables of proportions for z scores are called tables of the **standard normal distribution;** the standard normal distribution is just a normal distribution for which $\mu = 0$ and $\sigma = 1$. It is the distribution you get when you transform all of the scores from any normal distribution into z scores. On the other hand, you could work out a table for any particular normal distribution. For example, a table for a normal distribution with $\mu = 60$ and $\sigma = 20$ would show that the proportion of scores above 60 is .5, and there would be entries for 61, 62, etc. Of course, it would be impractical to have a table for every possible normal distribution, and it is easy enough to convert scores to z scores (or vice versa) when necessary, so you only need to have a table of the standard normal distribution (as shown in Table A.1 in the Appendix). It is also unnecessary to include negative z scores in the table. Because of the symmetry of the normal distribution, the proportion of scores above a particular positive z score is the same as the proportion below the corresponding negative z score (e.g., the proportion above $z = +1.5$ equals the proportion below $z = -1.5$; see Figure 5.3).

Suppose you want to know the proportion of the normal distribution that falls between the mean and one standard deviation above the mean (i.e., between $z = 0$ and $z = +1$). This portion of the distribution corresponds to the shaded area in Figure 5.4. Assume that all of the scores in the distribution fall between $z = -4$ and $z = +4$. (The fraction of scores not included in this region of the normal distribution is so tiny that you can ignore it without fear of making a noticeable

Figure 5.3

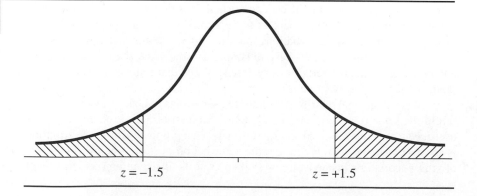

$z = -1.5$ $z = +1.5$

Figure 5.4

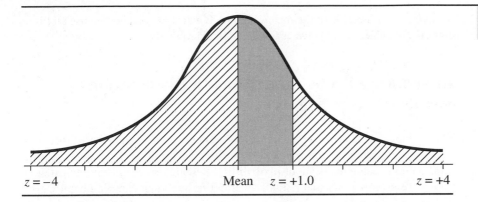

error.) Thus for the moment, assume that the shaded area plus the cross-hatched areas of Figure 5.4 represent 100% of the distribution, or 1.0 in terms of proportions. The question about proportions can now be translated into areas of the normal distribution. If you knew what proportion of the area of Figure 5.4 is shaded, you would know what proportion of the scores in the entire distribution were between $z = 0$ and $z = +1$. (Recall from Chapter 2 that the size of an **area under the curve** represents a proportion of the scores.) The shaded area looks like it is about one-third of the entire distribution, so you can guess that in any normal distribution about one-third of the scores will fall between the mean and $z = +1$.

Table of the Standard Normal Distribution

Fortunately, you do not have to guess about the relative size of the shaded area in Figure 5.4; Appendix A can tell you the exact proportion. A small section of Table A.1 has been reproduced in Table 5.1. The column labeled Mean to z tells you what you need to know. First, go down the column labeled z until you get to 1.00. The column next to it contains the entry .3413, which is the proportion of the normal distribution enclosed between the mean and z, when $z = 1.00$. This tells you that the shaded area in Figure 5.4 contains a bit more than one-third of the scores in the distribution—it contains .3413 (one-third would be .3333). The proportion between the mean and $z = -1.00$ is the same: .3413. Thus about 68% (a little over two-thirds) of any normal distribution is within one standard

z	**Mean to *z***	**Beyond *z***	**Table 5.1**
.98	.3365	.1635	
.99	.3389	.1611	
1.00	.3413	.1587	
1.01	.3438	.1562	
1.02	.3461	.1539	

deviation on either side of the mean. Section B will show you how to use all three columns of Table A.1 to solve various practical problems.

Introducing Probability: Smooth Distributions Versus Discrete Events

The main reason that finding areas for different portions of the normal distribution is so important to the psychological researcher is that these areas can be translated into statements about **probability.** Researchers are often interested in knowing the probability that a totally ineffective treatment can accidentally produce results as promising as the results they have just obtained in their own experiment. The next two chapters will show how the normal distribution can be used to answer such abstract questions about probability rather easily.

Before we can get to that point, it is important to lay out some of the basic rules of probability. These rules can be applied either to discrete events or to a smooth, mathematical distribution. An example of a discrete event is picking one card from a deck of 52 playing cards. Predicting which five cards might be in an ordinary poker hand is a more complex event, but it is composed of simple, discrete events (i.e., the probability of each card). Applying the rules of probability to discrete events is useful in figuring out the likely outcomes of games of chance (e.g., playing cards, rolling dice, etc.) and in dealing with certain types of nonparametric statistics. I will postpone any discussion of discrete probability until Part VI, in which nonparametric statistics are introduced. Until Part VI, I will be dealing with parametric statistics, which are based on variables that have smooth distributions. Therefore, at this point, I will describe the rules of probability only as they apply to smooth, continuous distributions, such as the normal curve.

A good example of a smooth distribution that resembles the normal curve is the distribution of height for adult females in a large population. Finding probabilities involving a continuous variable like height is very different from dealing with discrete events like selecting playing cards. With a deck of cards there are only 52 distinct possibilities. On the other hand, how many different measurements for height are there? It depends on how precisely height is measured. With enough precision, everyone in the population can be determined to have a slightly different height from everyone else. With an infinite population (which is assumed when dealing with the true normal distribution), there are infinitely many different height measurements. Therefore, instead of trying to determine the probability of any particular height being selected from the population, it is only feasible to consider the probability associated with a range of heights (e.g., 60 to 68 inches or 61.5 to 62.5 inches).

Probability as Area under the Curve

In the context of a continuous variable measured on an infinite population, an "event" can be defined as selecting a value within a particular range of a distribution (e.g., picking an adult female whose height is between 60 and 68 inches). Having defined an event in this way, we can next define the probability that a particular event will occur if we select one adult female at random. The probability of some event can be defined as the proportion of times this event occurs out of an infinite number of random selections from the distribution. This proportion is equal to the area of the distribution under the curve that is enclosed by the range in which the event occurs. This brings us back to finding areas of the distribution. If you want to know the probability that the height of the next adult female selected at random will be between one standard deviation below the mean and one standard deviation above the mean, you must find the proportion of the normal distribution enclosed by $z = -1$ and $z = +1$. I have already pointed out that this proportion is equal to about .68, so the probability is .68 (roughly two chances out of three) that the height of the next woman selected at random will fall in this range. If you wanted to know whether the next randomly selected adult female would be between 60 and 68 inches tall, you would need to convert both of these heights to z scores so that you could use Table A.1 to find the enclosed area according to procedures described in Section B.

Real Distributions Versus the Normal Distribution

Before I proceed, let me offer a couple of warnings concerning the use of Table A.1. First, none of the variables studied by psychologists matches the normal distribution perfectly. I can state this with confidence because the normal distribution never ends; no matter what the mean and standard deviation, the normal distribution extends infinitely in both directions. On the other hand, the measurements psychologists deal with have limits. For instance, if a psychophysiologist is studying the resting heart rates of humans, the distribution will have a lowest and a highest value, and therefore will differ from a true normal distribution. This means that the proportions found in Table A.1 will not apply exactly to the variables and populations in the problems of Section B. However, for many real-life distributions the deviations from Table A.1 tend to be small, and the approximation involved can be a very useful tool. Moreover, the error tends to be even less when dealing with a series of samples, as you will see in Chapter 6.

Finally, because z scores are usually used only when a normal distribution can be assumed, some students get the false impression that converting to z scores somehow makes any distribution more like the normal distribution. In fact, as I pointed out earlier, converting to z scores does not change the shape of a distribution at all. Certain transformations *will* change the shape of a distribution (as described in Section C of Chapter 4) and in some cases will "normalize" the distribution, but converting to z scores is not one of them.

Exercises

*1. If you convert each score in a set of scores to a z score, which of the following will be true about the resulting set of z scores?
   a) The mean will equal 1.
   b) The variance will equal 1.
   c) The distribution will be normal in shape.
   d) All of the above
   e) None of the above

* 2. Assume that the mean height for adult women (μ) is 65 inches and that the standard deviation (σ) is 3 inches.

    a) What is the z score for a woman who is exactly 5 feet tall?

    b) What is the z score for a woman who is 5 feet 5 inches tall?

    c) What is the z score for a woman who is 70 inches tall? 75 inches tall? 64 inches tall?

    d) How tall is a woman whose z score for height is -3? -1.33? -0.3? -2.1?

    e) How tall is a woman whose z score for height is $+3$? $+2.33$? $+1.7$? $+.9$?

3. a) Calculate μ and σ for the following set of scores, and then convert each score to a z score: 64, 45, 58, 51, 53, 60, 52, 49.

    b) Calculate the mean and standard deviation of these z scores. Did you obtain the values you expected? Explain.

* 4. What is the SAT score corresponding to a) $z = -0.2$? b) $z = +1.3$? c) $z = -3.1$? d) $z = +1.9$?

* 5. What is the z score that corresponds to an SAT score of a) 520? b) 680? c) 250? d) 410?

* 6. Suppose that the verbal part of the SAT contains 30 questions and that $\mu = 18$ correct responses with $\sigma = 3$. What SAT score corresponds to a) 15 correct? b) 10 correct? c) 20 correct? d) 27 correct?

* 7. Suppose the mean for a psychological test is 24 with $\sigma = 6$. What is the T score that corresponds to a raw score of a) 0? b) 14? c) 24? d) 35?

* 8. What is the z score that corresponds to each of the following IQ scores? a) 70 b) 90 c) 105 d) 120

* 9. Use Table A.1 to find the area of the normal distribution between the mean and z when z equals a) .18 b) .50 c) .88 d) 1.25 e) 2.11

* 10. Use Table A.1 to find the area of the normal distribution beyond z when z equals a) .09 b) .75 c) 1.05 d) 1.96 e) 2.57

BASIC STATISTICAL PROCEDURES

Chapter 2 pointed out that one of the most informative ways of locating a score in a distribution is by finding the percentile rank (PR) of that score (i.e., what percentage of the distribution is below that score). To find the PR of a score within a small set of scores, the techniques described in Chapter 2 are appropriate. However, if you want to find the PR of a score with respect to a very large group of scores whose distribution resembles the normal distribution (and you know both the mean and standard deviation of this reference group), you can use the procedure described below.

Finding Percentile Ranks

It is particularly easy to find the PR of a score that is above the mean of a normal distribution. The variable we will use for the examples in this section is the IQ of adults, which has a fairly normal distribution and is already expressed as a standardized score with $\mu = 100$ and (for the Stanford-Binet test) $\sigma = 16$. To use Table A.1, however, IQ scores will have to be converted back to z scores. I will illustrate this procedure by finding the PR for an IQ score of 116, using Formula 5.1:

$$z = \frac{116 - 100}{16} = \frac{16}{16} = +1.0$$

Figure 5.5

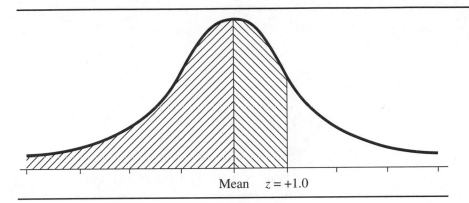

Mean $z = +1.0$

Next, we draw a picture of the normal distribution, always placing a vertical line at the mean ($z = 0$), and at the z score in question ($z = +1$, for this example). The area of interest, as shown by the cross-hatching in Figure 5.5, is the portion of the normal distribution to the left of $z = +1.0$. The entire cross-hatched area does not appear as an entry in Table A.1 (though some versions of Table A.1 include a column that would correspond to the shaded area). Notice that the cross-hatched area is divided in two portions by the mean of the distribution. The area to the left of the mean is always half of the normal distribution and therefore corresponds to a proportion of .5. The area between the mean and $z = +1.0$ can be found in Table A.1 (under Mean to z), as demonstrated in Section A. This proportion is .3413. Adding .5 to .3413 we get .8413, which is the proportion represented by the cross-hatched area in Figure 5.5. To convert a proportion to a percentage, we need only multiply by 100. Thus the proportion .8413 corresponds to a PR of 84.13%. Now we know that 84.13% of the population have IQ scores lower than 116. I emphasize the importance of drawing a picture of the normal distribution to solve these problems. In the problem above, it would have been easy to forget the .5 area to the left of the mean without a picture to refer to.

It is even easier to find the PR of a score below the mean, but you must know which column of Table A.1 to use. Suppose you want to find the PR for an IQ of 84. Begin by finding z:

$$z = \frac{84 - 100}{16} = \frac{-16}{16} = -1.0$$

Next, draw a picture and shade the area to the left of $z = -1.0$ (see Figure 5.6). Unlike the previous problem, the shaded area this time consists of only one section, which *does* correspond to an entry in Table A.1. First, you must temporarily ignore the minus sign of the z score and find 1.00 in the first column of Table A.1. Then look at the corresponding entry in the column labeled Beyond z, which is .1587. This is the proportion represented by the shaded area in Figure 5.6 (i.e., the area to the left of $z = -1.0$). The PR of 84 = .1587 · 100 = 15.87%; only about 15% of the population have IQ scores less than 84.

Figure 5.6

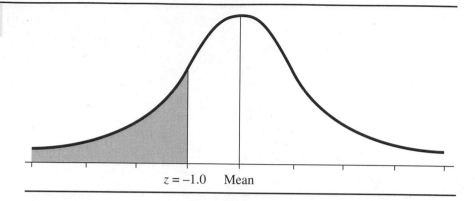

$z = -1.0$ Mean

The area referred to as Beyond z (in the third column of Table A.1) is the area that begins at z and extends *away* from the mean in the direction of the closest tail. In Figure 5.6, the area between the mean and $z = -1.0$ is .3413 (the same as between the mean and $z = +1.0$), and the area beyond $z = -1$ is .1587. Notice that these two areas add up to .5000. In fact, for any particular z score, the entries for Mean to z and Beyond z will add up to .5000. You can see why by looking at Figure 5.6. The z score divides one half of the distribution into two sections; together those two sections add up to half the distribution, which equals .5.

Finding the Area Between Two z Scores

Now we are ready to tackle more complex problems involving two different z scores. Suppose you have devised a teaching technique that is not accessible to someone with an IQ below 76 and would be too boring for someone with an IQ over 132. To find the proportion of the population for whom your technique would be appropriate, you must first find the two z scores and locate them in a drawing.

$$z = \frac{76 - 100}{16} = \frac{-24}{16} = -1.5$$

$$z = \frac{132 - 100}{16} = \frac{32}{16} = +2.0$$

From Figure 5.7 you can see that you must find two areas of the normal distribution, both of which can be found under the column Mean to z. For $z = -1.5$ you ignore the minus sign and find that the area from the mean to z is .4332. The corresponding area for $z = +2.0$ is .4772. Adding these two areas together gives a total proportion of .9104. Your teaching technique would be appropriate for 91.04% of the population.

Finding the area enclosed between two z scores becomes a bit trickier when both of the z scores are on the same side of the mean (i.e., both are positive or

Figure 5.7

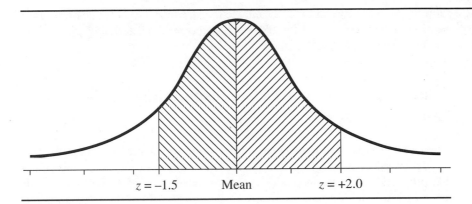

$z = -1.5$ Mean $z = +2.0$

both are negative). Suppose that you have designed a remedial teaching program that is only appropriate for those whose IQs are below 80 but would unfortunately be useless for someone with an IQ below 68. As in the problem above, you can find the proportion of people for whom your remedial program is appropriate by first finding the two z scores and locating them in your drawing.

$$z = \frac{80 - 100}{16} = \frac{-20}{16} = -1.25$$

$$z = \frac{68 - 100}{16} = \frac{-32}{16} = -2.0$$

The shaded area in Figure 5.8 is the proportion you are looking for, but it does not correspond to any entry in Table A.1. The trick is to notice that if you take the area from $z = -2$ to the mean and remove the section from $z = -1.25$ to the mean, then you are left with the shaded area. (You could also find the area beyond $z = -1.25$ and then remove the area beyond $z = -2.0$.) The area between $z = 2$ and the mean was found in the previous problem to be .4772. From this we

Figure 5.8

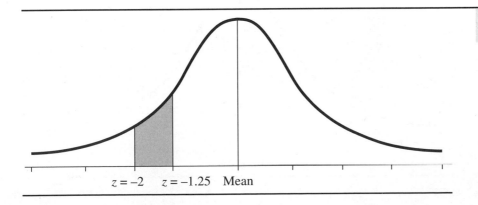

$z = -2$ $z = -1.25$ Mean

subtract the area between $z = 1.25$ and the mean, which is .3944. The proportion we want is $.4772 - .3944 = .0828$. Thus the remedial teaching program is suitable for use with 8.28% of the population.

Finding the Raw Scores Corresponding to a Given Area

Often a problem involving the normal distribution will be presented in terms of a given proportion, and it is necessary to find the range of raw scores that represents that proportion. For instance, a national organization called MENSA is a club for people with high IQs. Only people in the top 2% of the IQ distribution are allowed to join. If you were interested in joining and you knew your own IQ, you would want to know the minimum IQ score required for membership. Using the IQ distribution from the problems above, this is an easy question to answer (even if you are not qualified for MENSA). However, because you are starting with an area and trying to find a raw score, the procedure is reversed. You begin by drawing a picture of the distribution and shading in the area of interest, as in Figure 5.9. (Notice that the score that cuts off the upper 2% is also the score that lands at the 98th percentile; that is, you are looking for the score whose PR is 98.) Given a particular area (2% corresponds to a proportion of .0200) you cannot find the corresponding IQ score directly, but you can find the z score using Table A.1. Instead of looking down the z column, you look for the area of interest (in this case, .0200) first, and then see which z score corresponds to it. From Figure 5.9, it should be clear that the shaded area is the *area beyond* some as yet unknown z score, so you look in the Beyond z column for .0200. You will not be able to find this exact entry, as is often the case, so look at the closest entry, which is .0202. The z score corresponding to this entry is 2.05, so $z = +2.05$ is the z score that cuts off (about) the top 2% of the distribution. In order to find the raw score that corresponds to $z = +2.05$, you can use Formula 5.2:

Figure 5.9

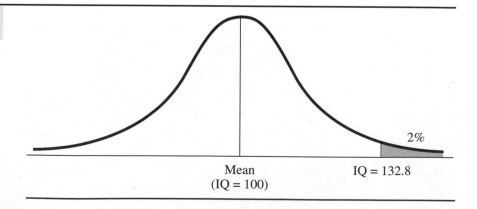

$$X = z\sigma + \mu = +2.05(16) + 100 = 32.8 + 100 = 132.8$$

Rounding off, you get an IQ of 133—so if your IQ is 133 or above you are eligible to join MENSA.

Areas in the Middle of a Distribution

One more very common type of problem involving normal distributions is to locate a given proportion in the middle of a distribution. Imagine an organization called MEZZA, which is designed for people in the middle range of intelligence. In particular, this organization will only accept those in the middle 80% of the distribution—those in the upper or lower 10% are not eligible. What is the range of IQ scores within which your IQ must fall if you are to be eligible to join MEZZA? The appropriate drawing is shown in Figure 5.10. From the drawing you can see that you must look for .1000 in the column labeled Beyond z. The closest entry is .1003, which corresponds to $z = 1.28$. Therefore, $z = +1.28$ cuts off (about) the upper 10%, and $z = -1.28$ the lower 10% of the distribution. Finally, both of these z scores must be transformed into raw scores, using Formula 5.2:

$$X = -1.28(16) + 100 = -20.48 + 100 = 79.52$$
$$X = +1.28(16) + 100 = +20.48 + 100 = 120.48$$

Thus (rounding off) the range of IQ scores that contain the middle 80% of the distribution extends from 80 to 120.

The above procedures relate raw scores to areas under the curve, and vice versa, by using z scores as the intermediate step, as shown below.

Raw score ⟷ (Formula) ⟷ z score ⟷ (Table A) ⟷ Area

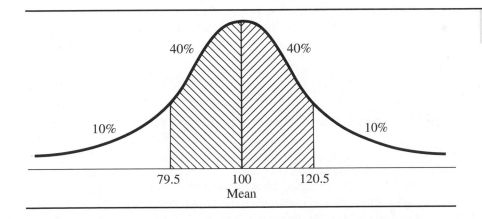

Figure 5.10

When given raw scores to start with, you move from left to right. A raw score can be converted to a z score using Formula 5.1. Then an area (or proportion) can be associated with that z score by looking down the appropriate column of Table A.1. Drawing a picture will make it clear which column is needed. When given an area or proportion to start with, you move from right to left. First, use Table A.1 backwards (look up the area in the appropriate column to find the z score), and then use Formula 5.2 to transform the z score into a corresponding raw score.

Exercises

* 1. Suppose that students in a large Introductory Psychology class have taken a midterm exam, and the scores are approximately normally distributed with $\mu = 75$ and $\sigma = 9$.

   a) What is the percentile rank (PR) for a student who scores 90?

   b) What is the PR for a score of 70?

   c) What PR corresponds to a score of 60?

   d) What is the PR for a score of 94?

* 2. Find the area between each of the following:

   a) $z = -0.5$ and $z = +1.0$

   b) $z = -1.5$ and $z = +0.75$

   c) $z = +0.75$ and $z = +1.5$

   d) $z = -0.5$ and $z = -1.5$

* 3. Assume that the resting heart rate in humans is normally distributed with $\mu = 72$ bpm (beats per minute) and $\sigma = 8$ bpm.

   a) What proportion of the population has resting heart rates above 82 bpm? Above 70?

   b) What proportion of the population has resting heart rates below 75 bpm? Below 50?

   c) What proportion of the population has resting heart rates between 80 and 85 bpm? Between 60 and 70? Between 55 and 75?

* 4. Refer again to the population of heart rates described in exercise 3.

   a) Above what heart rate do you find the upper 25% of the people? (That is, what heart rate is at the 75th percentile, or third quartile?)

   b) Below what heart rate do you find the lowest 15% of the people? (What heart rate is at the 15th percentile?)

   c) Between which two heart rates do you find the middle 75% of the people?

* 5. Imagine a new organization called DENSA, designed for people with a lower than average IQ. If to be eligible one must have an IQ between 75 and 90, what proportion of the population is eligible?

6. A new preparation course for the math SAT is open to those who have already taken the test once and scored in the middle 90% of the population. In what range must a test taker's previous score have fallen for the test taker to be eligible for the new course?

OPTIONAL MATERIAL

The Mathematics of the Normal Distribution

The true normal curve is determined by a mathematical equation, just as a straight line or a perfect circle is determined by an equation. The equation for the normal curve is a mathematical function into which you can insert any X value (usually represented on the horizontal axis) to find one corresponding Y value (usually plotted along the vertical axis). Because Y, the height of the curve, is a function of X, Y can be symbolized as $f(X)$. The equation for the normal curve can be stated as follows,

$$f(X) = \frac{1}{\sqrt{2\pi\sigma^2}} e^{-(X-\mu)^2/2\sigma^2}$$

where π is a familiar mathematical constant, and so is e ($e = 2.7183$, approximately). The symbols μ and σ^2 are called the *parameters* of the normal distribution, and they stand for the ordinary mean and variance. These two parameters determine which normal distribution is being graphed.

The above equation is a fairly complex one, but it can be simplified by expressing it in terms of z scores. This gives us the equation for the standard normal distribution and shows the intimate connection between the normal distribution and z scores:

$$f(z) = \frac{1}{\sqrt{2\pi}} e^{-z^2/2}$$

Because the variance of the standard normal distribution equals 1.0, $2\pi\sigma^2 = 2\pi$, and the power that e is raised to is just minus one-half times the z score squared. The fact that z is being squared tells us that the curve is symmetric around zero. The height of the curve depends on the magnitude of the z score but not on its direction. The height for $z = -2$ is the same as the height for $z = +2$, because in both cases, $z^2 = +4$. Because the exponent of e has a minus sign, the function is largest (i.e., the curve is highest) when z is smallest, namely zero. Thus the mode of the normal curve (along with the mean and median) occurs in the center, at $z = 0$. Note that the height of the curve is never zero; that is, the curve never touches the X axis. Instead, the curve extends infinitely in both directions, always getting closer to the X axis. In mathematical terms, the X axis is the *asymptote* of the function, and the function touches the asymptote only at infinity.

The height of the curve [$f(X)$, or Y] is called the density of the function, so the normal equation above is often referred to as a *probability density function*. In more concrete terms, the height of the curve can be thought of as representing the relative likelihood of each X value; the higher the curve, the more likely is the X value at that point. However, as I pointed out in Section A, the probability of any *exact* value occurring is infinitely small, so when we talk about the probability of some X value being selected, we talk in terms of a range of X values (i.e., an interval along the horizontal axis). The probability that the next random selection will come from that interval is equal to the proportion of the total distribution that is contained in that interval. This is the area under the curve corresponding to the given interval, and this area can be found mathematically by *integrating* the function over the interval using calculus. Fortunately, the areas between the mean and various z scores have already been calculated and entered into tables (see Table A.1), and other areas not in a table can be found by using the techniques described in Section B.

Probability

As I pointed out in Section A, statements about areas under the normal curve translate directly to statements about probability. For instance, if you select one person at random, the probability that that person will have an IQ between 76 and 132 is about .91, because that is the amount of area enclosed between those two IQ scores, as we found in Section B (see Figure 5.7). To give you a deeper understanding of probability and its relation to problems involving distributions, I will lay out some specific rules. To represent the probability of an event symbolically, I will write $p(A)$, where A stands for some event. For example, $p(IQ > 110)$ stands for the probability of selecting someone with an IQ greater than 110.

Rule 1

Probabilities range from zero (the event is certain *not* to occur) to 1 (the event is *certain* to occur), or from zero to 100 if probability is expressed as a percentage instead of a proportion. As an example of $p = 0$, consider the case of adult height. The distribution ends somewhere around $z = -15$ on the low end and $z = +15$ on the high end (unlike the true normal distribution, which extends infinitely in both directions). So for height, the probability of selecting someone for whom z is greater than $+20$ (or less than -20) is truly zero. An example of $p = 1$ is the probability that a person's height will be between $z = -20$ and $z = +20$ [i.e, $p(-20 < z < +20) = 1$].

Rule 2: The Addition Rule

If two events are **mutually exclusive,** the probability that either one event *or* the other will occur is equal to the sum of the two individual probabilities. Stated as Formula 5.5, the addition rule for mutually exclusive events is

$$p(A \text{ or } B) = p(A) + p(B) \qquad \text{Formula 5.5}$$

Two events are mutually exclusive if the occurrence of one rules out the occurrence of the other. For instance, if we select one individual from the IQ distribution, this person cannot have an IQ that is both above 120.5 and also below 79.5—these are mutually exclusive events. As I demonstrated in the discussion of the hypothetical MEZZA organization in Section B, the probability of each of these events is .10. We can now ask: What is the probability that a randomly selected individual will have an IQ above 120.5 *or* below 79.5? Using Formula 5.5, we simply add the two individual probabilities: $.1 + .1 = .2$. In terms of a single distribution, two mutually exclusive events are represented by two areas under the curve that do *not* overlap. (In contrast, the area from $z = -1$ to $z = +1$ and the area above $z = 0$ are *not* mutually exclusive, because they *do* overlap.) If the areas do not overlap we can simply add the two areas to find the probability that an event will be in one area *or* the other. The addition rule can be extended easily to any number of events, if all of the events are mutually exclusive (i.e., no event overlaps with any other). For a set of mutually exclusive events the probability that one of them will occur [p(A or B or C, etc.)] is the sum of the probabilities for each event [i.e., p(A) + p(B) + p(C), etc.].

The Addition Rule for Overlapping Events

The addition rule must be modified if events are not mutually exclusive. If there is some overlap between two events, the overlap must be subtracted after the two probabilities have been added. Stated as Formula 5.6, the addition rule for events that are *not* mutually exclusive is

$$p(A \text{ or } B) = p(A) + p(B) - p(A \text{ and } B) \qquad \text{Formula 5.6}$$

where p(A and B) represents the overlap (the region where A and B are both true simultaneously). For example, what is the probability that a single selection from the normal distribution will be either within one standard deviation of the mean *or* above the mean? The probability of the first event is the area between $z = -1$ and $z = +1$, which is about .68. The probability of the second event is the area above $z = 0$, which is .5. Adding these we get .68 + .5 = 1.18, which is more than 1.0, and therefore impossible. However, as you can see from Figure 5.11, these events are not mutually exclusive; the area of overlap corresponds to the interval from $z = 0$ to $z = +1$. The area of overlap [i.e., p(A and B)] equals about .34, and because it is actually being added in twice (once for each event), it must be subtracted once from the total. Using Formula 5.6, we find that p(A or B) = .68 + .5 − .34 = 1.18 − .34 = .84 (rounded off).

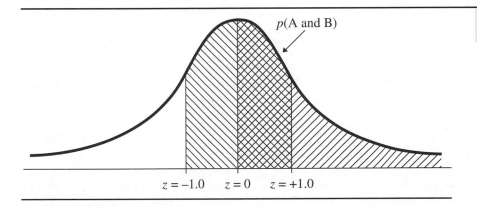

p(A and B)

$z = -1.0$ $z = 0$ $z = +1.0$

Figure 5.11

A Note about Exhaustive Events

Besides being mutually exclusive, two events can also be **exhaustive** if one or the other *must* occur (together they "exhaust" all the possible events). For instance, consider the event of being above the mean and the event of being below the

Figure 5.12

Exhaustive but Not Mutually Exclusive Events

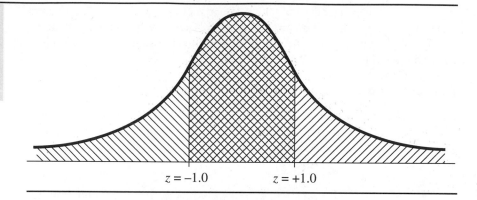

$z = -1.0$ $z = +1.0$

mean; these two events are not only mutually exclusive, they are exhaustive, as well. The same is true of being within one standard deviation from the mean and being at least one standard deviation away from the mean. When two events are both mutually exclusive and exhaustive, one event is considered the *complement* of the other, and the probabilities of the two events must add up to 1.0. If the events A and B are mutually exclusive and exhaustive, we can state that $p(B) = 1.0 - p(A)$.

Just as two events can be mutually exclusive but not exhaustive (e.g., $z > +1.0$ and $z < -1.0$), two events can be exhaustive without being mutually exclusive. For example, the two events $z > -1.0$ and $z < +1.0$ are exhaustive (there is no location in the normal distribution that is not covered by one event or the other), but they are not mutually exclusive; the area of overlap is shown in Figure 5.12. Therefore, the two areas represented by these events will not add up to 1.0, but rather somewhat more than 1.0.

Rule 3: The Multiplication Rule

If two events are **independent,** the probability that both will occur (i.e., A *and* B) is equal to the two individual probabilities multiplied together. Stated as Formula 5.7, the multiplication rule for independent events is

$$p(A \text{ and } B) = p(A)\,p(B) \qquad \text{Formula 5.7}$$

Two events are said to be independent if the occurrence of one in no way affects the probability of the other. The most common example of independent events is two flips of a coin. As long as the first flip does not damage or change the coin in some way—and it's hard to imagine how flipping a coin could change it—the second flip will have the same probability of coming up heads as the first flip. [If the coin is unbiased, $p(H) = p(T) = .5$.] Even if you have flipped a fair coin and have gotten ten heads in a row, the coin will not be altered by the flipping; the chance of getting a head on the eleventh flip is still .5. It may seem that after ten heads, a tail would become more likely than usual, so as to "even out" the total number of heads and tails. This belief is a version of the *gambler's fallacy;* in reality, the coin has no memory—it doesn't keep track of the previous ten heads in a row. The multiplication rule can be extended easily to any number of events that are all independent of each other [p(A and B and C, etc.) = p(A)p(B)p(C), etc.].

Consider two independent selections from the IQ distribution. What is the probability that if we choose two people at random, their IQs will both be within one standard deviation of the mean? In this case, the probability of both individual events is the same, about .68 (assuming we replace the first person in the pool of possible choices before selecting the second; see below). Formula 5.7 tells us that the probability of both events occurring jointly [i.e., p(A and B)] equals (.68)(.68) = .46. When the two events are *not* independent, the multiplication rule must be modified. If the probability of an event changes because of the occurrence of another event, we are dealing with a **conditional probability.**

Conditional Probability

A common example of events that are *not* independent are those that involve successive samplings from a finite population *without replacement.* Let us take the simplest possible case: A bag contains three marbles; two are white and one is black. If you grab marbles from the bag without looking, what is the probability of picking two white marbles in a row? The answer depends on whether you select marbles *with replacement* or *without replacement.* In selecting with replacement, you take out a marble, look at it, and then replace it in the bag before picking a second marble. In this case, the two selections are independent; the probability of picking a white marble is the same for both selections: 2/3. The multiplication rule tells us that when the two events are independent (e.g., sampling *with* replacement), we can multiply their probabilities. Therefore, the probability of picking two white marbles in a row is (2/3)(2/3) = 4/9, or about .44.

On the other hand, if you are sampling *without* replacement, the two events will *not* be independent, because the first selection will alter the probabilities for the second. The probability of selecting a white marble on the first pick is still 2/3, but if the white marble is *not* replaced in the bag, the probability of selecting a white marble on the second pick is only 1/2 (there is one white marble and one black marble left in the bag). Thus the conditional probability of selecting a white marble, *given that* a white marble has already been selected and not replaced [i.e., p(W|W)] is 1/2. To find the probability of selecting two white marbles in a row when not sampling with replacement, we need to use Formula 5.8 (the multiplication rule for dependent events):

$$p(\text{A and B}) = p(\text{A})p(\text{B|A}) \qquad \text{Formula 5.8}$$

In this case, both A and B can be symbolized by W (picking a white marble): p(W)p(W|W) = (2/3)(1/2) = 1/3, or .33 (less than the probability of picking two white marbles when sampling with replacement).

The larger the population, the less difference it will make whether you sample with replacement or not. (With an infinite population, the difference is infinitesimally small.) For the remainder of this text I will assume that the population from which a sample is taken is so large that sampling without replacement will not change the probabilities enough to have any practical consequences.

Exercises

* 1. Consider a normally distributed population of resting heart rates with $\mu = 72$ bpm and $\sigma = 8$ bpm.

a) What is the probability of randomly selecting someone whose heart rate is either below 58 or above 82 bpm?

b) What is the probability of randomly selecting someone whose heart rate is either between 67 and 75 bpm, above 80, or below 60?

c) What is the probability of randomly selecting someone whose heart rate is either between 66 and 77 or above 74?

* 2. Refer again to the population of heart rates described in the previous exercise.

a) What is the probability of randomly selecting two people in a row whose resting heart rates are both above 78?

b) What is the probability of randomly selecting three people in a row whose resting heart rates are all below 68?

c) What is the probability of randomly selecting two people, one of whom has a resting heart rate above 70 and one of whom has a resting heart rate below 76?

3. What is the probability of selecting each of the following at random from the population (assume $\sigma = 15$)?

a) One person whose IQ is either above 110 or below 95

b) One person whose IQ is either between 95 and 110 or above 105

c) Two people with IQs above 90

d) One person with an IQ above 90 and one person with an IQ below 115

* 4. An ordinary deck of playing cards consists of 52 different cards: 13 in each of four suits (hearts, diamonds, clubs, and spades).

a) What is the probability of randomly drawing two hearts in a row, if you replace the first card before picking the second?

b) What is the probability of randomly drawing two hearts in a row, if you draw *without* replacement?

c) What is the probability of randomly drawing one heart and then one spade in two picks *without* replacement?

* 5. A third-grade class contains 12 boys and 8 girls. Two blackboard monitors are chosen at random.

a) What is the probability that both will be girls?

b) Both will be boys?

c) There will be one of each gender?

d) Sum the probabilities found in parts a, b, and c. Can you say the events described in parts a, b, and c are all mutually exclusive and exhaustive?

SUMMARY

The Important Points of Section A

1. To localize a score within a distribution or compare scores from different distributions, *standardized scores* can be used.

2. The most common standardized score is the *z score.* The *z* score expresses a raw score in terms of the mean and standard deviation of the distribution of raw scores. The magnitude of the *z* score tells you how many standard deviations away from the mean the raw score is, and the sign of the *z* score tells you whether the raw score is above ($+$) or below ($-$) the mean.

3. If you take a set of raw scores and convert each one to a *z* score, the mean of the *z* scores will be zero and the standard deviation will be 1. The shape of the distribution of *z* scores, however, will be exactly the same as the shape of the distribution of raw scores.

4. *z* scores can be converted to *SAT scores* by multiplying by 100 and then adding 500. SAT scores have the advantages of not requiring minus signs or decimals to be sufficiently accurate. *T scores* are similar, but involve multiplication by 10 and the addition of 50.

5. The *normal distribution* is a symmetrical, bell-shaped mathematical distribution whose shape is precisely determined by an equation (see Section C). The normal distribution is actually a family of distributions, the members of which differ according to their means and/or standard deviations.

6. If all the scores in a normal distribution are converted to *z* scores, the resulting distribution of *z* scores is called the *standard normal distribution,* which is a normal distribution that has a mean of zero and a standard deviation of 1.

7. The proportion of the scores in a normal distribution that falls between a particular *z* score and the mean is given by the amount of area under the curve between the mean and *z* as compared to the total area of the distribution (defined as 1.0). These areas are given for all *z* scores (to the nearest hundredth) in a table of the standard normal distribution, such as Table A.1.

8. The probability that one random selection from a distribution will have a value between the mean and some *z* score is equal to the area under the curve between the mean and that *z* score, as given in Table A.1. The probability that the selection will have some *exact* value cannot be calculated; in a perfectly smooth distribution, the probability of any exact value is infinitely small.

9. Distributions based on real variables measured in populations of real subjects (whether people or not) can be similar to, but not exactly the same as, the normal distribution. This is because the true normal distribution extends infinitely in both the negative and positive directions.

The Important Points of Section B

If a variable is normally distributed and you know both the mean and standard deviation of the population, it is easy to find the proportion of the distribution that falls above or below any raw score, or between any two raw scores. Conversely, for a given proportion at the top, bottom, or middle of the distribution, you can find the raw score or scores that form the boundary of that proportion. I will illustrate these operations using the height of adult females as the variable, assuming $\mu = 65$ inches, and $\sigma = 3$ inches.

The Proportion (and PR) below a Given Raw Score

I will not describe a separate procedure for finding proportions above a given raw score, because once you have a drawing, the procedure should be clear. In any case, you can always find the proportion above a score by first finding the proportion below and then subtracting that proportion from 1.0.

Example A: What proportion of adult women are less than 63 inches tall?

Step 1. Find the z score (using Formula 5.1):

$$z = \frac{X - \mu}{\sigma} = \frac{63 - 65}{3} = \frac{-2}{3} = -.67$$

Step 2. Draw the picture (including vertical lines at the approximate z score and at the mean). As you can see from Figure 5.13, when the z score is negative, the area below (i.e., to the left of) the z score is the area beyond z (i.e., toward the tail of the distribution).

Step 3. Look in Table A.1. Look up .67 (ignoring the minus sign) in the z column, and then look at the entry under Beyond z. The entry is .2514, which is the proportion you are looking for.

Step 4. Multiply the proportion below the score by 100 if you want to find the PR for that score. In this example, PR $= .2514 \cdot 100 = 25.14\%$.

Figure 5.13

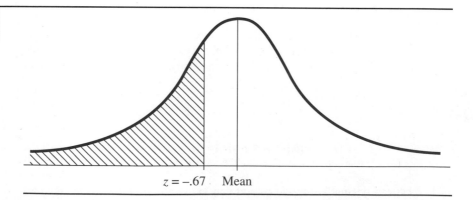

$z = -.67$ Mean

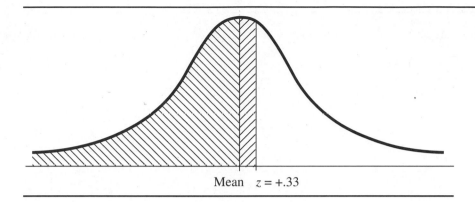

Figure 5.14

Mean $z = +.33$

Example B: What proportion of adult women are less than 66 inches tall?

Step 1. Find the z score:

$$z = \frac{66 - 65}{3} = \frac{1}{3} = +.33$$

Step 2. Draw the picture. As you can see from Figure 5.14, when the z score is positive, the area below the z score consists of two sections: the area between the mean and z, and the area below the mean (the latter always equals .5).

Step 3. Look in Table A.1. The area between the mean and z for $z = .33$ is .1293. Adding the area below the mean yields .1293 + .5 = .6293, which is the proportion you are looking for.

Step 4. Multiply by 100 to find the PR. In this case, the PR for 66 inches = .6293 · 100 = 62.93%.

The Proportion Between Two Raw Scores

Example A: What is the proportion of adult women between 64 inches and 67 inches in height?

Step 1. Find the z scores:

$$z = \frac{64 - 65}{3} = \frac{-1}{3} = -.33 \quad \text{and} \quad z = \frac{67 - 65}{3} = \frac{2}{3} = +.67$$

Step 2. Draw the picture. From Figure 5.15 you can see that when the z scores are opposite in sign (i.e., one is above the mean and the other is below), you must find two areas: the area between the mean and z for each of the two z scores.

Figure 5.15

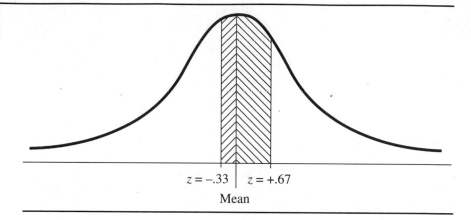

$z = -.33$ $z = +.67$
Mean

Step 3. Look in Table A.1. The area between the mean and z is .1293 for $z =$ .33 and .2486 for $z = .67$. Adding these gives $.1293 + .2486 = .3779$, which is the proportion you are looking for.

Example B: What proportion of adult women is between 67 inches and 68 inches tall?

Step 1. Find the z scores:

$$z = \frac{67 - 65}{3} = \frac{2}{3} = +.67 \quad \text{and} \quad z = \frac{68 - 65}{3} = \frac{3}{3} = +1.0$$

Step 2. Draw the picture. You can see from Figure 5.16 that both z scores are on the same side of the mean, which means that you must find the area between the mean and z for each and then subtract the two areas.

Figure 5.16

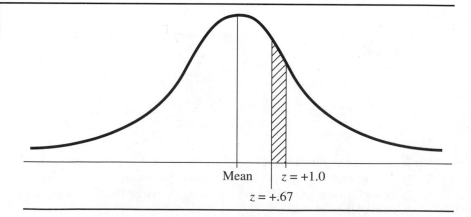

Mean $z = +1.0$
$z = +.67$

Figure 5.17

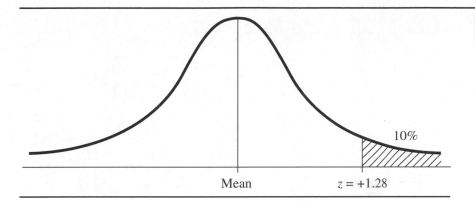

10%

Mean $z = +1.28$

Step 3. Look in Table A.1. The area between the mean and z is .2486 for $z = .67$ and .3413 for $z = 1.0$. Subtracting gives $.3413 - .2486 = .0927$, which is the proportion you are looking for. (Note: You cannot subtract the two z scores first and then look for the area; this will give you the wrong area.)

The Raw Score(s) Corresponding to a Given Proportion

Because you are starting with a proportion and looking for a raw score, the order of the steps is different.

Example A: Above what height are the tallest 10% of adult women?

Step 1. Draw the picture. Shade in the top 10% of the distribution, as shown in Figure 5.17. From the figure you can see that you want to know the z score for which the area beyond is 10%, or .1000.

Step 2. Look in Table A.1. Looking in the Beyond z column for .1000, you see that the closest entry is .1003, which corresponds to $z = +1.28$. (If you were looking for the bottom 10%, z would equal -1.28.)

Step 3. Find the raw score (using Formula 5.2):

$$X = z\sigma + \mu = +1.28(3) + 65 = 3.84 + 65 = 68.84$$

The tallest 10% of adult women are 68.84 inches tall or taller.

Example B: Between which two heights are the middle 90% of adult women?

Step 1. Draw the picture. Subtract 90% from 100%, which equals 10%, to find the percentage in the tails. Then divide this percentage in half to find the percentage that should be shaded in each tail (in this case, 5% of the distribution is shaded in each tail, as shown in Figure 5.18). Thus you need the z score for which the area beyond is 5%, or .0500.

Figure 5.18

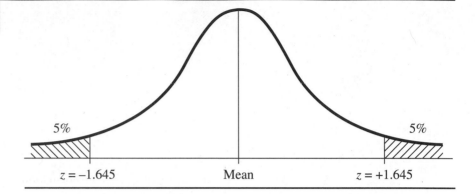

5% $z = -1.645$ Mean $z = +1.645$ 5%

Step 2. Look in Table A.1. You cannot find .0500 in the Beyond z column, but it is halfway between 1.64 (.0505) and 1.65 (.0495), so you can say that $z = 1.645$. From Figure 5.18, you can see that you want both $z = +1.645$ and $z = -1.645$.

Step 3. Find the raw scores:

$$X = +1.645(3) + 65 = +4.94 + 65 = 69.94$$
$$X = -1.645(3) + 65 = -4.94 + 65 = 60.06$$

The middle 90% of adult women are between 60.06 inches and 69.94 inches tall.

The Important Points of Section C

1. The equation of the normal distribution depends on two parameters: the mean and the variance (or standard deviation) of the population. If the equation is expressed in terms of z scores, the result is the standard normal distribution.

2. The highest point of the curve (i.e., the mode) is at $z = 0$, which is also the mean and the median, because the normal curve is perfectly symmetrical around $z = 0$.

The Rules of Probability

Rule 1: The probability of an event [e.g., $p(A)$] is usually expressed as a proportion, in which case $p(A)$ can range from zero (the event A is certain *not* to occur) to 1.0 (the event A *is* certain to occur).

   Rule 2: The addition rule for mutually exclusive events. If the occurrence of event A precludes the occurrence of event B, the probability of either A *or* B occurring, $p(A \text{ or } B)$, equals $p(A) + p(B)$.

The addition rule must be modified as follows if the events are *not* mutually exclusive (Formula 5.5): $p(A \text{ or } B) = p(A) + p(B) - p(A \text{ and } B)$. Also, if two events are both mutually exclusive and *exhaustive* (one of the two events must occur), then $p(A) + p(B) = 1.0$, and therefore, $p(B) = 1.0 - p(A)$.

Rule 3: The multiplication rule for independent events. The probability that two independent events will both occur, $p(A \text{ and } B)$, equals $p(A)p(B)$.

Two events are *not* independent if the occurrence of one event changes the probability of the other. The probability of one event, given that another has occurred, is called a *conditional probability*. The probability of two dependent events *both* occurring is given by a modified multiplication rule: The probability of one event is multiplied by the conditional probability of the other event, *given that the first* event has occurred. When you are sampling from a finite population without replacement, successive selections will not be independent. However, if the population is very large, *sampling without replacement* is barely distinguishable from *sampling with replacement* (any individual in the population has an exceedingly tiny probability of being selected twice), so successive selections can be considered independent even without replacement.

Definitions of Key Terms

z **score** A standardized score designed to have a mean of zero and a standard deviation of 1. The magnitude of the *z* score tells you how many standard deviations it is from the mean, and the sign of the *z* score tells you whether it is above or below the mean.

Standardized score A raw score that has been transformed by a simple formula, so that the mean and standard deviation are convenient numbers and the shape of the distribution is not changed.

Raw score A number that comes directly from measuring a variable, without being transformed in any way.

SAT scores Standardized scores that have a mean of 500 and a standard deviation of 100. They are often used to report the results of standardized tests, because they avoid the use of the minus signs and decimals that are necessary with *z* scores.

Standard normal distribution A normal distribution with a mean of zero and a standard deviation of 1. Any normal distribution can be transformed into the standard normal distribution by converting all the scores to *z* scores.

Area under the curve An area within a distribution bounded by vertical lines at each end of a range of values, the horizontal axis, and the curve of the distribution. The ratio of the area of this space to the area of the entire distribution (defined as having an area of 1.0) tells you the proportion of scores that fall within the given range of values.

Probability A number that describes the likelihood of the occurrence of some event, usually expressed as a proportion between zero (no chance of the event occurring) and 1 (certainty that the event will occur). In terms of a distribution, the probability that an event will be from a particular range of values is given by the proportion of area under the curve corresponding to that range of values.

Mutually exclusive events Two events are mutually exclusive if the occurrence of one event is incompatible with the occurrence of the other event. Areas in a distribution that do not overlap represent mutually exclusive events.

Exhaustive events A set of events is exhaustive if all possibilities are represented, such that one of the events must occur. When a coin is flipped, the events "heads" and "tails" are exhaustive (unless there is a chance the coin will land and remain on its edge). Heads and tails are also mutually exclusive.

Independent events Two events are independent if the occurrence of one event does not change the probability of the occurrence of the other event. Under ordinary circumstances, two flips of a coin are independent events.

Conditional probability The probability of one event, given that another event has already occurred. For example, the probability that a child will be above average in IQ is .5, but the probability of a child being above average, given that both of the child's parents are above average, is higher than .5.

Key Formulas

The z score corresponding to a raw score:

$$z = \frac{X - \mu}{\sigma}$$

Formula 5.1

The raw score that corresponds to a given z score:

$$X = z\sigma + \mu$$

Formula 5.2

The SAT score corresponding to a raw score, if the z score has already been calculated:

$$\text{SAT} = 100z + 500$$

Formula 5.3

The T score corresponding to a raw score, if the z score has already been calculated:

$$T = 10z + 50$$

Formula 5.4

The addition rule for mutually exclusive events:

$$p(A \text{ or } B) = p(A) + p(B) \qquad \text{Formula 5.5}$$

The addition rule for events that are *not* mutually exclusive:

$$p(A \text{ or } B) = p(A) + p(B) - p(A \text{ and } B) \qquad \text{Formula 5.6}$$

The multiplication rule for independent events:

$$p(A \text{ and } B) = p(A)p(B) \qquad \text{Formula 5.7}$$

The multiplication rule for events that are *not* independent (based on *conditional probability*):

$$p(A \text{ and } B) = p(A)p(B|A) \qquad \text{Formula 5.8}$$

Hypothesis Tests Involving One or Two Group Means

Part III introduces the concept of null hypothesis testing and culminates with an explanation of how the means of two samples can be used to draw an inference about whether two populations have the same or different means. When you want to test some new treatment (e.g., a drug or a therapy), you give the new treatment to one group of subjects and not to another. You will very likely observe *some* difference (at least a tiny bit) between the means of the two groups (on some outcome variable), even if your new treatment is totally ineffective. The practical problem addressed in Part III is, how tiny does the difference have to be before you conclude "This treatment is just not working"? Or, how large must the difference be before you can say with confidence "This difference is too large to be accidental, the treatment must be working"? In order to explain thoroughly a two-group experiment and its analysis, it is helpful to begin with a discussion of one-sample experiments and the concept of a sampling distribution. This progression is reflected in the chapters of Part III.

The effectiveness of a treatment in a psychology experiment is usually evaluated in terms of the mean of a group of subjects getting that treatment, as measured on some relevant variable (e.g., self-esteem, problem-solving performance). Because the focus of Part III is the means of groups (i.e., sample means), you will need to apply what you learned about the distributions of individual scores in Part II to distributions formed by many (ideally, infinitely many) sample means. The distribution of sample means is called the sampling distribution of the mean and is described in Section A of Chapter 6. In Section B, the concept of standardized scores, introduced in Chapter 5, is extended from individuals to the means of groups. Section C briefly describes the Central Limit Theorem, an important statistical law that explains why you can expect the sampling distribution of the mean to look like the normal distribution even when dealing with a variable whose distribution in the population looks nothing like the normal distribution.

Section A of Chapter 7 introduces the basic concepts of null hypothesis testing that will be used throughout the remainder of the text: test statistics, significance levels, one- and two-tailed tests, and type I and type II errors. In Section B I outline a six-step procedure for null hypothesis testing that is used in almost all of the remaining chapters of the text. Section C delves into some of the complexities of the logic of null hypothesis testing. Although the concepts discussed in Chapter 7 are vital to understanding the material in subsequent chapters, the one-sample hypothesis test described therein is rarely used. It is unusual to test an experimental condition on one sample without using a second sample for comparison purposes (i.e., a control group). Even when the sample represents some preexisting condition or population (e.g., vegetarians, left-handers), it is common to draw a comparison sample.

Chapter 8 continues the focus on one-sample experiments. In Section A, I deal with the realistic situation in which your sample is fairly small and you don't know any of the parameters of the population distribution of the variable you are measuring. In addressing this problem, I introduce the t distribution as a minor modification of the normal distribution, which must be used when your samples are small. Section B presents a very reasonable way to use the data from one sample. Instead of asking whether the population mean for a special group or treatment is different or the same as the mean of the general population, you can use the procedure in Section B to estimate the actual value of the population mean of interest by constructing a confidence interval. Section C adds some detail about estimation in general and about the estimation of a population's variance, rather than its mean.

Chapter 9 combines information about null hypothesis testing from Chapter 7 and information about the t distribution from Chapter 8 and applies these concepts to the analysis of a two-group experiment. Section A deals with the complexities of combining the variances of two samples into a single estimate of population variance to be used in the denominator of the t-test. Section B demonstrates how to apply both null hypothesis testing and confidence interval construction to the two-sample case. Section C tackles the problems of determining when two populations can be assumed to have the same variance and performing the two-group t-test when the two population variances cannot be assumed to be equal. Chapter 9 reiterates that the two-group experiment consists of an independent variable (which distinguishes the two groups) and a dependent variable (which must be measured on an interval or ratio scale). If subjects are assigned at random by the experimenter to the levels of the independent variable (e.g., experimental drug versus placebo), then the experiment is a true experiment. On the other hand, if the samples represent preexisting populations (e.g., males versus females), the experiment is called a quasi-experiment.

Finally, Chapter 10 completes the picture of null hypothesis testing by discussing factors that influence the type II error rate (beta), and therefore the power (1 − beta) of a two-group experiment. Section A explains how power depends on the t value you expect for your experiment and your alpha level. Your expected t value depends, in turn, on your specific alternative hypothesis expressed as the effect size (the true separation of the two populations expressed in standard deviations), as well as on the size of the samples you are using. Section B shows how to use tables and formulas to estimate power for a given effect size and sample size, or to estimate the sample size needed to attain the desired power level for a given effect size. Section C discusses some practical issues and limitations involved in null hypothesis testing.

By the end of Part III you should have a thorough understanding of two-group experiments: how to test for a significant difference in population means; what conclusions to draw; what to do if the population variances cannot be assumed equal; and how to estimate your chances of attaining statistical significance in terms of the population effect size, sample size, and significance level.

THE SAMPLING DISTRIBUTION OF THE MEAN

CHAPTER

You will need to use the following from previous chapters:

Symbols
μ: Mean of a population
$\overline{X}$: Mean of a sample
σ: Standard deviation of a population
N: Number of subjects (or observations) in a sample

Formulas
Formula 5.1: The z score

Procedures
Finding areas under the normal distribution (Chapter 5, Section B)

A

CONCEPTUAL FOUNDATION

Chapter 5 showed that if you are dealing with a variable that has a nearly normal distribution in the population, you can answer interesting questions easily, provided that you know the mean and standard deviation. Given any range of scores on the variable of interest, you can find the proportion of the population that falls in that range, and therefore the probability that someone chosen at random from the population will be in that range. Conversely, you can find the value of the variable that cuts off the upper 10% of the distribution, or the values surrounding the middle 50%, etc. Besides answering simple questions about a population, the areas under the normal distribution can be used to draw indirect inferences about the effects of experimental manipulations, as in the following example.

Suppose you know that heart rate at rest is approximately normally distributed, with a mean of 72 beats per minute (bpm) and a standard deviation of 10 bpm. You also know that a friend of yours, who drinks an unusual amount of coffee every day—five cups—has a resting heart rate of 95 bpm. Naturally, you suspect that the coffee is related to the high heart rate, but then you realize that some people in the ordinary population have resting heart rates just as high as your friend's. Coffee isn't necessary as an explanation of the high heart rate, because there is plenty of variability within the population based on genetic and other factors. Still, it may seem like quite a coincidence that your friend drinks so much coffee *and* has such a high heart rate. How much of a coincidence this really is depends in part on just how unusual your friend's heart rate is. If a large proportion of the population has heart rates as high as your friend's, you are not dealing with such a big coincidence; it would be very reasonable to suppose that your friend was just one of the many with high heart rates that have nothing to do with coffee consumption. On the other hand, if a very small segment of the population has heart rates as high as your friend's, you must believe either that your friend happens to be one of those rare individuals who naturally have a high heart rate or that the coffee is elevating his heart rate. The more unusual your friend's heart rate, the harder it is to believe that the coffee is not to blame. You can use your knowledge of the normal distribution to determine just how unusual your friend's heart rate is. Calculating your friend's z score (Formula 5.1), we find

$$z = \frac{X - \mu}{\sigma} = \frac{95 - 72}{10} = \frac{23}{10} = 2.3$$

The area beyond a z score of 2.3 is only about .011, so this is quite an unusual heart rate; only a little more than 1% of the population has heart rates that are as high or higher. Of course, it is still possible that your friend is one of those rare individuals who naturally has such a high heart rate, but now it does seem like more than just coincidence that he also drinks so much coffee.

The above example has applications for psychological research. However, a researcher would not be interested in finding out whether coffee has raised the heart rate of one particular individual. The more important question is whether coffee raises the heart rates of humans in general. One way to answer this question is to look at a series of individuals who are heavy coffee drinkers, and, in each case, find out how unusually high the heart rate is. Somehow all of these individual probabilities would have to be combined in order to decide whether these heart rates are just too unusual to believe that the coffee is uninvolved. There is a simpler way to attack the problem. Instead of focusing on one individual at a time, psychological researchers usually look at a group of subjects as a whole. This is certainly not the only way to conduct research, but because of its simplicity and widespread use, the group approach is the basis of statistics in introductory texts, including this one.

Sampling Distribution of the Mean

In this chapter I will shift focus from individuals to groups of individuals. Instead of the heart rate of an individual, we need to talk about the heart rate of a group. To do so we have to find a single heart rate to characterize an entire group. Chapter 3 showed that the mean, median, and mode are all possible ways to describe the central tendency of a group, but the mean has the most convenient mathematical properties and leads to the simplest statistical procedures. Therefore, for most of this text, the mean will be used to characterize a group; that is, when I want to refer to a group by a single number, I will use the mean.

A researcher who wanted to explore the effects of coffee on resting heart rate would assemble a group of heavy coffee drinkers and find the mean of their heart rates. Then the researcher could see if the group mean was unusual or not. However, to evaluate how unusual a group mean is, you cannot compare the group mean to a distribution of individuals. You need, instead, a distribution of groups (all the same size). This is a more abstract concept than a population distribution that consists of individuals, but it is a critical concept for understanding the statistical procedures in the remainder of this text.

If we know that heart rate has a nearly normal distribution with a mean of 72 and a standard deviation of 10, what can we expect for the mean heart rate of a small group? There is a very concrete way to approach this question. First you have to decide on the size of the groups you want to deal with; this makes quite a difference, as you will soon see. For our first example, let us say that we are interested in studying groups that have 25 subjects each. So we take 25 people from the general population and find the mean heart rate for that group. Then we do this again and again, each time recording the mean heart rate. If we do this many times, the mean heart rates will start to pile up into a distribution. As we approach an infinite number of group means, the distribution becomes smooth and continuous. One convenient property of this distribution of means is that it will be a normal distribution, provided that the variable has a normal distribution for the individuals in the population.

Because the groups that we have been hypothetically gathering are considered samples of the population, the group means are called *sample means,* and are symbolized by $\overline{X}$ (read as "*X* bar"). The distribution of group means is called a *sampling distribution.* More specifically, it is called the **sampling distribution of the mean.** (Had we been taking the median of each group of 25 and piling up these medians into a distribution, that would be called the sampling distribution of the *median.*) Just as the population distribution gives us a picture of how the individuals are spread out on a particular variable, the sampling distribution shows us how the sample means (or medians, or whatever is being used to summarize each sample) would be spread out if we grouped the population into very many samples. To make things simple, I will assume for the moment that we are always dealing with variables that have a normal distribution in the population. Therefore, the sampling distribution of the mean will always be a normal distribution, which implies that we need only know its mean and standard deviation to know everything about it.

First, consider the mean of the sampling distribution of the mean. This term may sound confusing, but it really is very simple. The mean of all the group means will always be the same as the mean of the individuals, i.e., the population mean (μ). This assumes that each group is a *random sample,* as discussed below. Otherwise there is no way to predict anything about the sampling distribution. It should make sense that if you have very many random samples from a population, there is no reason for the sample means to be more often above or below the population mean. For instance, if you are looking at the average heights for groups of men, why should the average heights of the groups be any different from the average height of individual men? However, finding the standard deviation of the sampling distribution is a more complicated matter. Whereas the standard deviation of the individuals within each sample should be roughly the same as the standard deviation of the individuals within the population as a whole, the standard deviation of the sample means is a very different kind of thing.

Standard Error of the Mean

The means of samples do not vary as much as the individuals in the population. To make this concrete, consider again a very familiar variable: the height of adult men. It is obvious that if you were to pick a man off the street at random, it is somewhat unlikely that the man would be over 6 feet tall, but not very unlikely (the chance would be better than .1). On the other hand, imagine selecting a group of 25 men *at random* and finding their average height. The probability that the 25 men would average over 6 feet in height is extremely small. Remember that the group was selected at random. It is not difficult to find 25 men whose average height is over 6 feet tall (you might start at the nearest basketball court), but if the selection is truly random, men below 5 feet 6 inches will be just as likely to be picked as men over 6 feet tall. The larger the group, the smaller the chance that the group mean will be far from the population mean (in this case, about 5 feet 9 inches). Imagine finding the average height of men in each of the 50 states of the United States. Could the average height of men in Wisconsin be much different from the average height of men in Pennsylvania or Alabama? Such extremely large groups will not vary much from each other or from the population mean. That sample means vary less from each other than do individuals is a critical concept for understanding the statistical procedures in Parts III, IV, and V of this book. The concept is critical because we will be judging whether groups are unusual, and the fact that groups vary less than individuals do implies that it takes a smaller deviation for a group to be unusual than for an individual to be unusual. Fortunately, there is a simple formula that can be used to find out just how much groups tend to vary.

Because sample means do not vary as much as individuals, the standard deviation for the sampling distribution will be less than the standard deviation for a population. As the samples get larger, the sample means are clustered more closely, and the standard deviation of the sample means therefore gets smaller. This characteristic can be expressed by a simple formula, but first I will introduce a new term. The standard deviation of the sampling distribution of the mean is called the **standard error of the mean** and is symbolized as $\sigma_{\bar{X}}$. For any particular sampling distribution all of the samples must be the same size, symbolized by N (for the number of observations in each sample). How many different random samples do you have to select to make a sampling distribution? The question is irrelevant, because nobody really creates a sampling distribution this way. The kinds of sampling distributions that I will be discussing are mathematical ideals that are based on drawing an infinite number of samples all of the same size. (This approach creates a sampling distribution that is analogous to the population distribution, which is based on an infinite number of individuals.)

Now I can show how the standard error of the mean decreases as the size of the samples increases. This relationship is expressed as Formula 6.1:

$$\sigma_{\bar{X}} = \frac{\sigma}{\sqrt{N}} \qquad\qquad \text{Formula 6.1}$$

To find the standard error, you start with the standard deviation of the population and then divide by the square root of the sample size. This means that for any given sample size, the more the individuals vary, the more the groups will vary (i.e., if σ gets larger, then $\sigma_{\overline{X}}$ gets larger). On the other hand, the larger the sample size, the *less* the sample variances will vary (i.e., as N increases, $\sigma_{\overline{X}}$ decreases). For example, if you make the sample size four times larger, the standard error is cut in half (e.g., σ is divided by 5 for a sample size of 25, but it is divided by 10 if the sample size is increased to 100).

Sampling Distribution versus Population Distribution

In Figure 6.1 you can see how the sampling distribution of the mean compares with the population distribution for a specific case. We begin with the population distribution for the heights of adult men. It is a nearly normal distribution with $\mu = 69$ inches and $\sigma = 3$ inches. For $N = 9$, the sampling distribution of the mean is also approximately normal. We know this because there is a statistical law that states that if the population distribution is normal, then the sampling distribution of the mean will also be normal. Moreover, there is a theorem that states that when the population distribution is not normal, the sampling distribution of the mean will be closer to the normal distribution than the population distribution (the Central Limit Theorem, discussed in Section B). So, if the population distribution is close to normal to begin with (as in the case of height for adults of a certain gender), we can be sure that the sampling distribution of the mean for this variable will be very similar to the normal distribution.

Compared to the population distribution, the sampling distribution of the mean will have the same mean but a smaller standard deviation (i.e., standard error). The standard error in this case is 1 inch ($\sigma_{\overline{X}} = \sigma/\sqrt{N} = 3/\sqrt{9} = 3/3 = 1$). For $N = 100$, the sampling distribution becomes even narrower; the standard error equals 0.3 inch. Notice that the means of groups tend to vary less from the

Figure 6.1

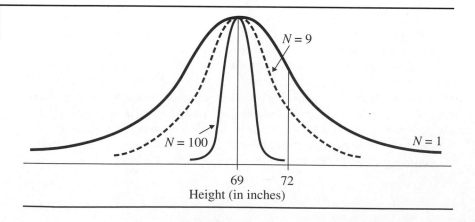

population mean than do the individuals, and that large groups vary less than small groups.

Referring to Figure 6.1, you can see that it is not very unusual to pick a man at random who is about 72 inches, or 6 feet, tall. This is just one standard deviation above the mean, so nearly one in six men is 6 feet tall or taller. On the other hand, to find a group of nine randomly selected men whose average height is over 6 feet is quite unusual; such a group would be three standard errors above the mean. This corresponds to a z score of 3, and the area beyond this z score is only about .0013. And to find a group of 100 randomly selected men who averaged 6 feet or more in height would be extremely rare indeed; the area beyond $z = 10$ is too small to appear in most standard tables of the normal distribution. Section B will illustrate the use of z scores when dealing with groups.

Exercises

* 1. The distribution of body weights for adults is somewhat positively skewed—there is much more room for people to be above average than below. If you take the mean weights for random groups of ten adults each and form a new distribution, how will this new distribution compare to the distribution of individuals?

a) The new distribution will be more symmetrical than the distribution of individuals.

b) The new distribution will more closely resemble the normal distribution.

c) The new distribution will be narrower (i.e., have a smaller standard deviation) than the distribution of individuals.

d) All of the above

e) None of the above

2. Each of the students in a statistics class rated his or her level of math anxiety on a ten-point scale (from 0 = no anxiety to 10 = paralyzing anxiety), as shown below.

4	3	6	9
5	3	2	8
2	4	6	2
8	5	4	5

a) How large is the unbiased sample standard deviation?

b) If the standard deviation for the entire population (σ) were equal to the value you found in part a, how large would the standard error of the mean be for samples of this size?

* 3. If the population standard deviation (σ) for some variable equals 17.5, what is the value of the standard error of the mean when a) $N = 5$; b) $N = 25$; c) $N = 125$; d) $N = 625$?

4. If the sample size is cut in half, what happens to the standard error of the mean for a particular variable?

* 5. If the standard deviation of variable A is three times as large as the standard deviation of variable B, how would the standard error of the mean ($\sigma_{\bar{X}}$) for variable A compare to $\sigma_{\bar{X}}$ for variable B (assuming N is the same in both cases)?

* 6. In one college, freshman English classes always contain exactly 20 students. An English teacher wonders how much students in these classes are likely to vary in terms of their verbal scores on the SAT. What would you expect for the standard deviation (i.e., standard error) of class means on the verbal SAT?

7. Assuming that IQ is normally distributed with a mean of 100 and a standard deviation of 15, describe completely the sampling distribution of the mean for a sample size (N) of 20.

*8. Suppose that a crew for the space shuttle consists of seven people, and we are interested in the average weights of all possible shuttle crews. If the standard deviation for weight is 30 pounds, what is the standard deviation for the mean weights of shuttle crews (i.e., the standard error of the mean)?

*9. If, for a particular sampling distribution of the mean, we know that the standard error is 4.6, and we also know that σ = 32.2, what is the sample size (*N*)?

10. A grocer has noticed that the weight of a dozen eggs varies slightly. After weighing many packages each containing a dozen eggs, she calculates a standard deviation of .41 ounce. What would be your guess for the standard deviation of eggs weighed individually (i.e., the population standard deviation)?

BASIC STATISTICAL PROCEDURES

In Chapter 5, I showed how to answer a variety of questions about a variable that has a nearly normal distribution, once you know the mean and standard deviation. For example, if you want to know the proportion of women between 65 and 70 inches tall, you can use *z* scores and the table for the standard normal distribution. Using the sampling distribution of the mean, you can find answers to questions about groups rather than individuals. Suppose there is a university that encourages women's basketball by assigning all of the female students to one basketball team or another at random. Assume that each team has exactly the required five players. Imagine that a particular woman wants to know the probability that the team she is assigned to will have an average height of over 67 inches. I will show how the sampling distribution of the mean can be used to answer that question.

We begin by assuming that the heights of women at this university form a normal distribution with a mean of 65 inches and a standard deviation of 3 inches. Next, we need to know what the distribution would look like if it were composed of the means from an infinite number of basketball teams, each with five players. In other words, we need to find the sampling distribution of the mean for *N* = 5. First, we can say that the sampling distribution will be a normal one, because we are assuming that the population distribution is normal. Given that the sampling distribution is normal, we need only specify its mean and standard deviation.

The *z* Score for Groups

The mean of the sampling distribution is the same as the mean of the population, that is, μ. For this example, μ = 65 inches. The standard deviation of the sampling distribution of the mean, called the standard error of the mean, is given by Formula 6.1. For this example the standard error, $\sigma_{\bar{X}}$, equals

$$\sigma_{\bar{X}} = \frac{\sigma}{\sqrt{N}} = \frac{3}{\sqrt{5}} = \frac{3}{2.24} = 1.34$$

Now that we know the parameters of the distribution of the means of groups of five (e.g., the basketball teams), we are prepared to answer questions about any particular group, such as the team that includes the inquisitive woman in our example. Because the sampling distribution is normal, we can use the standard normal table to determine, for example, the probability of a particular team having an average height greater than 67 inches. However, we first have to convert the particular group mean of interest to a z score—in particular, a z score with respect to the sampling distribution of the mean, or more informally, a z score for groups. The z score for groups closely resembles the z score for individuals and is given by Formula 6.2,

$$z = \frac{\overline{X} - \mu}{\sigma_{\overline{X}}} \qquad \text{Formula 6.2}$$

in which $\sigma_{\overline{X}}$ is a value found using Formula 6.1.

Formula 6.2 can also be written with a subscript on μ, as shown below:

$$z = \frac{\overline{X} - \mu_{\overline{X}}}{\sigma_{\overline{X}}} \qquad \text{Formula 6.2A}$$

Because $\mu_{\overline{X}}$, the mean of the sample means, is always equal to μ, the mean of the population, the subscript on μ is usually left off.

To show the parallel structures of the z score for individuals and the z score for groups, Formula 5.1 for the individual z score can be rewritten with subscripts:

$$z = \frac{X - \mu_{X}}{\sigma_{X}} \qquad \text{Formula 5.1A}$$

Comparing the formula above with Formula 6.2A, you can see that in both cases we start with a particular score (or the sample mean), subtract the mean of those scores (or of the sample means), and then divide by the standard deviation of those scores (or of the sample means). When dealing with a single variable, it is common to leave off the subscript X, and write Formula 5.1A as Formula 5.1. However, such subscripts will become useful in Part IV, when we deal with two variables at a time.

In the present example, if we want to find the probability that a randomly selected basketball team will have an average height over 67 inches, it is necessary to convert 67 inches to a z score for groups, as follows:

$$z = \frac{67 - 65}{1.34} = 1.49$$

The final step is to find the area beyond $z = 1.49$ in Table A.1; this area is approximately .068. As Figure 6.2 shows, most of the basketball teams have mean heights that are less than 67 inches; an area of .068 corresponds to fewer than

Figure 6.2

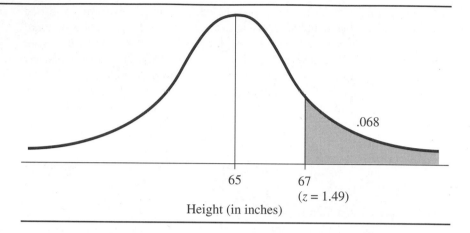

7 chances out of 100 (or about 1 out of 15) that the woman in our example will be on a team whose average height is at least 67 inches.

Using the *z* score for groups, you can answer a variety of questions about how common or unusual a particular group is. For the present example, because the standard error is 1.34 and the mean is 65, we know immediately that a little more than two-thirds of the basketball teams will average between 63.66 inches (i.e., 65 − 1.34) and 66.34 inches (i.e., 65 + 1.34) in height. Teams with average heights in this range would be considered fairly common, whereas teams averaging more than 67 inches or less than 63 inches in height would be relatively uncommon.

You could make similar determinations about the example introduced in Section A, which involved the possible effects of caffeine on heart rate. After gathering a group of heavy coffee drinkers and finding the average heart rate for that group, you could use the procedures described above to determine how unusual it would be to find a random group (the same size) with an average heart rate just as high. The more unusual the heart rate of the coffee-drinking group turns out to be, the more inclined you would be to suspect a link between coffee consumption and heart rate. Remember, however, that the observation that heavy coffee drinkers do indeed have higher heart rates does not imply that drinking coffee *causes* an increase in heart rate—although it may suggest that. It may be that people with high heart rates are more inclined to drink coffee. There are many possible alternative explanations that can only be ruled out by a *true* experiment, in which the experimenter decides (randomly) who will be drinking coffee and who will not. I will have more to say about the design of experiments and the kinds of conclusions that can be drawn from statistical tests in the chapters ahead. In Chapter 7, for instance, I will detail how the *z* score for groups can be used to answer questions that arise from psychological research. But first you need to know more about when it is appropriate to use the sampling distribution of the mean.

Assumptions

The use of the sampling distribution of the mean in the example of the basketball teams was based on assuming that the sampling distribution would be a normal distribution and that its standard deviation would be given by Formula 6.1. These assumptions are justified only if the following conditions apply.

The Variable Follows a Normal Distribution in the Population

Strictly speaking, the variable must be normally distributed in the population before you can be certain that the sampling distribution will also be normal. However, as discussed in the previous chapter, the variables that psychological researchers measure do not form perfectly normal distributions, although many of the variables come close. In fact, many variables that are important to psychology, such as reaction time, often form distributions that are quite skewed or are otherwise considerably different from the normal distribution. Fortunately there is a statistical law, known as the **Central Limit Theorem,** that states that for nonnormal population distributions, the sampling distribution of the mean will be closer to the normal distribution than the original population distribution. Moreover, as the samples get larger, the sampling distribution more closely approaches the shape of the normal distribution.

If the variable is close to being normally distributed in the population, the sampling distribution can be assumed to be normal, with a negligible amount of error, even for very small samples. On the other hand, if the population distribution is extremely skewed or is very different from the normal distribution in some other way, the sampling distribution won't be very close to normal, unless the samples are fairly large. (Usually sample sizes of 30 or 40 are adequate even if the population distribution is not very close to being normal.) It is, therefore, important *not* to assume that the sampling distribution of the mean is normal, if you know that the population distribution is far from normal *and* you are using small samples. The Central Limit Theorem and its implications for inferential statistics will be discussed in greater detail in Section C.

The Samples Are Selected Randomly

Many statistical laws, such as the Central Limit Theorem, that lead to the kinds of simple statistical procedures described in this text, assume that groups are being created by *simple random sampling*. The easiest way to define this method of sampling is to stipulate that if the sample size is equal to N, then every possible sample of size N that could be formed from the population must have the same probability of being selected. It is important to understand that this rule goes beyond stating that every individual in the population must have an equal chance of being selected. In addition, simple random sampling also demands that each selection must be independent of all the others.

For instance, imagine that you are collecting a sample from your neighborhood, but you refuse to include both members of a married couple; if one member of a couple has already been selected, and that person's spouse comes up at random as a later selection, the spouse is excluded and another selection is drawn. Adopting such a rule means that your groups will not be simple random samples. Even though everyone in the population has an equal chance at the outset, the probabilities change as you go along. After a particular married individual has been selected, the spouse's probability of being selected drops to zero. The selections are *not* independent of each other.

To state the above condition more formally: In order for all samples of the same size to have the same chance of being selected, two conditions must be satisfied. (1) All individuals in the population must have the same probability of being selected, and (2) each selection must be independent of all others. Because of the second condition, this type of sampling is sometimes called **independent random sampling.** Not including both members of a married couple is only one way to violate the independence rule. Another is to deliberately include friends and/or family members of someone who had been randomly selected. Technically, sampling should occur *with replacement,* which means that after a subject is selected his or her name is put back in the population so that the same subject could be selected more than once for the same group! Of course, in experiments involving real people sampling is always done *without* replacement. Fortunately, when dealing with large populations the chance of randomly selecting the same subject twice for the same group is so tiny that the error introduced by not sampling with replacement is negligible. The statistical laws that govern sampling from relatively small populations are too complicated for this text, but fortunately they are rarely used.

Random Samples Versus Convenient Samples

It is only when groups are chosen by simple random sampling that the sampling distribution of the mean follows simple laws that make it easy to use. For a psychologist to perform this kind of sampling, the names of everyone in the population of interest (e.g., adults in the United States) would have to be available for selection (each with an equal probability of being chosen) and then somehow the people randomly selected would have to be persuaded to cooperate. Whereas large, national organizations can sometimes approximate this kind of sampling, individual psychological researchers almost never do. First, the researcher usually draws subjects from a subpopulation, such as one particular college, hospital, or neighborhood. Second, the subjects are usually not selected at random from the subpopulation but are those who volunteer, or are most available, or can most easily be persuaded to participate. In other words, *a sample of convenience* (introduced in Chapter 1) is usually used. The fact that psychologists almost never engage in true random sampling can sometimes lead to limitations on the conclusions that can be drawn from such research; fortunately, the random assignment of subjects to experimental conditions can preserve the validity of a study based on a sample of convenience. These issues will be discussed where appropriate in the next few chapters. Section C will deal in more detail with the consequences of studying a variable that is not normally distributed in the population.

Exercises

* 1. A teacher thinks her class has an unusually high IQ because her 36 students have an average IQ (X) of 108.

  a) If the population mean is 100 and $\sigma = 15$, what is the z score for this class?

  b) What percentage of classes ($N = 36$, randomly selected) would have even higher IQs?

* 2. An aerobics instructor thinks that his class has an unusually low resting heart rate. If $\mu = 72$ bpm and $\sigma = 8$ bpm and his class of 14 pupils has a mean heart rate ($\overline{X}$) of 66, a) what is the z score for the aerobics class?

  b) What is the probability of randomly selecting a group of 14 people with a mean resting heart rate lower than the mean for the aerobics class?

3. Imagine that a test for spatial ability produces scores that are normally distributed in the population with $\mu = 60$ and $\sigma = 20$.

  a) Between which two scores will you find the middle 80% of the people?

  b) Considering the means of groups, all of which have 25 subjects, between what two scores will the middle 80% of these means be?

* 4. Suppose that the mean weight of adults (μ) is 150 pounds with $\sigma = 30$ pounds. Consider the mean weights of all possible space shuttle crews ($N = 7$). If the space shuttle cannot carry a crew that weighs more than a total of 1190 pounds, what is the probability that a randomly selected crew will be too heavy? (Assume that the sampling distribution of the mean is approximately normal.)

5. If freshman English classes always contain 20 students each, a) what percentage of these classes will have a mean verbal SAT score above 530?

  b) The lowest 15% of these classes will have a mean verbal SAT score below what value?

* 6. Suppose that the average person sleeps 8 hours each night and that $\sigma = .7$ hour.

  a) A group of 50 joggers is found to sleep an average of 7.6 hours per night. What is the z score for this group?

  b) If a group of 200 joggers also has a mean of 7.6, what is the z score for this larger group?

  c) Comparing your answers to parts a and b, can you determine the mathematical relation between sample size and z (when $\overline{X}$ remains constant)?

7. Refer back to exercise 2 above. If an aerobics class had a mean heart rate ($\overline{X}$) of 62, and this resulted in a group z score of 7.1, how large must the class have been?

* 8. Suppose that the mean height for a group of 40 women who had been breast-fed for at least the first 6 months of life was 66.8 inches.

  a) If $\mu - 65.5$ inches and $\sigma = 2.6$ inches, what is the z score for this group?

  b) If height had been measured in centimeters, what would the z score be? (Hint: Multiply $\overline{X}$, μ, and σ by 2.54 to convert inches to centimeters.)

  c) Comparing your answers to parts a and b, what can you say about the effect on z scores of changing units? Can you explain the significance of this principle?

9. Suppose that you want to study adults who were abused as children with regard to several psychological variables. What obstacles might you encounter in trying to obtain an independent random sample?

10. A social psychologist recruits several subjects for his experiment and then asks each of these subjects to bring in as many friends as possible to also be in the experiment. What is wrong with this kind of sampling?

The Central Limit Theorem

It is important to remember that the term *population* is used in a special way in mathematics. To a statistician a population is not a collection of people, but rather an infinite collection of numbers that can be described by some distribution that shows how frequently any particular number comes up compared to the others. A normally distributed population follows exactly the mathematical formula of the normal distribution, extending infinitely in both the positive and negative directions. To a psychological researcher, a population is usually a collection of numbers that comes from measuring each subject in an actual population on some variable. For instance, such a population could be the heights of all adults in the United States. Of course, it is rare to have all these measurements available, which is why samples are drawn and guesses are made about the population parameters based on sample statistics. However, a population of people is not infinite, and will not follow any simple mathematical distribution perfectly. (Psychologists may instead deal with populations of, for instance, animals, hospitals, or experimental trials. However, all of these populations tend to fall short of mathematical perfection as well.) The problem is that statistical procedures only work out simply when you are dealing with exact mathematical (probability) distributions, such as the normal distribution. To take statistical procedures that were worked out for perfect mathematical distributions and use them to try to make inferences about populations of people—or animals, hospitals, experiments, or whatever the population might be—involves a certain amount of error.

Strictly speaking, the sampling distribution of the mean will follow the normal distribution perfectly only if the population does, and we know that populations do not perfectly follow a normal distribution, no matter what variable is measured. Fortunately, the Central Limit Theorem (CLT) says that even for population distributions that are very far from normal, the sampling distribution of the mean can be assumed to be normal, provided that the sample size is large enough. It is because of the Central Limit Theorem that psychologists do not worry much about the shape of the population distributions of the variables they use, and the process of drawing statistical inferences is greatly simplified. Recognizing the importance of the Central Limit Theorem, I will state it formally and then discuss its implications.

> *Central Limit Theorem:* For any population that has mean μ and a finite variance σ^2, the distribution of sample means (each based on N independent observations) will approach a normal distribution with mean μ and variance σ^2/N, as N approaches infinity.

The statement about variance translates easily into a statement about standard deviation; if the population has standard deviation σ, the standard deviation of the sampling distribution (called the standard error of the mean) will be $\sigma/\sqrt{N}$. (This principle was presented as Formula 6.1.) However, the statement about N approaching infinity may sound strange. If each sample must be nearly infinite in size to guarantee that the sampling distribution will have a normal shape, why

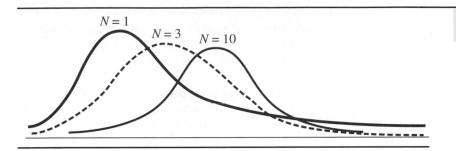

Figure 6.3

is the CLT so helpful? It is fortunate that the sampling distribution becomes very nearly normal long before the sample size becomes infinitely large. How quickly the *sampling distribution* becomes nearly normal as sample size increases depends on how close to normal the *population distribution* is to begin with. If the population distribution is normal to begin with, then the sampling distribution will be normal for any sample size. If the population distribution is fairly close to being normal, then the sampling distribution will be very nearly normal even for small sample sizes. On the other hand, if the population distribution is very far from normal, the sample size must be fairly large before the sampling distribution becomes nearly normal. The CLT guarantees us that no matter how bizarre the population distribution (provided that its variance is finite), the sample size can always be made large enough so that the sampling distribution approximates the normal distribution. The condition that the population variance be finite is never a problem in psychological research; none of the variables psychologists use would ever have an infinite variance.

I mentioned above that fairly large sample sizes are needed if the population distribution is very far from normal. It has been generally found for a wide range of population distribution shapes that when the sample size exceeds about 30, the sampling distribution closely approximates a normal distribution. There is nothing magical about the number 30, however. When $N = 29$ the sampling distribution is also very close to normal (only slightly less so than with $N = 30$), and when $N = 31$ the approximation to the normal distribution becomes better, but only slightly. For a distribution that is only moderately skewed, the sampling distribution may approximate the normal distribution with $N = 20$ or even less. However, if you are unsure about the shape of the population distribution, the safest approach is to use groups that are at least about 30 to 40 in size.

Figure 6.3 depicts the type of sampling distribution of the mean that is obtained for different sample sizes, when the population distribution is strongly skewed. Notice that in each case the sampling distribution is narrower, and more symmetrical, than the population distribution ($N = 1$), and that the sampling distribution becomes increasingly narrower and more symmetrical (and, of course, more normal) as the sample size increases.

Psychologists often deal with variables that are based on new measures for which little is known about the population distribution. Nonetheless, thanks to the Central Limit Theorem, they can assume that the sampling distribution of the mean will be nearly normal, as long as sufficiently large samples (N of at least 30 or 40) are used. If, however, a researcher has good reason to suspect that the population distribution of the measured variable will be extremely far from normal, it may be safer to use sample sizes of at least 100. Finally, if practical considerations prevent a researcher from obtaining a sample that he or she feels is sufficiently large to ensure a nearly normal sampling distribution, the use of nonparametric, or distribution-free, statistical tests should be considered (see Part VI).

Using Skewness and Kurtosis

Some variables that psychologists work with have been measured so extensively (e.g., simple reaction time) that there is little doubt about what shape the distribution would take in the entire population. Other variables are so new (e.g., a person's level of AIDS awareness) that you may have no basis for guessing about the shape of the population distribution other than the sample data you have collected. If you want to test the likelihood that a normal population distribution will produce the distribution in your sample, one possibility is a "goodness-of-fit" test (which will be described in Chapter 20). Another possibility is to measure the skewness and kurtosis of your sample data as described in Chapter 4, Section C.

Unless your sample distribution is perfectly symmetric, like the normal distribution, you will not come up with zero when you calculate skewness. However, small departures from zero are not inconsistent with samples from a normally distributed population. Just as the means of samples will usually cluster around, but differ somewhat from, the population mean, the skewness of samples from a normal distribution will cluster around zero, but usually will be somewhat positive or negative. Moreover, just as the standard error of the mean tells us how large a departure from the population mean to expect for a sample mean (within one standard error either way is likely, more than two standard errors away is not likely), the standard error of the skewness gives us similar information about the likely skewness of our sample.

The standard error of skewness depends only on sample size; as N increases, large departures from the skewness of the population become less likely (i.e., the standard error decreases). As it turns out, rather large sample sizes are required to draw reliable inferences about the skewness of the population distribution. When dealing with the relatively small samples often used in psychological research (e.g., 20 to 30), only extreme skewness in the sample data is cause for alarm. With skewness several standard errors from zero, you may have to consider a data transformation (see Chapter 3) or a nonparametric test (see Part VI).

A measure of kurtosis provides additional information for deciding whether a particular sample is likely to arise from a normal distribution. Even if the data in a sample are symmetric (i.e., skewness is zero), the kurtosis of the sample may suggest that the population distribution it comes from is not normal in shape. For instance, a large number of extreme outliers on both ends of the sample may balance out to yield a near-zero skewness but may result in a high kurtosis compared to the normal distribution (for which the kurtosis is zero) because of the heavy tails in the sample distribution. (Similarly, a total lack of extreme scores could lead to a negative kurtosis.) As with skewness, the standard error of kurtosis decreases as sample size increases; with small samples the kurtosis measure is generally thought to be too unreliable to be of much use. Only a very extreme value for kurtosis would be cause for concern when dealing with fairly small samples.

Information about skewness and kurtosis has been included here not because these measures are frequently used in psychological research, but because you will likely encounter them when using statistical software for data analysis. These distributional measures may be supplied automatically as part of a package of descriptive statistics, or offered as options. In either case, this brief introduction may keep you from being mystified when you encounter these terms.

Exercises

* 1. If the skewness in a sample is zero, which of the following distributional measures can still provide evidence that the sample was not randomly drawn from a normal distribution? a) Mean; b) Variance; c) Standard deviation; d) Kurtosis

2. The sampling distribution of the mean will tend toward the normal distribution with increasing N only if
   a) the samples are randomly drawn.
   b) the samples are all the same size.
   c) the variance is finite.
   d) all of the above
   e) none of the above

* 3. When drawing a sample from a normal distribution, as N increases, which of the following will be true?
   a) It becomes less likely that the sample will have a very negative kurtosis.
   b) It becomes more likely that the sample will have a very positive kurtosis.
   c) It becomes less likely that the sample will have zero skewness.
   d) It becomes more likely that the sample mean will be larger than the population mean.

SUMMARY

The Important Points of Section A

1. Just as it is sometimes useful to determine if an individual is unusual with respect to a population, it can also be useful to determine how unusual a group is compared to other groups that could be randomly selected.

2. Groups are usually thought of as *samples* from a population. The group mean (more often called the *sample mean*) is used to summarize the group with a single number.

3. In order to find out how unusual a sample is, the sample mean ($\overline{X}$) must be compared to a distribution, not of individuals, but of sample means, called, appropriately, the *sampling distribution of the mean*.

4. The sampling distribution could be found by taking very many samples from a population and gathering the sample means into a distribution, but there are statistical laws that tell you just what the sampling distribution of the mean will look like if certain conditions are met. The conditions are that the population distribution is normal and that the samples, all of which are the same size, are *independent random samples*.

5. Given the above conditions, the sampling distribution of the mean will be a normal distribution with a mean of μ (the same mean as the population). The only other measure of the sampling distribution that we need to know is its standard deviation.

6. The standard deviation of the sampling distribution of the mean is called the *standard error of the mean,* and it is less than the standard deviation of the population because group means vary less than individuals.

7. The larger the sample size, N, the smaller the standard error of the mean, which is equal to the population standard deviation divided by the square root of N (see Formula 6.1).

The Important Points of Section B

The basic statistical procedures of this chapter can be summarized with the following example. A fifth-grade teacher thinks that his class is unusually good at reading. On a standardized reading test, his class of $N = 36$ students scored a mean ($\overline{X}$) of 5.1. Assume that the national average (μ) for fifth-graders is 5.0 with a standard deviation (σ) of .5. How unusual is the class in question?

To determine how unusual the class is, we must find the percentage of all possible (random) classes that would have an even higher sample mean. This is the same as finding the probability of selecting 36 students at random from the population and finding that the mean of this group is higher than the mean of the class in question. Because this question concerns comparing a group mean to other group means, it must be answered in terms of the sampling distribution of the mean. The following steps must be performed.

1. *Find the appropriate sampling distribution of the mean.* It can be assumed that the sampling distribution is normal (see Assumptions below). The mean of the sampling distribution is always the same as the mean (μ) for the population. For this problem, $\mu = 5.0$. The standard deviation of the sampling distribution, called the standard error of the mean ($\sigma_{\overline{X}}$), is found by Formula 6.1:

$$\sigma_{\bar{X}} = \frac{\sigma}{\sqrt{N}} = \frac{.5}{\sqrt{36}} = \frac{.5}{6} = .083$$

2. *Find the z score of the group mean in question with respect to the sampling distribution of the mean.* The *z* score is found by using Formula 6.2:

$$z = \frac{\bar{X} - \mu}{\sigma_{\bar{X}}} = \frac{5.1 - 5.0}{.083} = \frac{.1}{.083} = 1.20$$

3. *Find the appropriate area.* To determine the probability that a randomly selected group of the same size would have a higher mean, you must look at the area beyond the *z* score found above. The area beyond *z* = 1.20 is .115 (see Table A.1).

From the above procedure you can see that the class in question is somewhat unusual in that only about 12% of randomly selected groups would have higher sample means. In the next chapter I will discuss how to make use of this determination.

Assumptions

When sampling distributions are discussed, it is assumed that all of the samples are the same size, *N,* and that an infinite number of samples are drawn. Matters are greatly simplified if we can also assume that the samples are obtained by a process known as *simple,* or *independent,* random sampling, in which each individual in the population has the same chance of being selected *and* each selection is *independent* of all others.

Furthermore, the sampling distribution of the mean can be assumed to be normal if we assume that the population distribution is normal or the sample size is at least about 30.

The Important Points of Section C

1. When the population distribution is normal, the sampling distribution of the mean is also normal, regardless of sample size.

2. If the population is *not* normally distributed (but the population variance is finite), the *Central Limit Theorem* states that the sampling distribution of the mean will become more like the normal distribution as sample size (*N*) increases—becoming normal when *N* reaches infinity.

3. For sample sizes of 30 to 40 or more, the sampling distribution of the mean will be approximately normal, regardless of the shape of the population distribution.

4. The skewness and kurtosis of the data in a sample can be used to judge whether it is reasonable to assume that the population follows a normal distribution.

5. Unfortunately, measures of skewness and kurtosis become very unreliable with small sample sizes. Only extreme values for skewness or kurtosis should make you question the normal distribution assumption when the sample is small.

Definitions of Key Terms

Sampling distribution of the mean A distribution composed of an infinite number of sample means, all from random samples of the same size.

Standard error of the mean The standard deviation of the sampling distribution of the mean.

Central Limit Theorem For any population that has mean μ and a finite variance σ^2, the distribution of sample means (each based on N independent observations) will approach a normal distribution with mean μ and variance σ^2/N, as N approaches infinity.

Independent random sample A sample collected so that every unit in the population has the same chance of being selected *and* every selection is independent of all the others.

Key Formulas

The standard error of the mean:

$$\sigma_{\bar{X}} = \frac{\sigma}{\sqrt{N}}$$

Formula 6.1

The z score for groups (the standard error of the mean must be found first):

$$z = \frac{\bar{X} - \mu}{\sigma_{\bar{X}}}$$

Formula 6.2

INTRODUCTION TO HYPOTHESIS TESTING: THE ONE-SAMPLE z-TEST

You will need to use the following from previous chapters:

Symbols
μ: Mean of a population
$\overline{X}$: Mean of a sample
σ: Standard deviation of a population
$\sigma_{\overline{X}}$: Standard error of the mean
N: Number of subjects (or observations) in a sample

Formulas
Formula 6.1: The standard error of the mean
Formula 6.2: The z score for groups

Concepts
Sampling distribution of the mean

Procedures
Finding areas under the normal distribution (Chapter 5, Section B)

CHAPTER 7

A

CONCEPTUAL FOUNDATION

The purpose of this chapter is to explain the concept of hypothesis testing in the simple case in which only one group of subjects is being compared to a population, and when we know the mean and standard deviation for that population. This situation is not common in psychological research. However, a description of the one-sample hypothesis test (with a known standard deviation) is useful as a simple context in which to introduce the logic of hypothesis testing, and, more specifically, as a stepping stone for understanding the very common two-sample t-test (see Chapter 9). In this overview I employ a rather fanciful example to help simplify the explanations. In Section B, I consider a more realistic example and suggest general situations in which the one-sample hypothesis test (with a known standard deviation) may serve a useful function.

Consider the following situation. You just met someone who tells you he can determine whether any person has a good aptitude for math just by looking at that person. (He gets "vibrations" and sees "auras," which are what clue him in to someone's math aptitude.) Suppose you are inclined to test the claim of this self-proclaimed psychic—maybe you want to show him he's wrong (or maybe you're excited by the possibility that he's right). Your first impulse might be to find a person whose math aptitude you know and ask the psychic to make a judgment. If the psychic were way off, you might decide at once that his powers could not be very impressive. If he were close, you'd probably repeat this test several times, and only if the psychic came close each time would you be impressed.

Selecting a Group of Subjects

Would you be interested if the psychic had only a slight ability to pick people with high math aptitude? Perhaps not. There wouldn't be much practical application of the psychic's ability in that case. But the possibility of even a slight degree of extrasensory perception is of such great theoretical interest that we should consider a more sensitive way to test our psychic friend. The approach that a psychological researcher might take is to have the psychic select from the population at large a group of people he believed to have a high mathematical aptitude. Then math SAT scores would be obtained for each person selected and the mean of these scores would be calculated. Assume that 500 is the mean math SAT score for the entire population. Note that in order for the mean of the sample group to be above average it is not necessary for each selected person to be above average. The mean is useful because it permits the selection of one person with a very high math SAT score to make up for picking several people slightly below average.

It is important to make clear that this experiment would not be fair if the psychic were roaming around the physics department of a university. If we are checking to see if the psychic can pick people above the average for a particular population (e.g., adults in the United States) then, theoretically, all of the people in that population should have an equal chance of being selected. Furthermore, each selection should be independent of all the other selections. For instance, it would not be fair if the psychic picked a particular woman and then decided it would be a good idea to include her sister as well.

The Need for Hypothesis Testing

Suppose we calculate the mean math SAT score of the people the psychic has selected and find that mean to be slightly higher than the mean of the population. Should we be convinced the psychic has some extrasensory ability—even though it appears to be only a slight ability? One reason for being cautious is that it is extremely unlikely that the mean math SAT score for the selected group of people will be *exactly* equal to the population mean. The group mean almost always will be slightly above or below the population mean even if the psychic has no special powers at all! And that, of course, implies that even if the psychic has zero amount of psychic ability, he will pick a group that is at least slightly above the population mean in approximately half of the experiments that you do. Knowing how easily chance fluctuations can produce positive results in cases in which there really is no reason for them, researchers are understandably cautious. This caution led to the creation of a statistical procedure known as *null hypothesis testing*. Null hypothesis testing involves what is called an indirect proof. You take the "worst-case scenario"—that is, you take the opposite of what you want to prove—and assume that it is true. Then you try to show that this assumption leads to ridiculous consequences and that therefore you can reject the assumption. Because the assumption was the opposite of what you wanted to prove, you are happy to reject the assumption.

It would not be surprising if the above explanation sounds a bit confusing. Cognitive psychologists have found that logical arguments are sometimes more understandable when they are cast in concrete and memorable terms (Griggs & Cox, 1982; Johnson-Laird, Legrenzi, & Legrenzi, 1972)—so I will take a concrete approach here. The indirect proof described above can be thought of as a "devil's advocate" procedure: You argue from the position of your opponent (the devil, so to speak) and try to show that your opponent's position has unreasonable implications. The idea of the "devil's advocate" inspired me to create an imaginary character who personifies scientific skepticism: a permanent opponent who is always skeptical of all the types of experiments I will describe in this text. This character is Dr. Null (the appropriateness of this name will soon become obvious). My intention in creating this character is not to be cute or amusing (though I will not be appalled if this is a by-product), and certainly not to be condescending. I'm simply trying to make the logic easier to comprehend by making it concrete. (If you are comfortable with the logic of indirect proofs, feel free to ignore my comments about Dr. Null.)

The Logic of Null Hypothesis Testing

The character of Dr. Null can be used in several ways. Imagine that whenever you run an experiment, like the psychic test described above, and write up the results and send them to a journal to be published, the first person to see your article (no matter which journal you send it to) is Dr. Null. Dr. Null has the same skeptical reaction to any results that are submitted. No matter how plausible the results may seem, Dr. Null insists that the experimental treatment was completely ineffective. In the case of the psychic tests, he would claim the psychic had no ability to pick people with high math aptitude at all. How does Dr. Null explain that the group selected by the psychic has a slightly higher mean than the overall population? Dr. Null is quick to point out the example of flipping a perfectly fair coin ten times. You don't always get five heads and five tails. You can easily get six of one and four of the other—which doesn't mean there's anything biased or "loaded" about the coin. In fact, in about one in a thousand sets of ten tosses, a perfectly fair coin will give you ten heads in a row. The same principle applies when sampling from a population, Dr. Null would explain. Sometimes, by chance, you'll get a few high scores and the mean will be above the population average; or, it's just as likely that some low scores will put you below the population mean. So regardless of how good your results look, Dr. Null offers the same simple hypothesis: Your experimental results merely come from sampling the original population; that is, your experiment does not involve a special population, or a different population, or a modified population. This hypothesis, about which I will have much more to say in a little while, is generally formalized into what is known as the **null hypothesis** (from which Dr. Null gets his name).

Imagine further that Dr. Null offers you a challenge. Dr. Null insists that before you proclaim that your experiment really "worked," he will go out and try to duplicate (or even surpass) your experimental results *without* adding the crucial experimental ingredient, just by taking perfectly random samples of the population. In the case of the psychic test, Dr. Null would select the same number of people but would make sure to do so randomly, to avoid any possibility of psychic ability entering into the process. Then if the mean of Dr. Null's group was as high or even higher than the mean of the psychic's group, Dr. Null would take the opportunity to ridicule your experiment. He might say, "How good could your psychic be if I just did even better (or just as well) without using any psychic ability at all?" You might want to counter by saying that Dr. Null got his results by chance and you got your results through the psychic's ability, but how could you be sure? And how could you convince the scientific community to ignore Dr. Null's arguments? Dr. Null's beating the psychic's results (or even just matching them) is an embarrassment to be avoided if possible. Your only chance to avoid such an embarrassment is to know your enemy, Dr. Null, as well as possible.

The Null Hypothesis Distribution

Is there ever a situation when experimental results are so good (i.e., so far from the population mean) that Dr. Null has no chance of beating them at all? Theoretically, the answer is no, but in some cases the chances of Dr. Null's beating your results are so ridiculously small that nobody would worry about it. The important thing is to know what the probability is that Dr. Null will beat your results, so you can make an informed decision and take a calculated risk (or decide not to take the risk).

How can you know Dr. Null's chances of beating your results? That's relatively easy. Dr. Null is only taking random samples, and the laws of statistics can tell us precisely what happens when you select random samples of a particular size from a particular population. These statistical laws are simple to work with only if we make certain assumptions (to be discussed in the next section), but fortunately these assumptions are reasonable to make in many of the situations that psychological researchers face. Most of the inferential statistics that psychologists use can be thought of, informally, as ways for researchers to figure out the probability of Dr. Null's beating their results and arbitrary rules to help them decide whether or not to risk embarrassment.

When you are comparing the mean of one sample to the mean of a population, the statistical laws mentioned above lead to a very simple formula (given some assumptions). To map out what Dr. Null can do in the one-sample case, imagine that he keeps drawing random samples of the same size, from the same population, and measuring the mean for each sample. If Dr. Null did this sampling many, many times (ideally, an infinite number of times), the sample means he would get would form a smooth distribution. This distribution is called the *null hypothesis distribution,* because it shows what can happen (what is likely and not so likely) when the null hypothesis is true.

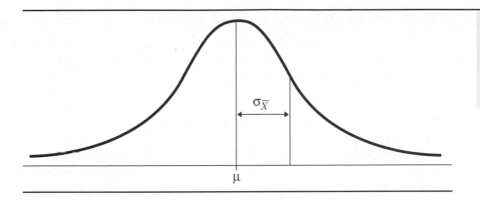

Figure 7.1

One-Sample Null Hypothesis Distribution

The Null Hypothesis Distribution for the One-Sample Case

You should recall from Chapter 6 that when you take many groups of the same size and form a distribution from the group means, you get the sampling distribution of the mean. If the population has a normal distribution for the variable of interest (in this case, math SAT scores), the sampling distribution will be normal and will have the same mean as the population from which the samples are being drawn (500 in the case of math SAT scores). But the sampling distribution will be narrower than the population distribution. As you saw in Chapter 6, the standard deviation for the sample means would equal the standard deviation of the population divided by the square root of the sample size. The null hypothesis distribution does not usually work out this simply, but in the simplest possible case—the one-sample test with a known population standard deviation—the null hypothesis distribution is just the sampling distribution of the mean, which you learned about in Chapter 6.

It should now be clear why I introduced the sampling distribution of the mean in the first place. Generally, experiments are conducted not by looking at one individual at a time but by selecting a group and looking at the mean of the group. To get our map of what Dr. Null can do in the one-sample case, we have to know what happens when many groups are selected and their means are calculated. So in this case our map of what Dr. Null can do is just the well-known sampling distribution of the mean (see Figure 7.1). We know the mean of Dr. Null's distribution because we know the mean of the population from which he is taking samples. Similarly, we can easily calculate the standard error of Dr. Null's distribution because we know the standard deviation of the population and the sample size. Because the sampling distribution of the mean tends to be a normal distribution, it can extend infinitely in each direction. This implies that, theoretically, there is no limit to what Dr. Null can do by random sampling. However, the tails of the distribution get very thin, which tells us that Dr. Null has little chance of drawing a sample whose mean is far from the population mean. If we assume that the sampling distribution *is* normal, then we can use the standard normal table to answer questions about group means, as described in Chapter 6.

z Scores and the Null Hypothesis Distribution

If we want to know the probability of Dr. Null's surpassing our psychic's results, there is a simple procedure for finding that out. We just have to see where our experimental results fall on Dr. Null's distribution. Then we look at how much area of the curve is beyond that point, because that represents the proportion of times Dr. Null will get a result equal to or better than ours. To find where our experimental result falls on Dr. Null's distribution, all we have to do is convert our result to a z score—a z score for groups, of course. We can use Formula 6.2 (which is presented below as Formula 7.1). Once we know the z score we can find the area beyond that z score, using the tables for the standard normal distribution as we did in Chapters 5 and 6. To do a one-sample hypothesis test, then, you begin by simply figuring out the z score of your results on the null hypothesis distribution and finding the area beyond that z score.

I will illustrate the procedure described above using actual numbers from a specific example. Suppose the psychic selects a group of 25 people and the mean for that group is 530. This is clearly above the population mean of 500, but remember that Dr. Null claims he can do just as well or better using only random sampling. To find the probability of Dr. Null's beating us, we must first find the z score of our experimental result (i.e., a mean of 530) with respect to the null hypothesis distribution, using Formula 7.1,

$$z = \frac{\overline{X} - \mu}{\sigma_{\overline{X}}}$$

Formula 7.1

where $\overline{X}$ is our experimental result, μ is the population mean, and $\sigma_{\overline{X}}$ is the standard deviation of the sampling distribution.

To get the standard deviation of the sampling distribution (i.e., the standard error of the mean) we need to know the standard deviation for the population. This is not something a researcher usually knows, but some variables (such as math SAT scores and IQ) have been standardized on such large numbers of people that we can say we know the standard deviation for the population. The standard deviation for math SAT scores is 100. Using Formula 6.1, we find that the standard error for groups of 25 is

$$\sigma_{\overline{X}} = \frac{\sigma}{\sqrt{N}} = \frac{100}{\sqrt{25}} = \frac{100}{5} = 20$$

The z score for the psychic's group therefore is

$$z = \frac{530 - 500}{20} = \frac{30}{20} = 1.5$$

How often can Dr. Null do better than a z score of $+1.5$? We need to find the area to the right of (beyond) $z = +1.5$; that is, the area shaded in Figure 7.2. Table A.1 tells us that the area we are looking for is .0668. This means that if

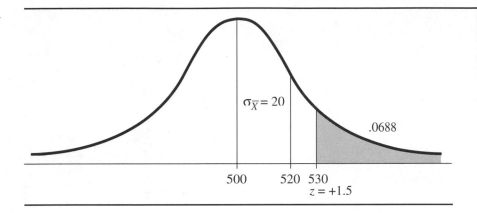

Figure 7.2

Null Hypothesis Distribution for Math SAT Scores

Dr. Null performs his "fake" psychic experiment many times (i.e., drawing random samples of 25 people each, using no psychic ability at all), he will beat us about 67 times out of a thousand, or nearly 7 times out of a hundred. The probability of Dr. Null's beating us is usually called the **p level** associated with our experimental results (*p* for *probability*). In this case the *p* level is .0668.

Statistical Decisions

Now that we know the probability of Dr. Null's embarrassing us by getting even better results than we did without using our experimental manipulation, we can make an informed decision about what we can conclude from our results. But we still must decide how much risk to take. If Dr. Null's chance of beating us is less than one in ten, should we take the risk of declaring our results statistically significant? The amount of risk you decide to take is called the **alpha (α) level,** and this should be decided *before* you do the experiment, before you see the results. Alpha is the chance you're willing to take that Dr. Null will beat you. When the chance of Dr. Null's beating you is low enough—that is, when it is less than alpha—you can say that you *reject* the null hypothesis. This is the same as saying your results have **statistical significance.** What you are telling your fellow scientists is that the chance of Dr. Null's beating you is too low to worry about, and that it is safe to ignore this possibility.

Does each researcher decide on his or her own alpha level? Not exactly. Among psychologists in particular, an alpha level of one chance in twenty (a probability corresponding to .05) is generally considered the largest amount of risk worth taking. In other words, by convention, other psychologists agree to ignore Dr. Null's claims and protests concerning your experiment as long as you can show that Dr. Null has less than a .05 chance of beating you (i.e., the *p* level for your experiment is less than .05). If you are using an alpha level of .05 and your *p* level is less than .05, then you can reject the null hypothesis at the .05 level of significance (see Figure 7.3).

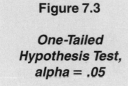

Figure 7.3

One-Tailed Hypothesis Test, alpha = .05

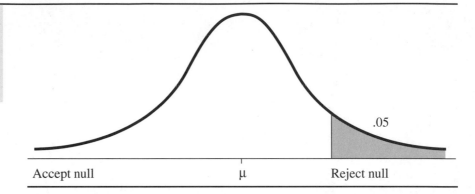

.05

Accept null   μ   Reject null

But what if your *p* level is greater than .05? What must you conclude then? Unless you have declared an alpha level greater than .05 (which is not likely to be accepted by your fellow psychologists without some special justification, which I will discuss below), you must *accept,* or *retain,* the null hypothesis. Does this mean you are actually agreeing with the infamous Dr. Null? Are you asserting that your experiment has absolutely no effect and that your experimental results are attributable entirely to chance (the luck involved in random sampling)? Generally, no. In fact, many psychologists and statisticians prefer to say not that they accept the null hypothesis in such cases, but that they *fail to reject it*—that they have insufficient evidence for rejecting it. When the *p* level for your experiment is greater than .05, you may still feel that your experiment really worked and that your results are not just due to luck. However, convention dictates that in this situation you must be cautious. Dr. Null's claims, though somewhat farfetched if your *p* level is only a little over .05, are not ridiculous or totally unreasonable. The null hypothesis in such a case must be considered a reasonable possibility for explaining your data, so you cannot declare your results statistically significant.

The *z* Score as Test Statistic

In describing math SAT scores earlier, I stated that the null hypothesis distribution was normal, with a mean of 500 and a standard error of 20. But there is no table for areas under this particular distribution, so we had to use *z* scores. When you convert to *z* scores, the *z* scores follow a standard normal distribution (mean = 0 and standard deviation = 1), for which there is a convenient table. Used this way, the *z* score can be called a **test statistic:** It is based on one or more sample statistics (in this case just the sample mean), and it follows a well-known distribution (in this case, the standard normal distribution). For convenience, it is the distribution of the test statistic that is viewed as the null hypothesis distribution. In later chapters, I will deal with more complicated test statistics and other well-known distributions (e.g., t and F), but the basic principle for making statistical decisions will remain the same.

The larger the z score, the lower the p level, so in general you would like the calculated z to be as large as possible. But how large can the z score in a one-sample test get? If you look at the standard normal table (Table A.1), you might be misled into thinking that the largest possible z score is 4.0, because that is where this particular table ends. (The area beyond $z = 4.0$ is listed as .0000, suggesting that there is no area left—we've come to the end of the normal distribution.) Actually, if there were more numbers beyond the decimal you would see that the area beyond $z = 4.0$ is .00003, but because the table is limited to four places beyond the decimal point, .00003 is rounded off to .0000. For most research purposes it is not important to distinguish among probabilities that are less than .0001, so the table does not bother to include them. So if your calculated z comes out to be 22, you can just say that p is less than .0001 and leave it at that. Even though statistical programs can find the area beyond even a very large z score, the p levels printed in the output are usually rounded off as in Table A.1, so that for large z scores you will see $p = .0000$. Of course, there is always *some* area beyond any z score, no matter how large, until you get to infinity.

Sometimes people are confused when they see a very large z score as the result of a one-sample test, because they think of such a z score as almost impossibly rare. They are forgetting that a large z score for a particular experiment is very rare only if the null hypothesis happens to be true for that experiment. The whole point of the one-sample test is to obtain a large z score, if possible, to show that we shouldn't believe the null hypothesis—precisely because believing that the null hypothesis is true leads to a very implausible conclusion, namely, that our results are a very rare fluke produced entirely by sampling error.

You should also keep in mind that the z score for groups can get very large even if the sample mean does not deviate much from the population mean (as compared to the population standard deviation). This is because the denominator of the formula for the z score for groups is the standard error of the mean, which gets smaller as the sample size gets larger. With a large enough sample size, even a slightly deviant sample mean can produce a large z score, so you should not be overly impressed by large z scores. Large z scores *do* lead to tiny p levels, which allow confident rejection of the null hypothesis—but that does not imply that the deviation of the sample mean is enough to be important or interesting. I will return to this point in greater detail in Chapter 10.

Type I and Type II Errors

Even if our p level is very much under .05 and we reject the null hypothesis, we are nonetheless taking some risk. But just what is it that we are risking? I've been discussing "risk" as the risk of Dr. Null's beating our results without using our experimental manipulation and thus making us look bad. This is just a convenient story I used to make the concept of null hypothesis testing as concrete as possible. In reality no psychologist worries that someone is going to do a "null experiment" (i.e., an experiment in which the null is really true) and beat his or her results. The real risk is that Dr. Null may indeed be right; that is, the null hypothesis may really be true and we got positive-looking results (i.e., $p < .05$) by luck. In other words, we may ourselves have done a "null experiment" without knowing it and obtained our positive results by accident. If we then reject the null hypothesis in a case when it is actually true (and our experimental manipulation was entirely ineffective), we are making an error. This kind of error is called a **Type I error.** The opposite kind of error occurs when you accept the null hypothesis even though it is false (in other words, you cautiously refrain from rejecting the null hypothesis even though in reality your experiment was at least somewhat effective). This kind of error is called a **Type II error.** The distinction between the two types of errors is usually illustrated as shown in Table 7.1. Notice the probabilities indicated in parentheses. When the null hypothesis is true, there are two decisions you can make (accept or reject), and the probabilities of these two decisions must add up to 1.0. Similarly, the probabilities for these two decisions add up to 1.0 when the null hypothesis is false. (Note that it makes no sense to add all four probabilities together, because they are *conditional* probabilities; see Section C.) The probability of making a Type II error is called β, but this value is not set by the researcher and can be quite difficult to determine. I will not discuss the probability of Type II errors in any detail until I deal with the topic of power in Chapter 10.

The main focus of null hypothesis testing is the avoidance of Type I errors. The chief reason for trying not to make a Type I error is that it is a kind of false alarm. By falsely claiming statistical significance for the results of your experiment, you are sending your fellow researchers off in the wrong direction. They may waste time, energy, and money trying to replicate or extend your conclusions. These subsequent attempts may produce occasional Type I errors as well, but if the null hypothesis is actually true—if your experiment was really totally ineffective—other experiments will most likely wind up *not* rejecting the null hypothesis. Several failures to replicate the statistical significance of your original results will suggest that you committed a Type I error in analyzing your results, and this can be somewhat embarrassing. But if you had used the conventional alpha of .05 and proper techniques of random sampling, etc., then your mistake

Table 7.1		
	Actual Situation	
Researcher's Decision	**Null Hypothesis Is True**	**Null Hypothesis Is False**
Accept the null hypothesis	Correct decision $(p = 1 - \alpha)$	Type II error $(p = \beta)$
Reject the null hypothesis	Type I error $(p = \alpha)$	Correct decision $(p = 1 - \beta)$

was an honest one that nobody could have prevented and, therefore, should not be a cause of embarrassment. You were simply a victim of chance factors.

The Trade-Off Between Type I and Type II Errors

To reiterate: The chief purpose of statistical hypothesis testing is to keep Type I errors to a low level, because false alarms can be quite disruptive, sending other researchers down blind alleys. One way to further reduce the number of Type I errors made by scientists in general would be to agree on an even lower alpha level, for instance, .01 or even .001. Since Type I errors are so disruptive and misleading, why not use a smaller alpha? As you will see in Chapter 10, changing the decision rule so as to make alpha smaller would at the same time result in more Type II errors being made. Scientists would become so cautious in trying to avoid a Type I error that they might have to ignore a lot of genuinely good experimental results because these results could very occasionally be produced by Dr. Null (i.e., produced when the null hypothesis is actually true). The rather arbitrary value of .05 that has been conventionally accepted as the minimal alpha level for psychology experiments represents a compromise between the risks of Type I errors and the risks of Type II errors.

In some situations Type II errors can be quite serious, so in those cases we might want to use a larger alpha. Consider the case of the researcher testing a cure for a terminal disease. A Type II error in this case would mean accepting the null hypothesis and failing to say that a particular cure has some effect— when in fact the cure does have some effect, but the researcher is too cautious to say so. In such a case the researcher might consider using an alpha of .1 instead of .05. This would increase Type I errors and, therefore, false alarms, but it would reduce Type II errors and therefore lower the chances that a potential cure would be overlooked. On the other hand, if a treatment for acne is being tested, the chief concern should be Type I errors. Because acne is not life-threatening, it would be unfortunate if many people wasted their money on a totally ineffective product simply because a study accidently obtained results that made the treatment look good.

Besides changing the alpha level, another way of reducing Type II errors in certain cases is to perform a one-tailed rather than a two-tailed test. I have ignored this distinction so far to keep things simple, but now it is time to explain the difference.

One-Tailed Versus Two-Tailed Tests

Suppose the group selected by the psychic had a mean math SAT score of only 400 or even 300. What could we conclude? Of course, we would not reject the null hypothesis, and we would be inclined to conclude that the psychic had no special powers for selecting people with high math aptitude. However, if the group mean were extremely low, could we conclude the psychic's powers were working in reverse, and he was picking people with low math aptitude more consistently than could be reasonably attributed to chance? In other words, could we test for statistical significance in the other direction—in the other tail of the distribution? We could if we had planned ahead of time to do a **two-tailed test** instead of a **one-tailed test.**

For the experiment I have been describing it may seem quite pointless to even think about testing significance in the other direction. What purpose could be served and what explanation could be found for the results? However, for many experiments the "other tail" cannot be so easily ignored, so you should know how to do a two-tailed as well as a one-tailed test.

Fortunately, there is no difference between one- and two-tailed tests when you calculate the *z* score for groups. The difference lies in the *p* level. The procedure I have been describing thus far applies only to a one-tailed test. To get the *p* level for a two-tailed test, you find the area beyond *z,* just as for a one-tailed test, but then you double the area. For example, suppose the psychic selected a group of 25 people whose mean math SAT turned out to be 450. The *z* score for this group would be

$$z = \frac{450 - 500}{100/\sqrt{25}} = \frac{-50}{20} = -2.5$$

If you had planned a one-tailed test, expecting the selected group to be above average, the rules of hypothesis testing would not allow you to test for statistical significance, and you would have to retain the null hypothesis. However, if you had planned a two-tailed test, you would be allowed to proceed by finding the two-tailed *p* level. First you have to look at the area beyond your calculated *z* score (see Figure 7.4). Because your *z* score is negative, take its absolute value first (remember there are no negative *z* scores in the normal curve table). The area beyond *z* = 2.5 is .006. To get the *p* level for a two-tailed test you simply double the area beyond *z* to get .012. If your alpha had been set to .05, you could then reject the null hypothesis. Of course, if you had decided to use alpha = .01, you could not then reject the null hypothesis. In our one-tailed test of the psychic's efforts, we found the *p* level to be .0668. The two-tailed *p* level would have been twice that, which is .1336. Since the one-tailed test was not significant at the .05 level, the two-tailed test would certainly not be significant either.

To understand why the *p* level is doubled when you perform a two-tailed test, think again in terms of what Dr. Null can do. When you plan a two-tailed test for one sample you expect the mean of your sample to be *different* from the population mean, but you are not committed to saying whether it will be higher or lower. This means that Dr. Null has *two* ways to embarrass you. He can embarrass you by beating (or tying) your results *in the same direction,* or he can get results in the opposite direction that are *just as different* from the population mean as your results. Since you didn't specify which direction your results would be in (only that they would be different from the population mean), Dr. Null can claim to have matched your results even if his results are in the other direction, because he has matched your results in terms of deviation from the population mean.

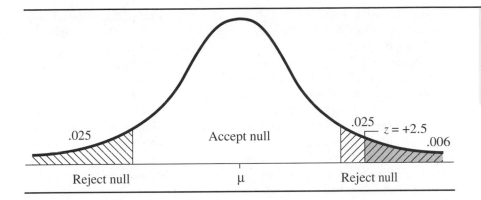

Figure 7.4

Two-Tailed Hypothesis Test, alpha = .05

That's why when finding the *p* level for a two-tailed test, we have to add the same amount of area for the other tail. Doing a two-tailed test makes it a bit harder to reach statistical significance, because a larger *z* score is required. This is the disadvantage of doing a two-tailed test. On the other hand, if you plan a one-tailed test and the results come out in the direction opposite to what you expected, it is "unfair" to test the significance of these unexpected results (see below). This is the disadvantage of a one-tailed test.

Selecting the Number of Tails

So when should you perform a one-tailed test and when a two-tailed test? Unfortunately, there is no simple rule. It's a matter of judgment. However, if you plan a one-tailed test, you must state the expected direction of the results before seeing them; if the results come out in the wrong direction, no test can be performed—not even a two-tailed test. You might at first think it would be alright to plan a one-tailed test and then switch to a two-tailed test if the results come out in the wrong direction. The problem is that if this were commonly done, the percentage of Type I errors would be greater than alpha. (If alpha is .05, then you are putting .05 in one tail of the null distribution and another .025 in the opposite tail, for a total alpha of .075.) Because of the potential abuse of one-tailed testing, the two-tailed test is generally considered more conservative (i.e., more likely to keep Type I errors at alpha) and is therefore less likely to be criticized. The other side of this argument is that a one-tailed test is more "powerful" if your prediction is correct (you are less likely to make a Type II error; I will explain more fully in Chapter 10). However, since researchers tend to be more openly concerned about Type I errors, you are usually safer from criticism if you always do a two-tailed test.

Perhaps the most justifiable case for performing a one-tailed test is when there is a strong basis for predicting results in a particular direction (e.g., the results of previous studies) and results in the "other" tail (i.e., in the unexpected direction) are really absurd, or are otherwise not worth testing. For example, suppose a medical researcher is testing an improved headache remedy, similar to others that have been effective in the past. Although it is not impossible that this new remedy could actually worsen headaches, the researcher is not interested in finding a drug that worsens headaches, and could legitimately plan not to test for such a possibility. Either reason—a strong basis for prediction (based on theory or previous results) or a total lack of interest in results in the other tail—can justify performing a one-tailed test, but bear in mind that psychologists tend to be very cautious about planning one-tailed tests

There are many cases when it may at first seem absurd to test the "other" tail, but not so absurd after more thought. For instance, it may seem obvious that a stimulant drug will allow subjects to produce more work in a given amount of time. But if the work requires some skill or concentration, the subjects may actually produce less because they are too stimulated to be patient or organized. Only experience in a particular field of research will tell you if a particular one-tailed test is likely to be considered acceptable by your colleagues. For psychologists in general, the two-tailed test is considered the "default"—the conventional thing to do, just as .05 is the conventional alpha level.

Exercises

* 1. a) If the calculated z for an experiment equals 1.35, what is the corresponding one-tailed p level? The two-tailed p level?

    b) Find the one- and two-tailed p levels corresponding to $z = -.7$.

    c) Find one- and two-tailed p levels for $z = 2.2$.

* 2. a) If alpha were set to the unusual value of .08, what would be the magnitude of the critical z for a one-tailed test? What would be the values for a two-tailed test?

    b) Find the one- and two-tailed critical z values for $\alpha = .03$.

    c) Find one- and two-tailed z values for $\alpha = .007$.

3. a) If the one-tailed p level for an experiment were .123, what would the value of z have to be?

    b) If the two-tailed p level for an experiment were .4532, what would the value of z have to be?

* 4. a) As alpha is made smaller (e.g., .01 instead of .05), what happens to the size of the critical z?

    b) As the calculated z gets larger, what happens to the corresponding p level?

* 5. An English professor suspects that her current class of 36 students is unusually good at verbal skills. She looks up the verbal SAT score for each student

and is pleased to find that the mean for the class is 540. Assuming that the general population of students has a mean verbal SAT score of 500 with a standard deviation of 100, what is the two-tailed p level corresponding to this class?

6. Consider a situation in which you have calculated the z score for a group of subjects and have obtained the unusually high value of 20. Which of the following statements would be true, and which would be false? Explain your answer in each case.

    a) You must have made a calculation error, because z scores cannot get so high.

    b) The null hypothesis cannot be true.

    c) The null hypothesis can be rejected, even if a very small alpha is used.

    d) The difference between the sample mean and the hypothesized population mean must have been quite large.

7. Suppose the z score mentioned in exercise 6 involved the measurement of height for a group of men. If $\mu = 69$ inches and $\sigma = 3$ inches, how can a group of men have a z score equal to 20? Give a numerical example illustrating how this can occur.

* 8. Compared to a one-tailed hypothesis test, a two-tailed test requires a) more calculation; b) more prior

planning; c) more integrity; d) a larger critical *t*-value; e) all of the above.

9. Describe a situation in which a one-tailed hypothesis test seems justified. Describe a situation in which a two-tailed test is clearly called for.

10. Describe a case in which it would probably be appropriate to use an alpha smaller than the conventional .05 (e.g., .01). Describe a case in which it might be appropriate to use an unusually large alpha (e.g., .1).

The example in Section A of the psychic selecting people high in math ability does not depict a situation that arises often in psychological research. In fact, one-sample hypothesis tests are relatively rare in psychology, for reasons which will be mentioned later in this chapter. However, there are circumstances that can reasonably lead to the use of a one-sample test. I have tried to make the following hypothetical experiment as clear and simple as possible, but it is similar in structure to real experiments that have been published.

Dr. Sara Tonin is a psychiatrist who specializes in helping women who are depressed and have been depressed since early childhood. She refers to these patients as life-long depressives (LLD). Over the years she has gotten the impression that these women tend to be shorter than average. The general research question that interests her is whether there may be some physiological effect of childhood depression on growth function. However, in order to design an experiment she must translate her general research question into a more specific research hypothesis. If her research hypothesis is framed in terms of population means (probably the most common way to proceed)—such as stating that the mean height of the LLD population is less than the mean height of the general population—she will need to use the procedures described below to perform a one-sample hypothesis test.

An hypothesis test can be described as a formal procedure with a series of specific steps. Although I will now describe the steps for a one-sample hypothesis test when the standard deviation for the population is known, the steps will be similar for the other hypothesis tests I will be describing in this text.

BASIC STATISTICAL PROCEDURES

Step 1. State the Hypotheses

Based on her observations and her speculations concerning physiology, Dr. Tonin has formed the *research hypothesis* that LLD women are shorter than women in the general population. Notice that she is not specifying how much shorter LLD women are. Her hypothesis is therefore not specific enough to be tested directly. This is common when dealing with research hypotheses. However, there is another, complementary hypothesis that is easy to test because it is quite specific. Dr. Tonin could hypothesize that LLD women are not shorter than average; rather, they are exactly the same height on average as the rest of the population. This hypothesis is easy to test only if you know the average height of all women in the population—but that measure happens to be fairly well known. Dr. Tonin thus tests her research hypothesis *indirectly*, by testing what she doesn't want to be true (i.e., the null hypothesis); she hopes the null hypothesis will be rejected, which would imply that the complementary hypothesis (i.e., the research hypothesis) is true. You will see shortly why specific hypotheses are easier to test. First we need to name these hypotheses.

What shall we call the specific hypothesis that states that LLD women are the same height as other women? This hypothesis essentially asserts that there is "nothing going on" with respect to the height of LLD women. So it is not surprising that this kind of hypothesis is called the "null hypothesis." (Recall from Section A that this is just the kind of hypothesis we would expect from Dr. Null.) Of course, this hypothesis does not imply that every LLD woman is average in height, or that you cannot find many LLD women who are below average. The null hypothesis concerns the mean of the population. Stated more formally, the null hypothesis in this case is that the mean height for the population of LLD women is exactly the same as the mean height for the population of all women. Dr. Tonin can also state this hypothesis symbolically. A capital H is used to represent a hypothesis. To indicate the null hypothesis, a subscript of zero is used: H_0. This is usually pronounced "H sub zero" or "H nought." To symbolize the mean of the population of LLD women, Dr. Tonin uses μ, as she would for any population mean. To symbolize the mean of the population she is using as the basis of comparison (in this case, all women) she uses μ with the subscript hyp or 0, to indicate that this mean is associated with the null hypothesis (μ_{hyp} or μ_0). Stated symbolically, the null hypothesis (that the mean height of LLD women is the same as the mean height for all women) is $H_0: \mu = \mu_0$. However, before she can test the null hypothesis, Dr. Tonin must get even more specific and fill in the value for μ_0. Although it is not likely that anyone has ever measured the height of every woman in the country and therefore we cannot know the exact value of the population mean, height is one of those variables for which enough data has been collected to make an excellent estimate of the true population mean (other examples include body temperature and blood pressure). For the purposes of this example we will assume that $\mu_0 = 65$ inches, so Dr. Tonin can state her null hypothesis formally as $H_0: \mu = 65$ inches.

So far we have paid a lot of attention to the null hypothesis and, as you will see, that is appropriate. But what of Dr. Tonin's research hypothesis? Isn't that what we should be more interested in? Now that we have defined the null hypothesis it will be easier to define the research hypothesis. In fact, the research hypothesis is often defined in terms of the null hypothesis; the former is the complement, or negation, of the latter. Thus defined, the research hypothesis is called the **alternative hypothesis** and is generally symbolized by the letter A (for *alternative*) or the number 1 (in contrast to 0, for the null hypothesis) as a subscript: H_A or H_1. Stated formally, the alternative hypothesis for the present example would read as follows: The mean height for the population of LLD women is *not* the same as the mean height for all women. Symbolically it would be written $H_A: \mu \neq \mu_0$, or in this case, $H_A: \mu \neq 65$ inches.

Either the null hypothesis or the alternative hypothesis must be true (either the two means are the same or they aren't). There is, however, one complication that you may have noticed. The alternative hypothesis just described is two-tailed, because there are two directions in which it could be correct. This implies that the alternative hypothesis could be true even if the LLD women were taller instead of shorter than average. This is the more conservative way to proceed (refer to Section A). However, if a one-tailed test is considered justified, the alternative hypothesis can be modified accordingly. If Dr. Tonin intends to test only the possibility that LLD women are shorter, the appropriate H_A would state that the mean height of the LLD women is *less* than the mean height of all women. Symbolically this is written H_A: $\mu < \mu_0$, or in this case, H_A: $\mu < 65$ inches. This is now a one-tailed alternative hypothesis. The other one-tailed alternative hypothesis would, of course, look the same except it would use the word *more* instead of *less* (or the greater than instead of the less than symbol). To summarize: Before proceeding to the next step, a researcher should state the null hypothesis, decide on a one- or two-tailed test, and state the appropriate alternative hypothesis. Bear in mind that the two-tailed test is far more common than the one-tailed test and is considered more conventional.

Step 2. Select the Statistical Test and the Significance Level

Because Dr. Tonin is comparing the mean of a single sample to a population mean, and the standard deviation of the population for the variable of interest (in this case, height) is known, the appropriate statistical test is the one-sample z-test. The convention in psychological research is to set alpha, the significance level, at .05. Only in unusual circumstances (e.g., testing a preliminary condition where the null must be accepted for the remaining tests to be valid) might a larger alpha such as .1 be justified. More common than setting a larger alpha is setting a stricter alpha, such as .01 or even .001, especially if many significance tests are being performed. (This process will be discussed in detail in Chapter 15.)

Step 3. Select the Sample and Collect the Data

If Dr. Tonin wants to know if LLD women are shorter than average, there is a definitive way to answer the question that avoids the necessity of performing an hypothesis test. She simply has to measure the height of every LLD woman in the population of interest (e.g., adult women in the United States) and compare the LLD mean with the mean of the comparison population. The difficulty with this approach is a practical one and should be obvious. The solution is to select a random sample of LLD women, find the mean for that sample, and continue with the hypothesis testing procedure. How large a sample? From the standpoint of hypothesis testing, the larger the better. The larger the sample, the more accurate will be the results of the hypothesis test. Specifically, there will be fewer Type II errors with a larger sample; the proportion of Type I errors depends only on alpha and therefore does not change with sample size. However, the same practical considerations that rule out measuring the entire population tend to limit the size of the sample. The size of the sample for any particular experiment will

Table 7.2			
65	59	68	63
61	62	60	66
63	63	59	67
58	66	64	64

depend on a compromise among several factors, such as the cost of measuring each subject and how large a difference between populations is expected (see Chapter 10). In the present example the sample size may be additionally limited by the number of LLD patients who can be found and their availability to the experimenter. For the purposes of illustration, let us suppose that Dr. Tonin could not find more than 16 LLD patients, so for this example $N = 16$. The heights (in inches) of the 16 LLD women appear in Table 7.2.

Step 4. Find the Region of Rejection

The test statistic Dr. Tonin will be using to evaluate the null hypothesis is just the ordinary z score for groups, and she is assuming that these z scores follow a normal distribution. More specifically, these z scores will follow the standard normal distribution (a normal distribution with $\mu = 0$ and $\sigma = 1$) if the null hypothesis is true, so the standard normal distribution is the null hypothesis distribution. Therefore, Dr. Tonin can calculate the z score for the data in Table 7.2 and look in Table A.1 to see if the area beyond this z score (i.e., the p level) is less than the alpha she set in Step 2. There is, however, an alternative procedure that is very convenient and commonly used, so I will introduce it now.

Given that she has already set alpha to .05, Dr. Tonin can determine beforehand which z score has exactly .05 area beyond it. It is not hard to find that z score. She just looks in Table A.1 at areas beyond z until she finds an area that equals .05 and then sees what z score corresponds to it. Unfortunately, the way the table is constructed, none of the areas comes out exactly to .05. However, a z score of 1.64 corresponds to .055, and a z score of 1.65 corresponds to .045, so by inter-polation Dr. Tonin can say that a z score of 1.645 should correspond closely to an "area beyond" (i.e., the area in the tail of the distribution) that equals .05. The value of the test statistic that corresponds exactly to alpha is called the **critical value** of the test statistic. In the case of z scores, the z score beyond which the area left over is equal to alpha is called a *critical z score*.

To get the critical value corresponding to alpha $= .05$ for the two-tailed test Dr. Tonin must divide alpha in half (half of alpha will go in each tail) and then find the z score for that area. So if alpha equals .05, she takes half of alpha—in this case, .025—and looks for that area beyond in the normal distribution table (Table A.1). If you look for .025 in the appropriate column, you will see that it corresponds to a z score of 1.96. So 1.96 is the critical z when you are doing a .05, two-tailed test. Actually, you have two critical values, $+1.96$ and -1.96, one in each tail (see Figure 7.5). In a similar manner you can find the critical z scores

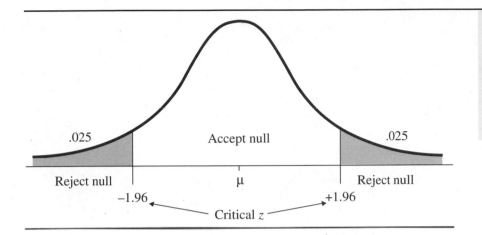

Figure 7.5

Two-Tailed Hypothesis Test, alpha = .05, Showing Critical Values for z

for the .01 one- and two-tailed tests; they are 2.33 and ±2.58, respectively (check the table to verify this for yourself).

Once you decide on the alpha level you will use and on your alternative hypothesis (one-tailed or two-tailed), the critical values of your z-test are determined. If the z score you calculate for your experimental group is larger than the critical value, then without having to look in the table you know your p level will be smaller than alpha, and your results can be declared statistically significant at the chosen alpha level. Consider the two-tailed test. Notice in Figure 7.5 that the area beyond +1.96 is .025. Any z score larger than +1.96 would correspond to a p level of less than .025, which when doubled would be less than .05 (similarly for any z score more negative than −1.96). Thus any z score larger than +1.96 would lead to the rejection of the null hypothesis at the .05, two-tailed level. That is why the area of the null hypothesis distribution above +1.96 is shaded and labeled "Reject null." For the same reason, the area below −1.96 is also labeled "Reject null." Each region of rejection has an area of .025, so when the two are added together the total area of rejection equals .05, which is the alpha set for this example.

For a one-tailed test, there would be only one region of rejection on the end of the null hypothesis distribution where the experimental result is predicted to fall. If alpha is set to .05 for a one-tailed test, then the region of rejection would have an area of .05 but would appear on only one side. Note that in Figure 7.6 this leads to a critical z score that is easier to beat than in the case of the two-tailed test (1.645 instead of 1.96). That is the advantage of the one-tailed test. But there is no region of rejection on the other side in case there is an unexpected finding in the opposite direction. This is the disadvantage of the one-tailed test, as discussed in Section A. Because the region of rejection for a z-test depends only on alpha and whether a one- or two-tailed test has been planned, the region of rejection can and should be specified before the z-test is calculated. It would certainly not be proper to change the region of rejection after you had seen the data.

Figure 7.6

One-Tailed Hypothesis Tests, alpha = .05

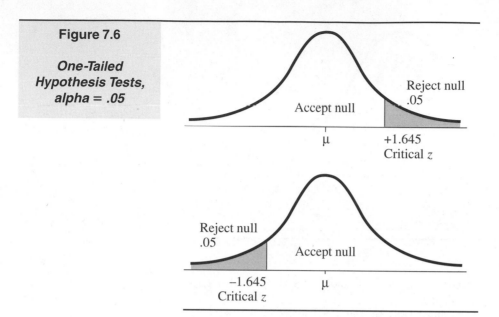

Step 5. Calculate the Test Statistic

Once Dr. Tonin has collected her random sample of LLD women she can measure their heights and calculate the sample mean ($\overline{X}$). If that sample mean happens to be exactly equal to the population mean specified by the null hypothesis, she does not need to do an hypothesis test; she has no choice but to accept (or retain, or fail to reject) the null hypothesis. If she specifies a one-tailed alternative hypothesis, and it turns out that the sample mean is actually in the opposite direction from the one she predicted (e.g., H_A states $\mu < 65$, but $\overline{X}$ turns out to be more than 65), again she must accept the null hypothesis. If, however, $\overline{X}$ is in the direction predicted by H_A, or if a two-tailed alternative has been specified and $\overline{X}$ is not equal to μ_0, an hypothesis test is needed to determine if $\overline{X}$ is so far from μ_0 (i.e., the value predicted by H_0) that the null hypothesis can be rejected.

The mean for the 16 measurements in Table 7.2 is 63 inches. (You should check this for yourself as practice.) Clearly, this mean is less than the average for all women. However, it is certainly possible to randomly select 16 non-LLD women, measure their heights, and get an average that is even lower than 63 inches. In order to find out the probability of randomly drawing a group of women with a mean height of 63 inches or less, we have to transform the 63 inches into a z score for groups, as described in Chapter 6. The z score of our experimental group of LLD patients with respect to the null hypothesis distribution is the *test statistic*.

When you have already calculated $\sigma_{\overline{X}}$, you can use Formula 7.1 to get the test statistic. However, to create a more convenient, one-step calculating formula,

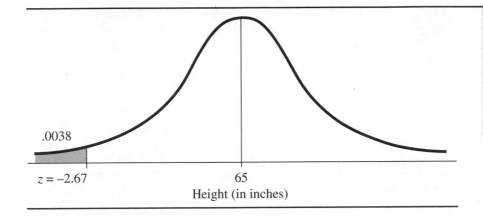

I will start with Formula 7.1 and replace the denominator ($\sigma_{\overline{X}}$) with Formula 6.1, to get Formula 7.2:

$$z = \frac{\overline{X} - \mu}{\sigma/\sqrt{N}}$$ Formula 7.2

Now you can plug in the values for the present example:

$$z = \frac{63 - 65}{3/\sqrt{16}} = \frac{-2}{3/4} = \frac{-2}{.75} = -2.67$$

Having found the z score that corresponds to the experimental group of LLD patients, you can proceed to the next step and compare the calculated z with the critical z. However, using Table A.1, you can also find the p level that is associated with the calculated z score (just ignore the minus sign when using Table A.1). From the table you can see that the area beyond this z score (2.67) is .0038. So the probability of drawing a group from the non-LLD population whose mean height is as short as (or shorter than) your LLD group is .0038. This is the p level Dr. Tonin would use if she had planned a one-tailed test with H_A: $\mu < 65$ (see Figure 7.7). Because we prefer to do a two-tailed test, the p level would be twice .0038, or .0076, which indicates the probability of drawing a group from the non-LLD population that is at least as *extreme* as the LLD group. Even in this case, the p level is clearly less than .05, so we can reject the null hypothesis that $\mu = 65$ inches and claim that our results are therefore significant at the .05 level.

Step 6. Make the Statistical Decision

The final step is to compare your calculated test statistic with the appropriate critical value. For a two-tailed test with alpha = .05, the critical value is ± 1.96.

Because -2.67 is less than -1.96, you can reject the null hypothesis. (Most often you just ignore the sign of your test statistic and work only with positive values—keeping in mind, of course, the direction of your results.) It turns out that even if you had planned a two-tailed test at the .01 alpha level, your calculated z would still be larger than the critical value (i.e., 2.58), and therefore you could have rejected the null hypothesis even at the .01 level with a two-tailed test.

Bear in mind that it is still possible that the two populations really have the same mean and the fact that your LLD sample is so short is just a fluke. In other words, by rejecting the null hypothesis you may be making a Type I error (i.e., rejecting a true null). But at least this is a calculated risk. When the null hypothesis is really true for your experiment, you will make a Type I error only 5% of the time. It is also important to realize that there is no way of knowing when, or even how often the null hypothesis is actually true, so you can never have any idea of just how many Type I errors are actually being made (see Section C). All you know is that when the null is true, you have a 95% chance of accepting the null (or as some prefer to say, failing to reject the null) if you use alpha = .05. And when you do accept the null hypothesis, you are not saying that you believe the null is actually true; rather, you are saying that the chance of making a Type I error is too large to reject the null in such a case.

The six steps detailed above can be used to describe any type of null hypothesis test, including all of the tests I will outline in the remainder of this text. That is why I took the time to describe these steps in detail. In subsequent chapters the steps can be described more briefly, and if you get confused about a step you can return to this chapter for a more complete explanation.

Assumptions Underlying the One-Sample *z*-Test

In order to do a one-sample z-test we had to know the null hypothesis distribution; that is, we had to know what group means are likely when you draw random samples of a specified size from a known population. That way we could judge whether it is likely to find a random sample of non-LLD women as short, on average, as our sample of LLD women. This procedure is fairly simple, but only if certain conditions apply. In fact, the results of our significance tests are strictly true only if these conditions are met. The following are the conditions that researchers try to meet, or must assume are met, in order for the one-sample z-test (as I have described it) to be valid.

The Sample Was Drawn Randomly

For instance, referring to the example involving LLD women, a random sample must be drawn in such a way that each LLD woman in the entire population of LLD women has an equal chance of being chosen, and each of these women is selected independently of all other selections. (Once you have selected a particular LLD woman you cannot then decide that you will include any of her close friends that also happen to be LLDs.) In reality, a researcher tries to draw a sample that is as representative as possible, but inevitably there will be limitations involving such factors as the subjects' geographical location, socioeconomic status, willingness to participate, etc. These factors should be kept in mind when you try to draw conclusions from a study or to generalize the results to the entire population.

In fact, any systematic bias could completely invalidate the results. For instance, if only poor LLD women can be attracted to the study (perhaps because of a small reward for participating), then these women may be shorter not because of depression, but because of inferior nutrition that results from being poor. Because the validity of one-sample tests hinges on the randomness of the sampling, you should be quite cautious about conducting or interpreting such studies.

The Variable Measured Has a Normal Distribution in the Population

Because of the Central Limit Theorem, this assumption is not critical when the size of the group is about 30 or more (see Chapter 6, especially Section C). However, if there is reason to believe that the population distribution is very far from a normal distribution, and fairly small sample sizes are being used, there is reason to be concerned. In such cases the use of nonparametric statistics should be considered (see Part VI). In all cases the researcher should look at the distribution of the sample data carefully to get an idea of just how far from normal the population distribution might be.

The Standard Deviation for the Sampled Population Is the Same as That of the Comparison Population

In the example of the psychic, I used the standard deviation of the general (i.e., comparison) population to find the standard error for the null hypothesis distribution. However, what if the population being sampled *does* have the same mean as the comparison population (as stated in the null hypothesis) but a very different variance? (Statisticians prefer to work with variances when dealing with normal distributions.) Then the probability of getting a particular sample mean would *not* follow the null hypothesis distribution, and it is even possible that in some situations we could too easily conclude that the sampled population has a different mean when it has only a different variance. Because it would be quite unusual to find a sampled population that has exactly the same mean as the comparison population but a rather different variance, this possibility is generally ignored, and the variances of the sampled and the comparison population are assumed to be equal. In a sense it is not a serious error if a researcher concludes that the mean of a sampled population is different from that of the general population, when it is only the variance that differs. In either case there is indeed something special about the population in question, which may be worthy of further exploration. I will return to this issue when I discuss the assumptions that underlie hypothesis tests involving two sample means.

When to Use the One-Sample Test

Testing a Preexisting Group

The one-sample *z*-test can be used whenever you know the mean and standard deviation of some variable of interest for an entire population and you want to know if your selected sample is likely to come from this population or from some population that has a different mean for that variable. The LLD example described above fits this pattern. The mean height and its standard deviation for the population of women in the United States are well known (or at least a very good estimate can be made). We can then ask if based on their mean height, the sample of LLD women is likely to come from the known population. If there is only a small chance of finding a random sample from the general population with a mean height as low as the LLD women, we can conclude that the mean for all LLD women is less than the mean for women, in general. Dr. Tonin would be happy to draw such a conclusion, because it lends support to her theory that the stress of LLD inhibits growth. It doesn't prove that her theory is correct—other theories might make the same prediction for very different reasons—but it is encouraging. If there were no difference in height, or if the difference was in the other direction, Dr. Tonin would be inclined to abandon or seriously modify her theory. (The amount of discouragement appropriate for negative results depends on power, which will be discussed in Chapter 10.)

The LLD "experiment" I have been describing is not a *true* experiment, because the experimenter did not decide which subjects in the population had been depressed as children and which had not. The LLD women represent a *preexisting group,* that is, a group that was not created by the experimenter; the experimenter only *selected* from a group that was already in existence. A true experiment would be more conclusive, but it would be virtually impossible to devise a true experiment in this case. The LLD study is an example of correlational research, as described in Chapter 1. The distinction between "true" and "correlational" (quasi) experiments is an important concept that I will be discussing in upcoming chapters.

Performing a One-Sample Experiment

The example I have been discussing deals with two somewhat distinct populations: LLD and non-LLD women. The one-sample *z*-test could also be used to test the effects of an experimental treatment on a sample from a population with a known mean and standard deviation for some variable. Consider the following example. You want to know if a new medication affects the heart rate of the people who take it. Heart rate is a variable for which we know the mean and standard deviation in the general population. Although it may not be obvious, testing the effect of some treatment (e.g., the effect of a drug on heart rate) can be viewed in terms of comparing populations. To understand how the heart rate experiment can be viewed as involving two different populations, you must think of the sample of people taking the new drug as representatives of a new population—not a population that actually exists, but one that would exist if everyone in the general population took the drug and their heart rates were measured. All these heart rates taken after drug ingestion would constitute a new population (not a preexisting population like the LLD women, but a population created by the experimenter's administration of the drug). Now we can ask if this new population has the same mean heart rate as the general population (those not taking this drug). If the sample of people taking the drug has a mean heart rate that is not likely to be found by choosing a random sample of people not taking the drug, you would conclude that the two populations have different means. This is the same as saying that the drug does indeed affect heart rate, or that if everyone in the population took the drug, the population mean for heart rate would change.

However, this one-sample procedure for testing the effects of the new drug on heart rate is not a very good way to do the experiment. The people taking the drug may be affected just by being in an experiment, even if the drug has no real effect (e.g., they may fear the injection or the side effects of the drug). The best way to do the experiment is to have a second group (a *control group*) that takes a fake drug and is compared to the group taking the real drug. Both groups should have the same fears, expectations, etc. I will discuss such two-group experiments in Chapter 9.

There are situations for which a one-group experiment is reasonable, but these are relatively rare. For instance, the SAT is usually administered in a classroom with many students taking the exam at the same time. If a researcher wanted to test the hypothesis that students score higher when taking the exam in a smaller room with just a few other students, only one random group of students would have to be tested, and they would be tested in the new condition (i.e., smaller room, etc.). A control group would not be necessary, since we already have a great deal of data about how students perform under the usual conditions. Although this experiment represents an appropriate use of the one-sample test, two-group experiments are far more common. In actual practice the one-sample test is used much more often with preexisting groups than in the context of a true experiment.

So far I have implied that the one-sample z-test can only be used when the mean and standard deviation of the comparison population are known. Actually if your sample is large enough, you need only have a population mean to compare to; the unbiased standard deviation of your sample can be used in place of the population standard deviation. How large does the sample have to be? When might you have a population mean, but not a corresponding population standard deviation? These questions will be answered at the beginning of Chapter 8.

Publishing the Results of One-Sample Tests

Let us suppose that Dr. Tonin has written a journal article about her LLD experiment, and has followed the guidelines in the *Publication Manual of the American Psychological Association,* fourth edition (1994). Somewhere in her results section she will have a statement such as the following: "As expected, the LLD women sampled were on average shorter ($\underline{M}$ = 63 inches) than the general population (μ = 65 inches); a one-sample test with alpha = .05 demonstrated that this difference was statistically significant, $\underline{z} = -2.67$, $\underline{p} < .05$, two-tailed." In this sentence, $\underline{M}$ stands for the sample mean, and it is often written instead of $\overline{X}$ (though both have the same meaning). The designation $\underline{p} < .05$ means that the p level associated with the test statistic (in this case, $\underline{z}$) turned out to be less than .05. The *M, z,* and *p* are all underlined to indicate that they should appear in italics in the published manuscript; according to APA style, all statistical symbols should be underlined in the manuscript. (Greek letters, such as μ, are generally not underlined.)

APA style requires that measurements originally made in nonmetric units (e.g., inches) also be expressed in terms of their metric equivalent. An alternative way of writing about the results of the LLD experiment would be "The mean height for the sample of LLD women was 63 inches (160 cm). A one-sample test with alpha = .05 showed that this is significantly less than the population mean of 65 inches (165 cm), $\underline{z} = -2.67$, $\underline{p} < .05$, two-tailed." Of course, if any particular test involved a different alpha or number of tails, they would have to be stated for the test in question.

The actual p level for Dr. Tonin's experiment was .0076, which is even less than .01. It is common to report p levels in terms of the lowest alpha level at which the results were significant, regardless of the original alpha set, so Dr. Tonin would probably have written $p < .01$, instead of $p < .05$. Generally, significance levels are viewed as values ending in 5 or 1 (e.g., $p < .005$ or $p < .0001$), but researchers sometimes prefer to report p levels more exactly, as they are given by statistical software (e.g., $p < .02$ or $p = .035$). It is especially common to report more accurate p levels when describing results that are close to but not below the conventional alpha of .05 (e.g., $p < .06$ or $p = .055$). However, if the p level is much larger than alpha, so that the result is not even close to statistically significant, it is common to state only that p is greater than alpha (e.g., $p > .05$).

Exercises

1. A psychiatrist is testing a new anti-anxiety drug, which seems to have the potentially harmful side effect of lowering the heart rate. For a sample of 50 medical students whose pulse was measured after 6 weeks of taking the drug, the mean heart rate was 70 beats per minute (bpm). If the mean heart rate for the population is 72 bpm with a standard deviation of 12, can the psychiatrist conclude that the new drug lowers heart rate significantly? (Set alpha = .05 and perform a one-tailed test.)

* 2. Can repressed anger lead to higher blood pressure? In a hypothetical study, 16 college students with very high repressed anger scores (derived from a series of questionnaires taken in an introductory psychology class) are called in to have their blood pressure measured. The mean systolic blood pressure for this sample ($\overline{X}$) is 124 mm Hg. (Millimeters of mercury are the standard units for measuring blood pressure.) If the mean systolic blood pressure for the population is 120 with a standard deviation of 10, can you conclude that repressed anger is associated with higher blood pressure? Use alpha = .05, two-tailed.

3. Suppose that the sample in exercise 2 had been four times as large (i.e., 64 students with very high repressed anger scores), but the same sample mean had been obtained.

   a) Can the null hypothesis now be rejected at the .05 level?

b) How does the calculated z for this exercise compare with that in exercise 2?

c) What happens to the calculated z if the size of the sample is multiplied by k but the sample mean remains the same?

* 4. A psychologist has measured the IQ for a group of 30 children, now in the third grade, who had been regularly exposed to a new interactive, computerized teaching device. The mean IQ for these children is $\overline{X} = 106$.

a) Test the null hypothesis that these children are no different from the general population of third-graders ($\mu = 100$, $\sigma = 16$) using alpha = .05.

b) Test the same hypothesis using alpha = .01. What happens to your chances of attaining statistical significance as alpha becomes smaller (all else being equal)?

5. Referring to exercise 4, imagine that you have read about a similar study of IQs of third-graders, in which the same sample mean (106) was obtained, but the z score reported was 3.0. Unfortunately, the article neglected to report the number of subjects that were measured for this study. Use the information just given to determine the sample size that must have been used.

* 6. The following are verbal SAT scores of hypothetical students who were forced to take the test under adverse conditions (e.g., construction noises, room too warm, etc.): 510, 550, 410, 530, 480, 500, 390, 420, 440. Do these scores suggest that the adverse conditions really made a difference (at the .05 level)?

7. Suppose that an anxiety scale is expressed as T scores, so that $\mu = 50$ and $\sigma = 10$. After an earth-quake hits their town, a random sample of the townspeople yields the following anxiety scores: 72, 59, 54, 56, 48, 52, 57, 51, 64, 67.

a) Test the null hypothesis that the earthquake did not affect the level of anxiety in that town (use alpha = .05).

b) Considering your decision in part a, which kind of error (Type I or Type II) could you be making?

* 8. Imagine that you are testing a new drug that seems to raise the number of T cells in the blood, and therefore has enormous potential for the treatment of disease. After treating 100 subjects, you find that their mean ($\overline{X}$) T cell count is 29.1. Assume that μ and σ (hypothetically) are 28 and 6, respectively.

a) Test the null hypothesis at the .05 level, two-tailed.

b) Test the same hypothesis at the .1 level, two-tailed.

c) Describe in practical terms what it would mean to commit a Type I error in this example.

d) Describe in practical terms what it would mean to commit a Type II error in this example.

e) How might you justify the use of .1 for alpha in similar experiments?

9. a) Assuming everything else in the previous problem stayed the same, what would happen to your calculated z if the population standard deviation (σ) were 3 instead of 6?

b) What general statement can you make about how changes in σ affect the calculated value of z?

* 10. Referring to exercise 8, suppose that $\overline{X}$ is equal to 29.1 regardless of the sample size. How large would N have to be for the calculated z to be statistically significant at the .01 level (two-tailed)?

The significance level, alpha, can be defined as the probability of rejecting the null hypothesis, *when the null hypothesis is true.* Alpha is called a conditional probability (as defined in Chapter 5, Section C) because it is the probability of rejecting the null *given that a particular condition is satisfied* (i.e., that the null hypothesis is true). Recall that in Table 7.1 the probabilities within each column added up to 1.0. That is because each column represented a different condition (status of H_0), and the probabilities were based on each condition separately.

OPTIONAL MATERIAL

Specifically, the first column takes all the times that the null hypothesis is true and indicates which proportion will lead to rejection of the null (α), and which proportion will not ($1 - \alpha$). The other column deals with those instances in which the null is not true; these latter probabilities will be discussed in detail in Chapter 10. For the remainder of this section, I will assume that alpha always equals .05, or 5%. This will make the discussion simpler and more concrete. It will then be easy to generalize to other choices for alpha.

You may find it easier to view alpha as a percentage than to view it as a probability. But if alpha is 5%, what is it 5% of? First, I will define a "null experiment" as one for which the null hypothesis is really true. Remember that if the null hypothesis is true, then either the experimental manipulation must be totally ineffective or the sampled population must be not at all different from the general population on the given variable—and we would not want such experiments to produce statistically significant results. Alpha can thus be defined as the percentage of null (i.e., *ineffective*) experiments that nonetheless attain statistical significance. Specifically, when alpha is 5%, it means that only 5% of the null experiments will be declared statistically significant.

The whole purpose of null hypothesis testing is to keep the null experiments from being viewed as effective. However, we cannot screen out all of the null experiments without screening out all of the effective ones, so by convention we allow some small percentage of the null experiments to be called statistically significant. That small percentage is called alpha. When you call a null experiment statistically significant you are making a Type I error, which is why we can say that alpha is the expected percentage of Type I errors.

Conditional probabilities can be difficult to understand and can lead to fallacies. For instance, it is common to forget that alpha is a conditional probability and instead to believe that 5% of *all* statistical tests will result in Type I errors. This is not true. It is the case that 5% of the statistical tests *involving ineffective experiments* will result in Type I errors, not 5% of *all* tests. Thus another way to view alpha is as the upper limit of Type I errors when considering a whole set of statistical tests. For example, suppose Dr. Johnson has performed 100 significance tests in her career so far (always using alpha = .05). What is the most likely number of Type I errors that she has made? The fallacy mentioned above might lead you to say five. But you really cannot answer this question without knowing how many of her experiments were ineffective. What you *can* say is that in the worst case, all 100 would have been ineffective. In that case, the most likely number of Type I errors would be five. But if any of her experiments were effective (the null hypothesis was *not* true), the expected number would be fewer than five, and if all her experiments were effective, she could not have made any Type I errors. (You can only make a Type I error when testing an ineffective treatment.) Thus although we do not know how many Type I errors to expect, we can say that the upper limit in this case is about five (for any given 100 null experiments, alpha is just an approximation), and the lower limit is, of course, zero.

Why do people have trouble with conditional probabilities? An interesting article by Pollard and Richardson (1987) discusses some of the reasons. In particular, they suggest that some of the blame for the difficulties psychology students face in understanding alpha can be attributed to the vagueness of discussions of alpha in most statistical textbooks. The discussion of alpha in this section was inspired in part by the Pollard and Richardson article.

Another fallacy concerning alpha arises when you ask psychologists to estimate the percentage of experiments that are truly ineffective—in other words, the probability of the null hypothesis being true. Many are tempted to say that the null hypothesis is true for about 5% of all experiments, but this too is a fallacy. In reality, there is no way to estimate the percentage of experiments that are ineffective, and it certainly does not depend on alpha. We don't even have a lower or upper limit (except, of course, for zero and 100%). It may be somewhat disturbing to realize that although a good deal of effort is spent to screen out null experiments with statistical tests, we have no idea how often null experiments are conducted, or even if any are conducted. All that we know is that however many ineffective experiments are performed, only about 5% of them will be declared statistically significant.

Finally, as Pollard and Richardson (1987) point out, there is a common tendency to believe that the reverse of a conditional probability will have the same value, and this seems to be true in the case of alpha. If you point out 100 statistically significant results in a journal, and ask researchers how many they suspect are Type I errors, the most likely response is 5% (assuming that was the alpha level for each of these tests). Again, this estimate is based on a fallacy and bears no resemblance to the truth. The fallacy arises from mentally reversing the conditional probability that alpha is based on. Recall that alpha is the percentage of ineffective experiments that attain statistical significance. If, on the other hand, you want to know what percentage of statistically significant results are Type I errors, you are really asking what percentage of results attaining significance are based on ineffective experiments. Or, what is the probability of the null hypothesis being true, given the condition that the results are significant? This is exactly the reverse of the conditional probability that is alpha, and there is no reason that the reverse should also equal alpha.

In fact, there is no way to determine the percentage of significant results that are really Type I errors. The problem can be expressed in a diagram (see Figure 7.8). As you can see from the upper part of the figure, significant results come from two sources: "effective" experiments (null is false) and "ineffective"— null— experiments (null is true). The two sources were drawn with wavy lines to indicate that in reality we have no idea how the number of null experiments compares to the number of effective ones. To add to the uncertainty, we know the percentage of null experiments that become significant (i.e., alpha), but we don't know what percentage of the effective experiments become significant. The latter percentage depends on just how effective the experiments are—that is, on the average power of the experiments—which is difficult to estimate (see Chapter 10). So we do not know what percentage of the total number of significant results comes from each source.

It is important to realize that it is quite possible that very few (much less than 5%) of the statistically significant results in any particular journal are Type I

Figure 7.8

Total Number of Statistically Significant Results

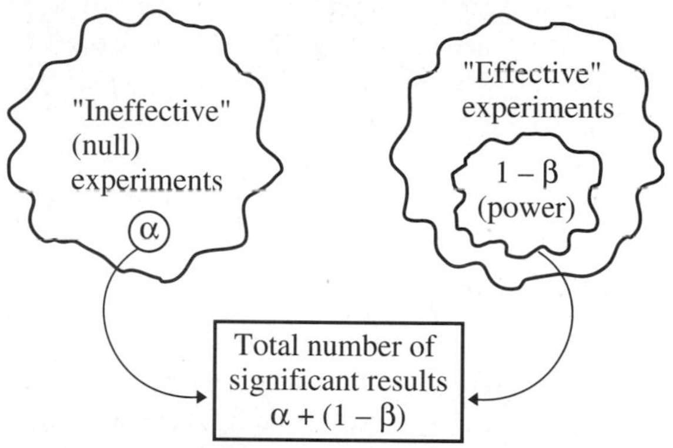

errors. (For instance, perhaps contributors to that journal almost always deal with powerful treatments.) On the other hand, you should be aware that a problem can arise from the fact that statistically significant results are much more likely to get published than nonsignificant results. The problem is that if 20 different researchers perform the same null experiment, the chances are that one of them will obtain significant results. Then, it is possible that the significant result (a Type I error) will be published but that the 19 "failures" (actually correct acceptances of the null) will never be heard of. If several of the "failures" were published, as well as the significant result, other researchers would be alerted to the Type I error; but nonsignificant results are usually not even submitted for publication, unless they have an impact on some important theory. Nonsignificant results are often filed away in a drawer somewhere, which is why our lack of knowledge of such results has been termed the "file drawer problem" (Rosenthal, 1979). Some way for journals briefly to note nonsignificant results might help lessen this problem. Otherwise, it is difficult to even guess at how many null experiments are being conducted in any given time period. On the other hand, even knowing for certain that the null hypothesis is not true may not be very informative in itself. I shall have more to say about the limitations of null hypothesis testing in Chapter 10; however, it will be easier to understand the shortcomings of null hypothesis testing and how they can be overcome after you have learned more about hypothesis testing in the next two chapters.

Exercises

* 1. Alpha stands for

a) the proportion of experiments that will attain statistical significance.

b) the proportion of experiments for which the null hypothesis is true that will attain statistical significance.

c) the proportion of statistically significant results for which the null hypothesis is true.

d) the proportion of experiments for which the null hypothesis is true.

2. Last year, Dr. Galton performed 50 experiments testing drugs for side effects. Suppose, however, that all of those drugs were obtained from the same fraudulent supplier, which was later revealed to have been sending only inert substances (e.g., distilled water,

sugar pills) instead of real drugs. If Dr. Galton used alpha = .05 for all of his hypothesis tests, how many of these 50 experiments would you expect to yield significant results? How many Type I errors would you expect? How many Type II errors would you expect?

*3. Since she arrived at the university, Dr. Pearson has been very productive and successful. She has already performed 20 experiments that have each attained the .05 level of statistical significance. What is your best guess for the number of Type I errors she has made so far? For the number of Type II errors?

4. Dr. Fisher is very ambitious and is planning to perform 100 experiments over the next two years and test each at the .05 level. How many Type I errors can Dr. Fisher be expected to make during that time? How many Type II errors?

SUMMARY

The Important Points of Section A

1. To perform a one-sample experiment, select a random sample of some population that interests you (or select a sample from the general population and subject it to some experimental condition), calculate the mean of the sample for the variable of interest, and compare it to the mean of the general population.

2. Even if the sample mean is different from the population mean in the direction you predicted, this result could be due to chance fluctuations in random sampling—that is, the null hypothesis could be true.

3. The null hypothesis distribution is a "map" of what results are or are not likely by chance. For the one-sample case, the null hypothesis distribution is the sampling distribution of the mean.

4. To test whether the null hypothesis can easily produce results that are just as impressive as the results you found in your experiment, find the z score of your sample mean with respect to the null hypothesis distribution. The area beyond that z score is the p value for your experiment (i.e., the probability that when the null hypothesis is true, results as good as yours will be produced).

5. When the p value for your experiment is less than the alpha level you set, you are prepared to take the risk of rejecting the null hypothesis and of declaring that your results are statistically significant.

6. The risk of rejecting the null hypothesis is that it may actually be true (even though your results look very good), in which case you are making a Type I error. The probability of making a Type I error when the null hypothesis is true is determined by the alpha level that you use (usually .05).

7. If you make alpha smaller to reduce the proportion of Type I errors, you will increase the proportion of Type II errors, which occur whenever the null hypothesis is *not* true but you fail to reject the null hypothesis (because you are being cautious). The probability of making a Type II error is not easily determined. Type II errors will be discussed thoroughly in Chapter 10.

8. One-tailed hypothesis tests make it easier to reach statistical significance in the predicted tail but rule out the possibility of testing results in the other tail. Because of the prevalence of unexpected results, the two-tailed test is more generally accepted.

The Important Points of Section B

You learned in Section B that hypothesis testing can be divided into a six-step process, as follows:

1. State the hypotheses.
2. Select the statistical test and the significance level.
3. Select the sample and collect the data.
4. Find the region of rejection.
5. Calculate the test statistic.
6. Make the statistical decision.

For review I will pose another example appropriate for one-group hypothesis testing and briefly work through the six steps. Do redheaded people have the same IQ as the rest of the population? Such a question can be answered definitively only by measuring the IQ of every redheaded individual in the population, calculating the mean IQ, and comparing that mean to the mean IQ of the general population. A more practical but less accurate way is to select a random sample of redheaded people and to find the mean IQ of that sample. It is almost certain that the sample mean will differ from the mean of the general population, but that doesn't automatically tell us that the mean IQ for the redheaded population differs from the mean IQ of the general population. However, if the mean IQ of the redheaded sample is far enough from the mean of the general population, we can then conclude that the two population means are different. To determine how far is far enough we need to follow the steps of a one-sample z-test.

Step 1. State the Hypotheses

The easiest way to proceed is to set up a specific null hypothesis that we wish to disprove. If the IQ of the general population is 100, then the null hypothesis would be expressed as $H_0: \mu = 100$. The complementary hypothesis is called the alternative hypothesis, and it is the hypothesis that we would like to be true. A two-tailed alternative would be written $H_A: \mu \neq 100$; a one-tailed alternative would be written either $H_A: \mu < 100$ or $H_A: \mu > 100$. Usually a two-tailed test is performed unless there is strong justification for a one-tailed test.

Step 2. Select the Statistical Test and the Significance Level

We are comparing one sample mean to a population mean, and we know the standard deviation of the population for the variable of interest, so it is appropriate to perform a one-sample z-test. Alpha is usually set at .05, unless some special situation requires a larger or smaller alpha.

Step 3. Select the Sample and Collect the Data

A random sample of the population of interest, in this case redheaded people, is selected. The larger the sample, the more accurate the results will be, but practical limitations will inevitably limit the size of the sample. Let us suppose that we have measured the IQ for each of ten randomly selected redheaded people, and the ten values are as follows: 106, 111, 97, 119, 88, 126, 104, 93, 115, 108.

Step 4. Find the Region of Rejection

The region of rejection can be found in terms of critical z scores—the z scores that cut off an area of the normal distribution that is exactly equal to alpha. (The region of rejection is equal to half of alpha in each tail of a two-tailed test.) Because we have set alpha = .05 and planned a two-tailed test, the critical z scores are $+1.96$ and -1.96. The regions of rejection are therefore the portions of the standard normal distribution that are above (i.e., to the right of) $+1.96$ or below (i.e., to the left of) -1.96.

Step 5. Calculate the Test Statistic

The first step is to calculate the mean of the sample, $\overline{X}$. We add up the ten numbers to get the sum of X, which is 1067. Then we divide the sum by the size of the sample, N, which is 10. So $\overline{X} = 1067/10 = 106.7$.

Immediately we see that the sample mean is above the population mean. Had we planned a one-tailed test and hypothesized that redheads are *less* intelligent than the general population, we could not ethically proceed with our statistical test. (To switch to a two-tailed test after you see the data implies that you are using an alpha of .05 in the hypothesized tail *and* .025 in the other tail, so your actual alpha would be the unconventionally high value of .075.)

Because we planned a two-tailed test we can proceed, but to perform the one-sample z-test we must know both the mean and the standard deviation of the comparison population. The mean IQ for the general population is set at 100, which is also the mean specified by the null hypothesis (Dr. Null thinks the redheads are no different from the rest of the population). The standard deviation for IQ (Wechsler test) is set at 15. Now we have all the values we need to use Formula 7.2 to get the z score:

$$z = \frac{\overline{X} - \mu}{\sigma/\sqrt{N}} = \frac{106.7 - 100}{15/\sqrt{10}} = \frac{6.7}{4.74} = 1.41$$

In calculating the z score for groups, there are only a few opportunities to make errors. A common error is to forget to take the square root of N. Perhaps an even more common error is to leave N out of the formula entirely and just divide by the population standard deviation instead of the standard error of the mean. Without the square root of N, the z score for groups is virtually identical to the ordinary z score for individuals, and sometimes people complain that they never know which type of z score to use. Keep the following rule in mind: If you are asking a question about a group, such as whether a particular group is extreme or unusual, then you are really asking a question about the *mean* of the group and should use the z score for groups (Formula 7.2). Only when you are concerned about one individual score, and where it falls in a distribution, should you use the simple z score for individuals.

Step 6. Make the Statistical Decision

If the z score you calculate is greater in magnitude than the critical z score, then you can reject the null hypothesis. In that case, as alpha had been set to .05, we would say that the results are significant at the .05 level. However, the calculated (or "obtained") z score is for this example less than the critical z, so we cannot reject the null hypothesis. Some would say that we therefore accept the null hypothesis, but most researchers prefer stating that there is insufficient evidence to reject the null hypothesis.

For our example, rejecting the null hypothesis would mean concluding that redheaded people have a mean IQ that is different from the general population. (Based on your data you would, of course, indicate in which direction the difference fell.) However, because our calculated z score was rather small and therefore fell too near the middle of the null hypothesis distribution, we would have to conclude that we do not have sufficient evidence to say that redheaded people differ in IQ from the rest of the population.

The Important Points of Section C

1. Alpha is a *conditional* probability; it is the probability that the null hypothesis (H_0) will be rejected *given* that H_0 is true, and it is also the probability that a statistical test involving a true H_0 will result in a Type I error.

2. Alpha is *not* the probability of *any* statistical test resulting in a Type I error— only the probability that a test for which H_0 is true will do so. If you are planning to perform a number of statistical tests, alpha represents the upper limit for the expected proportion of Type I errors in the "worst-case scenario"—that is, when the H_0 is true for all of the tests.

3. Alpha is also *not* the proportion of statistical tests for which the H_0 is true. Actually, there is no way to even guess at how often the null hypothesis is true when statistical tests are being performed.

4. There is a tendency to believe (erroneously) that the conditional probability represented by alpha can be reversed and still have the same value. The reverse of alpha is the probability that H_0 is true *given* that H_0 has been rejected, or the proportion of significant results that are really Type I errors. There is a tendency to believe that this probability is also equal to alpha, but in reality the proportion of significant results that are actually Type I errors depends on other probabilities that are not known.

Definitions of Key Terms

Null hypothesis A specific hypothesis that a researcher would like to disprove, usually one that implies that the experimental results are actually due entirely to chance factors involved in random sampling. In the one-sample case the null hypothesis is that the mean of the sampled or experimental population is the same as that of the comparison population.

***p* level** The probability that *when the null hypothesis is true* you will get a result that is at least as extreme as the one you found for your selected sample (or experimental group).

Alpha (α) level The percentage of Type I errors you are willing to tolerate. When the *p* level is less than the alpha level, you are willing to take the chance of making a Type I error by rejecting the null hypothesis.

Statistical significance The *p* level is low enough (i.e., below alpha) that the null hypothesis can be ignored as an explanation for the experimental results. (Statistical significance should not be confused with practical significance or importance. Ruling out chance as an explanation for your results does not imply that your results are strong enough to be useful or meaningful.)

Test statistic A sample statistic that follows a known mathematical distribution and can therefore be used to test a statistical hypothesis. The *z* score for groups is a test statistic that follows the standard normal distribution.

Type I error Rejecting the null hypothesis when it is actually true. This kind of error acts like a false alarm and can lure other experimenters into wasting time exploring an area where they are unlikely to find significant results.

Type II error Accepting (or failing to reject) the null hypothesis when the null hypothesis is actually false. This kind of error is a "miss"; a real effect has been missed, and other experimenters may mistakenly abandon exploring an area that in fact might be worth investigating.

Two-tailed test A test in which the alternative hypothesis is stated so that sample results can be tested in either direction. Alpha is divided in half; half of the probability is placed in each tail of the null hypothesis distribution.

One-tailed test A test in which the alternative hypothesis is stated so that only results that deviate in one particular direction from the null hypothesis can be tested. For instance, in the one-sample case, a one-tailed alternative hypothesis might state that the sampled (or experimentally treated) population mean is larger than the comparison population mean. The sample mean can then be tested for significance only if it is larger than the comparison population mean; no test can be performed if the sample mean is smaller than the comparison population mean.

Alternative hypothesis An hypothesis that is usually not stated specifically enough to be tested directly but rather in such a way that rejecting the null hypothesis implies that the alternative hypothesis is true. The alternative hypothesis is a formal statement of the research hypothesis that a researcher wants to prove.

Critical value The value of the test statistic for which the *p* level is exactly equal to the alpha level. Any test statistic larger than the critical value will have a *p* level less than alpha, leading to rejection of the null hypothesis.

Key Formulas

The *z* score for groups, used when the standard error of the mean ($\sigma_{\bar{X}}$) has already been calculated:

$$z = \frac{\bar{X} - \mu}{\sigma_{\bar{X}}}$$

Formula 7.1

The *z* score for groups, used when the standard error of the mean ($\sigma_{\bar{X}}$) has *not* already been calculated:

$$z = \frac{\bar{X} - \mu}{\sigma/\sqrt{N}}$$

Formula 7.2

THE ONE-SAMPLE *t*-TEST

You will need to use the following from previous chapters:

Symbols
μ: Mean of a population
$\overline{X}$: Mean of a sample
SS: Sum of squared deviations from the mean
σ: Standard deviation of a population
s: Unbiased standard deviation of a sample
$\sigma_{\overline{X}}$: Standard error of the mean
N: Number of subjects (or observations) in a sample

Formulas
Formula 6.1: The standard error of the mean
Formula 6.2 (or 7.1): The z score for groups
Formula 5.2: For finding X given a z score

Concepts
The null hypothesis distribution
Critical values of a test statistic

8

C H A P T E R

A

**CONCEPTUAL
FOUNDATION**

The previous chapter focused on variables that have been measured extensively in the general population, such as height and IQ. We have enough information about these variables to say that we know the population mean and standard deviation. However, suppose you are interested in a variable for which there is little information concerning the population, for instance, the number of hours per month each American family spends watching rented videotapes. The procedures you will learn in this chapter will make it possible to estimate the mean and standard deviation of the population from the data contained in one sample. You will also learn how to deal with the case in which you know the population mean but not its standard deviation, and you would like to know if a particular sample is likely to come from that population. This is not a common situation, but by studying it you will develop the tools you will need to handle the more common statistical procedures in psychological research. To illustrate a one-sample hypothesis test in which we know the population mean, but not its standard deviation, I have constructed the following example.

Suppose a friend of yours is considering psychotherapy but is complaining of the high cost. (I will ignore the possibility of therapy fees based on a "sliding scale" in order to simplify the example.) Your friend says, "The cost per hour keeps going up every year. I bet therapy wasn't so expensive back in the old days." You reply that because of inflation the cost of everything keeps rising, but your friend insists that even after adjusting for inflation it will be clear that psychotherapy is more expensive now then it was back in, for instance, 1960. The first step toward resolving this question is to find out the hourly cost of therapy in 1960. Suppose that after some library research your friend finds the results of an extensive survey of psychotherapy fees in 1960, which shows that the average hourly fee back then was $32. For the sake of the example, I will assume that no such survey has been conducted in recent years, so we will have to conduct our own survey to help settle the question.

The Mean of the Null Hypothesis Distribution

Suppose that the 1960 survey in our example was so thorough and comprehensive (it was conducted by some large national organization) that we can use the $32 average fee as the population mean for 1960. If we could conduct a current survey just as complete as the one done in 1960, we would have the answer to our question. We would only have to convert the 1960 fee to current dollars and compare the current average with the adjusted 1960 average. (For this example I will assume that 32 dollars in 1960 correspond to 63 current dollars, after adjusting for inflation.) However, considering that our resources are rather limited (as is often the case in academic research) and that there are quite a few more psychotherapists to survey today than there were in 1960, suppose that the best we can do is to survey a random sample of 100 current psychotherapists. If the mean hourly fee of this sample were $72, it would look like hourly fees had increased beyond the adjusted 1960 average (i.e., $63), but because our current figure is based on a limited sample rather than a survey of the entire population, we cannot settle the question with absolute certainty. If we were to announce our conclusion that psychotherapy is more expensive now than it was in 1960, Dr. Null would have something to say.

Dr. Null would say that the mean hourly fee is the same now as it was in 1960: $63 (in current dollars). He would say that our sample mean was just a bit of a fluke based on the chance fluctuations involved in sampling a population, and that by sampling a population with a mean of $63, he could beat our $72 mean on his first try! As you learned in Chapter 7, the chance of Dr. Null's beating us can be found by describing the null hypothesis distribution.

For this example, the null hypothesis distribution is what you get when you keep drawing samples of 100 psychotherapists and recording the mean hourly fee of each sample. The null hypothesis distribution will have the same mean as the population from which you are drawing the samples, in this case, $63. But the standard deviation of the null hypothesis distribution, called the standard error of the mean, will be smaller than the population standard deviation, because groups of 100 do not vary from each other as much as individuals do. In fact, Formula 6.1 presented a simple formula for the standard error:

$$\sigma_{\bar{X}} = \frac{\sigma}{\sqrt{N}}$$

When the Population Standard Deviation Is Not Known

If we try to apply Formula 6.1 to the present example, we immediately run into a problem. The hypothetical 1960 survey did not publish a standard deviation along with its average hourly fee. We have a population mean, but no population standard deviation, and therefore no σ to put into Formula 6.1. If there is no way to obtain the raw data from the 1960 survey, we cannot calculate σ. How can we find the null hypothesis distribution and make a decision about the null hypothesis?

The answer is that we can use the unbiased standard deviation (s) of our sample of 100 hourly fees in place of σ. (We assume that the variability has not changed from 1960 to the present; this assumption will be discussed further in Section B.) By making this substitution, we convert Formula 6.1 into Formula 8.1,

$$s_{\overline{X}} = \frac{s}{\sqrt{N}}$$

Formula 8.1

where $s_{\overline{X}}$ is our estimate of $\sigma_{\overline{X}}$. By substituting Formula 8.1 into Formula 7.2 we get a modified formula for the one-sample z-test, Formula 8.2:

$$z = \frac{\overline{X} - \mu}{s/\sqrt{N}}$$

Formula 8.2

This formula is called the **large-sample z-test,** and we can use it just as we used Formulas 7.1 or 7.2 to do a one-sample hypothesis test. The large-sample z-test works the same way as the one-sample z-test discussed in previous chapters and has the same assumptions, plus one added assumption: The large-sample test, which involves the use of s when σ is unknown, is only valid when the sample size is large enough. How large the sample must be is a matter of judgment, but statisticians generally agree that the lower limit for using the large-sample test is around 30 to 40 subjects. To be conservative (i.e., to avoid the Type I error rate being even slightly higher than the alpha you set), you may wish to set the lower limit higher than 40. However, most statisticians would agree that with a sample size of 100 or more, the large-sample test is quite accurate.

Calculating a Simple Example

Let us apply the large-sample z-test to our example of psychotherapy fees. We can plug the following values into Formula 8.2: $\mu = 63$ (mean adjusted hourly fee in 1960), $\overline{X} = 72$ (mean hourly fee for our sample), $N = 100$ (number of therapists in the sample), and $s = 22.5$ (the unbiased standard deviation of the

100 therapists in our sample). (Normally you would have to calculate *s* yourself from the raw data, but to simplify matters I will give you *s* for this example.)

$$z = \frac{72 - 63}{22.5/\sqrt{100}} = \frac{9}{22.5/10} = \frac{9}{2.25} = 4.0$$

As you should recall from Chapter 7, a *z* score this large is significant even at the .01 level with a two-tailed test. Therefore, the null hypothesis that $\mu = 63$ can be rejected. We can conclude that the mean hourly fee for the current population of therapists is greater than the mean was in 1960 even after adjusting for inflation. Dr. Null has very little chance of taking a random sample of 100 therapists from the 1960 survey and finding an adjusted mean hourly fee greater than the $72 we found for our current sample.

The *t* Distribution

So far all I have really done is reviewed what you learned in the previous chapter. But I have set the stage for a slightly more complicated problem. What would happen if we could only find 25 therapists for our current sample? I've already said that we could not use the one-sample *z*-test. It is time to tell you why and to show how we can modify this test for dealing with small samples. The problem involves the use of *s* as a substitute for σ. When the sample is large, *s* is a pretty good reflection of σ, and there is little error involved in using *s* as an estimate of σ. However, the smaller the sample gets, the greater is the possibility of *s* being pretty far from σ. As the sample size gets below about 30, the possible error involved in using *s* gets too large to ignore.

This was the problem confronting William Gosset in the early 1900s when he was working as a scientist for the Guinness Brewery Company (Cowles, 1989). It was important to test samples of the beer and draw conclusions about the entire batch, but practical considerations limited the amount of sampling that could be done. Gosset worked out a solution to the problem, which he published under the pseudonym "Student" (because of company rules about publishing). The solution involves a distribution that is similar to the normal distribution and is referred to as the **t distribution** (sometimes as Student's *t* distribution). The *t* distribution resembles the standard normal distribution, because it is bell-shaped, symmetrical, continues infinitely in either direction, and has a mean of zero. (The variance of the *t* distribution is a more complicated matter, but fortunately you will not have to deal with that matter directly.) Like the normal distribution, the *t* distribution is a mathematical abstraction that follows an exact mathematical formula but can be used in an approximate way as a model for sampling situations relevant to psychology experiments.

In Chapter 7, I made use of the fact that when you are dealing with a normal distribution, the *z* scores also have a normal distribution, and they can therefore be found in the standard normal table. The *z* scores have a normal distribution because all you are doing to the original, normally distributed, scores is subtracting a constant (μ) and dividing by a constant (σ or $\sigma_{\overline{X}}$). Now look at Formula 8.2 again. Instead of σ, you see s in the denominator. Unlike σ, s is *not* a constant. When the sample is large, s stays close enough to σ that it is almost a constant, and not much error is introduced. But with small samples s can fluctuate quite a bit from sample to sample, so we are no longer dividing by a constant. This means that z, as given by Formula 8.2, does not follow a normal distribution when N is small. In fact, it follows a t distribution. So when N is small we change Formula 8.2 to Formula 8.3:

$$t = \frac{\overline{X} - \mu}{s/\sqrt{N}} \qquad\qquad \text{Formula 8.3}$$

The only change, of course, is replacing z with t to indicate that a t distribution is now being followed.

Degrees of Freedom and the *t* Distribution

If you are drawing a series of samples, each of which contains only six subjects (or observations), the sample standard deviations will vary a good deal from the population standard deviation. However, when N is only 4 there is even more variation. So it makes sense that one t distribution cannot correspond to all possible sample sizes. In fact, there is a different t distribution for each sample size, so we say that there is a family of t distributions. However, we do not distinguish distributions by sample size. Instead, we refer to t distributions according to degrees of freedom (df). In the case of one sample, df $= N - 1$. This corresponds to the fact that s, the unbiased sample standard deviation, is calculated using $N - 1$ (see Formula 4.8). Therefore, when we have a sample size of 6, Formula 8.3 follows a t distribution with df $= 5$.

Before you start to think that these t distributions are complicated, take a look at Figure 8.1, which shows the t distributions for df $= 3$, df $= 9$, and df $= 20$. For comparison, the normal distribution is also included. Notice that as the df increase, the t distribution looks more like the normal distribution. By the time df equal 20, there is very little difference between the t and the normal distributions. So it should be understandable that for N above 30 or 40, some statisticians disregard the difference between the two distributions and just use the normal

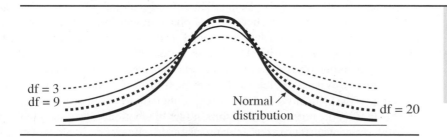

Figure 8.1

Comparing the t Distribution to the Normal Distribution

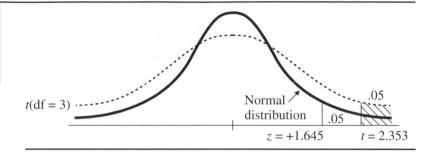

distribution. That's what the large-sample test is all about. As you look at Figure 8.1, notice that the important difference between the t distribution and the normal distribution is that the former has fatter tails. In terms of kurtosis, this means that the t distribution is leptokurtic (see Section C in Chapters 4 and 6). This is important because the t distribution is going to be used as our null hypothesis distribution representing what Dr. Null can do.

Figure 8.2 compares one of the t distributions (df = 3) with the normal distribution. Notice that for any z score in the tail area, the p value (the amount of area in the tail beyond that z score) is larger for the t than for the normal distribution, because the t distribution has more area in its tails. On the other hand, consider a particular z score that cuts off a particular amount of area in the tail; for instance, $z = 1.645$ has 5% (.05) of the area beyond it. A t value of 1.645 cuts off more than .05. To find the t value that cuts off exactly .05, we have to go further out in the tail. As shown in Figure 8.2, $t = 2.353$ has .05 area beyond it (but only if df = 3). As mentioned in Chapter 7, $z = 1.645$ is the critical z for alpha = .05, one-tailed. It should be clear from Figure 8.2 that $t = 2.353$ is the critical t for alpha = .05, one-tailed, df = 3. The degrees of freedom must be specified, of course, because as the degrees of freedom increase the tails of the t distribution get skinnier, and the critical t values therefore get smaller. You can see this in Table A.2 in Appendix A, as you look down any of the columns of critical t values. Part of Table A.2 is reproduced as Table 8.1.

Critical Values of the t Distribution

Table A.2 lists the critical values of the t distribution for the commonly used alpha levels and for different degrees of freedom. It should be no surprise that a .025, one-tailed test corresponds to the same critical values as the .05, two-tailed test, since the latter involves placing .025 area in each tail. Of course, critical t gets higher (just as critical z does) as the alpha level gets lower (look across each row in Table 8.1). And as I just mentioned, critical t values *decrease* as df increase. How small do the critical values get? Look at the bottom row of Table 8.1, where the number of degrees of freedom is indicated by the symbol ∞, which means infinity. When the df are as high as possible, the critical t values become the same as the critical values for z, because the t distribution becomes indistinguishable from the normal distribution. In fact, the bottom row of the t table can be used as a convenient reference for the critical z values.

	Area in One Tail			Table 8.1
df	.05	.025	.01	
3	2.353	3.182	4.541	
4	2.132	2.776	3.747	
5	2.015	2.571	3.365	
6	1.943	2.447	3.143	
7	1.895	2.365	2.998	
8	1.860	2.306	2.896	
9	1.833	2.262	2.821	
⋮	⋮	⋮	⋮	
∞	1.645	1.960	2.326	

You may wonder why the *t* table looks so different from the standard normal distribution table (Table A.1). You could make a *t* table that looks like Table A.1, but a different table would have to be created for each possible number of degrees of freedom. Then you could look up the *p* level corresponding to any particular *t* value. But there would have to be at least 30 tables, each the size of Table A.1. (After about 30 df, it could be argued that the *t* tables are getting so similar to each other and to the normal distribution that you don't have to include any more tables.) By creating just one *t* table containing only critical values, we sacrifice the possibility of looking up the *p* level corresponding to any particular *t* value, but researchers are often concerned only with rejecting the null hypothesis at a particular alpha level. However, now that computers are so readily available to perform *t*-tests, we no longer have to sacrifice exact *p* levels for the sake of convenience. Most computer programs that perform *t*-tests also calculate and print the exact *p* level that corresponds to the calculated *t*.

Calculating the One-Sample *t*-Test

A one-sample test that is based on the *t* distribution is called a **one-sample *t*-test.** I will illustrate this test by returning to the example of psychotherapy fees, but this time assume that only 25 current psychotherapists can be surveyed. I will use Formula 8.3 and assume that μ, $\overline{X}$, and s are the same as before, but that N has changed from 100 to 25.

$$t = \frac{72 - 63}{22.5/\sqrt{25}} = \frac{9}{22.5/5} = \frac{9}{4.5} = 2.00$$

If the 2.00 calculated above were a *z* score, it would be significant at the .05 level (critical *z* = 1.96). Because the sample size is so small, we call the calculated value a *t* value, and we must find the appropriate critical value from the *t* table. The df for this problem equal $N - 1 = 25 - 1 = 24$. Looking in Table A.2 down the column for alpha = .05, two-tailed (or .025, one-tailed), we find that the critical *t* is 2.064. The calculated *t* of 2.00 is *less* than the critical *t* of 2.064, so the calculated *t* does not fall in the rejection area—therefore the null hypothesis cannot be rejected. You can see that the *t* distribution forces a researcher to be more cautious when dealing with small sample sizes; it is more difficult to reach significance. Using the normal distribution when the *t* distribution is called for would result in too many Type I errors—more than the alpha level would indicate. Had we planned a one-tailed *t*-test at the same alpha level, the critical *t* would have been 1.711 (look down the appropriate column of Table A.2), and we could have rejected the null hypothesis. However, it would be difficult to justify a one-tailed test in this case.

Sample Size and the One-Sample *t*-Test

It is also important to notice that the *t* value for $N = 25$ was only half as large as the *z* score for $N = 100$ ($t = 2.00$ compared to $z = 4.00$), even though the two calculating formulas (8.2 and 8.3) are really the same. This difference is due entirely to the change in *N*. To see why an increase in *N* produces an increase in the value calculated from these formulas, you must look carefully at the structure of the formulas. The denominator of each formula is actually a fraction, and *N* is in the denominator of that fraction. So you can say that *N* is in the *denominator of the denominator* of these formulas; making *N* larger makes the whole denominator smaller, which in turn makes the whole ratio larger. Algebraically, for a number to be in the denominator of a denominator is the same as being in the numerator. Since we are taking the square root of *N,* increasing *N* by a factor of 4 (e.g., from $N = 25$ to $N = 100$) increases the whole ratio by the square root of 4, which is 2 (e.g., from $t = 2.00$ to $z = 4.00$). All other things being equal, increasing the sample size will increase the *t* value (or *z* score), and make it easier to attain statistical significance.

   You have seen that increasing the sample size can help you attain statistical significance in two ways. First, a larger sample size means more df, and therefore a smaller critical *t* value that you have to beat (until the sample size is large enough to use the normal distribution, after which further increases in sample size do not change the critical value). Second, increasing the sample size tends to increase the calculated *t* or *z* (no matter how large the sample is to start with). So why not always use a very large sample size? As mentioned previously, there are often practical circumstances that limit the sample size. In addition, very large sample sizes make it possible to achieve statistical significance even when there is a very small, uninteresting experimental effect taking place. This aspect of sampling will be discussed at length in Chapter 10.

Cautions Concerning the One-Sample *t*-Test

Remember that even if statistical significance is attained, your statistical conclusion is valid only to the extent that the assumptions of your hypothesis test have been met. Probably the biggest problem in conducting any one-sample test is ensuring the randomness of the sample that is drawn. This is especially true when dealing with a preexisting population. In the example of psychotherapy fees, all psychotherapists in the population being investigated must have an equal chance of being selected and reporting their fee, or the sample mean may not accurately represent the population. However, it is easy for sampling biases to creep into such a study. Therapists working at a clinic may be easier to find; therapists practicing a certain style of therapy may be more cooperative; therapists charging unusually low or high fees may be less willing to disclose their fees. Any such bias in the sampling could easily lead to a sample mean that misrepresents the true population mean and therefore results in an invalid conclusion. This situation certainly does not improve when small sample sizes are used, so this is one reason why one-sample *t*-tests are usually undesirable.

The main problem that arises when trying to use the one-sample *t*-test to evaluate the results of an experiment is that even if a truly random sample has been obtained, the lack of a comparison group can prevent you from ruling out various alternative explanations (i.e., "confounding variables"). For instance, if you take a random sample of office workers, change their source of lighting from ordinary incandescent or fluorescent bulbs to bulbs that simulate natural sunlight, and then find that these workers have fewer illnesses than the national average, you cannot be certain that the "sunlight" had an effect on health. It is possible that the workers felt "special" just because they were in an experiment and felt an increased motivation toward their work, which in turn benefited their health (this is a form of placebo effect). You need a second group that is treated and measured the same way but does not get natural sunlight bulbs (that is, both groups get new lamps and are told that the bulbs are different, but only one group gets the new bulbs). Then you can conclude that the difference in health between the two groups is due to the type of lighting. Two-group experiments will be discussed in the next chapter.

The example I used to illustrate the one-sample *t*-test, where we knew the population mean for the 1960 fees but not the population standard deviation, does not arise often in the real world. A much more common situation is one in which you would like to know the population mean for some variable, and there is no previous information at all. In terms of our previous example, even if there were no prior survey of psychotherapy fees, we might have a good reason to want to know the average fee right now. Some variables pertain to phenomena that are so new that there is little information to begin with—for example, the number of hours per month spent watching rented videotapes or using a personal computer. This kind of information can be of great interest to market researchers.

Estimating the Population Mean

A psychologist might want to know how many hours married couples converse about marital problems each week, or how many close friends the average person has. Or a psychologist might be interested in ordinary variables in a particular subpopulation (e.g., the average blood pressure of African Americans or the average IQ of left-handed people). To find the population mean for any of these variables the procedure is not complicated or difficult to understand. All you need to do is to measure every individual in the population of interest, and then take the mean of all these measures. The problem, of course, is a practical one. The most practical solution involves taking a random sample of the population and measuring all the individuals in the *sample.*

We can use a random sample to estimate the mean of the population (μ). Not surprisingly, the best estimate of the population mean that you can get from a sample is the mean of that sample ($\overline{X}$). As you learned in Chapter 6, the larger the sample, the closer the sample mean is likely to be to the population mean. This implies that larger samples give better (i.e., more accurate) estimates of the population mean.

Interval Estimation and the Confidence Interval

When using the sample mean as an estimate of the population mean, we are making a **point estimate,** suggesting a single value or number—a point—where the population mean is expected to be. In the example concerning psychotherapy fees, the point estimate of the current hourly fee was the mean of our sample—that is, $72. Because larger samples give more accurate point estimates, the larger the sample, the greater our confidence that the estimate will be near the actual population mean. However, the point estimate alone cannot tell us how much confidence to invest in it. A more informative way to estimate the population mean is through **interval estimation.** By using an interval (a range of values instead of just one point) to estimate the population mean, it becomes easy to express how much confidence we have in the accuracy of that interval. Such an interval is therefore called a **confidence interval.**

The common way of constructing a confidence interval (and the only one I will consider) is to place the point estimate in the center and then mark off the same distance below and above the point estimate. How much distance is involved depends on the amount of confidence we want to have in our interval estimate. Selecting a confidence level is similar to selecting an alpha (i.e., significance) level. For example, one of the most common confidence levels is 95%. After constructing a 95% confidence interval, you can feel 95% certain that the population mean lies within the interval specified. To be more precise, suppose that Ms. Clare Inez constructs 95% confidence intervals for a living. She specifies hundreds of such intervals each year. If all the necessary assumptions are met (see Section B), the laws of statistics tell us that about 95% of her intervals will be "hits"; that is, in 95% of the cases, the population mean will really be in the interval Ms. Inez specified. On the other hand, 5% of her intervals will be "misses"; the population mean will not be in those intervals at all.

The confidence intervals that are "misses" are something like the Type I errors in null hypothesis testing; in ordinary practice one never knows which interval is a "miss," but the overall percentage of misses can be controlled by selecting the degree of confidence. The 95% confidence interval is popular because in many circumstances a 5% miss rate is considered tolerable, just as a 5% Type I error rate is considered tolerable in null hypothesis testing. If a 5% miss rate is considered too large, you can construct a 99% confidence interval in order to lower the miss rate to 1%. The drawback of the 99% confidence interval, however, is that it is larger and thus identifies the probable location of the population mean less precisely.

Confidence intervals based on large samples tend to be smaller than those based on small samples. If we wanted to estimate the current hourly fee for psychotherapy by constructing a confidence interval, the interval would be centered on our point estimate of $72. In the case where $N = 100$, the interval would be smaller (the limits of the interval would be closer to the point estimate) than in the case where $N = 25$. In fact, any confidence interval can be made as small as desired by using a large enough sample. This concept should be more understandable when I demonstrate the procedure for calculating the confidence interval in Section B.

Exercises

* 1. The unbiased variance (s^2) for a sample of 200 subjects is 55.

   a) What is the value of the estimated standard error of the mean ($s_{\overline{X}}$)?

   b) If the variance were the same but the sample were increased to 1800 subjects, what would be the new value of $s_{\overline{X}}$?

2. A survey of 144 college students reveals a mean beer consumption rate of 8.4 beers per week, with a standard deviation of 5.6.

   a) If the national average is 7 beers per week, what is the z score for the college students? What p level does this correspond to?

   b) If the national average were 4 beers per week, what would the z score be? What can you say about the p level in this case?

* 3. a) In a one-group t-test based on a sample of ten subjects, what is the value for df?

   b) What are the two-tailed critical t values for alpha $= .05$? For alpha $= .01$?

   c) What is the one-tailed critical t for alpha $= .05$? For alpha $= .01$?

4. a) In a one-group t-test based on a sample of 20 subjects, what is the value for df?

   b) What are the two-tailed critical t values for alpha $= .05$? For alpha $= .01$?

   c) What is the one-tailed critical t for alpha $= .05$? For alpha $= .01$?

* 5. a) Twenty-two stroke patients performed a maze task. The mean number of trials ($\overline{X}$) was 14.7 with $s = 6.2$. If the population mean (μ) for this task is 6.5, what is the calculated value for t? What is the critical t for a .05, two-tailed test?

   b) If only eleven patients had been run but the data were the same as in part a, what would be the calculated value for t? How does this value compare with the t value calculated in part a?

6. a) Referring to part a of exercise 5, what would the calculated t value be if $s = 3.1$ (all else remaining the same)?

   b) Comparing the t values you calculated for exercises 5a and 6a, what can you say about the relation between t and the sample standard deviation?

* 7. Imagine that the t value has been calculated for a one-sample experiment. If a second experiment used a larger sample but resulted in the same calculated t, how would the p level for the second experiment compare to the p level for the first experiment? Explain.

8. The calculated t for a one-sample experiment was 1.1. Which of the following can you conclude?

　a) The sample mean must have been close to the mean of the null hypothesis distribution.

　b) The sample variance must have been quite large.

　c) The sample size (N) could *not* have been large.

　d) The null hypothesis *cannot* be rejected at the .05 level.

　e) None of the above can be concluded without further information.

BASIC STATISTICAL PROCEDURES

The formal step-by-step procedure for conducting a large-sample z-test is identical to the procedure for conducting a one-sample z-test described in Section B of Chapter 7. The only difference is that z is calculated using s instead of σ. The procedure for the one-sample t-test is identical to the procedure for conducting a large-sample z-test, except for the use of the t table to find the critical values. Therefore, I will not repeat the steps of null hypothesis testing as applied to the one-sample t-test in this section. The statistical procedure on which I will concentrate in this section is the calculation of a confidence interval (CI) for the population mean. (This procedure is common, but probably more so in fields other than psychology, such as business and politics.) For the sake of simplicity I will continue with the example of psychotherapy fees. Rather than comparing current fees with those of preceding years, however, I will modify the psychotherapy example and assume that a national organization of therapists is trying to determine the current mean hourly fee for psychotherapy in the United States. The following four-step procedure can be used to find the CI for the population mean.

Step 1. Select the Sample Size

The first step is to decide the size of the random sample to be drawn. For any given level of confidence, the larger the sample, the smaller (and more precise) will be the confidence interval. (Of course, if your sample was as large as the entire population, the sample mean would equal the population mean and the confidence interval would shrink to a single point, giving an exact answer.) However, at some point, the goal of increasing the accuracy of the confidence interval will not justify the added expense in time or money of increasing the sample size. If a particular size confidence interval is desired and the population standard deviation is known, the required sample size can be determined (as I will show later in this section). For now, in keeping with the previous example, I will assume that our sample size has been set to 100.

Step 2. Select the Level of Confidence

Next, the level of confidence must be determined for the interval estimate. This is analogous to choosing an alpha level for hypothesis testing. As I stated before, the most common choice is 95%, so I will begin with this level.

Step 3. Select the Random Sample and Collect the Data

Once you have some idea of the desired sample size, a random sample must be collected. The accuracy of the confidence interval depends upon the degree to which this sample is truly random. With a vast, diversified population some form of *stratified* sampling may be used to help ensure that each segment of the population is proportionally represented. This form of sampling, which is often used for public opinion polls and marketing research, requires that the population be divided and sampled according to mutually exclusive strata (e.g., men versus women, levels of socioeconomic status). Stratified sampling and related methods, which are discussed more extensively in texts devoted to research design, often require the resources of a large organization. On the other hand, individual psychological researchers are more likely to rely on samples based on convenience (e.g., samples of college students), which do not readily allow generalization to the larger population. However, psychologists are usually less interested in estimating population means than in studying the differences between groups or the effects of treatments. Therefore, psychologists more often draw two or more samples and rely on the random assignment of subjects, as will be discussed in subsequent chapters.

Step 4. Calculate the Limits of the Interval

Formula for Large Samples

The formula that is used to find the limits of a confidence interval should be quite familiar, as it is just a variation of the one used in null hypothesis testing. In fact, the two procedures are closely related. Recall that in null hypothesis testing, we start with a particular population mean (e.g., the 1960 adjusted mean of $63) and want to know what kind of sample means are likely, so we find the appropriate sampling distribution and center it on the population mean (see Figure 8.3). To

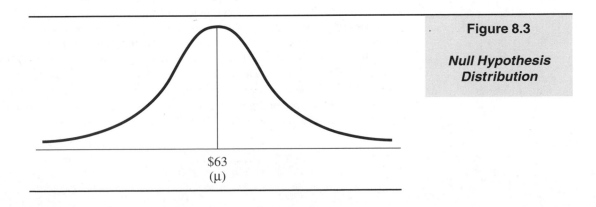

Figure 8.3

Null Hypothesis Distribution

$63
(μ)

Figure 8.4

*95% Confidence
Interval for
Population Mean*

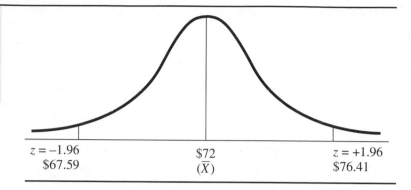

$z = -1.96$ $\$72$ $z = +1.96$
$\$67.59$ $(\overline{X})$ $\$76.41$

find a confidence interval the problem must be reversed. Starting with a particular *sample* mean, the task is to find which *population* means are likely to produce that sample mean and which are not. Therefore, we take a normal distribution just like the null hypothesis distribution and center it on the *sample mean,* which is our point estimate of the population mean.

In our example, the mean for the random sample of 100 psychotherapists was $72. A normal distribution centered on the sample mean can then indicate which population means are likely and which are not (as in Figure 8.4). For instance, if we want to find the most likely 95% of the possible population means, we just have to find the z scores that mark off the middle 95% of the normal distribution. Hopefully, you recall that $z = -1.96$ and $z = +1.96$ enclose the middle 95% of a normal distribution. These are the limits of the 95% CI in terms of z scores, as Figure 8.4 shows. Although Figure 8.4 can be a useful teaching device to clarify the procedure for constructing a CI, it must not be taken literally. Population means do *not* vary around a sample mean—in fact, they do not vary at all! For any given problem, there is just one population mean, and it stays the same no matter what we do to look for it. However, we do not know the value of the population mean for most problems and never will. Therefore, it can be useful to think in terms of "possible" population means as distributed around the sample mean; the closer the possible population mean is to the sample mean, the more likely a candidate it is to be the actual population mean.

Knowing that the limits of the CI are $z = \pm 1.96$, we know that we have to go about two standard deviations above and below the point estimate of 72. If we knew the size of the standard deviation of this distribution, we could easily find the limits of the CI. Fortunately, the standard deviation of this imaginary distribution around the sample mean is the same as for the null hypothesis distribution; it is the standard error of the mean given by Formula 8.1, which we computed as part of our calculations for the null hypothesis test in Section A:

$$s_{\overline{X}} = \frac{s}{\sqrt{N}} = \frac{22.5}{\sqrt{100}} = \frac{22.5}{10} = 2.25$$

Now that we know the standard error, we can go about two standard errors (1.96 to be more exact) above and below our point estimate of 72, in order to find the limits of our confidence interval. Two standard errors = $2 \cdot 2.25 = 4.5$, so the limits are approximately $72 - 4.5 = 67.5$ and $72 + 4.5 = 76.5$.

We have just found, in an informal way, approximate limits for the 95% confidence interval for the population mean. There is a formula for this procedure, but I wanted you to get the general idea first. The formula is really the same as the one we used for null hypothesis testing (Formula 8.2), but it has been algebraically manipulated to make it easy to find the confidence interval.

The purpose of Formula 8.2 is to find the z score corresponding to a sample mean with respect to the null hypothesis distribution. When constructing a CI you start by knowing the z scores (because once the confidence percentage is picked the z score limits are determined). Then those z scores must be converted back to raw scores, which in this case are limits on the population mean (see Figure 8.4). Formula 8.2 can be used for this purpose by filling in the z score and solving for μ. (You must perform this calculation twice, once with the positive z score and again with the corresponding negative z score.) Or you can solve Formula 8.2 for μ *before* inserting any particular values. Multiplying both sides of Formula 8.2 by the standard error, then adding the sample mean to both sides and adjusting the signs to indicate both positive and negative z scores will result in the following formula for the CI:

$$\mu = \overline{X} \pm z_{crit} s_{\overline{X}}$$

If this formula has a familiar look, that's because it resembles Formula 5.2, which is used when you know the z score and want to find the corresponding raw score. The above formula is also used when you know the z score, except in this case it is the z score for groups, and you are looking for the population mean instead of for a raw score. This formula gives two values for μ: one above $\overline{X}$ and the other the same distance below $\overline{X}$. These two values are the upper and lower limits of μ, respectively, and together they define the confidence interval. So the formula above can be rewritten in two parts, which together will be designated Formula 8.4:

$$\mu_{lower} = \overline{X} - z_{crit} s_{\overline{X}}$$
$$\mu_{upper} = \overline{X} + z_{crit} s_{\overline{X}}$$

Formula 8.4

The term μ_{lower} in the above formula is just a shorthand way of indicating the lower limit of the confidence interval for the population mean. Any value for μ that is even lower than μ_{lower} would not be considered as being in the confidence interval; an analogous interpretation applies to μ_{upper}. Formula 8.4 can be used to find the exact 95% CI for the example of psychotherapy fees, as follows:

$$\mu_{lower} = 72 - (1.96)(2.25) = 72 - 4.41 = 67.59$$
$$\mu_{upper} = 72 + (1.96)(2.25) = 72 + 4.41 = 76.41$$

Based on the above calculations we can state with a confidence level of 95% that the mean hourly fee for the entire population of current psychotherapists is somewhere between $67.59 and $76.41. Any particular 95% CI that is calculated from real data may be incorrect (i.e., the population mean may not actually be in the specified interval), but on the average, 95% of the intervals constructed in this manner will contain the population mean.

The 99% Confidence Interval

For confidence intervals to be correct more often, say, 99% of the time, the appropriate z_{crit} must be used in Formula 8.4. For the 99% CI, z_{crit} is the z that corresponds to alpha = .01, two-tailed, so $z_{crit} = \pm 2.58$. Plugging the appropriate z_{crit} into the formula, we get

$$\mu_{lower} = 72 - (2.58)(2.25) = 72 - 5.80 = 66.20$$
$$\mu_{upper} = 72 + (2.58)(2.25) = 72 + 5.80 = 77.80$$

Based on the above calculations we can say with 99% confidence that the mean hourly fee for psychotherapy is currently between $66.20 and $77.80. Note that this interval is larger than the 95% CI, but there is also more confidence associated with this interval estimate. It should be clear that greater confidence requires a larger z score, which in turn results in a larger interval. Whereas 95% confidence is the most conventional level, practical considerations can dictate either a larger or smaller level of confidence. If you are designing a piece of equipment (e.g., a protective helmet) that must be able to accommodate nearly everyone in the population, you may want a 99% or even a 99.9% confidence interval for the population mean (e.g., head circumference). On the other hand, if a marketing researcher needs a rough estimate of the average yearly per capita beer consumption in the United States, a 90% CI may suffice.

Interval Width as a Function of Sample Size

There is a way to make the interval estimate smaller and more precise without sacrificing confidence, and that is to increase the sample size. Enlarging N has the effect of reducing $s_{\overline{X}}$, and according to Formula 8.4, the reduction of $s_{\overline{X}}$ will reduce the size of the interval proportionally (e.g., cutting $s_{\overline{X}}$ in half will halve the size of the interval). A glance at Formula 8.1 confirms that the multiplication of N by some factor results in $s_{\overline{X}}$ being divided by the square root of that factor (e.g., multiplying N by 4 divides $s_{\overline{X}}$ by 2). Therefore, the confidence interval is reduced by the square root of whatever factor N is multiplied by (assuming that the sample standard deviation stays the same, which is a reasonable approximation when dealing with large samples).

Because the size of the CI depends on $s_{\overline{X}}$, which depends on N, you can specify any desired width for the interval and figure out the approximate N required to obtain that width, assuming that the population standard deviation can be reliably estimated. When dealing with large sample sizes, the width of the CI will be about four times the standard error of the mean (1.96 in each direction). If the width of the CI is represented by W, then $W = 4 s_{\bar{X}} = 4s/\sqrt{N}$. Therefore, $\sqrt{N} = 4s/W$, and $N = (4s/W)^2$. If you want to estimate the mean IQ for left-handed students (assuming $\sigma = 16$), and you want the width of the interval to be 4 IQ points, then the required $N = (4 \cdot 16/4)^2 = 16^2 = 256$.

Confidence Intervals Based on Small Samples

Formula 8.4 is only valid when using large samples, or when you know the standard deviation of the population (σ). (In the latter case, $\sigma_{\bar{X}}$ is used in place of $s_{\bar{X}}$.) I did not consider the situation in which you know σ but you are trying to estimate μ, because the situation is uncommon in psychological research. Generally, if you know enough about a population to know its standard deviation, you also know its mean. (In the case of physical measurement, the standard deviation may be based entirely on errors of measurement that are well known, but this is not likely in psychology, where individual differences among subjects can be quite unpredictable.) The reverse situation, wherein you know the population mean but not the population standard deviation, is not very common either, but our example involving psychotherapy fees represents one possibility, and others will be mentioned later in the chapter and in the exercises.

I discussed the problem of using s to estimate σ when sample size is not large in connection with one-sample hypothesis tests; the same discussion applies to interval estimation as well. For smaller sample sizes (especially those below 30), we must substitute t_{crit} for z_{crit} in Formula 8.4 to create Formula 8.5:

$$\mu_{\text{lower}} = \bar{X} - t_{\text{crit}}s_{\bar{X}}$$

$$\mu_{\text{upper}} = \bar{X} + t_{\text{crit}}s_{\bar{X}}$$

Formula 8.5

The appropriate t_{crit} corresponds to the level of confidence in the same way that z_{crit} does: You subtract the confidence percentage from 1 to get the corresponding alpha (e.g., $1 - 95\% = 1 - .95 = .05$) and find the two-tailed critical value. In the case of t_{crit} you also must know the df, which equals $N - 1$, just as for one-sample hypothesis testing. For the small-sample example from Section A, in which only 25 therapists could be polled, df $= N - 1 = 25 - 1 = 24$. Therefore, t_{crit} for a 95% CI would be 2.064—the same as the critical value used for the .05, two-tailed hypothesis test. For the small sample, $s_{\bar{X}}$ is $22.5/\sqrt{25} = 22.5/5 = 4.5$. To find the 95% CI for the population mean for this example we use Formula 8.5:

$$\mu_{\text{lower}} = 72 - (2.064)(4.5) = 72 - 9.29 = 62.71$$

$$\mu_{\text{upper}} = 72 + (2.064)(4.5) = 72 + 9.29 = 81.29$$

Notice that the 95% CI in this case—62.71 to 81.29—is considerably larger than the 95% CI found earlier for $N = 100$. This occurs for two reasons. First, t_{crit} is larger than the corresponding z_{crit}, reflecting the fact that there is more error in estimating σ from s when the sample is small, and therefore the CI should be larger and less precise. Second, the standard error ($s_{\bar{X}}$) is larger with a small sample size, because s is being divided by a smaller N.

Relationship Between Interval Estimation and Null Hypothesis Testing

By now I hope the similarities between the procedures for interval estimation and for null hypothesis testing have become obvious. Both procedures involve the same critical values and the same standard error. The major difference is that $\overline{X}$ and μ exchange roles, in terms of which is the "center of attention." Interestingly, it turns out that once you have constructed a confidence interval, you do not have to perform a separate procedure in order to conduct null hypothesis tests; in a sense, you get null hypothesis testing "for free."

To see how this works, let's look again at the 95% CI we found for the $N = 100$ psychotherapists example. The interval ranged from \$67.59 to \$76.41. Now recall the null hypothesis test in Section A. The adjusted 1960 population mean we were testing was \$63. Notice that \$63 does not fall within the 95% CI. Therefore, \$63 would not be considered a likely possibility for the population mean, given that the sample mean was \$72 (and given the values of s and N in the problem). In fact, the 1960 population mean of \$63 was rejected in the null hypothesis test in Section A (for $N = 100$). The 95% CI tells us that \$67 or \$77 would also have been rejected, but \$68 or \$76 would have been accepted as null hypotheses. So interval estimation provides a shortcut to null hypothesis testing by allowing you to see at a glance which population means would be accepted as null hypotheses and which would not.

Because the 95% CI gives the range of possible population means that has a 95% chance of containing the population mean, any population mean outside of that range can be rejected as *not* likely at the $1 - 95\% = 5\%$ significance level. In general, if you want to perform a two-tailed hypothesis test with alpha (α) as your criterion, you can construct a CI with confidence equal to $(1 - \alpha) \cdot 100\%$ and check to see whether the hypothesized μ is within the CI (accept H_0) or not (reject H_0). Thus a 99% CI shows which population means would be rejected at the .01 level. Looking at the 99% CI for $N = 100$ psychotherapists, which ranged from \$66.20 to \$77.80, we see that \$63 falls outside the 99% CI; therefore, that population mean would be rejected not only at the .05 level but at the .01 level as well. Finally, we turn to the 95% CI for the $N = 25$ example, which ranged from \$62.71 to \$81.29. In this case, \$63 falls *within* the 95% CI, implying that a population mean of \$63 would have to be *accepted* at the .05 level when dealing with the smaller sample. This decision agrees with the results of the corresponding null hypothesis test performed in Section A.

Assumptions Underlying the Large-Sample Test and Confidence Interval

The assumptions that justify the use of the large-sample z-test are the same as those for the one-sample z-test described in Chapter 7, with the obvious addition that the sample must be large.

Independent Random Sampling

Remember that any test involving one sample is only valid to the extent that the sample is truly random. If psychotherapists with a certain orientation, or those who work at a clinic rather than in a private practice, are more likely to be sampled, a misleading view will emerge. The results would also be biased if groups of psychotherapists who work together were included in the sampling. This would violate the principle of independence; each psychotherapist should be sampled independently of any others. If you are planning to perform a one-sample experiment, you should draw a random sample from the population.

Normal Distribution

When a large sample is being used, there is little need to worry about whether the variable of interest is normally distributed in the population. The Central Limit Theorem (Chapter 6, Sections B and C) implies that unless the population distribution has a very extreme skew or other bizarre form, the null hypothesis distribution will be close enough to a normal distribution that we can look up the calculated z scores in the standard normal table. For the case of psychotherapy fees, the distribution is likely to have a moderately positive skew, with the bulk of fees somewhat near the low end and a few fees considerably above the median. (Fees cannot get lower than zero, so there is a limit on the negative end of the distribution.) But with large-sample experiments, this skew should produce a negligible amount of error.

Standard Deviation of the Sampled Population Equals That of the Comparison Population

In the formula for the large-sample test and the formula for the large-sample CI, the standard deviation of the sample is used as an estimate of the standard deviation for the comparison population. This only makes sense if we assume that the two populations have the same standard deviation, even though they may or may not have different means. In terms of the psychotherapy example, we are assuming that the standard deviation in 1960 was the same as it is now, and we are only testing to see if the mean is different. As Chapter 7 mentioned, a problem arises only if the standard deviation has changed but the mean has not. This is a rather unlikely possibility that is usually ignored.

The Sample Is Large

Whereas some statisticians consider 30 a large enough sample to justify large-sample z-tests, the more cautious approach is to use the t distribution until the sample size is over about 100. When in doubt, use the t distribution; no matter how large the sample is, the t distribution will always yield the appropriate critical value.

Assumptions Underlying the One-Sample *t*-Test

The first three assumptions detailed above apply equally to the one-sample *t*-test. The only difference is that because the *t*-test can be performed with very small sample sizes, you should be a bit more concerned about having a normal distribution in the population. Fortunately, statisticians have determined that the *t*-test is "robust" with respect to the normality assumption, which means that the population distribution can be quite far from normal before the *t*-test loses its validity. However, if you are dealing with a very small sample *and* a very nonnormal population distribution, you should consider the use of nonparametric hypothesis tests (see Part VI).

Use of One-Sample Tests and the Confidence Interval for the Population Mean

In Chapter 7, I assumed that one-sample tests could be performed only for variables whose mean and standard deviation in the population were known. In this chapter you have seen that when the population standard deviation is not known, the sample standard deviation can be used instead. This can expand the possibilities of one-sample tests to variables for which there is enough information to make a good estimate of the population mean, but no clear idea of the standard deviation. For instance, the soap industry may be able to tell us the total amount of soap used in the United States in any given year, and from that we could calculate the mean consumption per person, without having any idea about the standard deviation. Or we may have an excellent idea of the average weight of one-year old babies but no good estimate of the corresponding standard deviation. In that case, it is not hard to imagine a "special" group of babies you might want to compare to the general population. Finally, a researcher might hypothesize a particular population mean based on a theoretical prediction and then test this value with a one-sample experiment. For instance, a task may require a musician to rehearse a particular piece of music mentally and signal when finished. The null hypothesis mean could be the amount of time normally taken to play that piece of music.

If one-sample *z*-tests are rare, one-sample *t*-tests are probably rarer—with one notable exception. If the subjects of two groups are matched in pairs on some basis that is relevant to the variable measured, the two-group *t*-test can be transformed into a one-group ("matched") *t*-test on the difference scores. Because this is a common and important test, I will devote an entire chapter to it. However, in order to understand matched *t*-tests, it is important to understand the concept of correlation, so the topic of matched *t*-tests will be postponed until Chapter 13.

One-sample experiments are not common mainly because comparison of the experimental group to some second (control) group is nearly always helpful, as discussed in Section A. On the other hand, constructing confidence intervals for the population mean is quite common. For example, various ratings services, such as the A. C. Nielsen Company, periodically estimate the number of televisions in use at a given moment, or the mean number of hours that each family has the television turned on during a particular week. These estimates come with "error bars" that create a confidence interval. Also common are opinion polls and election polls. Here the results are usually given as proportions or percentages, again with error bars that establish a confidence interval.

When finding a confidence interval for the mean it is preferable to use a sample size of at least 100, but there are circumstances that limit sample size so that the *t*-distribution must be used. An example would be the estimation of jaw size for a particular prehistoric ape. The anthropologist wishing to make the estimate may have only a dozen jawbones of that ape available for measurement. Or, a medical researcher may have access to only 20 patients with a particular genetic abnormality, but would like to estimate that group's population mean for some physiological variable, such as red blood cell count.

Publishing the Results of One-Sample *t*-Tests

In Chapter 7, I described the reporting of one-sample tests when the population standard deviation is known, and that description applies here as well, with a couple of simple additions. In the case of the large-sample *z*-test when the population standard deviation is *not* known, it is a good idea to include the sample size in parentheses to help the reader judge to what extent the large-sample test is appropriate. For instance: "The hourly fee (M = \$72) for our sample of current psychotherapists is significantly greater, $\underline{z}$ (100) = 4.0, $\underline{p}$ < .001, than the 1960 hourly rate (μ = \$63, in current dollars)."

In reporting the results of a *t*-test, the number of degrees of freedom should always be put in parentheses after the *t*, because that information is required to determine the critical value for *t*. Also, the sample standard deviation should be included along with the mean, in accord with the guidelines listed in the fourth edition of the APA *Publication Manual*. For instance: "Although the mean hourly fee for our sample of current psychotherapists was considerably higher ($\underline{M}$ = \$72, SD = 22.5) than the 1960 population mean (μ = \$63, in current dollars), this difference only approached statistical significance, $\underline{t}$ (24) = 2.00, $\underline{p}$ <. 06."

Exercises

* 1. A high school is proud of its advanced chemistry class, in which its 16 students scored an average of 89.3 on the statewide exam, with $s = 9$.

a) Test the null hypothesis that the advanced class is just a random selection from the state population ($\mu = 84.7$), using alpha = .05 (two-tailed).

b) Test the same hypothesis at the .01 level (two-tailed). Considering your decision with respect to the null hypothesis, what type of error (Type I or Type II) could you be making?

2. Are serial killers more introverted than the general population? A sample of 14 serial killers serving life sentences was tested and found to have a mean introversion score ($\overline{X}$) of 42 with $s = 6.8$. If the population mean (μ) is 36, are the serial killers significantly more

introverted at the .05 level? (Perform the appropriate one-tailed test, though normally it would not be justified.)

* 3. A researcher is trying to estimate the average number of children born to couples who do not practice birth control. The mean for a random sample of 100 such couples is 5.2 with $s = 2.1$.

a) Find the 95% confidence interval for the mean (μ) of all couples not practicing birth control.

b) If the researcher had sampled 400 couples and found the same sample mean and standard deviation as in part a, what would be the limits of the 95% CI for the population mean?

c) Compare the width of the CI in part a with the width of the CI in part b. What is the general principle relating changes in sample size to changes in the width of the CI?

* 4. A psychologist studying the dynamics of marriage wanted to know how many hours per week the average American couple spends discussing marital problems. The sample mean ($\overline{X}$) of 155 randomly selected couples turned out to be 2.6 hours with $s = 1.8$.

a) Find the 95% confidence interval for the mean (μ) of the population.

b) A European study had already estimated the population mean to be 3 hours per week for European couples. Are the American couples significantly different from the European couples at the .05 level? Show how your answer to part a makes it easy to answer part b.

5. If the psychologist in exercise 4 wanted the width of the confidence interval to be only half an hour, how many subjects would have to be sampled?

6. A market researcher needed to know how many blank videotapes are purchased by the average American family each year. To find out, 22 families were sampled at random. The researcher found that $\overline{X} = 5.7$ with $s = 4.5$.

a) Find the 95% confidence interval for the population mean.

b) Find the 99% CI.

c) How does increasing the amount of confidence affect the width of the confidence interval?

* 7. A study is being conducted to investigate the possibility that autistic children differ from other children in the number of digits from a list that they can correctly repeat back to the experimenter (i.e., digit retention). A sample of ten twelve-year-old autistic children yielded the following number of digits for each child: 10, 15, 14, 8, 9, 14, 6, 7, 11, 13.

a) If a great deal of previous data suggest that the population mean for twelve-year-old children is 7.8 digits, can the researcher conclude that autistic children are significantly different on this measure at the .05 level (two-tailed)? At the .01 level (two-tailed)?

b) Find the 95% confidence interval for the mean digit retention of all twelve-year-old autistic children.

8. A psychologist would like to know how many casual friends are in the average person's social network. She interviews a random sample of people and determines for each the number of friends or social acquaintances they see or talk to at least once a year. The data are as follows: 5, 11, 15, 9, 7, 13, 23, 8, 12, 7, 10, 11, 21, 20, 13.

a) Find the 90% confidence interval for the mean number of friends for the entire population.

b) Find the 95% CI.

c) If a previous researcher had predicted a population mean of ten casual friends per person, could that prediction be rejected as an hypothesis at the .05 level, one-tailed? At the .05 level, two-tailed?

* 9. Which of the following will make your confidence interval for the population mean smaller (assuming everything else stays the same)? a) Increasing the size of the sample; b) Increasing the amount of confidence; c) Increasing the sample variance; d) All of the above; e) None of the above

10. A study of 100 pet owners revealed that this group has an average heart rate ($\overline{X}$) of 70 beats per minute. By constructing confidence intervals (CI) around this sample mean a researcher discovered that the population mean of 72 was contained in the 99% CI but not in the 95% CI. This implies that

a) pet owners differ from the population at the .05 level, but not at the .01 level.

b) pet owners differ from the population at the .01 level, but not at the .05 level.

c) pet owners differ from the population at both the .05 and the .01 level.

d) pet owners differ from the population at neither the .05 nor the .01 level.

In Section A you learned that the mean of a sample ($\overline{X}$) can be used as a point estimate of the population mean (μ). It should also be clear from the discussion of the sampling distribution of the mean that as the sample size increases, the sample mean becomes a better estimate of the population mean, in that the probability of $\overline{X}$ being near μ increases. I have also mentioned in previous chapters that the sample variance can be used to estimate the population variance. In this section I will discuss the estimation of the population variance in greater detail, and in so doing add to your understanding of the statistical principles involved in the estimation of any population parameter.

OPTIONAL MATERIAL

The variance of a population can be calculated according to Formula 4.4:

$$\sigma^2 = \frac{\Sigma(X - \mu)^2}{N}$$

It would seem logical then to use the same formula (substituting $\overline{X}$ for μ) with the data from a sample in order to obtain a point estimate of the population variance. However, as Chapter 4 pointed out, using this formula to calculate the sample variance provides a point estimate that is biased—it tends to underestimate the population variance. I will explain this concept more fully by discussing the sampling distribution of the variance.

The Sampling Distribution of the Variance

If you draw very many random samples, all the same size (and all from the same population, of course), and calculate the variance of each sample according to Formula 4.1, these sample variances will form a distribution, which will become smooth and continuous as the number of samples approaches infinity. If the individuals in the population are normally distributed, the distribution of the sample variances will have a characteristic shape but will generally *not* resemble the normal distribution. In fact, the sampling distribution of the variance will be positively skewed (as shown in Figure 8.5); whereas there is no limit to how large the sample variances can get, the sample variances can never be less than zero. Actually, the shape of the sampling distribution of the variance changes as the samples get larger (the skewing decreases), and as the sample size approaches infinity, this sampling distribution becomes indistinguishable from the normal distribution. However, our concern at this point is not with the shape of the distribution, but rather with its mean.

Figure 8.5

Sampling Distribution of the Variance

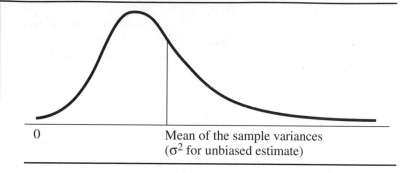

0   Mean of the sample variances
                     (σ^2 for unbiased estimate)

Mean of the Sampling Distribution of the Variance

Recall that the mean of the sampling distribution of the mean (i.e., the mean of all the sample means) was equal to the population mean. It would seem reasonable to expect a similar result for the sampling distribution of the variance: that its mean would be the population variance (σ^2). However, this is not true when the sample variances are calculated according to Formula 4.4. When this formula is used, the mean of all the sample variances is somewhat *less* than σ^2; it is actually

$$\frac{N - 1}{N} \sigma^2$$

Thus the mean of the sampling distribution of the variance (the location of which is indicated in Figure 8.5) is not σ^2. That is why we say that the sample variance, as calculated by Formula 4.4, is a *biased* estimator of the population variance. On the average, it underestimates σ^2. Fortunately, it is easy to create an *unbiased* estimator of σ^2. As you have seen, all we need to do to Formula 4.4 is to subtract 1 in the denominator (and substitute $\overline{X}$ for μ) to create Formula 4.7:

$$s^2 = \frac{\Sigma(X - \overline{X})^2}{N - 1}$$

The sample variance, as calculated by Formula 4.7, is labeled s^2, and it is an unbiased estimator of σ^2. The sampling distribution of s^2 has a mean that is equal to σ^2. Bear in mind, however, that the difference between the value given by Formula 4.4 and that given by Formula 4.7 becomes negligible for large sample sizes, because $N - 1$ does not differ much from N when N is large.

Properties of Estimators

An *estimator* is a formula, applied to sample data, that is used to estimate a population parameter. The formula for the sample mean is an estimator, as are Formulas 4.4 and 4.7. One property of an estimator is whether or not it is biased.

The sample mean ($\sum X^2/N$) and Formula 4.7 are unbiased, whereas Formula 4.4 is biased. In general, estimators that are unbiased are more desirable and are considered better estimators than those that are biased. However, bias is not the only important property of estimators. In fact, a biased estimator can be better than an unbiased one if it is superior in certain other properties. I will briefly mention two other properties of estimators here; you can consult a more advanced text (e.g., Hays, 1994) for a fuller treatment of point estimators and their properties.

In addition to being unbiased, another desirable property of an estimator is *consistency*. An estimator is consistent if it has a greater chance of coming close to the population parameter as the sample size (N) increases. We know that the sample mean is consistent because the sampling distribution of the mean gets narrower as N gets larger. Both the biased and the unbiased sample variances are consistent estimators, as well.

One reasonable way of determining which estimator of a particular population value is better is based on *relative efficiency*. If there are two different unbiased estimators of the same population parameter, the estimator that has the smaller variance is more efficient (assuming the same sample size for the two estimators).

More about Sampling Distributions

The shape of the distribution in Figure 8.5 is related to the shape of a well-known distribution called the chi-square distribution (pronounced "kie square"; "kie" rhymes with "pie"), which is symbolized by the Greek letter chi squared (χ^2). Actually, the chi-square distribution is a family of distributions that vary according to the number of degrees of freedom, just as the t distribution does. Any particular chi-square distribution can be symbolized as $\chi^2(\text{df})$; the same is true of the t distribution: $t(\text{df})$. When mathematicians refer to distributions, the number of degrees of freedom is symbolized by the Greek letter nu (ν), so $\chi^2(\nu)$ may be written instead of $\chi^2(\text{df})$. This also applies to the t distribution, which can be referred to as $t(\nu)$. Thus both types of distributions are said to depend on a single parameter, ν. Like the t distribution, the chi-square distribution looks more like the normal distribution as ν increases; when ν is large enough the two distributions are indistinguishable.

The connection between the χ^2 and t distributions can be seen by looking at the formula for the one-sample t-test:

$$t = \frac{\overline{X} - \mu}{s/\sqrt{N}} \qquad \text{Formula 8.3}$$

Note that the numerator follows a normal distribution because the sample mean is normally distributed (given the usual assumptions), and subtracting a constant (i.e., the population mean) does not change the shape of the distribution. The denominator, however, follows a different distribution; it contains the sample standard deviation, which follows a distribution that is related to the chi-square distribution. If the population standard deviation were known, the denominator would instead be a constant, and we would have a normal distribution being divided by a constant, which is still a normal distribution. (That is why you can use the normal distribution table when you know σ.) It is the fact that the normal distribution is being divided by a distribution related to the chi-square distribution that gives rise to the *t* distribution.

Although the sampling distribution of the standard deviation is related (by a square-root transformation) to the sampling distribution of the variance, it turns out that the square root of the unbiased sample variance (*s,* which is often referred to as the unbiased sample standard deviation) is not actually an unbiased estimate of the population standard deviation (σ), though it is less biased than the square root of the biased sample variance. Fortunately, this is a very minor point, because the bias involved in using *s* is so small that it is routinely ignored. Further discussion of the sampling distribution of *s,* as well as the many other interesting sampling distributions that have been studied (e.g., the sampling distribution of the median), is beyond the scope of this text.

The chi-square distribution can be used to test hypotheses about the population variance based on the variance of a single sample. However, this procedure is so rarely used it will not be discussed in this text. A more common procedure is to draw inferences based on comparing two sample variances; this topic will be discussed further in Chapter 9 (Section C) and again in Chapter 14 (Section C). The most common use of the chi-square distribution in psychological research, however, is to evaluate the outcome of nonparametric tests on categorical data (see Chapter 20).

Exercises

* 1. The sampling distribution of the variance will resemble the normal distribution when a) the number of samples is very large; b) the size of the samples is very large; c) the population variance is very large; d) the unbiased formula is used; e) none of the above

* 2. When deciding whether to use a particular estimator to estimate a population parameter, which of the following should be considered? a) Relative efficiency; b) Consistency; c) Amount of bias; d) All of the above; e) None of the above

SUMMARY

The Important Points of Section A

1. If you have a sample (e.g., hourly fees of current psychotherapists) and you want to know if it could reasonably have come from a population with a particular mean (1960 population of psychotherapists), but you don't know the population standard deviation (σ), you can still conduct a one-sample *z*-test.

2. To conduct the one-sample z-test when you don't know σ, you must have a large enough sample (preferably 100 or more). In that case you can use the unbiased sample standard deviation (s) in place of σ.

3. If the sample size is fairly small (and you still don't know the population standard deviation), you can nonetheless conduct a one-sample test, but you must use the t distribution to find your critical values.

4. The t distribution is actually a family of distributions that differ depending on the number of degrees of freedom (df equals the sample size minus 1).

5. The t distribution has fatter tails than the normal distribution, which means that the critical value for t will be larger than the critical value for z for the same alpha level (and, of course, the same number of tails).

6. As the df increase, the t distribution more closely resembles the normal distribution. By the time df reach about 100, the difference between the two types of distributions is negligible.

7. One-sample t-tests are not common (except for the matched t-test; see Chapter 13); it is much more common to compare two sample means (see Chapter 9).

8. The sample mean can be used as a *point estimate* of the population mean.

9. *Interval estimation* provides more information than point estimation. A *confidence interval* (CI) for the population mean can be constructed with the point estimate in the center.

10. The greater the degree of confidence, the larger the confidence interval must be. The 95% CI is the most common. For a given level of confidence, increasing the sample size will tend to make the CI smaller.

The Important Points of Section B

I will use a hypothical example to review the calculation of a one-group t-test and the construction of a confidence interval. Imagine the following situation: An anthropologist has just discovered a previously unknown group of humans native to Antarctica. These "Antarcticans" have adapted to their cold climate and apparently function quite well at body temperatures somewhat below what is considered normal for the rest of the human population. The anthropologist wants to know whether the mean body temperature for the population of Antarcticans is really less than the mean body temperature of other humans. Let us suppose that 98.6°F is the population mean (μ) for humans in general, but we do not have enough information to establish the population standard deviation (σ). Suppose also that the anthropologist was able to take the temperatures of only nine Antarcticans. The nine temperature measurements are as follows: 97.5, 98.6, 99.0, 98.0, 97.2, 97.4, 98.5, 97.0, 98.8. Because the anthropologist has a small sample

Figure 8.6

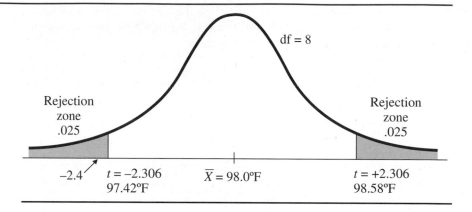

size and does not know the population standard deviation, he will have to perform a one-sample *t*-test.

The appropriate formula for this test is Formula 8.3, which requires that we calculate $\overline{X}$ and *s* before proceeding. The methods of Chapters 3 and 4 can be used to find that $\overline{X} = 98.0$ and $s = .75$. Because the numbers are high and fairly close together, I recommend using the definitional, rather than the computational, formula for *s*. Now we have all the values needed to compute the *t* statistic, according to Formula 8.3:

$$t = \frac{\overline{X} - \mu}{s/\sqrt{N}} = \frac{98.0 - 98.6}{.75/\sqrt{9}} = \frac{-.6}{.25} = -2.4$$

To complete the *t*-test, we need to know the critical value from the *t* distribution (see Table A.2), and to look up the critical value we need to know the number of degrees of freedom. For a one-group test, df = $N - 1$; in this case df = $9 - 1 = 8$. For a two-tailed test with alpha = .05, the critical $t = 2.306$. Because this is a two-tailed test, the region of rejection has two parts, one in each tail, which encompass *t* values greater than $+2.306$ or less than -2.306. The calculated *t* of -2.4 is less than -2.306, so the calculated *t* falls in the rejection zone (see Figure 8.6). Therefore we can conclude that the difference in body temperatures between Antarcticans and other humans is statistically significant (at the .05 level).

It cannot be overemphasized that in order for this test to be valid, the sample must truly be an independent random sample. If only the younger Antarcticans or the friendlier ones (could the friendlier ones be "warmer"?) were included, for instance, the conclusions from this test would not be valid.

Because body temperature is so tightly constrained by physiological requirements it would be interesting to find any group of people that maintained a different body temperature, even if the difference was fairly small. So the conclusion of the hypothesis test above would be of some interest in itself. But scientists in several disciplines would probably want more information; many would want to know the mean body temperature of all Antarcticans (μ_A). Based on the study above, the point estimate for μ_A would be $\overline{X} = 98.0$. However, there is a certain amount of error associated with that estimate, and the best way to convey the uncertainty involved is to construct a confidence interval.

The sample size being small, and the population standard deviation unknown, the appropriate formula is 8.5. This formula calls for $\overline{X}$ and $s_{\overline{X}}$, which have already been calculated for the hypothesis test above ($\overline{X} = 98.0$ and $s_{\overline{X}} = .25$). If we choose to construct a 95% CI, we must also find t_{crit} for df $= 8$ and alpha $= .05$, two-tailed. (In general, for an X% CI, alpha $= 1 - .X$, two-tailed.) This is the same t_{crit} we used for the hypothesis test above: $t_{crit} = 2.306$. The upper and lower limits for μ are calculated below.

$$\mu_{lower} = \overline{X} - t_{crit}s_{\overline{X}} = 98.0 - (2.306)(.25) = 98.0 - .577 = 97.42$$
$$\mu_{upper} = \overline{X} + t_{crit}s_{\overline{X}} = 98.0 + (2.306)(.25) = 98.0 + .577 = 98.58$$

Based on the above calculations we can say with 95% confidence that the mean body temperature for the entire population of Antarcticans is between 97.42°F and 98.58°F (see Figure 8.6). Note that the mean for the general population, 98.6°F, does not fall within the 95% confidence interval, which is consistent with the results of our null hypothesis test.

Of course, any particular confidence interval is either right or wrong—that is, it either contains the population mean or it does not. We never know which of our confidence intervals are right or wrong, but of all the 95% CIs we construct, we know that about 95% of them will be right (similarly, 99% of our 99% CIs will be right).

The Important Points of Section C

1. If an infinite number of random samples (all the same size) are drawn from a normal population, and the *biased* sample variance is calculated for each, these sample variances will form a positively skewed distribution. However, the mean of that distribution will be somewhat less than the population variance, which is why those sample variances are considered biased.

2. If the unbiased sample variance is calculated instead (using $N - 1$ in place of N in the denominator of the formula), the mean of the sampling distribution will be *equal* to the population variance.

3. Being unbiased rather than biased is just one of the properties that renders one estimator better than another in estimating a particular population parameter from sample data. Consistency and relative efficiency are two other desirable properties.

4. Put briefly, a *consistent* estimator is one that tends to get closer to the population parameter as sample size increases (the sample mean is one example), and a more *efficient* estimator exhibits less variance for a given sample size than a less efficient estimator.

Definitions of Key Terms

Large-sample *z*-test A test identical to the one-sample *z*-test, except that it is performed when the population standard deviation is not known but the sample is large enough that the sample standard deviation can be used to estimate the population standard deviation with little error.

***t* distribution** A family of distributions that differ according to the number of degrees of freedom. Each *t* distribution looks like the normal distribution except with fatter tails. As df increase, the *t* distribution more closely approximates the normal distribution.

One-sample *t*-test A test similar to the one-sample *z*-test, except that the sample standard deviation is used in place of the population standard deviation and the critical values are found from the *t* distribution. It is used when the sample size is not very large.

Point estimation Using a single value to estimate a population parameter, such as using the sample mean as an estimate of the population mean.

Interval estimation Using a range of values to estimate a population parameter.

Confidence interval An interval constructed so that a certain percentage of the time the interval will contain the specified population parameter. For instance, a particular 95% CI for the population mean may not contain the population mean, but 95% of such intervals will.

Key Formulas

Estimate of the standard error of the mean (when the population standard deviation is not known):

$$s_{\bar{X}} = \frac{s}{\sqrt{N}}$$

Formula 8.1

The *z*-score for groups, or the large-sample test (this formula can be used to conduct a one-sample *z*-test when the population standard deviation is not known but the sample size is sufficiently large):

$$z = \frac{\bar{X} - \mu}{s\sqrt{N}} \qquad \text{Formula 8.2}$$

One-sample *t*-test (population standard deviation is not known):

$$t = \frac{\bar{X} - \mu}{s/\sqrt{N}} \qquad \text{Formula 8.3}$$

One-sample *t*-test (estimate of standard error has already been calculated):

$$t = \frac{\bar{X} - \mu}{s_{\bar{X}}} \qquad \text{Formula 8.3A}$$

One-sample *t*-test (raw-score version; SS is the sum of squared deviations, as calculated by Formula 4.3):

$$t = \frac{\bar{X} - \mu}{\sqrt{SS/N(N-1)}} \qquad \text{Formula 8.3B}$$

Confidence interval for the population mean (sample size is sufficiently large):

$$\mu_{lower} = \bar{X} - z_{crit}s_{\bar{X}} \qquad \text{Formula 8.4}$$
$$\mu_{upper} = \bar{X} + z_{crit}s_{\bar{X}}$$

Confidence interval for the population mean (using *t* distribution):

$$\mu_{lower} = \bar{X} - t_{crit}s_{\bar{X}} \qquad \text{Formula 8.5}$$
$$\mu_{upper} = \bar{X} + t_{crit}s_{\bar{X}}$$

THE *t*-TEST FOR TWO INDEPENDENT SAMPLE MEANS

You will need to use the following from previous chapters:

Symbols
SS: Sum of squared deviations from the mean
σ: Standard deviation of a population
$\sigma_{\overline{X}}$: Standard error of the mean
s: Unbiased standard deviation of a sample
$s_{\overline{X}}$: Sample estimate of the standard error of the mean

Formulas
Formula 6.1: The standard error of the mean
Formula 8.3: The one-sample *t*-test
Formula 8.5: The confidence interval for the population mean

Concepts
The *t* distribution
Degrees of freedom (df)

Procedures
Finding critical values of the *t* distribution in Table A.2

CONCEPTUAL FOUNDATION

The purpose of this chapter is to explain how to apply null hypothesis testing in the case when you have two samples (treated differently), but do not know any of the population means or standard deviations involved. For the procedures of this chapter to be appropriate, the two samples must be completely separate, independent random samples, such that the individuals in one sample are not in any way connected or related to individuals in the other sample. It is sometimes desirable to use two samples that are somehow related, but I will not deal with this case until Chapter 13. As usual, I begin with an imaginary example in order to explain the concepts involved.

Suppose you have a friend who loves to exercise and is convinced that people who exercise regularly are not sick as often, as long, or as severely as those who do not exercise at all. She asks for your advice in designing a study to test her hypothesis, and the two of you agree that a relatively easy way to test the hypothesis would be to select at random a group of regular exercisers and an equal-sized group of nonexercisers and follow the individuals in both groups for a year, adding up the days each person would be considered sick. (You both agree that it would be too difficult to measure the severity of sickness in this informal study.) Now suppose that the year is over, and your friend is delighted to find that the mean number of sick days for the exercisers is somewhat lower than the mean for the nonexercisers. She starts to talk about the possibility of publishing the findings. It's time to tell your friend about Dr. Null.

If your friend tries to publish her results, Dr. Null is sure to know about it. He'll say that if the yearly sick days of the entire population of exercisers were measured, the mean number of sick days would be exactly the same as the mean for the entire population of nonexercisers. This, of course, is the null hypothesis applied to the two-group case. Dr. Null is also prepared to humiliate your friend in the following manner: He is ready to conduct his own experiment using two random samples (the same size as your friend's samples), but he will take both random samples from the nonexerciser population. He arbitrarily labels one of the samples exercisers and the other nonexercisers. (The subjects in Dr. Null's experiments wouldn't know this, of course.) Dr. Null claims that in his bogus experiment he will find that his phony exercise group has fewer sick days than the nonexercise group, and that the difference will be just as large as, if not larger than, the difference your friend found.

Your friend admits that it would be quite humiliating if Dr. Null could really do that. "But," she insists, "this Dr. Null is relying solely on chance, on sampling fluctuations. It's very unlikely that he will find as large a difference as I found, and half the time his results will be in the wrong direction." At this point (having read Chapters 7 and 8 of this text) you must explain that researchers cannot rely on their subjective estimates of probability; the only way to have any confidence about what Dr. Null can and cannot do is to draw a map of what he can do—a probability map—called the null hypothesis distribution. Once you know the parameters of this distribution, you truly know your enemy; you have Dr. Null's "number." You know when you can reject him with confidence and when you are better off "playing it safe." Finding the null hypothesis distribution in the two-group case is a bit trickier than in the one-group case. But considering that hypothesis tests involving two sample means are extremely common, it is well worth the effort to understand how these tests work.

Null Hypothesis Distribution for the Differences of Two Sample Means

In order to find the null hypothesis distribution in the two-group case, we have to consider what Dr. Null is doing when he performs his bogus experiment. All he is doing is selecting two random samples from the same population (actually, from two populations that have the same mean, but the distinction is not important at this point) and measuring the difference in their means, which can be negative or positive, depending on whether the first or second sample has the higher mean. If he conducts this experiment very many times, these difference scores will begin to pile up and form a distribution. If the variable being measured happens to have a normal distribution, then our task is fairly easy. A statistical theorem states that the differences (from subtracting the two sample means) will also form a normal distribution.

Finding the mean of this distribution is particularly easy in the case we've been describing. If Dr. Null's hypothesis is that the exercisers and the non-exercisers have the same population mean, then sometimes the exerciser sample

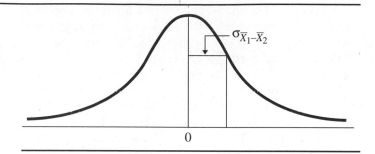

Figure 9.1

Normal Distribution of Differences of Sample Means

will have a higher mean and sometimes the nonexerciser sample mean will be higher, so some of the differences will be positive and some negative. However, there's no reason for the positives to outweigh the negatives (or vice versa) if the two population means are indeed equal, so Dr. Null's distribution will be a normal distribution centered at zero (see Figure 9.1).

Small differences will be the most common, with larger differences becoming increasingly less likely. (Differences that are *exactly* zero will not be common; see Section C.) The only thing we still need to know about the null hypothesis distribution is its standard deviation. Given the standard deviation, we will know everything we need to know to make judgments about the null hypothesis in the case of two independent groups.

Standard Error of the Difference

The standard deviation of Dr. Null's difference scores is called the **standard error of the difference** and is symbolized as $\sigma_{\bar{X}_1 - \bar{X}_2}$; the subscript shows that each difference score is the subtraction of two sample means. You may recall that finding the standard error of the mean ($\sigma_{\bar{X}}$), as we did in Chapter 6, was rather easy—it is just the standard deviation of the individual scores divided by the square root of N. The formula for finding the standard error of the differences is a little more complicated, but not much. In order to show the similarity between $\sigma_{\bar{X}}$ and $\sigma_{\bar{X}_1 - \bar{X}_2}$, I will rewrite Formula 6.1 in a different but equivalent form. Formula 6.1 looks like this:

$$\sigma_{\bar{X}} = \frac{\sigma}{\sqrt{N}}$$

Now I square both sides to get

$$\sigma_{\bar{X}}^2 = \frac{\sigma^2}{N}$$

Finally, I take the square root of both sides to get an alternate Formula 6.1 that looks like this:

$$\sigma_{\bar{X}} = \sqrt{\frac{\sigma^2}{N}}$$

Note that this modified version of Formula 6.1 gives you the same value as the original version if you plug in numbers, but it is expressed in terms of the population variance instead of the standard deviation. The reason I have bothered to change the appearance of Formula 6.1 is that this alternate version looks a lot like the formula for $\sigma_{\bar{X}_1 - \bar{X}_2}$, which I will refer to as Formula 9.1:

$$\sigma_{\bar{X}_1 - \bar{X}_2} = \sqrt{\frac{\sigma_1^2}{N_1} + \frac{\sigma_2^2}{N_2}} \qquad \text{Formula 9.1}$$

Formula 9.1 gives us just what we want. This formula tells you how large a difference is "typical" when you draw two samples from populations with the same mean and subtract the two sample means. (One sample has N_1 subjects and comes from a population with variance equal to σ_1^2; the other has N_2 subjects and is from a population with variance equal to σ_2^2.) Of course, *on the average* you would expect *no* difference, because the two samples come from populations with the same mean. However, you know that because of chance fluctuations the sample means will virtually never be the same. When you calculate the formula above for $\sigma_{\bar{X}_1 - \bar{X}_2}$ you know that about two-thirds of the differences of sample means will be less (in magnitude) than the value you get from the formula. (Recall that about two-thirds of any normal distribution falls within one standard deviation on either side of the mean.)

Because Formula 9.1 refers to what happens when Dr. Null is right, and the two populations (e.g., exercisers and nonexercisers) have the same mean, you may wonder why the formula allows for the possibility that the two populations have different variances. This is a somewhat strange possibility, so I will be returning to it later in this section and again in Section C. You may also wonder why it is that to get the standard error of the differences we *add* the two variances in the formula (after dividing each by its corresponding N) instead of subtracting them. Section C will give you some intuitive feeling for why the variances are added.

Formula for Comparing the Means of Two Samples

Now that we know the mean and standard deviation of Dr. Null's distribution, and we know that it is a normal distribution, we have a map of what Dr. Null can do, and we can use it to get a p value and make a decision about whether we can reject the null hypothesis and be reasonably confident about doing it. (We are assuming that the variable of interest has a normal distribution in both populations.) As in Chapter 7, we must take our experimental results and see where they fall on Dr. Null's distribution. We need to find the z score of our results with respect to the null hypothesis distribution, because that will allow us to find the p level, which in turn tells us the chance of Dr. Null's beating our results without actually doing our experiment—that is, just by performing random sampling. The z score for the case of two independent groups is very similar to the z score for one group. Examine the one-group formula (Formula 7.1), which appears below for comparison:

$$z = \frac{\overline{X} - \mu_0}{\sigma_{\overline{X}}}$$

Now consider the formula for the two-group case, Formula 9.2:

$$z = \frac{(\overline{X}_1 - \overline{X}_2) - (\mu_1 - \mu_2)}{\sigma_{\overline{X}_1 - \overline{X}_2}}$$ Formula 9.2

Note that the denominator in Formula 9.2 is given by Formula 9.1.

Notice that in both cases the z score formula follows the same basic structure: The numerator is the difference between your actual experimental results and what Dr. Null expects your experimental results to be. The denominator is the amount of difference that is "typical" when the null hypothesis is true, the kind of difference that can arise from chance fairly easily. If the difference you found for your experiment (i.e., the numerator) is less than or equal to the standard error (i.e., the denominator), then the z score will be less than or equal to 1, and your results will fall near the middle of the null hypothesis distribution, indicating that Dr. Null can beat your results rather easily. If the experimental difference you found is at least twice the standard error, then you can conclude (using $\alpha = .05$) that results as extreme as yours do not come up often enough by chance to worry.

Null Hypothesis for the Two-Sample Case

I will soon present numerical examples to make the use of these formulas more concrete, but for the moment take a closer look at the numerator of Formula 9.2. The first part, $\overline{X}_1 - \overline{X}_2$, represents your experimental results—for instance, the difference in mean sick days you found between exercisers and nonexercisers. The second part, $\mu_1 - \mu_2$, represents the difference Dr. Null expects, that is, the null hypothesis. You might expect this term always to be zero, so that you needn't bother to include it in the formula. After all, doesn't Dr. Null always expect there to be no difference between the two populations? Actually, the second part of the numerator often *is* left out—it is almost always zero. I have left the term $\mu_1 - \mu_2$ in Formula 9.2 for the sake of completeness, because it is possible (even though extremely rare) to consider a null hypothesis in which $\mu_1 - \mu_2$ does not equal zero. The following is an example of a plausible case in which the null hypothesis would not involve a zero difference.

Suppose a group of boys and a group of girls are selected and each child is given a growth hormone between the ages of 5 and 7. Then the mean height for each group is measured at age 18. The purpose of the experiment may be to show that the growth hormone is more effective on boys than on girls, but even if the hormone has the same effect on boys and girls, the boys will be taller on average than the girls at age 18. Knowing this, Dr. Null doesn't expect the mean heights for boys and girls to be equal. Rather, because he expects the growth hormone to have the same effect on both groups, he expects the difference in the heights of boys and girls to be the difference normally found at 18 when no growth hormone is administered. In this unusual case, the null hypothesis is not $\mu_1 - \mu_2 = 0$, but rather $\mu_1 - \mu_2 = d$, where d equals the normal height difference at 18. So in this special case, the experimenter is hoping that the $\overline{X}_1 - \overline{X}_2$ difference found in the experiment is not only greater than zero, but greater than d (or less than d if the experimenter expects the hormone to be more effective with girls).

Now let's return to the use of Formula 9.2, using our example about exercise and sick days. First I will substitute Formula 9.1 into the denominator of Formula 9.2 to obtain Formula 9.3:

$$z = \frac{(\overline{X}_1 - \overline{X}_2) - (\mu_1 - \mu_2)}{\sqrt{\sigma_1^2/N_1 + \sigma_2^2/N_2}} \qquad \text{Formula 9.3}$$

For the example involving exercise and sick days, the null hypothesis would be that $\mu_1 - \mu_2 = 0$—that the population mean for exercisers is the same as the population mean for nonexercisers. So when we use Formula 9.3, the second part of the numerator will drop out. Suppose the mean annual sick days for the exerciser sample is 5, and for the nonexercisers the mean is 7. Thus the numerator of Formula 9.3, $\overline{X}_1 - \overline{X}_2$, is $5 - 7 = -2$. This result is in the predicted direction—so far so good. Next assume that the two samples are the same size, $N_1 = N_2 = 15$. Now all we need to know is the variance for each of the two populations.

Actually, that's quite a bit to know. If we knew the variance of each population, wouldn't we also know the mean of each population? Yes—generally, if you have enough information to find the variance for a population, you have enough information to find the mean. (It is possible to devise an example in which the amount of variance is known from past experience but the population mean is not, but such situations are very rare in psychological research.) However, if we knew the mean of each population, there would be no reason to select samples or do a statistical test. We would have our answer. If the population mean for exercisers turned out to be less than that for nonexercisers, that could be reported as a fact, and not as a decision with some chance of being in error. (For instance, we do not need to perform a statistical test to state that women in the United States are shorter than men, because we know the means for both populations.)

The *z*-Test for Two Large Samples

In Chapter 7 we considered a case in which we knew the mean and variance for one population (i.e., the height of adult women in the United States) and wanted to compare the known population to a second, unknown population (i.e., LLD women). In that situation a one-sample test was appropriate. However, when you want to compare two populations on a particular variable and you don't know the mean of either population, it is appropriate to conduct a two-sample test. That is the case with the exercise and sick days example. But for any case in which you don't know either population mean, it is safe to say that you won't know the population variances either. Therefore you won't be able to use Formula 9.3. In fact, it looks like any time you have enough information to use Formula 9.3 you won't need to, and if you don't have the information you won't be able to! At this point you must be wondering why I introduced Formula 9.3 at all. Fortunately, there is a fairly common situation in which it is appropriate to use Formula 9.3: when you are dealing with large samples.

If your samples are very large the variances of the samples can be considered excellent estimates of the corresponding population variances—such good estimates, in fact, that the sample variances can be used in Formula 9.3 where population variances are indicated. Then the z score can be calculated and tested against the appropriate critical z, as described in Chapter 7. Therefore, Formula 9.3 can be called the **large-sample test for two independent means.** How large do the samples have to be? There is no exact answer. The error involved in estimating the population variances from the sample variances continues to decrease as sample size increases. However, once each sample contains more than 100 subjects the decrease in error due to a further increase in sample size becomes negligible. Some statisticians would suggest using Formula 9.3 even when you have only about 30 to 40 subjects per group, but the conservative approach (i.e., an approach that emphasizes minimizing Type I errors) would be to consider only samples of 100 or more as large. What can you do if your sample sizes cannot be considered large?

Separate-Variances *t*-Test

When sample sizes are not large there can be considerable error in estimating population variances from sample variances. The smaller the samples, the worse the error is likely to get. In Chapter 8 you learned how to compensate for this kind of error. For the current example we must again use the t distribution. If we take Formula 9.3 and substitute the sample variance (s^2) for the population variance (σ^2), we get a formula that we would expect to follow the t distribution, Formula 9.4:

$$t = \frac{(\overline{X}_1 - \overline{X}_2) - (\mu_1 - \mu_2)}{\sqrt{s_1^2/N_1 + s_2^2/N_2}}$$ Formula 9.4

This formula is called the separate-variances *t*-test, for reasons that will be made clear later. Unfortunately, this formula does not follow the t distribution in a simple way, and its use is somewhat controversial. Therefore, I will postpone the discussion of this formula until Section C. In the meantime, by assuming homogeneity of variance (i.e., assuming that the two population variances are equal, which will be explained in Section B), we can modify Formula 9.4 and make it easy to use. This modification concerns only the denominator of the formula: We need to describe a different way of estimating the standard error of the difference, $\sigma_{\overline{X}_1 - \overline{X}_2}$. Because this modified way of estimating $\sigma_{\overline{X}_1 - \overline{X}_2}$ involves pooling together the two sample variances, the resulting formula is referred to as the **pooled-variances *t*-test.** This test is easier to perform without a computer and much more commonly used than the separate-variances *t*-test.

The Pooled-Variances Estimate

The two sample variances can be pooled together to form a single estimate of the population variance (σ^2). The result of this pooling is called the **pooled variance,** s_p^2, and it is inserted into Formula 9.4 as a substitute for both sample variances, which produces Formula 9.5A. The pooled variance can be factored out of each fraction in the denominator of Formula 9.5A to produce the algebraically equivalent Formula 9.5B.

$$t = \frac{(\overline{X}_1 - \overline{X}_2) - (\mu_1 - \mu_2)}{\sqrt{s_p^2/N_1 + s_p^2/N_2}} \qquad \text{Formula 9.5A}$$

$$t = \frac{(\overline{X}_1 - \overline{X}_2) - (\mu_1 - \mu_2)}{\sqrt{s_p^2(1/N_1 + 1/N_2)}} \qquad \text{Formula 9.5B}$$

The rationale for pooling the two sample variances is based on the assumption that the two populations have the same variance, so that both sample variances are estimates of the same value. However, when the sample sizes are unequal, it is the larger sample that is the better estimate of the population variance, and it should have more influence on the pooled-variance estimate. We ensure that this is the case by taking a weighted average (as defined in Chapter 5) of the two sample variances.

When dealing with sample variances, the weight of a particular variance is just one less than the number of observations in the sample (i.e., $N - 1$)—in other words, the weight is the degrees of freedom (df) associated with that variance. Recall Formula 4.7: $s^2 = SS/(N - 1)$. Notice that the weight associated with each variance is actually the denominator that was used to calculate the variance in the first place. To take the weighted average of the two sample variances, multiply each variance by its weight (i.e., its df), add the two results together, and then divide by the total weight (the df of the two variances added together). The resulting weighted average is the pooled variance, as shown in Formula 9.6A:

$$s_p^2 = \frac{(N_1 - 1)s_1^2 + (N_2 - 1)s_2^2}{N_1 + N_2 - 2} \qquad \text{Formula 9.6A}$$

Formula 4.7 provides a way of rewriting the numerator of Formula 9.6A. If we multiply both sides of Formula 4.7 by $N - 1$, we get $SS = (N - 1)s^2$. This information allows us to rewrite Formula 9.6A as Formula 9.6B:

$$s_p^2 = \frac{SS_1 + SS_2}{N_1 + N_2 - 2} \qquad \text{Formula 9.6B}$$

Formula 9.6B gives us another way to think of the pooled variance. You simply add the SS (the top part of the variance formula) for the two groups together and then divide by the total degrees of freedom (the df for both groups added together).

The Pooled-Variances *t*-Test

We are now ready to put the pieces together to create one complete formula for the pooled-variances *t*-test. By inserting Formula 9.6A into Formula 9.5B, we get a formula that looks somewhat complicated but is easy to calculate, Formula 9.7A:

$$t = \frac{(\overline{X}_1 - \overline{X}_2) - (\mu_1 - \mu_2)}{\sqrt{\dfrac{(N_1 - 1)s_1^2 + (N_2 - 1)s_2^2}{N_1 + N_2 - 2}\left(\dfrac{1}{N_1} + \dfrac{1}{N_2}\right)}} \qquad \text{Formula 9.7A}$$

As an alternative you can instead insert Formula 9.6B into Formula 9.5B to get Formula 9.7B:

$$t = \frac{(\overline{X}_1 - \overline{X}_2) - (\mu_1 - \mu_2)}{\sqrt{\dfrac{SS_1 + SS_2}{N_1 + N_2 - 2}\left(\dfrac{1}{N_1} + \dfrac{1}{N_2}\right)}} \qquad \text{Formula 9.7B}$$

Of course, both Formula 9.7A and Formula 9.7B will give you exactly the same *t* value. Formula 9.7A is more convenient to use when you are given, or have already calculated, either the standard deviation or the variance for each sample. Formula 9.7B may be more convenient when you want to go straight from the raw data to the *t*-test. Because I recommend always calculating the means and standard deviations of both groups and looking at them carefully before proceeding, I favor Formula 9.7A when performing a pooled-variances *t*-test.

Formula for Equal Sample Sizes

If both samples are the same size, that is, $N_1 = N_2$, then the weighted average becomes a simple average, so

$$s_p^2 = \frac{s_1^2 + s_2^2}{2}$$

Instead of writing N_1 and N_2, you can use N without a subscript to represent the number of subjects (or more generally speaking, the number of observations) in each sample. In this case, Formula 9.5B becomes

$$t = \frac{(\bar{X}_1 - \bar{X}_2) - (\mu_1 - \mu_2)}{\sqrt{s_p^2(2/N)}}$$

After substituting the simple average for s_p^2 in the above formula, we can cancel out the factor of 2 and get the formula for the pooled-variance t-test for two groups of equal size, Formula 9.8:

$$t = \frac{(\bar{X}_1 - \bar{X}_2) - (\mu_1 - \mu_2)}{\sqrt{(s_1^2 + s_2^2)/N}} \qquad \text{Formula 9.8}$$

Formula 9.8 is also the separate-variances t-test for two groups of equal size. If you replace both N_1 and N_2 with N in Formula 9.4, then you have a common denominator in the bottom part of the formula, as in Formula 9.8. This means that when the two groups are the same size, it doesn't matter whether a pooled or a separate-variances t-test is more appropriate; in either case you use Formula 9.8 to calculate your t value. The critical t, however, will depend on whether you choose to perform a separate- or a pooled-variances t-test (see Section C).

Calculating the Two-Sample *t*-Test

Let us return to our example involving exercisers and nonexercisers. I deliberately mentioned that both samples contained 15 subjects because I want to deal numerically with the simpler case first, in which the two samples are the same size. (In Section B, I will tackle the general case.) Because the two samples are the same size we can use Formula 9.8. As I mentioned above, the difference between the two sample means for this example is -2, and the null hypothesis is that the two population means are equal, so the value of the numerator of Formula 9.8 is -2. To calculate the denominator we need to know the variances of both samples. This would be easy to calculate if you had the raw data (i.e., 15 numbers for each group). For the sake of simplicity, assume that the standard deviation for the 15 exercisers is 4 and the standard deviation for the 15 nonexercisers is 5. Now to get the sample variances all you have to do is square the standard deviations: $s_1^2 = 16$ and $s_2^2 = 25$. Plugging these values into Formula 9.8 gives a t ratio, as follows:

$$t = \frac{-2}{\sqrt{(16 + 25)/15}} = \frac{-2}{\sqrt{2.733}} = \frac{-2}{1.65} = -1.21$$

Note that the N in Formula 9.8 refers to the size of each group, not the total of both groups, which is why in this case $N = 15$. Also, the t value above is a negative number, but normally you ignore the sign of the calculated t when testing for significance. In fact, it is common to subtract the means in the numerator in whatever order makes the t value come out positive. Of course, the very first step would be to compare the two sample means to determine the direction of the results. (In the case of a one-tailed test you are at the same time determining whether a statistical test will be performed at all.) Then when you calculate the t value, you need only be concerned with its magnitude, to see if it is extreme enough that you can reject the null hypothesis. The magnitude of the t value in our example is $t = 1.21$. To determine whether this value is statistically significant, we have to find the appropriate critical value for t.

Interpreting the Calculated t

When we attempt to use the t table (Table A.2 in Appendix A), we are reminded that we must first know the number of degrees of freedom that apply. For the pooled-variances t-test of two independent groups, the degrees of freedom are equal to the denominator of the formula that was used to calculate the pooled variance (Formula 9.6A or Formula 9.6B)—that is, $N_1 + N_2 - 2$. If the two groups are the same size, this formula simplifies to $N + N - 2$, which equals $2N - 2$. For the present example, then, df $= 2N - 2 = 30 - 2 = 28$. If we use the conventional alpha $= .05$ and perform a two-tailed test, under df $= 28$ in the t table we find that the critical t equals 2.048. Thus our calculated t of 1.21 is smaller than the critical t of 2.048. What can we conclude?

Because our calculated t is smaller than the critical t we cannot reject the null hypothesis; we cannot say that our results are statistically significant at the .05 level. Our calculated t has landed too near the middle of the null hypothesis distribution. This implies that the difference we found between exercisers and nonexercisers is the kind of difference that occurs fairly often by chance (certainly more than .05, or 1 time out of 20, which is the largest risk we are willing to take). So we must be cautious and concede that Dr. Null has a reasonable case when he argues that there is really no difference at all between the two populations being sampled. We may not be completely convinced that Dr. Null is right about our experiment, but we have to admit that the null hypothesis is not farfetched enough to be dismissed. If the null hypothesis is *not* really true for our experiment, then by failing to reject the null hypothesis we are committing a Type II error. But we have no choice—we are following a system (i.e., null hypothesis testing) that is designed to keep Type I errors down to a reasonably low level (i.e., alpha).

Limitations of Statistical Conclusions

What if our calculated t were larger than the critical t, and we could therefore reject the null hypothesis? What could we conclude in that case? First, we should keep in the back of our minds that in rejecting the null hypothesis we could be making a Type I error. Perhaps the null hypothesis is true, but we got "lucky." Fortunately, when the calculated t is larger than the critical t, we are permitted to ignore this possibility and conclude that the two population means are not the same—that in fact the mean number of sick days for all exercisers in the population is lower than the mean for all the nonexercisers. However, rejecting the null hypothesis does not imply that there is a large or even a meaningful difference in the population means. Rejecting the null hypothesis, regardless of the significance level attained, says nothing about the size of the difference in the population means. That is why it can be very informative to construct a confidence interval for the difference of the means. This topic will be covered in detail in Section B.

An important limitation of the experiment described above should be made clear. Had the null hypothesis been rejected, the temptation would have been to conclude that regular exercise somehow made the exercisers less susceptible to getting sick. Isn't that why the experiment was performed? However, what we have been calling an experiment—comparing exercisers to nonexercisers—is not a true experiment, and therefore we cannot conclude that exercise *causes* the reduction in sick days. The only way to be sure that exercise is the factor that is causing the difference in health is to assign subjects randomly to either a group that exercises regularly or one that does not. The example we have been discussing is a *quasi-experiment* (as discussed in Chapter 1), because it was the subjects themselves who decided whether to exercise regularly or not. The problem with this design is that we cannot be sure that exercising is the only difference between the two groups. It is not unreasonable to suppose that healthy people (perhaps because of genetic factors or diet) are more likely to become exercisers, and that they would be healthier even if they didn't exercise. Or there could be personality differences between those who choose to exercise and those who do not, and it could be those personality factors, rather than the exercise itself, that is responsible for their having fewer sick days. It is important to know whether it is the exercise or other factors that is reducing illness. If it is the exercise that is responsible, we can encourage nonexercisers to begin exercising, because we have reason to believe that the exercise improves health. But if exercisers are healthier because of genetic or personality factors, there is no reason for nonexercisers to begin exercising; there may be no evidence that exercise will change their genetic constitution or their personality. To prove that it is the exercise that reduces the number of sick days, you would have to conduct a true experiment, randomly assigning subjects to either a group that must exercise regularly or a group that does not.

Exercises

* 1. Hypothetical College is worried about its attrition (i.e., drop-out) rate, so it measured the entire freshman and sophomore classes on a social adjustment scale to test the hypothesis that the better adjusted freshmen are more likely to continue on as sophomores. The mean for the 150 freshmen tested was 45.8, with $s = 5.5$; the mean for the 100 sophomores was 47.5, with $s = 4$.

a) Use the appropriate formula to calculate a test statistic to compare the means of the freshman and sophomore classes.

b) What is the *p* level associated with your answer to part a?

2. a) For a two-group experiment, $N_1 = 14$ and $N_2 = 9$. How many degrees of freedom would be associated with the pooled-variances *t*-test?

b) What are the two-tailed critical *t* values, if alpha = .05? If alpha = .01?

* 3. A group of 101 subjects has a variance $(s_1{}^2)$ of 12, and a second group of 51 subjects has a variance $(s_2{}^2)$ of 8. What is the pooled variance, $s_p{}^2$, of these two groups?

4. The weights of 20 men have a standard deviation $s_M = 30$ lbs., and the weights of 40 women have $s_W = 20$ lbs. What is the pooled variance, $s_p{}^2$, of these two groups?

* 5. A particular psychology experiment involves two equal-sized groups, one of which has a variance of 120, the other of which has a variance of 180. What is the pooled variance? (Note: It doesn't matter how

large the groups are, as long as they are both the same size.)

6. The two groups in a psychology experiment both have the same variance, $s^2 = 135$. What is the pooled variance? (Note: It doesn't matter how large the groups are, or even whether they are the same size.)

* 7. In a study of a new treatment for phobia, the data for the experimental group were $\overline{X}_1 = 27.2$, $s_1 = 4$, and $N_1 = 15$. The data for the control group were $\overline{X}_2 = 34.4$, $s_2 = 14$, and $N_2 = 15$.

a) Calculate the separate-variances *t* value.

b) Calculate the pooled-variances *t* value.

c) Compare your answers in parts a and b. Can you determine the general principle that is being illustrated?

8. a) Design a true experiment involving two groups (i.e., the experimenter decides, at random, in which group each subject will be placed).

b) Design a quasi-experiment (i.e., an observational or correlational experiment) involving groups not created, but only selected, by the experimenter. How are your conclusions from this experiment limited, even if the results are statistically significant?

BASIC STATISTICAL PROCEDURES

The previous section described an experiment that was observational or correlational in nature; the subjects were sampled from groups that were already in existence. In this section I will describe a true experiment, employing the same dependent variable as in the previous section but a different independent variable. Another difference in the example in this section is that in this true experiment, the two groups are *not* the same size. This gives an opportunity to demonstrate the calculation of the two-group *t*-test when $N_1 \neq N_2$. There is absolutely no difference between the statistical analysis of a true experiment and the corresponding analysis of an observational experiment. I illustrate both types of experimental designs so that I can comment on the different conclusions that can be drawn *after* the statistical analysis.

Another factor that may affect the number of days an individual is sick, and one that has received much publicity in recent years, is taking very large doses (i.e., megadoses) of vitamin C daily. Although quite a few studies have already been conducted in this area, the conclusions remain controversial. However, this topic serves our purposes well at this point because it is relatively easy to design a true experiment to test the effects of vitamin C. All we need to do is select a group of volunteers who are not already taking large amounts of vitamin C, and form two groups by randomly assigning half the subjects to the vitamin C group and the rest to the placebo group. (A placebo is some harmless substance that we can be sure won't affect the subjects at all.)

To prevent the biases of the subjects or the experimenters from affecting the results, the study should follow a "double-blind" design—that is, the vitamin C and placebo capsules should be indistinguishable by the subjects and coded in such a way that the experimenters who interact with the subjects also have no idea which subjects are taking vitamin C and which the placebo. (The code is not broken until the end of the experiment.) Because the subjects have been randomly assigned to groups, and the effects of expectations are the same for both, any statistically significant difference between the two groups can be attributed specifically to the action of vitamin C. If we tried to perform this experiment with only one sample, we would not have a placebo group for comparison. In that case, any reduction in sick days for the vitamin C group could be either from vitamin C or from the expectations of the subjects (i.e., the "placebo effect").

To set up and analyze a two-sample experiment we can use the same six steps that we used in Chapter 7 to test hypotheses involving one sample. In fact, having learned and practiced the use of these six steps for one-sample problems, you should find the steps below quite familiar; it is only the calculation formula that looks different. And I hope that Section A helped you to see that whereas the two-group *t* formula can look rather complicated, it is conceptually very similar to the one-sample *t* and one-sample *z* formulas.

Step 1. State the Hypotheses

The research hypothesis is that subjects who take the vitamin C will be sick fewer days than the subjects who take the placebo. However, this hypothesis is not specific enough to be tested directly. The common solution to this problem, as you saw in the previous two chapters, is to set up a specific null hypothesis, which the researcher hopes to disprove. The appropriate null hypothesis in this case is that it makes no difference whether the subjects take vitamin C or the placebo—the average number of sick days will be the same. Next we need to translate this idea into a statement about populations. If everyone in the population of interest had been in the vitamin C group, the mean of that distribution could be designated μ_C. Similarly, if the entire population had been in the placebo group, the mean of the resulting distribution could be designated μ_P. The null hypothesis (H_0) for this experiment is that these two hypothetical population means are equal: $\mu_C = \mu_P$, or $\mu_C - \mu_P = 0$.

The alternative hypothesis is the complement of the null hypothesis: H_A: $\mu_C \neq \mu_P$, or $\mu_C - \mu_P \neq 0$. If a one-tailed alternative hypothesis is considered justified, there are, of course, two possibilities: either H_A: $\mu_C < \mu_P$ (or $\mu_C - \mu_P < 0$), or H_A: $\mu_C > \mu_P$ (or $\mu_C - \mu_P > 0$). According to the research hypothesis that inspired the present example, if a one-tailed hypothesis were to be used, it would be H_A: $\mu_C < \mu_P$, because with vitamin C we expect *fewer* sick days. Whether it is appropriate to conduct a one-tailed test in a two-group experiment depends on the same issues that were already discussed for one-sample experiments (see Chapter 7, Section A). For the present example I will follow the more conservative approach and use the two-tailed alternative hypothesis.

Step 2. Select the Statistical Test and the Significance Level

We want to draw an inference about two population means from the data in two fairly small samples. Assuming that the populations have the same variance, we can perform the pooled-variance *t*-test for two independent samples. There is nothing different about setting alpha for a two-group test as compared to the one-group test. The most common alpha level is .05, and only unusual circumstances would justify a larger alpha.

Step 3. Select the Samples and Collect the Data

If the populations already exist, as in the study of exercisers and nonexercisers in Section A, then a random sample of each should be taken. Of course, truly random samples are virtually impossible, but the validity of the exercise study would be completely destroyed if, for instance, any attempt were made to exclude sickly exercisers or to encourage especially robust exercisers to participate. On the other hand, for the vitamin C study, one sample would be drawn from the general population (excluding people who are already vitamin C users). The researcher would try to make the sample as representative of the population as feasible, but most likely would use a sample of convenience. Then the subjects would be *randomly assigned* to one condition or the other. The random assignment of subjects would ensure that there were no systematic differences between the two groups before the experimental conditions could be imposed; the only differences between the groups would be those due to chance.

The larger the samples the more accurate will be the conclusions. Samples large enough to allow the use of the large-sample *z*-test of two independent means (Formula 9.3) are preferable, as in that case there is little reason to worry about the shapes of the population distributions. Such large samples may not be available or affordable, and if you are looking for a rather large experimental effect they may not be necessary. (This latter point will be explained in the next chapter.) When using fairly small samples (each less than about 30 or 40) an effort should be made to ensure that the two groups are the same size, so that the more complicated separate-variances *t*-test need not be considered (see Section C). For the present example, suppose that 12 subjects were assigned to each group, but at the end of the experiment two subjects confessed that they frequently forgot to take their pills, and therefore their data had to be eliminated. Imagine that, coincidentally, both of these subjects were from the placebo group, so that $N_C = 12$ and $N_P = 10$. The data for the vitamin C experiment would thus consist of 22 numbers, each of which would represent the number of sick days for a particular subject during the course of the experiment. To streamline the calculations in this section I will give only the means and standard deviations for each group; in Section D, I will illustrate the calculation of the two-group *t*-test directly from the raw data.

Step 4. Find the Region of Rejection

Because the sample sizes in our example are not large enough to justify a large-sample *z*-test, we will have to use the *t* distribution. Having selected alpha = .05 and a two-tailed test, we only need to know the degrees of freedom to look up the critical *t* value. The df for the two-group test is $N_1 + N_2 - 2$. Therefore, the df for the present example is $12 + 10 - 2 = 22 - 2 = 20$. Table A.2 shows that the critical $t = 2.086$. Because we are planning for a two-tailed test there is a rejection region in each tail of the *t* distribution, at -2.086 and at $+2.086$, as shown in Figure 9.2.

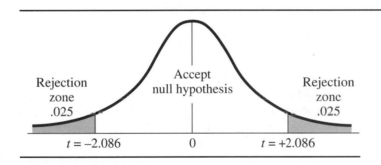

Figure 9.2

t **Distribution Based on 20 Degrees of Freedom**

Step 5. Calculate the Test Statistic

Suppose that $\overline{X}_C = 4.25$ sick days, $s_C = 3$, $\overline{X}_P = 7.75$ sick days, and $s_P = 4$. First you would check to see which mean is higher. Even though a two-tailed test is planned, it is clear that the designer of this experiment is hoping that the vitamin C mean is lower so that the results come out in the desired direction. In fact, had the means come out in the direction opposite the expected one, the researcher might well have decided not to even bother testing the results for significance. (If a one-tailed test had been planned and the results had come out in the direction opposite what he expected, the researcher would have been ethically bound to refrain from testing the results.) Because the samples are not large and not equal in size, the appropriate formula to use is either Formula 9.5A or Formula 9.5B (but Formula 9.5B is easier to look at):

$$t = \frac{(\overline{X}_1 - \overline{X}_2) - (\mu_1 - \mu_2)}{\sqrt{s_p^2(1/N_1 + 1/N_2)}}$$

I will start by calculating the numerator:

$$(\bar{X}_1 - \bar{X}_2) - (\mu_1 - \mu_2) = (7.75 - 4.25) - 0 = 3.5 - 0 = 3.5$$

Note that I deliberately arranged the sample means in the formula so the smaller would be subtracted from the larger, giving a positive number. This is often done to avoid the added complexity of negative numbers. Since we have already taken note of which mean was larger, we do not have to rely on the sign of the t value to tell us the direction of our results. Also note that because our null hypothesis is that $\mu_1 - \mu_2 = 0$, that term becomes zero in the numerator.

The next step is to pool the two sample variances together to get s_p^2. Because we already have the standard deviations and can therefore easily get the variances, we will use Formula 9.6A. First we square each of the standard deviations to get $s_C^2 = 3^2 = 9$ and $s_P^2 = 4^2 = 16$, and then we plug these values into the formula:

$$s_p^2 = \frac{(N_1 - 1)s_1^2 + (N_2 - 1)s_2^2}{N_1 + N_2 - 2} = \frac{11(9) - 9(16)}{12 + 10 - 2}$$
$$= \frac{99 + 144}{20} = \frac{243}{20} = 12.15$$

Note that the pooled variance falls between the two sample variances, but not exactly midway between. Had the two samples been the same size, the pooled variance would have been 12.5, which is halfway between 9 and 16. When the two samples are not the same size, the pooled variance will be closer to the variance of the larger sample; in this case the larger sample has variance $= 9$, so the pooled variance is closer to 9 than it is to 16. This is a consequence of taking a weighted average. Now we are ready to finish the calculation of Formula 9.5B:

$$t = \frac{(\bar{X}_1 - \bar{X}_2) - (\mu_1 - \mu_2)}{\sqrt{s_p^2(1/N_1 + 1/N_2)}} = \frac{3.5}{\sqrt{12.15(1/12 + 1/10)}}$$
$$= \frac{3.5}{\sqrt{12.15(.1833)}} = \frac{3.5}{\sqrt{2.23}} = \frac{3.5}{1.49} = 2.345$$

Step 6. Make the Statistical Decision

The calculated t equals 2.345, which is larger than the critical t (2.086), so the null hypothesis can be rejected. By looking at Figure 9.2 you can see that $t = 2.345$ falls in the region of rejection. We can say that the difference between our two samples is statistically significant at the .05 level, allowing us to conclude that the means of the two populations (i.e., the vitamin C population and the placebo population) are not exactly the same.

Interpreting the Results

Because our vitamin C experiment was a true experiment, involving the random assignment of subjects to the two groups, we can conclude that it is the vitamin C that is responsible for the difference in mean number of sick days. We can rule out various alternative explanations, such as placebo effects (subjects in both groups thought they were taking vitamin C), personality differences, etc. However, bear in mind that no experiment is perfect; there can be factors at play that the experimenters are unaware of. For example, researchers conducting one vitamin C experiment found that some subjects had opened their capsules and tasted the contents to try to find out if they had the vitamin or the placebo (these subjects knew that vitamin C tastes sour). But even if the experiment *were* perfect, declaring statistical significance is not in itself very informative. By rejecting the null hypothesis, all we are saying is that the effect of vitamin C is not identical to the effect of a *placebo*, a totally inactive substance. We *are* saying that vitamin C produces a reduction in sick days (relative to a placebo), but we are *not* saying by how much. With sufficiently large groups of subjects, statistical significance can be obtained with mean differences too small to be of any practical concern.

If you are trying to decide whether to take large amounts of vitamin C to prevent or shorten the common cold or other illnesses, knowing that the null hypothesis was rejected in the study described above will not help you much. What you need to know is the *size* of the reduction due to vitamin C; in other words, how many sick days can you expect to avoid if you go to the trouble and expense of taking large daily doses of vitamin C? For an average reduction of half a day per year, you may decide not to bother. However, if the expected reduction is several days, it may well be worth considering. There are many cases when rejecting the null hypothesis in a two-group experiment is not in itself very informative and a researcher needs additional information concerning the difference between the means. In such cases, a confidence interval can be very helpful.

Confidence Intervals for the Difference Between Two Population Means

If an experiment is conducted to decide between two competing theories that make opposite predictions, simply determining which population mean is larger may be more important than knowing the magnitude of the difference. On the other hand, in many practical cases, it would be useful to know just how large a difference there is between the two population means. In Chapter 8, when we wanted to estimate the mean of a population, we used the sample mean, $\overline{X}$, as a *point estimate* of μ. A similar strategy is used to estimate the difference of two population means, $\mu_1 - \mu_2$; the best point estimate from our data would be $\overline{X}_1 - \overline{X}_2$. For the vitamin C experiment, our point estimate of $\mu_C - \mu_P$ is $\overline{X}_C - \overline{X}_P = 7.75 - 4.25 = 3.5$ sick days. If this experiment were real, a reduction of 3.5 sick days per year would be worthy of serious consideration. However, even if the experiment were real, this would be just an estimate. As you know, there is a certain amount of error associated with this estimate, and it would be helpful to know just how much error is involved before we take any such estimate too seriously.

As you learned in Chapter 8, a point estimate can be supplemented by an interval estimate, called a *confidence interval*. We can use the same procedure described in that chapter, modified slightly for the two-group case, to construct a confidence interval for the difference between two population means. First, recall that in testing the null hypothesis in a two-group experiment, we center the null hypothesis distribution on the value of $\mu_1 - \mu_2$ that is predicted by the null hypothesis, usually zero. Then, in a two-tailed test, we mark off the critical values (of t or z, whichever is appropriate) symmetrically on either side of the expected value (as in Figure 9.2). To find the confidence interval, the process is similar, except that we center the distribution on the point estimate for $\mu_1 - \mu_2$, that is, $\overline{X}_1 - \overline{X}_2$. Critical values are then marked off symmetrically just as in a null hypothesis test. The final step is to translate those critical values back into upper and lower estimates for $\mu_1 - \mu_2$. I will illustrate this procedure by constructing a confidence interval around the difference in sample means that we found in the vitamin C experiment.

Constructing a 95% Confidence Interval

We start by using the same distribution we used to test the null hypothesis: a t distribution with 20 degrees of freedom. However, the value at the center of the distribution is not zero, but $\overline{X}_C - \overline{X}_P = 3.5$. Then we must decide what level of confidence is desired. If we choose to construct a 95% confidence interval (the most common), the critical values are the same as in the .05, two-tailed test: -2.086 and $+2.086$ (see Figure 9.3). To convert these critical values into upper and lower estimates of $\mu_1 - \mu_2$ we need to turn the t formula around as we did in the one-group case. This time we start with a generic version of the two-group t formula, as follows:

$$t = \frac{(\overline{X}_1 - \overline{X}_2) - (\mu_1 - \mu_2)}{s_{\overline{X}_1 - \overline{X}_2}}$$

We already know the value of t—it is equal to the critical value. What we want to do is turn the formula around in order to solve for $\mu_1 - \mu_2$:

$$(\overline{X}_1 - \overline{X}_2) - (\mu_1 - \mu_2) = t_{\text{crit}} s_{\overline{X}_1 - \overline{X}_2}$$

Note that t_{crit} can be positive or negative. A final rearrangement to isolate $\mu_1 - \mu_2$ gives Formula 9.9:

$$\mu_1 - \mu_2 = (\overline{X}_1 - \overline{X}_2) \pm t_{\text{crit}} s_{\overline{X}_1 - \overline{X}_2} \qquad \text{Formula 9.9}$$

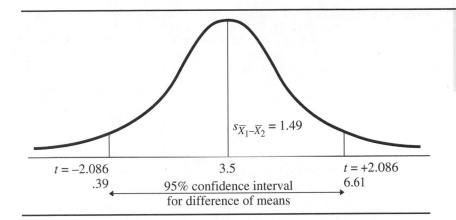

Figure 9.3

t Distribution (df = 20)

$s_{\overline{X}_1-\overline{X}_2} = 1.49$

$t = -2.086$ 3.5 $t = +2.086$
.39   95% confidence interval   6.61
for difference of means

Note the resemblance between Formula 9.9 and Formula 8.5. By plugging in values that we have already calculated for the t-test ($\overline{X}_1 - \overline{X}_2 = 3.5$ and $t_{crit} = 2.086$) we can find the upper and lower boundaries for the 95% confidence interval for $\mu_C - \mu_P$. The only other value we need in Formula 9.9 is $s_{\overline{X}_1-\overline{X}_2}$, which is the denominator of the t value we calculated in step 5; the denominator was equal to 1.49. Plugging these values into Formula 9.9, we get

$$\mu - \mu_2 = 3.5 \pm 2.086(1.49) = 3.5 \pm 3.11$$

So the upper estimate for the population mean difference is $3.5 + 3.11 = 6.61$, and the lower estimate is $3.5 - 3.11 = 0.39$. What does this tell us? First, notice that because both boundaries are positive numbers, zero is not included in the confidence interval. This means that based on our sample data, we can say that the chance of $\mu_C - \mu_P$ being zero is less than 5% (because zero is not in the interval that has a 95% chance of containing $\mu_C - \mu_P$). You should notice that this is the same as rejecting the null hypothesis that $\mu_C - \mu_P = 0$, at the .05 level in a two-tailed test. As I pointed out in Chapter 8, whatever the value predicted by the null hypothesis, if it is included in the 95% confidence interval, then the null hypothesis cannot be rejected at the .05 level (two-tailed).

Something else you may have noticed about the confidence interval we just calculated is that it is so wide that it is not very helpful. In order to be 95% certain, we must concede that the true reduction in sick days due to vitamin C may be as low as .39 or as high as 6.61. Of course, if we are willing to be less certain of being right, we can provide a narrower confidence interval (e.g., 90% or even 80%). The only way to be sure of making the confidence interval narrower without reducing confidence is to increase the number of subjects (unless you can find some way to reduce subject-to-subject variability). Increasing the number of subjects reduces the standard error of the difference, and in general provides more accurate information concerning population parameters.

Constructing a 99% Confidence Interval

A 99% confidence interval will be even larger than the 95% CI, but I will construct one to further illustrate the use of Formula 9.9. The critical t is the one that corresponds to an alpha of .01 (two-tailed); the df are still 20. Table A.2 shows that $t_{crit} = 2.845$. The rest of the values in Formula 9.9 remain the same:

$$\mu - \mu_2 = 3.5 \pm 2.845(1.49) = 3.5 \pm 4.24$$

The 99% confidence interval ranges from $3.5 - 4.24 = -.74$ to $3.5 + 4.24 = 7.74$. Notice that zero *is* included in this interval, which tells us that the null hypothesis of zero difference between population means cannot be rejected at the .01 level, two-tailed. To be 99% certain of our statement, we would have to concede that whereas vitamin C may reduce sick days by more than a week, it may actually *increase* sick days slightly. The latter possibility cannot be ruled out if we are to have 99% confidence.

Assumptions of the *t*-Test for Two Independent Samples

Independent Random Sampling

Ideally, both groups should be simple random samples, which implies that in the selection of each group, each member of the appropriate population has an equal chance of being selected, and each individual is selected independently of all the others in the same group (see Chapter 6). The concept of independence is involved in a second way: Each individual selected for one sample should be selected independently of all the individuals in the other sample. For example, in a study comparing men and women, independence would be violated if a group of married men was selected and then all of their wives were selected to be the group of women. The *t*-test formulas in this chapter apply only when the two samples are independent of each other. (Sometimes it is advantageous to create a systematic relationship or dependence between the two samples; the analysis of that type of experiment is the topic of Chapter 13.)

In psychology experiments, true random sampling is virtually impossible, so in order for the experimental conclusions to have any validity, the subjects must be *randomly assigned* to the two experimental conditions. For example, the results of the vitamin C experiment would have no validity if the experimenter had had information about the general health of each subject when deciding in which group to put each subject. In the exercise experiment described in Section A, subjects were not randomly assigned to groups, so for the study to have any validity, each sample must be as representative as possible of its population (or, at the least, the two samples should be as similar as possible in every regard except for the variable that distinguishes them). In addition, sampling biases must be avoided; the experimenter should not deliberately select sickly nonexercisers and/or healthy exercisers.

Normal Distributions

Ideally, the variable being measured (e.g., number of sick days) should be normally distributed in both populations. This assures that the differences of sample means will also follow a normal distribution, so that we can perform a valid two-group hypothesis test using z scores (for large samples) or the t distribution (when the population variances must be estimated from small samples). However, as I have mentioned, though many variables follow the normal distribution pretty closely, many of the variables in psychology experiments do not. In fact, in the example I have been using, the variable (sick days) is not likely to have a normal distribution; its distribution is likely to be positively skewed, with most people not very far from zero and a few people far above the others.

Fortunately, it is a consequence of the Central Limit Theorem that even when two populations are not normally distributed, the distribution of sample mean differences will approach the normal distribution as the sample sizes increase. In fact, for most population distributions, the sample sizes do not have to be very large before the methods of this chapter can be used with acceptable accuracy. As with the one-sample t-test, we can say that the two-group t-test is *robust* with respect to the assumption of normal distributions. Nonetheless, you should observe how the data are distributed in each sample (using the techniques described in Chapter 2). If the data are extremely skewed or otherwise strangely distributed, and sample sizes are not large, you should consider using a data transformation (see Chapter 3) or a nonparametric procedure (such as the Mann-Whitney test, described in Chapter 21).

Homogeneity of Variance

The pooled-variances t-test described in this chapter is strictly valid only if you assume that the two populations (e.g., exercisers and nonexercisers) have the same variance—a property called **homogeneity of variance**—whether or not they have the same mean. This assumption is reasonable because we expect that exposing a population to a treatment (e.g., regular exercise, regular doses of vitamin C) is likely to change the mean of the population (shifting everyone in the population by about the same amount) without appreciably changing its variance. However, there are situations in which a treatment could increase the variance of a population with or without affecting the mean, so the possibility of heterogeneity of variance must always be considered (see Section C).

Fortunately, there are three situations in which you can perform a two-group hypothesis test without having to worry about homogeneity of variance. The first situation is when both samples are quite large (at least about 100 subjects in each). Then you can use Formula 9.3 and the critical values for z. The second situation is when the two samples are the same size; then you can use Formula 9.8. Statisticians have demonstrated that even large discrepancies in population variances have little influence on the distribution of calculated *t* when the two samples are the same size. The third situation is when your two sample variances are very similar; in that case, it is reasonable to assume homogeneity of variance without testing for it and to proceed with the pooled-variances test. A popular rule of thumb is that if one variance is no more than four times larger than the other, you can assume homogeneity of variance.

If your samples are not very large, not equal in size, and one sample variance is more than four times the other, you must consider the possibility of performing a separate-variances *t*-test. These tests are difficult to do by hand, requiring special tables and complex mathematical formulas. Because they are now performed automatically by commonly available, easy-to-use computerized statistical packages (such as SPSS), it is becoming increasingly important to understand them. Thus most of Section C is devoted to this topic.

When to Use the Two-Sample *t*-Test

The two-sample *t*-test can be used to analyze a study of samples from two preexisting populations—a quasi-experiment—or to analyze the results of subjecting two randomly assigned samples to two different experimental conditions—a true experiment. However, a *t*-test is appropriate only when the *dependent variable* in a two-group experiment has been measured on an interval or ratio scale (e.g., number of sick days per year, weight in pounds). If the dependent variable has been measured on an ordinal or categorical scale, the nonparametric methods described in Part VI of this text will be needed. On the other hand, the *independent variable,* which has only two values, is usually measured on a categorical scale (e.g., exercisers and nonexercisers; psychotics and normals). In some cases, the independent variable may be measured originally using an interval or ratio scale, which is later converted into a dichotomy (the simplest form of categorical scale) by the researcher.

For example, a teacher may give out a social anxiety questionnaire at the beginning of a semester and then keep track of how much time each student speaks during class. In order to analyze the data, the teacher could use the anxiety measure to divide the class into two equal groups: those "high" in anxiety and those "low" in anxiety. (This arrangement is called a median split; the quote marks indicate that these are just relative distinctions within this one class.) A *t*-test can be performed to see if the two groups differ significantly in class speaking time. However, if the social anxiety scores form a nearly normal distribution, it is probably better to use a correlational analysis (to be described in Chapter 11) than to throw away information by merely classifying students as either high or low in anxiety. In fact, the exercise example in Section A might be better designed as a correlation between amount of exercise and the number of sick days in a sample of subjects varying widely in their degree of daily or weekly exercise. It is when the two groups are distinctly different, perhaps clustered at the two opposite extremes of a continuum or representing two qualitatively different categories, that the *t*-test is particularly appropriate.

When to Construct Confidence Intervals

If I have done a good job of selling the importance of constructing confidence intervals for a two-group experiment, you may feel that they should always be constructed. However, for many two-group experiments, confidence intervals for the difference in means may not be very meaningful. Consider, for example, an experiment in which one group of subjects watches a sad movie, while the other group watches a happy movie. The dependent variable is the number of "sad" words (e.g., "funeral") subjects can recall from a long list of words studied right after the movie. A difference of two more words recalled by those who watched a sad movie could turn out to be statistically significant, but a confidence interval would not be easy to interpret. The problem is that the number of words recalled is not a universal measure that can easily be compared from one experiment to another; it depends on the specifics of the experiment: the number of words in each list, the time spent studying the list, etc. On the other hand, days lost from work each year because of illness is meaningful in itself. When the units of the dependent variable are not universally meaningful, the confidence interval for the difference of means may not be helpful; a standardized measure of effect size or strength of association may be preferred. One such measure will be described in the next chapter.

Publishing the Results of the Two-Sample *t*-Test

If we were to try to publish the results of our vitamin C experiment, we would need to include a sentence like the following in the results section: "Consistent with our predictions, the vitamin C group averaged fewer days sick ($\underline{M} = 4.25$, $\underline{SD} = 3$) than did the placebo group ($\underline{M} = 7.75$, $\underline{SD} = 4$), $\underline{t}\,(20) = 2.34$, $\underline{p} < .05$, two-tailed." The number in parentheses following *t* is the number of degrees of freedom associated with the two-group test ($N_1 + N_2 - 2$).

The following is an excerpt from the results section of a published journal article titled "Group decision making under stress" (Driskell & Salas, 1991), which adheres to APA style rules. For this experiment, a two-group *t*-test was the most appropriate way to test two preliminary hypotheses: "Results also indicated that subjects in the stress conditions were more likely than subjects in the no-stress conditions to report that they were excited (Ms = 3.86 vs. 4.78), $t(72) = 2.85$, $p < .01$, and that they felt panicky (Ms = 4.27 vs. 5.08), $t(72) = 2.64$, $p = .01$" (p. 475). Note that the second *p* value stated was exactly .01 and was therefore expressed that way. Also, note that the hypotheses being tested were of a preliminary nature. In order to study the effects of stress on decision making, the authors had to induce stress in one random group but not the other. Before comparing the two groups on decision-making variables, the authors checked on the effectiveness of the stress manipulation by comparing the responses of subjects in the two groups on a self-report questionnaire (hence, this is called a manipulation check). The significant differences on the questionnaire (as noted in the quote above) suggest that the stress manipulation was at least somewhat effective and that stress could be the cause of the group differences in the main (dependent) variables being measured.

Exercises

*1. Seven acute schizophrenics and ten chronic schizophrenics have been measured on a clarity of thinking scale. The mean for the acute sample was 52, with $s = 12$, and the mean for the chronic sample was 44, with $s = 11$. Perform a pooled-variance t-test with alpha $= .05$ (two-tailed), and state your statistical conclusion.

*2. A group of 30 subjects is divided in half based on their self-rating of the vividness of their visual imagery. Each subject is tested on how many colors of objects they can correctly recall from a briefly seen display. The "more vivid" visual imagers recall an average ($\overline{X}_1$) of 12 colors with $s_1 = 4$; the "less vivid" visual imagers recall an average ($\overline{X}_2$) of 8 colors with $s_2 = 5$.

   a) Perform a two-group t-test of the null hypothesis that vividness of visual imagery does not affect the recall of colors; use alpha $= .01$, two-tailed. What is your statistical conclusion?

   b) What are some of the limitations that prevent you from concluding that vivid visual imagery *causes* improved color recall in this type of experiment?

*3. On the first day of class, a third-grade teacher is told that 12 of his students are "gifted," as determined by IQ tests, and the remaining 12 are not. In reality, the two groups have been carefully matched on IQ and previous school performance. At the end of the school year, the "gifted" students have a grade average of 87.2 with $s = 5.3$, whereas the other students have an average of 82.9, with $s = 4.4$. Perform a t-test to decide whether you can conclude from these data that false expectations can affect student performance; use alpha $= .05$, two-tailed.

*4. A researcher tested the diastolic blood pressure of 60 marathon runners and 60 nonrunners. The mean for the runners was 75.9 mm Hg with $s = 10$, and the mean for the nonrunners was 80.3 mm Hg with $s = 8$.

   a) Find the 95% confidence interval for the difference of the population means.

   b) Find the 99% confidence interval for the difference of the population means.

   c) Use the confidence intervals you found in parts a and b to test the null hypothesis that running has no effect on blood pressure at the .05 and .01 levels, two-tailed.

5. Imagine that the study described in exercise 4 was conducted with four times as many subjects.

   a) Find both the 95% and 99% confidence intervals.

   b) Compare the widths of these intervals to their counterparts in exercise 4, and state the general principle illustrated by the comparison. (Your comparison should be an approximate one, because in addition to rounding error, a slight change in the critical t from exercise 4 to exercise 5 will influence the comparison.)

*6. A psychologist is studying the concentration of a certain enzyme in saliva as a possible indicator of chronic anxiety level. A sample of 12 anxiety neurotics yields a mean enzyme concentration of 3.2 with $s = .7$. For comparison purposes, a sample of 20 subjects reporting low levels of anxiety is measured and yields a mean enzyme concentration of 2.3, with $s = .4$. Perform a t-test (alpha $= .05$, two-tailed) to determine whether the two populations sampled differ with respect to their mean saliva concentration of this enzyme.

*7. Will students wait longer for the arrival of an instructor who is a full professor than for one who is a graduate student? This question was investigated by counting how many minutes students waited in two small seminar classes, one taught by a full professor and one taught by a graduate student. The data (in minutes) are as follows:

Graduate Student Instructor: 9, 11, 14, 14, 16, 19, 37
Full Professor: 13, 15, 15, 16, 18, 23, 28, 31, 31

a) Use the pooled-variances *t*-test to test the null hypothesis at the .05 level, two-tailed.

b) Find the limits of the 95% confidence interval for the difference of the two population means.

* 8. Suppose that the subject in exercise 7 who waited 37 minutes for the graduate student wasn't really waiting but simply had fallen asleep. Eliminate the measurement for that particular subject.

a) Retest the null hypothesis. What can you say about the susceptibility of the *t*-test to outliers?

b) On the average, how much more time did students wait for the professor than for the graduate student? Construct the 95% confidence interval for the difference in mean waiting times.

c) Would a separate-variances *t*-test have been appropriate to answer part a? Explain.

* 9. An industrial psychologist is investigating the effects of different types of motivation on the performance of simulated clerical tasks. The ten subjects in

the "individual motivation" sample are told that they will be rewarded according to how many tasks they successfully complete. The ten subjects in the "group motivation" sample are told that they will be rewarded according to the average number of tasks completed by all the subjects in their sample. The number of tasks completed by each subject are as follows:

Individual Motivation: 11, 17, 14, 10, 11, 15,
10, 8, 12, 15
Group Motivation: 10, 15, 14, 8, 9, 14, 6, 7, 11, 13

a) Perform a two-sample *t*-test. Can you reject the null hypothesis at the .05 level?

b) Based on your answer in part a, what type of error (Type I or Type II) might you be making?

10. Suppose that a second industrial psychologist performed the same experiment described in exercise 9 but used considerably larger samples. If the second experimenter obtained the same sample variances and the same calculated *t* value, which experimenter obtained the larger difference in sample means? Explain your answer.

Zero Differences Between Sample Means

OPTIONAL MATERIAL

The virtual impossibility of obtaining a zero difference between two sample means involves a mathematical paradox. If you look at the null hypothesis distribution for the study of exercisers versus nonexercisers (see Figure 9.1), it is easy to see that the mode (the highest vertical point of the distribution) is zero—implying that zero is the most common difference that Dr. Null will find. But it is extremely unlikely to get a zero difference between two sample means, even when doing the bogus experiments that Dr. Null does. (Similarly, recall that in describing one-sample tests I mentioned that it is very unlikely that the sample mean selected by Dr. Null will be exactly equal to the population mean.) The problem is that when dealing with a smooth mathematical distribution, picking any one number from the distribution *exactly,* including exactly zero, has an infinitely small probability. This is why probability was defined in terms of intervals or ranges of a smooth distribution in Chapter 5. So, although numbers around zero are the most common means for Dr. Null's experiments, getting *exactly* zero (i.e., an infinite number of zeros past the decimal point) can be thought of as impossible.

Adding Variances to Find the Variance of the Difference

To get some feeling for why you add the variances of two samples when calculating the standard error of the differences, first imagine randomly selecting one score at a time from a normal distribution. Selecting a score within one standard deviation from the mean would be fairly likely. On the other hand, selecting a score about two standard deviations from the mean would be quite unusual. Next, imagine randomly selecting two scores at a time. It would not be unusual to select two scores, each of which is about one standard deviation from the mean; however, it would be just as likely for both scores to be on the same side of the mean as it would be for the two scores to be on opposite sides. If both are on the same side of the mean, the difference between them will be about zero. But if the scores are on opposite sides, the difference between them would be about two standard deviations. The point of this demonstration is to suggest that it is easier to get a *difference of two scores* that is equal to two standard deviations than it is to get a single score that is two standard deviations from the mean. The reason is that when you have two scores, even though each could be a fairly typical score, they could easily be from opposite sides of the distribution, and that increases the difference between them. Although this explanation is not very precise, it is meant merely to give you some feeling for why you could expect difference scores to have more, rather than less, variability from the mean than single scores. The statistical law in this case states that when you subtract (or add) two random scores from two normal distributions, the variance of the difference (or sum) is the sum of the two variances. (Note that you can add variances, but not standard deviations.)

When to Use the Separate-Variances *t*-Test

The values calculated from the pooled-variance formula (Formula 9.5B) follow the *t* distribution with $N_1 + N_2 - 2$ degrees of freedom only if the null hypothesis is true *and* the two population variances are equal. If the null hypothesis concerning the population means is true but the population variances are not equal, the pooled-variance *t* will not follow the *t* distribution we expect it to, and, therefore, the calculated *t* may exceed the critical *t* more or less often than alpha. The major concern is that we may not be screening out ineffective experiments as much as our alpha level would suggest. If you have a good reason to believe that the population variances are not equal, you should not be doing a pooled-variance *t*-test. The pooling of the sample variances is only justified if each is representing the same value (i.e., the same population variance). Before continuing, however, I remind you that there are two common cases in which you do not have to worry about pooling the variances: when both samples are large (in which case use Formula 9.3), or when the two samples are the same size (in which case use Formula 9.8). On the other hand, if your samples are fairly small and not equal in size, you do need to be concerned about whether homogeneity of variance is a reasonable assumption to make.

How can we decide whether the two population variances are equal? All we have to go on are the two sample variances. Of course, we don't expect the two sample variances to be exactly the same even if the two population variances are equal. However, if the two sample variances are very similar, it is safe to assume that the two population variances are equal. If, on the other hand, one of the sample variances is more than four times as large as the other, we have cause for concern. When the sample variances are not very similar, we can perform a statistical test of the null hypothesis that the population variances are equal. If the test leads to statistical significance, we conclude that the population variances are not equal (i.e., there is **heterogeneity of variance**), and we must perform a separate-variances *t*-test. If, on the other hand, the test is not significant, we should be able to conclude that we have homogeneity of variance and that pooling the sample variances is justified.

Several procedures have been suggested for testing homogeneity of variance. The simplest is the *F*-test, in which the two sample variances are divided to obtain an *F* ratio; I will return to the *F*-test in greater detail after I have introduced the *F* distribution in Chapter 14. Although it is the simplest, the *F*-test is not the best test of homogeneity of variance, and several alternatives have been proposed—some of which are calculated automatically when a statistical computer package is used to calculate a *t*-test. Unfortunately, homogeneity of variance tests tend to lose their power and validity with small sample sizes, which is when they are most needed. Moreover, such tests are not as robust as the *t*-test when the population distributions differ from the normal distribution. Indeed, the whole matter of testing homogeneity of variance and deciding when to apply the separate-variances *t*-test remains controversial. Perhaps the safest thing to do when your sample sizes are fairly small and unequal and one sample variance is at least four times as large as the other is to perform the separate-variances *t*-test, as described below.

The **separate-variances *t*-test** is so named because each sample variance is separately divided by its own sample size, and the variances are not pooled together. The test begins with the calculation of *t* according to Formula 9.4, shown below:

$$t = \frac{(\overline{X}_1 - \overline{X}_2) - (\mu_1 - \mu_2)}{\sqrt{s_1^2/N_1 + s_2^2/N_2}} \qquad \text{Formula 9.4}$$

The calculation of this *t* value for any particular experiment is not difficult; the problem is to find the proper critical *t* that corresponds to the desired alpha. Unfortunately, the *t* calculated from the above formula does not follow the *t* distribution with $N_1 + N_2 - 2$ degrees of freedom. Finding the null hypothesis distribution that corresponds to the separate-variances *t* has come to be known as the *Behrens-Fisher problem*, after the two statisticians who pioneered the study of this topic.

Fortunately, the separate-variances t follows *some* t distribution fairly well, and one way to view the problem boils down to finding the best estimate for the df of that t distribution. Both Welch and Satterthwaite independently devised the same rather complicated formula for estimating df for the separate-variances t, known as the *Welch-Satterthwaite solution*. I will not reproduce that formula here; most statistical packages calculate it for you and present the fractional value for df that often results. Instead, I will illustrate a procedure by which you may be able to avoid the formula entirely.

The following simple procedure is suggested by Howell (1992). Based on the Welch-Satterthwaite solution, the lowest that df will ever be for the separate-variances t-test is either $N_1 - 1$ or $N_2 - 1$, whichever is lower for the experiment in question. That is the "worst-case scenario" no matter how unequal the population variances are. The highest that df can be, of course, is the df for the pooled-variances test, $N_1 + N_2 - 2$. By now you should be familiar with the idea that as df gets lower, critical t gets higher, and therefore harder to beat. All this leads to a simple strategy.

Once you have made the decision to do a separate variances t-test, use Formula 9.4 to calculate the t value. Then compare this value to the critical t for $N_1 + N_2 - 2$ degrees of freedom. If you cannot beat that critical t, you will not be able to reject the null hypothesis; no further procedure is required. Any procedure for the separate-variances t is going to produce a lower df and higher critical t and make it even harder to reject the null. If you do beat the critical t for df $= N_1 + N_2 - 2$, then see if you can also beat the critical t at $N_1 - 1$ or $N_2 - 1$ (whichever df is lower). If you can beat this new critical t, you have beaten the highest critical t that can result from the separate-variances t-test, so there is no need for any additional procedure. You can reject the null hypothesis without worry.

The only difficult case arises when your calculated t falls between these two extremes; that is, when you beat the critical t at $N_1 + N_2 - 2$ df, but not at the lower of $N_1 - 1$ or $N_2 - 1$. Then you must use the Welch-Satterthwaite formula (or an equally complex alternative solution) to determine the closest df, and, therefore, the most appropriate critical t for your data. Fortunately, most computerized statistical packages provide an exact separate-variances test, along with the pooled-variances test, whenever a two-group t-test is requested, regardless of the circumstances. A test for homogeneity of variance is also provided automatically by most statistical packages. When the sample sizes are large and/or equal, the pooled- and separate-variances tests should produce very similar p values and lead to the same conclusion with respect to the null hypothesis. Otherwise, the results of the homogeneity of variance test can dictate which t-test to use and which conclusion to draw. (If the p value associated with the homogeneity of variance test is less than .05, the assumption of homogeneity of variance is usually rejected, along with the possibility of performing a pooled-variances t-test.) As with the decision to plan a one- or two-tailed test, there is no escaping the fact that some judgment is involved in deciding which homogeneity of variance test to rely on, and with which alpha, and whether that test can be trusted in your particular situation.

Heterogeneity of Variance as an Experimental Result

Besides testing an assumption of the pooled-variances t-test, the homogeneity of variance test can sometimes reveal an important experimental result. For instance, one study found that filling out a depression questionnaire by itself changed the mood of most of the subjects. Subjects who were fairly depressed to begin with were made even more depressed by reading the depressing statements in the questionnaire. On the other hand, nondepressed subjects were actually happier after completing the questionnaire because of a contrast effect: They were happy to realize that the depressing statements did not apply to them. The subjects made happier balanced out the subjects made sadder, so that the mean mood of the experimental group was virtually the same as that of a control group that filled out a neutral questionnaire. Because these two sample means were so similar, the numerator of the t-test comparing the experimental and control groups was near zero, and therefore the null hypothesis concerning the population means was *not* rejected. However, the *variance* of the experimental group increased compared to that of the control group, as a result of subjects becoming either more sad or more happy than usual. In this case, accepting the null hypothesis about the population means is appropriate, but it does not indicate that the experimental treatment *did* do *something*. However, the homogeneity of variance test can reveal that the experimental treatment had some effect worthy of further exploration. Although researchers usually hope that the homogeneity of variance test will fail to reach significance so that they can conclude that their samples *do* have homogeneity of variance and proceed with the pooled-variance t-test, the above example shows that sometimes a significant difference in variance can be an interesting and meaningful result in itself.

Exercises

Note: For purposes of comparison, I sometimes refer back to exercises from previous sections of a chapter or from earlier chapters. A shorthand notation, consisting of the chapter number and section letter followed by the problem number, will be used to refer to exercises. For example, below, exercise 9B1 refers to exercise 1 at the end of Section B of this chapter.

*1. Calculate the separate-variances t-test for exercise 9B1. Use the simple procedure suggested by Howell to determine which of the following decisions to make: retain H_0, reject H_0, or test more precisely (e.g., use the Welch-Satterthwaite formula to calculate the df).

2. a) Repeat the steps of the previous exercise with the data from exercise 9B6.

b) Repeat the steps of the previous exercise with the data from exercise 9B8.

*3. Suppose the results of a study were as follows: $\overline{X}_1 = 24$, $s_1 = 4.5$, $N_1 = 5$; $\overline{X}_2 = 30$, $s_2 = 9.4$, $N_2 = 15$.
   a) Calculate the pooled-variances t-test.
   b) Calculate the separate-variances t-test, and compare the result to your answer for part a.

4. Imagine that the standard deviations from exercise 3 are reversed, so that the data are $\overline{X}_1 = 24$, $s_1 = 9.4$, $N_1 = 5$; $\overline{X}_2 = 30$, $s_2 = 4.5$, $N_2 = 15$.
   a) Calculate the pooled-variances t-test.
   b) Calculate the separate-variances t-test, and compare the result to your answer for part a.
   c) When the larger group has the larger standard deviation, which t-test yields the larger value? When

the larger group has the smaller standard deviation, which *t*-test yields the larger value?

5. Describe as realistic a two-group study as you can think of in which the variance of the two populations could be expected to differ though the population means are virtually the same.

SUMMARY

The Important Points of Section A

1. In a two-sample experiment, the two sample means may differ in the predicted direction, but this difference may be due to chance factors involved in random sampling.

2. To find the null hypothesis distribution that corresponds to the two-sample case, we could draw two random samples (of appropriate size) from populations with the same mean and find the difference between the two sample means. If we did this many times, the differences would pile up into a normal distribution (provided that the variable being measured had a normal distribution in both populations).

3. If the null hypothesis is that the two populations have the same mean, then the null hypothesis distribution will have a mean of zero. The standard deviation for this distribution is called the standard error of the difference.

4. If both samples are large, a *z* score can be used to make a decision about the null hypothesis. The denominator of the formula for the *z* score is the separate-variances formula for the standard error of the difference.

5. If the samples are not large and the population variances can be assumed equal (i.e., you can assume homogeneity of variance), then a pooled-variance *t*-test is recommended. The standard error of the difference is based on a weighted average of the two sample variances. The critical values come from the *t* distribution with df $= N_1 + N_2 - 2$. If the population variances cannot be assumed equal, a separate-variances *t*-test should be considered (see Section C).

6. If the two sample sizes are equal, you need not worry about homogeneity of variance—a simplified formula for *t* can be applied.

7. If the two populations being sampled were not created by the experimenter, the possible conclusions are limited. Also, concluding that the population means differ significantly does not imply that the difference is large enough to be interesting or important. In some cases, a confidence interval for the difference in population means can provide helpful information.

The Important Points of Section B

To review the statistical analysis of a two-group experiment, I will describe a hypothetical study from the field of neuropsychology. A researcher believes that a region in the middle of the right hemisphere of the brain is critical for solving paper-and-pencil mazes, and that the corresponding region of the left hemisphere is not involved. She is lucky enough to find six patients with just the kind of brain damage that she thinks will disrupt maze learning. For comparison purposes, she is only able to find four patients with similar damage to the left hemisphere. Each of the ten patients is tested with the same maze. The variable measured is how many trials it takes for each patient to learn the maze perfectly (i.e., execute an errorless run). The data collected for this hypothetical experiment appear in the table below:

Left Damage	Right Damage
5	9
3	13
8	8
6	7
	11
	6
$\bar{X}_L = 22/4 = 5.5$	$\bar{X}_R = 54/6 = 9$

The results are in the predicted direction: The mean number of trials required to learn the maze was considerably higher for the group with damage to the right side of the brain. Fortunately, the amount of variability in the two groups is similar, so it is appropriate to perform a pooled-variance *t*-test. Working directly from the data in the table and using one of the formulas from Chapter 4, you could calculate SS for each group and plug these values into Formula 9.7B, as shown below.

$$t = \frac{(\bar{X}_1 - \bar{X}_2) - (\mu_1 - \mu_2)}{\sqrt{\dfrac{SS_1 + SS_2}{N_1 + N_2 - 2}\left(\dfrac{1}{N_1} + \dfrac{1}{N_2}\right)}} = \frac{(9 - 5.5) - 0}{\sqrt{\dfrac{13 + 34}{8}\left(\dfrac{1}{4} + \dfrac{1}{6}\right)}}$$

$$= \frac{3.5}{\sqrt{5.875(.4167)}} = \frac{3.5}{\sqrt{2.45}} = \frac{3.5}{1.56} = 2.24$$

Watch out for the following errors that students commonly make when using Formulas 9.7A and B:

1. Don't confuse SS with the variance, s^2; remember that SS is only the numerator when you calculate the variance. Also, don't forget that the variance in Formula 9.7A is the *unbiased* variance, calculated with $N - 1$ in the denominator.

2. If given the standard deviation, s, don't forget to square it to get the variance for Formula 9.7A.

3. After obtaining the pooled variance, remember that you must still multiply by $(1/N_1 + 1/N_2)$; and don't forget to take the square root of the denominator.

Note that I deliberately subtracted the sample means in the numerator in the order that would make the t value come out positive. This is frequently done to avoid the awkwardness of working with negative numbers. We already know from looking at the two sample means which is larger; we don't need the sign of the t value to tell us in which direction the results fell.

To find the critical t values we must first calculate the number of degrees of freedom. For this example, $df = 4 + 6 - 2 = 10 - 2 = 8$. For alpha $= .05$, two-tailed, the critical values are -2.306 and $+2.306$. Because the calculated t (2.24) is smaller than the critical t, we cannot reject the null hypothesis, and we cannot say that our results are significant at the .05 level. The statistical test in this example probably had very little power, and therefore little chance of rejecting the null even if the population means were not exactly the same. The important topic of power is discussed in the next chapter.

Finally, although you would probably not bother to calculate a confidence interval for the above experiment because of the small sample sizes and lack of statistical significance, I will find the 95% confidence interval just for review. I will use Formula 9.9 and plug in the values found earlier:

$$\mu_1 - \mu_2 = (\overline{X}_1 - \overline{X}_2) \pm t_{\text{crit}} s_{\overline{X}_1 - \overline{X}_2} = 3.5 \pm (2.306)(1.56) = 3.5 \pm 3.6$$

Note that for the 95% confidence interval the critical t values are the same ones used for the .05, two-tailed hypothesis test. Also note that the value for $s_{\overline{X}_1 - \overline{X}_2}$ in the above formula is the value of the denominator from the calculation of t above. The 95% confidence interval extends from $-.1$ to 7.1 maze-learning trials. The fact that the 95% confidence interval contains zero tells us that we would not be able to reject zero as the null hypothesis, using alpha $= .05$, two-tailed. This is consistent with the result of our hypothesis test.

The Important Points of Section C

1. If the null hypothesis ($\mu_1 = \mu_2$) is true but the population variances are *not* equal, the pooled-variances t-test may produce significant results more or less often than the alpha level used; in the worst case, there will be more Type I errors than expected. The separate-variances *t*-test is needed in this situation to keep the proportion of Type I errors as close as possible to the alpha being used.

2. There is no need to perform the separate-variances *t*-test if (1) the sample sizes are equal; (2) both samples are very large; or (3) the two sample variances are very similar (one is less than four times the other). If, however, the two sample sizes are fairly small and unequal and the two sample variances are quite dissimilar, a separate-variances *t*-test should be performed.

3. A test for homogeneity of variance can be performed to determine whether it is valid to pool the variances. If the test is significant (i.e., $p < .05$), the equality of the population variances cannot be assumed, and a separate-variances *t*-test should be performed. However, note that there is more than one legitimate test for homogeneity of variance, and there is some debate about the validity of these tests with small samples (which is when they are needed most).

4. Calculating the separate-variances *t* value is easy, but finding the appropriate critical *t* value to compare it to is not. The problem can be framed as finding the appropriate df for determining the critical *t*. Most computer packages find the df automatically whenever a two-group *t*-test is conducted.

5. The text presented a simplified solution for determining the significance of the separate-variances *t*-test that should work for most cases. It is based on determining the smallest and largest df that may be associated with this test, and making the appropriate comparisons.

6. In certain experimental situations, differences in variance may be more dramatic or interesting than the difference in the means. For such situations, the homogeneity of variance test may have important practical or theoretical implications.

Definitions of Key Terms

Standard error of the difference This phrase is short for "standard error of the difference between means." It is the value you would get if you kept selecting two random samples at a time (from populations with equal means), finding the difference between the means of the two samples, and after piling up a huge (preferably infinite) number of these difference scores, calculated the standard deviation.

Large-sample test for two independent means This test is valid only when the two samples are quite large (30 or 40 subjects per group, at the very least). The two sample variances are not pooled. The critical values are found from the standard normal distribution.

Pooled-variances *t*-test A *t*-test based on the use of the pooled variance to estimate the population variance, which, strictly speaking, is only valid when the two populations have the same variance.

Pooled variance This phrase is short for "pooled-variance estimate of the population variance." It is a weighted average of the two sample variances, which produces the best estimate of the population variance when the two populations being sampled have the same variance.

Homogeneity of variance The assumption that two populations have the same variance, which underlies the use of the pooled-variances *t*-test.

Heterogeneity of variance The assumption that two populations do *not* have the same variance, which we conclude if a test for homogeneity of variance is statistically significant.

Separate-variances *t*-test This *t*-test is appropriate when the samples are small, of unequal size, and it cannot be assumed that the two populations have the same variance. The sample variances are not pooled for this test, but rather each is separately divided by its own sample size. The critical values must be found through special procedures.

Key Formulas

The *z*-test for two samples (use only when you know both of the population variances or when the samples are sufficiently large):

$$z = \frac{(\overline{X}_1 - \overline{X}_2) - (\mu_1 - \mu_2)}{\sqrt{\sigma_1^2/N_1 + \sigma_2^2/N_2}}$$ Formula 9.3

The separate-variances *t*-test (may *not* follow a *t* distribution with df $= N_1 + N_2 - 2$; see Section C):

$$t = \frac{(\overline{X}_1 - \overline{X}_2) - (\mu_1 - \mu_2)}{\sqrt{s_1^2/N_1 + s_2^2/N_2}}$$ Formula 9.4

Pooled-variances *t*-test (pooled-variance estimate has already been calculated):

$$t = \frac{(\overline{X}_1 - \overline{X}_2) - (\mu_1 - \mu_2)}{\sqrt{s_p^2(1/N_1 + 1/N_2)}}$$ Formula 9.5B

Pooled-variance estimate of the standard error of the difference (when the variances of both samples have already been calculated):

$$s_p^2 = \frac{(N_1 - 1)s_1^2 + (N_2 - 1)s_2^2}{N_1 + N_2 - 2}$$ Formula 9.6A

Pooled-variances *t*-test (when the variances of both samples have already been calculated):

$$t = \frac{(\overline{X}_1 - \overline{X}_2) - (\mu_1 - \mu_2)}{\sqrt{\dfrac{(N_1 - 1)s_1^2 + (N_2 - 1)s_2^2}{N_1 + N_2 - 2}\left(\dfrac{1}{N_1} + \dfrac{1}{N_2}\right)}} \qquad \text{Formula 9.7A}$$

Pooled-variances *t*-test (raw-score version; SS has already been calculated for each sample):

$$t = \frac{(\overline{X}_1 - \overline{X}_2) - (\mu_1 - \mu_2)}{\sqrt{\dfrac{SS_1 + SS_2}{N_1 + N_2 - 2}\left(\dfrac{1}{N_1} + \dfrac{1}{N_2}\right)}} \qquad \text{Formula 9.7B}$$

The *t*-test for equal-sized samples (note that *N* in the formula is the number of subjects in *each* group, and not the total in both groups combined):

$$t = \frac{(\overline{X}_1 - \overline{X}_2) - (\mu_1 - \mu_2)}{\sqrt{(s_1^2 + s_2^2)/N}} \qquad \text{Formula 9.8}$$

The confidence interval for the difference of two population means ($s_{\overline{X}_1 - \overline{X}_2}$ has already been calculated using the denominator of Formula 9.4, 9.7A, 9.7B, or 9.8, as appropriate):

$$\mu_1 - \mu_2 = (\overline{X}_1 - \overline{X}_2) \pm t_{\text{crit}} s_{\overline{X}_1 - \overline{X}_2} \qquad \text{Formula 9.9}$$

STATISTICAL POWER

You will need to use the following from previous chapters:

Symbols
μ: Mean of a population
σ: Standard deviation of a population

Formulas
Formula 9.8: *t*-Test for two equal-sized groups

Concepts
The null hypothesis distribution

Procedures
Finding areas under the normal curve

10

C H A P T E R

A

**CONCEPTUAL
FOUNDATION**

In this chapter I will discuss how to estimate the probability of making a Type II error, and thus the probability of *not* making a Type II error, which is called the *power* of a statistical test. I will also introduce the concept of effect size, and show how effect size, sample size, and alpha combine to affect power. Although the usual purpose of power analysis is to plan an experiment, a thorough understanding of this topic will enable you to be much more astute as an interpreter of experimental results already obtained. Most importantly, the discussion of power and effect size makes the difference between statistical significance and practical importance clear. It will be easiest to introduce power in the context of the two-sample *t*-test described in the previous chapter; the topic of power will be revisited in subsequent chapters.

The Alternative Hypothesis Distribution

In the previous three chapters we looked carefully at the null hypothesis distribution (NHD), which helped clarify how Type I errors are controlled. However, Type II errors cannot occur when the null hypothesis is true; a Type II error occurs only when the null hypothesis is *not* true but it is accepted anyway. (Take another look at Table 7.1 on page 200.) To understand the factors that determine the rate of Type II errors, you need a picture of what can happen when the null hypothesis (H_0) is false and the alternative hypothesis (H_A) is true. Unfortunately, this approach is a bit more complicated than it sounds. The null hypothesis distribution was fairly easy to find because the null hypothesis is stated specifically (e.g., $\mu_0 = 65$ inches, or $\mu_1 = \mu_2$). If we try to find the **alternative hypothesis distribution (AHD),** the problem we encounter is that the alternative hypothesis is usually stated as the complement of the null hypothesis and is therefore *not* specific (e.g., $\mu_0 \neq 65$ inches, or $\mu_1 \neq \mu_2$). If the H_A states that the population mean is *not* 65 inches, there are many possibilities left to choose from (e.g., $\mu_0 = 66$ inches, $\mu_0 = 60$ inches, etc.). In order to introduce the study of Type II errors, and therefore the topic of power analysis, I will need to begin with an alternative hypothesis that is stated specifically.

Suppose that a Martian scientist has been studying the earth for some time and has noticed that adult humans come in two varieties: men and women. The Martian has also noticed that the women seem to be shorter than the men, but being a scientist, the Martian does not trust its (I don't know what gender to use to refer to Martians) own judgment. So the Martian decides to perform an experiment. The Martian's null hypothesis is that the mean height for all men is the same as the mean height for all women. But we on earth know that the null hypothesis is not true, and moreover, we know which of the possible alternatives *is* true. For the sake of simplicity I will assume that the mean height for men is exactly 69 inches, and for women it is 65 inches, with a standard deviation of 3 inches for both genders.

Because in this case the H_0 is not true, the Martian cannot make a Type I error; it will either reject H_0 correctly or retain H_0 and make a Type II error. To find the probability of committing a Type II error, we need to draw the alternative hypothesis distribution. Suppose that the Martian plans to select at random four men and four women and then perform a *t*-test for two sample means. It would be logical to assume that the AHD will be a *t* distribution with $4 + 4 - 2 = 6$ degrees of freedom, just as the NHD would be. However, the difference is that the NHD would be centered on zero, which is the *t* value that Dr. Null expects to get on the average, but the AHD is centered on some value other than zero. A *t* distribution that is not centered on zero is called a *noncentral t distribution*. This noncentral *t* distribution is rather complicated, so to simplify matters AHD is often assumed to be a normal distribution. This assumption involves some error, especially for small sample sizes, but the concept is basically the same whether we use a normal distribution or a *t* distribution—and as you will see, power analysis is a matter of approximation anyway.

In order to find the value upon which the AHD is centered, we need to know the *t* value that the Martian would get, on the average, for its experiment. (I will continue to refer to *t* instead of *z*, even though I will use the normal distribution to simplify the calculations.) It will be helpful to look at the formula for a two-sample test when the sample sizes are equal (Formula 9.8):

$$ t = \frac{(\overline{X}_1 - \overline{X}_2)}{\sqrt{(s_1^2 + s_2^2)/N}} $$

To get the average *t* value, we replace each variable in the above formula with its average value; for the present example, the average for the group of men will be 69 inches, for the women 65 inches, and the two variances would each average $3^2 = 9$.

$$ \text{Average } t = \frac{69 - 65}{\sqrt{(9 + 9)/4}} = \frac{4}{\sqrt{4.5}} = \frac{4}{2.12} = 1.89 $$

Figure 10.1

Alternative Hypothesis Distribution (alpha = .05)

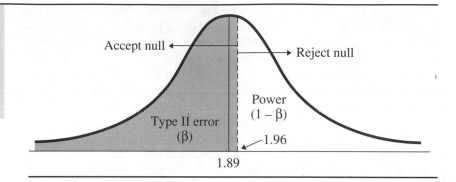

Accept null ← → Reject null

Type II error (β)

Power (1 − β)

1.96

1.89

From the above calculation you can see that, on average, the Martian will get a *t* value of 1.89. However, if we assume that the AHD is normal, the critical value needed for significance (alpha = .05, two-tailed) is 1.96, so most of the time the Martian will not be able to reject the null hypothesis, and therefore it will commit a Type II error. From Figure 10.1 it appears that the Martian will make a Type II error (falsely concluding no height difference between men and women) about 53% of the time, and correctly reject the null hypothesis about 47% of the time. The proportion of the AHD that results in Type II errors (i.e., the Type II error rate) is symbolized by the Greek letter **beta (β);** in this case β equals about .53. The proportion of the AHD that results in rejecting the null hypothesis is 1 − β (as shown in Table 7.1), and it is called the **power** of the test; in this case, power is about 1 − .53 = .47.

It is important to realize that even though there is a considerable height difference (on average) between men and women, the Martian's experiment most often will not provide sufficient evidence for it. This is the price of null hypothesis testing. The critical *t* in this case is determined by the need to prevent 95% of "null experiments" from being significant. But that also means that many experiments for which the null is *not* true (like the Martian's experiment) will nonetheless fail to beat the critical *t* and be judged not significant. In fact, the power of the Martian's experiment is so low (because of the small sample sizes) that it hardly pays to perform such an experiment. One simple way to increase the power would be to increase the sample sizes, as will be explained below. Another way to increase power is to raise alpha from .05 to some larger value such as .1. This would change the critical value to 1.645. As you can see from Figure 10.2 (on page 297), power is now close to 60%, and the proportion of Type II errors has been reduced accordingly. However, this increase in power comes at the price of increasing Type I errors from 5% to 10%, which is usually considered too high a price.

The Expected *t* Value (Delta)

To find the percentage of Figure 10.1 that was below the critical value and the percentage above, I made a rough approximation based on visual inspection of the distribution. Section B will present a method involving the use of tables to

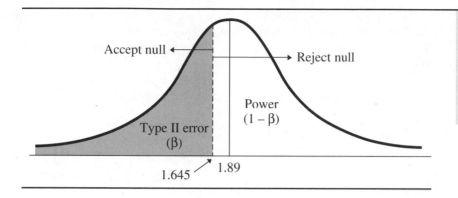

Figure 10.2

Alternative Hypothesis Distribution (alpha = .10)

determine the percentages more accurately. For now, the concept of power needs further explanation. Thus far, power has been described as dependent on the average t value for some specific AHD. This average t is generally symbolized by the Greek letter **delta (δ).** The delta corresponding to the null hypothesis equals zero (when H_0 is $\mu_1 = \mu_2$). If delta happens to just equal the critical value needed to attain statistical significance, then power (as well as β) will equal 50%; half the time the t value will be above its average and therefore significant, and half the time below.

Delta can also be thought of as the "expected" t value that corresponds to a particular AHD. The expected t can be calculated for any specific alternative hypothesis (provided that the population standard deviation is also specified); if that alternative hypothesis is true, the actual t values from experiments would fluctuate around, but would average out to, the expected t. Based on the formula for a two-sample t-test with equal-sized groups (Formula 9.8), a general formula can be derived for calculating delta. First, we replace each term in Formula 9.8 by its expected or average value, which is the value for that statistic in the whole population. If we make the common assumption that $\sigma_1^2 = \sigma_2^2$ (homogeneity of variance), the result is as follows:

$$\delta = \frac{\mu_1 - \mu_2}{\sqrt{(\sigma_1^2 + \sigma_2^2)/N}} = \frac{\mu_1 - \mu_2}{\sqrt{2\sigma^2/N}}$$

Next, it will be useful to use the laws of algebra to rearrange this formula a bit. The denominator can be separated into two square roots, as follows:

$$\delta = \frac{\mu_1 - \mu_2}{\sqrt{2/N}\sqrt{\sigma^2}} = \frac{\mu_1 - \mu_2}{\sqrt{2/N}\,\sigma}$$

Finally, the remaining square root can be moved to the numerator, which causes it to be turned upside down, yielding Formula 10.1:

$$\delta = \frac{\sqrt{N/2}\,(\mu_1 - \mu_2)}{\sigma} = \sqrt{\frac{N}{2}}\frac{(\mu_1 - \mu_2)}{\sigma} \qquad \text{Formula 10.1}$$

Formula 10.1 shows that delta can be conceptualized as the product of two easily understood terms. The first term depends only on the size of the samples. The second term is the separation of the two population means, in terms of standard deviations; it is like the *z* score of one population mean with respect to the other. This term is called the **effect size** and is sometimes symbolized by the Greek letter gamma (γ). The formula for gamma will be designated Formula 10.2:

$$\gamma = \frac{\mu_1 - \mu_2}{\sigma} \qquad \text{Formula 10.2}$$

Expressing Formula 10.1 in terms of gamma gives Formula 10.3:

$$\delta = \sqrt{\frac{N}{2}}\,\gamma \qquad \text{Formula 10.3}$$

The separation of the expected *t* value into the effect size and a term that depends only on sample size was pioneered by Dr. Jacob Cohen (1988) and is both useful and instructive. However, the concept of effect size requires some explanation.

The Effect Size

The term *effect size* suggests that the difference in two populations is the effect of something; for instance, the height difference between males and females can be thought of as just one of the effects of gender. The effect size for the male-female height difference can be found by plugging the appropriate values into Formula 10.2:

$$\gamma = \frac{\mu_1 - \mu_2}{\sigma} = \frac{69 - 65}{3} = \frac{4}{3} = 1.33$$

An effect size of 1.33 is considered quite large, for reasons soon to be made clear. One way to get a feeling for the concept of effect size is to think of effect size in terms of the amount of overlap between two population distributions. Figure 10.3 depicts four pairs of overlapping population distributions, each pair corresponding to a different effect size. Notice that when gamma = .2 there is a great deal of overlap. If, for instance, the two populations are subjects who were instructed to use imagery for memorization and subjects who received no such instruction, and the variable being measured is amount of recall, then when gamma = .2, many subjects who received no instructions recalled more than subjects who *were* instructed. On the other hand, when gamma = 4.0 there is

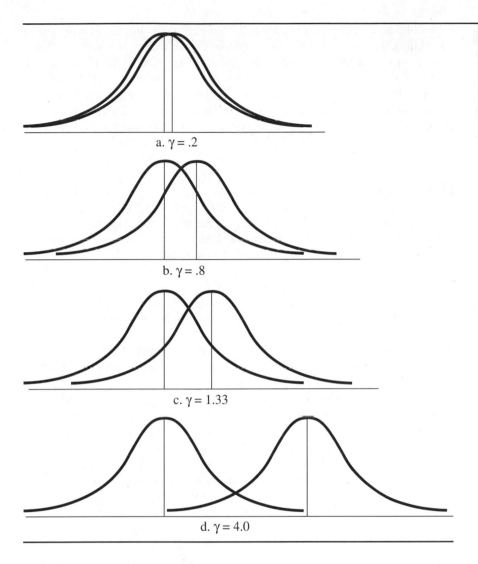

a. γ = .2

b. γ = .8

c. γ = 1.33

d. γ = 4.0

Figure 10.3

*Overlap of
Populations As a
Function of Effect
Size*

very little overlap. In this case, very few of the subjects who received no instructions performed better than even the worst of the subjects who did receive instructions. It is rare to find an experiment in which gamma = 4.0, because the effect being measured would be so obvious it would be noticed without the need for an experiment. (For example, in comparing the reading skills of first-graders and fourth-graders the effect of age on reading ability is clear.) In fact, even gamma = 1.33 is generally too large to require an experiment. Although there is a fair amount of overlap between the heights of men and women, the difference is quite noticeable. An experiment in which gamma = 1.33 would most likely be performed only in a situation in which there had been little opportunity to observe the effect (such as a test of a new and effective form of therapy).

According to guidelines suggested by Cohen (1988), gamma = .8 represents a large effect size: not so large as to be obvious from casual observation, but large enough to have a good chance of being found statistically significant with a modest number of subjects. By contrast, gamma = .2 is considered a small effect; effect sizes much smaller would usually not be worth the trouble of investigating. The difference in overlap between gamma = .2 and gamma = .8 can be seen in Figure 10.3. Finally, gamma − .5 is considered to be a medium effect size.

Power Analysis

Power analysis is the technique of predicting the power of a statistical test before an experiment is run. If the predicted power is too low, it would not be worthwhile to conduct the experiment, unless some change could be made to increase the power. How low is too low? Most researchers would agree that .5 is too low; before investing the time, money, and effort required to perform an experiment one would prefer to have better than a 50% chance of attaining significant results. A power of .7, which corresponds to a Type II error rate of .3, is often considered minimal. On the other hand, it is not generally considered important to keep the Type II error rate as low as the Type I error rate. Keeping the Type II error rate as low as the usual .05 Type I error rate would result in a power of .95, which is generally considered higher than necessary. A power of about .8 is probably the most reasonable compromise, in that it yields a high enough chance of success to be worth the effort without the drawbacks involved in raising power any higher. (The problems associated with increasing power will be discussed shortly.)

Because the experimenter can choose the alpha and sample size to be used, power analysis boils down to predicting the effect size of the intended experiment. You cannot always predict effect size, but with power analysis you can find the power associated with each possible effect size, for a given combination of alpha and sample size. This kind of analysis can help the experimenter decide whether to run the experiment as planned, try to increase the power, or abandon the whole enterprise. Before we consider the difficult task of predicting effect sizes, let us look further at how sample size combines with effect size to determine power.

Once alpha and the number of tails are set, power depends only on delta, the expected *t* value. Looking at Formula 10.3, you see that delta is determined by multiplying the effect size by a term that depends on sample size. This means that power can be made larger either by making the effect size larger or by making the sample size larger. Because it is often not possible to influence effect size (e.g., the gender difference in height), power is usually manipulated by changing the sample size. Theoretically, power can always be made as high as desired, *regardless of the effect size,* by sufficiently increasing the sample size (unless the effect size is exactly zero). This fact has very important implications; probably the most important involves the interpretation of statistical results.

The Interpretation of *t* Values

Suppose that a particular two-sample experiment produces a very large *t* value, such as 17. The *p* value associated with that *t* would be extremely low; the results would be significant even with alpha as low as .0001. What does this imply about effect size? Actually, it tells us nothing about effect size. The large *t* *does* tell us something about delta, however. It is extremely unlikely that delta could be zero, as the null hypothesis would imply; that is why we can reject the null hypothesis with such a high degree of confidence. In fact, delta is likely to be somewhere near 17; if the expected *t* were as low as, say, 5 or as high as about 30, there would be little chance of obtaining a *t* of 17. So, although there is always some chance that the obtained *t* of 17 is a fluke that is either much higher or lower than delta, the chances are that delta is indeed in the vicinity of 17. The reason this tells us nothing about the probable effect size is that we have not yet considered the sample size involved. It is possible for delta to be 17 when the effect size is only .1 (in which case there would have to be about 58,000 subjects in each group!) or when the effect size is about 10 (in which case there would need to be only about 6 in each group).

What does a very large *t* value indicate? The *t* value is a test of the null hypothesis. The usual H_0 is $\mu_1 - \mu_2 = 0$, which implies that the effect size also equals zero [$\gamma = (\mu_1 - \mu_2)/\sigma = 0/\sigma = 0$)]. A very large *t* value leads to a very small *p* value, which means that it would be very unlikely to get such results if the null were true—that is, if the effect size were zero. Therefore, when a very large *t* is obtained we can feel very sure that the effect size is not zero. (And unless sample sizes are extremely small, even *t* = 5 can be considered very large.) However, no matter how certain we are that the effect size is not zero, this does not imply that the effect size must be fairly large. As discussed above, even a very tiny effect size can lead to a large expected *t* if very large samples are used. And it is important to remember that even if the *p* value is very small (e.g., *p* < .0001), statistical significance does not imply that the effect size is large enough to be interesting or of any practical importance. Of course, if the effect size is really zero, it doesn't matter how large the samples are: the expected *t* will still be zero (zero times any number equals zero).

On the other hand, a small obtained *t* value does not allow the rejection of the null hypothesis; an effect size of zero cannot be ruled out and must be considered a reasonable possibility. Again, nothing definitive can be said about the effect size simply because the *t* value is small, but a small obtained *t* implies that delta is probably small. If a small *t* is obtained using large sample sizes, a smaller effect size is implied than if the same *t* value were obtained with small samples. This principle has implications for the interpretation of negative (i.e., not significant) results. When testing a new drug, therapy, or any experimental manipulation, negative results suggest that the treatment is not at all effective. However, a nonsignificant result could actually be a Type II error. The smaller the samples being used, the lower the power, and the more likely is a Type II error. Negative results based on small samples are less conclusive and less trustworthy than negative results based on large samples.

Estimating Gamma

I began this chapter with an example for which gamma is well known (the male-female height difference), so power could be estimated easily. Of course, if you actually knew the value of gamma for a particular experimental situation, there would be no reason to conduct an experiment. In fact, most experiments are conducted in order to decide whether gamma may be zero (i.e., whether the null hypothesis is true). Usually the experimenter believes gamma is not zero and has a prediction regarding the direction of the effect but no exact prediction about the size of gamma. However, in order to estimate power and determine whether the experiment has a good chance of producing positive (i.e., significant) results, we must make some guess about gamma. One way to estimate gamma is to use previous research or theoretical models to arrive at estimates of the two population means and the population standard deviation that you are dealing with, and then calculate an estimated gamma according to Formula 10.2. The estimate of power so derived would only be true to the extent that the estimates of the population parameters were reasonable.

It can be especially difficult to estimate population parameters if you are dealing with newly created measures. For instance, you might want to compare memory during a happy mood with memory during a sad mood, measuring memory in terms of the number of words recalled from a list devised for the experiment. As this word list may never have been used before, there would be no way to predict the number of words that would be recalled in each condition. However, even when a previous study cannot help you estimate particular population parameters, the sample statistics from a previous study can be combined to provide an overall estimate of gamma. A simple estimate of gamma that could be derived from a previous two-group study is based on the difference of the two sample means divided by the square root of the pooled variance (s_p), as shown in Formula 10.4:

$$g = \frac{\overline{X}_1 - \overline{X}_2}{s_p}$$
                                                        Formula 10.4

Note the similarity to Formula 10.2; each population parameter from Formula 10.2 has been replaced by a corresponding sample statistic. My use of the letter *g* to represent a point estimate of γ is consistent with Hedges (1981) and is in keeping with the usual convention of using English letters for sample statistics (and their combinations), and corresponding Greek letters for characteristics of populations. The value of *g* from a similar study performed previously can then be used to estimate your value of gamma in your proposed study. [As calculated above, *g* is a somewhat biased estimator of gamma, but because the correction factor is quite complicated and the bias becomes very slight for large samples (Hedges, 1981), the value you obtain from Formula 10.4 will suffice as an estimate of gamma.]

Of course, there may not be any previous studies similar enough to provide an estimate of gamma for your proposed study. If you must take a guess at the value of gamma, you can use the guidelines established by Cohen (1988), in which .2, .5 and .8 represent small, medium, and large effect sizes, respectively. [Note that Cohen (1988) uses the letter d to represent the population effect size that I am calling gamma, d_s to represent what I am calling g, and g to represent an effect size measure for use with proportions.] To make such an estimate, you would have to know the subject area well, in terms of the variables being measured, experimental conditions involved, etc.

Manipulating Power

The most common way of manipulating power is the regulation of sample size. Though power can be altered by changing the alpha level, this approach is not common because of widespread concern about keeping Type I errors to a fixed, low level. A third way to manipulate power is to change the effect size. This last possibility may seem absurd at first. If we consider the male-female height example, it *is* absurd; we cannot change the difference between these populations. However, if the two populations represent the results of some experimental effect, there is the possibility of manipulation. For example, if a drug raises the heart rate, on average, by 5 beats per minute (bpm), and the standard deviation is 10 bpm, the effect size will be 5/10 = .5. It is possible that a larger dose of the drug could raise the heart rate even more, without much change in the standard deviation; in that case, the larger dose would be associated with a larger effect size. Other treatments or therapies could be intensified in one way or another to increase the relative separation of the population means, and thus increase the effect size. Of course, it is not always possible to intensify an experimental manipulation, and sometimes doing so could be unpleasant, if not actually dangerous, to the subjects. From a practical standpoint it is often desirable to keep the effect in the experiment at a level that is normally encountered in the real world (or would be if the new treatment were to be adopted).

There is another potential way to increase the effect size without changing the difference between the population means: the standard deviation (σ) could be decreased. (Because σ is in the denominator of the formula for gamma, lowering σ will increase gamma.) Normally, researchers try to keep σ as low as possible by maintaining the same experimental conditions for each subject. But even if very little random error is introduced by the experiment, there will always be the individual differences of subjects contributing to σ. For some experiments, the subject-to-subject variability can be quite high, making it difficult to have sufficient power without using a lot of subjects. A very useful way to avoid much of the subject-to-subject variability in any experiment is to measure each subject in more than one condition, or to match similar subjects and then place them in different experimental groups. This method can greatly increase the power without intensifying the experimental manipulation or increasing the sample size. However, techniques for matching subjects will be better understood after correlation is explained, so this topic will be delayed until Chapter 13.

Exercises

1. What is the Type II error rate (β) when power is a) .65? b) .96? What is the power when β is c) .12? d) .45?

* 2. Suppose that the mean heart rate of all pregnant women (μ_P) is 75 bpm, and the mean heart rate for nonpregnant women (μ_N) is 72. If the standard deviation for both groups is 10 bpm, what is the effect size (γ)?

3. If the mean verbal SAT score is 510 for women and 490 for men, what is the effect size (γ)?

* 4. In exercise 2, if a *t*-test were performed to compare 28 pregnant women with an equal number of nonpregnant women, what would be the expected *t* value (δ)? If a two-tailed test were planned with alpha = .05, would the results come out statistically significant more than half the time, or not?

5. Suppose the experiment in exercise 4 above were performed with four times as many subjects in each group. What would be the new expected *t*? How does this compare with the answer you found in exercise 2? Can you state the general rule that applies?

* 6. If two population means differ by one and a half standard deviations, what is the value of γ? If a *t*-test is performed using 20 subjects from each population, what will delta be?

7. If two population means differ by three standard deviations, and a *t*-test is performed with 20 subjects in each group, what will the delta be? Compare this value to your answer for the previous exercise. What general rule is being demonstrated?

* 8. Calculate *g,* the sample estimate of effect size, a) for exercise 9B1; b) for exercise 9B6.

9. The *t* value calculated for a particular two-group experiment turned out to be −23. Which of the following can you conclude? Explain your choice.
   a) A calculation error must have been made.
   b) The number of subjects must have been large.
   c) The effect size must have been large.
   d) The delta was probably large.
   e) The alpha level was probably large.

* 10. Suppose you are in a situation in which it is more important to reduce Type II errors than to worry about Type I errors. Which of the following could be helpful in reducing Type II errors? Explain your choice.
   a) Make alpha unusually large (e.g., .1).
   b) Use a larger number of subjects.
   c) Try to increase the effect size.
   d) All of the above.
   e) None of the above.

BASIC STATISTICAL PROCEDURES

In Section A, I mentioned that the alternative hypothesis distribution (AHD) for a *t*-test involving two small groups is actually a *noncentral t distribution*. The value that it is centered on, delta, is therefore called the *noncentrality parameter*. To find power accurately, we need to find the proportion of the appropriate noncentral *t* distribution that is in the rejection zone of the test. Unfortunately, finding the exact proportions requires tables for the various noncentral *t* distributions.

Using Power Tables

Because delta can take on a large range of possible values, you need a whole series of noncentral *t* tables, as presented by Cohen (1988). As an alternative, you might consult a graph containing a series of "power curves" that depict power as a function of effect size for different possible sample sizes. A simpler (albeit less accurate) alternative, and the method used in Section A, is to use the normal distribution as an approximation, regardless of sample size. This approach serves the purpose of teaching the concepts of power, and fortunately requires only one simple table, which has been included in Appendix A as Table A.3. (Also, the reverse of Table A.3 has been included as Table A.4.) The following examples will demonstrate how the tables work.

Assume that the delta for a particular experiment is 3.0. The possible experimental outcomes can be approximated as a normal distribution centered on the value 3.0, as shown in Figure 10.4. The critical value for alpha = .05, two-tailed is 1.96, so you'll notice a vertical line at this value. The area to the left of the critical value contains those experimental results for which the null hypothesis must be accepted (i.e., Type II errors). Because the critical value is about one standard deviation below the mean (3.0 − 1.96 = 1.04), the area to the left of the critical value is about 16% (i.e., the area beyond $z = 1.0$). Therefore, for this problem, the Type II error rate (β) = .16, and power = 1 − .16 = .84 (see Figure 10.4). Table A.3 can be used to obtain the same result. Look down the column for the .05, two-tailed test until you get to $\delta = 3.0$. The entry for power is .85, which is close to the value of .84 approximated above (1.96 is a little *more* than one standard deviation below the mean, leaving a bit more area to the right). Also notice that the power associated with $\delta = 2.0$ is just slightly more than .50 (because 2.0 is slightly greater than 1.96). The table makes the determination of power easier, because the subtraction of the critical value from delta and the calculation of the area have been done for you.

Table A.3 can also be used to find the delta that is required to yield a particular level of power. If, for instance you'd like a power of .80 at alpha = .05,

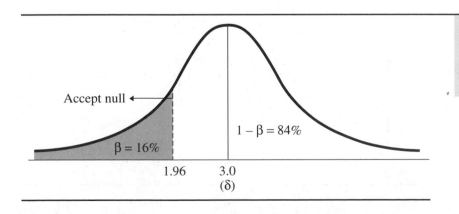

Figure 10.4

Power for delta = 3.0

two-tailed, you could look down the appropriate column within the table to find that level of power (or the closest value in the table), and then look across for the corresponding delta; in this case, the exact value of .80 happens to appear in the table, corresponding to delta = 2.8. Table A.4 makes this task easier, by displaying the values of delta that correspond to the most convenient levels of power. Notice that next to power = .8 under the appropriate column heading is, of course, delta = 2.8. Also notice that if you read across from power = .50, the delta in each column is the corresponding critical *z*. This again shows that when the critical value is the expected value, half the time you will obtain significant results (power = .5) and half the time you will not (β = .5).

The Relationship Between Alpha and Power

Looking again at Table A.3, you can see the effect of changing alpha level. Pick any value for delta, and the row of power values corresponding to that delta will show how power changes as a function of alpha. (For now we will consider only the two-tailed values.) If delta = 2.6, for example, power is a rather high .83 for alpha = .1. However, if alpha is lowered to the more conventional .05 level, power is lowered as well, to .74. And if alpha is lowered further, to .01, power is reduced to only .5. Figure 10.5 shows graphically that as alpha is decreased, the critical value moves further to the right on the alternative hypothesis distribution, so that there is less area above (i.e., to the right of) the critical value, representing fewer statistically significant results (i.e., lower power).

Considering the greater power associated with alpha = .05, one-tailed, as compared to alpha = .05, two-tailed, the advantage of the one-tailed test should be obvious. However, the one-tailed test, as discussed earlier, involves a "promise" on the part of the researcher not to test results in the direction opposite what is predicted. Whether such a "promise" is considered appropriate or acceptable depends on circumstances and conventions (see Chapter 7). Note also that if alpha is reduced in order to be more conservative (i.e., more cautious about Type

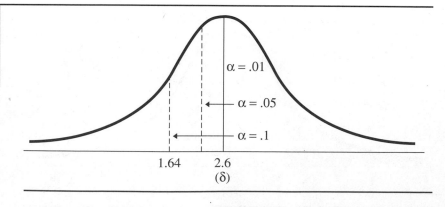

Figure 10.5

Power As a Function of Alpha

I errors), power will be reduced as well. On the other hand, in situations in which a Type II error may have serious consequences, and there is less than the usual concern about Type I errors, alpha can be increased in order to raise power. Such situations, however, are quite rare. Typically, researchers try to manipulate power by manipulating factors that affect delta, most commonly the sample sizes. However, in some situations, the researcher may have little control over the sizes of his or her samples.

Power Analysis with Fixed Sample Sizes

Let us return to the example of the Martian scientist trying to prove that men and women differ in height. If the Martian were to compare eight randomly selected men with eight randomly selected women, and it knew that gamma = 1.33, it could calculate delta using Formula 10.3:

$$\delta = \sqrt{\frac{N}{2}}\, \gamma = \sqrt{\frac{8}{2}}\, 1.33 = \sqrt{4}\, (1.33) = (2)(1.33) = 2.66$$

The Martian rounds off the calculated delta, 2.66, to 2.7 in order to use Table A.3. For alpha = .05, two-tailed, the power is .77.

Unequal Sample Sizes

Although researchers try to ensure that both groups in a two-group experiment will be the same size (to alleviate any worry about homogeneity of variance), there are occasions when this isn't possible. Suppose a researcher has available, for example, ten patients with Alzheimer's disease, but can find only five patients with Korsakoff syndrome for comparison. Eliminating five of the Alzheimer's patients in order to have equal-sized groups would lower the power considerably; the advantage of not having to worry about homogeneity of variance would not be worth the loss of power. However, if you wanted to calculate the power for an experiment in which $N_1 = 10$ and $N_2 = 5$, you would soon notice that Formula 10.3 for delta contains only N, which assumes that the two groups are the same size. The obvious solution would be to take the ordinary (i.e., arithmetic) mean of the sample sizes [in this case, $(10 + 5)/2 = 15/2 = 7.5$] and use that mean in place of N in the formula. As it turns out, there is a more accurate procedure, but as this procedure involves an additional formula and is not often needed, discussion of the case of unequal Ns will be delayed until Section C.

Finding Gamma for a Desired Level of Power

For a situation in which the sample sizes are to some extent fixed by circumstances, the researcher may find it useful to estimate the power for a range of effect sizes, in order to determine the smallest effect size for which the power is reasonable (i.e., worth performing the experiment). If the smallest effect size that yields adequate power is considerably larger than could be optimistically expected, the researcher would probably decide not to conduct the experiment as planned.

Instead of finding power for a range of effect sizes, we can find the smallest effect size that yields adequate power by solving Formula 10.3 for gamma to create Formula 10.5:

$$\delta = \sqrt{\frac{N}{2}}\,\gamma$$

$$\gamma = \frac{\delta}{\sqrt{N/2}}$$

$$\gamma = \sqrt{\frac{2}{N}}\,\delta \qquad\qquad \text{Formula 10.5}$$

Find delta by looking in Table A.4 under the lowest value of power considered worthwhile (this is a matter of judgment, and depends on the cost of doing the experiment, etc.). Then use the value of delta found in the table along with the fixed sample sizes in Formula 10.5. For instance, suppose .6 is the lowest power you are willing to consider. This corresponds to $\delta = 2.21$ for alpha $= .05$, two-tailed. If you are stuck with only ten subjects in each group, then $N = 10$. Plugging these values into Formula 10.5, we get

$$\gamma = \sqrt{\frac{2}{10}}\,(2.21) = (.447)(2.21) = .99$$

An effect size of .99 is needed to attain power $= .6$. Any smaller effect size would yield less than adequate power in this example. An effect size of .99 is fairly large, and if such a large effect size cannot be expected for this hypothetical experiment, then the experiment is not worth doing.

Sample Size Determination

Usually a researcher has a fair amount of flexibility in determining the size of the groups to be used in an experiment. However, there is invariably some cost per subject that must be considered. The cost can be quite high if the subject is being tested with a sophisticated medical procedure (e.g., PET scan of blood flow in the brain), or quite low if the subject is filling out a brief questionnaire in a psychology class. But there is always some cost in time and effort to run the subject, analyze the data, etc. Even if the subjects are volunteers, remember that the supply is not endless, and often the larger the number of subjects used for one experiment, the fewer that are available for future experiments. With this in mind, most researchers try to use the smallest number of subjects necessary to have a reasonable chance of obtaining significant results with a meaningful effect size. Power analysis can aid this endeavor.

First, you must decide on the lowest acceptable level for power. As mentioned before, this is based on each researcher's own judgment. However, power = .8 is often considered desirable, so this value will be used for our example. Deciding on power, as well as alpha, allows us to find delta from Table A.4. Next, an estimate for gamma must be provided. For our first example, we will use the male-female height difference, for which gamma = 1.33. The values for delta and gamma can be inserted into Formula 10.3 to find the required sample size, but it is a bit more convenient to solve Formula 10.3 for N to create Formula 10.6:

$$\delta = \sqrt{\frac{N}{2}}\,\gamma$$

$$\sqrt{\frac{N}{2}} = \frac{\delta}{\gamma}$$

$$\frac{N}{2} = \left(\frac{\delta}{\gamma}\right)^2$$

$$N = 2\left(\frac{\delta}{\gamma}\right)^2 \qquad\qquad \text{Formula 10.6}$$

Inserting the values for gamma and delta into Formula 10.6, we get

$$N = 2\left(\frac{2.8}{1.33}\right)^2 = 2(2.1)^2 = (2)(4.41) = 8.82$$

The above calculation indicates that nine subjects are needed in each of the two groups in order to have acceptable power. The required sample size in this example is unusually low, because gamma is unusually high.

Determining the sample size is very straightforward if gamma can be estimated. However, the above procedure can still be useful even if no estimate for gamma is at hand. One approach is to decide on the smallest effect size worth dealing with. There is no point in using enough subjects to obtain statistical significance with an effect size too small to be important. How small is too small when it comes to gamma? One way to decide this question involves focusing on the numerator of the formula for gamma—that is, $\mu_1 - \mu_2$. In many situations it is not difficult to determine when the difference between the population means is too small to be of any consequence. For instance, if a new reading technique increases reading speed on average by only one word per minute, or a new therapy for overeating results in the loss of only three pounds a year, these differences may be too small to be worth finding statistically significant. Having decided on the smallest $\mu_1 - \mu_2$ that is worth testing, you must then divide this number by an estimate of σ, in order to derive a minimal value for gamma (γ). For most variables, there is usually enough prior information to make at least a crude guess about σ. However, if there is insufficient basis to speculate about

$\mu_1 - \mu_2$ or σ individually, it is still possible to decide when γ is too low to be important. Although the minimal gamma depends on the specific situation, $\gamma = .2$ is generally considered a small effect size. For the following example, I will suppose that an effect size less than .2 is not worth the trouble of testing. To determine the required sample size, insert the minimal value of gamma in Formula 10.6, along with the value of delta that corresponds to what is considered an adequate level of power. Using $\gamma = .2$ and $\delta = 2.8$, wc find the following:

$$N = 2\left(\frac{\delta}{\gamma}\right)^2 = 2\left(\frac{2.8}{.2}\right)^2 = (2)(14)^2 = (2)(196) = 392$$

The calculation above demonstrates that in order to have a reasonable chance (.8) of obtaining significant results when the effect size is actually .2, nearly 400 subjects are needed in each group. It should also be clear that there would be little point to using more than 400 subjects per group, because that would result in a large chance of obtaining significant results when gamma is so tiny that it is trivial. It is usually fortunate that the sample sizes typically used in psychology experiments (30 to 80 per group) yield little power at small effect sizes.

The Power of a One-Sample Test

Because one-sample experiments are so much less common than two-sample experiments, power analysis thus far has been described in terms of the latter rather than the former. Fortunately, there is very little difference between the two cases. The chief difference is in the formula for delta and its variations. For the one-sample case,

$$\delta = \frac{\mu_1 - \mu_2}{\sigma/\sqrt{N}}$$

which leads to Formula 10.7:

$$\delta = \sqrt{N}\,\gamma \qquad\qquad \text{Formula 10.7}$$

Formula 10.7 can be solved for N to create Formula 10.8 for finding the sample size in the one-sample case:

$$N = \left(\frac{\delta}{\gamma}\right)^2 \qquad\qquad \text{Formula 10.8}$$

Notice that the one-sample formulas differ from the two-sample formulas only by a factor of 2; all of the procedures and concepts are otherwise the same. In later chapters, I will show how to calculate power for other statistical procedures.

Exercises

* 1. What is the Type II error rate (β) and the power associated with δ = 1.5 (alpha = .05, two-tailed)? With δ = 2.5?

2. Repeat exercise 1 for alpha = .01, two-tailed. What is the effect on the Type II error rate of reducing alpha?

* 3. What delta is required to have power = .4? To have power = .9? (Assume alpha = .05, two-tailed for this exercise.)

4. a) A researcher has two sections of a psychology class, each with 30 students, to use as subjects. The same puzzle is presented to both classes, but the students in one of the classes are given a helpful strategy for solving the puzzle. If the gamma is .9 in this situation, what will be the power at alpha = .05, two-tailed?

   b) If you had no estimate of gamma, but considered power less than .75 to be unacceptable, how high would gamma have to be for you to run the experiment?

* 5. a) To attain power = .7 with an effect size that also equals .7, how many subjects are required in each group of a two-group experiment (use alpha = .01, two-tailed)?

   b) How many subjects are required per group to attain power = .85 (all else equal)?

6. In exercise 5, how many subjects would be required in a one-group experiment? How does this compare to your previous answer?

* 7. If the smallest gamma for which it is worth showing significance in a particular experiment is .3, and power = .8 is desired, a) what is the largest number of subjects per group that should be used, when alpha = .05, one-tailed)? b) When alpha = .01, one-tailed?

8. A drug for treating headaches has a side effect of lowering diastolic blood pressure by 8 mm Hg compared to a placebo. If the population standard deviation is known to be 6 mm Hg, a) what would be the power of an experiment (alpha = .01, two-tailed) comparing the drug to a placebo using 15 subjects per group?

   b) How many subjects would you need per group to attain power = .95, with alpha = .01, two-tailed?

* 9. In exercise 9B3, if the effect size in the experiment were equal to the value of *g* obtained, a) how many subjects would have been needed to attain power = .7 with alpha = .05, two-tailed?

   b) Given the number of subjects in exercise 9B3, how large an effect size would be needed to have a power of .9, with alpha = .05, two-tailed?

10. a) If gamma less than .2 were always considered too small to be of interest, power = .7 were always considered acceptable, and you always tested with alpha = .05, two-tailed, what is the largest number of subjects in a one-sample experiment that you would ever need?

   b) If you never expect gamma to be more than .8, you never want power to be less than .8, and you always test at alpha = .05, two-tailed, what is the smallest number of subjects in a one-sample experiment that you would ever use?

The Case of Unequal Sample Sizes

All of the power formulas in Section B that dealt with two groups assume that the two groups are the same size: the *N* that appears in each formula is the *N* of each group. When the two samples are not the same size, the formulas of Section B can still be used by averaging the two different *N*s together. However, the ordinary average, or mean, does not provide an accurate answer. The power

**OPTIONAL
MATERIAL**

formulas are most accurate when N is found by calculating the *harmonic mean* of N_1 and N_2. Formula 10.9 for the harmonic mean of two numbers is as follows:

$$\text{Harmonic mean} = \frac{2N_1N_2}{N_1 + N_2} \qquad \text{Formula 10.9}$$

Let us return to the example described in Section B, in which there were initially ten Alzheimer's and five Korsakoff patients. The harmonic mean in this case would be

$$N = \frac{(2)(10)(5)}{10 + 5} = \frac{100}{15} = 6.67$$

Again assuming that $\gamma = 1.0$, we find the value of δ from Formula 10.3:

$$\delta = \sqrt{\frac{N}{2}}\,\gamma = \sqrt{\frac{6.67}{2}}\,(1.0) = \sqrt{3.33} = 1.83$$

If we round off δ to 1.8 and look in Table A.3 (alpha = .05, two-tailed), we find that power is only .44. However, this is nonetheless an improvement over the power that would remain (i.e., .36) if we threw away half the Alzheimer's patients to attain equal-sized groups.

The Limitations of Null Hypothesis Testing

Although null hypothesis testing is not the only approach that psychologists use to analyze data, it plays a major role in many areas of research, especially in social, industrial, and clinical psychology. The chief purpose of null hypothesis testing is to prevent the results of experiments for which the null hypothesis is actually true from being judged statistically significant. However, as mentioned before (see Chapter 7, Section C), it is not possible to estimate the proportion of experiments for which the null hypothesis is true, and it is possible that the null hypothesis is so rarely true that we need not worry about it. One reason that the null hypothesis is not likely to be true is that it is so specific. In the two-group case, the H_0 states that the two population means are equal—not just similar, but identical. In reality, no one would insist that the null hypothesis requires the population means to be exactly equal with infinite accuracy. It is more reasonable to think of the null hypothesis as representing a tiny range of values around zero, whose extent depends on the accuracy of measurements in the domain of interest.

The Improbability of the Null Hypothesis

Even given this more practical notion of the null hypothesis, some psychologists have argued that the many complex relationships among human variables virtually guarantee that changing any one variable will have at least some tiny effect on any other variable that is observed, so that ". . . the strict null hypothesis must always be assumed to be false" (Lykken, 1968, p. 152). These psychologists would likely prefer a null hypothesis defined by a small range around zero, wherein the effect sizes are too tiny to be of any interest or consequence. Moreover, it could be argued that psychologists rarely perform experiments in which the effect size is so tiny that it is barely distinguishable from zero. Psychologists are usually building on the results of previous research and using manipulations and population differences similar to those already shown to have some consequence. Only when a psychologist is breaking new ground or going against convention (e.g., research in "paranormal" phenomena) is there a considerable danger that the effect size may be within a tiny range surrounding zero. Should all the results of psychological studies be subjected to the same null hypothesis testing procedure, just to screen out 95% (i.e., $1 - \alpha$) of the studies for which the null hypothesis is indeed true, when such studies may be quite rare to begin with? Some psychologists (e.g., Cohen, 1994) argue for the greater use of *Bayesian statistics* (probabilities determined by Bayes's Theorem), in which the initial probabilities of the null and various alternative hypotheses being true are taken into account in determining the final *p* level. In this framework, research on paranormal phenomena would require greater evidence than research with familiar, well-understood manipulations. [For a recent and spirited critique of null hypothesis testing, the article by Cohen (1994) is highly recommended.]

Supplementing the Null Hypothesis Test

Rejecting the null hypothesis, regardless of how tiny our *p* level may be (e.g., $p < .00001$), means only that we feel safe in ignoring the possibility that the effect size is virtually indistinguishable from zero. By itself, this is not a very informative conclusion. To begin with, although null hypothesis testing is framed in terms of making a *decision* about the null hypothesis, no decision is really being made, except perhaps whether to publish the result. In reality, results that lead to $p = .051$ are not treated very differently from those that lead to $p = .049$. No psychologist gives up a line of inquiry or pet theory because of a study in which $p = .051$. And no one feels complete confidence in a result that leads to $p = .049$. In addition to deciding whether the effect size is likely to be zero or not, it is usually very helpful to have some estimate of just how large the effect size is likely to be. This purpose can be served by a confidence interval for the difference of means, if the variable being measured is well known. Whereas it may be of some interest to know that the difference in mean annual sick days between those who take vitamin C and those who take a placebo is not zero, it would be of greater interest to know that vitamin C was associated with a reduction of somewhere between two and five annual sick days, with 95% confidence. As I pointed out in Chapter 9, once you have found a $(1 - \alpha)\%$ confidence interval, you do not have to perform a two-tailed null hypothesis test at the level of α—you get that for free! The potential usefulness and importance of the confidence interval cannot be overemphasized; as Rozeboom (1960) put it, "Whenever possible, the basic statistical report should be in the form of a *confidence interval*" (p. 426, emphasis in the original).

In cases in which the variable measured is not well known and may be specific to the experiment being analyzed, a confidence interval for the difference of means may not be very meaningful. In such cases, a useful alternative would be to construct a confidence interval for gamma. In a two group experiment, the CI for gamma would be centered on *g*, the point estimate of gamma as found by Formula 10.4. In the vitamin C experiment mentioned above, the CI for gamma could be found by dividing the limits of the CI for the population mean by an estimate of σ. If the CI for the mean were between two and five days, and σ is estimated to be about 5 days, then the effect size would be estimated to be between .4 (2/5) and 1 (5/5) with 95% confidence.

Uses of the Null Hypothesis Test

If the only function served by null hypothesis testing were to screen out the results of experiments for which the null hypothesis were literally true, the need for this procedure would be quite limited, and it would be hard to understand its pervasiveness in the psychological literature. However, lest you feel quite discouraged about learning all the procedures in this text, I should point out that null hypothesis testing serves at least two fairly important functions.

Determining the Direction of the Effect Sometimes it appears rather obvious that the results of an experiment will come out in a particular direction (e.g., a new relaxation treatment will *reduce,* rather than *raise,* anxiety levels), but in other cases competing theories may make opposite predictions. For example, does watching violent movies make people more or less aggressive? A "catharsis" theory would suggest less aggressiveness afterwards, because watching the violent movies could allow aggression to be released vicariously, and therefore harmlessly. On the other hand, another theory might predict that aggressive impulses will be stimulated by watching the violent movies, producing more aggressive behavior afterwards. If the results of an appropriate experiment demonstrate significantly more aggression on the part of subjects who watch a violent movie (compared, perhaps, to subjects who watch a neutral movie), it is not just the null hypothesis (i.e., no effect) that is ruled out—the opposite prediction (and thus the competing theory) is ruled out as well. The finding of significantly less aggression would be just as helpful. (Of course, no single experiment would be considered the last word on the issue, but if the experiment were well designed *and* statistically significant, its conclusions would be taken seriously.) Nonsignificant findings would probably not convince anyone that violent movies have no effect on aggression either way, and would likely lead to the design of experiments that are more sensitive, in order to settle the issue.

Even in less controversial cases that do not involve sharply opposing theories, there is often a reasonable possibility of finding results in either direction. In such cases, finding significant results in one direction can be useful in either confirming expectations or providing somewhat paradoxical conclusions that must be explained. For instance, it might be predicted that subjects with higher achievement motivation will finish a particular experimental task more quickly. Significant results in the expected direction would tend to validate the method used to assess achievement motivation. Significant results in the opposite direction could lead to speculation that the subjects with high achievement motivation thought the particular experimental task was a waste of their time and energy. Further experiments involving a variety of tasks would be necessary to confirm such speculation.

If the direction of the results is so obvious that a one-tailed test seems justified (e.g., the group with more practice performs more skillfully), the chief purpose of null hypothesis testing is reduced to determining whether the difference between the groups is indeed zero. Thus you might think that when the direction of the results is not in question, null hypothesis testing is not very important. However, there is another practical (though incidental) function of null hypothesis testing that needs to be discussed.

Screening Out Small Effect Sizes If the null hypothesis were actually true for a particular experiment, we would not want the null hypothesis to be rejected, and using alpha as our criterion there is a good chance (i.e., $1 - \alpha$) that it would *not* be rejected. However, even if the null hypothesis were not literally true, but rather almost true (i.e., the effect size were rather tiny), it would still be more useful to the scientific community to retain rather than reject the null hypothesis. As you have learned in this chapter, small effect sizes will lead to retention instead of rejection of H_0 (a Type II error) when power is low, which will be the case when the sample sizes are not large. It is fortunate, then, that psychologists often use modest sample sizes. This is usually a matter of economic constraints rather than choice, but it is a happy by-product of limited resources. If psychologists had unlimited access to subjects, they would be tempted to use very large samples in each study in an attempt to virtually guarantee the statistical significance usually required for publishing results (which, in turn, is necessary for gaining grants, tenure, promotions, etc.). But if very large samples were routinely used, small effect sizes as well as large ones would lead to statistical significance, and it could become difficult to see the forest for all the trees. That is, the literature would soon be cluttered with "significant" results representing utterly trivial effects. (For example, is it important that reading a book printed on green paper may increase reading speed by an average of one word per minute as compared to reading the same book printed on plain white paper?) This would be less of a problem if researchers routinely reported measures of effect size. Although this practice is increasing, encouraged by the APA *Manual of Style,* it is still far from universal. Psychologists generally rely on the likelihood that a significant result also represents a reasonably large effect size, because the sample sizes used are usually small. The use of unusually large sample sizes raises a "red flag" that any significant results may not be large enough to be of practical importance. Thus because of the common use of small samples, null hypothesis testing serves (incidentally) to screen out not only experiments for which the null hypothesis is true but many other experiments for which the effect size would be too small to be interesting anyway. Of course, the down side of small samples is that when the null hypothesis is not rejected, there is a good chance that this is just a Type II error. A combination of larger sample sizes and greater emphasis on the reporting of effect sizes and confidence intervals would be optimal.

Meta-Analysis

Another application of the principles studied in this chapter is a relatively new statistical procedure called meta-analysis. The concept of meta-analysis is that different studies involving similar variables can be compared or combined by estimating the effect size in each study; for a two-group study, *g* or a slight modification of *g* (see Hedges, 1982) can be used to estimate effect size. For example, even though two studies may use different word lists to measure recall, it may be reasonable to compare the effect sizes in the two studies. Statistical procedures have been devised to decide whether the effect sizes of the two studies are significantly different from each other (Rosenthal, 1993). On the other hand, if the effect sizes of several studies (employing essentially the same variables) are similar, averaging these effect sizes together (perhaps giving proportionally more weight to studies based on larger samples) can give a more reliable estimate of the true effect of the variable in question than any one study (Rosenthal, 1993). If a statistical test is performed on the combined effect size, power may be greatly increased because it is based on the total number of subjects in all of the combined studies. Note, however, that it is often not obvious which separate studies can be legitimately combined, and how to test the "combined" study statistically. Meta-analysis is relatively new, and its procedures are still being worked out. It is a promising technique, nonetheless, which offers the economical advantage of taking several studies, each with low power, and creating a much more powerful "combined" study.

Exercises

1. What are the ordinary (i.e., arithmetic) and harmonic means of 20 and 10?

* 2. You are comparing 20 schizophrenics to 10 manic depressives on some physiological measure for which gamma is expected to be about .8. What would be the power of a *t*-test between these two groups, using alpha = .05, two-tailed? Would this experiment be worth doing?

3. Assume that for the experiment described in exer-

cise 2 above, power = .75 is considered the lowest acceptable level. How large would γ have to be to reach this level of power?

* 4. In exercise 9B1, if gamma were 1.1, what would be the power of this experiment when tested with alpha = .05, two-tailed?

5. In exercise 9B6, how large would gamma have to be to attain power = .7, with alpha = .05, two-tailed?

SUMMARY

The Important Points of Section A

1. To estimate the Type II error rate (β) for a particular experiment, it is helpful to choose a specific alternative hypothesis and then to draw the alternative hypothesis distribution (AHD).

2. For the two-group case, the null hypothesis distribution is usually centered on zero, but the AHD is centered on the expected t value, which depends on the specific alternative hypothesis.

3. Once the AHD is determined, the critical value is found with respect to it. The proportion of the AHD below (to the left of) the critical value is β, and the proportion above (to the right) is $1 - \beta$, which is called power.

4. Changing alpha changes the critical value, which, in turn, changes the power. A smaller alpha is associated with fewer Type I errors but more Type II errors, and therefore lower power.

5. The expected t value is called delta, and it can be calculated as the product of two terms: one that depends on the sample size, and one that is called the effect size, or gamma (γ).

6. Gamma is related to the separation (or conversely, the degree of overlap) of the two population distributions; for many purposes $\gamma = .2$ is considered a small effect size, .5 is medium, and .8 is large.

7. If gamma is known or can be accurately estimated (and both alpha and the sample size have been chosen), it is easy to determine power. However, if you knew gamma there would be no reason to do the experiment. Often gamma must be estimated roughly from previous results or theoretical considerations.

8. Once gamma has been estimated, you can find the sample sizes necessary to achieve a reasonable amount of power; .8 is usually considered reasonable.

9. A large obtained t value implies that the expected t is probably large but does not say anything about effect size until sample size is taken into account.

10. Nonsignificant results associated with small sample sizes are less informative than nonsignificant results based on large sample sizes.

The Important Points of Section B

Power analysis can be separated into two categories: fixed and flexible sample sizes.

1. When the sample sizes are fixed by circumstance (e.g., a certain number of patients with a particular condition are available), the effect size is estimated in order to find the power. If the power is too low, the experiment is not performed as designed. Conversely, you can decide on the lowest acceptable power, find the corresponding delta, and then put the fixed sample size into the equation to solve for gamma. The calculated gamma is the lowest gamma that yields an acceptable level of power with the sample sizes available. If the actual effect size in the proposed experiment is expected to be less than this gamma, power would be too low to justify performing the experiment.

2. When there is a fair amount of flexibility in determining sample size, gamma is estimated and the delta corresponding to the desired level of power is found. The appropriate equation can then be used to find the sample sizes needed to attain the desired power level with the effect size as estimated. This procedure can also be used to set limits on the sample size. You find the smallest gamma that is worth testing and put that gamma into the equation along with the delta corresponding to the lowest acceptable power level. The calculated sample size is the *largest* that can be justified. Using any more subjects would produce too great a chance of obtaining statistically significant results with a trivial effect size. Conversely, the largest gamma that can reasonably be expected for the proposed experiment is put into the same equation. The sample size that is found is the bare minimum that is worth employing. Using any fewer subjects would mean that the power would be less than acceptable even in the most optimistic case (gamma as large as can be expected).

The Use of Power Tables and Formulas

A psychologist is studying the relationship between type of crime and sociopathy in prison inmates. For a comparison of arsonists to burglars, there are unfortunately only 15 of each available for testing. What is the power for this experiment? First, gamma must be estimated. Suppose that $\gamma = .3$ can be expected for the sociopathy difference between arsonists and burglars. Now, delta can be found from Formula 10.3:

$$\delta = \sqrt{\frac{N}{2}}\,\gamma = \sqrt{\frac{15}{2}}\,.3 = (2.74)(.3) = .82$$

Finally, an alpha level must be selected in order to use Table A.3. Assuming alpha $= .05$, two-tailed, a delta of .8 (rounding off) corresponds to a power level of .13. This is a very low power level, and it would certainly not be worth running the experiment. If gamma were really only .3, the sample sizes being used would only produce significant results about 13 times in 100 experiments.

If the lowest acceptable power were judged to be .7, Table A.4 tells us that we need delta = 2.48 (given α = .05, two-tailed). Using the sample size we are stuck with, we can use Formula 10.5 to find the minimal effect size.

$$\gamma = \sqrt{\frac{2}{15}}\,(2.48) = (.365)(2.48) = .91$$

The calculated gamma, .91, is the lowest gamma that will yield acceptable power with the available sample sizes at the chosen alpha level. If there is little chance that the sociopathy difference could lead to an effect size this large, then there is little point to conducting the experiment as planned.

Another psychologist is studying the difference between men and women in recalling the colors of objects briefly seen. A very large number of college students of both genders is readily available for testing. How many subjects should be used? First, we need to decide on a level of power. Let us say that power = .85 is desired. Table A.4 shows that this level of power for a test involving alpha = .05, two-tailed, requires that delta = 3.0. Next, gamma must be estimated. Assume that from previous experiments, gamma is expected to be of medium size, so γ = .5. Finally we apply Formula 10.6:

$$N = 2\left(\frac{\delta}{\gamma}\right)^2 = 2\left(\frac{3.0}{.5}\right)^2 = (2)6^2 = (2)(36) = 72$$

Seventy-two males and seventy-two females are required in order to have power equal to .85, with an effect size that equals .5.

If we have no estimate for gamma, another strategy is to decide on the smallest gamma worth testing. For the present example, you might decide that if the effect size is less than .1, there is no point to finding statistical significance. In this case, any effect size less than .1 might be considered about as unimportant as an effect size of zero. If power = .85 is desired (and therefore delta is still 3.0) and the minimum gamma = .1, you again find the number of subjects from Formula 10.6:

$$N = 2\left(\frac{3.0}{.1}\right)^2 = (2)(30)^2 = (2)(900) = 1800$$

A total of 1800 males and 1800 females would be required to attain a power equal to .85 if gamma was only equal to .1. It is not likely that so many subjects would be used for such an experiment, but the above calculation demonstrates that there would certainly be no point to using more than that number of subjects. Using more subjects would lead to a high level of power for effect sizes so small that they would be considered trivial.

Finally, we could ask: What is the largest that gamma might be for the difference in color memory? Suppose that the answer is $\gamma = .7$. Assuming the same desired level of power as above, the required number of subjects would be

$$N = 2\left(\frac{3.0}{.7}\right)^2 = (2)(18.4) = 36.7$$

The above calculation tells us that there would be no point in using fewer than 37 subjects in each of the two groups. Using fewer subjects would mean that the power would be less than desirable even for the largest effect size expected, and therefore less than desirable for the somewhat smaller effect sizes that are likely to be true.

The Important Points of Section C

1. Power formulas based on having two equal-sized groups can be used when the Ns are different, if the *harmonic* mean of the two Ns is calculated (Formula 10.9) and used in place of N in the usual power formulas.

2. The ostensible purpose of null hypothesis testing is to prevent a large percentage (i.e., $1 - \alpha$) of experiments for which the null hypothesis is true from producing statistically significant results. However, the actual percentage of psychology experiments for which the null hypothesis is true or very nearly true may be rather small. Some psychologists have gone so far as to declare that the null hypothesis is never strictly true in psychological research.

3. Because null hypothesis testing is directed only at ruling out a possibility that some consider quite rare (i.e., zero effect size), it is useful to supplement a significance test with the additional information provided by a confidence interval.

4. Beyond attempting to decide whether the effect size is zero, a useful purpose for null hypothesis testing is determining the direction of the difference between two population means, when that direction is the object of some debate. A statistically significant difference in one direction provides evidence that it is unlikely that the population means are actually different in the opposite direction.

5. Another coincidental purpose served by null hypothesis testing is to screen out experiments for which the effect size is too small to be important. Because of the relatively small sample sizes used in psychology experiments, power is fairly low when the effect size is small, so small effect sizes usually do not lead to statistical significance, and therefore do not distract researchers from noticing more important results.

6. Meta-analysis offers the possibility of using effect size estimation to compare studies that use different measures to assess the same variable, or even to combine several small studies into one much more powerful "composite" study. This technique is new but rapidly gaining in popularity.

Definitions of Key Terms

Alternative hypothesis distribution (AHD) The distribution of the test statistic when the null hypothesis is *not* true. When dealing with one or two sample means, the AHD is usually a *noncentral t distribution* whose mean depends on the specific alternative hypothesis being dealt with.

Beta (β) The probability of making a Type II error (i.e., accepting the null hypothesis when the null hypothesis is *not* true).

Power The probability of rejecting the null hypothesis when the null hypothesis is *not* true. Power equals $1 - β$.

Delta (δ) The *t* value that can be expected for a particular alternative hypothesis. Delta, referred to as the "expected *t*," is more formally known as the *noncentrality parameter,* because it is the value on which a noncentral *t* distribution is centered.

Effect size, or gamma (γ) The separation of two population means in terms of standard deviations (assuming homogeneity of variance). The effect size determines the amount of overlap between two population distributions.

Key Formulas

Formulas 10.1 through 10.6 and 10.9 are for the two-group case.

Delta (expected *z* or *t*), when gamma has not already been calculated or estimated:

$$\delta = \sqrt{\frac{N}{2}} \frac{(\mu_1 - \mu_2)}{\sigma} \qquad \text{Formula 10.1}$$

Gamma, when the appropriate population parameters are known or can be estimated:

$$\gamma = \frac{\mu_1 - \mu_2}{\sigma} \qquad \text{Formula 10.2}$$

Delta, when gamma has already been calculated or estimated:

$$\delta = \sqrt{\frac{N}{2}}\,\gamma \qquad\qquad \text{Formula 10.3}$$

Estimate of gamma based on sample statistics from a two-group experiment:

$$g = \frac{\overline{X}_1 - \overline{X}_2}{s_\text{p}} \qquad\qquad \text{Formula 10.4}$$

Gamma, when delta and the sample sizes have already been determined (the formula tells you how large gamma needs to be to attain a particular level of power with a given sample size):

$$\gamma = \sqrt{\frac{2}{N}}\,\delta \qquad\qquad \text{Formula 10.5}$$

The required sample size (of each group in a two-group experiment) for a given level of power (in terms of delta) and a particular value for gamma:

$$N = 2\left(\frac{\delta}{\gamma}\right)^2 \qquad\qquad \text{Formula 10.6}$$

Delta for the one-group case (corresponds to Formula 10.3):

$$\delta = \sqrt{N}\,\gamma \qquad\qquad \text{Formula 10.7}$$

Required sample size for the one-group case (corresponds to Formula 10.6).

$$N = \left(\frac{\delta}{\gamma}\right)^2 \qquad\qquad \text{Formula 10.8}$$

Harmonic mean of two numbers (allows two-group power formulas to be used when the two sample sizes are unequal):

$$\text{Harmonic mean} = \frac{2N_1N_2}{N_1 + N_2} \qquad\qquad \text{Formula 10.9}$$

Hypothesis Tests Involving Two Measures on Each Subject

The statistical procedures described thus far deal with interval/ratio data from only one variable at a time. The three chapters in Part IV all involve situations in which each subject has been measured on an interval/ratio scale twice, and the data from both variables are to be included in the same analysis.

Chapter 11 deals with the case in which both variables are viewed as dependent variables, and our goal is to assess the degree of linear relationship. Section A introduces the concept of the correlation coefficient and emphasizes graphing the relation between two variables in a scatterplot, in order to spot patterns for which a measure of linear correlation would be misleading. Section B is concerned with the actual calculation of the correlation coefficient and the test for its statistical significance. Three main topics are covered in Section C: estimating the power of a test for linear correlation; constructing confidence intervals for the population correlation coefficient (ρ); and testing null hypotheses other than $\rho = 0$.

Chapter 12 also deals with situations for which it is appropriate to compute a correlation coefficient, but the emphasis is different. Chapter 12 focuses on linear regression, in which one variable is viewed as the independent variable and the other as the dependent variable. In reality, the independent variable is usually *not* manipulated by the experimenter; more often, the independent variable is just another dependent variable—but a variable that the experimenter wishes to use in order to make predictions about the second variable. Section A lays out the main concepts of linear regression, including the proportion of variance in one variable accounted for by another. Section B works through the mechanics of making predictions and constructing confidence intervals for those predictions. Section C is devoted to two important topics. First, the principles of correlation and regression are applied to the two-group experiment, in order to show that the concept of "proportion of variance accounted for" can be usefully applied to two-group experiments as well as to correlational studies. Second, the principal concepts of multiple regression, in which several independent variables are combined to predict a single dependent variable, are introduced.

Chapter 13 deals with the situation in which the same variable is being measured twice, but under different conditions. Although a correlation coefficient could be computed, in this case the researcher may be interested in comparing the means of the two variables with a *matched* t-test. Because the matched t-test described in this chapter is closely related to the two-group t-test presented in Chapter 9, some instructors may choose to cover Chapter 13 immediately after Chapter 9 or 10. Others may assign Chapter 13 immediately after Chapter 8, because the mechanics of the matched t-test so closely resemble the mechanics of the one-sample t-test. I have placed the discussion of the matched t-test after the chapters on correlation and regression, because I feel that an understanding of these concepts is essential to appreciating the difference between the matched t-test and the t-test for two independent samples. As usual, the basic structure of the matched t-test is described in Section A, and the step-by-step mechanics of applying the test are detailed in Section B. Section C is concerned with the power of the matched t-test, and how that power varies as a function of the correlation between the two variables in the population.

An example will summarize the distinction among the three chapters of Part IV. Consider a study in which a student's self-esteem is measured both before and after the student attends college. If we were interested in the stability of self-esteem over time (i.e., the test-retest reliability of the self-esteem measure), we would calculate the correlation between the two measures (Chapter 11). If our concern were predicting self-esteem levels after college from earlier self-esteem levels, a regression approach would be appropriate (Chapter 12). Finally, if we wanted to test whether self-esteem levels are significantly higher after college than before, we would perform a matched t-test (Chapter 13).

LINEAR CORRELATION

You will need to use the following from previous chapters:

Symbols
μ: Mean of a population
$\overline{X}$: Mean of a sample
σ: Standard deviation of a population
s: Unbiased standard deviation of a sample
SS: Sum of squared deviations from the mean

Formulas
Formula 5.1: The z score
Formula 8.3: The t-test for one sample

Concepts
The normal distribution

A

**CONCEPTUAL
FOUNDATION**

In this chapter, I will take a concept that is used in everyday life, namely, correlation, and show how to quantify it. You will also learn how to test hypotheses concerning correlation and how to draw conclusions about correlation, knowing the limitations of these methods.

People without formal knowledge of statistics make use of correlations in many aspects of life. For instance, if you are driving to a supermarket and see many cars in the parking lot, you automatically expect the store to be crowded, whereas a nearly empty parking lot would lead you to expect few people in the store. Although the correlation between the number of cars and the number of people will not be perfect (sometimes there will be more families, and at other times more single people driving to the store, and the number of people walking to the store may vary, as well), you know intuitively that the correlation will be very high.

We also use the concept of correlation in dealing with individual people. If a student scores better than most others in the class on a midterm exam, we expect similar performance on the final. Usually midterm scores correlate highly with final exam scores (unless the exams are very different—for example, multiple choice and essays). However, we do not expect the correlation to be perfect; many factors will affect each student's performance on each exam, and some of these factors can change unexpectedly. In order to understand how to quantify the degree of correlation, I will begin by describing perfect correlation.

Perfect Correlation

Consider the correlation between midterm and final exam scores for a hypothetical course. The simplest way to attain **perfect correlation** would be if each student's score on the final was identical to that student's score on the midterm (student A gets 80 on both, student B gets 87 on both, etc.). If you know a student's midterm score, then you know his or her final exam score, as well. But this is not the only way to attain perfect correlation. Suppose that each student's final score is exactly 5 points less than his or her midterm score. The correlation is still perfect; knowing a student's midterm score means that you also know that student's final score.

It is less obvious that two variables can be perfectly correlated when they are measured in different units and the two numbers are not so obviously related. For instance, theoretically, height measured in inches can be perfectly correlated with weight measured in pounds, though the numbers for each will be quite different. Of course, height and weight are not perfectly correlated in a typical group of people, but we could find a group (or make one up for an example) in which these two variables *were* perfectly correlated. However, height and weight would not be the same number for anyone in the group, nor would they differ by some constant. Correlation is not about a subject's having the same *number* on both variables, it is about a subject's being in the same position on both variables relative to the rest of the group. In other words, in order to have perfect correlation someone slightly above average in height should also be slightly above average in weight, for instance, and someone far below average in weight should also be far below average in height.

In order to quantify correlation, we need to transform the original score on some variable to a number that represents that score's position with respect to the group. Does this sound familiar? It should, because the z score is just right for this job. Using z scores, perfect correlation can be defined in a very simple way: If both variables are normally distributed, then perfect **positive correlation** can be defined as each person in the group having the same z score on both variables (e.g., subject A has $z = +1.2$ for both height and weight; subject B has $z = -.3$ for both height and weight, etc.).

Negative Correlation

I have not yet mentioned **negative correlation,** but perfect negative correlation can be defined just as simply: Each person in the group has the same z score in magnitude for both variables, but the z scores are opposite in sign (e.g., someone with $z = +2.5$ on one variable must have $z = -2.5$ on the other variable, etc.). An example of perfect negative correlation is the correlation between the score on an exam and the number of points taken off: If there are 100 points on the exam, a student with 85 has 15 points taken off, whereas a student with 97 has only 3 points taken off. Note that a correlation does not have to be based on individual people, each measured twice. The subject being measured can be a school, a city, or even an entire country. As an example of less-than-perfect negative correlation, consider the following. If each country in the world is measured twice, once in terms of average yearly income, and then in terms of the rate of infant deaths, we have an example of a negative correlation (more income, fewer infant deaths, and vice versa) that is *not* perfect. For example, the United States has one of the highest average incomes, but not one of the lowest infant death rates.

The Correlation Coefficient

In reality, the correlation between height and weight, or between midterm and final scores, will be far from perfect, but certainly not zero. To measure the amount of correlation, a *correlation coefficient* is generally used. A coefficient of $+1$ represents perfect positive correlation, -1 represents perfect negative correlation, and 0 represents a total lack of correlation. Numbers between 0 and 1 represent the relative amount of correlation, with the sign (i.e., $+$ or $-$) representing the direction of the correlation. The correlation coefficient that is universally used for the kinds of variables dealt with in this chapter is the one devised by Karl Pearson (Cowles, 1989); **Pearson's correlation coefficient** is symbolized by the letter r and is often referred to as "Pearson's r." (It is also sometimes referred to as Pearson's "product-moment" correlation coefficient, for reasons too obscure to clarify here.) In terms of z scores, the formula for Pearson's r is remarkably simple (assuming that the z scores are calculated according to Formula 5.1). It is given below as Formula 11.1:

$$r = \frac{\Sigma z_X z_Y}{N} \qquad \text{Formula 11.1}$$

This is not a convenient formula for calculating r. You would first have to convert all of the scores on each of the two variables to z scores, then find the cross product for each subject (i.e., the z score for the X variable multiplied by the z score for the Y variable), and finally find the mean of all of these cross products. There are a variety of alternative formulas that give the same value for r but are easier to calculate. These formulas will be presented in Section B. For now we will use Formula 11.1 to help explain how correlation is quantified.

First, notice what happens to Formula 11.1 when correlation is perfect. In that case, z_X always equals z_Y, so the formula can be rewritten as:

$$r = \frac{\Sigma z_X z_X}{N} = \frac{\Sigma z_X^2}{N}$$

The latter expression is the variance for a set of z scores, and therefore it always equals $+1$, which is just what r should equal when the correlation is perfect. If you didn't recognize that $\Sigma z_X^2/N$ is the variance of z scores, consider the definitional formula for variance (Formula 4.1):

$$\sigma^2 = \frac{\Sigma(X - \mu)^2}{N}$$

Expressed in terms of z scores, the formula is

$$\frac{\Sigma(z - \mu_z)^2}{N}$$

Because the mean of the z scores is always zero, the μ_z term drops out of the second expression, leaving $\Sigma z^2/N$. For perfect negative correlation, the two z scores are always equal but opposite in sign, leading to the expression

$$r = \frac{\Sigma(-z)^2}{N} = \frac{-\Sigma z^2}{N} = -1$$

When the two z scores are always equal, the largest z scores get multiplied by the largest z scores, which more than makes up for the fact that the smallest z scores are being multiplied together. Any other pairing of z scores will lead to a smaller sum of cross products, and therefore a coefficient less than 1.0 in magnitude. Finally, consider what happens if the z scores are randomly paired (as in the case when the two variables are not correlated at all). For some of the cross products the two z scores will have the same sign, so the cross product will be positive. For just about as many cross products, the two z scores would have opposite signs, producing negative cross products. The positive and negative cross products would cancel each other out, leading to a coefficient near zero.

Linear Transformations

Going back to our first example of perfect correlation—having the same score on both the midterm and the final—it is easy to see that the z score would also be the same for both exams, thus leading to perfect correlation. What if each final exam score is 5 points less than the corresponding midterm score? You may recall that subtracting a constant from every score in a group changes the mean but not the z scores, so in this case the z scores for the two exams are still the same, which is why the correlation is still perfect. The midterm scores were converted into the final scores by subtracting 5 points, a conversion that does not change the z scores. In fact, there are other ways that the midterm scores could be changed into final scores without changing the z scores—in particular, multiplying and dividing. It turns out that you can do any combination of adding, subtracting, multiplying, and dividing by constants and the original scores will be perfectly correlated with the transformed scores. This kind of transformation, which does not change the z scores, is called a **linear transformation,** for reasons that will soon be made "graphically" clear.

A good example of a linear transformation is the rule that is used to convert Celsius (also called centigrade) temperatures into Fahrenheit temperatures. The formula is °F $= 9/5$ °C $+ 32$. If you measure the high temperature of the day in both F and C degrees for a series of days, the two temperatures will, of course, be perfectly correlated. Similarly, if you calculate the correlation between height in inches and weight in pounds, and then recalculate the correlation for the same people but measure height in centimeters and weight in kilograms, the correlation will come out exactly the same. It is the relative positions of the measures of the two variables, and not the absolute numbers, that are important. The relative positions are reflected in the z scores, which do not change with simple (i.e., linear) changes in the scale of measurement. Any time one variable is a linear transformation of the other, the two variables will be perfectly correlated.

Graphing the Correlation

The correlation coefficient is a very useful number for characterizing the relationship between two variables, but using a *single* number to describe a potentially complex relationship can be misleading. Recall that describing a group of numbers in terms of the mean for the group can be misleading if the numbers are very spread out. Even adding the standard deviation still does not tell us if the distribution is strongly skewed or not. Only by drawing the distribution could we be sure if the mean and standard deviations were good ways to describe the distribution. Analogously, if we want to inspect the relationship between two variables we need to look at some kind of bivariate distribution. The simplest way to picture this relationship is to draw what is called a **scatterplot,** or scatter diagram (**scattergram,** for short).

A scatterplot is a graph in which one of the variables is plotted on the X axis and the other variable is plotted on the Y axis. Each subject is represented as a single dot on the graph. For instance, Figure 11.1 depicts a scatterplot for height versus weight in a group of ten subjects. Because there are ten subjects, there are ten dots on the graph. If an eleventh subject were 65 inches tall and 130 pounds,

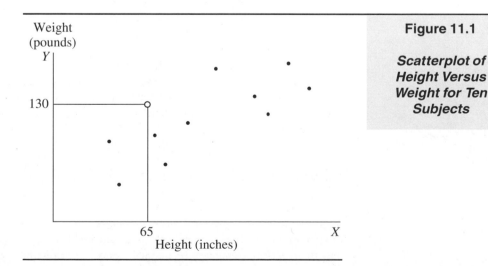

Figure 11.1

Scatterplot of Height Versus Weight for Ten Subjects

Figure 11.2

Scatterplot Depicting Perfect Positive Correlation

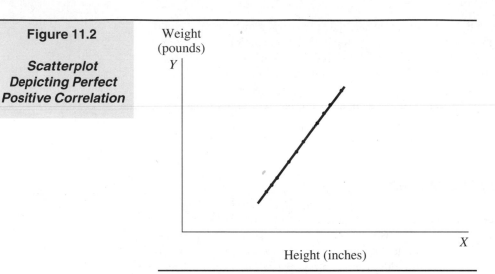

you would go along the X axis to 65 inches, and then up from that point to 130 pounds on the Y axis, putting an eleventh dot at that spot. Now look at Figure 11.2. This scatterplot represents ten people for whom height and weight are perfectly correlated. Notice that all of the dots fall on a single straight line. This is the way a scatterplot always looks when there is perfect (linear) correlation (for perfect negative correlation, the line slants the other way). As mentioned above, if the scores on one variable are a linear transformation of the scores on the other variable, the two variables will be perfectly correlated. You can see from Figure 11.2, why this is called a *linear* transformation.

The straight line formed by the dots in Figure 11.2 tells us that a linear transformation will convert the heights into the weights (and vice versa). That means the same simple formula can give you the weight of any subject once you put in his or her height (or the formula could be used in reverse to get height from weight). In this case the formula is $W = 4H - 120$, where W represents weight in pounds and H is height in inches. Of course, for any ten people it is very unlikely that the height-weight correlation will be perfect. In the next chapter you will learn how to find the best linear formula (i.e., the closest straight line) even when you don't have perfect correlation. At this point, however, we can use scatterplots to illustrate the characteristics and the limitations of the correlation coefficient.

Dealing with Curvilinear Relationships

An important property of Pearson's r that must be stressed is that this coefficient measures only the degree of *linear* correlation. Only scatterplots in which all of the points fall on the same straight line will yield perfect correlation, and for less-than-perfect correlation Pearson's r only measures the tendency for all of

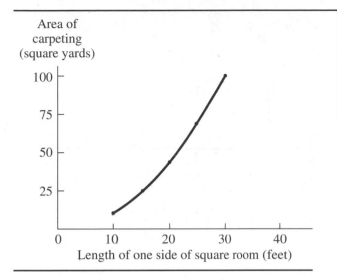

Figure 11.3

Scatterplot Depicting Perfect Curvilinear Correlation

the points to fall on the same straight line. This characteristic of Pearson's r can be viewed as a limitation in that the relation between two variables can be a simple one, with all of the points of the scatterplot falling on a smooth curve, and yet r can be considerably less than 1.0. An example of such a case is depicted in Figure 11.3. The curve depicted in Figure 11.3 is the scatterplot that would be obtained for the relationship between the length (in feet) of one side of a square room and the amount of carpeting (in square yards) that it would take to cover the floor. These two variables are perfectly correlated in a way, but the formula that relates Y to X is not a linear one (in fact, $Y = X^2/9$). To demonstrate the perfection of the relationship between X and Y in Figure 11.3 would require a measure of **curvilinear correlation,** which is a topic beyond the scope of this text. Fortunately, linear relationships are common in nature (including in psychology), and for many other cases a simple transformation of one or both of the variables can make the relationship linear. (In Figure 11.3, taking the square root of Y would make the relation linear.)

A more extreme example of a curvilinear relation between two variables is pictured in Figure 11.4 (page 332). In this figure, age is plotted against running speed for a group of people who range in age from 5 to 75 years old (assuming none of these people are regular runners). If Pearson's r were calculated for this group it would be near zero. Notice that if r were calculated only for the range from 5 years old to the optimum running age (about 20 years old), it would be positive, and quite high. However, r for the range past the optimum point would be of the same magnitude, but negative. Over the whole range, the positive part cancels out the negative part to produce a near-zero linear correlation. Unfortunately, the near-zero r would not do justice to the degree of relationship evident in Figure 11.4. Again, we would need a measure of curvilinear correlation to show the high degree of predictability depicted in Figure 11.4. If you preferred

Figure 11.4

Scatterplot Depicting Curvilinear Correlation (and Near-Zero Linear Correlation)

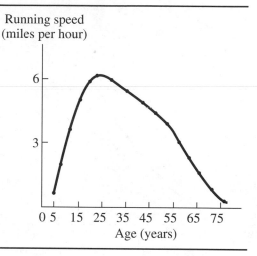

to work with measures of linear relationship, it might be appropriate to restrict the range of one of the variables, so that you were dealing with a region of the scatterplot in which the relationship of the two variables was fairly linear. Bear in mind that when psychologists use the term "correlation" without specifying linear or curvilinear, you can assume that linear correlation is being discussed; I will adhere to this convention, as well.

Problems in Generalizing from Sample Correlations

When psychologists wonder whether IQ is correlated with income, or whether income is correlated with self-esteem, they are speculating about the correlation coefficient you would get by measuring everyone in the population of interest. The Pearson's *r* that would be calculated if an entire population had been measured is called a **population correlation coefficient** and is symbolized by ρ, the Greek letter rho. Of course, it is almost never practical to calculate this value directly. Instead, psychologists must use the *r* that is calculated for a sample in order to draw some inference about ρ. A truly random sample should yield a sample *r* that reflects the ρ for the population. Any kind of biased sample could lead to a sample *r* that is misleading. One of the most common types of biased samples involves a narrow (i.e., truncated) range on one (and often both) of the variables, as described next.

Restricted, or Truncated, Ranges

Data from a sample that is certainly not random are pictured in Figure 11.5. This is a scatterplot of math anxiety versus number of careless errors made on a math

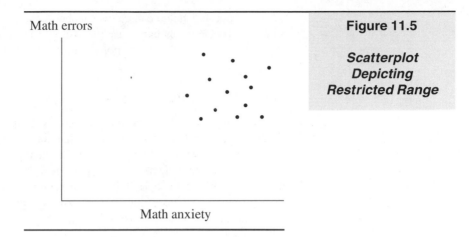

Math errors

Math anxiety

Figure 11.5

Scatterplot Depicting Restricted Range

test by a group of "math phobics." Most students, when asked to guess at the value of *r* that would be calculated for the data in Figure 11.5, say that the correlation will be highly positive; actually, *r* will be near zero. The reason *r* will be near zero is that the points in this figure do not fall on a straight line, or even nearly so. The reason students often expect a high positive correlation is that they see that for each subject a high number on one variable is paired with a high number on the other variable. The problem is that they are thinking in terms of absolute numbers when they say that all subjects are "high" on both variables. However, correlation is based not on absolute numbers, but on relative numbers (i.e., *z* scores) within a particular group. If you calculated Pearson's *r* you would find that all of the points in Figure 11.5 do *not* have high positive *z* scores on both variables. The math phobic with the lowest anxiety in the group (the point furthest left on the graph) will have a negative *z* score because he or she is *relatively* low in anxiety in *that* group, despite being highly anxious in a more absolute sense.

In order for the group of math phobics to produce a high value for *r,* the least anxious one should also have the fewest errors (and vice versa). In fact, all of the math phobics should form a straight, or nearly straight, line. Intuitively, you would be correct in thinking that small differences in anxiety are not going to be reliably associated with small changes in the number of errors. But you would expect large differences in one variable to be associated with fairly large differences in the other variable. Unfortunately, there are no large differences within the group depicted in Figure 11.5. This situation is often referred to as a **truncated (or restricted) range.** If some people low in anxiety were added to Figure 11.5, and these people were also low in errors committed, then all of the math phobics *would* have high positive *z* scores on both variables (the newcomers would have negative *z* scores on both variables), and the calculated *r* would become highly positive.

A truncated range usually leads to a sample *r* that is lower than the population ρ. However, in some cases, such as the curvilinear relation depicted in Figure

11.4, a truncated range can cause r to be considerably higher than ρ. Another sampling problem that can cause the sample r to be much higher or lower than the correlation for the entire population is the one created by outliers, as described next.

Bivariate Outliers

A potential problem with the use of Pearson's r is its sensitivity to outliers (sometimes called outriders). You may recall that both the mean and standard deviation are quite sensitive to outliers; methods for dealing with this problem were discussed in Chapter 3, Section C. Because the measurement of correlation depends on *pairs* of numbers, correlation is especially sensitive to **bivariate outliers.** A bivariate outlier need not have an extreme value on either variable, but the combination it represents must be extreme (e.g., a 6 ft 2 in. tall man who weighs 140 pounds, or a 5 ft 7 in. tall man who weighs 220 pounds). A graphical example of a bivariate outlier is shown in Figure 11.6. Notice that without the outlier there is a strong negative correlation; with the outlier the correlation actually reverses direction and becomes slightly positive.

Where do outliers come from? Sometimes their origins can be pretty obvious. If you are measuring performance on a task, an outlier may arise because a subject may have forgotten the instructions or applied them incorrectly, or may even have fallen asleep. If a self-report questionnaire is involved, it is possible that a subject was not being honest or didn't quite understand what was being asked. If you can find an independent basis for eliminating an outlier from the data (other than the fact that it is an outlier), then you should remove the outlier before calculating Pearson's r. On the other hand, sometimes you simply have to live with an outlier. Once in a while the outlier represents a very unlucky event that is not likely to appear in the next sample. (This possibility highlights the importance of replication.) Or the outlier may represent the influence of as yet unknown factors that are worthy of further study.

Figure 11.6

Scatterplot Containing One Bivariate Outlier

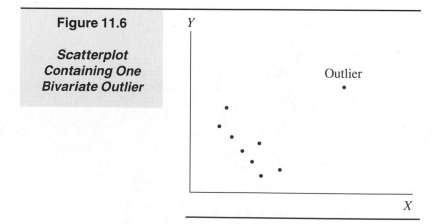

In any case, the scatterplot should always be inspected before any attempt is made to interpret the correlation. It is always possible that your Pearson's *r* has been raised or lowered in magnitude, or even reversed in direction, by a truncated range or a single outlier. An unexpected curvilinear trend can be discovered. Finally, the spread of the points may not be even as you go across the scatterplot. (The importance of this last problem—heteroscedasticity—will be discussed in Chapter 12.)

Correlation Does Not Imply Causation

A large sample *r* suggests that ρ for the population may be large, as well, or at least larger than zero. In Section B you will see how the sample *r* can be used to draw an inference about ρ. Once you have concluded that two variables are indeed closely related, it can be tempting to believe that there must be a causal relationship between the two variables. But it is much more tempting in some situations than others. If you take a sample of pupils from an elementary school, you will find a high correlation between shoe size and vocabulary size, but it is safe to say that you would *not* be tempted to think that having larger feet makes it easier to learn new words (or vice versa). And, of course, you would be right in this case. The high correlation most likely results from having a range of ages in your sample. The older the pupil, the larger his or her feet tend to be, and the larger his or her vocabulary tends to be, as well. Shoe size and vocabulary size are not affecting each other directly; both are affected by a third variable—namely, age.

On the other hand, if you observed a high negative correlation between number of hours spent per week on exercise and number of days spent sick each year, you probably would be tempted to believe that exercise was *causing* good health. However, it is possible that the flow of causation is in the opposite direction; healthier people may have a greater inclination to exercise. It is also possible that, as in the example above, some unnoticed third variable is responsible for both the increased exercise and the decreased illness. For instance, people who are more optimistic may have a greater inclination to exercise as well as better psychological defenses against stress that protect them from illness. In other words, the optimism may be causing both the inclination to exercise and the improved health. If this latter explanation were true, then encouraging people to exercise is not the way to improve their health; you should be trying to find a way to increase their optimism. Although sometimes the causal link underlying a strong correlation seems too obvious to require experimental evidence (e.g., the correlation between number of cigarettes smoked and likelihood of contracting lung cancer), the true scientist must resist the temptation of relying solely on theory and logic. He or she should make every effort to obtain confirmatory evidence by means of true experiments.

If the above discussion sounds familiar, it is because the same point was made with respect to the two-group experiment. If a group of exercisers is compared to a group of nonexercisers with respect to sick days, this is a correlational design that is open to the "third variable" (e.g., optimism) explanation used above. It is by randomly assigning subjects to an exercise or no-exercise group that the experimenter can definitively rule out third variable explanations and conclude that exercise directly affects health.

True Experiments Involving Correlation

If instead of forming two distinct groups, an experimenter randomly assigned each subject a different number of exercise hours per week, it would then not be appropriate to calculate a t-test; it would make more sense to calculate the correlation coefficient between amount of exercise and number of sick days. However, in this special case, a significant correlation could be attributed to a causal link between exercise and health. Third variables, such as optimism, could be ruled out, because the *experimenter* determined the amount of exercise; personality and constitutional variables were not allowed to determine the subjects' amount of exercise. Such experiments, which lead directly to a correlation coefficient instead of a t value, are rare, but they will be discussed in more detail in the context of linear regression in Chapter 12.

Exercises

1. Describe a realistic situation in which two variables would have a high positive correlation. Describe another situation for which the correlation would be highly negative.

2. A recent medical study found that the moderate consumption of alcoholic beverages is associated with the fewest heart attacks (as compared to heavy drinking or no drinking). It was suggested that the alcohol *caused* the beneficial effects. Devise an explanation for this relationship that assumes there is no direct causal link between drinking alcohol and having a heart attack. (Hint: Consider personality.)

* 3. A study compared the number of years a person has worked for the same company (X) with the person's salary in thousands of dollars per year (Y). The data for nine subjects appear in the table below. Draw the scatterplot for these data and use it to answer the following questions.

Years (X)	Annual Salary (Y)
5	24
8	40
3	20
6	30
4	50
9	40
7	35
10	50
2	22

a) Considering the general trend of the data points, what direction do you expect the correlation to be in (positive or negative)?

b) Do you think the correlation coefficient for these data would be meaningful or misleading? Explain.

*4. A psychologist is studying the relation between anxiety and memory for unpleasant events. Subjects are measured with an anxiety questionnaire (X) and with a test of their recall of details (Y) from a horror movie shown during the experimental session. Draw the scatterplot for the data in the following table and use it to answer the following questions.

Anxiety (X)	Number of Details Recalled (Y)
39	3
25	7
15	5
12	2
34	4
39	3
22	6
27	7

a) Would you expect the magnitude of the correlation coefficient to be high or low? Explain.

b) Do you think the correlation coefficient for these data would be meaningful or misleading? How might you explain the results of this study?

*5. A clinical psychologist believes that depressed people speak more slowly than others. He measures the speaking rate (words per minute) of six depressed patients who had already been measured on a depression inventory. Draw the scatterplot for the data in the following table and use it to answer the following questions.

Depression (X)	Speaking Rate (Y)
52	50
54	30
51	39
55	42
53	40
56	31

a) What direction do you expect the correlation to be in (i.e., what sign do you expect the coefficient to have)?

b) What expectations do you have about the magnitude of the correlation coefficient (very low, very high, or moderate)?

6. Everyone in an *entire population* has been measured on two different variables, and the correlation coefficient (ρ) has been calculated. You are disappointed at how low ρ has turned out to be, because you thought the two variables were closely related. Which of the following can explain why ρ was not higher?

a) One or both of the variables has a restricted range.

b) The relationship between the two variables is curvilinear.

c) The number of degrees of freedom is too small.

d) One variable was just a linear transformation of the other.

*7. A psychologist is studying the relationship between the reported vividness of visual imagery and the ability to rotate objects mentally. A sample of graduate students at a leading school for architecture is tested on both variables, but the Pearson's *r* turns out to be disappointingly low. Which of the following is the most likely explanation for why Pearson's *r* was not higher?

a) One or both of the variables has a restricted range.

b) The relationship between the two variables is curvilinear.

c) The number of degrees of freedom is too small.

d) One variable was just a linear transformation of the other.

8. A psychologist conducted a small pilot study involving the correlation of two variables and was surprised to find a Pearson's *r* much higher than anticipated. The most likely explanation for this surprise is that

a) sampling error produced the high correlation.

b) the variability was smaller than expected.

c) an unknown third variable was playing a strong role.

d) the relationship of the two variables was not as linear as expected.

* 9. The correlation between scores on the midterm and scores on the final exam for students in a hypothetical psychology class is .45.

a) What is the correlation between the midterm scores and the number of points taken off for each midterm?

b) What is the correlation between the number of points taken off for each midterm and the final exam score?

10. Suppose there is a perfect negative correlation between the amount of time spent (on average) answering each question on a test and the total score on the test. Assume that $\overline{X}_1 = 30$ seconds and $s_1 = 10$ for the time per item, and that $\overline{X}_2 = 70$, with $s_2 = 14$ for the total score.

a) If someone spent 10 seconds per item, what total score would they receive?

b) If someone's score on the test was 49, how much time did they spend on each item?

BASIC STATISTICAL PROCEDURES

We now turn to the practical problem of calculating Pearson's correlation coefficient. Although Formula 11.1 is easy to follow, the necessity to convert all scores to z scores before proceeding makes it unnecessarily tedious. A formula that is easier to use can be created by plugging Formula 5.1 for the z score into Formula 11.1 and rearranging the terms algebraically until the formula assumes a convenient form. The steps below lead to Formula 11.2:

$$r = \frac{\sum z_X z_Y}{N} = \frac{\sum[(X - \mu_X)/\sigma_X][(Y - \mu_Y)/\sigma_Y]}{N} = \frac{\sum(X - \mu_X)(Y - \mu_Y)}{N\sigma_X\sigma_Y}$$

$$= \frac{\sum XY - N\mu_X\mu_Y}{N\sigma_X\sigma_Y}$$

$$r = \frac{\dfrac{\sum XY}{N} - \mu_X\mu_Y}{\sigma_X\sigma_Y}$$

Formula 11.2

Notice that the first term in Formula 11.2, $\sum XY/N$, is in itself similar to Formula 11.1, except that it is the average cross product for the original scores, rather than for the cross products of z scores. Instead of subtracting the mean and dividing by the standard deviation for each score to obtain the corresponding z score, with Formula 11.2 we need only subtract once ($\mu_X\mu_Y$) and divide once ($\sigma_X\sigma_Y$). Note also that we are using population values for the mean and standard deviation, because we are assuming for the moment that Pearson's r is being used only as a means of describing a set of scores (actually pairs of scores) at hand, so for our purposes that set of scores is a population. Of course, for the mean it does not matter whether we refer to the population mean (μ) or the sample mean ($\overline{X}$); the calculation is the same. There is a difference, however, between σ and s—the former is calculated with N in the denominator, whereas the latter is calculated with $N - 1$ (see Chapter 4). For Formula 11.2 to yield the correct answer, σ must be calculated rather than s.

The Covariance

The numerator of Formula 11.2 has a name of its own; it is called the **covariance.** This part gets larger in magnitude as the two variables show a greater tendency to "covary"—that is, to vary together, either positively or negatively. Dividing by the product of the two standard deviations ensures that the correlation coefficient will never get larger than $+1.0$ or smaller than -1.0, no matter how the two variables covary. Formula 11.2 assumes that you have calculated the biased standard deviations, but if you are dealing with a sample and planning to draw inferences about a population, it is more likely that you have calculated the unbiased standard deviations. It is just as easy to calculate the correlation coefficient in terms of s rather than σ, but then the numerator of Formula 11.2 must be adjusted accordingly. Just as the denominator of Formula 11.2 can be considered biased (when extrapolating to the larger population), the numerator in this formula has a corresponding bias, as well. These two biases cancel each other out so that the value of r is not affected, but if you wish to use unbiased standard deviations in the denominator you need to calculate an unbiased covariance in the numerator.

The Unbiased Covariance

The bias in the numerator of Formula 11.2 is removed in the same way we removed bias from the formula for the standard deviation: We divide by $N - 1$, instead of N. Adjusting the covariance accordingly, and using the unbiased standard deviations in the denominator gives Formula 11.3, as shown below:

$$r = \frac{\frac{1}{N-1}(\Sigma XY - N\overline{X}\overline{Y})}{s_X s_Y} \qquad \text{Formula 11.3}$$

I used the symbol for the sample mean in Formula 11.3 in order to be consistent with using the sample standard deviation. It is very important to realize that Formula 11.3 always produces the same value for r as Formula 11.2, so it does not matter which formula you use. If you have already calculated σ for each variable, or you plan to anyway, you should use Formula 11.2. If you have instead calculated s for each variable, or plan to, then it makes sense to use Formula 11.3.

An Example of Calculating Pearson's r

To make the computation of these formulas as clear as possible, I will illustrate their use with the following example. Suppose that a researcher has noticed a trend for women with more years of higher education (i.e., beyond high school) to have fewer children. To investigate this trend, she selects six women at random

Table 11.1

X (Years of Higher Education)	Y (Number of Children)	XY
0	4	0
9	1	9
5	0	0
2	2	4
4	3	12
1	5	5
$\overline{X} = 3.5$	$\overline{Y} = 2.5$	$\sum XY = 30$
$\sigma_X = 2.99$	$\sigma_Y = 1.71$	
$s_X = 3.27$	$s_Y = 1.87$	$\sum XY/N = 30/6 = 5$

and records the years of higher education and number of children for each. The data appear in Table 11.1, along with the means and standard deviations for each variable and the sum of the cross products.

We can apply Formula 11.2 to the data in Table 11.1 as follows. (Note the use of σ instead of s.)

$$r = \frac{\sum XY/N - \mu_X \mu_Y}{\sigma_X \sigma_Y} = \frac{30/6 - (3.5)(2.5)}{(2.99)(1.71)} = \frac{5 - 8.75}{5.1} = \frac{-3.75}{5.1} = -.735$$

Notice that r is highly negative, indicating that there is a strong tendency for *more* years of education to be associated with *fewer* children in this particular sample. Notice that we get exactly the same answer for r if we use Formula 11.3 instead of Formula 11.2. (Note the use of s rather than σ.)

$$r = \frac{\frac{1}{N-1}(\sum XY - N\overline{X}\overline{Y})}{s_X s_Y} = \frac{1/5[30 - (6)(3.5)(2.5)]}{(3.27)(1.87)} = \frac{-4.5}{6.115} = -.735$$

Alternative Formulas

There are other formulas for Pearson's r that are algebraically equivalent to the formulas above and therefore give exactly the same answer. I will show some of these below because you may run into them elsewhere, but I do not recommend their use. My dislike of these formulas may be more understandable after you have seen them.

A traditional variation of Formula 11.2 is shown as Formula 11.4 below:

$$r = \frac{\sum(X - \overline{X})(Y - \overline{Y})}{\sqrt{SS_X SS_Y}} = \frac{SP}{\sqrt{SS_X SS_Y}} \qquad \text{Formula 11.4}$$

The numerator of Formula 11.4 is sometimes referred to as the sum of products of deviations, shortened to the sum of products, and symbolized by SP. The SS terms simply represent the numerators of the appropriate variance terms. [You'll recall that Formula 4.9 states that $s^2 = SS/(N - 1) = SS/df$.] The chief drawback of Formula 11.4 is that it requires calculating a deviation score for each and every raw score—which is almost as tedious as calculating all of the z scores. To simplify the calculations, Formula 11.4 can be algebraically manipulated into the following variation:

$$r = \frac{\Sigma XY - (\Sigma X)(\Sigma Y)/N}{\sqrt{[\Sigma X^2 - (\Sigma X)^2/N][\Sigma Y^2 - (\Sigma Y)^2/N]}}$$

The above formula can be made even more convenient for calculation by multiplying both the numerator and the denominator by N to yield Formula 11.5:

$$r = \frac{N\Sigma XY - (\Sigma X)(\Sigma Y)}{\sqrt{[N\Sigma X^2 - (\Sigma X)^2][N\Sigma Y^2 - (\Sigma Y)^2]}} \qquad \text{Formula 11.5}$$

This formula is often referred to as a "raw-score" computing formula because it calculates r directly from raw scores without first producing intermediate statistics, such as the mean or SS.

Which Formula To Use

At this point I want to emphasize that all of the formulas for r in this chapter will give exactly the same answer (except for slight deviations involved in rounding off numbers before obtaining the final answer), so which formula you use is merely a matter of convenience. In the days before electronic calculators or computers, it was of greater importance to use a formula that minimized the steps of calculation, so Formula 11.5 was generally favored. I dislike Formula 11.5, however, because it is easy to make a mistake on one of the sums or sums of squares without noticing it, and it can be hard to track down the mistake once you realize that the value for r is incorrect (for example, when you obtain an r greater than 1.0).

In contrast, the means and standard deviations that go into Formula 11.2 or Formula 11.3 can each be checked to see that they are reasonable before proceeding. Just be sure to keep at least four digits past the decimal point for each of these statistics when inserting them in Formula 11.2 or Formula 11.3; otherwise the error introduced by rounding off these statistics can seriously affect the final value of r you obtain. The only term whose value can be way off without looking obviously wrong is $\Sigma XY/N$, so this term must be checked very carefully. Of course, in actual research, the correlation will be most often calculated by a computer. However, for teaching the concepts of correlation, I prefer Formula 11.2 or Formula 11.3, because their structures reveal something about the quantity being calculated.

Testing Pearson's *r* for Significance

Using the *t* Distribution

The *r* that we calculated for the example about women's education and number of children was quite high ($r = -.735$), and it would appear that the hypothesis $\rho = 0$ must be quite far-fetched. On the other hand, recall that our correlation was based on only six randomly selected women, and you should be aware that fairly large correlations can occur easily by chance in such a tiny sample. We need to know the null hypothesis distribution that corresponds to our experiment. That is, we need to know, in a population where $\rho = 0$ for these two variables, how the sample *r*s will be distributed when randomly drawing six pairs of numbers at a time.

Given some assumptions that will be detailed shortly, the laws of statistics tell us that when $\rho = 0$ and the sample size is quite large, the sample *r*s will be normally distributed with a mean of 0 and a standard error of about $1/\sqrt{N}$. For sample sizes that are not large, the standard error can be estimated by using the following expression:

$$\sqrt{\frac{1 - r^2}{N - 2}}$$

The significance of *r* can then be tested with a formula that resembles the one for the one-group *t*-test,

$$t = \frac{r - \rho_0}{\sqrt{\dfrac{1 - r^2}{N - 2}}}$$

where ρ_0 represents the value of the population correlation coefficient according to the null hypothesis. In the most common case, the null hypothesis specifies that $\rho_0 = 0$, and this simplifies the formula. Also, it is common to rearrange the formula algebraically into the following form (Formula 11.6):

$$t = \frac{\sqrt{N - 2}\, r}{\sqrt{1 - r^2}} \qquad \text{Formula 11.6}$$

To test a null hypothesis other than $\rho = 0$ or to construct a confidence interval around a sample *r* requires special methods, which are discussed in Section C.

Using the Table of Critical Values for Pearson's *r*

There is a simpler way to test *r* for statistical significance. Statisticians have already performed the *t*-tests for the common alpha levels and put the results in a convenient table (see Table A.5 in Appendix A) that allows you to look up the critical value for *r* for each combination of alpha and df. For instance, to find the critical r corresponding to the education/children study, we need to know alpha (we will use .05, two-tailed, for this example) and the degrees of freedom. For correlation problems, df $= N - 2$, where N is the number of subjects (or *pairs* of numbers). In the education/children study, $N = 6$, so df $= 6 - 2 = 4$. If we look this up in Table A.5, we see that the critical $r = .811$. Because the magnitude of our calculated r is .735 (the sign of the correlation does not matter when testing for significance), which is less than .811, we must retain (i.e., we *cannot* reject) the null hypothesis that $\rho = 0$.

The Critical Value for r as a Function of Sample Size

The above statistical conclusion may be surprising, considering how high the sample r was, but Table A.5 reveals that rather high rs can be commonly found with samples of only six subjects, even when there is no correlation at all in the population. By looking down the column for the .05, two-tailed alpha in Table A.5, you can see how the critical r changes with df (and therefore with the sample size). With a sample of only four subjects (df $= 2$), the sample r must be over .95 in magnitude to be statistically significant. With ten subjects (df $= 8$) the critical value reduces to .632, whereas with 102 subjects (df $= 100$), a sample r can be significant just by being larger than .195. Also, it should not be surprising to see that the critical values for r become larger as alpha gets smaller (looking across each row toward the right).

An important point to remember is that, unlike z scores or t values, any sample $r,$ no matter how small (unless it is exactly zero) can be statistically significant, if the sample size is large enough. Conversely, even a correlation coefficient close to 1.0 can fail to attain significance, if the sample is too small. One way to understand the dependence of the critical r on sample size is to realize that the sample rs are clustered more tightly around the population ρ as sample size increases—it becomes more and more unlikely to obtain a sample r far from ρ when N gets larger.

Table A.5 assumes that $\rho = 0$, which is why we can ignore the sign of the sample when testing for significance; the distribution of sample rs around $\rho = 0$ will be symmetric. This will not be the case, however, when the null hypothesis specifies any ρ other than zero. This latter case is more complex, and therefore I will postpone that discussion until Section C.

Assumptions Associated with Pearson's r

Independent Random Sampling

This assumption applies to all of the hypothesis tests in this text. In the case of correlation, it means that even though a relation may exist between the two numbers of a pair (e.g., a large number for height may be consistently associated with a large number for weight), each pair should be independent of the other pairs, and all of the pairs in the population should have an equal chance of being selected.

Normal Distribution

Each of the two variables should be measured on an interval or ratio scale and be normally distributed in the population.

Bivariate Normal Distribution

If the assumption of a bivariate normal distribution is satisfied, then you can be certain that the assumption above will also be satisfied, but it is possible for each variable to be normally distributed separately without the two variables jointly following a bivariate normal distribution (which is a stricter assumption than the one discussed above).

In order to picture a **bivariate distribution,** consider the univariate distribution. In the univariate distribution (the kind that was described in Chapter 2), one variable is placed along the horizontal axis, and the height of the distribution represents the likelihood of each value. In the bivariate distribution, both the vertical and horizontal axes are needed to represent the two variables; each spot on the plane represents a possible pair of values (just as in the scatterplot). How can the likelihood of each pair be represented? One way would be to use the third dimension to represent the height at each point (as in a topographical, or relief, map). A common bivariate distribution would come out looking like a wide-brimmed hat, as shown in Figure 11.7. Values near the middle are more common, whereas values that are far away in any direction, including diagonally, are less common.

A bivariate normal distribution can involve any degree of linear relationship between the two variables from $\rho = 0$ to $\rho = +1.0$ or -1.0, but a curvilinear relationship would not be consistent with a bivariate normal distribution. If there is good reason to suspect a nonlinear relationship, Pearson's r should not be used;

Figure 11.7

Bivariate Normal Distribution (Three Dimensions Are Indicated)

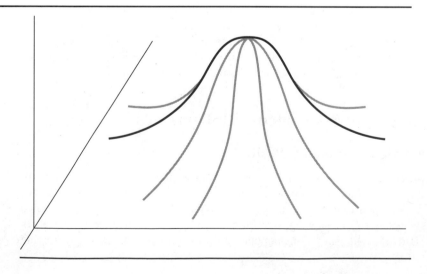

instead, you should consult advanced texts for alternative techniques. If the bivariate distribution in your sample is very strange, or if one of the variables has been measured on an ordinal scale, then the data should be converted to ranks before applying the Pearson correlation formula (see Part VI, Chapter 21). When the correlation is calculated for ranked data, the *Spearman rank-order* correlation formula is commonly used, and the resulting coefficient is often called the Spearman rho (r_S). This correlation coefficient is interpreted in the same way as any other Pearson's *r*.

As with the assumption of a normal distribution, the assumption of a bivariate normal distribution becomes less important as sample size increases. For very large sample sizes, the assumption can be grossly violated with very little error. Generally, if the sample size is above about 30 or 40, and the bivariate distribution does not deviate a great deal from bivariate normality, the assumption is usually ignored.

Uses of the Pearson Correlation Coefficient

Reliability and Validity

One of the most common uses of Pearson's *r* is in the measurement of reliability, which occurs in a variety of contexts. For instance, to determine whether a questionnaire is assessing a personality trait that is stable over time, each subject is measured twice (with a specified time interval in between), and the correlation of the two scores is calculated to determine the *test-retest reliability*. There should be a strong tendency for subjects to have the same score both times; otherwise, we may not be measuring some stable aspect of individuals. The internal consistency of the questionnaire can also be checked; separate subscores for the odd- and even-numbered items can be correlated to quantify the tendency for all items in the questionnaire to measure the same trait. This is called *split-half reliability*. (Usually more complicated procedures are preferred for measuring internal reliability.)

Sometimes a variable is measured by having someone act as a judge to rate the behavior of a subject (e.g., how aggressive or cooperative a particular child is in a playground session). In order to feel confident that these ratings are not peculiar to the judge used in the study, a researcher may have two judges rate the same behavior, so that the correlation of these ratings can be assessed. It is important to have high *inter-rater reliability* in order to trust these ratings. In general, correlation coefficients for reliability that are below .7 lead to a good deal of caution and rethinking.

Another frequent use of correlation is to establish the *criterion validity* of a self-report measure. For instance, subjects might fill out a questionnaire containing a generosity scale, and then later, in a seemingly unrelated experiment, for which they are paid, be told that the experimenter is running low on funds and wants them to give back as much of their payment as they are comfortable giving. A high correlation between the self-reported generosity score and the actual amount of money subsequently donated would help to validate the self-report measure. It is also common to measure the degree of correlation between two questionnaires that are supposed to be measuring the same variable, such as two measures of anxiety or two measures of depression.

Relationships Between Variables

The most interesting use for correlation is to measure the degree of association between two variables that are not obviously related but are predicted by some theory or past research to have an important connection. For instance, one aspect of Freudian theory might give rise to the prediction that stinginess will be positively correlated with stubbornness, because both traits are associated with the anal retentive personality. As another example, the correlation that has been found between mathematical ability and the ability to imagine how objects would look if rotated in three-dimensional space supports some notions about the cognitive basis for mathematical operations. On the other hand, some correlations are not predicted but can provide the basis for future theories, such as the correlations that are sometimes found between various eating and drinking habits and particular health problems.

Finally, correlation can be used to evaluate the results of an experiment when the levels of the manipulated variable come from an interval or ratio scale. For example, the experimenter may vary the number of times particular words are repeated in a list to be memorized, and then find the correlation between number of repetitons and the probability of recall. A social psychologist may vary the number of "bystanders" (actually, confederates of the experimenter) at an accident scene, to see if a subject's response time for helping someone who seems to have fainted (another confederate, of course) is correlated with the number of other onlookers.

Publishing the Results of Correlational Studies

If the results of the study of years of education versus number of children were published, the results section of the article might contain a sentence such as the following: "Although a linear trend was observed between the number of years of education and the number of children that a woman had, the correlation coefficient failed to reach significance, r (6) $= -.735$, $p > .05$." The number in parentheses following r is the sample size. Of course, a considerably larger sample would usually be used, and such a high correlation would normally be statistically significant.

An example of the use of Pearson's r in the psychological literature is found in the following excerpt from a study of the subjective responses of undergraduate psychology students who reported marijuana use (Davidson & Schenk, 1994). Subjects completed two scales concerning their first experience with marijuana, indicating amount of agreement with each of several statements like the following: "Marijuana made small things seem intensely interesting" (Global Positive 1) and "Marijuana caused me to lose control and become careless" (Global Negative 1). Among other results, the authors found that "Global Positive 1 and Global Negative 1 scores were related to each other [$r(197) = .31, p < .01$]." Notice the large sample size, which makes it relatively easy to attain statistical significance without a large correlation coefficient.

This is a good example of a situation in which a one-tailed significance test would not be justified. You might have expected a negative correlation (more positive features of marijuana use associated with fewer negative features, and vice versa), but it is also understandable that some students would be generally affected more by their first use of marijuana and would experience more of both the positive and negative features, whereas other students would be relatively unaffected.

Exercises

1. A professor has noticed that in her class a student's score on the midterm is a good indicator of the student's performance on the final exam. She has already calculated the means and (biased) standard deviations for each exam for a class of 20 students: $\mu_M = 75$, $\sigma_M = 10$; $\mu_F = 80$, $\sigma_F = 12$; therefore, she only needs to calculate ΣXY to find Pearson's r.

a) Find the correlation coefficient using Formula 11.2, given that $\Sigma XY = 122,000$.

b) Can you reject the null hypothesis (i.e., $\rho = 0$) at the .01 level (two-tailed)?

c) What can you say about the degree to which the midterm and final are linearly related?

2. Calculate Pearson's r for the data in exercise 11A4, using Formula 11.2. Test the significance of this correlation coefficient by using Formula 11.6.

3. a) Calculate Pearson's r for the data in exercise 11A5, using Formula 11.3.

b) Recalculate the correlation you found in part a above, using Formula 11.5 (Hint: The calculation can be made easier by subtracting a constant from all of the scores on the first variable.)

c) Are the two answers exactly the same? If not, explain the discrepancy.

*4. a) Calculate Pearson's r for the data in exercise 11A3 and test for significance with alpha = .05 (two-tailed).

b) Delete the subject with 4 years at the company and recalculate Pearson's r. Test for significance again.

c) Describe a situation in which it would be legitimate to make the deletion indicated in part b above.

*5. A psychiatrist has noticed that the schizophrenics who have been in the hospital the longest score the lowest on a mental orientation test. The data for ten schizophrenics are listed in the table below:

Years of Hospitalization (X)	Orientation Score (Y)
5	22
7	26
12	16
5	20
11	18
3	30
7	14
2	24
9	15
6	19

a) Calculate Pearson's r for the data.

b) Test for statistical significance at the .05 level (two-tailed).

*6. If a test is reliable, each subject will tend to get the same score each time he or she takes the test. Therefore, the correlation between two administrations of the test (test-retest reliability) should be high. The reliability of the verbal GRE score was tested using five subjects, as shown in the table below:

Verbal GRE (1)	Verbal GRE (2)
540	570
510	520
580	600
550	530
520	520

a) Calculate Pearson's *r* for the test-retest reliability of the verbal GRE score.

b) Test the significance of this correlation with alpha = .05 (one-tailed). Would this correlation be significant with a two-tailed test?

7. A psychologist wants to know if a new self-esteem questionnaire is internally consistent. For each of the nine subjects who filled out the questionnaire, two separate scores were created: one for the odd-numbered items and one for the even-numbered items. The data appear below. Calculate the split-half reliability for the self-esteem questionnaire using Pearson's *r*.

Subject	Odd Items	Even Items
1	10	11
2	9	15
3	4	5
4	10	6
5	9	11
6	8	12
7	5	7
8	6	11
9	7	7

* 8. A psychologist is preparing stimuli for an experiment on the effects of watching violent cartoons on the play behavior of children. Each of six cartoon segments is rated on a scale from 0 (peaceful) to 10 (ex-

tremely violent) by two different judges, one male and one female. The ratings are shown below.

Segment	Male Rater	Female Rater
1	2	4
2	1	3
3	8	7
4	0	1
5	2	5
6	7	9

a) Calculate the interrater reliability using Pearson's *r*.

b) Test the significance of the correlation coefficient at the .01 level (one-tailed).

* 9. Does aerobic exercise reduce blood serum cholesterol levels? To find out, a medical researcher assigned subjects who were not already exercising regularly to do a randomly selected number of exercise hours per week. After 6 months of exercising the prescribed number of hours, each subject's cholesterol level was measured, yielding the data in the table below.

Subject	Hours of Exercise per Week	Serum Cholesterol Level
1	4	220
2	7	180
3	2	210
4	11	170
5	5	190
6	1	230
7	10	200
8	8	210

a) Calculate Pearson's correlation coefficient for the data in the table.

b) Test the significance of the correlation coefficient at the .05 level (two-tailed).

c) What conclusions can you draw from your answer to part b?

* 10. One of the most common tools of the cognitive psychologist is the lexical decision task, in which a string of letters is flashed on a screen and a subject must decide as quickly as possible whether those letters form a word. This task is often used as part of a more complex experiment, but this exercise considers reaction time as a function of the number of letters in a string for a single subject. The data below represent the subject's reaction times in response to three strings of each of four lengths: 3, 4, 5, or 6 letters. Calculate Pearson's correlation coefficient for the data.

Trial	Number of Letters in String	Reaction Time (in milliseconds)
1	6	930
2	5	900
3	3	740
4	5	820
5	4	850
6	4	720
7	3	690
8	6	990
9	4	810
10	3	830
11	6	880
12	5	950

The Power Associated with Correlational Tests

It is not difficult to apply the procedures of power analysis that were described in the previous chapter to statistical tests involving Pearson's r. Just as the null hypothesis for the two-group t-test is almost always $\mu_1 - \mu_2 = 0$, the null hypothesis for a correlational study is almost always $\rho_0 = 0$. The alternative hypothesis is usually stated as $\rho_A \neq 0$ (for a one-tailed test, $\rho_A < 0$ or $\rho_A > 0$), but in order to study power it is necessary to hypothesize a particular value for the population correlation coefficient. Given that ρ_A does not equal 0, the sample rs will be distributed around whatever value ρ_A does equal. The value for ρ_A specified by the alternative hypothesis can be thought of as an "expected" r. How narrowly the sample rs are distributed around the expected r depends on the sample size.

However, to understand power analysis for correlation, it is important to appreciate a fundamental difference between Pearson's r and the t value for a two-group test. The expected r (ρ_A) is a measure similar to γ (the effect size associated with a t-test); it does not depend on sample size. Rather, the expected r describes the size of an effect in the population, and its size does not tell you whether a particular test of the null hypothesis is likely to be statistically significant. By contrast, the t-value is a reflection of *both* γ *and* sample size, and it *does* give you a way to determine the likelihood of attaining statistical significance. The point is that when performing power analysis for correlation, the expected r plays the same role as γ in the power analysis of a t-test, and not the role of expected t. We still need to transform ρ_A (i.e., the expected r) into a delta value that can be looked up in the power table. This transformation is done with Formula 11.7:

$$\delta = \sqrt{N - 1}\, \rho_A \qquad \qquad \text{Formula 11.7}$$

Notice the similarity to the formula for delta in the one-group t-test, and recall that ρ_A plays the same role as γ. For instance, if we have reason to expect a correlation of .35, and we have 50 subjects available,

$$\delta = \sqrt{50 - 1}\,(.35) = \sqrt{49}\,(.35) = 7(.35) = 2.45$$

Assuming alpha = .05, two-tailed, from Table A.3 we find that for $\delta = 2.45$, power is between .67 and .71. Chances are considerably better than 50% that the sample of 50 subjects will produce a statistically significant Pearson's r (if we are right that the true $\rho = .35$), but the chances may not be high enough to justify the expense and effort of conducting the study.

If we still expect that $\rho = .35$, but we desire power to be .85, we can calculate the required N by solving Formula 11.7 for N to create Formula 11.8:

$$N = \left(\frac{\delta}{\rho_A}\right)^2 + 1 \qquad \qquad \text{Formula 11.8}$$

From Table A.4 we see that to obtain power = .85, we need $\delta = 3.0$ (alpha = .05, two-tailed). Plugging this value into the formula above, we find that

$$N = (3/.35)^2 + 1 = (8.57)^2 + 1 = 73.5 + 1 = 74.5$$

Therefore, a sample of 75 subjects is required to have power = .85, if $\rho = .35$.

As demonstrated in Chapter 10, power analysis can be used to determine the maximum number of subjects that should be used. First determine δ based on the desired levels for power and α. Then plug the smallest correlation of interest for the variables you are dealing with into Formula 11.8 in place of ρ_A. The N given by the calculation is the largest sample size you should use. Any larger N will have too high a chance of giving you statistical significance when the true correlation is so low that you don't care about it. Similarly, in place of ρ_A you can plug in the largest correlation that can reasonably be expected. The N that is thus calculated is the minimum sample size you should employ; any smaller sample size will not give you a strong enough chance of obtaining statistical significance.

Sometimes the magnitude of the correlation expected in a new study can be predicted based on previous studies (as with gamma). At other times the expected correlation is characterized roughly as small, medium, or large. The conventional guideline for Pearson's r (Cohen, 1988) is that .1 = small, .3 = medium, and .5 = large. Correlations much larger than .5 usually involve two variables that are measuring the same thing, such as the various types of reliability described earlier, or for instance, two different questionnaires both designed to assess a subject's current level of depression.

You may have noticed that whereas $r = .5$ is considered a "large" correlation, it is only a "medium" value for gamma. This difference arises from the fact that even though ρ and γ are similar in terms of what they are assessing in the population, they are measured on very different scales. Correlation can only range between 0 and 1 in magnitude, whereas γ is like a z score, with no limit to how large it can get. Although it is a measure of effect size, correlation is more often referred to as a measure of the *strength of association* between two variables.

Once you have determined that a particular sample r is statistically significant, you can rule out (with a degree of confidence that depends on the alpha level used) that ρ = 0—that the two variables have *no* linear relationship in the population. More than that, your sample r provides a point estimate of ρ; if $r = .4$, then .4 is a good guess for ρ. However, as I first discussed in Chapter 8, the accuracy of the point estimate depends on the sample size. You can be more confident that ρ is close to .4 if the r of .4 comes from a sample of 100 subjects than if the .4 was calculated for only 10 subjects. An interval estimate would be more informative than the point estimate, providing a clearer idea of what values are likely for the population correlation coefficient. Interval estimation for ρ is not performed as often as it is for μ, probably because it has fewer practical implications—but it should be performed more often than it is. Unfortunately, constructing a confidence interval for ρ involves a complication that does not arise when dealing with μ, as you will see next.

Fisher *Z* Transformation

The complication in constructing a confidence interval for ρ concerns the distribution of sample rs around ρ. When ρ = 0, the sample rs form a symmetrical distribution around 0, which can be approximated by a normal distribution (especially as the sample size gets fairly large). However, when ρ equals, for example, $+.8$, the sample rs are not going to be distributed symmetrically around $+.8$. There is a "ceiling" effect; r cannot get higher than $+1.0$, so there is much more room for r to be lower than $+.8$ (as low as -1.0) than to be higher than $+.8$. The distribution will be negatively skewed (it would be positively skewed for ρ = $-.8$); the closer ρ gets to $+1.0$ or -1.0, the more skewed the distribution becomes. Fortunately, Fisher (1970) found a way to transform the sample rs, so that they will follow an approximate normal distribution, regardless of ρ. This method allows us once again to use the simplicity of z scores in conjunction with the standard normal table (Table A.1). But first we have to transform our sample r. Although the transformation is based on a formula that requires finding logarithms, the transformations have already been done by statisticians and listed in a convenient table (see Table A.6). The transformed rs are called Zs (the capital Z is usually used to avoid confusion with z scores), and the process is referred to as the Fisher Z transformation. I will make use of this transformation below.

The Confidence Interval for ρ

To construct a confidence interval for ρ, we would expect to begin by putting a point estimate for ρ (e.g., the sample r) in the middle of the interval. But to ensure a normal distribution, we first transform r by finding the corresponding Z_r in Table A.6. We must also select a level of confidence. We'll begin with the usual 95% CI. Assuming a normal distribution around the transformed point estimate, we know which z scores will mark off the middle 95%: +1.96 and −1.96. This means that we need to go about two standard errors above and below Z_r. Finally, we need to know the standard error for Z_r. Fortunately, this standard error is expressed in an easy formula:

$$\sqrt{\frac{1}{N-3}} = \frac{1}{\sqrt{N-3}}$$

We can now work out a numerical example. The Pearson's r for the study of years of education and number of children was −.735. Looking in Table A.6 we find that the corresponding Z_r for +.735 is .94, so we know that for −.735, Z_r equals −.94. Next we find the standard error. Because $N = 6$ in our example, the standard error $= 1/\sqrt{N-3} = 1/\sqrt{6-3} = 1/\sqrt{3} = 1/1.732 = .577$. Having chosen to construct a 95% CI, we add and subtract 1.96 · .577 from −.94. So the upper boundary is −.94 + 1.13 = +.19 and the lower boundary is −.94 − 1.13 = −2.07.

Perhaps you noticed something peculiar about these boundaries. Correlation cannot be less than −1, so how can the lower boundary be −2.07? But remember that these boundaries are for the *transformed* correlations, not for ρ. To get the upper and lower limits for ρ, we have to use Table A.6 in reverse. We look up the values of the boundaries calculated above in the Z_r column to find the corresponding rs. Although we cannot find .19 as an entry for Z_r, there are entries for .187 and .192, corresponding to rs of .185 and .190, respectively. Therefore, we can estimate that $Z_r = .190$ would correspond approximately to $r = .188$. That is the upper limit for ρ. To find the lower limit we look for 2.07 under the Z_r column. (As in the normal distribution table, there is no need for negative entries—the distribution is symmetrical around zero.) There are entries for 2.014 and 2.092, corresponding to rs of .965 and .970, respectively. Therefore, we estimate that the lower limit for ρ is approximately −.968. For this example, we can state with 95% confidence that the population correlation coefficient for these two variables is somewhere between −.968 and +.188. Note that ρ = 0 is included in the 95% CI. That tells us that the null hypothesis (i.e., ρ = 0) cannot be rejected at the .05, two-tailed level, confirming the hypothesis test that we conducted in Section B.

I hope you noticed that the CI we just found is so large as to be virtually useless. This result is due to the ridiculously small sample size. Normally, you would not bother to calculate a CI when dealing with such a small N. In fact, the use of $1/\sqrt{N-3}$ to represent the standard error of Z is not accurate for small sample sizes, and the inaccuracy gets worse as the sample r deviates more from zero. It is also assumed that the variable in question has a bivariate normal distribution in the population. If your distribution seems strange, and/or the sample r is quite high in magnitude, it is important to use a fairly large sample size. As usual, 30 to 40 would be considered minimal in most situations, but an N of more than 100 may be required for accuracy in more extreme situations.

The process outlined above for constructing a confidence interval for ρ can be formalized in the following formula:

$$Z_\rho = Z_r \pm z_{\text{crit}} \, \sigma_r$$

Note how similar this formula is to Formula 8.4 for the population mean. If we insert the equation for the standard error into the above formula and separate the formula into upper and lower limits, we obtain Formula 11.9.

$$\text{Upper } Z_\rho = Z_r + z_{\text{crit}} \frac{1}{\sqrt{N-3}}$$

$$\text{Lower } Z_\rho = Z_r - z_{\text{crit}} \frac{1}{\sqrt{N-3}} \qquad \text{Formula 11.9}$$

Don't forget that Formula 11.9 gives the confidence interval for the transformed correlation. The limits found with Formula 11.9 must be converted back to ordinary correlation coefficients by using Table A.6 in reverse.

Testing a Null Hypothesis Other Than $\rho = 0$

I mentioned in the previous section that the null hypothesis for correlation problems is usually $\rho = 0$, and that matters get tricky if $\rho \neq 0$. The problem is that when $\rho \neq 0$, the sample rs are not distributed symmetrically around ρ, making it difficult to describe the null hypothesis distribution. This is really the same problem as finding the confidence interval around a sample r that is not zero, and it is also solved by the Fisher Z transformation. Let us once more consider the correlation between education and number of children. Suppose that a national survey 50 years ago showed the correlation (ρ) to be $-.335$, and we want to show that the present correlation, $r = -.735$, is significantly more negative. Also suppose for this problem that the present correlation is based on 19 subjects instead of only 6.

Now the null hypothesis is that $\rho_0 = -.335$. To test the difference between the present sample r and the null hypothesis, we use the formula for a one-group z-test, modified for the transformed correlation coefficients:

$$z = \frac{Z_r - Z_\rho}{\sigma_r}$$

If we insert the equation for the standard error (the same equation used for the standard error in the confidence interval) into the above equation, we obtain Formula 11.10:

$$z = \frac{Z_r - Z_\rho}{\sqrt{1/(N - 3)}}$$ Formula 11.10

Before we can apply this formula to our example, both the sample r and ρ_0 must be transformed by Table A.6. As we found before, $-.735$ corresponds to $-.94$. For $-.335$ (ρ_0) the transformed value is $-.348$. Inserting these values into Formula 11.10, we get

$$z = \frac{-.94 - (-.348)}{\sqrt{1/16}} = \frac{-.592}{.25} = -2.37$$

Because $z = -2.37$ is less than -1.96, the null hypothesis, that $\rho_0 = -.335$, can be rejected at the .05 level. We can conclude that the education/children correlation is more negative in today's population than it was 50 years ago.

Testing the Difference of Two Independent Sample *r*s

The procedures described above can be modified slightly to handle one more interesting test involving correlations. If a sample r can be tested in comparison to a population ρ that is not zero, it is just a small step to creating a formula that can test the difference between two sample *r*s. For example, the education/children correlation we have been dealing with was calculated on a hypothetical random group of six women. Suppose that for the purpose of comparison, a group of nine men is randomly selected and measured on the same variables. If the correlation for men were in the reverse direction and equal to $+.4$, it would seem likely that the correlation for men would be significantly different from the correlation for women. However, a hypothesis test is required to demonstrate this.

The formula for comparing two sample *r*s is similar to the formula for a two-group z-test:

$$z = \frac{Z_{r_1} - Z_{r_2}}{\sigma_{r_1 - r_2}}$$

The standard error of the difference of two sample *r*s is a natural extension of the standard error for one correlation coefficient, as the following formula shows:

$$\sigma_{r_1 - r_2} = \sqrt{\frac{1}{N_1 - 3} + \frac{1}{N_2 - 3}}$$

Combining this formula with the one above, we obtain a formula for testing the difference of two sample rs, Formula 11.11:

$$z = \frac{Z_{r_1} - Z_{r_2}}{\sqrt{1/(N_1 - 3) + 1/(N_2 - 3)}} \qquad \text{Formula 11.11}$$

To apply Formula 11.11 to our present example, we must first transform each of the sample rs, and then insert the values, as follows:

$$z = \frac{-.94 - (+.424)}{\sqrt{1/(6 - 3) + 1/(9 - 3)}} = \frac{-1.364}{\sqrt{1/2}} = \frac{-1.364}{.707} = -1.93$$

Surprisingly, the null hypothesis cannot be rejected at the .05 level (two-tailed); -1.93 is not less than -1.96. The reason that two sample rs that are so discrepant are not significantly different is that the sample sizes are so small, making the power quite low.

It is sometimes interesting to compare two sample rs calculated on the *same* group of people, but with different variables. For instance, you may wish to test whether the correlation between annual income and number of children is higher than the correlation between years of education and number of children for a particular group of women. Because the two correlation coefficients involve the same people, they are not independent, and you cannot use Formula 11.11. A statistical solution has been worked out to test the difference of two non-independent rs, and the interested reader can find this solution in a more advanced text, such as the one by Howell (1992).

Exercises

* 1. What is the value of Fisher's Z transformation for the following Pearson correlation coefficients? a) .05 b) .1 c) .3 d) .5 e) .7 f) .9 g) .95 h) .99 What is the value of Pearson's r that corresponds most closely with each of the following values for Fisher's Z transformation? i) .25 j) .50 k) .95 l) 1.20 m) 1.60 n) 2.00 o) 2.50

2. a) As Pearson's r approaches 1.0, what happens to the discrepancy between r and Fisher's Z?
b) For which values of Pearson's r does Fisher's Z transformation seem unnecessary?

* 3. In exercise 11B5,
a) what would the power of the test have been if the correlation for the population (ρ) were .5?
b) What would the power of the test have been

if ρ were equal to the sample r found for that problem?
c) How many schizophrenics would have to be tested if ρ were equal to the sample r and you wanted power to equal .90?

4. For exercise 11B5, a) find the 95% confidence interval (CI) for the population correlation coefficient relating years of hospitalization to orientation score.
b) Find the 99% CI for the same problem. Is the sample r significantly different from zero, if alpha = .01 (two-tailed)? Explain how you can use the 99% CI to answer this question.

* 5. In exercise 11B5, suppose the same two variables were correlated for a sample of 15 prison inmates, and the sample r were equal to .2. Is the sample r for the

schizophrenics significantly different (alpha = .05, two-tailed) from the sample r for the inmates?

6. a) If a correlation less than .1 is considered too small to be worth finding statistically significant, and power = .8 is considered adequate, what is the largest sample size you should use if a .05 two-tailed test is planned?

b) If you have available a sample of 32 subjects, how highly would two variables have to be correlated in order to have power = .7 with a .01, two-tailed test?

*7. Suppose that the population correlation for the quantitative and verbal portions of the SAT equals .4.

a) A sample of 80 psychology majors is found to have a quantitative/verbal correlation of .5. Is this cor-relation significantly different from the population at the .05 level, two-tailed?

b) A sample of 120 English majors has a quantitative/verbal correlation of .3. Is the correlation for the psychology majors significantly different from the correlation for the English majors (α = .05, two-tailed)?

8. For a random sample of 50 adults the correlation between two well-known IQ tests is .8.

a) Find the limits of the 95% confidence interval for the population correlation coefficient.

b) If a sample of 200 subjects had the same corre-lation, what would be the limits of the 95% CI for this sample?

c) Compare the widths of the CIs in parts a and b. What can you say about the effects of sample size on the width of the confidence interval for the population correlation?

SUMMARY

The Important Points of Section A

1. If each subject has the same score on two different variables (e.g., midterm and final exams), the two variables will be perfectly correlated. However, this condition is not necessary. Perfect correlation can be defined as all subjects having the same z score (i.e., the same relative position in the distribution) on both variables.

2. Negative correlation is the tendency for high scores on one variable to be associated with low scores on a second variable (and vice versa). Perfect negative correlation occurs when each subject has the same magnitude z score on the two variables, but the z scores are *opposite in sign*.

3. Pearson's correlation coefficient, $r,$ can be defined as the average of the cross products of the z scores on two variables. Pearson's r ranges from -1.0 for perfect negative correlation, to 0 when there is no linear relationship between the vari-ables, to $+1.0$ when the correlation is perfectly positive.

4. A *linear transformation* is the conversion of one variable into another by only arithmetic operations (i.e., adding, subtracting, multiplying, or dividing) involv-ing constants. If one variable is a linear transformation of another, each subject will have the same z score on both variables, and the two variables will be perfectly correlated. Changing the units of measurement on either or both of the variables will not change the correlation, as long as the change is a linear one (as is usually the case).

5. A *scatterplot,* or *scattergram,* is a graph of one variable plotted on the X axis versus a second variable plotted on the Y axis. A scatterplot of perfect correlation will be a straight line that slopes up to the right for positive correlation and down to the right for negative correlation.

6. One important property of Pearson's r is that it assesses only the degree of *linear* relationship between two variables. Two variables can be closely related by a very simple curve, and yet produce a Pearson's r near zero.

7. Problems can occur when Pearson's r is measured on a subset of the population but you wish to extrapolate these results to estimate the correlation for the entire population (ρ). The most common problem is having a *truncated,* or *restricted,* range on one or both of the variables. This problem usually causes the r for the sample to be considerably less than ρ, although in rare cases the opposite can occur (e.g., when you are measuring one portion of a curvilinear relationship).

8. Another potential problem with Pearson's r is that a few *bivariate outliers* in a sample can drastically change the magnitude (and, in rare instances, even the sign) of the correlation coefficient. It is very important to inspect a scatterplot for curvilinearity, outliers, and other aberrations before interpreting the meaning of a correlation coefficient.

9. Correlation, even if high in magnitude and statistically significant, does not prove that there is any causal link between two variables. There is always the possibility that some third variable is separately affecting each of the two variables being studied. An experimental design, with random assignment of subjects to conditions, is required to determine whether one particular variable is *causing* changes in a second variable.

The Important Points of Section B

In order to review the statistical procedures discussed in Section B, I will describe a hypothetical study to determine whether people who come from large immediate families (i.e., people who have many siblings) tend to create large immediate families (i.e., produce many children). In this example, each "subject" is actually an entire immediate family (parents and their children). The X variable is the average number of siblings for the two parents (e.g., if the mother has one brother and one sister, and the father has two brothers and two sisters, $X = 3$); the Y variable is the number of children in the selected family. The families

Table 11.2

X^2	X	XY	Y	Y^2
9	3	9	3	9
1	1	2	2	4
16	4	20	5	25
4	2	4	2	4
2.25	1.5	1.5	1	1
4	2	8	4	16
16	4	12	3	9
6.25	2.5	10	4	16
4	2	4	2	4
1	1	1	1	1
Totals 63.5	23	71.5	27	89

should be selected independently and randomly (each family being selected as a single subject). For this hypothetical example, ten families were selected, so $N = 10$. The data in Table 11.2 consist of the mean number of parental siblings, *X,* and the number of children for each of the ten selected families, *Y,* as well as the squared values and cross products required for some of the computational formulas.

As long as we are satisfied with assessing the degree of linear correlation, the appropriate test statistic is Pearson's *r,* and it can be calculated with any of the formulas presented in Section B. I will illustrate the use of both Formula 11.2 and Formula 11.5, to demonstrate that two formulas can look very different and yet (if they are algebraically equivalent) always yield the same answer (except for error due to rounding off). First, I will insert the various sums in the last line of the table into Formula 11.5:

$$r = \frac{N\Sigma XY - (\Sigma X)(\Sigma Y)}{\sqrt{[N\Sigma X^2 - (\Sigma X)^2][N\Sigma Y^2 - (\Sigma Y)^2]}} = \frac{715 - (23)(27)}{\sqrt{(635 - 23^2)(890 - 27^2)}}$$
$$= \frac{715 - 621}{\sqrt{(106)(161)}} = \frac{94}{\sqrt{17066}} = \frac{94}{130.64} = +.71955$$

Having found Pearson's *r* using the raw-score formula, I will calculate the means and standard deviations, so that we can use Formula 11.2. The mean of *X* equals $23/10 = 2.3$, and the mean of *Y* equals $27/10 = 2.7$. Using the appropriate formula from Chapter 4, you can verify that $\sigma_X = 1.0296$ and $\sigma_Y = 1.2688$. Plugging these values into Formula 11.2, we get

$$r = \frac{\Sigma XY/N - \mu_X\mu_Y}{\sigma_X\sigma_Y} = \frac{71.5/10 - (2.3)(2.7)}{(1.0296)(1.2688)} = \frac{7.15 - 6.21}{1.3064} = \frac{.94}{1.3064} = .71955$$

As you can see, both formulas produce exactly the same value for Pearson's r. I prefer to use Formula 11.2, because the means and standard deviations can be checked for accuracy before proceeding. However, because the numerator and denominator are smaller in magnitude (as compared to those in Formula 11.5), you must be careful not to round off too much on the intermediate steps; for instance, in calculating the standard deviation, retain at least four digits *after* the decimal point (as in the example above).

The r we calculated can be tested for statistical significance by comparing it to a critical r in Table A.5. For correlation problems the number of degrees of freedom equals $N - 2$; for this example, df $= 10 - 2 = 8$. Looking in Table A.5 under alpha $= .05$, two-tailed, we find that the critical r is .632. Therefore, the region of rejection consists of sample rs that are above $+.632$ or below $-.632$. Because the sample r calculated above, $+.72$, is higher than the positive critical r, the null hypothesis (i.e., $\rho = 0$) can be rejected. To test a null hypothesis other than $\rho = 0$, you would use the methods of Section C.

Our statistical conclusion, it must be noted, is only valid if we have met the assumptions of the test, which are as follows:

1. *Independent random sampling.* Any restriction in the range of values sampled can threaten the accuracy of our estimate of ρ.
2. *Normal distributions.* Each variable should be inspected separately to see that it follows an approximately normal distribution, especially if sample sizes are small.
3. *Bivariate normal distribution.* The scatterplot of the data should be inspected for unusual characteristics, such as bivariate outliers or curvilinearity.

The Important Points of Section C

1. The calculation of power for testing the significance of a correlation coefficient was discussed. The hypothesized population correlation coefficient (ρ_A) plays the same role played by gamma in finding the power of a two-group t-test; ρ_A must be combined with the proposed sample size to find a value for delta, which can then be used to look up power.

2. A confidence interval, centered on the sample r, can be constructed to estimate ρ. To construct CIs around a sample r, Fisher's Z transformation must be used (Table A.6). The CI formula is otherwise similar to the one for the population mean.

3. The Fisher Z transformation is also needed for significance testing when the null hypothesis specifies that ρ is some particular value other than zero, and for testing the significance of a difference between two sample rs.

Definitions of Key Terms

Perfect correlation Perfect positive correlation means each subject has the same z score on two variables. Perfect negative correlation means each subject has the same magnitude z score on two variables, but the z scores are opposite in sign.

Positive correlation As the magnitude of one variable increases, the second variable tends to increase, as well; as one decreases, the other tends to decrease.

Negative correlation As the magnitude of one variable increases, the second variable tends to decrease; decreases in the first variable tend to be associated with increases in the second variable.

Pearson's correlation coefficient (r) A measure of the degree of linear relationship between two variables. Ranges from -1 for perfect negative correlation, to 0 for a total lack of linear relationship, to $+1$ for perfect positive correlation; also called "Pearson's product-moment correlation coefficient."

Linear transformation The transformation of one variable into another variable by adding, subtracting, multiplying by, and/or dividing by constants (but using no other mathematical operations). After a linear transformation, each subject has the same z score as before the transformation.

Scatterplot (scatter diagram or scattergram) A graph in which one variable is plotted on the X axis and the other variable is plotted on the Y axis. Each subject (or observation) is represented by a single dot on the graph.

Curvilinear correlation A relationship between two variables such that the scatterplot appears as a curve instead of a straight line. Coefficients of curvilinear correlation measure such relationships.

Population correlation coefficient (ρ) The Pearson correlation coefficient calculated on an entire population.

Truncated (or restricted) range A problem that occurs when a sample fails to include the full range of values of some variable that is represented in the population. This usually reduces the magnitude of the sample r as compared to ρ.

Bivariate outliers Data points that need not be extreme on either variable separately, but rather represent an unusual combination of values of the two variables. Even one bivariate outlier can greatly influence the correlation coefficient.

Covariance The tendency of two variables either to vary together or to vary consistently in opposite directions; must be divided by the product of the standard deviations of the two variables in order to yield a correlation coefficient that varies in magnitude between 0 and 1.

Bivariate distribution A distribution that represents the relative likelihood of each possible *pair* of values for two variables. In order to test Pearson's r for significance, it must be assumed that the two variables follow a *bivariate normal distribution.*

Key Formulas

Pearson's product-moment correlation coefficient (definitional form; not convenient for calculating, unless *z* scores are readily available):

$$r = \frac{\sum z_X z_Y}{N} \qquad \text{Formula 11.1}$$

Pearson's product-moment correlation coefficient (convenient when the *biased* sample variances have already been calculated):

$$r = \frac{\dfrac{\sum XY}{N} - \mu_X \mu_Y}{\sigma_X \sigma_Y} \qquad \text{Formula 11.2}$$

Pearson's product-moment correlation coefficient (convenient when the *unbiased* sample variances have already been calculated):

$$r = \frac{\dfrac{1}{N-1}(\sum XY - N\bar{X}\bar{Y})}{s_X s_Y} \qquad \text{Formula 11.3}$$

Pearson's product-moment correlation coefficient (not a convenient formula for calculation; it is included here only because it appears in various other texts):

$$r = \frac{\sum(X - \bar{X})(Y - \bar{Y})}{\sqrt{SS_X SS_Y}} = \frac{SP}{\sqrt{SS_X SS_Y}} \qquad \text{Formula 11.4}$$

Pearson's product-moment correlation coefficient, raw-score version:

$$r = \frac{N\sum XY - (\sum X)(\sum Y)}{\sqrt{[N\sum X^2 - (\sum X)^2][N\sum Y^2 - (\sum Y)^2]}} \qquad \text{Formula 11.5}$$

A *t* value that can be used to test the statistical significance of Pearson's *r* against the H_0 that $\rho = 0$ (as an alternative, the critical *r* can be found directly from Table A.5):

$$t = \frac{\sqrt{N-2}\, r}{\sqrt{1 - r^2}} \qquad \text{Formula 11.6}$$

Delta, to be used in the determination of power for a particular hypothesized value of rho and a given sample size:

$$\delta = \sqrt{N - 1}\, \rho_A$$

Formula 11.7

The required sample size to attain a given level of power (in terms of delta) for a particular hypothesized value of rho:

$$N = \left(\frac{\delta}{\rho_A}\right)^2 + 1$$

Formula 11.8

Confidence interval for the population correlation coefficient (ρ) in terms of Fisher's Z transformation; the limits found by this formula must then be transformed back into ordinary rs using Table A.6:

$$\text{Upper } Z_\rho = Z_r + z_{\text{crit}} \frac{1}{\sqrt{N - 3}}$$

$$\text{Lower } Z_\rho = Z_r - z_{\text{crit}} \frac{1}{\sqrt{N - 3}}$$

Formula 11.9

z-test to determine the significance of a sample r against a specific null hypothesis (other than $\rho_0 = 0$):

$$z = \frac{Z_r - Z_\rho}{\sqrt{1/(N - 3)}}$$

Formula 11.10

z-test to determine whether two independent sample rs are significantly different from each other (the null hypothesis is that the samples were drawn from populations with the same ρ):

$$z = \frac{Z_{r_1} - Z_{r_2}}{\sqrt{1/(N_1 - 3) + 1/(N_2 - 3)}}$$

Formula 11.11

LINEAR REGRESSION

You will need to use the following from previous chapters:

Symbols
μ: Mean of a population
$\overline{X}$: Mean of a sample
σ: Standard deviation of a population
s: Unbiased standard deviation of a sample
r: Pearson's product-moment correlation coefficient

Formulas
Formula 5.1: The z score
Formula 8.3: The t-test for one sample
Formula 11.5: Pearson correlation (raw-score version)

Concepts
The normal distribution
Linear transformations

12

CHAPTER

A

CONCEPTUAL FOUNDATION

In the previous chapter I demonstrated that two variables could be perfectly correlated, even if each subject did not have the same number on both variables. For instance, height and weight could be perfectly correlated for a particular group of people, although the units for the two variables are very different. The most useful property of perfect correlation is the perfect predictability that it entails. For a group in which height and weight are perfectly correlated, you can use a particular subject's height to predict, without error, that subject's weight (with a simple formula). Of course, the predictability would be just as perfect for two variables that were negatively correlated, as long as the negative correlation were perfect. It should also come as no surprise that when correlation is nearly perfect, prediction is nearly perfect as well, and therefore very useful. Unfortunately, when predictability is needed most in real-life situations (e.g., trying to predict job performance based on a test), correlation is not likely to be extremely high, and certainly not perfect. Fortunately, however, even when correlation is not very high, the predictability may be of some practical use. In this chapter you will learn how to use the linear relationship between two variables to make predictions about either one.

Perfect Predictions

The prediction rule that is used when linear correlation is perfect could not be simpler, especially when it is expressed in terms of z scores (as calculated by Formula 5.1). The rule is that the z score you predict for the Y variable is the same as the z score for the X variable. (It is conventional to plot the variable you wish to predict on the Y axis of a scatterplot and the variable you are predicting *from* on the X axis.) As a formula, this rule would be written as $z_{Y'} = z_X$, where the prime following the subscript Y signifies that it is the Y value that is being predicted. (Y' is pronounced "Y prime"; other symbols are sometimes used to indicate that Y is being predicted, but Y' seems to be the most popular convention.) For perfect negative correlation, the formula is the same except for the minus sign: $z_{Y'} = -z_X$. For nearly perfect correlation it might seem reasonable to follow the same rule, but whenever the correlation is less than perfect, the above rule does not give the best predictions. As the correlation becomes smaller in magnitude, there is a greater need for a modified prediction rule, as shown below.

Predicting with z Scores

Imagine trying to predict a student's math SAT score from his or her verbal SAT score. If we use the simple prediction rule described above, then a student two deviations above the mean in the verbal SAT score ($z = +2$) would be predicted to have a math SAT score two standard deviations above the mean, as well. However, because the correlation between these two variables is always far from perfect, there is plenty of room for error, so predicting $z = +2$ for the math SAT would be going out on a limb. The usual procedure is to "hedge your bet," knowing that the lower the correlation, the greater room there is for error. A modified rule has been devised to minimize the error of predictions when the correlation is less than perfect. The rule is given by Formula 12.1,

$$z_{Y'} = r \, z_X \qquad \qquad \text{Formula 12.1}$$

where r is the Pearson correlation coefficient described in Chapter 11.

Note that when r is $+1$ or -1, the formula reverts to the simple rule for perfect correlation that was discussed above. Note also that when there is no linear correlation between the two variables (i.e., $r = 0$), the prediction is always that the z score is zero, which implies that we are predicting that the Y variable will be at its mean. This strategy makes sense. If you have to predict the weight of each person in a group, and you have no information at all, error is minimized by predicting the mean weight in each case. Why this minimizes the error has to do with how we measure the error of predictions, which will be discussed later in this section. When correlation is less than perfect but greater than zero (in magnitude), Formula 12.1 represents a compromise between predicting the same z score as the first variable and predicting the mean of the second variable. As the correlation becomes lower, there is less of a tendency to expect an extreme score on one variable to be associated with an equally extreme score on the other. On the other hand, as long as the correlation is not zero, the first variable is taken into account in the prediction of the second variable; the first variable clearly has an influence, but that influence lessens as the correlation between the variables is reduced.

Calculating an Example

To make the discussion more concrete, let us apply Formula 12.1 to the prediction of math SAT scores from verbal SAT scores. If we assume that the correlation between these two variables is $+.4$, Formula 12.1 becomes $z_{Y'} = .4z_X$. A person with a verbal SAT z score of $+1.0$ would be predicted to have a math SAT z score of $+.4$. A person with a verbal SAT of $z = -2.0$ would be predicted to have $z = -.8$ in math. Of course, in most applications the data will not be in the form of z scores; in order to use Formula 12.1 you would first have to convert each score into a z score. Rather than converting to z scores, it is easier to use a prediction formula that is designed to deal with the original scores. The derivation of this formula will be presented shortly.

Regression Toward the Mean

I have been calling Formula 12.1 a prediction formula, which it is, but it is more commonly called a *regression formula*. The term comes from the work of Sir Francis Galton, who, among other projects, studied the relation between the height of a subject and the heights of the subject's parents. He found that Formula 12.1 applied to his data with a correlation coefficient of about .67. In fact, Karl Pearson's derivation of the formula we now use for the correlation coefficient was motivated by Galton's work (Cowles, 1989). Galton noted the tendency for unusually tall people to have children shorter than themselves; the children were usually taller than average, of course, but not as tall as the parents. Similarly, unusually short people generally had children closer to the mean than themselves. Galton referred to this tendency as "regression toward mediocrity," but it is now referred to as **regression toward the mean.** At one point, this phenomenon was seen as some sort of natural pressure pushing toward mediocrity, but now scientists recognize that regression toward the mean is just a consequence of the laws of probability, when correlation is less than perfect. Looking at Formula 12.1, you can see that as the correlation gets lower, the prediction gets closer to $z = 0$; that is, it regresses toward the mean. Because it is based on r, which measures linear correlation, Formula 12.1 is a formula for *linear* regression. Other forms of regression have been devised to handle more complex relationships among variables, but only linear regression will be covered in this text.

Graphing Regression in Terms of z Scores

The use of Formula 12.1 can be clarified by means of a scatterplot. When a scatterplot is used in conjunction with z scores, the zero point (called the *origin*) is in the middle, so that the horizontal and vertical axes can extend in the negative direction (to the left and down, respectively; see Figure 12.1). In terms of z scores, the scatterplot for perfect positive correlation is just a diagonal line (at an angle of 45 degrees) that passes through the origin, as in Figure 12.1. Notice that each point along the line corresponds to the same z score on both the X and the

Figure 12.1

Scatterplot for Perfect Correlation in Terms of z-Scores

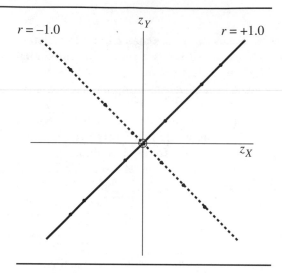

Y axis. Perfect negative correlation is represented by a similar diagonal line slanted in the opposite direction.

If correlation is not perfect, for example, if $r = +.5$, Formula 12.1 becomes $z_{Y'} = .5z_X$, which corresponds to the prediction line shown in Figure 12.2. In keeping with the traditional terminology, the prediction line is called the **regression line.** For any z_X, you can use this line to find the best prediction for z_Y (in this case, half of z_X). Notice that the regression line for $r = +.5$ makes a smaller angle with the *X* axis than $r = +1$. (Compare Figure 12.2 to Figure 12.1.)

Figure 12.2

Scatterplot for $r = .5$ in Terms of z-Scores

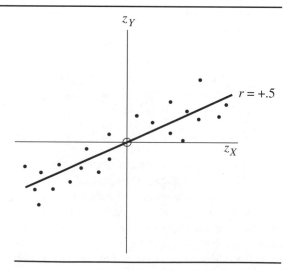

The angle of the regression line (as with any straight line on a graph) is called the **slope of the regression line.** The slope can be measured in degrees, but it is usually measured as the change in the Y axis divided by the change in the X axis. When $r = +1$, the line goes *up* one unit (change in Y axis) for each unit it moves to the *right* (change in X axis). However, when $r = +.5$, the line goes up only $\frac{1}{2}$ unit each time it moves one unit to the right, so the slope is .5.

The slope of the regression line, plotted in terms of z scores, always equals the correlation coefficient. Figure 12.2 illustrates the scatter of the data points around the regression line when $r = .5$. When correlation is perfect, all of the data points fall on the regression line, but as the correlation coefficient gets lower, the data points are more widely scattered around the regression line (and the slope of the regression line gets smaller, too). Regression lines are particularly easy to draw when dealing with z scores, but this is not the way regression is commonly done. However, in order to deal directly with the original scores, we need to modify the regression formula as shown below.

The Raw-Score Regression Formula

In order to transform Formula 12.1 into a formula that can accommodate raw scores, Formula 5.1 for the z score must be substituted for z_X and z_Y. (This is very similar to the way Formula 11.1 was transformed into Formula 11.2 in the previous chapter.) The formula that results must then be solved for Y' in order to be useful.

$$z_{Y'} = r\, z_X$$

$$\frac{Y' - \mu_Y}{\sigma_Y} = r\left(\frac{X - \mu_X}{\sigma_X}\right)$$

$$Y' - \mu_Y = r\left(\frac{X - \mu_X}{\sigma_X}\right)\sigma_Y$$

$$Y' - \mu_Y = \frac{\sigma_Y}{\sigma_X}\, r\,(X - \mu_X)$$

$$Y' = \frac{\sigma_Y}{\sigma_X}\, r\,(X - \mu_X) + \mu_Y \qquad \text{Formula 12.2}$$

Note that Formula 12.2 is expressed in terms of population parameters (i.e., μ and σ). This is particularly appropriate when regression is being used for descriptive purposes only. In that case, whatever scores you have are treated as a population; it is assumed you have no interest in making inferences about a larger, more inclusive set of scores. In Section B, I will express these two formulas in terms of sample statistics and take up the matter of inference, but fortunately the formulas will change very little. In the meantime, Formula 12.2 can be put into an even simpler form with a bit more algebra. First I will create Formula 12.3 and define a new symbol, as follows:

$$b_{YX} = \frac{\sigma_Y}{\sigma_X} r$$
Formula 12.3

Formula 12.2 can be rewritten in terms of this new symbol:

$$Y' = b_{YX}(X - \mu_X) + \mu_Y$$

Multiplying to get rid of the parentheses yields

$$Y' = b_{YX}X - b_{YX}\mu_X + \mu_Y$$

One final simplification can be made by defining one more symbol, using Formula 12.4:

$$a_{YX} = \mu_Y - b_{YX}\mu_X$$
Formula 12.4

If you realize that $-a_{YX} = b_{YX}\mu_X - \mu_Y$, then you can see that Formula 12.2 can be written in the following form, which will be designated Formula 12.5:

$$Y' = b_{YX}X + a_{YX}$$
Formula 12.5

The Slope and the *Y* Intercept

Formula 12.5 is a very convenient formula for making predictions. You start with an X value to predict from, you multiply it by b_{YX}, and then you add a_{YX}; the result is your prediction for the Y value. You may also recognize Formula 12.5 as the formula for *any* straight line. (Usually in mathematics the formula for a straight line is given by the equation $y = mx + b$, but unfortunately the letters used for the constants conflict with the notation conventionally used for regression, so I will stick with the regression notation.) The term b_{YX} is the slope of the regression line using raw scores ($b_{YX} = r$, when regression is calculated in terms of z scores). If the line goes through the origin, all you need is the slope to describe the line. However, to describe lines that do not pass through the origin, you need to indicate at what point the line hits the Y axis, i.e., the value of Y when X is zero. This is called the **Y intercept,** and it is represented by a_{YX} in Formula 12.5. Again, these descriptions can be made more concrete by drawing a graph. Imagine that height and weight are perfectly correlated for a group of people. The scatterplot would form a straight line, as shown in Figure 12.3.

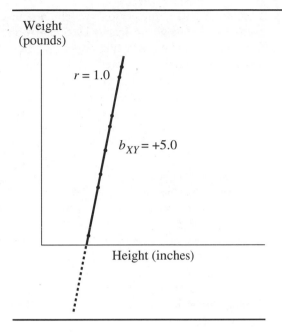

Figure 12.3

Predictions Based on Raw Scores

When correlation is perfect, as in Figure 12.3, the slope of the line is $b_{YX} = \sigma_Y/\sigma_X$ (or $b_{YX} = -\sigma_Y/\sigma_X$ for perfect negative correlation). The mean height is 69 inches; the standard deviation for height is about 3 inches. The mean weight is 155 pounds; the standard deviation for weight is about 15 pounds for this group. (σ_Y for this group is unusually small because all the people have the same "build"; it is only height that is producing the difference in weight.) Therefore, the slope for this graph is 15/3 = 5. This means that whenever two people differ by 1 inch in height, we know they will differ by 5 pounds in weight. Even though the correlation is perfect, you would not expect a slope of 1.0, because you would not expect a 1-inch change in height to be associated with a 1-*pound* gain in weight. The slope *is* 1.0 for perfect correlation when the data are plotted in terms of *z* scores, because you would expect a change of one *standard deviation* in height to be associated with a change of one *standard deviation* in weight.

It is important to point out that if the regression line in Figure 12.3 were extended it would not pass through the origin. That is why we cannot predict weight in pounds by taking height in inches and multiplying by 5. If it were extended (as shown by the dotted portion of the line), the regression line would hit the *Y* axis at −190 pounds, as determined by Formula 12.4:

$$a_{YX} = \mu_Y - b_{YX}\mu_X = 155 - 5(69) = 155 - 345 = -190$$

So a_{YX} would be −190 pounds, and the full regression equation would be

$$Y' = 5X - 190$$

Within this group, any person's weight in pounds can be predicted exactly: Just multiply their height in inches by 5 and subtract 190. For example, someone 6 feet tall would be predicted to weigh $5 \cdot 72 - 190 = 360 - 190 = 170$ lb. Of course, there is no need to predict the weight of anyone in this group, because we already know both the height *and* the weight for these particular people—that is how we were able to find the regression equation in the first place. However, the regression equation can be a very useful way to describe the relationship between two variables and can be applied to cases in which prediction is really needed. But first I must describe how the regression line looks when correlation is not perfect.

In the previous example, the slope of the regression line was 5. However, if the correlation were not perfect, but rather $+.5$, the slope would be 2.5 [$b_{YX} = (\sigma_Y/\sigma_X)r = (15/3)(.5) = 5(.5) = 2.5$]. The Y intercept would be -17.5 [$a_{YX} = \mu_Y - b_{YX}\mu_X = 155 - 2.5 (69) = 155 - 172.5 = -17.5$]. Therefore, the regression equation would be $Y' = 2.5X - 17.5$. A person 6 feet tall would be predicted to weigh $2.5 \cdot 72 - 17.5 = 180 - 17.5 = 162.5$ pounds. Notice that this prediction is midway between the prediction of 170 lb. when the correlation is perfect and the mean weight of 155 lb. The prediction is exactly in the middle because the correlation in this example is midway between perfect and zero.

Interpreting the Y Intercept

The Y intercept for the above example does not make sense; weight in pounds cannot take on negative values. The problem is that the height-weight relationship is not linear for the entire range of weights down to zero. For many regression examples, either it does not make sense to extend the regression line down to zero or the relationship does not remain linear all the way down to zero. For instance, performance on a mental task may correlate highly with IQ over a wide range, but trying to predict performance as IQ approaches zero would not be meaningful. On the other hand, if vocabulary size were being predicted from the number of years of formal education, it would be meaningful to estimate vocabulary size for people with no formal schooling. As another example, you would expect a negative correlation between heart rate and average amount of time spent in aerobic exercise each week. When predicting heart rate from exercise time, the Y intercept would meaningfully represent the heart rate for individuals who do not exercise at all. If, instead, heart rate were being used as the predictor of, say, reaction time in a mental task, the Y intercept would not make sense, because it would be the reaction time associated with a zero heart rate!

Quantifying the Errors Around the Regression Line

When correlation is less than perfect, the data points do not all fall on the same straight line, and predictions based on any one straight line will often be in error. The regression equation (Formula 12.5) gives us the straight line that minimizes

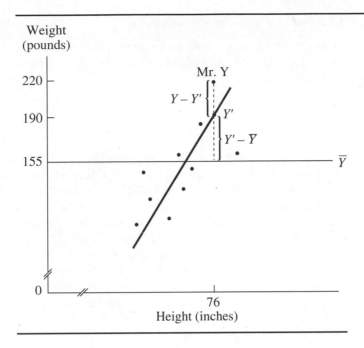

Figure 12.4

the error involved in making predictions. This will be easier to understand after I explain how error around the regression line is measured. Suppose we label as Y the variable that is to be predicted. (The equations can be reversed to predict the X variable from the Y variable, as I will show in Section B.) Then we can measure the difference between an actual Y value and the Y value predicted for it by the regression equation. This difference, $Y - Y'$, is called a **residual,** because it is the amount of the original value that is left over after the prediction is subtracted out. The use of residuals to measure error from the regression line can be illustrated in a graph.

Figure 12.4 is a graph of height versus weight for a group of 12 men. Notice in particular the point on the graph labeled Mr. Y. This point represents a man who is quite tall (6 ft 4 in.), and also very heavy (220 lb). The fact that this point is *above* the regression line indicates that this man is even heavier than would be expected for such a tall man. Using the regression line, you would predict that any man who is 76 inches tall would weigh 190 lb (this point is labeled Y' on the graph). The difference between Mr. Y's actual weight (220) and his predicted weight (190), his residual weight, is $Y - Y'$, which equals 30. This is the amount of error for one data point. To find the total amount of error, we must calculate $Y - Y'$ for each data point. However, if all of these errors were simply added, the result would be zero, because some of the data points are below the regression line, producing negative errors. The regression line functions like an "average," in that the amount of error above the line always balances out the amount of error below.

The Variance of the Estimate

To quantify the total amount of error in the predictions, all of the residuals (i.e., $Y - Y'$) are squared, then added together and divided by the total number, as indicated in Formula 12.6:

$$\sigma^2_{\text{est } Y} = \frac{\Sigma(Y - Y')^2}{N} \qquad \text{Formula 12.6}$$

If the above formula looks like a variance, it should. It is the variance of the data points around the regression line, called the **variance of the estimate,** or, the *residual variance* (literally, the variance of the residuals). The closer the data points are to the regression line, the smaller will be the error involved in the predictions, and the smaller $\sigma^2_{\text{est } Y}$ will be. Lower correlations are associated with more error in the predictions, and therefore a larger $\sigma^2_{\text{est } Y}$. When correlation is at its lowest possible value (i.e., zero), $\sigma^2_{\text{est } Y}$ is at its maximum. What happens to $\sigma^2_{\text{est } Y}$ when $r = 0$ is particularly interesting. When $r = 0$, the regression line becomes horizontal; its slope is therefore zero. The same prediction is therefore being made for all X values; for any X value the prediction for Y is the mean of Y. Because Y' is always $\overline{Y}$ (when $r = 0$), Formula 12.6 in this special case becomes:

$$\sigma^2_{\text{est } Y} = \frac{\Sigma(Y - \overline{Y})^2}{N}$$

This means that the variance of the predictions around the regression line is just the ordinary variance of the Y values. Thus the regression line isn't helping us at all! Without the regression line, there is a certain amount of variability in the Y variable (i.e., σ_Y^2). If the X variable is not correlated with the Y variable, the regression line will be flat, and the variance around the regression line ($\sigma^2_{\text{est } Y}$) will be the same as the variance around the mean (σ_Y^2). However, for any correlation greater than zero, $\sigma^2_{\text{est } Y}$ will be less than σ_Y^2, and that represents the advantage of performing regression.

Explained and Unexplained Variance

The difference between the variance of the estimate and the total variance is the amount of variance "explained" by the regression equation. To understand the concept of **explained variance,** it will help to look again at Figure 12.4. Mr. Y's total deviation from the mean weight ($Y - \overline{Y}$) is $220 - 155 = 65$ lb. This total deviation can be broken into two pieces. One piece is Mr. Y's deviation from the regression line ($Y - Y'$), which equals 30, and the other piece is the difference between the prediction and the mean ($Y' - \overline{Y}$), which equals 35. The two pieces, 30 and 35, add up to the total deviation. In general, $(Y - Y') + (Y' - \overline{Y}) = (Y - \overline{Y})$. The first part, $Y - Y'$, which I have been referring to as the error of the prediction, is sometimes thought of as the "unexplained" part of the variance, in contrast to the second part, $(Y' - \overline{Y})$, which is the "explained" part. In terms of

Mr. Y's unusually high weight, part of his weight is "explained" (or predicted) by his height—he is expected to be above average in weight. But he is even heavier than someone his height is expected to be; this extra weight is not explained by his height, so in this context it is "unexplained."

If all the "unexplained" pieces were squared and added up $[\Sigma(Y - Y')^2]$, we would get the unexplained sum of squares, or unexplained SS; dividing the unexplained SS by N yields the **unexplained variance,** which is the same as the variance of the estimate discussed above. Similarly, we could find the explained SS based on the $Y' - \overline{Y}$ pieces, and divide by N to find the amount of variance "explained." Together the explained and unexplained variances would add up to the total variance. The important concept is that whenever r is not zero, the unexplained variance is less than the total variance, so error or uncertainty has been reduced. We can guess a person's weight more accurately if we know his or her height than if we know nothing about the person at all. In terms of a scatterplot, when there is a linear trend to the data, the points tend to get higher (or lower, in the case of negative correlation) as you move to the right. The regression line slopes in order to follow the points, and thus it leads to better predictions and less error than a horizontal line (i.e., predicting the mean for everybody), which doesn't slope at all.

The Coefficient of Determination

If you want to know how well your regression line is doing in terms of predicting one variable from the other, you can divide the explained variance by the total variance. This ratio is called the **coefficient of determination** because it represents the proportion of the total variance that is explained (or determined) by the predictor variable. You might think that it would take a good deal of calculation to find the variances that form this ratio—and it would. Fortunately, the coefficient of determination can be found much more easily; it is always equal to r^2. If $r = .5$, then the coefficient of determination is $.5^2 = .25$. If this were the value of the coefficient of determination in the case of height predicting weight, it would mean that 25% of the variation in weight is being accounted for by variations in height. It is common to say that r^2 gives the "proportion of variance accounted for." Because of the squaring involved in finding this proportion, small correlations account for less variance than you might expect. For instance, a low correlation of .1 accounts for only $.1^2 = .01$, or just 1%, of the variance.

The Coefficient of Nondetermination

Should you want to know the proportion of variance not accounted for, you can find this by dividing the unexplained variance (i.e., $\sigma^2_{\text{est }Y}$) by the total variance. Not surprisingly, this ratio is called the **coefficient of nondetermination,** and it is simply equal to $1 - r^2$, as shown in Formula 12.7A:

$$\frac{\sigma^2_{\text{est }Y}}{\sigma_Y^2} = 1 - r^2 \qquad\qquad \text{Formula 12.7A}$$

The coefficient of nondetermination is sometimes symbolized as k^2. For most regression problems, we'd like k^2 to be as small as possible and r^2 to be as large as possible. When $r = +1$ or -1, $k^2 = 1 - 1 = 0$, which is the most desirable situation. In the worst case, $r = 0$ and $k^2 = 1$, implying that the variance of the regression is just as large as the ordinary variance. For the example above in which $r = +.5$, $k^2 = 1 - .5^2 = 1 - .25 = .75$, indicating that the variance around the regression line is 75% of the total amount of variance. Because $k^2 = 1 - r^2$, the coefficient of determination added to the coefficient of nondetermination will always equal 1.0 for a particular regression problem.

Calculating the Variance of the Estimate

By rearranging Formula 12.7A, we can obtain a convenient formula for calculating the variance of the estimate. If we multiply both sides of Formula 12.7A by the population variance of Y, the result is Formula 12.7B:

$$\sigma^2_{\text{est } Y} = \sigma_Y^2 (1 - r^2) \qquad \text{Formula 12.7B}$$

The above formula is a much easier alternative to Formula 12.6, once you have calculated the population variance of Y and the correlation between the two variables. (It is likely that you would want to calculate these two statistics anyway, before proceeding with the regression analysis.)

Bear in mind that I am only describing linear regression in this text; if two variables have a curvilinear relationship, other forms of regression will account for even more of the variance. The assumptions and limitations of linear regression are related to those of linear correlation and will be discussed in greater detail in the next section.

Exercises

1. Consider a math exam for which the highest score is 100 points. There will be a perfect negative correlation between a student's score on the exam and the number of points the student loses because of errors.

a) If the number of points Student A loses is half a standard deviation below the mean (i.e., he has $z = -.5$ for points lost), what z score would correspond to Student A's score on the exam?

b) If Student B attains a z score of -1.8 on the exam, what z score would correspond to the number of points Student B lost on the exam?

*2. Suppose that the Pearson correlation between a measure of shyness and a measure of trait anxiety is $+.4$.

a) If a subject is one and a half standard deviations above the mean on shyness, what would be the best prediction of that subject's z score on trait anxiety?

b) What would be the predicted z score for trait anxiety if a subject's z score for shyness were $-.9$?

3. In exercise 2, a) what proportion of the variance in shyness is accounted for by trait anxiety?

b) If the variance for shyness is 29, what would be the variance of the estimate when shyness is predicted by trait anxiety?

*4. On a particular regression line predicting heart rate in beats per minute (bpm) from number of milligrams of caffeine ingested, heart rate goes up 2 bpm for every 25 milligrams of caffeine. If heart rate is predicted to be 70 bpm with no caffeine, what is the raw score equation for this regression line?

* 5. For a hypothetical population of men, waist size is positively correlated with height, such that Pearson's $r = +.6$. The mean height (μ_X) for this group is 69 inches with $\sigma_X = 3$; mean waist measurement (μ_Y) is 32 inches with $\sigma_Y = 4$.

a) What is the slope of the regression line predicting waist size from height?

b) What is the value of the Y intercept?

c) Does the value found in part b above make any sense?

d) Write the raw-score regression equation predicting waist size from height.

* 6. Based on the regression equation found in exercise 5 above, a) what waist size would you predict for a man who is 6 feet tall?

b) What waist size would you predict for a man who is 62 inches tall?

c) How tall would a man have to be for his predicted waist size to be 34 inches?

7. a) In exercise 5, what is the value of the coefficient of determination?

b) How large is the coefficient of non-determination?

c) How large is the variance of the estimate (residual variance)?

* 8. What is the magnitude of Pearson's r when the amount of unexplained variance is equal to the amount of explained variance?

9. Describe a regression example in which the Y intercept has a meaningful interpretation.

* 10. In a hypothetical example, the slope of the regression line predicting Y from X is -12. This means that

a) a calculation error must have been made.

b) the correlation coefficient must be negative.

c) the magnitude of the correlation coefficient must be large.

d) the coefficient of determination equals 144.

e) none of the above.

Whenever two variables are correlated, one of them can be used to predict the other. However, if the correlation is very low, it is not likely that these predictions will be very useful; a low correlation means that there will be a good deal of error in the predictions. For instance, if $r = .1$, the variance of the data from the predictions (i.e., around the regression line) is 99% as large as it would be if you simply used the mean as the prediction in all cases. Even if the correlation is high there may be no purpose served by making predictions. It might be of great theoretical interest to find a high correlation between the amount of repressed anger a person has, as measured by a projective test, and the amount of depression the person experiences, as measured by a self-report questionnaire, but it is not likely that anyone will want to make predictions about either variable from the other. On the other hand, a high correlation between scores on an aptitude test and actual job performance can lead to very useful predictions. I will use the following example concerning the prediction of life expectancy to illustrate how linear regression can be used to make useful predictions.

BASIC STATISTICAL PROCEDURES

Life Insurance Rates

In order to decide on their rates, life insurance companies must use statistical information to estimate how long an individual will live. The rates are usually

Table 12.1	LQ Score (X)		Number of Years Lived (Y)	
	$\overline{X}$	36	$\overline{Y}$	74
	s_X	14	s_Y	10
	$r = +.6$			
	$N = 40$			

based on life expectancies averaged over large groups of people, but the rates can be adjusted for subgroups (e.g., women live longer than men, nonsmokers live longer than smokers, etc.). Imagine that in an attempt to individualize its rates, an insurance company has devised a lifestyle questionnaire (LQ), that can help predict a person's life expectancy based on his or her habits (amount of exercise, typical levels of stress, alcohol or caffeine consumption, smoking, etc.). LQ scores can range from 0 to 100, with 100 representing the healthiest lifestyle possible. A long-term study is conducted in which each person fills out the LQ on his or her fiftieth birthday, and eventually his or her age at death is recorded. Because we want to predict the total number of years a person will live based on the LQ score, number of years will be the Y variable, and LQ score will be the X variable.

In Section D, I will show how to calculate the regression equation from scratch (i.e., using the raw data). To streamline the procedures in this section, I will assume that the means and standard deviations for both variables, as well as the Pearson correlation coefficient, have already been calculated. Table 12.1 shows these values for a hypothetical sample of 40 subjects.

Regression in Terms of Sample Statistics

Note that Table 12.1 gives the *unbiased* standard deviations. The regression formulas in Section A were given in terms of population parameters, as though we would have no desire to go beyond the actual data already collected. Actually, it is more common to try to extend the results of a regression analysis beyond the data given. In this section, the regression formulas will be recast in terms of sample statistics, which can serve as unbiased estimators of population parameters. Except for a few minor changes, however, the formulas will look exactly the same. For instance, the slope of the regression line can be written as $b_{YX} = (s_Y/s_X)r$ instead of $(\sigma_Y/\sigma_Y)r$. Both formulas always give exactly the same value for b_{YX}. Similarly, the formula for a_{YX} becomes $\overline{Y} - b_{YX}\overline{X}$.

You must continue to use the subscript YX on the variable for both the slope and the intercept, because the subscript indicates that "Y is being regressed on X," which is another way of saying that Y is being *predicted from X*. The slope and intercept will usually be different when X is being regressed on Y, as I will show later. We need not use the subscript on Pearson's r, because there is no distinction between X correlated with Y and Y correlated with X.

Finding the Regression Equation

Once the means, standard deviations, and Pearson's *r* have been found, the next step in the regression analysis is to calculate the slope and the *Y* intercept of the regression line using Formulas 12.3 and 12.4 (substituting sample statistics for population parameters), as shown below:

$$b_{YX} = \frac{s_Y}{s_X} r = \frac{10}{14}(.6) = .714(.6) = .43$$

$$a_{YX} = \bar{Y} - b_{YX}\bar{X} = 74 - .43(36) = 74 - 15.43 = 58.6$$

Finally, we insert the values for b_{YX} and a_{YX} into Formula 12.5:

$$Y' = b_{YX}X + a_{YX} = .43X + 58.6$$

Making Predictions

The regression formula can now be used to predict a person's life expectancy based on his or her LQ score. For instance, the regression formula would give the following prediction for someone with an average LQ score (see Table 12.1):

$$Y' = .43X + 58.6 = .43(36) + 58.6 = 15.48 + 58.6 = 74.1$$

It is not surprising that someone with the average LQ score is predicted to have the average life expectancy. (The slight error is due to rounding off the slope to only two digits.) This will always be the case. Thinking in terms of *z* scores, the average of *X* corresponds to $z_X = 0$, which leads to a prediction of $z_Y = 0$, regardless of *r*.

Consider the life expectancy predicted for the person with the healthiest possible lifestyle (i.e., LQ = 100):

$$Y' = .43(100) + 58.6 = 43 + 58.6 = 101.6$$

On the other hand, the prediction for the least healthy lifestyle (LQ = 0) has already been found; it is the *Y* intercept (the point where the regression line hits the *Y* axis when LQ = 0):

$$Y' = .43(0) + 58.6 = 58.6$$

The process is the same for any value in between. If LQ equals 50, the prediction is

$$Y' = .43(50) + 58.6 = 21.5 + 58.6 = 80.1$$

For LQ = 10,

$$Y' = .43(10) + 58.6 = 4.3 + 58.6 = 62.9$$

These predictions can have very practical implications for the life insurance company. Although a correlation of .6 leaves plenty of room for error in the predictions, it would be possible to lower insurance rates somewhat for individuals with high LQ scores, while raising the rates proportionally for low LQ scorers. A correlation of .6 means that $.6^2$, or 36%, of the variance in life expectancy can be accounted for by the LQ scores; much of the remaining variance would be connected to, for instance, genetic factors. The life insurance company could increase its profits by drawing in people with a healthy lifestyle with the promise of lower rates. (Of course, the company would not lower its rates too much; it would want to leave plenty of room for bad luck—that is, it would want to charge rates high enough to cover instances when, for example, a person with a high LQ score was killed by lightning.)

Using Sample Statistics to Estimate the Variance of the Estimate

When correlation is less than perfect, every prediction has a "margin" for error. With high correlation, it is unlikely that any of the predictions will be way off, so the margin is relatively small. However, the margin for error increases as correlation decreases. The margin for error is based on the degree to which the data points are scattered around the regression line, and this scatter is measured by the variance of the estimate, as described in the previous section. It would be useful for the insurance company to be able not only to generate life-expectancy predictions but also to specify the margin for error around each prediction in terms of a confidence interval. Certain assumptions are required if these confidence intervals are to be valid; the most important of these is **homoscedasticity.** This term means that the variance around the regression line is the same at every part of the line. In other words, we can calculate the variance in Y for any particular X value, and we will always find the same amount of spread no matter which X value we choose; this condition is illustrated in Figure 12.5. This assumption justifies our using the same value for the variance of the estimate as our margin for error, regardless of which part of the regression line we are looking at.

In most situations in which we would like to make predictions, it is not realistic to use a variance of estimate formula that assumes we have data for the entire population. The variance of the estimate for the entire population, $\sigma^2_{\text{est } Y}$, must be estimated from the available sample data according to the following formula:

$$s^2_{\text{est } Y} = \frac{\Sigma(Y - Y')^2}{N - 2}$$

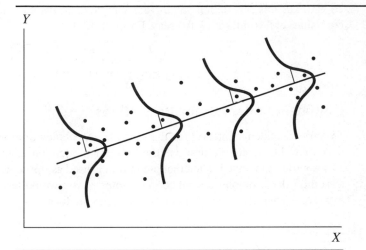

Figure 12.5

Scatterplot Depicting Homoscedasticity

The distribution of the *Y* values has the same spread (i.e., variance) at every value for *X*.

If you compare the above formula to Formula 12.6, you will notice that the only difference is that the denominator is $N - 2$ instead of N. Dividing by N would result in a biased estimate of $\sigma^2_{\text{est } Y}$. (As mentioned in Chapter 11, the degrees of freedom for correlational problems is $N - 2$ rather than $N - 1$.)

There is a much easier way to calculate $s^2_{\text{est } Y}$, if you have already found s_Y^2 and r; you can use Formula 12.8, which is very similar to Formula 12.7B.

$$s^2_{\text{est } Y} = \left(\frac{N - 1}{N - 2}\right) s_Y^2 (1 - r^2) \qquad \text{Formula 12.8}$$

Even though we are using the unbiased sample variance, we need the factor of $(N - 1)/(N - 2)$ to ensure that we have an unbiased estimator of the variance of the estimate.

Standard Error of the Estimate

Because confidence intervals are based on standard deviations, rather than variances, we need to introduce one new term: the square root of the variance of the estimate, $\sigma_{\text{est } Y}$, which is called the **standard error of the estimate.** This is the standard deviation of points—in this case, in the vertical direction—from the regression line, and gives an idea of how scattered points are from the line (about two thirds of the points should be within one standard error above or below the line). To find the standard error of estimate in the population, we take the square root of both sides of Formula 12.7B:

$$\sigma_{\text{est } Y} = \sigma_Y \sqrt{(1 - r^2)}$$

To estimate this population value from sample data, we take the square root of both sides of Formula 12.8, to create Formula 12.9:

$$s_{\text{est } Y} = s_Y \sqrt{\frac{N-1}{N-2}(1 - r^2)}$$ Formula 12.9

Confidence Intervals for Predictions

To find a confidence interval for the *true* life expectancy of an individual with a particular LQ score, we first use the regression equation to make a prediction (Y') for life expectancy. Then that prediction is used as the center of the interval. Based on the principles described in Chapter 8, you would have good reason to expect the confidence interval to take the following form,

$$Y' \pm t_{\text{crit}} \, s_{\text{est } Y}$$

where t_{crit} depends on the level of confidence and $s_{\text{est } Y}$ is the estimated standard error of estimate. (For a 95% CI, t_{crit} corresponds to alpha = .05, two-tailed.) Unfortunately there is one more complication. The actual formula for the confidence interval contains an additional factor, and therefore looks like this (Formula 12.10):

$$Y' \pm t_{\text{crit}} s_{\text{est } Y} \sqrt{1 + \frac{1}{N} + \frac{(X - \overline{X})^2}{(N-1)s_X^2}}$$ Formula 12.10

The reason for the additional factor is that the regression line, which is based on a relatively small sample ($N = 40$), is probably wrong—that is, it does not perfectly represent the actual relationship in the population. So there are really two very different sources of error involved in our prediction. One source of error is the fact that our correlation is less than perfect; in the present example, unknown factors account for 64% of the variability in life expectancy. The second source (which necessitates the additional factor in the confidence interval formula) is the fact that different samples would lead to different regression lines, and therefore different predictions. To demonstrate the use of this complex-looking formula, let us find a 95% confidence interval for a particular prediction.

An Example of a Confidence Interval

Earlier in this section, I used the regression equation to show that a person with an LQ score of 50 would be predicted to live 80.1 years. Now I will use interval estimation to supplement that prediction. The t_{crit} for the 95% CI with 38 degrees of freedom is about 2.02. The standard error of estimate for Y is found by using Formula 12.9:

$$s_{\text{est } Y} = s_Y \sqrt{\frac{N-1}{N-2}(1 - r^2)} = 10 \sqrt{\frac{39}{38}(1 - .6^2)} = 10\sqrt{.657} = 10(.81) = 8.1$$

Inserting the appropriate values into Formula 12.10, we get

$$Y' \pm 2.02(8.1)\sqrt{1 + \frac{1}{40} + \frac{(50 - 36)^2}{(39)(196)}}$$

$$= 80.1 \pm 16.36\sqrt{1 + .025 + .0256}$$

$$= 80.1 \pm 16.36\sqrt{1.0506} = 80.1 \pm 16.36(1.025) = 80.1 \pm 16.77$$

Our confidence interval predicts that a person with LQ = 50 will live somewhere between 63.33 and 96.87 years. Of course, 5% of these 95% CIs will turn out to be wrong, but that is a reasonable risk to take. If Pearson's r were higher, $s_{\text{est } Y}$ would get smaller (because $1 - r^2$ gets smaller), and therefore the 95% CI would get narrower. Increasing the sample size also tends to make the CI smaller by reducing both t_{crit} and the additional factor (in which N is in the denominator), but this influence is limited. With an extremely large N, t_{crit} approaches z_{crit}, and the additional factor approaches 1, but $s_{\text{est } Y}$ approaches $s_Y(1 - r^2)$, which is *not* affected by sample size. If Pearson's r is small, $s_{\text{est } Y}$ will not be much smaller than s_Y, no matter how large the sample size is. The bottom line is that unless the correlation is high, there is plenty of room for error in the prediction. The confidence interval found above may not look very precise, but without the LQ score, our prediction for the same individual would have been the mean (74), and the 95% confidence interval would have been from about 54 to 94 (about two standard deviations in either direction). Given the LQ information, the insurance company can be rather confident that the individual will not die before the age of 63, instead of having to use 54 as the lower limit.

Assumptions Underlying Linear Regression

As with linear correlation, linear regression can be used purely for descriptive purposes—to show the relationship between two variables for a particular group of subjects. In that case the assumptions described below need not be made. However, it is much more common to want to generalize your regression analysis to a larger group (i.e., a population) and to scores that were not found in your original sample, but might be found in future samples. The confidence intervals discussed above are valid only if certain assumptions are actually true.

Independent Random Sampling

This is the same assumption described for Pearson's r in Chapter 11, namely, that each subject (i.e., pair of scores) should be independent of the others and should have an equal chance of being selected.

Linearity

The results of a linear regression analysis will be misleading if the two variables have a curvilinear relationship in the population.

Normal Distribution

At each possible value of the X variable, the Y variable must follow a normal distribution in the population.

Homoscedasticity

For each possible value of the X variable, the Y variable has the same population variance. This property is analogous to homogeneity of variance in the two-group t-test.

Regressing X on Y

What happens if you wish to predict the X variable from scores on the Y variable, instead of the other way around? The obvious answer is that you can switch the way the two variables are labeled and use the equations already presented in this chapter. However, if you would like to try the regression analysis in both directions for the same problem (e.g., predict a subject's maximum running speed on a treadmill from his or her reaction time in a laboratory *and* predict reaction time from running speed), it would be confusing to relabel the variables. It is easier to switch the X and Y subscripts in the formulas to create a new set of formulas for regressing X on Y. For instance, the regression formula becomes

$$X' = b_{XY}Y + a_{XY}$$

where

$$b_{XY} = \frac{s_X}{s_Y}r \quad \text{and} \quad a_{XY} = \overline{X} - b_{XY}\overline{Y}$$

Unless the two variables have the same standard deviation, the two slopes will be different. If b_{YX} is less than 1, b_{XY} will be greater than 1, and vice versa.

Raw Score Formulas

The formula that I recommend for the calculation of the regression slope [$b_{YX} = (s_Y/s_X)r$] is simple, but it does require that both standard deviations, as well as Pearson's r, be calculated first. Because it is hard to imagine a situation in which those latter statistics would not be computed for other reasons anyway, the formula for b_{YX} does not really require extra work. However, it is also true that b_{YX} can be calculated directly from the raw scores, without finding any other statistics. By taking the raw-score formula for Pearson's r (Formula 11.5) and multiplying by the raw-score formula for the ratio of the two standard deviations, we derive Formula 12.11:

$$b_{YX} = \frac{N(\sum XY) - (\sum X)(\sum Y)}{N\sum X^2 - (\sum X)^2}$$ Formula 12.11

You may recognize that the numerator in Formula 12.11 is the same as the numerator for calculating Pearson's r (Formula 11.5); it is the biased covariance. The denominator is just the biased variance of X (σ_X^2). So another way to view b_{YX} is that it is the covariance of X and Y divided by the variance of X. Although a similar raw-score formula could be derived for a_{YX}, it would not be efficient to use it. Once b_{YX} has already been calculated, Formula 12.4 is the most sensible way to calculate a_{YX}. You may encounter other formulas for b_{YX} and a_{YX} in terms of sums of squares and sums of products (SS and SP), but these are algebraically equivalent to the formulas presented here, and therefore produce the same answers (SS divided by N yields a variance, whereas SP divided by N results in a covariance).

When to Use Linear Regression

Prediction

The most obvious application of linear regression is in predicting the future performance of something or someone—for example, a person's performance on a job or in college based on some kind of aptitude test. Hiring or admissions decisions can be made by finding a cutoff score on the aptitude test, above which an individual can be expected to perform adequately on the performance measure (e.g., college grades or job skill evaluation). The example used thus far in this section is another appropriate application of regression. Insurance rates can be individualized based on life-expectancy predictions. In addition to its practical applications, regression can also be used in testing theories. A regression equation could be devised based on a theoretical model, and the model's predictions could be confirmed or disconfirmed by the results of an experiment.

Statistical Control

Regression analysis can be used to adjust statistically for the effects of a confounding variable. For instance, if you are studying factors that affect vocabulary size (VS) in school-aged children, age can be an extraneous variable. You can try to keep age constant within your study, but if VS varies linearly with age, you can control for the effects of age by using it to predict VS. An age-adjusted VS can be created by subtracting each subject's VS from the VS predicted for his or her age. The result is a residual that should not be correlated with age; it should indicate whether a subject's VS is high or low for his or her age. This procedure is referred to as "partialing out" the effects of age, and it will be discussed in greater detail in the latter half of Section C, and in the context of the analysis of covariance in Chapter 18, Section C.

Regression with Manipulated Variables

So far I have been describing regression in the situation in which neither of the variables is being controlled by the experimenter. Although it is much less common, regression can also be used to analyze the results of a genuine experiment in which subjects are randomly assigned to different values of the *X* variable. For instance, prospective psychotherapy patients could be randomly assigned to one, two, three, four, or five sessions per week (*X* variable), and then after a year of such therapy the patients could be tested for severity of neurotic symptoms (*Y* variable). If the regression equation had a slope significantly different from zero, there would be evidence that the number of sessions per week makes a difference in the effectiveness of psychotherapy. However, such experimental designs are usually evaluated with the methods of analysis of variance, as described in Chapter 14.

For an experimental example in which the use of linear regression is especially appropriate, I turn to a classic experiment in cognitive psychology conducted by Saul Sternberg (1966). The mental task he assigned subjects was to keep a small list of digits in mind and then answer as quickly as possible whether a test digit was included in that list. For instance, if one of the memory lists was 1, 3, 7, 8 and the test item were 7, the answer would be "yes," but if the test item were 4 the answer would be "no." For half the test items the correct answer was "no." The memory lists ranged in length from one to six digits, and, as you might expect, reaction time (RT) was longer for the longer lists, because subjects had to scan more digits. In fact, if you plot the average RT for each length list for the "no" trials (as Sternberg did), you can see that the relation is linear (See Figure 12.6, which is derived from Sternberg's published data.) The slope of the line is about 38, which means that increasing the memory list by one digit lengthens RT by 38 milliseconds (ms). The slope of the regression line provides an estimate of how long it takes to scan each digit in the memory list. The *Y* intercept

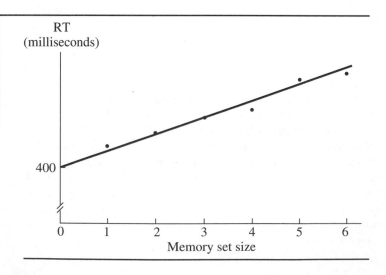

Figure 12.6

Regression Line for Predicting Reaction Time from Size of Memory Set

of the regression line is also meaningful. It can be thought of as a kind of "over-head"; it depends on the speed of motor commands, how quickly the test item can be recognized, and other fixed amounts that are not related to the size of the memory list.

The Sternberg regression model has practical applications in the exploration of cognitive functioning. For instance, one study found that the intake of marijuana (as compared with a placebo) increased the Y intercept of the regression line but did not alter the slope. There seemed to be a general slowing of the motor response, but no slowing of the rate at which the digits were being mentally scanned. On the other hand, elderly or mentally retarded subjects exhibited a steeper regression slope (indicating a slower rate of mental scanning) without much difference in the Y intercept. The regression approach allows the researcher to separate factors that change with the manipulated variable from factors that remain relatively constant.

Exercises

*1. The data from exercise 11A3, with the outlier eliminated, are reproduced below.

Years (X)	Annual Salary (Y)
5	24
8	40
3	20
6	30
9	40
7	35
10	50
2	22

a) Find the regression equation for predicting an employee's annual salary from his or her number of years with the company.

b) What salary would you predict for someone who's been working at the same company for four years?

c) How many years would you have to work at the same company to have a predicted salary of $60,000 per year?

2. A statistics professor has devised a diagnostic quiz, which, when given during the first class, can accurately predict a student's performance on the final exam. So far data are available for ten students, as shown below:

Student	Quiz Score	Final Exam Score
1	5	82
2	8	80
3	3	75
4	1	60
5	10	92
6	6	85
7	7	86
8	4	70
9	2	42
10	6	78

a) Find the regression equation for predicting the final exam score from the quiz.

b) Find the (unbiased) standard error of the estimate.

c) What final exam score would be predicted for a student who scored a 9 on the quiz?

d) Find the 95% confidence interval for your prediction in part c.

*3. Refer to exercise 11B5. a) What proportion of

the variance in orientation scores is accounted for by years of hospitalization?

b) Find the regression equation for predicting orientation scores from years of hospitalization.

c) What is the value of the Y intercept for the regression line? What is the meaning of the Y intercept in this problem?

d) How many years does someone have to be hospitalized before they are predicted to have an orientation score as low as 10?

4. A cognitive psychologist is interested in the relationship between spatial ability (e.g., ability to rotate objects mentally) and mathematical ability, so she measures 12 subjects on both variables. The data appear in the table below.

Subject	Spatial Ability Score	Math Score
1	13	19
2	32	25
3	41	31
4	26	18
5	28	37
6	12	16
7	19	14
8	33	28
9	24	20
10	46	39
11	22	21
12	17	15

a) Find the regression equation for predicting the math score from the spatial ability score.

b) Find the regression equation for predicting the spatial ability score from the math score.

c) According to your answer to part a, what math score is predicted from a spatial ability score of 20?

d) According to your answer to part b, what spatial ability score is predicted from a math score of 20?

* 5. For the data in exercise 11B9, a) use the correlation coefficient you calculated in part a of exercise 11B9

to find the regression equation for predicting serum cholesterol levels from the number of hours exercised each week.

b) Find the regression slope using the raw-score formula. Did you obtain the same value for b_{YX} as in part a?

c) What cholesterol level would you predict for someone who does not exercise at all?

d) What cholesterol level would you predict for someone who exercises 14 hours per week?

e) Find the 99% confidence interval for your prediction in part d.

6) For the data in exercise 11B10, a) use the correlation coefficient calculated in exercise 11B10 to find the regression equation for predicting reaction time from the number of letters in the string.

b) How many milliseconds are added to the reaction time for each letter added to the string?

c) What reaction time would you predict for a string with seven letters?

d) Find the 95% confidence interval for your prediction in part c.

* 7. If you calculate the correlation between shoe size and reading level in a group of elementary school children, the correlation will turn out to be quite large, provided that you have a large range of ages in your sample. The fact that each variable is correlated with age means that they will be somewhat correlated with each other. The following table illustrates this point. Shoe size is measured in inches, for this example, reading level is by grade (4.0 is average for the fourth grade), and age is measured in years.

Child	Shoe size	Reading level	Age
1	5.2	1.7	5
2	4.7	1.5	6
3	7.0	2.7	7
4	5.8	3.1	8
5	7.2	3.9	9
6	6.9	4.5	10
7	7.7	5.1	11
8	8.0	7.4	12

a) Find the regression equation for predicting shoe size from age.

b) Find the regression equation for predicting reading level from age.

c) Use the equations from parts a and b to make shoe size and reading level predictions for each child. Subtract each prediction from its actual value to find the residual.

*8. a) Calculate Pearson's *r* for shoe size and reading level using the data from exercise 7.

b) Calculate Pearson's *r* for the two sets of residuals you found in part c of exercise 7.

c) Compare your answer in part b with your answer to part a above. The correlation in part b is the partial correlation between shoe size and reading level after the confounding effect of age has been removed from each variable (see Section C).

The Point-Biserial Correlation Coefficient

**OPTIONAL
MATERIAL**

It is certainly not obvious that a two-group experiment can be analyzed by linear regression; however, not only is it possible, it is instructive to see how it works. As an example of a two-group experiment, suppose that subjects in one group receive a caffeine pill, while subjects in the other receive a placebo. All subjects are then measured on a simulated truck-driving task. The results of this experiment can be displayed on a scatterplot, although it is a rather strange scatterplot, because the *X* variable (drug condition) has only two values (see Figure 12.7). Drawing the regression line in this case is relatively easy. The regression line passes through the mean of the *Y* values at each level of the *X* variable; that is, the regression line connects the means of the two groups, as shown in Figure 12.7. The two group means serve as the predictions.

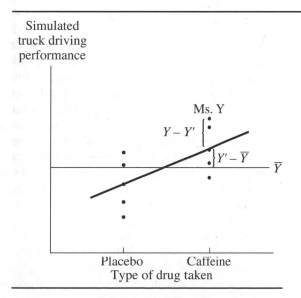

Figure 12.7

Remember that a prediction based on the regression line is better than predicting the mean of Y for everyone. The same is true for our two-group example. If we didn't know which group each subject was in, we would have to use the overall mean (i.e., the mean of all subjects combined regardless of group—the mean of the two group means when the groups are the same size) as the prediction for each subject. Therefore, we can divide each subject's score into an explained (i.e., predicted), and an unexplained (deviation from prediction) portion. We focus on Ms. Y, who has the highest score in the caffeine group. The part labeled $Y - Y'$ in Figure 12.7 is unexplained; we do not know why she scored so much better than the other subjects in her group, all of whom had the same dosage of caffeine. The part labeled $Y' - \overline{Y}$ is the difference between using her own group mean to predict Ms. Y's score and using the mean of all subjects to predict it. This part is explained; we could expect Ms. Y to perform better than the overall mean, because she had taken a caffeine pill. Together $Y - Y'$ and $Y' - \overline{Y}$ add up to $Y - \overline{Y}$, the total deviation.

If all the unexplained pieces are squared and averaged, we get the variance of the estimate, but in this case the variance of the estimate is based on the variance around each group mean. In the two-group example, $s^2_{\text{est } Y}$ is just the average of s^2_{caffeine} and s^2_{placebo}. (It would have to be a weighted average—a pooled variance— if the groups were not equal in size.) Clearly, the variances around each group mean are smaller than the variance of all subjects around the overall mean (i.e., the total variance) would be. This reduction in variance is equal to the variance of the explained pieces. From Section A you know that the ratio of the explained variance to the total variance is equal to r^2. But what is r for this strange type of regression? If you divide the explained variance by the total variance, and then take the square root in a two-group problem, the result is a correlation coefficient that is often called the **point-biserial r** (symbolized as r_{pb}).

Calculating r_{pb}

Strange as it seems, we could have calculated Pearson's r directly for our two-group example. It is strange because to calculate r, each subject must have both an X and a Y value. Each subject has a Y value—his or her simulated truck driving score—but what is the X value for each subject? The solution is to assign the same arbitrary number to all of the subjects in the caffeine group, and a different number to all of the placebo subjects. The simplest way is to use the numbers zero and 1; the X value for all subjects in the first group is zero, and for the second group it is 1. (The designations "first" and "second" are totally arbitrary.) Once each subject has both an X and a Y value, any of the formulas for Pearson's r can be used. However, the sign of r_{pb} is not important, because it depends on which group has the higher X value, and that is chosen arbitrarily. Therefore, r_{pb} can always be reported as a positive number as long as the means are given, or it is otherwise made clear in which direction the effect is going.

The repetition of the X values in the two-group case results in a simplified computational formula for r_{pb}, but it has become so easy to calculate a correlation with an electronic calculator or computer that I will not bother to show that formula here. Moreover, the specialized formula tends to obscure the fact that r_{pb} is like any other Pearson's r; the name "point-biserial r" is only to remind you that the correlation is being used in a special situation, in that one of the variables has only two values. Probably the most useful function of the point-biserial r is to provide supplementary information after a two-group t-test has been conducted. I turn to this point next.

Deriving r_{pb} from a t Value

For a two-group experiment, you could calculate r_{pb} as an alternative to the t value and then test r_{pb} for statistical significance. To test r_{pb} you could use Table A.5 in the Appendix or Formula 11.6, which gives a t value for testing Pearson's r. One version of Formula 11.6 is shown below because it will provide a convenient stepping stone to a formula for r_{pb}:

$$t = \frac{r}{\sqrt{(1 - r^2)/(N - 2)}}$$

In the case of the two-group experiment, N is the total number of subjects in both groups: $N_1 + N_2$. Because $N_1 + N_2 - 2$ is the df for the two-group case, the formula above can be rewritten for r_{pb} as follows to give Formula 12.12:

$$t = \frac{r_{pb}}{\sqrt{(1 - r_{pb}^2)/df}} \qquad \text{Formula 12.12}$$

The important connection is that the t value for testing the significance of r_{pb} is the same t value you would get by using the pooled-variance t test formula. So, instead of using the pooled-variance t-test formula, you could arrive at the same result by calculating r_{pb} and then finding the t value to test it for significance. However, this would probably not be any easier. What is more interesting is that you can find the pooled-variance t-test first, plug it into the formula above, and solve for r_{pb}. You would probably want to find the t value anyway to test for significance, and r_{pb} is easily found using that t value. For convenience the above formula can be solved for r_{pb} ahead of time, yielding the very instructive and useful Formula 12.13:

$$r_{pb} = \sqrt{\frac{t^2}{t^2 + df}} \qquad \text{Formula 12.13}$$

Interpreting r_{pb}

The reason for spending this whole section on the point-biserial r is that this statistic provides information that the t value alone does not. Unfortunately, the size of the t value by itself tells you nothing about the effect size in your samples.

As Chapters 9 and 10 emphasized, a large t value for a two-group experiment does not imply that the difference of the means was large, or even that this difference was large compared to the standard deviations of the groups. There could be almost a total overlap between the two samples, and yet the t value could be very large and highly significant—as long as the df were very large. But r_{pb} *does* tell you something about the overlap of the two samples in your experiment; r_{pb}^2 tells you the proportion by which the variance is reduced by knowing which group each subject is in (as compared to the variance around the overall mean). The point-biserial r tells you the *strength of association* between group membership and the dependent variable—that is, the strength of the tendency for scores in one group to be consistently higher than scores in the other group. If $r_{pb} = .9$, you know that the scores of the two groups are well separated; if $r_{pb} = .1$, you can expect a good deal of overlap. What r_{pb} does not tell you is anything about statistical significance. Even $r_{pb} = .9$ is not significant if there are only two people in each group, and $r_{pb} = .1$ *can* be significant, but only if there are at least about 200 people per group. That is why the t value is also needed—to test for statistical significance.

Looking at Formula 12.13, you can see the relationship between t and r_{pb}. Suppose that the t value for a two-group experiment is 4 and the df is 16:

$$r_{pb} = \sqrt{\frac{4^2}{4^2 + 16}} = \sqrt{\frac{16}{32}} = \sqrt{.5} = .71$$

The point-biserial r is quite high, indicating a strong differentiation of scores between the two groups. On the other hand, suppose that t is once again 4, but for a much larger experiment in which df = 84:

$$r_{pb} = \sqrt{\frac{4^2}{4^2 + 84}} = \sqrt{\frac{16}{100}} = \sqrt{.16} = .4$$

The strength of association is still moderately high, but considerably less than in the previous experiment. Finally, if $t = 4$ and df = 1000,

$$r_{pb} = \sqrt{\frac{4^2}{4^2 + 1000}} = \sqrt{\frac{16}{1016}} = \sqrt{.016} = .125$$

Now the strength of association is rather low and unimpressive. A t of 4.0 indicates a strong effect in a small experiment, but not in a very large one.

Strength of Association in the Population (Omega Squared)

In its relationship to the t value, r_{pb} may remind you of g, the effect size in the sample (discussed in Chapter 10). There is a connection. Instead of r_{pb}, you could use $(\overline{X}_1 - \overline{X}_2)/s_p$ to describe the strength of association. However, the values of

r_{pb}, which range from 0 to 1, are more readily interpretable, and r_{pb}^2 indicates directly the proportion of variance explained by group membership. If you calculated r_{pb} for an entire population, the result could be referred to as ρ_{pb}. However, the square of ρ_{pb}, which would measure the proportion of variance accounted for in the population is not generally referred to as such. For historical reasons, the term **omega squared** (symbolized as ω^2) is used instead to stand for the same quantity. Assuming that we are talking about equal-sized groups, the relation between ω^2 and γ^2 (i.e., the population effect size squared) is a simple one, as given in Formula 12.14 below:

$$\omega^2 = \frac{\gamma^2}{\gamma^2 + 4} \qquad \text{Formula 12.14}$$

Note that γ can take on any value, but ω^2, like r_{pb}^2, is a proportion that cannot exceed 1.0. An unusually high γ of 4 corresponds to

$$\omega^2 = \frac{4^2}{4^2 + 4} = \frac{16}{20} = .8$$

This means that the r_{pb} for a particular sample should come out to be somewhere near $\sqrt{.8}$, which is about .9. A small effect size, $\gamma = .2$, corresponds to

$$\omega^2 = \frac{.2^2}{.2^2 + 4} = \frac{.04}{4.04} = .0099$$

Only about 1% of the variance in the population is accounted for by group membership, and the r_{pb} for a particular sample would be expected to be around $\sqrt{.01} = .1$.

If you wished to estimate ω^2 from r_{pb}, you could square r_{pb}, but the estimate would be somewhat biased. A better, though not perfect, estimate of ω^2 is given by Formula 12.15:

$$\text{Estimated } \omega^2 = \frac{t^2 - 1}{t^2 + df + 1} \qquad \text{Formula 12.15}$$

Notice that Formula 12.15 is just a slight modification of the formula you would get by squaring Formula 12.13.

The term ω^2 gives us an alternative to gamma (γ) for describing population differences. For instance, in Chapter 10 I suggested that γ for the difference in height between men and women was about $(69 - 65)/3 = 4/3 = 1.33$. For this γ, the corresponding ω^2 (using Formula 12.14) is

$$\omega^2 = \frac{1.33^2}{1.33^2 + 4} = \frac{1.78}{4.78} = .37$$

Therefore, we can say that about 37% of the variance in height among adults is accounted for by gender. One important advantage of ω^2 is that whereas gamma is a measure that is specific to a two-group comparison, ω^2 can be used in many situations, including correlations and multigroup comparisons (see Chapter 14).

Biserial *r*

The point-biserial *r* should only be used when the two groups represent a true dichotomy, such as male and female. On the other hand, in some studies two groups are formed rather arbitrarily according to differences on some continuous variable. For instance, a group of tall people and a group of short people can be compared on some dependent variable, such as self-esteem. If the variable used for dividing the two groups follows something like a normal distribution, as height does, then assigning zeroes and ones to the two groups and applying the Pearson formula will lead to a misleading result. It is more accurate to use a similar, but adjusted, correlation coefficient, known simply as the **biserial *r.*** Because it is not often used, I will not complicate matters any further by presenting the formula for the biserial *r* in this text; the interested reader can find the formula in a more advanced statistics textbook (e.g., Howell, 1992).

Multiple Regression

Suppose that a statistics teacher wants to know who will perform well in her class. She obtains math SAT scores for each student and finds a correlation of .5 between math SAT and grades on her midterm exam. This is a pretty good correlation; the math SATs account for $.5^2 = .25 = 25\%$ of the variance in midterm grades. But there is plenty of variance (75%) that is apparently unrelated to math aptitude, which remains to be accounted for. Some of this remaining variance may be related to how much each student studied for the exam (let us hope!).

Suppose that the teacher finds out the total number of hours each student studied for the midterm and obtains a correlation of .4 between number of hours of studying and midterm grades. Another 16% of the midterm grade variance has been accounted for! Suppose the correlation between math SAT scores and hours of studying is zero. (It is hard to find zero correlations in real life, but the correlation between these two variables could be quite small.) Then these two variables account for different parts of the variance in midterm grades, as shown in Figure 12.8. This type of figure is called a **Venn diagram.** Together, math SAT scores and hours of studying account for $25 + 16 = 41\%$ of the variance in midterm grades. The total amount of variance accounted for in a dependent (predicted) variable by a combination of two or more independent (predictor) variables is symbolized by R^2, where R is called the **multiple correlation coefficient.** In this example, $R^2 = .41$, so $R = \sqrt{.41} = .64$.

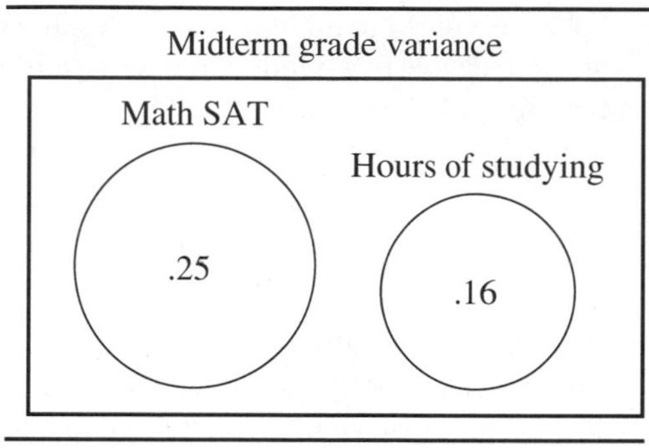

Midterm grade variance

Figure 12.8

Venn Diagram

The square represents the total variance of the criterion, and the circles represent the proportions of variance accounted for by different predictor variables.

Uncorrelated Predictors

In order to explain what R really means I have to move from simple linear regression to multiple linear regression (usually just referred to as multiple regression). Because it will be easier to explain multiple regression in terms of z scores, I will be dealing mostly with **standardized regression equations.** As you may recall from Section A, the standardized regression equation for simple linear regression is $z_Y = rz_X$, so for math SAT scores the equation would be $z_Y = .5z_X$, and for hours of studying, $z_Y = .4z_X$. However, if we know both math SAT and the number of hours studied for each subject, we can make a better prediction about midterm grades than we can using either predictor variable by itself. The way multiple regression works is by using a weighted combination of the two predictor variables to make the prediction. When the two predictors are uncorrelated with each other, the weights are just the ordinary Pearson's rs of each predictor separately with the criterion (i.e., the predicted variable):

$$z_{Y'} = r_{YX_1}z_{X_1} + r_{YX_2}z_{X_2} \qquad \text{Formula 12.16}$$

For the midterm grades example, $z_{Y'} = .5z_{X_1} + .4z_{X_2}$, where X_1 refers to math SAT scores and X_2 refers to hours of studying.

In simple linear regression, the correlation between your predictions for Y and the actual values for Y is always just r. For multiple regression this correlation is R. For our example the correlation between the values from the multiple regression equation (i.e., $.5z_{X_1} + .4z_{X_2}$) and the actual z scores for Y is .64, which is higher than the r for either predictor variable by itself. For the general case of two uncorrelated predictors, the following simple equation, Formula 12.17, applies:

$$R = \sqrt{r_{YX_1}{}^2 + r_{YX_2}{}^2} \qquad \text{Formula 12.17}$$

The case of three predictor variables in which $r = 0$ for each pair is a simple extension of the two-predictor case described above. The multiple regression equation would be

$$z_{Y'} = r_{YX_1}z_{X_1} + r_{YX_2}z_{X_2} + r_{YX_3}z_{X_3}$$

However, it is relatively rare to find two virtually uncorrelated predictors, and finding three mutually uncorrelated predictors, each having a meaningful correlation with the criterion, is so rare that there is little value in discussing this case. Correlations among predictors greatly complicate the process of finding the multiple regression equation, so I will return to the two-predictor case to begin to explain this problem.

Two Correlated Predictors

A common example of multiple regression is the prediction of graduate school success (GSS—I will dodge the issue of how this can best be measured) from both a student's total GRE score (TGRE) and the student's undergraduate grade point average (UGPA). It is highly unlikely that these two predictors would be uncorrelated. Let us assume that we know the correlation between TGRE and GSS is $r_{1Y} = .4$, and that between UGPA and GSS is $r_{2Y} = .5$. To make the subscripts easier to read I will refer to the first predictor as 1 instead of X_1, the second predictor as 2, and place these numbers before the criterion, Y. I will assume that the correlation between the two predictors is $r_{12} = .3$. The correlation between the two predictors means that there is some overlap in the variance that each accounts for in graduate school success. This overlap can be pictured as the intersection of two circles in a Venn diagram, as shown in Figure 12.9. To find the weights (the **regression coefficients**) for the multiple regression equation when there are only two predictors and the predictors are correlated with each other, we use a simple formula, Formula 12.18. If the equation is in terms of z scores, the two regression coefficients are called **standardized regression**

Figure 12.9

Venn Diagram of Two Correlated Predictors

Note that R^2 = Area A + Area B + Area C, and $1 - R^2$ = Area D.

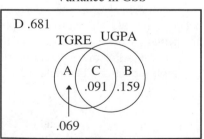

Variance in GSS

D .681

TGRE   UGPA

A   C   B
.091   .159

.069

coefficients, and are designated β_1 and β_2. (These should not be confused with the β that stands for the Type II error rate.)

$$\beta_1 = \frac{r_{1Y} - r_{2Y}r_{12}}{1 - r_{12}{}^2} \quad \beta_2 = \frac{r_{2Y} - r_{1Y}r_{12}}{1 - r_{12}{}^2} \qquad \text{Formula 12.18}$$

The multiple regression equation becomes

$$z_{Y'} = \beta_1 z_{X_1} + \beta_2 z_{X_2} \qquad \text{Formula 12.19}$$

Note that if $r_{1Y} = r_{2Y}$, then $\beta_1 = \beta_2$, which means that both predictors will have equal weight in the regression equation, which makes sense. Also note that when the predictors are not correlated (i.e., $r_{12} = 0$), $\beta_1 = r_{1Y}$ and $\beta_2 = r_{2Y}$, and Formula 12.19 becomes Formula 12.16. In any case the *beta weights,* as they are frequently called, are chosen in such a way that the predictions they produce have the highest possible correlation with the actual values of Y; no other weights would produce a higher multiple R.

For the example above, $r_{1Y} = .4$, $r_{2Y} = .5$, and $r_{12} = .3$. Therefore,

$$\beta_1 = \frac{.4 - .15}{1 \quad .09} = \frac{.25}{.91} = .275 \quad \beta_2 = \frac{.5 - .12}{1 - .09} = \frac{.38}{.91} = .418$$

$$z_{Y'} = .275 z_{X_1} + .418 z_{X_2}$$

With two correlated predictors the multiple R becomes

$$R = \sqrt{\beta_1 r_{1Y} + \beta_2 r_{2Y}} \qquad \text{Formula 12.20}$$

This formula can be extended easily to accommodate any number of variables. For the above example,

$$R = \sqrt{.275 \cdot .4 + .418 \cdot .5} = \sqrt{.319} = .5648$$

R^2, the amount of variance in GSS accounted for by the combination of TGRE and UGPA, equals $.5648^2 = .319$. The amount by which R^2 is less than $r_{1Y}{}^2 + r_{2Y}{}^2 = .16 + .25 = .41$ is equal to the overlapping area of variance in GSS (labeled C in Figure 12.9) that is predicted by both TGRE and UGPA, which equals .091 (overlap $= r_{1Y}{}^2 + r_{2Y}{}^2 - R^2 = .16 + .25 - .319 = .091$). In most cases, greater correlation between the predictors means more overlap and a reduction in R and R^2. However, keep in mind that the relationships among three variables cannot always be conveniently displayed in a Venn diagram. Some variables, known as **suppressor variables,** may have no correlation with the criterion (and therefore take up no area in the Venn diagram), but nonetheless have a considerable correlation with one of the predictors. A suppressor variable can actually increase R^2 by removing unwanted variance from one of the predictors, and thus improve predictability. For instance, if we had a measure of a subject's speed in filling out little circles with a number 2 pencil, this variable might correlate somewhat with TGRE (if that is how the GRE is administered) but not with GSS. If TGRE could be "corrected" to remove the effects of circle-filling speed, we would have a more accurate measure of aptitude, and therefore a better predictor. Although suppressor variables occasionally can play a useful role in multiple regression, their effects are too complex to be dealt with in this brief introduction to the topic. I will deal only with situations that can be represented by simple Venn diagrams.

When r_{YX} and therefore b_{YX} are not statistically significant in a simple linear regression problem, there is little point in reporting the regression equation or using it to generate predictions. Multiple R can also be tested for statistical significance, and if the null hypothesis (i.e., $R = 0$) cannot be rejected, there is probably no point in reporting the multiple regression equation. (Actually it is R^2 that is tested with an F ratio; the methods will be covered in Chapter 14.) However, if R is found to be significant in a two-predictor problem, it doesn't mean that both predictors are contributing significantly to the multiple correlation; it is quite possible that one predictor is doing nearly all the work and that the other predictor is contributing very little. I will return to this important point shortly.

Raw-Score Prediction Equation

If both of your predictors prove to be statistically significant, you may want to use your multiple regression equation to generate predictions. In that case, it would probably be more convenient to use a raw-score equation, rather than the z score equation we have been working with (just as we switched from $z_{Y'} = rz_X$ to $Y' = b_{YX}X + a_{YX}$ for simple regression). For two predictors, the raw-score formula would take the following form,

$$Y' = b_0 + b_1X_1 + b_2X_2 \qquad \text{Formula 12.21}$$

where b_0 is the Y intercept (rather than a) and is found by the following formula:

$$b_0 = \overline{Y} - b_1X_1 - b_2X_2$$

The **unstandardized regression coefficients** can be found from the βs using Formula 12.22:

$$b_1 = \frac{\sigma_Y}{\sigma_{X_1}}\beta_1 \qquad b_2 = \frac{\sigma_Y}{\sigma_{X_2}}\beta_2 \qquad \text{Formula 12.22}$$

(Note the resemblance to Formula 12.3 for the one-predictor case, with β in place of r).

Recall that in the raw-score formula for simple linear regression, the magnitude of the slope tells us nothing about the strength of the relationship between X and Y; r tells us what we want to know, but b comes from multiplying r by a ratio of standard deviations, so b can be small when r is large, and vice versa. Similarly, the unstandardized regression weights in multiple regression (i.e., b_1, b_2, etc.) give us no idea how much each predictor variable is contributing to the final prediction. When the focus of multiple regression is on the relative contributions of each variable in the set of predictors (which is more often the case), rather than on actually making predictions, it is the standardized regression coefficients (the βs) that should be inspected.

The beta weights have a rather straightforward interpretation. For the example above, the fact that TGRE had a beta of .275 in the standardized regression equation means that, if all else remains equal, an increase in one standard deviation in TGRE should be associated with an increase of .275 standard deviation in graduate school success. Although there is some debate as to whether the beta weights give the best measure of the relative importance of different predictors in a set, and alternative measures have been proposed, in most cases the beta weights give a reasonably accurate picture of the relative contributions of the different predictors.

Highly Correlated Predictors

Before a researcher would use a variable in a multiple regression situation, he or she would make sure that the variable did not have a very low correlation with the criterion to begin with (unless that variable was expected to serve as a suppressor variable). However, just because a particular variable is highly correlated with the variable being predicted does not mean that it will contribute much in a multiple regression equation. Its contribution will depend on how highly correlated it is with the other predictor variables—that is, the extent to which it is accounting for variance in the criterion not already accounted for by one or more other predictors. The case of two highly correlated predictors will be explored in the following example.

A psychiatrist wants to predict how long newly admitted schizophrenics will stay in the hospital before being released. Two measures taken on admission are available: a symptom checklist (SC), and a global assessment (GA) of psychoticism. Suppose that the data collected so far yield a correlation of .42 between SC and length of stay (r_{1Y}), whereas GA and length of stay (r_{2Y}) have a correlation of .45. However, it turns out that these two measures are quite redundant, as the correlation between them (r_{12}) equals .9. Using Formula 12.18, you will find that the beta weight for SC (β_1) is .079 and for GA (β_2) it is .379. The beta weights suggest that SC is not contributing much to the prediction. Figure 12.10 (on page 398) will help to show why this is so. (Note that the areas in Figure 12.10 are found by first calculating R^2 using Formula 12.20 and then subtracting either r_1^2 or r_2^2 to find the two non-overlapping areas.)

You can see from Figure 12.10 that GA does not account for much additional

Figure 12.10	Variance in length of stay

Figure 12.10

Venn Diagram of Two Highly Correlated Predictors

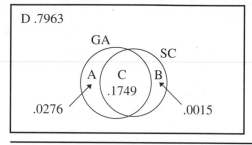

The square root of area A is the semipartial correlation of the first predictor with respect to the second, and the square root of area B is the semipartial correlation of the second predictor with respect to the first.

variance beyond what is accounted for by SC—less than 3% (area A)—but SC contributes even less unique variance accounted for—only about .15%. Because it makes sense to include in your regression equation the variable with the higher initial correlation, you would accept GA and then look at how much additional variance could be accounted for by adding SC to the equation. Adding SC to the equation adds so little variance accounted for that it is not likely the increment would be statistically significant, and therefore GA would probably be used by itself.

Semipartial Correlations

The proportion of variance in the criterion that is accounted for by one predictor but not any of the others—the predictor's unique contribution to the multiple correlation—plays a special role in multiple regression. The square root of this proportion is a correlation coefficient called the **semipartial correlation coefficient** (or "part" correlation). To find the semipartial correlation of GA with length of stay, take the square root of area A in Figure 12.10 (the area accounted for by GA but not SC), which equals $\sqrt{.0276} = .166$. This is the correlation that you would obtain if you followed the following three steps. First, perform a simple linear regression predicting GA from SC (i.e., predicting one predictor from the other). Second, find the residual for each subject's GA—that is, subtract from each GA the value predicted on the basis of SC. Finally, find the correlation between length of stay and the residuals found in the previous step. In other words, the semipartial correlation of GA with the criterion, length of stay (with SC "partialed out" of GA), is the correlation of the criterion with that part of GA that is not related to SC.

To decide on whether a variable should be added to the multiple regression equation, we can focus on its semipartial correlation with all the other predictors "partialed out," and test that value for statistical significance. The semipartial correlation is tested against the null hypothesis of zero like any other Pearson's r: you can use a t-test (Formula 11.6) or a critical value from Table A.5. However, you must use df $= N - k - 1$, where k is the number of predictors. For the two-predictor case, df $= N - 2 - 1 = N - 3$. For the example above, SC has a semipartial correlation of $\sqrt{.0015} = .039$, which is so tiny that it would probably not justify its being added to the equation (even if it were statistically significant).

An easy way to calculate the semipartial correlation is to find R^2 with the variable in question included in the regression equation, and then to find R^2 again without that variable. The increment in R^2 due to that variable is the square of its semipartial correlation. Alternatively, if you have already calculated the beta weights for a two-predictor problem, the semipartial correlations can be found from Formula 12.23:

$$r_{Y(1.2)} = \beta_1 \sqrt{1 - r_{12}^2} \quad r_{Y(2.1)} = \beta_2 \sqrt{1 - r_{12}^2} \qquad \text{Formula 12.23}$$

The above formula shows just how closely related β is to the semipartial correlation, and why it makes sense to refer to the βs as standardized *partial* regression coefficients. The notation $r_{Y(1.2)}$ stands for the semipartial correlation between the criterion (Y) and the first predictor, partialing out the effects of the second predictor on the first. Similarly, $r_{Y(2.1)}$ indicates the semipartial correlation between Y and the second predictor, with the effects of the first predictor partialed out.

Partial Correlation

The reason the correlation coefficient described above is called *semi*partial will be made clear as I discuss a related measure, the **partial correlation coefficient.** Compared to the semipartial correlation, the partial correlation is less directly linked to multiple regression and more likely to be used on its own. A common use for partial correlation is when the correlation between two variables of interest is misleadingly high because of the effects of some uninteresting third variable, and you want to remove the effects of that third variable statistically. For instance, suppose you are a gerontologist interested in the relation between memory and social skills in people over 80 years old, and you find a correlation between the two variables (r_{1Y}) that equals .6. The problem is that this correlation may not be due to any connection between memory and social skills but may be due instead to a general decline in the values of both variables with increasing age over 80.

That both variables show a decline with age was not what the researcher wanted to show; the point of the study was to show that for any particular age over 80, the subjects with the better memories also have the better social skills. Unfortunately, it may not be practical to hold age constant; there may not be enough subjects available who are both over 80 and all the same age to have a reasonable chance of statistical significance. The usual alternative to direct control of a "nuisance" variable, such as age in this study, is statistical control through regression. If you can assume that both variables show a linear decline with age, then the solution to this problem is to "remove" the effects of age from both variables through linear regression. This is the same process used in semi-

Figure 12.11

Venn Diagram in Which the Second Predictor Is a "Nuisance" Variable

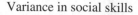

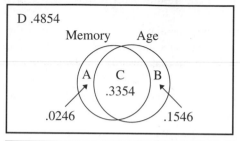

The partial correlation between the first predictor and the criterion, partialing out the second predictor, equals the square root of the ratio between area A and the sum of areas A and D.

partial correlation—making predictions and finding residuals—only this time you find residuals for *both* variables and then compute the correlation between the two sets of residuals. Now you can see where the term semipartial comes from. If age is partialed out of both variables, the correlation between the two sets of residuals is called a partial correlation. If age were partialed out of only the other predictor, but not the criterion, you would have a semipartial correlation.

   The partial correlation between memory and social skills in our example can often be found by manipulating the areas of the Venn diagram we have been using to represent multiple regression problems. For the example above, memory is the first predictor, and social skills is the criterion; the nuisance variable, age, becomes the second predictor. We know that r_{1Y}, the correlation between memory and social skills, is .6. Now, let us suppose that the correlations between age and memory and between age and social skills are both .7; that is, $r_{12} = r_{2Y} = .7$. For these correlations the overlapping areas of variance accounted for would appear as in Figure 12.11. The first area we are interested in is A: the variance in social skills that is explained by memory but not age. This area is the square of the semipartial correlation, so $r_{Y(1.2)} = \sqrt{.0246} = .157$. But this time we are interested in the partial correlation, $r_{Y1.2}$, between predictor 1 and the criterion (Y), partialing out predictor 2 from both. To get $r_{Y1.2}^2$ we divide area A by the total of areas A + D; that is, we divide by the proportion of variance *not* accounted for by age, because we are partialing age out of the criterion, as well as out of the other predictor. From Figure 12.11 we find that

$$r_{Y1.2}^2 = \frac{\text{Area A}}{\text{Area A} + \text{Area D}} = \frac{.0246}{.0246 + .4854} = .048$$

Therefore, $r_{Y1.2}^2 = \sqrt{.048} = .22$.

   The correlation between memory and social skills is reduced from .6 to .22 by partialing out the linear influence of age. With a sample of 20 subjects the correlation would have gone from significant at the .01 level (two-tailed) to not significant at the .05 level (two-tailed). The partial correlation can be found directly from the original correlations by Formula 12.24:

$$r_{Y1.2} = \frac{r_{1Y} - r_{2Y}r_{12}}{\sqrt{(1 - r_{2Y}^2)(1 - r_{12}^2)}}$$ Formula 12.24

It is also possible to find R^2 (the square of the multiple correlation coefficient) directly from the three r values by means of Formula 12.25:

$$R^2 = \frac{r_{1Y}^2 + r_{2Y}^2 - 2r_{1Y}r_{2Y}r_{12}}{1 - r_{12}^2}$$ Formula 12.25

Multiple Regression with More Than Two Predictors

When there are three predictors of the criterion, the calculation of the beta weights gets quite complicated, and with even more predictors the calculation becomes a very serious task. Fortunately, statistical software makes it easy to use a computer to find the multiple regression equation, even with quite a few predictors. However, when there are more than two predictors, deciding which predictors to include in the equation can be difficult. With only two predictors the decision is usually quite simple: Include the predictor that has the larger correlation with the criterion (if that correlation is statistically significant), and include the second predictor, as well, if it adds significantly to the variance accounted for by the first (i.e., if its semipartial correlation is significant). Even with only three predictors this decision process becomes complex and leads to various alternative methods.

You might think that the strategy mentioned for two predictors could easily be extended to any number of predictors: Start with the predictor that has the highest correlation with the criterion, then include the predictor that adds the most variance accounted for to the first (if its contribution is significant), then, of the remaining predictors, include the one that adds the most variance to the combination of the first two, and continue like this until none of the remaining predictors adds a statistically significant amount. This is a legitimate multiple regression procedure, called **forward selection,** but it has a serious drawback that is illustrated by the following example.

Imagine that you have three predictors of graduate school success (GSS): IQ, achievement motivation (AM), and undergraduate GPA. Further imagine that IQ and achievement motivation are each moderately correlated with GSS but barely correlated with each other, whereas UGPA is more highly correlated with GSS than either IQ or AM. (You would expect UGPA to have a reasonably high correlation with GSS.) Using forward selection, UGPA would be entered first. Assuming AM and IQ qualify to be added as well, the final situation can be represented as shown in Figure 12.12 (on page 402). The particular situation being depicted in Figure 12.12 is one in which IQ and AM together predict UGPA very well; UGPA adds very little variance accounted for beyond the combination of IQ and AM. However, with forward selection it is possible that neither IQ nor AM by itself would add enough variance to UGPA, so you could be stuck with UGPA as your only predictor, even when the combination of IQ

Figure 12.12

Venn Diagram of Three Correlated Predictors

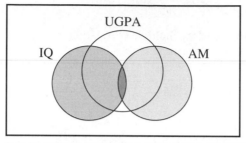

The combination of two predictors (IQ and AM) accounts for most of the variance in GSS; the third predictor, UGPA, adds little variance beyond that.

and AM would lead to a significantly higher R^2 than just UGPA by itself. The forward selection procedure does not account for the fact that *combinations* of subsequent predictors may be almost entirely redundant with one of the predictors included earlier.

Forward selection is one form of a procedure known as **stepwise regression.** In its more general form stepwise regression avoids the problem described above by retesting predictors entered earlier to see if they are still adding a significant amount of variance after other predictors have been entered. For instance, in the example described above, if IQ and AM were entered into the multiple regression, UGPA would be tested again to determine its semipartial correlation (partialing out both IQ and AM). At this point, UGPA might not be adding enough unique variance, and if that were the case it would be dropped, leaving only IQ and AM.

One way you could have known UGPA had a good chance of being dropped in a stepwise regression would have been to measure its **tolerance.** Tolerance is the amount of variance in a predictor that is not explained by all of the other predictors. To the extent that a predictor can itself be predicted by the other predictors (as UGPA could be predicted quite well by a combination of IQ and AM), it will have low tolerance. Even if a predictor has a high correlation with the criterion, if its tolerance is low it is not explaining much that isn't explained by the other predictors, and it should, perhaps, be dropped from the set. The extreme case, in which the tolerance of a predictor is zero, means that that predictor can be *perfectly* predicted by some combination of other predictors. In that case, the correlation matrix is said to exhibit **singularity,** and the multiple regression equation cannot be solved until the redundant predictor is dropped from the set. This extreme case is most likely to occur when the various subscales of a test are being used as predictors, and either the total score is being included as well or the subscales add up to some fixed number.

The principles I have been describing for choosing which predictors to include in a multiple regression equation are generally accepted as useful when the purpose is to predict some criterion as efficiently as possible. Even then there is considerable debate about the best way to apply stepwise regression, and several alternative methods have been proposed. A course of at least an entire semester is appropriate for teaching the mechanics of multiple regression. The subject becomes even more complicated when multiple regression is used to formulate or help confirm some theoretical model for how a set of variables interact to influence some criterion. The order in which you enter variables becomes not just a matter of how much variance each explains; it also depends in part on the causal directions between variables (e.g., parental income may affect a student's performance in school, but we don't expect an influence in the reverse direction) and on how directly one variable affects another (e.g., parental IQ may affect a student's performance, but there are a number of indirect ways this can occur). The practice of using a theoretical model or other a priori basis for determining the order for entering variables in multiple regression is generally referred to as **hierarchical regression.** Unfortunately, it is beyond the scope of this text to even introduce the issues involved in the hierarchical method of multiple regression. Therefore, I will conclude this section with a few warnings about common pitfalls you may encounter when using multiple regression.

Cautions Concerning Multiple Regression

The assumptions that apply to bivariate correlation and simple linear regression apply as well to significance tests involving multiple linear regression. In addition, some tests assume that all of the variables in the multiple regression jointly follow a multivariate normal distribution. As usual, this assumption is routinely ignored when sample sizes are large. If any curvilinear relationships occur among the possible pairs of variables, data transformations or curvilinear regression techniques may be needed, or more simply, those variables causing the problem can be dropped.

There are a few other problems that you must be especially cautious about when dealing with multiple regression. One problem that has already been alluded to above occurs when there are high correlations among a set of predictors. This problem, known as **multicollinearity,** usually leads to low tolerances and misleading beta weights. One way this problem can arise is when a researcher who wants to measure a trait such as anxiety includes several similar measures, not knowing which will work the best. Including all the related measures in the final regression analysis is likely to lead to regression coefficients that change considerably from one sample to another. After a preliminary analysis you should select the measure that works best and delete the others that are closely related.

Another common problem is using too many predictors relative to the size of your samples. No one would think of using only two subjects in a bivariate problem. The degrees of freedom ($N - 2$) would be zero, and you would be guaranteed to obtain perfect correlation (either $+1$ or -1). It is somewhat less obvious that with ten subjects and nine predictors you are guaranteed a perfect multiple R. The df in multiple regression equals $N - p - 1$, where p is the number of predictors. In general, the value of R found for a particular sample is an overestimate of the R that would be found in the population, and this problem gets worse as the number of predictors gets closer to the number of subjects. To better estimate the population R, the sample R can be adjusted according to Formula 12.26:

$$\text{Adjusted } R^2 = 1 - \frac{(1 - R^2)(N - 1)}{N - p - 1} \qquad \text{Formula 12.26}$$

Notice that when N is much larger than p not much adjustment occurs.

Related to the problem described above is the fact that R is likely to become smaller if you use the regression equation found from one sample to make predictions about values in a completely independent sample. The general problem is called **shrinkage,** and it occurs because the regression coefficients found for one sample are tailored to fit the data in that sample, including chance fluctuations that are not likely to appear in other samples. Some amount of shrinkage is virtually inevitable when using the regression coefficients obtained from one sample to make predictions in another sample or the population, but to keep shrinkage to a reasonable level a good rule is to use at least 20 subjects for each predictor—that is, when choosing a sample size, aim for $N > 20p$.

Some regression methods lead to Rs that fluctuate more from sample to sample and some methods lead to more stable Rs, but regardless of how a regression equation is obtained from a sample, it must be subjected to **cross-validation** on an independent sample in order to be taken seriously. Any of the statistical procedures described in this text can be affected by chance fluctuations; multiple regression just seems to provide more opportunities for chance to influence the results. That is why the need for replication (i.e., cross-validation) is especially strong when dealing with many, possibly related, predictors.

Exercises

*1. Section D of Chapter 9 included a table (page 289) showing data for a two-group experiment comparing patients with brain damage to the left or right cerebral hemisphere on their ability to solve paper-and-pencil mazes.

a) Calculate r_{pb} for those data, using any of the formulas for Pearson's r. (Note: It is convenient to assign the values 0 and 1 to the two groups.)

b) Use Formula 11.6 to find the t value to test the correlation coefficient you found in part a.

c) Compare the t value you calculated in part b to the t value found in Chapter 9, Section D.

2. a) Based on your answer to part a of exercise 1, determine the proportion of variance in number of mazes solved that is accounted for by location of brain damage.

b) Estimate the proportion of variance accounted for in the population (i.e., ω^2).

*3. If t for an experiment with two equal-sized groups equals 10, what proportion of variance in the dependent variable is accounted for a) when the sample size is 20?

b) When the sample size is 200?

c) What are your estimates for omega squared for parts a and b?

4. Suppose that you are planning a two-group experiment. How high would your t value have to be for

half the variance of your dependent variable to be accounted for a) if you use 25 subjects per group?

b) If you use 81 subjects per group?

* 5. According to the guidelines suggested by Cohen (1988), gamma = .8 is a large effect size; any effect size much larger would probably be too obvious to require an experiment.

a) What proportion of population variance is accounted for when gamma reaches this value?

b) What proportion of population variance is accounted for when gamma is moderate in size, i.e., gamma = .5?

c) How high does gamma have to be for half of the population variance to be accounted for?

6. a) If a rating based on reference letters from previous employers has a .2 correlation with future job performance, and a rating based on a job interview has a .4 correlation, and these two predictors have no correlation with each other, what is the multiple correlation coefficient, and what proportion of the variance in job performance is accounted for by a combination of the predictors?

b) Answer the questions in part a again, assuming the two correlations are both .6.

c) Can you answer the same questions if the two correlations are .7 and .8? Why not?

d) If the correlation between the job interview and job performance is .8, what is the largest correlation the reference letter rating can have with job performance, if there is no correlation between the two predictors?

7. Three predictors have correlations with a criterion variable that are .28, −.43, and .61, respectively. Assume that the correlation between any two of the predictors is zero.

a) Write the multiple regression equation.

b) How much variance in the criterion is accounted for by the predictions made by the regression equation you wrote for part a?

c) How large is the correlation between the predictions from your regression equation and the actual values of the criterion?

8. Suppose that the "concreteness" of a word has a .4 correlation with the probability of recall and the "imageability" of a word has a .5 correlation with the same criterion. Write the regression equation, calculate R, and find the semipartial correlation of each predictor with respect to the other, if the correlation between concreteness and imageability is a) .3 b) .6 c) .8.

d) Explain anything strange about your regression equation for part c.

9. The average resting heart rate ($\overline{Y}$) for the members of a large running club is 62 bpm with $s_Y = 5$. The average number of minutes spent running each day ($\overline{X}_1$) is 72 with $s_1 = 15$, and the average number of miles run per day ($\overline{X}_2$) is 9.5 with $s_2 = 2.5$. Suppose both running time and running distance have correlations of −.3 with resting heart rate, and the two predictors have a correlation of +.7 with each other.

a) Write the standardized multiple regression equation for predicting heart rate.

b) How much unique explained variance does each predictor add to the other?

c) Write the unstandardized multiple regression equation.

d) If a club member runs 5 miles in 50 minutes each day, what would you predict for his heart rate?

e) If a club member runs 9.5 miles in 72 minutes each day, what would you predict for her heart rate? Why did you not need the regression equation to make this prediction?

10. a) In a study of her own patients, a psychoanalyst found a correlation of .6 between a measure of "insight" and a measure of symptom improvement (relative to the start of therapy). She then realized, however, that because both symptom improvement and insight tend to increase with more years of therapy, the correlation she found might have resulted, in part, from using patients who differ considerably as to how much therapy they have had. Because she wants to know the relation between insight and symptom improvement when number of years in therapy is held constant, she calculates the partial correlation between insight and symptom improvement, partial-

ing out number of therapy years. How high would that partial correlation be if the correlation between symptom improvement and number of therapy years is .8, and the correlation between insight and number of therapy years is .5?

b) Use Formula 12.24 to calculate the partial correlation between shoe size and reading level, partialing out age, in the data for exercise 12B7. How does your answer compare to your answer to part b of exercise 12B8?

SUMMARY

The Important Points of Section A

1. For perfect linear positive correlation, the z score predicted for Y is the same as the z score for X. For perfect *negative* correlation, the z score predicted for Y is the same *in magnitude* as the z score for X, but opposite in sign.

2. If correlation is less than perfect, the z score predicted for Y is just r (i.e., Pearson's correlation coefficient) times the z score for X. If $r = 0$, the prediction for Y is always the mean of Y, regardless of the value of X.

3. The regression equation in terms of raw scores is $Y' = b_{YX}X + a_{YX}$, in which b_{YX} is the *slope* of the line and a_{YX} is the Y *intercept*. For regression in terms of z scores, the slope is just r, and the Y intercept is always zero.

4. For raw scores the slope is r times the ratio of the two standard deviations. The Y intercept is just the mean of Y minus the slope times the mean of X. The Y intercept is not always meaningful; it may not make sense to extend the regression line to values of X near zero.

5. The variance around the regression line is called the *variance of the estimate,* or the *residual variance,* symbolized by $\sigma^2_{\text{est }Y}$. When correlation is perfect (i.e., $r = +1$ or -1), $\sigma^2_{\text{est }Y} = 0$, and there is no error involved in the predictions. When $r = 0$, $\sigma^2_{\text{est }Y}$ equals σ_Y^2, which is the total variance of the Y scores.

6. The total variance can be divided into two portions: the unexplained variance (the variance of the estimate) and the explained variance. The ratio of the explained variance to the total variance is called the *coefficient of determination,* and it is symbolized by r^2.

7. The ratio of the variance of the estimate to the total variance is called the *coefficient of nondetermination* and is symbolized by k^2. In terms of Pearson's r, $k^2 = 1 - r^2$.

The Important Points of Section B

I will review the procedures presented in Section B by presenting a set of raw scores and showing how each step in the regression process can be calculated. Imagine that a career counselor is interested in how well college grade point averages (GPA) predict annual salaries five years after graduation. From her records, she selects at random six individuals who graduated five years ago, and finds out their final cumulative GPA (X), as well as their current annual salaries (Y). The data are shown in Table 12.2 along with the squared values and cross-products. (Highest possible GPA is 4.0, which means As in all courses; annual salary is expressed in thousands of dollars.)

Assuming that the two variables are linearly related, it can be useful to find the linear regression equation that predicts annual salary based on a subject's college GPA. The first step is to find the slope, which can be found either from the raw scores or from Pearson's r and the standard deviations. I will illustrate both ways, beginning with the raw-score method. Inserting the appropriate sums from the last line of the table into Formula 12.11 yields the following value:

$$b_{YX} = \frac{N(\Sigma XY) - (\Sigma X)(\Sigma Y)}{N\Sigma X^2 - (\Sigma X)^2} = \frac{6(479.3) - (17.4)(163)}{6(51.84) - 17.4^2}$$

$$= \frac{2875.8 - 2836.2}{311.04 - 302.76} = \frac{39.6}{8.28} = 4.7826$$

To find the regression slope with Formula 12.3, we need the standard deviations of both variables. Assuming that we had already calculated the unbiased standard deviations ($s_X = .5254$ and $s_Y = 3.7103$), the next step is to calculate Pearson's r using Formula 11.3:

$$r = \frac{\frac{1}{N-1}(\Sigma XY - N\bar{X}\bar{Y})}{s_X s_Y} = \frac{\frac{1}{5}[479.3 - 6(2.9)(27.17)]}{(.5254)(3.7103)} = \frac{1.32}{1.949} = .6771$$

	X^2	X	XY	Y	Y^2	
	9.61	3.1	99.2	32	1024	**Table 12.2**
	6.25	2.5	67.5	27	729	
	12.96	3.6	108.0	30	900	
	4.84	2.2	52.8	24	576	
	10.89	3.3	92.4	28	784	
	7.29	2.7	59.4	22	484	
Totals	51.84	17.4	479.3	163	4497	

Now we can use Formula 12.3, to find that the slope equals

$$b_{YX} = r\frac{s_Y}{s_X} = .6771\left(\frac{3.7103}{.5254}\right) = .6771\,(7.062) = 4.782$$

which, as you see, is the same (within rounding error) as the value obtained from Formula 12.11.

The second step in determining the regression equation is to find the Y intercept using Formula 12.4. To use this formula we need the slope and the means of both variables, which are $\overline{X} = \Sigma X/N = 17.4/6 = 2.9$, and $\overline{Y} = \Sigma Y/N = 163/6 = 27.17$. Then

$$a_{YX} = \overline{Y} - b_{YX}\overline{X} = 27.17 - 4.7826\,(2.9) = 27.17 - 13.87 = 13.3$$

Finally, we find the regression equation by plugging the slope and Y intercept into Formula 12.5, as shown below:

$$Y' = b_{YX}X + a_{YX} = 4.78X + 13.3$$

The equation above can be used to predict the salary for any GPA, even those not in our original sample. For instance, consider a GPA less than any in the sample, one that would barely allow graduation: $X = 1.8$. The person's predicted annual salary five years after graduation would be

$$Y' = 4.786\,(1.8) + 13.3 = 8.615 + 13.3 = 21.9 \text{ (i.e., \$21,900)}$$

However, given the small size of our sample, this prediction could not be considered very accurate. A confidence interval surrounding this prediction would give a much clearer picture of its accuracy. However, before we can construct the CI, we need to calculate the sample standard error of the estimate, $s_{\text{est } Y}$. Because we have already computed the unbiased standard deviation and Pearson's r, we can use Formula 12.9:

$$s_{\text{est } Y} = s_Y\sqrt{\frac{N-1}{N-2}(1 - r^2)}$$

$$= 3.71\sqrt{\frac{5}{4}(1 - .677^2)} = 3.71\sqrt{.677} = 3.71(.823) = 3.053$$

If we choose to find the 95% CI, we need to look up t_{crit} for alpha $= .05$ (two-tailed), and df $= N - 2 = 6 - 2 = 4$, which is 2.776 (from Table A.2). Now we can use Formula 12.10 to find the CI for our prediction:

$$Y' \pm t_{crit}s_{est\ Y} \sqrt{1 + \frac{1}{N} + \frac{(X - \overline{X})^2}{(N-1)s_X^2}}$$

$$= 21.9 \pm 2.776(3.053) \sqrt{1 + \frac{1}{6} + \frac{(1.8 - 2.9)^2}{(5).5254^2}}$$

$$= 21.9 \pm 8.47 \sqrt{2.04} = 21.9 \pm 8.47(1.43) = 21.9 \pm 12.12$$

The interval extends from 9.8 to 34.0. The small sample size is what makes the interval required for 95% confidence so large. The confidence interval would have been a bit smaller, however, for a prediction closer to the mean, where there is less likelihood of error.

Assumptions

The confidence interval calculated for the prediction is accurate only if we can make the following assumptions: (1) random sampling; (2) linear relationship; (3) normal distributions (of Y at each value of X); and (4) homoscedasticity (the same variance of Y at each value of X, i.e., the scatter around the regression line is uniform along the whole length of the line).

The most common uses for linear regression are:

1. Predicting future performance from measures taken previously.
2. Statistically removing the effects of a confounding or unwanted variable.
3. Evaluating the linear relationship between the quantitative levels of a truly independent (i.e., manipulated) variable and a continuous dependent variable.
4. Testing a theoretical model that predicts values for the slope and Y intercept of the regression line.

The Important Points of Section C

The Point-Biserial Correlation Coefficient

1. If the results of a two-group experiment are described in the form of linear regression, the square root of the ratio of the explained variance to the total variance yields a correlation coefficient called the *point-biserial r,* symbolized as r_{pb}. The point-biserial r can be calculated directly, using any formula for Pearson's r, by assigning X values to all the subjects, such that the subjects in one group are all assigned the same value (e.g, zero) and the subjects of the other group are all assigned some different value (e.g., 1).

2. The point-biserial r can also be found from the t value for a pooled-variance test. For a given t value, the larger the degrees of freedom, the smaller is r_{pb}. If the df stay the same, r_{pb} goes up as the t value increases.

3. The point-biserial r provides an alternative to g (the effect size in a sample) for assessing the strength of association between the independent and dependent variables in a two-group experiment.

4. A slight modification of the formula for r_{pb} squared can be used to estimate omega squared (ω^2), the proportion of variance accounted for in the dependent variable by group membership in the population. There is a simple relationship between ω^2 and the square of gamma, the population effect size.

5. The point-biserial r should only be used when dealing with two groups that represent a true dichotomy (e.g., male and female). If the two groups represent opposite sides of an underlying normal distribution (e.g., tall people and short people), a different correlation coefficient, the biserial r, should be used.

Multiple Linear Regression

6. When two predictors of a criterion variable are not correlated with each other, the proportion of variance they account for together is equal to the sum of the two separate proportions of variance accounted for, and R (the *multiple correlation coefficient*) is the square root of that sum. The coefficients of the multiple regression equation in this case are just the two sample rs, when dealing with standardized scores. These principles can be extended to deal with any number of mutually uncorrelated predictors, but the possibility of more than two or three such predictors is highly unlikely.

7. When two predictors are correlated, R^2 will usually be less than the sum of the two r^2, because of the overlap in variance accounted for (unless one of them is a *suppressor variable*). The *standardized regression coefficient*, β, for each variable will reflect (to some extent) that variable's contribution to the prediction made by the multiple regression equation.

8. The *semipartial correlation* is the square root of the unique variance accounted for by one predictor but not any of the others, and it can be tested for significance to determine whether that variable should be added to the regression equation. The *partial correlation* is the correlation between two sets of residuals obtained by removing the linear effects of some confounding (or "nuisance") variable from each. It is often used apart from multiple regression.

9. When there are more than two predictors, there are alternative ways to decide which predictors will be included in the final multiple regression equation. *Step-wise regression* includes variables according to the amount of variance each adds to that accounted for by the predictors already included and involves retesting variables previously included to determine whether they still make large enough unique contributions to be kept (unless forward selection is being used). In *hierarchical regression*, the order in which variables are entered into the equation is determined on theoretical grounds (e.g., which variables are likely to *cause* changes in which other variables), rather than on the basis of amount of variance accounted for.

10. An important problem to watch for in multiple regression is *multicollinearity*. This problem occurs when some of the predictors are highly correlated with each other, and it leads to low *tolerances* (some predictors explain very little variance that is not explained by other predictors in the set). If some of the closely related predictors are not eliminated, the regression coefficients will not be reliable and may change considerably when you attempt to *cross-validate* your prediction equation on an independent sample. Another important problem is *shrinkage,* which is the reduction in R^2 that occurs when using regression coefficients from one sample to make predictions in another sample or the population. To minimize the problem, try to have at least 20 cases for each predictor.

Definitions of Key Terms

Regression toward the mean The statistical tendency for extreme values of one variable to be associated with less extreme values (i.e., values closer to the mean) on a second less-than-perfectly correlated variable.

Regression line When regressing Y on X, the straight line that minimizes the squared differences between the actual Y values and the Y values predicted by the line. Regressing X on Y leads to a different line (unless correlation is perfect) that minimizes the squared differences in the direction of the X axis.

Slope of the regression line When regressing Y on X, the amount by which the Y variable changes when the X variable changes by one unit.

Y intercept of the regression line When regressing Y on X, the Y intercept is the value of Y when X is zero.

Residual In linear regression, the portion of an original score that is left over after a prediction has been subtracted from it.

Variance of the estimate ($\sigma^2_{est\ Y}$) Also called residual variance; when regressing Y on X, the variance of the Y values from the regression line, or equivalently, the variance of the residuals after each Y value has been subtracted from its predicted value.

Explained variance The variance which, when added to the unexplained variance, yields the total variance; stated another way, it is the amount by which the variance around the regression line is less than the total variance.

Unexplained variance Another way to refer to the variance of the estimate.

Coefficient of determination The proportion of the total variance that is "explained" by linear regression; it is the ratio of explained variance to total variance, and it is symbolized by r^2.

Coefficient of nondetermination The proportion of the total variance that is "unexplained" by linear regression; it is the ratio of unexplained variance to total variance, and it equals $1 - r^2$, which is sometimes symbolized as k^2.

Homoscedasticity The condition that exists when the variance around the regression line at one location (i.e., one particular X value) is the same as at any other location.

Standard error of the estimate A standard deviation that is the square root of the variance of the estimate.

Point-biserial r A Pearson's correlation coefficient that is calculated when one of the variables has only two possible values. It can be used to measure the strength of association between the independent and dependent variables in a two-group experiment.

Omega squared (ω^2) The proportion of variance accounted for in one variable by another (usually discrete) variable in a population.

Biserial r A correlation coefficient related to the point-biserial r, except that the two values for one of the variables represent an arbitrary division of an underlying normal distribution.

Venn diagram A graphical method of depicting sets and their interactions. A Venn diagram is sometimes used to represent the portions of criterion variance that are accounted for by several predictors in a multiple regression.

Multiple correlation coefficient (R) A coefficient that measures the correlation between the predictions from a weighted combination of predictor variables and the actual values of the variable being predicted. R^2 tells you the proportion of variance in the predicted variable explained by the combination of predictors.

Standardized regression equation A regression equation in which all of the predictor variables have been standardized and are expressed in terms of z scores.

Regression coefficient The weight given to a predictor variable in a regression equation.

Standardized regression coefficient (β) The weight given to a particular predictor variable in a standardized regression equation. It is more formally known as the standardized partial regression coefficient, or informally as the beta weight.

Suppressor variable A variable that may have little or no correlation with the criterion, but increases R^2 by removing extraneous (i.e., error) variance from one or more predictor variables.

Unstandardized regression coefficient (b) The weight given to a particular predictor variable in an unstandardized (raw-score) regression equation; also known as the unstandardized partial regression coefficient.

Semipartial correlation coefficient A coefficient that measures the correlation between one variable and the residuals of a second variable, after a third variable (or any number of other variables) has been "partialed out" of the second. In multiple regression, the square of this coefficient tells you how much explained variance is added (i.e., the increment in R^2) by one particular predictor.

Partial correlation coefficient A coefficient that measures the correlation between the residuals of two variables after the same third variable (or set of variables) has been "partialed out" of both variables. This type of correlation is often used to remove statistically the effects of some extraneous or nuisance variable (or variables) from the bivariate relation of interest.

Forward selection A multiple regression method in which the variable entered next is the one that adds the most explained variance to the variance already accounted for by previous predictors. The adding of variables stops when the next variable fails to contribute a significant amount of variance, according to some preset criterion.

Stepwise regression A general multiple regression method that begins with forward selection, but at each step includes a retesting of predictors previously entered to determine whether each previous predictor still contributes a sufficient amount of unique explained variance to meet a preset criterion for significance.

Tolerance The amount of variance in a predictor variable that is left over after all the other predictors in the set have been used to predict its values. A predictor variable that overlaps a great deal with other predictors will have low tolerance, and therefore it will not be able to explain much of the criterion variance that isn't already explained by other predictors, regardless of how well the variable correlates with the criterion by itself.

Singularity The situation that exists when at least one of the predictor variables can be perfectly predicted by one or a combination of other predictors (i.e., at least one predictor has zero tolerance). In this case, the regression coefficients cannot be determined until the redundant predictor is eliminated.

Hierarchical regression A multiple regression method in which the order for entering variables is determined to some extent by theoretical constraints.

Multicollinearity A multiple regression situation in which one or more predictor variables are highly correlated with other predictor variables, leading to several low tolerance levels.

Shrinkage The degree to which R^2 is reduced when the regression coefficients that fit the data for one sample are used to generate predictions in an independent sample or the entire population. Shrinkage becomes more of a problem as the number of predictors approaches the number of subjects or cases.

Cross-validation Using the regression equation derived from one sample to generate predictions in a completely independent sample.

Key Formulas

Regression equation for predicting Y from X (in terms of z scores):

$$z_{Y'} = r\, z_X \qquad\qquad \text{Formula 12.1}$$

Regression equation for predicting Y from X (in terms of population parameters):

$$Y' = \frac{\sigma_Y}{\sigma_X} r\,(X - \mu_X) + \mu_Y \qquad\qquad \text{Formula 12.2}$$

The slope of the regression line for predicting Y from X:

$$b_{YX} = \frac{\sigma_Y}{\sigma_X} r \qquad\qquad \text{Formula 12.3}$$

The Y intercept of the regression line for predicting Y from X:

$$a_{YX} = \mu_Y - b_{YX}\mu_X \qquad\qquad \text{Formula 12.4}$$

The regression equation for predicting Y from X, as a function of the slope and Y intercept of the regression line:

$$Y' = b_{YX}X + a_{YX} \qquad\qquad \text{Formula 12.5}$$

The variance of the estimate for Y (definitional formula, not convenient for calculating):

$$\sigma^2_{\text{est } Y} = \frac{\Sigma(Y - Y')^2}{N} \qquad\qquad \text{Formula 12.6}$$

The coefficient of nondetermination:

$$\frac{\sigma^2_{\text{est } Y}}{\sigma_Y^2} = 1 - r^2 \qquad\qquad \text{Formula 12.7A}$$

The variance of the estimate for Y; convenient for calculating. (Note: This formula is a rearrangement of Formula 12.7A):

$$\sigma^2_{\text{est } Y} = \sigma_Y^2\,(1 - r^2) \qquad\qquad \text{Formula 12.7B}$$

Estimate of the population variance of the estimate for Y (based on the *unbiased* estimate of total variance):

$$s^2_{\text{est } Y} = \left(\frac{N-1}{N-2}\right) s_Y^2 (1 - r^2) \qquad \text{Formula 12.8}$$

Estimate of the population standard error of the estimate for Y. (This formula is just the square root of Formula 12.8):

$$s_{\text{est } Y} = s_Y \sqrt{\frac{N-1}{N-2}(1 - r^2)} \qquad \text{Formula 12.9}$$

The confidence interval for a prediction based on sample data:

$$Y' \pm t_{\text{crit}}\, s_{\text{est } Y} \sqrt{1 + \frac{1}{N} + \frac{(X - \overline{X})^2}{(N-1)s_X^2}} \qquad \text{Formula 12.10}$$

Regression equation for predicting Y from X (raw-score version; not recommended for calculations because errors are difficult to locate):

$$b_{YX} = \frac{N(\Sigma XY) - (\Sigma X)(\Sigma Y)}{N\Sigma X^2 - (\Sigma X)^2} \qquad \text{Formula 12.11}$$

The t-test for determining the significance of r_{pb}:

$$t = \frac{r_{\text{pb}}}{\sqrt{(1 - r_{\text{pb}}^2)/\text{df}}} \qquad \text{Formula 12.12}$$

The point-biserial correlation, based on the two-group t-value and the degrees of freedom:

$$r_{\text{pb}} = \sqrt{\frac{t^2}{t^2 + \text{df}}} \qquad \text{Formula 12.13}$$

Omega squared (the proportion of variance accounted for in the population) expressed in terms of gamma (the population effect size):

$$\omega^2 = \frac{\gamma^2}{\gamma^2 + 4} \qquad \text{Formula 12.14}$$

Estimate of omega squared, in terms of the two-group *t*-value and the degrees of freedom:

$$\text{Estimated } \omega^2 = \frac{t^2 - 1}{t^2 + df + 1} \qquad \text{Formula 12.15}$$

Standardized multiple regression formula for two uncorrelated predictors:

$$z_{Y'} = r_{YX_1} z_{X_1} + r_{YX_2} z_{X_2} \qquad \text{Formula 12.16}$$

Multiple correlation coefficient for two uncorrelated predictors:

$$R = \sqrt{r_{YX_1}^2 + r_{YX_2}^2} \qquad \text{Formula 12.17}$$

Standardized regression coefficients (beta weights) for two correlated predictors:

$$\beta_1 = \frac{r_{1Y} - r_{2Y} r_{12}}{1 - r_{12}^2} \quad \beta_2 = \frac{r_{2Y} - r_{1Y} r_{12}}{1 - r_{12}^2} \qquad \text{Formula 12.18}$$

Standardized multiple regression formula for two correlated predictors:

$$z_{Y'} = \beta_1 z_{X_1} + \beta_2 z_{X_2} \qquad \text{Formula 12.19}$$

Multiple correlation coefficient for two correlated predictors (in terms of the beta weights):

$$R = \sqrt{\beta_1 r_{1Y} + \beta_2 r_{2Y}} \qquad \text{Formula 12.20}$$

Unstandardized (raw-score) multiple regression formula for two correlated predictors:

$$Y' = b_0 + b_1 X_1 + b_2 X_2 \qquad \text{Formula 12.21}$$

Unstandardized regression coefficients for two correlated predictors (in terms of the standardized coefficients):

$$b_1 = \frac{\sigma_Y}{\sigma_{X_1}} \beta_1 \quad b_2 = \frac{\sigma_Y}{\sigma_{X_2}} \beta_2 \qquad \text{Formula 12.22}$$

Semipartial correlation coefficients (in terms of the beta weights):

$$r_{Y(1.2)} = \beta_1 \sqrt{1 - r_{12}^2} \quad r_{Y(2.1)} = \beta_2 \sqrt{1 - r_{12}^2} \qquad \text{Formula 12.23}$$

Partial correlation coefficient (in terms of the pairwise correlation coefficients):

$$r_{Y1.2} = \frac{r_{1Y} - r_{2Y}r_{12}}{\sqrt{(1 - r_{2Y}^2)(1 - r_{12}^2)}} \qquad \text{Formula 12.24}$$

The squared multiple correlation coefficient for two correlated predictors (in terms of the pairwise correlation coefficients):

$$R^2 = \frac{r_{1Y}^2 + r_{2Y}^2 - 2r_{1Y}r_{2Y}r_{12}}{1 - r_{12}^2} \qquad \text{Formula 12.25}$$

Estimate of R^2 in the population (sample R^2 is adjusted for shrinkage):

$$\text{Adjusted } R^2 = 1 - \frac{(1 - R^2)(N - 1)}{N - p - 1} \qquad \text{Formula 12.26}$$

THE MATCHED *t*-TEST

You will need to use the following from previous chapters:

Symbols
$\overline{X}$: Mean of a sample
s: Unbiased standard deviation of a sample
r: Pearson's correlation coefficient

Formulas
Formula 4.15: Computational formula for *s*
Formula 8.3: The *t*-test for one sample
Formula 9.8: The *t*-test for two equal-sized samples

Concepts
The *t* distribution
Linear correlation

Procedures
Finding critical values of the *t* statistic in Table A.2

13

C H A P T E R

A

CONCEPTUAL FOUNDATION

The previous two chapters dealt with the situation in which each subject has been measured on two variables. In the simplest case, subjects are measured twice by the same instrument with some time in between, in order to assess test-retest reliability. A large positive correlation coefficient indicates that whatever is being measured about the subject is relatively stable over time (i.e., high scorers tend to remain high scorers, low scorers are still low scorers). However, the correlation would be just as high if all the subjects scored a few points higher the second time (or all scored lower). This would make the mean score for the second testing higher (or lower), and this difference in means can be interesting in itself. The passage of time alone can make a difference—for example, people may tend to become happier (or less happy) as they get older—or the experimenter may apply some treatment between the two measurements—for example, subjects may receive psychotherapy and thus show lower depression scores. If the focus is on the difference in means between the two testings, then a *t*-test, rather than a correlation coefficient, is needed to assess statistical significance. But the *t*-test for independent groups (described in Chapter 9) would not be the optimal choice. The correlation between the two sets of scores can be used to advantage by a special kind of *t*-test, called the *repeated-measures,* or **matched *t*-test,** which is the topic of this chapter.

Before-After Design

Imagine that you are testing a new weight-loss program to see if it is effective. Subjects are weighed before starting the program and then again after 6 months

Before	After	Table 13.1
280	278	
160	159	
220	218	
180	179	
200	198	
$\overline{X}_{before} = 208$	$\overline{X}_{after} = 206.4$	
$s_{before} = 46.043$	$s_{after} = 45.632$	

of adhering to the program. The data for five subjects are given in Table 13.1. Although no subject lost much weight, the mean weight after the program was 1.6 pounds less than before, which suggests that the program was at least a little bit effective. However, we must again deal with the skeptical Dr. Null, who will say that a person's weight tends to fluctuate over time, and that we were just lucky that all our subjects happened to be a little lighter after the program. The null hypothesis is, of course, that the program is not effective at all, so that in the entire population, people will average out to be exactly the same weight after the program as before. Someone who didn't know about the matched *t*-test might be tempted to answer Dr. Null by conducting a *t*-test for two independent groups, hoping to show that $\overline{X}_{after}$ is significantly less than $\overline{X}_{before}$. If you were to conduct an independent-groups *t*-test on the data in Table 13.1, it would be easiest to use Formula 9.8, because the sample size before is the same as the sample size after. I will apply Formula 9.8 to the data in the table to show what happens:

$$t = \frac{(\overline{X}_1 - \overline{X}_2)}{\sqrt{(s_1^2 + s_2^2)/N}} = \frac{208 - 206.4}{\sqrt{(46.04^2 + 45.63^2)/5}} = \frac{1.6}{\sqrt{840.5}} = \frac{1.6}{29} = .055$$

We don't have to look up the critical *t* value to know that $t = .055$ is much too small to reject the null hypothesis, regardless of the alpha level. It is also easy to see why the calculated *t* value is so low. The average amount of weight loss is very small compared to the variation from subject to subject.

If you have the feeling that the *t*-test above does not do justice to the data, you are right. When Dr. Null says that the before-after differences are random, you might be inclined to point out that although the weight losses are small, they are remarkably consistent; all five subjects lost about the same amount of weight. This is really not very likely to happen by accident. To demonstrate the consistency in our data, we need a different kind of *t*-test; one that is based on the before-after difference scores and is sensitive to their similarities. That is what the matched *t*-test is all about.

The Direct-Difference Method

The procedure described in this section requires the addition of a column of difference scores to Table 13.1, as shown in Table 13.2. The simplest way to calculate the matched *t*-test is to deal only with the difference scores, using a

Table 13.2	Before	After	Difference
	280	278	2
	160	159	1
	220	218	2
	180	179	1
	200	198	2
			$\overline{D} = 1.6$
			$s_D = .5477$

procedure called the **direct-difference method.** To understand the logic of the direct-difference method, it is important to understand the null hypothesis that applies to this case. If the weight-loss program were totally ineffective, what could we expect of the difference scores? Because there would be no reason to expect the difference scores to be more positive than negative, over the long run the negative differences would be expected to balance out the positive differences. Thus the null hypothesis would predict the mean of the differences to be zero. In order to reject this null hypothesis, the t-test must show that the mean of the difference scores ($\overline{D}$) for our sample is so far from zero that the probability of beating $\overline{D}$ when the null hypothesis is true is too low (i.e., less than alpha) to worry about. To test $\overline{D}$ against zero requires nothing more than the one-group t-test you learned about in Chapter 8. To review the one-group t-test, consider again Formula 8.3:

$$t = \frac{\overline{X} - \mu_0}{s/\sqrt{N}}$$

In the present case, $\overline{X}$ (the sample mean) is the mean of the difference scores, and we can therefore relabel it as $\overline{D}$. For the matched t-test, the population mean predicted by the null hypothesis, μ_0, is the expected mean of the difference scores, so it can be called μ_D. Although it is possible to hypothesize some value other than zero for the mean of the difference scores, it is so rarely done that I will only consider situations in which $\mu_D = 0$. The unbiased standard deviation of the sample, s, is now the standard deviation of the difference scores, which can therefore be symbolized by adding a subscript as follows: s_D. Finally, Formula 8.3 can be rewritten in a form that is convenient to use for the matched t-test; this revised form will be referred to as Formula 13.1,

$$t = \frac{\overline{D}}{s_D/\sqrt{N}} \qquad \qquad \text{Formula 13.1}$$

where N is the number of difference scores. Let us apply this formula to our weight-loss problem using $\overline{D}$ and s_D from Table 13.2 and compare the result to that of our independent-groups t-test:

$$t = \frac{1.6}{.5477/\sqrt{5}} = \frac{1.6}{.5477/2.236} = \frac{1.6}{.245} = 6.53$$

Again, we don't need to look up the critical t value. Because by now we are somewhat familiar with Table A.2, we know that $t = 6.53$ is significant at the .05 level (two-tailed); the only time a calculated t value over 6 is not significant at the .05 level is in the extremely unusual case in which there is only 1 degree of freedom (i.e., two difference scores). Notice that the matched t-test value is over 100 times larger than the t value from the independent groups test; this is because I set up an extreme case to make a point. However, when the matched t-test is appropriate it almost always yields a value higher than would an independent-groups test; this will be easier to see when I present an alternative to the direct-differences formula (Formula 13.1). Notice also that the numerators for both t-tests are the same, 1.6. This is not a coincidence. The mean of the differences, $\overline{D}$, will always equal the difference of the means, $\overline{X}_1 - \overline{X}_2$, because it doesn't matter whether you take the difference first and then average, or average first and then take the difference. It is in the denominator that the two types of t-tests differ.

Because the matched t value in the above example is statistically significant, we can reject the null hypothesis that the before and after measurements are really the same in the entire population; it seems that the new weight-loss program may have some effectiveness. However, your intuition may tell you that in this situation it is difficult to be sure that it is the weight-loss program that is responsible for the before-after difference, and not some confounding factor. The drawbacks of the before-after design will be discussed later in this section.

The Matched *t*-Test As a Function of Linear Correlation

The degree to which the matched t value exceeds the independent-groups t value for the same data depends on how highly correlated the two samples (e.g., before and after) are. The reason the matched t value was more than 100 times larger than the independent groups t value for the weight-loss example is that the before and after values are very highly correlated; in fact, $r = .99997$. In order to see how the correlation coefficient affects the value of the matched t-test, let's look at a matched t-test formula that gives the same answer as the direct-difference method (Formula 13.1) but is calculated in terms of Pearson's r. That formula will be designated Formula 13.2:

$$t = \frac{(\overline{X}_1 - \overline{X}_2)}{\sqrt{(s_1{}^2 + s_2{}^2)/N - 2rs_1s_2/N}} \qquad \text{Formula 13.2}$$

Note the resemblance between Formula 13.2 and Formula 9.8; the difference is that a term involving Pearson's r is subtracted in the denominator of the matched t-test formula. However, when the two samples are *not* correlated (i.e., the groups are independent), $r = 0$, and the entire subtracted term in Formula 13.2 becomes zero. So when $r = 0$, Formula 13.2 is identical to Formula 9.8—as it should be, since the groups are not really matched in that case. As r becomes more positive, a larger amount is subtracted in the denominator of the *t*-test, making the denominator smaller. The smaller the denominator gets, the larger the *t* value gets, which means that if all other values remain equal, increasing the correlation of the samples (i.e., the match between them) will increase the *t* value. On the other hand, if *r* were negative between the two samples, variance would actually be added to the denominator (but fortunately when an attempt has been made to match the samples there is very little danger of obtaining a negative correlation).

Comparing Formulas 13.1 and 13.2, you can see how increasing the correlation influences the variability of the difference scores. Because the numerators of the two formulas, $\overline{D}$ and $\overline{X}_1 - \overline{X}_2$, will always be equal (as mentioned above), it follows that the denominators of the two formulas must also be equal (because both formulas always produce the same *t* value). If the two denominators are equal, then increasing the correlation (which decreases the denominator of Formula 13.2) must decrease the variability of the difference scores. This fits with the concept of correlation. If the same constant is added (or subtracted) from all of the scores in the first set in order to get the scores in the second set (e.g., everyone loses the same amount of weight), the correlation will be perfect, and the variability of the difference scores will be zero. To the extent that subjects tend to stay in the same relative position in the second set of scores as they did in the first, the correlation will be high, and the variability of the difference scores will be low.

Reduction in Degrees of Freedom

It may seem that any amount of positive correlation between two sets of scores will make the matched *t*-test superior to the independent-samples *t*-test. However, there is a relatively minor disadvantage to the matched *t*-test that must be mentioned: The number of degrees of freedom is only half as large as for the independent groups *t*-test. For the independent *t*-test for the weight-loss example, the df we would have used to look for the critical *t* was $N_1 + N_2 - 2 = 5 + 5 - 2 = 10 - 2 = 8$. For the matched *t*-test the df is equal to $N - 1$, where N is the number of difference scores (this is also the same as the number of pairs of scores). The critical *t* would have been df $= N - 1 = 5 - 1 = 4$, which is only half as large as the df for the independent test. Because critical *t* gets higher when the df are reduced, the critical *t* for the matched *t*-test is higher and therefore harder to beat. This disadvantage is usually more than offset by the increase in the calculated *t* due to the correlation of the two samples, unless the sample size is rather small or the correlation is fairly low. When the sample size is very large, the critical *t* will not increase noticeably (in fact, the normal distribution can be used) even if the df are cut in half, and even a rather small correlation between the two sets of scores can be helpful.

Drawback of the Before-After Design

Although the results of a before-after *t*-test are often statistically significant, you may feel uneasy about drawing conclusions from such a design. If you feel that something is missing, you are right. What is missing is a control group. Even though the weight loss in our example could not easily be attributed to chance factors, it is quite possible that the same effect could have occurred without the new weight-loss program. Using a bogus diet pill that is really a placebo might have had just as large an effect. A well-designed experiment would include an appropriate control group, and sometimes more than one type of control group. (For example, one control group might get a placebo pill and another might simply be measured twice without any manipulation in between measurements.) However, even in a well-designed experiment, a before-after matched t-test might be performed on each group. A significant t for the experimental group would not be conclusive in itself without comparison to a control group, but a *lack* of significance for the experimental group would indeed be disappointing. Moreover, a significant difference for the control group would alert you to the presence of extraneous factors affecting your dependent variable.

Other Repeated-Measures Designs

The before-after experiment is just one version of a **repeated-measures** (or *within-subjects*) **design.** It is possible, for instance, to measure each subject twice on the same variable under different conditions within the same experimental session (practically simultaneously). Suppose that a researcher wants to test the hypothesis that subjects recall "emotional" words (i.e., words that tend to evoke affect, like "funeral" or "birthday") more easily than they recall neutral words. Each subject is given a list of words to memorize, in which emotional and neutral words are mixed randomly. Based on a recall test, each subject receives two scores: one for emotional word recall and one for neutral word recall. Given that each subject has two scores, the researcher can find difference scores just as in the before-after example and can calculate the matched t-test.

The researcher can expect that the emotional recall scores will be positively correlated with the neutral recall scores. That is, subjects who recall more emotional words than other subjects are also more likely to recall more neutral words; some subjects just have better memories, in general, than other subjects. This correlation will tend to increase the matched t value by decreasing the denominator. In addition to a reasonably high correlation, however, it is also important that recall scores be generally higher (or lower, if that is the direction hypothesized) for the emotional than for the neutral words. This latter effect is reflected in the numerator of the matched t-test. As with any t-test, too small a difference in the numerator can prevent a t value from being large enough to attain statistical significance, even if the denominator has been greatly reduced by matching.

The simultaneous repeated-measures design described above does not require a control group. However, there is always the potential danger of confounding variables. Before the researcher can confidently conclude that it is the emotional content of the words that leads to better recall (assuming the matched t-test is significant and in the predicted direction), he or she must be sure that the emotional and neutral words do not differ from each other in other ways, such as familiarity, concreteness, imageability, etc. In general, simultaneous repeated measures are desirable as an efficient way to gather data. Unfortunately, there are many experimental situations in which it would not be possible to present different conditions simultaneously (or mixed together randomly in a single session). For instance, you might wish to compare performance on a tracking task in a cold room as compared to performance in a hot room. In this case the conditions would have to be presented *successively*. However, it would not be fair if the cold condition (or the hot condition) were always presented first. Therefore, a *counterbalanced* design would be used; this design is discussed further in Section B.

Matched-Pairs Design

Sometimes a researcher wants to compare two conditions, but it is not possible to test both conditions on the same subject. For instance, it may be of interest to compare two very different methods for teaching long division to children, in order to see which method results in better performance after a fixed amount of time. However, after a child has been taught by one method for 6 months, it would be counterproductive to then try to teach the child the same skill by a different method, especially as the child may have already learned the skill quite well. It would seem logical for the researcher to give up on the matched *t*-test and its advantages and to resign herself to the fact that the matched *t*-test is not appropriate for this type of situation. Although it is true that a repeated-measures design does not seem appropriate for this example, the researcher can still gain the advantage of a matched *t*-test by using a **matched-pairs design.** In fact, the term "matched *t*-test" is derived from the use of a *t*-test in analyzing this type of design.

The strategy of the matched-pairs design can be thought of in the following way. If it is not appropriate to use the same subject twice, the next best thing is to find two people who are as similar as possible. Then each member of the pair is randomly assigned to one of the two different conditions. This is done with all of the pairs. After all of the measurements are made, the differences can be found for each pair, and Formula 13.1 can be used to find the matched *t* value. The similarities within each pair make it likely that there will be a high correlation for the two sets of scores. For some experiments, the ideal pairs are sets of identical twins, because they have the same genetic makeup. This, of course, is not the easiest subject pool to find, and fortunately the use of twins is not critical for most experiments. In the long division experiment, the students can be matched into pairs based on their previous performance on arithmetic exams. It is not crucial that the two students in a pair be similar in every way, as long as they are similar on whatever characteristic is relevant to the variable being measured in the study. The researcher hopes that the two students in each pair will attain similar performance on long division, ensuring a high correlation between the two teaching methods. However, it is also very important that one method consistently produce better performance. For instance, if method A works better than method B, the A member of each pair should perform at least a little better than the B member, and these differences should be about the same for each pair to maximize the matched *t* value.

Correlated, or Dependent, Samples

The *t*-test based on Formula 13.1 (or Formula 13.2) can be called a repeated-measures *t*-test or a matched *t*-test, because it is the appropriate statistical procedure for either design. For purposes of calculation it doesn't matter whether the pairs of scores represent two measurements of the same subject or measurements of two similar subjects. In the above example, the sample of long division performance scores for students taught with method B will likely be correlated with the sample of scores for students taught with method A, because each score in one sample corresponds to a fairly similar score (the other member of the pair) in the other sample. Therefore, the matched *t*-test is often called a *t*-test for correlated samples (or sometimes just "related samples"). Another way of saying that two samples are correlated is to say that they are dependent. So the *t*-test in this chapter is frequently referred to as a *t*-test for two dependent samples. Because it is the means of the two dependent samples that are being compared, this *t*-test can also be called a *t*-test for the difference of two dependent means—in contrast to the *t*-test for the difference of two independent means (or samples), which was the subject of Chapter 9.

When Not to Use the Matched *t*-Test

When matching works—that is, when it produces a reasonably high positive correlation between the two samples—it can greatly raise the *t* value (compared to what would be obtained with independent samples) and thus eliminate the need to increase the number of subjects in the study. Matching (or using the same subjects twice, when appropriate) can be an economical way to attain a reasonable chance of statistical significance (i.e., to attain adequate power; see Section C) while using a fairly small sample. However, in some situations in which it would be impossible to use the same subject twice, it is also not possible to match the subjects. In such a situation, you must give up the increased power of the matched *t*-test and settle for the *t*-test for independent samples. An example of just such a situation is presented below.

A researcher wants to know whether subjects will more quickly come to the aid of a child or an adult when all they hear is a voice crying for help. To test this hypothesis, each subject is asked to sit in a waiting room while the "experimental" room is prepared. The experiment actually takes place in the waiting room, because the researcher plays a recorded voice of someone crying for help so that the subject can hear it. The dependent variable is the amount of time that elapses between the start of the recording and the moment the subject opens the door of the waiting room to leave. The independent variable is the type of voice on tape: adult or child. It should be obvious that having tested a particular subject with either the child or adult voice, the researcher could not run the same subject again using the other voice.

If you wished to obtain the power of a matched *t*-test with the above experiment, you would have to find some basis for matching subjects into pairs; then one member of the pair would hear the child's voice and the other would hear the adult's voice. But what characteristic could we use as a basis for matching? If we had access to personality questionnaires the subjects had previously filled out, we might match together subjects who were similar in traits of altruism, heroism, tendency to get involved with others, etc. However, it is not likely that we would have such information, nor could we be confident in using that information for matching unless there were a large body of literature to guide us. For the type of experiment just described, it would be very reasonable to give up on the matched design and simply perform a *t*-test for independent groups. In the next section, I present an example for which the matched design is appropriate and demonstrate the use of the direct-difference method for calculating the matched *t* value.

Exercises

1. For each of the following experimental designs, how many degrees of freedom are there, and how large is the appropriate critical *t* ($\alpha = .05$, two-tailed)?

a) Twenty-five subjects measured before and after some treatment

b) Two independent groups of 13 subjects each

c) Two groups of 30 subjects each, matched so that every subject in one group has a "twin" in the other group

d) Seventeen brother-sister pairs of subjects

* 2. Can the depression of psychotherapy patients be reduced by treating them in a therapy room painted in bright primary colors, as compared to a room with a more conservative look with wood paneling? Ten patients answered depression questionnaires after receiving therapy in a primary-colored room, and ten patients answered the same questionnaire after receiving therapy in a traditional room. Mean depression was lower for patients treated in the "primary room" ($\overline{X}_p = 35$) than for those treated in the traditional room ($\overline{X}_t = 39$); the standard deviations were $s_p = 7$ and $s_t = 5$, respectively.

a) Calculate the *t* value for the test of two independent means (Formula 9.8).

b) Is this *t* value significant at the .05 (two-tailed) level? (Make sure you base your critical *t* value on the appropriate degrees of freedom for this test.)

* 3. Suppose that the patients in the exercise above had been matched in pairs, based on general depression level, before being assigned to groups.

a) If the correlation were only .1, how high would the matched *t* value be?

b) Is this matched *t* value significant at the .05 (two-tailed) level? Explain any discrepancy between this result and the conclusion you drew in part b of exercise 2.

c) How high would the matched *t* value be if the correlation were .3?

d) If the correlation were .5?

4. Exercise 9B3 described an experiment in which 12 students arbitrarily labeled "gifted" obtained a grade average of 87.2 with $s = 5.3$, as compared to 12 other students not so labeled, who had an average of 82.9 with $s = 4.4$. Suppose now that each "gifted" student was matched with a particular student in the other group, and that the correlation between the two sets of scores was .4.

a) Calculate the matched *t* and compare it with your answer to the *t* value you found in exercise 9B3.

b) Calculate the matched *t* if the correlation were .8 and compare that with the matched *t* you found in part a above.

* 5. Calculate the mean and unbiased standard deviation for the following sets of difference scores.

a) $-6, +2, +3, 0, -1, -7, +3, -4, +2, +8$

b) $+5, -11, +1, +9, +6, -2, 0, -2, +7$

6. a) Design an experiment in which the repeated measures are simultaneous.

b) Design an experiment in which counterbalanced repeated measures would be appropriate.

7. a) Design an experiment for which it would be appropriate for the researcher to match the subjects into pairs.

b) Design an experiment in which the subjects have already been matched "naturally" (by circumstance).

* 8. Suppose that the matched *t* value for a before-after experiment turns out to be 15.2. Which of the following can you conclude?

a) The before and after scores must be highly correlated.

b) A large number of subjects must have been involved.

c) The before and after means must be quite different (as compared to the standard deviation of the difference scores).

d) The null hypothesis can be rejected at the .05 level.

e) No conclusion is possible without more information.

BASIC STATISTICAL PROCEDURES

In order to make the weight-loss example in Section A as simple as possible, I arranged the numbers so that all of the differences were in the same direction, i.e., everybody lost weight. In reality, the treatments we are usually interested in testing do not work on every individual, so some of the differences could go in the opposite direction. This makes calculation of the direct-difference formula a little trickier, as I will demonstrate in this section. For the example below I will use a matched-pairs design.

Suppose that the progressive Sunny Day elementary school wants to conduct its own experiment to compare two very different methods for teaching children to read. I will refer to these two methods as the "visual" and the "phonic." The simplest way to conduct such a study would be to select two random samples of children who are ready to start learning to read, teach each sample with a different method, measure the reading ability of each child after one year, and conduct a *t*-test for independent means to see if the groups differ significantly. However, as you learned in the previous section, we have a better chance of attaining statistical significance with a matched *t*-test. Of course, a repeated-measures design is out of the question if we are interested in the initial acquisition of reading skills. So we are left to consider the matched-pairs design. It seems reasonable to suppose that a battery of tests measuring various prereading skills could be administered to all of the students, and that we could find some composite measure for matching children that would tend to predict their ability to learn reading by either method. I will illustrate how the six steps of null hypothesis testing can be applied to this situation.

Step 1. State the Hypotheses

The research hypothesis that motivates this study is that the two methods differ in their effect on reading acquisition. The appropriate null hypothesis is that the population means representing the two methods are equal, H_0: $\mu_V = \mu_P$; or H_0: $\mu_V - \mu_P = 0$. In terms of difference scores, H_0: $\mu_D = 0$. The appropriate alternative hypothesis in this case is two-tailed. Although the researcher may expect a particular method to be superior, there would be no justification for ignoring results in the unexpected direction. The alternative hypothesis is expressed as H_A: $\mu_V \neq \mu_P$; or H_A: $\mu_V - \mu_P \neq 0$. In terms of difference scores, H_A: $\mu_D \neq 0$. A one-tailed version would be either H_A: $\mu_D > 0$ or H_A: $\mu_D < 0$.

Step 2. Select the Statistical Test and the Significance Level

Because each subject in one group is matched with a subject in the other group, the appropriate test is the *t*-test for correlated or dependent samples (i.e., the matched *t*-test). There is no justification for using a significance level that is larger or smaller than .05, so I will use the conventional .05 level.

Step 3. Select the Samples and Collect the Data

First, one large sample is selected at random (or as randomly as possible within the usual practical constraints). Then subjects are matched into pairs according to their similarity on relevant variables (for this example, the composite score on the battery of prereading skill tests). Finally, each child in a pair is randomly assigned to either the visual or the phonic group. (For instance, the flip of a fair coin could determine which member of the pair is assigned to the visual method and which to the phonic method.) Because effective matching can help lead to a high *t* value, researchers tend to use smaller samples for a matched design than for an independent-groups design. In fact, a major reason for matching is that usually fewer subjects are needed to attain statistical significance (i.e., matching increases power without having to increase sample size; see Section C). To reduce the amount of calculation, this example will be based on a total of 20 subjects, which means only 10 pairs, so $N = 10$. The reading levels after one year in the experimental program are recorded for each subject in Table 13.3. A reading level of 2.0 is considered average for a second grader and would be the level expected for these subjects had they not been in this experiment.

Table 13.3	Pair	Visual	Phonic	D	D^2
	1	2.3	2.5	−.2	.04
	2	2.0	1.9	+.1	.01
	3	2.1	2.6	−.5	.25
	4	2.4	2.2	+.2	.04
	5	1.9	2.1	−.2	.04
	6	2.2	2.5	−.3	.09
	7	1.8	2.2	−.4	.16
	8	2.4	2.7	−.3	.09
	9	1.6	1.9	−.3	.09
	10	1.7	1.6	+.1	.01
				$\Sigma D = -1.8$	$\Sigma D^2 = .82$

Step 4. Find the Region of Rejection

If the number of pairs were quite large, especially if N were greater than 100, we could safely use the normal distribution to test our null hypothesis. However, because N is small and we do not know the standard deviation for either population (or for the population of difference scores), we must use the appropriate t distribution. The number of degrees of freedom is $N - 1$ (where N is the number of *pairs*), so for this example, df $= N - 1 = 10 - 1 = 9$. Looking at Table A.2 in the .05, two-tailed column, we find that the critical t for df $= 9$ is 2.262. If our calculated matched t value is greater than $+2.262$ or less than -2.262, we can reject the null hypothesis (see Figure 13.1).

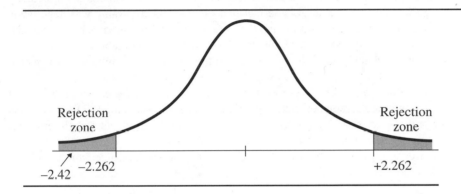

Figure 13.1

Step 5. Calculate the Test Statistic

I will use Formula 13.1 to calculate the matched t value. The first step is to calculate the mean of the difference scores, $\overline{D}$, which equals $\Sigma D/N$. It is important when finding the sum of the difference scores to keep track of which differences are positive and which negative. In our example, $\Sigma D = -1.8$, so $\overline{D} = -1.8/10 = -.18$. We can check this result by comparing it to $\overline{X}_V - \overline{X}_P$, to which it should be equal ($\overline{X}_V - \overline{X}_P = 2.04 - 2.22 = -.18$).

The next step is to find the unbiased standard deviation of the difference scores. Table 13.3 includes a column for the squared difference scores, so we can use Formula 4.15 (expressed in terms of difference scores) as follows:

$$s_D = \sqrt{\frac{1}{N-1}\left[\Sigma D^2 - \frac{(\Sigma D)^2}{N}\right]} = \sqrt{\frac{1}{9}\left[.82 - \frac{(-1.8)^2}{10}\right]}$$

$$= \sqrt{\frac{1}{9}(.496)} = \sqrt{.0551} = .235$$

Now we have the values needed for Formula 13.1:

$$t = \frac{\overline{D}}{s_D/\sqrt{N}} = \frac{-.18}{.235/\sqrt{10}} = \frac{-.18}{.0743} = -2.42$$

Step 6. Make the Statistical Decision

Because the calculated t (-2.42) is less (i.e., more negative) than the critical t of -2.262, we can reject the null hypothesis (see Figure 13.1) and conclude that μ_V does not equal μ_P—that is, the two teaching methods do not have the same effect on reading acquisition. We can say that the phonic method is better than the visual method, and that the difference between the two methods is statistically significant at the .05 level. The fact that our calculated t value was negative is just a consequence of subtracting the phonic score from the visual score instead of the other way around. It would have been perfectly acceptable to have reversed the order of subtraction just so we could have reduced the number of negative differences.

In order to see just how well matched our pairs of subjects were, we could calculate the Pearson correlation coefficient for the visual and phonic scores. As a preliminary step, I have computed all of the cross-products and the relevant summary statistics, as shown in Table 13.4.

Because we have already calculated the *unbiased* standard deviations, we use Formula 11.3 to calculate Pearson's r:

$$r = \frac{1/(N-1)(\sum XY - N\overline{X}\,\overline{Y})}{s_X s_Y} = \frac{(1/9)[45.98 - 10(2.04)(2.22)]}{(.2875)(.3553)} = \frac{.0769}{.1021} = .753$$

Table 13.4

Visual (X)	Phonic (Y)	XY
2.3	2.5	5.75
2.0	1.9	3.8
2.1	2.6	5.46
2.4	2.2	5.28
1.9	2.1	3.99
2.2	2.5	5.5
1.8	2.2	3.96
2.4	2.7	6.48
1.6	1.9	3.04
1.7	1.6	2.72
20.4	22.2	45.98
$s_X = .2875$	$s_Y = .3553$	

Having calculated $r = .753$, we can see that the pairs of subjects were indeed well matched in their second-grade reading levels, suggesting that our matching criterion (the composite score of prereading skills) was appropriate. Now that we know Pearson's r, it is easy to demonstrate that Formula 13.2 gives exactly the same t value (within error due to rounding) as the direct-difference method (Formula 13.1):

$$t = \frac{(\overline{X}_1 - \overline{X}_2)}{\sqrt{(s_1{}^2 + s_2{}^2)/N - 2rs_1s_2/N}}$$

$$= \frac{2.04 - 2.22}{\sqrt{[(.2875)^2 + (.3553)^2]/10 - 2(.753)(.2875)(.3553)/10}}$$

$$= \frac{-.18}{\sqrt{.0209 - .0154}} = \frac{-.18}{\sqrt{.0055}} = \frac{-.18}{.0743} = -2.42$$

Notice that both the numerator and the denominator are the same in the above calculations as they were when we used Formula 13.1—but the denominator is found in a very different way. With Formula 13.2 it is easy to see that a large portion of the variance in the denominator is being "subtracted out," and that if Pearson's r were even higher, even more would have been subtracted out. If Pearson's r had turned out to be zero, or if an independent-groups test had been conducted instead, the denominator would have been larger, resulting in a smaller t value, as shown below:

$$t = \frac{(\overline{X}_1 - \overline{X}_2)}{\sqrt{(s_1{}^2 + s_2{}^2)/N}} = \frac{2.04 - 2.22}{\sqrt{[(.2875)^2 + (.3553)^2]/10}} = \frac{-.18}{\sqrt{.0209}} = \frac{-.18}{.1445} = -1.245$$

Without the matching, the t value would have been only half as large, and the null hypothesis could not have been rejected.

Raw-Score Formula for the Matched *t*-Test

Chapter 11 presented a formula for Pearson's r (Formula 11.5) that was based directly on sums and sums of squared values and did not require the intermediate steps of calculating means or standard deviations. There is also a raw-score formula for the matched t-test. Although this raw-score formula (Formula 13.3) reduces the amount of calculation by leading directly from the raw scores to the matched t value, the reduction is slight and may not be worth the decreased chance of noticing a calculation error during the process. Formulas 13.1 and 13.2 have an advantage in that if any of the intermediate statistics they are based on (such as the mean of the difference scores) is drastically wrong, you will probably notice immediately. On the other hand, these formulas involve an increased risk of rounding-off error (compared to the raw-score formula) if not enough digits are retained in the intermediate steps. The worst thing about Formula 13.3 is that it looks confusing; its separate pieces do not correspond to commonly used statistics. However, in order to be comprehensive, and because you may encounter this formula elsewhere, I present the raw-score formula (Formula 13.3) below:

$$t = \frac{\Sigma D}{\sqrt{[1/(N-1)]/[N\Sigma D^2 - (\Sigma D)^2]}}$$
Formula 13.3

Of course, Formula 13.3 always produces the same *t* value as Formula 13.1 or Formula 13.2 (within rounding error), but I will postpone a demonstration of this formula until Section D.

Assumptions of the Matched *t*-Test

Because the matched *t*-test can be viewed as a one-sample hypothesis test of the difference scores, the assumptions behind it can be framed in terms of the population of difference scores. There are two major assumptions:

Normality

The population of difference scores follows a normal distribution. As with other *t*-tests, the normality assumption is not critical for relatively large samples. However, if the sample size is small (less than about 30), and the distribution seems to be very far from normal in shape, nonparametric statistics may be more valid (see Part VI).

Independent Random Sampling

Although there is a relationship between the two members of a pair of scores, the pairs should be independent of each other and, ideally, should be selected at random from all possible pairs.

Experimental Designs That Call for the Matched *t*-Test

The matched *t*-test procedure described in this chapter can be used with a variety of repeated-measures and matched-pairs designs. I will consider the most common of these designs.

Repeated Measures

The two basic types of repeated-measures design are *simultaneous* and *successive*. The experiment described in Section A, involving the recall of a mixed list of emotional and neutral words, is an example of the simultaneous design. It is not important that the subject literally see an emotional word and a neutral word simultaneously. The fact that the two types of words are randomly mixed throughout the session ensures that neither type has the temporal advantage of coming when the subject has more energy or more practice.

The successive repeated-measures experiment has two major subtypes: the *before-after design* and the *counterbalanced design.* The before-after design has already been discussed in connection with the weight-loss experiment. Recall that whereas this design can establish a significant change in scores from before to after, a control group is often required to rule out such alternative explanations as practice or a placebo effect.

As an example of a **counterbalanced design,** imagine that a researcher proposes that subjects can solve more arithmetic problems while listening to happy music than while listening to sad music. A simultaneous design is not feasible. But if subjects always hear the same type of music first, their scores for problem solving while listening to the second type may reflect the advantage of a practice effect (or possibly the disadvantage of a fatigue effect). To eliminate **order effects,** half the subjects would hear the happy music first, and the other half would hear the sad music first; the order effects should then cancel out when the data from all subjects are considered. This procedure is known as counterbalancing and will be dealt with again in Chapter 17 when more than two repeated measures are discussed.

Counterbalancing does not always work. If we use strong stimuli (e.g., selected movie segments) to make our subjects very sad or very happy, the emotion induced first may linger on and affect the induction of the second emotion. A sufficient period of time between conditions may avoid the effects of "carry-over." On the other hand, the two experimental conditions being compared may be such that one involves special instructions or hints (e.g., use visual imagery to remember items) that should not be used in the other condition. If subjects are not able to forget or ignore those instructions in the second condition, a repeated-measures design is not recommended. A matched-pairs design may be appropriate.

Matched Pairs

The matched-pairs design also has two main subtypes, *experimental* and *natural.* In the first type, the experimenter creates the pairs, based either on a relevant pretest or on other available data (e.g., gender, age, IQ, etc.). The experiment described in Section B, comparing the two methods for teaching reading, is an example of an experimental matched-pairs design. Another way to create pairs is to have two different subjects act as "judges" and rate the same stimulus or set of stimuli. For instance, the judges may be paired at random, with each member of the pair rating the same yearbook photo, but one member is told that the student in the photo is intelligent, whereas the other is told the opposite. The members of each pair are now matched in that they are exposed to the same stimulus (i.e., photo), and difference scores can be taken to see if the judge who is told the person in the photo is intelligent gives consistently higher (or lower) attractiveness ratings than the other member of the pair.

In contrast, in a natural matched-pairs design, the pairs occur "naturally" and are used by the experimenter. For example, husbands and wives can be compared with respect to their economic aspirations; daughters can be compared to their mothers to see if the former have higher educational goals. Because the pairs are not determined by the experiment, researchers must exercise extra caution in drawing conclusions from the results.

Basically, if a repeated-measures design is appropriate (e.g., there are no carryover effects), it is usually the best choice, because it yields more power (i.e., a better chance of statistical significance if the null hypothesis is not true) with fewer subjects. When a repeated-measures design is not appropriate, a matched-pairs design offers much of the same advantage, as long as there is some reasonable basis for matching (e.g., a pretest or relevant demographic or background data on each subject). If an independent *t*-test must be performed, it is helpful to use the principles described in Chapter 10 to estimate the sample size required for adequate power.

Publishing the Results of a Matched *t*-Test

Reporting the results of the matched *t*-test is virtually the same as reporting on an independent groups *t*-test. To report the results of the hypothetical reading acquisition experiment, we might use the following sentence: "A *t*-test for correlated samples revealed that the phonic method produced significantly better reading performance ($\underline{M}$ = 2.22) than the visual method ($\underline{M}$ = 2.04) when the pupils were tested at the end of 6 months, $\underline{t}(9)$ = −2.42, $\underline{p}$ < .05 (two-tailed)." Of course, the df reported in parentheses after the *t* value is the df that is appropriate for the matched *t*-test.

I will illustrate the use of the matched *t*-test in the psychological literature with an excerpt from an article by Kaye and Bower (1994). They report the results of a study of 12 newborns (less than two days old) that demonstrate their ability to match the shape of a pacifier in their mouths with a shape that they see on a screen. One of the main findings was that "the mean first-look duration at the image of the pacifier-in-mouth was 10.24 s (SD = 8.20). The mean first-look duration at the image of the other pacifier was 4.66 s (SD = 4.02). The difference was significant (t = 3.25, df = 11, p < .01). . . . There was no order effect" (p. 287). Note that the SD given in parentheses are the standard deviations for each set of scores separately; the standard deviation of the difference scores is not given but can easily be found by multiplying the difference in means by the square root of df and then dividing by the given *t* value. (This is the same as solving Formula 13.1 for s_D).

Exercises

1. The stress levels of 30 unemployed subjects were measured by a questionnaire before and after a real job interview. The stress level rose from a mean of 63 points to a mean of 71 points. The (unbiased) standard deviation of the difference scores was 18.

    a) What is the appropriate null hypothesis for this example?

    b) What is the critical value of *t* for a .05, two-tailed test?

c) What is the observed (i.e., calculated) value of t?

d) What is your statistical decision with respect to the null hypothesis?

e) Given your conclusion in part d, could you be making a Type I or Type II error?

* 2. In exercise 9B9, subjects in an individual motivation condition were compared to others in a group motivation condition in terms of task performance. Now assume that the subjects were matched in pairs based on some pretest. The data from exercise 9B9 are reproduced below, showing the pairing of the subjects.

Individual Motivation	Group Motivation
11	10
17	15
14	14
10	8
11	9
15	14
10	6
8	7
12	11
15	13

a) Perform a matched t-test on these data ($\alpha = .05$, two-tailed).

b) Compare the matched t with the independent t that you found for exercise 9B9.

3. a) Using the data from exercise 11B6, reproduced below, determine whether there is a significant tendency for verbal GRE scores to improve on the second testing. Calculate the matched t in terms of the Pearson correlation coefficient already calculated for exercise 11B6.

b) Recalculate the matched t-test according to the direct-difference method and compare the result to your answer for part a.

c) What would be the drawback to using the raw-score formula (Formula 13.3) to find the matched t for this problem?

Verbal GRE (1)	Verbal GRE (2)
540	570
510	520
580	600
550	530
520	520

* 4. An educator has invented a new way to teach geometry to high school students. To test this new teaching method, sixteen tenth-graders are matched into eight pairs based on their grades in previous math courses. Then each student in a pair is randomly assigned to either the new method or the traditional method. At the end of a full semester of geometry training, all students take the same standard high school geometry test. The scores for each student in this hypothetical experiment are given in the table below.

Traditional Method	New Method
65	67
73	79
70	83
85	80
93	99
88	95
72	80
69	100

a) Perform a matched t-test for this experiment ($\alpha = .01$, two-tailed). Is there a significant difference between the two teaching methods?

b) Use Formula 8.5 (substituting $\overline{D}$ for $\overline{X}$ and $s_{\overline{D}}$ for $s_{\overline{X}}$) to find the 99% confidence interval for the population difference of the two teaching methods.

* 5. In exercise 11B8, a male and a female judge rated the same cartoon segments for violent content. Using the data from that exercise, reproduced below, perform a matched t-test to determine whether there is a significant tendency for one of the judges to give

higher ratings (use whichever formula and significance level you prefer).

Segment	Male Rater	Female Rater
1	2	4
2	1	3
3	8	7
4	0	1
5	2	5
6	7	9

* 6. Do teenage boys tend to date teenage girls who have a lower IQ than they do? To try to answer this question, ten teenage couples (i.e., people who are dating regularly) are randomly selected, and each member of each couple is given an IQ test. The results are given in the table below. Perform a one-tailed matched *t*-test ($\alpha = .05$) to determine whether the boys have higher IQs than their girlfriends. What can you conclude?

Boy	Girl
110	105
100	108
120	110
90	95
108	105
115	125
122	118
110	116
127	118
118	126

7. A neuropsychologist believes that right-handed people will recognize objects placed in their right hands more quickly than objects placed in their left hands when they are blindfolded. The scores below represent how many objects each subject could identify in two minutes with each hand.

a) Use Formula 13.1 to test the null hypothesis of no difference between the two hands ($\alpha = .05$, two-tailed).

b) Recalculate the matched *t* using Formula 13.3

and compare the result to the *t* value you obtained in part a.

Subject	Object in Left Hand	Object in Right Hand
1	8	10
2	5	9
3	11	14
4	9	7
5	7	10
6	8	5
7	10	15
8	7	7
9	12	11
10	6	12
11	11	11
12	9	10

* 8. A cognitive psychologist is testing the theory that short-term memory is mediated by subvocal rehearsal. This theory can be tested by reading aloud a string of letters to a subject, who must repeat the string correctly after a brief delay. If the theory is correct, there will be more errors when the list contains letters that sound alike (e.g., G and T) than when the list contains letters that look alike (e.g., P and R). Each subject gets both types of letter strings, which are randomly mixed in the same experimental session. The numbers of errors for each type of letter string for each subject are shown in the table below. Perform a matched *t*-test ($\alpha = .05$, one-tailed) on the data and state your conclusions.

Subject	Letters That Sound Alike	Letters That Look Alike
1	8	4
2	5	5
3	6	3
4	10	11
5	3	2
6	4	6
7	7	4
8	11	6
9	9	7

In Section A I demonstrated that the matched *t*-test will produce a much higher *t* value than the independent-groups test when the two sets of scores are well matched. All other things being equal, the better the matching (i.e., the higher the correlation), the higher the *t* value. You may recall from Chapter 10 that a higher expected *t* value (i.e., δ) means higher power. Thus matching can increase power and free a researcher from having to increase the sample size. To illustrate how this works I will start with Formula 13.2 and replace each sample statistic with its expected value in the population. (Note that if we assume homogeneity of variance, σ^2 is the expected value for both s_1^2 and s_2^2.)

$$\delta = \frac{\mu_1 - \mu_2}{\sqrt{[(\sigma^2 + \sigma^2)/N] - (2\rho\sigma^2)/N}} = \frac{\mu_1 - \mu_2}{\sqrt{[2\sigma^2 - 2\rho\sigma^2]/N}} = \frac{\mu_1 - \mu_2}{\sqrt{[2(1 - \rho)\sigma^2]/N}}$$

The denominator can then be separated and part of it can be flipped over and put in the numerator, as follows:

$$\delta = \frac{\mu_1 - \mu_2}{\sigma\sqrt{2(1 - \rho)/N}} = \sqrt{\frac{N}{2(1 - \rho)}} \frac{\mu_1 - \mu_2}{\sigma}$$

Finally, we can separate the term involving *N* from the term involving ρ to produce Formula 13.4, which finds δ for a matched *t*-test:

$$\delta_{\text{matched}} = \sqrt{\frac{1}{1 - \rho}} \sqrt{\frac{N}{2}} \frac{\mu_1 - \mu_2}{\sigma} \qquad \text{Formula 13.4}$$

We can look up the value of δ in Table A.3 to find power, as discussed in Chapter 10.

Compare Formula 13.4 above to Formula 10.1, below:

$$\delta_{\text{ind}} = \sqrt{\frac{N}{2}} \frac{(\mu_1 - \mu_2)}{\sigma}$$

The expected matched *t* (δ_{matched}) is the same as the expected independent *t* (δ_{ind}), except for the term $\sqrt{1/(1 - \rho)}$, which is the same as $1/\sqrt{1 - \rho}$. This relationship can be expressed in the following manner:

$$\delta_{\text{matched}} = \left(\frac{1}{\sqrt{1 - \rho}} \right)(\delta_{\text{ind}})$$

For instance, a correlation of .5 would cause δ_{ind} to be multiplied by

$$\frac{1}{\sqrt{1 - \rho}} = \frac{1}{\sqrt{1 - .5}} = \frac{1}{\sqrt{.5}} = \frac{1}{.707} = 1.41$$

A correlation of .9 would result in a multiplication factor of

$$\frac{1}{\sqrt{1-.9}} = \frac{1}{\sqrt{.1}} = \frac{1}{.316} = 3.16$$

Suppose that, according to a particular alternative hypothesis, δ for an independent t-test is only 2.0. This corresponds to a power of 52% (from Table A.3 with $\alpha = .05$, two-tailed). If the experiment could be changed to a matched design for which a correlation (ρ) of .5 could be expected, the new δ would be 1.41 times the old delta (see calculation above for $\rho = .5$), or 2.82. This new delta corresponds to a much more reasonable level of power—about .8. To get the same increase in power *without* matching, the number of subjects in each group would have to be *doubled* in this particular case. The relationship between changes in ρ and changes in power is not a simple one in mathematical terms, but generally speaking, increasing ρ will increase power. Of course, it is not always possible to increase ρ, and there are limits to the amount of increase in power that is possible, due to the variability of individuals (and within individuals over time). Nevertheless, it can be worth the trouble to match subjects more closely (or to keep conditions as constant as possible for each subject being measured twice).

The trickiest part of a power analysis for a matched-pairs or repeated-measures design is that in addition to estimating gamma, as in an independent-groups test, you must estimate ρ as well. As usual, you are often forced to rely on data from relevant studies already published. There is also a potential danger involved in increasing the power by increasing ρ; it is the same danger inherent in using large sample sizes, as discussed in Chapter 10: Too much power allows effect sizes that are otherwise trivial to become statistically significant. From a practical point of view, we would be better off ignoring some of these tiny effects. Notice that, like increasing the sample size, increasing ρ yields greater power without changing the separation of the two sample means.

For instance, a weight-loss program that causes every subject to lose either 3 or 4 *ounces* of weight over a 6-month period will lead to a highly significant matched t value. Although the numerator of the t ratio will be small (the difference in means before and after will be about 3.5 ounces), the denominator will be even smaller (the standard deviation of the difference scores will be about half an ounce, which is then divided by the square root of N). The high t value, and corresponding low p value in this case are not a result of the effectiveness of the weight-loss program (in that people lose a lot of weight) but rather of the *consistency* of the weight-loss program (in that everyone loses about the same amount of weight).

In general, the higher the t value, the more confident we can be in rejecting the null hypothesis. In the case of the weight-loss program, a higher t value gives us more confidence that the weight-loss program is not totally ineffective (i.e., the before and after means are not identical). However, being sure that the effect is not zero does not imply that the effect is large. In the example above, the consistent small weight loss gives us much confidence that the effect is *not* zero but does not convince us that the effect is large—or even large enough to be of any practical significance. It is important to remember that matching, like increasing the sample size, helps you detect small differences in population means—and this can be either a blessing or a curse, depending on the situation.

Exercises

* 1. Imagine that a researcher plans an experiment in which there are two groups, each containing 25 subjects. The effect size (gamma) is estimated to be about .4.

a) If the groups are to be matched and the correlation is expected to be .5, what is the power of the matched *t*-test being planned, with alpha = .05, two-tailed?

b) If the correlation in the above example were .7 and all else remained the same, what would the power be?

c) Recalculate the power for part b above for alpha = .01 (two-tailed).

2. If a before-after *t*-test is planned with 35 subjects who are undergoing a new experimental treatment, and the after scores are expected to be one half of a standard deviation higher than the before scores, how high would the correlation need to be for the test to have power = .8, with alpha = .05, two-tailed?

* 3. A matched *t*-test is being planned to evaluate a new method for learning foreign languages. From previous research, an effect size of .3 and a correlation of .6 are expected.

a) How many subjects would be needed in each

matched group, in order to have power = .75, with a two-tailed test at alpha = .05?

b) What would your answer to part a be if alpha were changed to .01?

4. The correlation expected for a particular matched *t*-test is .5.

a) If it is considered pointless to have more than .7 power to detect a difference in population means as small as one tenth of a standard deviation, with an alpha = .05, two-tailed test, what is the largest sample size that should ever be used?

b) What would your answer to part a be if the expected correlation were only .3?

* 5. Which of the following can decrease the Type II error rate associated with a matched *t*-test?

a) Increasing the sample size

b) Increasing the correlation between the two sets of scores (i.e., improving the matching)

c) Using a larger alpha level (e.g., .1 instead of .05)

d) All of the above

e) None of the above

SUMMARY

The Important Points of Section A

1. In an experiment in which each subject is measured twice (before and after some treatment) the variability of the *difference scores* will probably be less than the variability of either the before scores or the after scores.

2. Because of this reduced variability, a one-group *t*-test comparing the mean of the difference scores to zero will yield a higher *t* value than an independent *t*-test comparing the before mean to the after mean.

3. The *t*-test on the difference scores is often called a *matched t-test* (or, if appropriate, a *repeated-measures t-test*), and it makes use of the consistency of the difference scores to show that the before-after differences are not likely to be merely chance fluctuations from a mean of zero.

4. The matched *t*-test can be expressed as a formula that contains a term that is subtracted in the denominator. The size of the subtracted term depends on the Pearson's *r* between the two sets of scores: The higher the correlation, the more that is subtracted (so the denominator gets smaller) and the larger the *t* value (all else staying equal).

5. A disadvantage of the matched *t*-test is that the degrees of freedom are reduced by half, resulting in a higher critical *t* value to beat. However, if the matching is reasonably good, the increase in the calculated *t* will outweigh the increase in the critical *t*.

6. The before-after design is often inconclusive if there is no control group for comparison. However, there are other *repeated-measures designs* that involve simultaneous measurement or counterbalancing in which a control group may not be needed (see Section B).

7. The difference scores in a matched *t*-test need not come from two measures on the same subject. A *matched-pairs design,* in which similar subjects undergo different treatments, can be analyzed in exactly the same way.

8. Another way of referring to the matched *t*-test is as a *t*-test for correlated (or dependent) samples (or means).

The Important Points of Section B

Two procedures for calculating the matched *t*-test were described in Section B. In this section, only the direct-difference method will be reviewed, as it is easier and far more commonly used than the method based on Pearson's *r*. Although the direct-difference method is relatively simple, calculations must be done carefully because of the nearly inevitable mixture of positive and negative numbers.

For the following example, imagine that each subject views the same videotape of a husband and wife arguing. Afterward, the subject is asked to rate (on a scale from 0 to 10) the likability of both the husband and the wife. This is an example of a simultaneous repeated-measures design. Each subject's ratings are sorted according to whether the subject is rating someone of the same or the opposite gender. In this example, six individuals (three of each gender) were chosen at random, and each provided two different ratings. The two ratings of

Table 13.5

Subject	Rating of Same Gender	Rating of Opposite Gender	D	D²
1	9	5	+4	16
2	5	5	0	0
3	8	3	+5	25
4	4	5	−1	1
5	6	3	+3	9
6	7	9	−2	4
			$\Sigma D = +9$	$\Sigma D^2 = 55$

each subject appear in Table 13.5. (I've used only six subjects in this example in order to minimize the amount of calculation.)

The first step is to create a column of difference scores, as shown in the table. The scores can be subtracted in either order (as long as you are consistent for every pair, of course), so you may want to choose the order that minimizes the minus signs. Adding a column for the squares of the difference scores allows the use of simple computational formulas that are based on ΣD^2. In finding the sum of the difference scores, you may want to add the positive and negative numbers separately and then add these two sums at the end. In this case the positive numbers are +4, +5, and +3, whose sum is +12. The negative numbers are −1 and −2, whose sum is −3. Finally, adding +12 and −3, we find that $\Sigma D = +9$. Dividing this sum by N, which is 6 for this example, we find that $\overline{D} = 1.5$. To obtain the (unbiased) standard deviation of the difference scores, s_D, we use Formula 4.15 (as modified for difference scores):

$$s_D = \sqrt{\frac{1}{N-1}\left[\Sigma D^2 - \frac{(\Sigma D)^2}{N}\right]} = \sqrt{\frac{1}{5}\left[55 - \frac{9^2}{6}\right]}$$
$$= \sqrt{\frac{1}{5}(41.5)} = \sqrt{8.3} = 2.88$$

The matched t value can now be found using Formula 13.1:

$$t = \frac{\overline{D}}{s_D/\sqrt{N}} = \frac{1.5}{2.88/\sqrt{6}} = \frac{1.5}{1.18} = 1.28$$

To demonstrate that Formula 13.3 yields the same t value (within rounding error) as Formula 13.1 (and, of course, Formula 13.2), I will use the sums from the table and calculate the t value again:

$$t = \frac{\sum D}{\sqrt{[1/(N-1)][N\sum D^2 - (\sum D)^2]}} = \frac{9}{\sqrt{(1/5)(330-81)}}$$

$$= \frac{9}{\sqrt{(1/5)(249)}} = \frac{9}{\sqrt{49.8}} = \frac{9}{7.06} = 1.27$$

Because the N is so small we must use the t distribution to represent our null hypothesis. The df $= N - 1$, which equals $6 - 1 = 5$, the critical t ($\alpha = .05$, two-tailed) is 2.571 (see Table A.2 in Appendix A). Because the calculated t (1.28) is well below this value, the null hypothesis—that the mean of the difference scores is zero—cannot be rejected.

The assumptions underlying the matched t-test are that the difference scores are normally distributed and random and independent of each other.

The matched t-test can be used with a number of experimental designs. The two basic types of repeated-measures design are *simultaneous* and *successive*. The successive repeated-measures experiment has two major subtypes: the *before-after design* and the *counterbalanced design*. The before-after design usually requires a control group to allow you to draw valid conclusions. Counterbalancing does not always get rid of effects that are due to the order of conditions; in such cases, repeated measures should not be used.

The two main subtypes of matched-pairs design are *experimental* and *natural*. The experimenter may create pairs based either on a relevant pretest or on other available data (e.g., gender, age, IQ, etc.). Another way to create pairs is to have two subjects rate or judge the same stimulus. Rather than creating pairs, the experimenter may use naturally occurring pairs.

The Important Points of Section C

1. Compared to an independent samples t-test, and assuming the means and standard deviations remain the same, the matched t-test is associated with a higher expected t (delta), as long as a positive correlation can be expected between the two sets of scores.

2. Increasing the matching (i.e., the correlation) between the two sets of scores increases delta (all else remaining equal) and therefore the power without the need to increase the number of subjects.

3. In addition to estimating gamma, as in the independent-samples t-test, the power analysis for a matched design requires estimating the population correlation (ρ) for the two sets of scores. Previous experimental results can provide a guideline.

4. Keep in mind that a well-matched design, like a study with very large sample sizes, can produce a high t value (and a correspondingly low p value), even though the separation of the means (relative to the standard deviation) is not very large, and in fact may be too small to be of any practical interest.

Definitions of Key Terms

Matched *t*-test Also called *t*-test for the difference of correlated (or dependent) means (or samples); a *t*-test that takes into account the degree of linear correlation between two sets of scores.

Direct-difference method A method for performing the matched *t*-test, whereby difference scores are computed for each pair of scores and then a one-group *t*-test is performed on the difference scores, usually against the null hypothesis that the mean of the difference scores in the population (μ_D) is zero.

Repeated-measures design Also called a within-subjects design; an experimental design in which each subject is measured twice on the same variable, either before and after some treatment (or period of time) or under different conditions (simultaneously or successively presented).

Matched-pairs design A design in which subjects are paired off based on their similarity on some relevant variable and then randomly assigned to two conditions. Naturally occurring pairs can be used, but this will limit the conclusions that can be drawn.

Counterbalanced design When a successive repeated-measures design involves two conditions, counterbalancing requires that half the subjects receive the conditions in one order, while the other half receive the conditions in the reverse order. This approach will average or balance out simple order effects.

Order effects Effects caused by the order in which subjects are exposed to conditions. When each subject is measured twice under successive conditions, order effects can increase or decrease the measurements, depending on whether a condition is administered first or second. Differences in scores caused by practice and fatigue are examples of order effects.

Key Formulas

The matched *t*-test (difference-score method):

$$t = \frac{\overline{D}}{s_D/\sqrt{N}}$$

Formula 13.1

The matched *t*-test, in terms of Pearson's correlation coefficient (definitional form):

$$t = \frac{(\overline{X}_1 - \overline{X}_2)}{\sqrt{(s_1{}^2 + s_2{}^2)/N - 2r\, s_1 s_2/N}} \qquad \text{Formula 13.2}$$

The matched *t*-test (raw-score form):

$$t = \frac{\Sigma D}{\sqrt{[1/(N-1)][N\Sigma D^2 - (\Sigma D)^2]}} \qquad \text{Formula 13.3}$$

Delta for the matched *t*-test (power can then be found by consulting Table A.3):

$$\delta = \sqrt{\frac{1}{1-\rho}} \sqrt{\frac{N}{2}} \frac{\mu_1 - \mu_2}{\sigma} \qquad \text{Formula 13.4}$$

One- and Two-Way Analysis of Variance

The main topic of Part V is a statistical method called the analysis of variance (ANOVA). This name can be a bit misleading because the object of this method is not to study the variances of populations, but rather to draw inferences about the means of populations. In this sense the methods of Part V are just an extension of those you learned in Part III—but applied to experiments in which there are more than two samples, and, in some cases, more than one independent variable. As in Part III, there will be only one dependent variable to deal with at a time. When an analysis of variance incorporates more than one dependent variable it becomes a multivariate analysis of variance (MANOVA); space permits only a brief introduction to this topic in Chapter 18.

In Chapter 14, the two-group analysis of Chapter 9 is extended to accommodate any number of different samples (i.e., levels of the independent variable). Section A introduces the basic concepts of analysis of variance (including the F ratio and the F distribution) that will be used throughout Part V. Section B presents a simple step-by-step procedure for the one-way (i.e., one independent variable) ANOVA. Section C applies the concepts of power, size of effect, and homogeneity of variance to the multigroup case.

Chapter 15 revisits the two-group t-test in the context of a multigroup experiment. There is more than one method for following up a significant one-way ANOVA, and several of these are introduced in this chapter. In Section A the basic issues of Type I error rate and power are discussed in the context of performing multiple t-tests, and two simple methods for controlling Type I errors, one by Fisher and one by Tukey, are introduced. In Section B the mechanics of calculating Fisher's LSD and Tukey's HSD are detailed, and several alternative methods are briefly described. Section C delves into the more complex topics of planned comparisons, complex comparisons, and orthogonal contrasts.

Chapter 16 introduces the topic of factorial designs, in which the effects of two or more independent variables (i.e., factors) are combined in a single experiment. This chapter describes only the two-way design, but the same principles apply to designs with three or more factors. Section A outlines the analysis of variance for a two-way design and explains the critical concept of interaction. Section B presents computational formulas for the two-way ANOVA within the context of a step-by-step procedure for null hypothesis testing. Section C considers the two-way ANOVA in a more complex case (unbalanced cells) and deals with the problem of follow-up (i.e., post hoc) tests.

Chapter 17, on the one-way repeated-measures ANOVA, combines concepts from Chapter 14 with the correlational concepts from the chapters in Part IV. Section A discusses this statistical procedure from several angles, comparing it conceptually to the matched *t*-test and computationally to the two-way ANOVA. The chapter discusses the special problems involved in repeating measures several times on the same subject, along with possible solutions. Section B provides the usual details concerning computational formulas and the steps of the null hypothesis testing procedure. Section C covers several advanced topics, including how to handle a severe violation of an important assumption (homogeneity of covariance), and how to neutralize order effects when the same subject receives several different treatments (counterbalancing).

Chapter 18 deals with a very common experimental design, the mixed design. Only the simplest form of this design—with one between-subjects factor and one repeated-measures factor—is described. As you might expect, this chapter draws from the concepts in the two chapters before it. Section A begins with a one-way repeated-measures ANOVA and shows what happens when a between-subjects factor is added. Section B contains its usual focus on computational formulas and formats for presenting results. Section C provides a brief introduction to the use of analysis of covariance (ANCOVA), a statistical procedure that uses linear regression to remove extraneous variance from a dependent variable. Finally, multivariate analysis of variance (MANOVA) is introduced, both as a means for enhancing the effect of an independent-samples ANOVA by including several dependent variables simultaneously and as an alternative to repeated-measures ANOVA that has less restrictive assumptions.

ONE-WAY INDEPENDENT ANOVA

You will need to use the following from previous chapters:

Symbols
$\overline{X}$: Mean of a sample
s: Unbiased standard deviation of a sample
s^2: Unbiased variance of a sample
SS: Sum of squared deviations from the mean
MS: Mean of squared deviations from the mean (same as variance)

Formulas
Formula 4.11B: Computational formula for SS
Formula 9.5B: The pooled-variance t-test

Concepts
Homogeneity of variance
The pooled variance, s_p^2
Null hypothesis distribution

14

C H A P T E R

A

**CONCEPTUAL
FOUNDATION**

In Chapter 9, I described a hypothetical experiment in which the effect of vitamin C on sick days taken off from work was compared to the effect of a placebo. Suppose the researcher was also interested in testing the claims of some vitamin enthusiasts who predict that vitamin C combined with B complex and other related vitamins is much more effective against illness than vitamin C alone. A third group of subjects, a multivitamin group, could be added to the previous experiment. But how can we test for statistically significant differences when there are three groups? The simplest answer is to perform t-tests for independent groups, taking two groups at a time. With three groups only three different t-tests are possible. First, the mean of the vitamin C group can be tested against the mean of the placebo group to determine whether vitamin C alone makes a significant difference in sick days. Next, the multivitamin group can be tested against vitamin C alone to determine whether the multivitamin approach really adds to the effectiveness of vitamin C. Finally, it may be of interest to test the multivitamin group against the placebo group, especially if vitamin C alone does not significantly differ from placebo.

There is a better procedure for testing differences among three or more independent means; it is called the **one-way analysis of variance,** and by the end of this chapter and the next one, you will understand its advantages as well as the details of its computation. The term *analysis of variance* is usually abbreviated as ANOVA. The term *one-way* refers to an experimental design in which there is only one independent variable. In the vitamin experiment described above, there are three experimental conditions, but they are considered to be three different *levels* of the same independent variable, which is "type of vitamin." In the context of ANOVA, an independent variable is called a **factor**; this terminology will be more useful in Chapter 16, when the two-way ANOVA is discussed.

In this chapter I will deal only with the one-way ANOVA of *independent* samples. If the samples are matched, you must use the procedures described in Chapter 17.

Before proceeding, I should point out that when an experiment involves only three groups, performing all of the possible *t*-tests (i.e., testing all three pairs), though not optimal, usually leads to the same conclusions that would be derived from the one-way analysis of variance. The drawbacks of performing multiple *t*-tests become more apparent, however, as the number of groups increases. Consider an experiment involving seven different groups of subjects. For example, a psychologist may be exploring the effects of culture on emotional expressivity by comparing the means of seven samples of subjects, each group drawn from a different cultural community. In such an experiment there are 21 possible two-group *t*-tests. (In the next chapter you will learn how to calculate the number of possible pairs given the number of groups.) If the .05 level is used for each *t*-test, the chances are better than 50% that one of these *t*-tests will attain significance even if all of the cultural groups are identical on the variable measured. Moreover, with so many groups, the psychologist's initial focus would likely be to see whether there are any differences at all among the groups, rather than to ask whether any one particular group mean is different from another group mean. In either case, Dr. Null would say that all seven cultural groups are identical to each other. To test this null hypothesis in the most valid and powerful way, we need to know how much seven sample means are likely to vary from each other when all seven samples are drawn from the same population (or different populations that all have the same mean). We need to develop the formula for the one-way ANOVA.

Transforming the *t*-Test into ANOVA

I will begin with the *t*-test for two independent groups and show that it can be modified to accommodate any number of groups. As a starting point, I will use the *t*-test formula as expressed in terms of the pooled variance (Formula 9.5B):

$$t = \frac{\overline{X}_1 - \overline{X}_2}{\sqrt{s_p^2(1/N_1 + 1/N_2)}}$$

We can simplify this formula by dealing with the case in which $N_1 = N_2$. The above formula becomes

$$t = \frac{\overline{X}_1 - \overline{X}_2}{\sqrt{s_p^2(1/N + 1/N)}} = \frac{\overline{X}_1 - \overline{X}_2}{\sqrt{s_p^2(2/N)}}$$

which leads to Formula 14.1:

$$t = \frac{\overline{X}_1 - \overline{X}_2}{\sqrt{2s_p^2/N}} \qquad\qquad \text{Formula 14.1}$$

When $N_1 = N_2$, the pooled variance is just the ordinary average of the two variances, and Formula 9.8 (the *t*-test for equal-sized groups) is easier to use than Formula 14.1. But it is informative to consider the formula in terms of the pooled variance. The next step is to square Formula 14.1, so that we are dealing more directly with the variances rather than with standard deviations (or standard errors). After squaring we multiply both the numerator and the denominator by the sample size (N) and then divide both by 2 to achieve the following expression, Formula 14.2:

$$t^2 = \frac{(\bar{X}_1 - \bar{X}_2)^2}{2s_p^2/N} = \frac{N(\bar{X}_1 - \bar{X}_2)^2}{2s_p^2} = \frac{N(\bar{X}_1 - \bar{X}_2)^2/2}{s_p^2} \qquad \text{Formula 14.2}$$

Bear in mind that as strange as Formula 14.2 looks, it can still be used to perform a *t*-test when the two groups are the same size, as long as you remember to take the square root at the end. If you do not take the square root, you have performed an analysis of variance on the two groups, and the result is called an *F* ratio (for reasons that will soon be made clear). Although Formula 14.2 works fine for the two-group experiment, it must be modified to accommodate three or more groups in order to create a general formula for the one-way analysis of variance.

Expanding the Denominator

Changing the denominator of Formula 14.2 to handle more than two sample variances is actually very easy. The procedure for pooling three or more sample variances is a natural extension of the procedure described in Chapter 9 for pooling two variances. However, it is customary to change the terminology for referring to the pooled variance when dealing with the analysis of variance. (I introduced this terminology in Chapter 4 to prepare you, when I showed that the variance, s^2, equals SS/df, where SS stands for the sum of squares, i.e., the sum of squared deviations.) Because the variance is really a mean of squares, it can be symbolized as MS. Therefore, the pooled variance can be referred to as MS. Specifically, the pooled variance is based on the variability *within* each group in the experiment, so the pooled variance is often referred to as the **mean-square-within,** or **MS$_w$.** When all of the samples are the same size, MS$_w$ is just the ordinary average of all the sample variances.

Expanding the Numerator

The numerator of the two-group *t*-test is simply the difference between the two sample means. If, however, you are dealing with *three* sample means and you wish to know how far apart they are from each other (i.e., how spread out they are), there is no simple difference score that you can take. Is there some way to measure how far apart three or more numbers are from each other? The answer is that the ordinary variance will serve this purpose nicely. (So would the standard deviation, but it will be easier to deal only with variances.) Although it is certainly not obvious from looking at the numerator of Formula 14.2, the term that follows

N is equal to the variance of the two group means. In order to accommodate three or more sample means, the numerator of Formula 14.2 must be modified so that it equals N times the variance of all the sample means.

For example, if the average heart rates of subjects in three groups taking different medications were 68, 70, and 72 bpm, the (unbiased) variance of the three sample means would be 4. (Taking the square root would give you a standard deviation of 2, which you could guess just from looking at the numbers.) If the sample means were more spread out, e.g., 66, 70, and 74, their variance would be greater (in this case, 16). To modify the numerator of Formula 14.2, the variance of the group means must be multiplied by the sample size. (The procedure gets a little more complicated when the sample sizes are not all equal.) Therefore, if each sample had eight subjects, the numerator would equal 32 in the first case (when the variance of the means was 4) and 128 in the second case (when the sample means were more spread out and the variance was 16).

Like the denominator of the formula for ANOVA, the numerator also involves a variance, so it too is referred to as MS. In this case, the variance is between groups, so the numerator is often called **mean-square-between,** $MS_{between\text{-}groups}$, or $\mathbf{MS_{bet}}$, for short.

The F Ratio

When MS_{bet}, as described above, is divided by MS_w, the ratio that results is called the F **ratio,** as shown in Formula 14.3:

$$F = \frac{MS_{bet}}{MS_w} \qquad \text{Formula 14.3}$$

When the null hypothesis is true, the F ratio will follow a well-known probability distribution, called the F **distribution,** in honor of R. A. Fisher, who pioneered the use of this distribution for null hypothesis testing. The letter F is used to represent a test statistic that follows the F distribution. As with the t value from a two-group test, the F ratio gets larger as the separation of the group means gets larger relative to the variability within groups. A researcher hopes that treatments will make MS_{bet} much larger than MS_w and thus produce an F ratio larger than what would easily occur by chance. However, before we can make any judgment about when F is considered large, we need to see what happens to the F ratio when Dr. Null is right—that is, when the treatments have no effect at all.

The F Ratio As a Ratio of Two Population Variance Estimates

In order to gain a deeper understanding of the F distribution and why it is the appropriate null hypothesis distribution for ANOVA, you need to look further at the structure of the F ratio and to consider what the parts represent. A ratio will follow the F distribution when both the numerator and the denominator are independent estimates of the same population variance. It should be relatively easy to see how the denominator of the F ratio, MS_w, is an estimate of the population variance, so I will begin my explanation of the F ratio at the bottom.

Pooling the sample variances in a t-test is based on the assumption that the two populations have the same variance (i.e., homogeneity of variance), and the same assumption is usually made about all of the populations that are sampled in a one-way ANOVA. Under this assumption there is just one population variance, σ^2, and pooling all the sample variances, MS_w, gives the best estimate of it. If the null hypothesis is true, the numerator of the F ratio also serves as an estimate of σ^2, but why this is so is not very obvious.

The numerator of the F ratio consists of the variance of the group means, multiplied by the sample size. (For simplicity I will continue to assume that all the samples are the same size.) To understand the general relationship between the variance of group means, $\sigma_{\bar{X}}^2$, and the variance of individuals in the population, σ^2, you have to go back to Formula 6.1,

$$\sigma_{\bar{X}} = \frac{\sigma}{\sqrt{N}}$$

and square both sides of the formula:

$$\sigma_{\bar{X}}^2 = \frac{\sigma^2}{N}$$

If we know the population variance and the size of the samples, we can use the formula above to determine the variance of the group means. On the other hand, if we have calculated the variance of the group means directly, as we do in the one-way ANOVA, and then multiplied this variance by N, the result should equal the population variance (i.e., $N\sigma_{\bar{X}}^2 = \sigma^2$). This is why the numerator of the F ratio in an ANOVA also serves as an estimate of σ^2. Calling the numerator of the F ratio $MS_{\text{between-groups}}$ can be a bit misleading if it suggests that it is just the variance among group means. MS_{bet} is actually an estimate of σ^2 based on multiplying the variance of group means by the sample size.

MS_{bet} is an estimate of σ^2 only when the null hypothesis is true; if the null hypothesis is false, the size of MS_{bet} reflects not only the population variance, but whatever treatment we are using to make the groups different. But, as usual, our focus is null hypothesis testing, which means we are interested in drawing a "map" of what can happen when the null hypothesis *is* true. When the null hypothesis is true, both MS_{bet} and MS_w are estimates of σ^2. Moreover, they are independent estimates of σ^2. MS_w estimates the population variance directly by pooling the sample variances. MS_{bet} provides an estimate of σ^2 that depends on the variance of the group means and is not affected by the size of the sample variances in that particular experiment. Either estimate can be larger, and the F ratio can be considerably greater or less than 1 (but never less than zero). If we want to know which values are common for our *F* ratio and which are unusual when the null hypothesis is true, we need to look at the appropriate *F* distribution, which will serve as our null hypothesis distribution.

Degrees of Freedom and the *F* Distribution

Like the *t* distribution, the *F* distribution is really a family of distributions. Because the *F* distribution is actually the ratio of two distributions, each of which changes shape according to its degrees of freedom, the *F* distribution changes shape depending on the number of groups as well as the total number of subjects. (Both the numerator and the denominator follow chi-square distributions; see Chapter 8, Section C for more details.) Therefore, two df components must be calculated. The df associated with MS_{bet} is 1 less than the number of groups; as the letter *k* is often used to symbolize the number of groups in a one-way ANOVA, we can say that

$$df_{bet} = k - 1 \qquad\qquad \text{Formula 14.4A}$$

The df associated with MS_w is equal to the total number of subjects (all groups combined) minus 1 for each group. Thus

$$df_w = N_T - k \qquad\qquad \text{Formula 14.4B}$$

When there are only two groups, $df_{bet} = 1$ and $df_w = N_T - 2$. Notice that in this case, df_w is equal to the df for a two-group *t*-test ($N_1 + N_2 - 2$). When df_{bet} is 1, the *F* ratio is just the *t* value squared.

The Shape of the *F* Distribution

Let us look at a typical *F* distribution; the one in Figure 14.1 corresponds to $df_{bet} = 2$ and $df_w = 27$. This means that there are three groups of ten subjects each, for a total of 30 subjects. The most obvious feature of the *F* distribution is that it is positively skewed. Whereas there is no limit to how high the *F* ratio can get, it cannot get lower than zero. *F* can never be negative because it is the ratio of two variances, and variances can never be negative. The mean of the distribution is

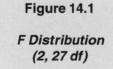

Figure 14.1

F Distribution (2, 27 df)

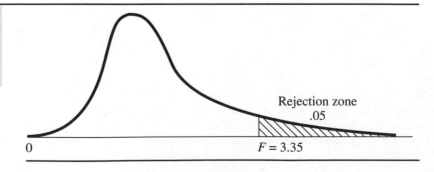

Rejection zone
.05

0   *F* = 3.35

near 1.0; to be more exact, it is $df_w/(df_w - 2)$, which gets very close to 1.0 when the sample sizes are large. Also, as the sample sizes get large, the F distribution becomes less skewed; as df_w approaches infinity, the F distribution becomes indistinguishable from a normal distribution with a mean of 1.0.

Notice that the upper 5% of the F distribution shown in Figure 14.1 is beyond $F = 3.35$; this is the critical value for this distribution when alpha equals .05. I will discuss below how to find these critical values in the F table, but first I must explain why ANOVA always involves a one-tailed test using the upper tail of the F distribution.

ANOVA As a One-Tailed Test

In the case of the t distribution, each tail represents a situation in which one of the two sample means is larger. But for the F distribution, one tail represents the situation in which all of the sample means are close together, whereas the other tail represents a spreading of the sample means. An F ratio larger than 1.0 indicates that the sample means are further apart than what we could expect (on average) as a result of chance factors. However, the size of F tells us nothing about *which* sample means are higher or lower than the others. On the other hand, an F ratio smaller than 1.0 indicates that the sample means are closer together than could be expected by chance. Because the null hypothesis is that the population means are all the same, a small F, even an F that is unusually small and close to zero, can only suggest that the null is really true.

Unfortunately, it can be confusing to refer to ANOVA as a one-tailed test, because from dealing with t-tests you get in the habit of thinking of a one-tailed test as a case in which you can change the critical value by specifying in advance which mean you expect to be larger. However, when you are dealing with more than two groups and performing the simple ANOVA described in this chapter, it makes no difference whether you make predictions about the relative sizes of the means. Thus some texts use the analogy of the t-test and refer to the ANOVA F-test as "two-tailed," because you are not predicting the order of the means. (In fact, when there are only two groups, a one-tailed F-test is equivalent to a two-tailed t-test at the same alpha, because regardless of which mean is larger, a big difference between the two means will lead to a large positive F.) Whichever way you look at it, you do not have a one- or two-tail choice with ANOVA, as you do with the t-test. The entire .05 area (or whatever alpha you are using) is always placed in the positive tail.

Using Tables of F Values

In order to look up a critical F value, we must know the df associated with the numerator (df_{bet}) as well as the denominator (df_w). If you look at Tables A.7, A.8, and A.9 in Appendix A, you will see that the numerator df determines which column to look in, and the denominator df determines how far down to look. In the t table (Table A.2), the columns represent different alpha levels. In an F table, both the columns and rows represent the df, so the entire F table usually corresponds to just one alpha level. Appendix A includes three F tables, one for each of the following alpha levels: .05, .025, and .01. Each table represents a one-tailed test; the alpha for the table tells you how much area in the positive tail lies above (to the right of) the critical value. The .05 and .01 tables are used for ANOVAs that are tested at those alpha levels. The .025 table applies when using both tails of the F distribution (even if the tail near zero is quite short), which is appropriate for testing homogeneity of variance (see Section C).

An Example with Three Equal-Sized Groups

To demonstrate how easy it can be to perform an analysis of variance, I will present a simple example based on the experiment described at the beginning of this chapter. The independent variable is vitamin treatment, and it has three levels: vitamin C, multivitamins, and placebo. The dependent variable is the number of sick days taken off work during the experimental period. Three samples of ten subjects each are selected at random, and each sample receives a different level of the vitamin treatment. At the end of the experiment, the means and standard deviations are as follows:

	Placebo	Vitamin C	Multivitamin
$\overline{X}$	9	7	5.5
s	3.5	3	2.5

In the special case when all of the samples are the same size, the formula for the one-way ANOVA becomes very simple. The numerator of the F ratio (MS_{bet}) is just the size of each sample (N) times the variance of the sample means,

$$MS_{bet} = N\frac{\Sigma(\overline{X}_i - \overline{X}_G)^2}{k - 1} = \frac{N\Sigma(\overline{X}_i - \overline{X}_G)^2}{k - 1}$$

where k is the number of groups in the experiment and $\overline{X}_G$ is the **grand mean** (i.e., the mean of all of the subjects in the whole experiment, regardless of group). An alternative notation, frequently used to represent the grand mean, is $\overline{\overline{X}}$ (pronounced "X double bar"). The two bars over the letter representing the variable suggest that the grand mean is a mean of means, which is one way to think of it. The subscript i on $\overline{X}$ reminds us that $\overline{X}$ is a variable (it usually has a different value for each sample) and that we must perform the subtraction and squaring for each of the sample means.

The denominator of the F ratio is just the average of all of the sample variances and can be symbolized as $MS_w = \Sigma s^2/k$. Combining this expression with the one above gives us Formula 14.5, which can only be used when all of the samples are the same size:

$$F = \frac{N\Sigma(\overline{X}_i - \overline{X}_G)^2/(k - 1)}{\Sigma s^2/k}$$
Formula 14.5

Calculating a Simple ANOVA

We will begin by calculating MS_{bet} for the vitamin example, using the formula for MS_{bet} above. First, we need to find $\overline{X}_G$, the grand mean. Because all the samples are the same size, the grand mean is just the average of the group means (otherwise we would have to take a weighted average): $\overline{X}_G = (7 + 5.5 + 9)/3 = 21.5/3 = 7.17$. There are three groups in this example, so $k = 3$. Now we can find the variance of the sample means:

$$\frac{\Sigma(\overline{X}_i - \overline{X}_G)^2}{k - 1} = \frac{(7 - 7.17)^2 + (5.5 - 7.17)^2 + (9 - 7.17)^2}{3 - 1}$$

$$= \frac{(-.17)^2 + (-1.67)^2 + (1.83)^2}{2} = \frac{.028 + 2.78 + 3.36}{2}$$

$$= \frac{6.17}{2} = 3.085$$

The variance of the sample means (7, 5.5, and 9) is 3.085. [You can check that the standard deviation ($\sqrt{3.085} = 1.76$) seems reasonable for these three numbers.] Finally, because there are ten subjects per group, $N = 10$, so that $MS_{bet} = 10 \cdot 3.085 = 30.85$.

When all the samples are the same size, MS_w is just the average of the three variances: $(3^2 + 2.5^2 + 3.5^2)/3 = (9 + 6.25 + 12.25)/3 = 27.5/3 = 9.17$. (If you are given standard deviations, don't forget to square them.) To complete the calculation of Formula 14.5, we form the following ratio: $F = MS_{bet}/MS_w = 30.85/9.17 = 3.36$.

Our calculated (or obtained) F ratio is well above 1.0, but to find out whether it is large enough to reject the null hypothesis, we need to look up the appropriate critical F in Table A.7, assuming that .05 is the alpha set for this ANOVA. Next, we find the df components. The df for the numerator is df_{bet}, which equals $k - 1$. For our example, $k - 1 = 3 - 1 = 2$. The df for the denominator is df_w, which equals $N_T - k = 30 - 3 = 27$. Therefore, we go to the second column of the .05 table and then down the column to the entry corresponding to 27, which is 3.35. Because our calculated F (3.36) is larger than the critical F (just barely), our result falls in the rejection zone of the F distribution (see Figure 14.1), and we can reject the null hypothesis that all three population means are equal.

Interpreting the F Ratio

The denominator of the F ratio reflects the variability within each group of scores in the experiment. The variability within a group can be due to individual differences, errors in the measurement of the dependent variable, or fluctuations in the conditions under which subjects are measured. Regardless of its origin, all of this variability is "unexplained," and it is generally labeled error variance. For this reason the denominator of the F ratio is often referred to as MS_{error}, rather than

MS_w, and in either case it is said to be the *error term* of the ANOVA. According to the logic of ANOVA, subjects in different groups should have different scores because they have been treated differently (i.e., given different experimental conditions) or because they belong to different populations, but subjects within the same group ought to have the same score. The variability within groups is not produced by the experiment, and that is why it is considered error variance.

The variability in the numerator is produced by the experimental manipulations (or preexisting population differences), but it is also increased by the error variance. Bear in mind that even if the experiment is totally ineffective and all populations really have the same mean, we do not expect all the *sample* means to be identical; we expect some variability among the sample means simply as a result of error variance. Thus there are two very different sources of variability in the numerator, only one of which affects the denominator. This idea can be summarized by the following equation:

$$F = \frac{\text{treatment effect} + \text{error variance}}{\text{error variance}}$$

If the null hypothesis is true, the treatment effect in the numerator will be zero, and both the top and the bottom of the F ratio will consist of error variance. You would expect the F ratio to equal 1.0 in this case—and on the average, it does equal about 1.0, but you must remember that the error variance in the numerator is estimated differently from the error variance in the denominator, so either one can be larger. Therefore, the F ratio will fluctuate above and below 1.0 (according to the F distribution) when the null hypothesis is true.

When the null hypothesis is *not* true, we expect the F ratio to be greater than 1.0; but even when there is some treatment effect, the F can turn out to be less than 1.0 through bad luck. (The effect may be rather weak, or it may not work well with the particular subjects selected, while the within-group variability can come out unusually high.) On the other hand, because we know the F ratio can be greater than 1.0 when there is no treatment effect at all, we must be cautious and reject the null only when our obtained F ratio is so large that only 5% (or whatever alpha we set) of the F distribution (i.e., the null hypothesis distribution) produces even larger F ratios. When we reject the null hypothesis, as we did for the vitamin example above, we are asserting our belief that the treatment effect is not zero—that the population means (in this case, placebo, vitamin C, and multivitamin) are not all equal.

We can picture the null hypothesis in the one-way ANOVA in terms of the population distributions; if three groups are involved, the null hypothesis states that the three population means are identical, as shown in the first part of Figure 14.2. (We also assume that all the populations are normally distributed with the same variance, so the three distributions should overlap perfectly.) The alternative hypothesis states that the three population means are not the same, but says nothing about the relative separation of the distributions; one possibility is shown in part b of Figure 14.2. The larger the separation of the population means

Figure 14.2

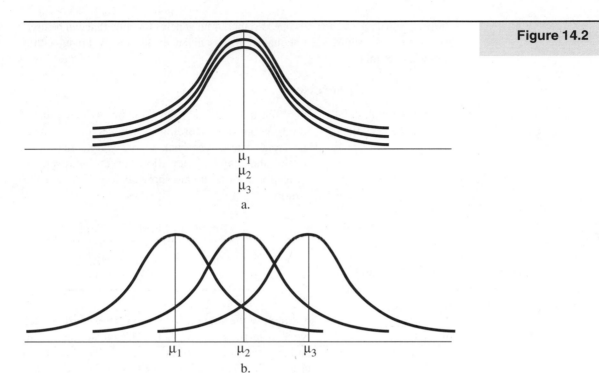

μ₁
μ₂
μ₃
a.

μ₁   μ₂   μ₃
b.

relative to the spread of the distributions, the larger the treatment effect, and the higher *F* tends to be. Section C describes a way to quantify the relative separation of the population means (i.e., the effect size of the experiment).

Advantages of ANOVA

Having rejected the null hypothesis for an experiment containing more than two groups, we would probably want to test each pair of group means to see more specifically where the significant differences are. (For instance, is the multivitamin condition significantly different from the vitamin C condition?) Why not go directly to performing the three *t*-tests and just skip the ANOVA entirely? There are two main reasons. One reason was mentioned at the beginning of this chapter: The chance of making a Type I error increases as more pairs of means are tested. The problem is minor when there are only three groups, but it becomes quite serious as groups are added to the design. The second reason is that the ANOVA can find a significant difference among several group means even when no two of the means are significantly different from each other. These advantages of the ANOVA will not be entirely obvious, however, until I discuss procedures for following the ANOVA with *t*-tests in Chapter 15.

Had our three groups not contained the same number of subjects, we could not have used Formula 14.5, and we might have been concerned about whether all three populations had the same variance. The general formula that can be used even when the groups are unequal in size is given in Section B, along with a numerical example.

Exercises

* 1. Consider a one-way ANOVA with five samples that are all the same size, but whose standard deviations are different: $s_1 = 10$, $s_2 = 15$, $s_3 = 12$, $s_4 = 11$, and $s_5 = 10$. Can you calculate MS_w from the information given? If so, what would be the value for MS_w?

2. a) If 120 subjects are divided equally among three groups, what are the df that you need to find the critical F, and what is the critical F for a .05 test?

    b) What is the critical F for a .01 test?

    c) If 120 subjects are divided equally among six groups, what are the df that you need to find the critical F, and what is the critical F for a .05 test?

    d) What is the critical F for a .01 test?

    e) Compare the critical Fs in parts a and c, and do the same for parts b and d. What is the effect on critical F of adding groups, if the total number of subjects remains the same?

* 3. If $df_{bet} = 4$, $df_w = 80$, and all groups are the same size, how many groups are there in the experiment, and how many subjects are in each group?

4. In exercise 9B2, a two-group t-test was performed to compare 15 "more vivid" visual imagers with 15 "less vivid" visual imagers on color recall. For the more vivid group, $\overline{X}_1 = 12$ colors with $s_1 = 4$; for the less vivid group, $\overline{X}_2 = 8$ colors with $s_2 = 5$.

    a) Calculate the F ratio for these data.

    b) How does your calculated F in part a compare to the t value you found for exercise 9B2? What is the general rule relating t and F in the two-group case?

    c) What is the appropriate critical F for testing your answer to part a? How does this value compare with the critical t value you used in exercise 9B2?

    d) Which statistical test, t or F, is more likely to lead to statistical significance when dealing with two equal-sized groups? Explain.

* 5. The 240 students in a large introductory psychology class are scored on an introversion scale that they filled out in class, and then they are divided equally into three groups according to whether they sit near the front, middle, or back of the lecture hall. The means and standard deviations of the introversion scores for each group are shown below. Calculate the F ratio.

	Front	Middle	Back
$\overline{X}$	28.7	34.3	37.2
s	11.2	12.0	13.5

* 6. Suppose the standard deviations in exercise 5 were twice as large, as shown below. Calculate the F ratio and compare it to the F ratio you calculated for exercise 5. What is the effect on the F ratio of doubling the standard deviations?

	Front	Middle	Back
s	22.4	24.0	27.0

* 7. A psychologist is studying the effects of various drugs on the speed of mental arithmetic. In an exploratory study, 32 subjects are divided equally into four drug conditions, and each subject solves as many problems as he or she can in 10 minutes. The mean number of problems solved is given below for each drug group, along with the standard deviations.

	Mari-juana	Ampheta-mine	Valium	Alcohol
$\overline{X}$	7	8	5	4
s	3.25	3.95	3.16	2.07

    a) Calculate the F ratio.

    b) Find the critical F ($\alpha = .05$).

c) What can you conclude with respect to the null hypothesis?

8. If the study in exercise 7 were repeated with a total of 64 subjects, a) what would be the new value for calculated F?

b) How does the F ratio calculated in part a compare to the F calculated in exercise 7? What general rule relates changes in the F ratio to changes in sample size (when all samples are the same size and all else remains unchanged)?

c) What is the new critical F ($\alpha = .05$)?

*9. Suppose that the F ratio you have calculated for a particular one-way ANOVA is .04. Which of the following can you conclude?

a) A calculation error has probably been made.

b) The null hypothesis can be rejected because $F < .05$.

c) There must have been a great deal of within-group variability.

d) The null hypothesis cannot be rejected.

e) No conclusions can be drawn without knowing the df.

10. Suppose that the F ratio you have calculated for another one-way ANOVA is 23. Which of the following can you conclude?

a) A calculation error has probably been made (F values this high are too unlikely to arise in real life).

b) The null hypothesis can be rejected at the .05 level (as long as all groups contain at least two subjects).

c) The group means must have been spread far apart.

d) There must have been very little within-group variability.

e) The sample size must have been large.

Just as in the case of the *t*-test, the levels of an independent variable in a one-way ANOVA can be created experimentally or can occur naturally. In the following ANOVA example, I will illustrate a situation in which the levels of the independent variable occur naturally, and the experimenter randomly selects subjects from these preexisting groups. At the same time, I will be illustrating the computation of a one-way ANOVA in which the samples are not all the same size.

BASIC STATISTICAL PROCEDURES

An ANOVA Example with Unequal Sample Sizes

A psychologist has hypothesized that the death of a parent, especially before a child is 12 years old, undermines the child's sense of optimism, and that this deficit is carried into adult life. She further hypothesizes that the loss of both parents before the age of 12 amplifies this effect. To test her hypothesis, the psychologist has found four young adults who were orphaned before the age of 12, five more who lost one parent each before age 12, and, for comparison, six young adults with both parents still alive. Each subject was tested on a 50-point optimism scale. The research hypothesis that parental death during childhood affects optimism later in life can be tested following the six steps of hypothesis testing that you have already learned.

Step 1. State the Hypotheses

In the case of a one-way ANOVA, the null hypothesis is very simple. The null hypothesis always states that all of the population means are equal; in symbols,

$\mu_1 = \mu_2 = \mu_3 = \mu_4 = \ldots$, etc. Because our example involves three groups, the null hypothesis is H_0: $\mu_1 = \mu_2 = \mu_3$.

For one-way ANOVA, the alternative hypothesis is simply that the null hypothesis is *not* true, i.e., that the population means are not all the same. However, this is not simple to state symbolically. The temptation is to write H_A: $\mu_1 \neq \mu_2 \neq \mu_3$, but this is *not* correct. Even if $\mu_1 = \mu_2$, the null hypothesis could be false if μ_3 were not equal to μ_1 and μ_2. In fact, in the case of three groups, there are four ways H_0 could be false: $\mu_1 = \mu_2 \neq \mu_3$; $\mu_1 \neq \mu_2 = \mu_3$; $\mu_1 \neq \mu_3 = \mu_2$; or $\mu_1 \neq \mu_2 \neq \mu_3$. The alternative hypothesis does not state which of these will be true, only that one of them will be true. Of course, with more groups in the experiment there would be even more ways that H_0 could be false and H_A could be true. Therefore, we do not worry about stating H_A symbolically, other than to say that H_0 is not true.

Step 2. Select the Statistical Test and the Significance Level

Because we want to draw an inference about more than two population means simultaneously, the one-way analysis of variance is appropriate—assuming that our optimism questionnaire can be considered an interval or ratio scale. As usual, alpha = .05.

Step 3. Select the Samples and Collect the Data

When you are dealing with preexisting groups and therefore not randomly assigning subjects to different levels of the independent variable, it is especially important to select subjects as randomly as possible. In addition, it is helpful to select samples that are all the same size and rather large, so that you need not worry about homogeneity of variance (see the discussion of assumptions later in this section) or low power. However, practical considerations may limit the size of your samples, and one group may be more limited than another. In the present example, it may have been difficult to find more than four young adults with the characteristics needed for the first group. Then, having found five subjects for

Table 14.1	Both Parents Deceased	One Parent Deceased	Both Parents Alive
	29	30	35
	35	37	38
	26	29	33
	22	32	41
	$\bar{X} = 28$	25	28
		$\bar{X} = 30.6$	40
			$\bar{X} = 35.83$

the second group, the psychologist probably decided it would not be worth the loss of power to throw away the data for one subject in the second group just to have equal-sized groups. The same considerations apply to having six subjects in the third group. The optimism rating for each subject is shown in Table 14.1 along with the mean of each group.

Step 4. Find the Region of Rejection

Because we are using the F ratio as our test statistic, and it is the ratio of two independent estimates of the same population variance, the appropriate null hypothesis distribution is one of the F distributions. To locate the critical F for our example we need to know the df for both the numerator and the denominator. For this example, k (the number of groups) = 3, and N_T (the total number of subjects) = $4 + 5 + 6 = 15$. Therefore, $df_{bet} = k - 1 = 3 - 1 = 2$, and $df_w = N_T - k = 15 - 3 = 12$. Looking at the .05 table (Table A.7), we start at the column labeled 2 and move down to the row labeled 12. The critical F listed in the table is 3.89. As you can see from Figure 14.3, the region of rejection is the portion of the F distribution that is above (i.e., to the right of) the critical F. The test is one-tailed; the .05 area is entirely in the upper tail, which represents sample means that are more spread out than typically occurs through random sampling.

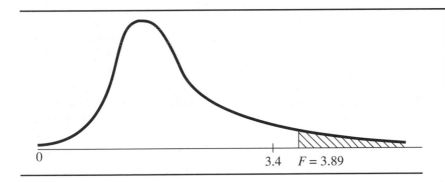

Figure 14.3

F Distribution (2, 12 df)

Step 5. Calculate the Test Statistic

In order to deal with the unequal sample sizes in our example, we must modify Formula 14.5. The denominator is easy to modify; instead of taking the simple average of the variances we go back to the concept of the pooled variance and take a weighted average, as we did for the t-test. We can expand Formula 9.6A to accommodate any number of groups. (Remember that s_p^2 is now being referred to as MS_w.) However, we will be making one minor change in notation. Because ANOVA can deal with many groups of different sizes, I will use the lowercase n (with a subscript when necessary) to represent the size of any one group, and the uppercase N to represent the total number of subjects in all the groups combined.

(I will use the subscript T with N to make it extra clear that this is a total N.) The formula for the denominator (MS_w) becomes

$$MS_w = \frac{(n_1 - 1)s_1^2 + (n_2 - 1)s_2^2 + (n_3 - 1)s_3^2 + \cdots + (n_k - 1)s_k^2}{N_T - k}$$

The above expression can be rewritten more compactly to create Formula 14.6:

$$MS_w = \frac{\sum(n_i - 1)s_i^2}{df_w} \qquad\qquad \text{Formula 14.6}$$

The numerator of Formula 14.5, written in long form, becomes

$$MS_{bet} = \frac{n_1(\overline{X}_1 - \overline{X}_G)^2 + n_2(\overline{X}_2 - \overline{X}_G)^2 + n_3(\overline{X}_3 - \overline{X}_G)^2 + \cdots + n_k(\overline{X}_k - \overline{X}_G)^2}{k - 1}$$

Written more compactly, the above expression becomes Formula 14.7:

$$MS_{bet} = \frac{\sum n_i(\overline{X}_i - \overline{X}_G)^2}{df_{bet}} \qquad\qquad \text{Formula 14.7}$$

The equations above represent by far the most common approach to calculating the one-way ANOVA when the sample sizes are unequal. This approach is called the *analysis of weighted means,* and it is the only method I will discuss in this chapter. An alternative approach, the *analysis of unweighted means,* will be discussed in Chapter 16, Section C, in the context of more complex ANOVA situations. I will expand on the data from the optimism example, so that I can illustrate the use of computational formulas to find the mean and variance of each group; the data are given in Table 14.2. Subscripts are used to distinguish the three groups.

The means for the three groups were shown in Table 14.1. The sum of squares (SS) for each group can be found by plugging the appropriate sum and sum of

Table 14.2	X_1	X_1^2	X_2	X_2^2	X_3	X_3^2
	29	841	30	900	35	1225
	35	1225	37	1369	38	1444
	26	676	29	841	33	1089
	22	484	32	1024	41	1681
	112	3226	25	625	28	784
			153	4759	40	1600
					215	7823

squared values (shown at the bottom of each column in Table 14.2) into Formula 4.5. With the addition of a subscript to represent the group number, the formula becomes

$$SS_i = \Sigma X^2 - \frac{(\Sigma X)^2}{n_i}$$

The SS for the three groups are as follows:

$$SS_1 = 3226 - 12{,}544/4 = 3226 - 3136 \quad = \quad 90$$

$$SS_2 = 4759 - 23{,}409/5 = 4759 - 4681.8 = \quad 77.2$$

$$SS_3 = 7823 - 46{,}225/6 = 7823 - 7704.2 = 118.8$$

Finally, the (unbiased) variance is given by Formula 4.7, which is shown below (including a subscript to distinguish groups):

$$s_i^2 = SS_i/(n_i - 1)$$

The three variances are:

$$s_1^2 = 90/3 \quad = 30$$

$$s_2^2 = 77.2/4 \quad = 19.3$$

$$s_3^2 = 118.83/5 = 23.77$$

By taking the square root of each variance, you can find the standard deviations ($s_1 = 5.48$, $s_2 = 4.39$, $s_3 = 4.88$) and check that they seem reasonable for the data for each group.

Now that we have the three variances it is easy to apply Formula 14.6:

$$MS_w = \frac{\Sigma(n_i - 1)s_i^2}{df_w} = \frac{3(30) + 4(19.3) + 5(23.8)}{12} = \frac{286.03}{12} = 23.84$$

Remember that MS_w is a weighted average of the sample variances, so it must be somewhere between the smallest and largest of your sample variances or you have made a calculation error. In this case, $MS_w = 23.84$ is right in the middle of the three variances (30, 19.3, 23.77).

Before we can apply Formula 14.7 to find MS_{bet}, we must find the grand mean, $\overline{X}_G$, which is the weighted average of all of the sample means. The easiest way to find the grand mean is just to add all of the scores from all of the groups together and then divide by the total number of scores (N_T). Summing the sums of the three groups, we get $112 + 153 + 215 = 480$; dividing by N_T, which is 15, we find that $\overline{X}_G = 480/15 = 32$. Now we can calculate Formula 14.7:

$$MS_{bet} = \frac{\sum n_i(\overline{X}_i - \overline{X}_G)^2}{df_{bet}} = \frac{4(28 - 32)^2 + 5(30.6 - 32)^2 + 6(35.83 - 32)^2}{2}$$

$$= \frac{4(16) + 5(1.96) + 6(14.69)}{2} = \frac{64 + 9.8 + 88.17}{2} = \frac{161.97}{2} = 80.98$$

Finally, we calculate the *F* ratio:

$$F = \frac{MS_{bet}}{MS_w} = \frac{80.98}{23.84} = 3.4$$

Step 6. Make the Statistical Decision

The calculated *F* (3.4) is not as large as the critical *F* (3.89), so our result does *not* land in the region of rejection (see Figure 14.3). We cannot reject the null hypothesis that all of the population means are equal. This is not surprising when you consider that with sample sizes as tiny as those in this experiment, the power of a one-way ANOVA is low for any effect size that is not very large. (See Section C for further discussion of power and effect size in relation to the one-way ANOVA.)

Had we rejected the null, we could have concluded that the population means are not all the same, but without additional *t*-tests we could not say which pairs of population means are significantly different (see Chapter 15). Moreover, because the three conditions were not created by the experimenter, rejecting the null would not mean that parental death *causes* changes in optimism. Alternative explanations are possible. For instance, pessimism may run in families, and pessimism may lead to poor health and early death. According to this hypothetical explanation, both the child's pessimism *and* the early parental death are caused by a third variable—that is, the parent's pessimism.

Raw-Score Formulas

Within-Group Sum of Squares (SS_w)

When calculating the one-way ANOVA directly from raw data (that is, when means and variances have not yet been calculated), it is customary to deal first with sums of squares, calculating variances only at the final step. To find MS_w it is not necessary to calculate the variance of each sample in the experiment. The SS can be calculated separately for each group, and then all of the SS can be added together; the sum of all the SS is called SS_w. When SS_w is divided by df_w, the result is MS_w. Actually, this is the same method I described as an alternative way to compute the pooled variance for the two-group *t*-test. If you look at Formula 9.6B, you will see that the SS for each group are added and then the sum is divided by the appropriate degrees of freedom. This formula can be generalized to accommodate any number of groups, as follows (Formula 14.8):

$$MS_w = \frac{SS_1 + SS_2 + SS_3 + \cdots + SS_k}{n_1 + n_2 + n_3 + \cdots + n_k - k} = \frac{\Sigma SS_i}{N_T - k}$$

$$MS_w = \frac{SS_w}{df_w} \qquad\qquad \text{Formula 14.8}$$

The SS for each group can be found either by using the definitional formula, Formula 4.3, $SS = \Sigma(X - \overline{X})^2$, or by using the computational formula, Formula 4.11B, $SS = \Sigma X^2 - (\Sigma X)^2/N$.

If the computational formula is used for each SS, the separate formulas can be summed over the groups to create one combined computational formula for SS_w. To simplify the notation, I define T (for total) as ΣX, and T_i as the sum for one particular group. Now the formula for SS_w can be written as Formula 14.9,

$$SS_w = \Sigma X^2 - \Sigma\left(\frac{T_i^2}{n_i}\right) \qquad\qquad \text{Formula 14.9}$$

where ΣX^2 is the sum of *all* the squared scores (from all of the groups). Note that the above formula gives the same answer as calculating the SS for each group separately and then adding them up. I will illustrate the use of this formula by applying it to the optimism example. To find ΣX^2 for the whole experiment we add the sum of squared values for each group (from Table 14.2). Therefore, $\Sigma X^2 = 3226 + 4759 + 7823 = 15,808$. Then we square the sum of each group, divide by the group size, and subtract the result from ΣX^2, as shown below:

$$\begin{aligned}
SS_w = \Sigma X^2 - \Sigma\left(\frac{T_i^2}{n_i}\right) &= 15,808 - \left(\frac{112^2}{4} + \frac{153^2}{5} + \frac{215^2}{6}\right) \\
&= 15,808 - \left(\frac{12,544}{4} + \frac{23,409}{5} + \frac{46,225}{6}\right) \\
&= 15,808 - (3136 + 4681.8 + 7704.17) \\
&= 15,808 - 15,521.97 = 286.03
\end{aligned}$$

Again, SS_w must be divided by df_w to yield MS_w. For the above example, $MS_w = SS_w/df_w = 286.03/12 = 23.84$, which, of course, is the same value for MS_w we obtained by using the definitional formula earlier in this section.

Between-Group Sum of Squares (SS_{bet})

Just as MS_w can be expressed as SS_w divided by df_w, a similar relation applies to MS_{bet}, as shown in Formula 14.10:

$$MS_{bet} = \frac{SS_{bet}}{df_{bet}} \qquad\qquad \text{Formula 14.10}$$

Like the formula for SS_w, the formula for SS_{bet} can be converted to a very convenient computational form, as shown below (I will not show the algebraic manipulations here):

$$SS_{bet} = \Sigma\left(\frac{T_i^2}{n_i}\right) - \frac{(\Sigma X)^2}{N_T}$$

Notice that the first term in the formula above is exactly the same as the second term in Formula 14.9. The second term in the above formula is the sum of all scores, which is squared and divided by the total number of scores. By expressing the ΣX for all scores (i.e., the **grand total**) as T_T, we can rewrite the above formula to create Formula 14.11. (I added the subscript T so that this sum won't be confused with the sum for any particular group in the experiment.)

$$SS_{bet} = \Sigma\left(\frac{T_i^2}{n_i}\right) - \frac{T_T^2}{N_T} \qquad \text{Formula 14.11}$$

To complete the example, we apply Formula 14.11 to the data in Table 14.2. First we need to find the total of all scores (T_T), which is also the sum of all the group totals, ΣT_i. Therefore, $T_T = 112 + 153 + 215 = 480$. Next, notice that we do not have to recalculate the first term of Formula 14.11; we already found this value in the calculation of Formula 14.9: $\Sigma(T_i^2/n_i) = 15{,}521.97$. Inserting these values into Formula 14.11, we obtain:

$$SS_{bet} = \Sigma\left(\frac{T_i^2}{n_i}\right) - \frac{T_T^2}{N_T} = 15{,}521.97 - \frac{480^2}{15} = 15{,}521.97 - 15{,}360 = 161.97$$

The value for SS_{bet} must be divided by df_{bet} to yield $MS_{bet} = 161.97/2 = 80.98$, which is the same answer we obtained using the definitional formula.

The Total Sum of Squares (SS_{tot})

If SS_{bet} and SS_w are added together the result is SS_{tot}, just as when we added the SS explained and SS unexplained for regression (see Chapter 12). Formula 14.12 represents this relationship:

$$SS_{tot} = SS_{bet} + SS_w \qquad \text{Formula 14.12}$$

It is also simple to find SS_{tot} directly; the formula for SS_{tot} is just $\Sigma(X - \overline{X}_G)^2$. This means that the grand mean is subtracted from every score and the results are squared and added. The groups are ignored; that is, the scores in the experiment are treated as though they were all from one big group. The computational formula for SS_{tot}, as with any SS, is $\Sigma X^2 - (\Sigma X)^2/N$. Using T_T to represent the total of all scores, we get the following computational formula (Formula 14.13) for SS_{tot}:

$$SS_{tot} = \Sigma X^2 - \frac{T_T^2}{N_T} \qquad\qquad \text{Formula 14.13}$$

Watch what happens when we add the computational formulas for SS_w and SS_{bet}:

$$\Sigma X^2 - \Sigma\left(\frac{T_i^2}{n_i}\right) + \Sigma\left(\frac{T_i^2}{n_i}\right) - \frac{T_T^2}{N_T}$$

The two middle terms cancel each other out, and we end up with formula 14.13.

Let us apply Formula 14.13 to our example. We already calculated the term ΣX^2 to find SS_w using Formula 14.9. The second term is also the second term of Formula 14.11, which we calculated in finding SS_{bet}. Therefore, Formula 14.13 gives us

$$SS_{tot} = \Sigma X^2 - \frac{T_T^2}{N_T} = 15{,}808 - 15{,}360 = 448$$

Note that using Formula 14.12—that is, simply adding SS_{bet} and SS_w—yields the same answer.

The concept of dividing SS_{tot} into SS_{bet} and SS_w parallels the concept behind the SS in regression and will be discussed further in Section C. For the moment I will point out that calculating SS_{tot} separately acts as a useful check on your calculations of SS_w and SS_{bet}. If SS_w and SS_{bet} do not add up to SS_{tot}, you know you have made a calculation error. (If the discrepancy is very slight it is probably due to some rounding off.)

If you do not want to perform the extra calculations required to check your SS, as described above, it is quicker to calculate SS_{tot} and SS_w and then obtain SS_{bet} by subtraction. Alternatively, you can calculate SS_{tot} and SS_{bet} and find SS_w by subtraction. (Generally SS_{tot} is less confusing to calculate from raw scores than either SS_w or SS_{bet}.) However, if the raw data are not available and you have only the means, standard deviations (or variances), and sample sizes, you will have to use the definitional formulas (Formulas 14.6 and 14.7) described in step 5 of the hypothesis testing procedure. Even if the raw data are available, if you have already calculated the means and standard deviations for other purposes, you may find it easier to use the definitional formulas (that is, if you are not using a computer to analyze your data).

Assumptions of the One-Way ANOVA for Independent Groups

The assumptions underlying the test described in this chapter are the same as the assumptions underlying the *t*-test of two independent means. The assumptions are briefly reviewed below, but for greater detail you should consult Section B of Chapter 9.

Independent Random Sampling

Ideally, all of the samples in the experiment should be drawn randomly, and individuals in one sample should have no relationship to individuals in any other sample. If you are not dealing with preexisting populations, it is critical that subjects be *randomly* assigned to the different experimental conditions. If all of the samples are in some way matched, as described for the matched *t*-test, the procedures of this chapter are not appropriate. Methods for dealing with more than two matched samples are presented in Chapter 17.

Normal Distributions

It is assumed that all of the populations are normally distributed. As in the case of the *t*-test, however, with large samples we need not worry about the shapes of our population distributions. In fact, even with fairly small samples, the *F*-test for ANOVA is not very sensitive to departures from the normal distribution—in other words, it is robust, especially when all the distributions are symmetric or skewed in the same way. However, when dealing with small samples, and distributions that look extremely different from the normal distribution and from each other, you should consider using nonparametric tests, such as the Kruskal-Wallis *H* test (see Chapter 21 in Part VI) or data transformations (see Chapter 3).

Homogeneity of Variance

It is assumed that all of the populations involved have the same variance. However, when the sample sizes are all equal, this assumption can be safely ignored. Even if the sample sizes and sample variances differ slightly there is no need to worry. Generally, if no sample variance is more than four times as large as another (i.e., no standard deviation is more than twice as large as another), and no sample is more than one and a half times as large as another, you can proceed with the ordinary ANOVA procedure, using the critical *F* from the table with negligible error. It is when the sample sizes are considerably different (and not very large) and the sample variances are not very similar, that there is some cause for concern. In addition to normalizing the shapes of the distributions, data transformations can also help to equate the variances. Otherwise, you may have to consider using a nonparametric test (see Section C for further discussion of this matter).

When to Use the One-Way ANOVA for Independent Groups

As you may have guessed, the analysis of variance can replace the *t*-test as a means of testing the significance of the results from a two-group experiment. Just bear in mind that the one-tailed critical *F* that you find for ANOVA corresponds to a two-tailed critical *t* in the two-group case. To perform a one-tailed *t*-test after calculating the *F* ratio, you would need to take the square root of *F* and compare it to the appropriate one-tailed critical *t* value. In case you are wondering why you learned about *t*-tests in the first place, it may help to know that the *t*-test is performed very frequently, not only for the two-group experiment but also as a method for following up on a significant ANOVA that involves more than two groups (see Chapter 15).

Two aspects of experimental design can affect how the ANOVA is calculated and/or interpreted. These two aspects are discussed below.

Fixed Versus Random Effects

One aspect of experimental design that may affect the interpretation of ANOVA concerns the manner in which the levels of the independent variable are chosen. For instance, for the vitamin experiment introduced in Section A, many different vitamins could have been tested. Vitamin C and the multivitamin were chosen specifically because we had a reason to be interested in their particular effects. Sometimes, however, there are so many possible treatments in which we are equally interested that we can only select a few levels at random to represent the entire set. For example, suppose a psychologist is creating a new questionnaire and is concerned that the order in which items are presented may affect the total score. There is a very large number of possible orders, and the psychologist may not be sure of which ones to focus on. Therefore, the psychologist selects a few of the orders at random to be the levels of a one-way ANOVA design. When the levels of the independent variable are selected *at random* from a larger set, the appropriate model for interpreting the data is a *random effects* ANOVA. The design that has been covered in this chapter thus far (in which the experimenter is specifically interested in the levels chosen and is not concerned with extrapolating the results to levels not actually included in the experiment) is called a *fixed effects* ANOVA. Effect size and power are estimated differently for the two models, and conclusions are drawn in a different manner. Because the use of the random effects model is relatively rare in psychological research, and the concept behind it is difficult to explain, I will continue to deal only with the fixed effects model in this chapter and the remainder of the text. To learn more about the random effects model of ANOVA, you may want to consult a more advanced text in statistics, such as Hays (1994).

Qualitative Versus Quantitative Levels

There is yet another way in which the levels of an independent variable in an ANOVA may differ, and this distinction can lead to a very different procedure for analyzing the data. The levels can represent different values of either a *qualitative* (nominal, categorical) or a *quantitative* (interval/ratio) scale. All the examples mentioned in this chapter thus far involve independent variables that have qualitative levels. Now consider an example of a one-way ANOVA in which the independent variable has quantitative levels. A researcher wants to know if more therapy sessions per week will speed a patient's progress, as measured by a well-being score. Five groups are randomly formed and subjects in each group are assigned to either one, two, three, four, or five sessions per week for the 6-month experimental period. Because both the independent variable (i.e., number of sessions per week) and the dependent variable (i.e., well-being score) are

Figure 14.4

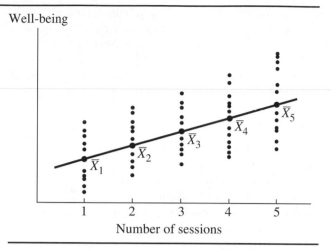

Well-being

Number of sessions

quantitative, a linear regression approach may be advantageous. ANOVA can be considered a special case of regression, as shown in Section C. (Linear regression can be used instead of ANOVA even when the independent variable has categorical levels, but in that case it has no special advantage.)

To get an idea of the advantage of linear regression over ANOVA when dealing with a quantitative independent variable, imagine that the mean well-being score is a perfect linear function of the number of sessions, as shown in Figure 14.4. Because all of the sample means fall on the same straight line, the SS explained by linear regression (called $SS_{regression}$, in this context) is the same as SS_{bet}, and the unexplained SS (called $SS_{residual}$) is equal to SS_w. The difference is in the degrees of freedom. With the linear regression approach, the numerator df is just 1, so unlike SS_{bet}, which would be divided by 4 for this example (i.e., $k - 1$), $SS_{regression}$ is only divided by 1. Thus the numerator for linear regression is four times as large as the numerator for the ANOVA. Because the denominator for regression will be a bit smaller than the denominator for ANOVA ($SS_{residual}$ is divided by $N - 1$, as compared to $N - k$ for SS_w), the F ratio for testing the linear regression will be somewhat more than four times the F ratio for ANOVA.

As the sample means deviate more and more from a straight line, $SS_{regression}$ becomes less than SS_{bet}, and the advantage of the regression approach decreases; but as long as the relation between the independent and dependent variables is somewhat linear, the regression approach will probably lead to a higher F than ANOVA. If the sample means fall on some kind of curve, a more sophisticated approach, such as multiple regression or ANOVA with trend components, would probably have more power. These procedures are described in more advanced statistics texts. At this point I just want you to know that a study involving, for example, several dosage levels of the same drug, or different amounts of training on some task, or some other quantitative independent variable should probably *not* be analyzed with the methods in this chapter; more advanced techniques will usually yield more useful information and more power.

Publishing the Results of a One-Way ANOVA

Because the results of the parental death/optimism experiment were not statistically significant, they would probably be described for publication as follows: "Although there was a trend in the predicted direction, with parental loss associated with decreased optimism, the results of a one-way ANOVA failed to attain significance, F (2, 12) = 3.4, MSE = 23.84, p > .05." Had the F ratio been as large as say, 4.1, significant results could have been reported in the following way: "Early parental death had an impact on the levels of optimism in young adults. Subjects who had lost both parents before the age of 12 were least optimistic (M = 28), those who had lost only one parent were somewhat more optimistic (M = 30.6), whereas those whose parents were still alive were the most optimistic (M = 35.83). A one-way ANOVA demonstrated that these differences were statistically reliable, F (2, 12) = 4.1, MSE = 23.84, p < .05." The numbers in parentheses after F are df_{bet} followed by df_w. MSE stands for MS_{error}, which is another way of referring to MS_w in a one-way ANOVA. Reporting MSE is in accord with the recommendation of the APA style manual to include measures of variability along with means and the test statistic (in this case, F), so that the reader can reconstruct your analysis. For instance, with the information given above, the reader could create a summary table. It is also becoming common to report a measure of effect size, such as omega squared, along with F. Effect size measures will be discussed in Section C.

Summary Table for One-Way ANOVA

In addition to presenting the means, the F ratio, and the significance level, the entire ANOVA analysis is sometimes presented in the form of a **summary table,** including the value of each SS component. Summary tables are more likely to be presented when dealing with complex ANOVAs, but I will introduce the format here in Table 14.3, which shows the summary for the one-way ANOVA performed on the parental death/optimism data. The heading of the first column, "Source," refers to the fact that there is more than one source of variability if we look at all scores in the experiment. One source is "between-groups," sometimes called "treatment"; if the group means were further apart there would be greater variability in this component. The other source is "within-groups"; increasing the variability within each group would lead to greater variability in this component. These two sources are independent, in that either one can be increased without increasing the other; together they add up to the total variability. Notice that each SS in the table is divided by the appropriate df to yield MS. (Some

Source	SS	df	MS	F	p	
Between-groups	161.97	2	80.98	3.4	>.05	**Table 14.3**
Within-groups	286.03	12	23.84			
Total	448.00	14				

statisticians prefer not to use the term MS to stand for variance; they use s_{bet}^2 and s_{within}^2 instead.) Finally, the two MS are divided to produce F. As usual, the p value can be given in terms of a significance level (in this case, $p > .05$), or it can be given exactly, which is the way it is given by most statistical packages (e.g., SPSS) that generate a summary table as part of the one-way ANOVA output.

Excerpt from the Psychological Literature

The one-way ANOVA for independent samples is commonly used in the psychological literature, most often with only three or four levels of the independent variable. The following excerpt illustrates the APA style for reporting the results of a typical one-way ANOVA. Brown, Wheeler, and Cash (1980) measured the duration of laughter and smiles of two- to four-year-olds listening to tape-recorded jokes under three different conditions: in the presence of a laughing model, a nonlaughing model, or no model. The results were reported as follows: "A one-way analysis of variance of smiling times indicated a significant difference ($F(2, 39) = 3.31$, $p < .05$) among the three groups of children." Note that you can determine the total number of subjects in the experiment by adding together the two numbers in parentheses after F ($df_{bet} + df_w = df_{tot}$) and then adding 1: thus $(2 + 39 + 1) = 42$, so there were three groups of 14 each.

Exercises

* 1. Are all antidepressant medications equally effective? To test this null hypothesis, a psychiatrist randomly assigns one of five different antidepressants to each of 15 depressed patients. At the end of the experiment, each patient's depression level is measured. Because some patients did not take their medication or dropped out of the experiment for other reasons, the final sample sizes are not equal. The means, standard deviations, and sample sizes are as follows: $\overline{X}_1 = 23$, $s_1 = 6.5$, $n_1 = 12$; $\overline{X}_2 = 30$, $s_2 = 7.2$, $n_2 = 15$; $\overline{X}_3 = 34$, $s_3 = 7$, $n_3 = 14$; $\overline{X}_4 = 29$, $s_4 = 5.8$, $n_4 = 12$; $\overline{X}_5 = 26$, $s_5 = 6$, $n_5 = 15$. Use the six-step one-way ANOVA procedure to test the null hypothesis at the .05 level.

2. Consider a one-way ANOVA with five samples that are not all the same size, but whose standard deviations are the same: $s = 15$. Can you calculate MS_w from the information given? If so, what is the value of MS_w?

* 3. A researcher suspects that schizophrenics have an abnormally low level of hormone X in their blood-streams. To test this hypothesis, the researcher measures the blood level of hormone X in five acute schizophrenics, six chronic schizophrenics, and seven normal control subjects. The measurements appear in the table below.

Acute	Chronic	Normal
86	75	49
23	42	28
47	35	68
51	56	52
63	70	63
	46	82
		36

a) Calculate the F ratio using the raw-score formulas (Formulas 14.9 and 14.11).

b) Recalculate the F ratio using the definitional formulas (Formulas 14.6 and 14.7).

c) Explain how an inspection of the means would have saved you a good deal of computational effort.

* 4. A social psychologist wants to know how long people will wait before responding to cries for help from an unknown person, and whether the gender or age of the person in need of help makes any difference. One at a time, subjects sit in a room waiting to be called for an experiment. After a few minutes they hear cries for help from the next room, which are actually on a tape recording. The cries are in either an adult male's, an adult female's, or a child's voice; seven subjects are randomly assigned to each condition. The dependent variable is the number of seconds from the time the cries begin until the subject gets up to investigate or help.

Child's Voice	Adult Female's Voice	Adult Male's Voice
10	17	20
12	13	25
15	16	14
11	12	17
5	7	12
7	8	18
2	3	7

a) Calculate the F ratio.
b) Find the critical F ($\alpha = .05$).
c) What is your statistical conclusion?
d) Present the results of the ANOVA in a summary table.

* 5. A psychologist is interested in the relationship between color of food and appetite. To explore this relationship, the researcher bakes small cookies with icing of one of three different colors (green, red, or blue). The researcher offers cookies to subjects while they are performing a boring task. Each subject is run individually under the same conditions, except for the color of the icing on the cookies that are available. Six subjects are randomly assigned to each color. The number of cookies consumed by each subject during the 30-minute session is shown in the table below.

Green	Red	Blue
3	3	2
7	4	0
1	5	4
0	6	6
9	4	4
2	6	1

a) Calculate the F ratio.
b) Find the critical F ($\alpha = .01$).
c) What is your statistical decision with respect to the null hypothesis?
d) Present your results in the form of a summary table.

6. Suppose that the data in exercise 5 had turned out differently. In particular, suppose that the number of cookies eaten by subjects in the green condition remains the same, but each subject in the red condition ate 10 more cookies than in the previous data set, and each subject in the blue condition ate 20 more. This modified data set is shown below.

Green	Red	Blue
3	13	22
7	14	20
1	15	24
0	16	26
9	14	24
2	16	21

a) Calculate the F ratio. Is the new F ratio significant at the .01 level?
b) Which part of the F ratio has changed from the previous exercise and which part has remained the same?
c) Put your results in a summary table to facilitate comparison with the results of exercise 5.

7. A college is studying whether there are differences in job satisfaction among faculty members from different academic disciplines. Faculty members rate their own job satisfaction on a scale from 1 to 10. The data collected so far are listed below by the academic area of the respondent.

Social Sciences	Natural Sciences	Humanities
6	8	7
7	10	5
10	7	9
8	8	4
8		
9		

a) State the null hypothesis in words and again in symbols.

b) Test the null hypothesis with a one-way ANOVA at the .05 level.

c) Based on your statistical conclusion in part b, what type of error (Type I or Type II) might you be making?

*8. A social psychologist is studying the effects of attitudes and persuasion on incidental and intentional memory. Subjects read a persuasive article containing ten distinct arguments that either coincide or clash with their own opinion (as determined by a prior questionnaire). Half the subjects are told that they will be asked later to recall the arguments (intentional condition), and half are tested without warning (incidental condition). Five subjects are randomly assigned to each of the four conditions; the number of arguments recalled by each subject is shown below.

Incidental		Intentional	
Agree	Disagree	Agree	Disagree
8	2	6	7
7	3	8	9
7	2	9	8
9	4	5	5
4	4	8	7

a) Test the null hypothesis at the .05 level.

b) Test the null hypothesis at the .01 level.

c) Present the results of the ANOVA in a summary table.

*9. A psychologist is investigating cultural differences in emotional expression among small children. Three-year-olds from six distinct cultural backgrounds are subjected to the same minor stressor (e.g., their parent leaves the room), and their emotional reactions are rated on a scale from 0 to 20. The means and sample sizes are as follows: $\overline{X}_1 = 14.5$, $n_1 = 8$; $\overline{X}_2 = 12.2$, $n_2 = 7$; $\overline{X}_3 = 17.8$, $n_3 = 8$; $\overline{X}_4 = 15.1$, $n_4 = 6$; $\overline{X}_5 = 13.4$, $n_5 = 10$; $\overline{X}_6 = 12.0$, $n_6 = 7$.

a) Given that $\Sigma X^2 = 9820$, use the raw-score formulas to calculate the F ratio. (Hint: $T_i^2/n_i = n_i\overline{X}_i^2$).

b) What is the critical F ($\alpha = .05$)?

c) What is your statistical decision with respect to the null hypothesis?

*10. A researcher is exploring the effects of strenuous physical exercise on the onset of puberty in girls. The age of menarche is determined for six young athletes who started serious training before the age of eight. For comparison, the same measure is taken on four girls involved in an equally serious and time-consuming training regimen starting at the same age, but not involving strenuous physical exercise—playing the violin. An additional six girls not involved in any such training are included as a control group. The data are as follows.

Controls	Athletes	Musicians
12	14	13
11	12	12
11	14	13
13	16	11
11	15	
12	13	
11		

a) Calculate the F ratio.

b) Find the critical F ($\alpha = .01$).

c) What is your statistical decision?

d) What are the limitations to the conclusions you can draw from this study?

Double Summation Notation

In Section B, I used the symbol T_i to represent the sum of any group, in order to avoid using a combination of two summation signs in the computational formulas for ANOVA. The advantage of using double summations is that new terms do not have to be introduced arbitrarily; the use of summation signs is more precise and universal. Like the authors of most introductory statistics texts, I am minimizing the use of summation notation for the sake of simplicity. However, I introduced double subscripts and double summations in Chapter 1 (Section C), and I will illustrate their use here, because you are likely to come across them in advanced texts. For instance, when all of the groups are the same size (n), the total of all scores (for which I use the symbol T_T) can be expressed as

$$\sum_{j=1}^{k} \sum_{i=1}^{n} X_{ij}$$

The subscript i represents any one of the n subjects within a group, whereas the subscript j stands for any one of the k different groups. The total sum of squares can therefore be expressed in the following manner:

$$SS_{tot} = \sum_{j=1}^{k} \sum_{i=1}^{n} X_{ij}^2 - \frac{\left(\sum_{j=1}^{k} \sum_{i=1}^{n} X_{ij} \right)^2}{N_T}$$

It should already be clear why I am not using this notation throughout the text. To complete the illustration, the computational formulas for SS_{bet} and SS_w (with equal n), using double summations, are shown below:

$$SS_{bet} = \frac{1}{n} \sum_{j=1}^{k} \left(\sum_{i=1}^{n} X_{ij} \right)^2 - \frac{\left(\sum_{j=1}^{k} \sum_{i=1}^{n} X_{ij} \right)^2}{N_T}$$

$$SS_w = \sum_{j=1}^{k} \sum_{i=1}^{n} X_{ij}^2 - \frac{1}{n} \sum_{j=1}^{k} \left(\sum_{i=1}^{n} X_{ij} \right)^2$$

Effect Size and Proportion of Variance Accounted For

Knowing that your F ratio is significant does not tell you which of the population means in your one-way ANOVA are different from which others; this problem will be dealt with in the next chapter. But there is something else important that the size of your F ratio alone cannot tell you: the proportion of variance accounted for by the independent variable in your experiment, as well as an estimate of that proportion in the population. This is the same problem we encountered with the *t*-test. The *t* value does not tell you how much overlap there is between the two samples, or the likely separation (in standard deviations) of the population means. In the two-group case, gamma provides a way of describing the separation of the population means, and as an alternative to *g*, the point-biserial *r* (symbolized as r_{pb}), described in Chapter 12, Section C, can serve as a useful description of the data in your experiment. A measure analogous to gamma or *g* is available for the multigroup case (Cohen, 1988, calls this effect size measure *f*), but I will take a more flexible and universal approach to the concept of effect size and power in the one-way ANOVA by focusing on a correlational measure that is an extension of r_{pb}^2.

In Chapter 12, to explain how a two-group experiment could be viewed in terms of linear regression and described by a correlation coefficient (i.e., r_{pb}), I graphed the results of an imaginary study involving the effects of caffeine and a placebo on a simulated truck driving task. It is a simple matter to add a third group to the study—for instance, a group that gets an even stronger stimulant, such as amphetamine. Figure 12.7 has been modified to include a third group and appears as Figure 14.5. If the three groups represented three levels of a quantitative independent variable (e.g., three dosage levels of caffeine), it would be appropriate to use a linear regression approach, as described briefly in Section B. In this section, however, I will show how a one-way ANOVA can be viewed in terms of regression even when the levels of the independent variable represent qualitative distinctions.

The breakdown of each score into separate components is exactly the same for three (or more) groups as it was for two groups. The grand mean of all the subjects in the experiment is shown in Figure 14.5, and the graph focuses on Mr. Y, the highest scoring subject in the amphetamine group. Mr. Y's deviation from

Figure 14.5

Simulated truck driving performance

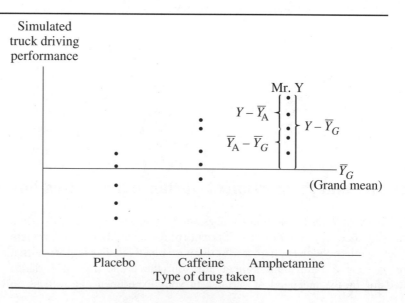

the grand mean $(Y - \overline{Y}_G)$ can be divided into his deviation from the mean of the amphetamine group $(Y - \overline{Y}_A)$ and the deviation of the amphetamine group mean from the grand mean $(\overline{Y}_A - \overline{Y}_G)$. In Chapter 12, I called the deviations from the group mean "unexplained," and when they were all squared and added, the result was called the unexplained SS. The unexplained SS corresponds exactly to the quantity that I have been calling SS_w in this chapter. Similarly, the explained SS corresponds to SS_{bet}.

The explained SS divided by the total SS gives the proportion of explained variance. In a correlational study, this proportion equals r_{XY}^2. In the two-group case, the same proportion is referred to as r_{pb}^2. In the multigroup case, this proportion, which becomes SS_{bet}/SS_{total}, is designated by a new term: η^2 (pronounced **eta squared**), as shown below:

$$\eta^2 = SS_{bet}/SS_{tot} \qquad \text{Formula 14.14}$$

The term η^2, which is the proportion of variance accounted for in the dependent variable by the independent variable in the results of a particular one-way ANOVA, is interpreted in the same way as r_{pb}^2. The use of η to denote a sample statistic is an exception to the common rule of using Greek letters to describe only population characteristics.) Of course, in the two-group case, η^2 is exactly equal to r_{pb}^2.

If you are given an F ratio and the degrees of freedom, perhaps in a published article, but you do not have access to a summary table of SS or the raw data, you can still find η^2 by using Formula 14.15:

$$\eta^2 = \frac{df_{bet}F}{df_{bet}F + df_w} \qquad \text{Formula 14.15}$$

In the two-group case, $F = t^2$ and $df_{bet} = 1$, so Formula 14.15 becomes the same as the square of Formula 12.13 [i.e., $\eta^2 = r_{pb}^2 = t^2/(t^2 + df)$].

Whereas η^2 is an excellent statistic for describing your data and a valuable supplement to the information provided by the F ratio, the major weakness of η^2 is that, like r_{pb}^2, it gives a rather biased estimate of omega squared (ω^2), the proportion of variance accounted for in the population. Nearly all of the bias in η^2 can be corrected by using Formula 14.16 to estimate ω^2:

$$\text{Estimated } \omega^2 = \frac{SS_{bet} - (k-1)MS_w}{SS_{tot} + MS_w} \qquad \text{Formula 14.16}$$

I will apply these formulas to the ANOVA described in Section B. First, I will use η^2 (Formula 14.14) to find the proportion of variance in optimism that can be accounted for by early parental death in the *samples* of our hypothetical experiment:

$$\eta^2 = \frac{SS_{bet}}{SS_{tot}} = \frac{161.97}{448} = .36$$

Then, to estimate the true proportion of variance accounted for in the entire population, I use Formula 14.16:

$$\text{Estimated } \omega^2 = \frac{SS_{bet} - (k-1)MS_w}{SS_{tot} + MS_w} = \frac{161.97 - 2(23.84)}{448 + 23.84} = \frac{114.29}{471.84} = .24$$

Eta squared shows us that 36% of the variance is accounted for in our data, which is really quite a bit compared to most psychological experiments. Normally, the F ratio would be significant with such a high η^2; it is because our sample sizes were so small (to simplify the calculations) that F fell short of significance for our ANOVA. Our estimate of ω^2 suggests that η^2 is overestimating the proportion of variance that would be accounted for in the population; however, $\omega^2 = .24$ is still a respectable proportion of explained variance in a three-group ANOVA. It is when ω^2 drops below .05 that psychologists generally begin to question the value of a study, though the results can be of some theoretical interest even in the neighborhood of only 1% variance accounted for.

Note that whenever the F ratio is less than 1.0, Formula 14.16 produces a negative number as an estimate of ω^2, which makes no sense. Although we cannot use Formula 14.16 when F is less than 1, we are usually not interested in estimating ω^2 in those cases, anyway. In fact, an estimate of ω^2 is usually reported only if the F is significant, in which case the estimate of ω^2 can replace MSE when you state your results (see "Publishing the Results . . ." in Section B). Whereas the significant F implies only that there are some differences among the population means, the estimate of ω^2 provides useful information concerning the probable effect size.

The Power of ANOVA

The concept of power as applied to the one-way independent ANOVA is essentially the same as the concept discussed in Chapter 10 with respect to the t-test. When the null hypothesis is true but the ANOVA is repeated many times anyway, the F ratios will follow one of the ordinary F distributions. However, if the null hypothesis is not true, the F ratios will follow a similar distribution called the **noncentral F distribution.** The average, or expected, F of a noncentral F distribution is higher than that of the ordinary F and depends on a *noncentrality parameter,* just as in the case of the noncentral t distribution. The important concept is that the noncentrality parameter, and therefore the expected $F,$ can be viewed as a function of both ω^2 and the sample sizes. Increasing either ω^2 or the sample sizes will lead to a larger expected $F,$ and therefore a greater chance that the calculated F will surpass the critical F and be declared significant (i.e., the power will be increased). Of course, power can also be increased by making critical F smaller by using a larger value for alpha. However, in this case the increased power is at the expense of increased Type I errors.

The power of the *F*-test for a one-way ANOVA can be found from Table A.10 (Appendix A) if ω^2 can be estimated (e.g., from previous studies), guessed at, or hypothesized about, and if the size and number of the groups in the experiment, as well as alpha, have been fixed. Table A.10 in this text is limited to the case of three, four, or five groups; for a similar table covering up to eight groups, see Hays (1994). Note also that Table A.10 does not include small effect sizes; the smallest nonzero value of ω^2 included is .1, which corresponds to a correlation coefficient of $\sqrt{.1} = .316$, or (in the two-group case) a gamma of 1.0. By the time you reach $\omega^2 = .5$, the equivalent correlation is .71 and the corresponding gamma is 4, so there is little need to look at higher values for ω^2.

For example, if you are planning a three-group experiment with 20 subjects per group, and you have good reason to expect that your independent variable will account for 10% of the variance in the dependent variable (i.e., $\omega^2 = .1$), power will be only .6 for a .05 test, and a very inadequate .34 for a .01 test. Raising the number of subjects to 30 per group yields a reasonable amount of power, .81, for the .05 test. Notice that a four-group experiment with 15 subjects per group ($\omega^2 = .1$, $\alpha = .05$) has less power (.53) than the corresponding experiment with three groups of 20 subjects each (.6), even though both experiments have the same total number of subjects. As you might expect, there is some loss of power in dividing a fixed number of subjects into a greater number of groups.

If your sample size is not fixed, you can use Table A.10 to determine the sample size you would need to attain adequate power for a given value of ω^2. For instance, if you expect that $\omega^2 = .2$ in a three-group experiment, and you require power to be at least .75 for a .05 test, you can look down the appropriate column of Table A.10 and see that you need at least 13 subjects per group ($n = 12$ yields power of only .73). On the other hand, if your experiment is fixed with four groups of ten subjects each, and you want to have power greater than .75 for a .01 test, but you don't know what value for ω^2 to expect, you can use Table A.10 to find that ω^2 must be about .3 or greater ($\omega^2 = .2$ is associated with a power of only .45 under these conditions, whereas $\omega^2 = .3$ leads to a power of .78).

I mentioned above that one way to increase power in an ANOVA is to increase ω^2, which involves the same problems discussed in Chapter 10 in terms of increasing gamma. Because ω^2 depends on the overlap of the populations, increasing the relative separation of the population means by increasing the effectiveness of your treatments and/or holding subject-to-subject variability to a minimum is your best hope for increasing ω^2, but of course, such manipulations are not possible in all studies. A powerful way to reduce variability is to match subjects, as described for the matched *t*-test in Chapter 13; the application of matching to more than two treatments or conditions is described in Chapter 17.

As you learned in Chapter 10, increasing the sample size will always increase power (unless the null hypothesis is actually true, in which case power is zero regardless of sample size), but it is not always worthwhile to do so. Using a very large number of subjects in an ANOVA makes it fairly likely you will obtain a significant *F* in situations where ω^2 (and, therefore, the effect size) is too small to be of any practical importance. Bear in mind, however, that even when ω^2 is quite small, the fact that it is not zero may be of some theoretical interest and therefore may be worth demonstrating.

Testing Homogeneity of Variance

I mentioned in Chapter 9 (Section C) that the F ratio could be used to test homogeneity of variance. Now that you have learned something about the F distribution, I can go into greater detail. When performing a **homogeneity of variance test** in the two-group case, whether to justify a pooled-variance t-test or to determine whether an experimental treatment increases the variance of a population compared to a control procedure, the null hypothesis is that the two populations have the same variance ($H_0 = \sigma_1^2 = \sigma_2^2$). Therefore, both sample variances are considered independent estimates of the same population variance. The ratio of two such estimates will follow the F distribution if the usual statistical assumptions are satisfied.

For example, suppose that the variance of a control group of 31 subjects is 135, and the variance of 21 experimental subjects is 315. Because our F table contains only F ratios greater than 1.0, we take the larger of the two variances and divide by the smaller to obtain $F = 315/135 = 2.33$. To find the critical F, we need to know the df for both the numerator and denominator of the F ratio. Because we put the experimental group's variance in the numerator, $df_{num} = N - 1 = 21 - 1 = 20$. The $df_{denom} = N - 1 = 31 - 1 = 30$, because the control group's smaller variance was used in the denominator. Finally, we must choose alpha. If we set $\alpha = .05$, however, we cannot use the F table for that alpha; we must use the F table for $\alpha/2 = .025$. The reason we must use $\alpha/2$ is that, in contrast to ANOVA, this time we are really performing a two-tailed test. The fact is that if the control group had had the larger variance, our F ratio would have been less than 1.0, and possibly in the left tail of the distribution. Of course, in that case, we would have turned the ratio upside down to make F greater than 1.0, but it would have been "cheating" not to allow for alpha in both tails, since either variance could be the larger. (In ANOVA we would never turn the ratio upside down, so it is reasonable to perform a one-tailed test.)

Looking up F (20, 30) at $\alpha = .025$, we find that the critical F is 2.20. Because our calculated F is higher than the critical F, the null hypothesis that the population variances are equal (i.e., homogeneity of variance) must be rejected. A pooled t-test is not justified, and you can conclude that your experimental treatment alters the variance of the population. However, these conclusions are justified only if it is reasonable to make the required assumptions. The most important assumption is that the two populations follow normal distributions. Whereas the t-test and the ANOVA F-test are robust with respect to violations of the normality assumption, the major drawback of the F-test for homogeneity of variance is that it *is* strongly affected by deviations from normal distributions, especially when the samples are not large. This problem makes the F ratio suspect as a test for homogeneity of variance in most psychological experiments, and therefore it is rarely used.

The F-test discussed above can be modified to test homogeneity of variance for an experiment that includes more than two groups. An F ratio is formed by dividing the largest of the sample variances by the smallest; the resulting ratio is called F_{max} (Hartley, 1950), and it must be compared to a critical value from a special table. Hartley's F_{max} test is also sensitive to violations of the normality assumption, which is why many statisticians argue against its use. Alternative procedures, less sensitive to violations of the normality assumption, have been devised for testing homogeneity of variance in the two-group as well as the multigroup case (e.g., Levene, 1960; O'Brien, 1981). These tests are too complex to be explained here, but fortunately, some of these tests are presented automatically by statistical computer packages (e.g., SPSS performs Levene's test) when any t-test or ANOVA is requested. Unfortunately, none of these tests has much power when sample sizes are small, which of course is when these tests are most often needed.

As a general rule, when the sample sizes are equal or nearly equal and the largest sample variance is no more than four times the smallest, you can proceed with the ANOVA without bothering to test for homogeneity of variance. However, the more the sample sizes differ, the more you need to be concerned about considerable differences in sample variances. If a test of homogeneity of variance is deemed appropriate, and it turns out to be significant, you have the choice of adjusting the calculation of F and the critical F (in a manner analogous to what must be done for the separate-variances t-test) or transforming the data by using some function, like the natural logarithm or the square root. The latter procedure is preferred if the distributions are not at all similar to the normal distribution, and the transformation has the effect of normalizing the distributions. However, the details of such procedures (mentioned briefly in Chapter 3) are beyond the scope of this text.

Exercises

* 1. a) If $F = 5$ in a three-group experiment, what proportion of variance is accounted for (i.e., what is η^2) when the total number of subjects is 30? 60? 90?

   b) If $F = 5$ in a six-group experiment, what proportion of variance is accounted for (i.e., what is η^2) when the total number of subjects is 30? 60? 90?

2. a) If $F = 10$ in a three-group experiment, what proportion of variance is accounted for (i.e., what is η^2) when the total number of subjects is 30? 60? 90?

   b) If $F = 10$ in a six-group experiment, what proportion of variance is accounted for (i.e., what is η^2) when the total number of subjects is 30? 60? 90?

* 3. a) For the data in exercise 14B5, find eta squared (the proportion of variance in cookie consumption that is accounted for by color). Estimate omega squared for the same data (or explain why you cannot).

   b) Answer part a for exercise 14B6.

4. a) Find eta squared and estimate omega squared for the data in exercise 14B1.

   b) Answer part a for exercise 14B3. Explain your answer.

* 5. a) What is the power associated with a five-group experiment that has six subjects in each group and

30% of the variance accounted for, if $\alpha = .05$? If $\alpha = .01$?

b) What is the power for the same experiment if only 20% of the variance is accounted for and $\alpha = .05$? If $\alpha = .01$?

6. a) Approximately how many subjects per group are needed in a four-group experiment, if ω^2 is expected to be .1 and power must be at least .7 for a .05 test?

b) How many subjects per group would be needed if $\omega^2 = .2$?

*7. a) If you have three groups of 15 subjects each, and you want power to be at least .85 for a .05 test, how large does ω^2 have to be?

b) If you have only five subjects in each of the three groups, and you want power to be at least .70 for a .05 test, how large does ω^2 have to be?

8. In exercise 9B4, the standard deviation in diastolic blood pressure for 60 marathon runners was $s = 10$, and for the nonrunners it was $s = 8$. Use an F ratio to test the homogeneity of variance assumption for that study at the .05 level.

*9. In exercise 9B6, a sample of 12 anxiety neurotics had a standard deviation in enzyme concentration of .7, whereas the standard deviation for the 20 control subjects was .4.

a) Use an F ratio to test the homogeneity of variance assumption at the .05 level.

b) Would a pooled-variances t-test be justified in this case?

10. The incentive of a very high reward can lead to improved performance for some people, but the same incentive can lead to decreased performance in others when the performance being rewarded involves complex motor behavior (e.g., playing a video game). If this is true, we could expect the variance of performance scores to be larger in a high-pressure condition than a low-pressure condition for some tasks. To test this hypothesis, 30 teenagers were assigned to play a video game with a small chance of a very large reward, while another 30 teenagers played the same game for a certain chance of a rather small reward. The variances were 65 for the high-pressure group and 41 for the low-pressure group. Using alpha = .05, test the null hypothesis that the two populations share the same variance.

SUMMARY

The Important Points of Section A

1. A *one-way* ANOVA has only one independent variable, which can have any number of *levels*. Each group in the experiment represents a different level of the independent variable.

2. The significance of the ANOVA is tested with an *F ratio*. The *F ratio* follows the *F distribution* when the null hypothesis is true. The null hypothesis is that all of the population means are equal (H_0: $\mu_1 = \mu_2 = \mu_3 = \ldots$).

3. The denominator of the F ratio, usually called MS_w (or MS_{error}), is just the average of the variances of all the groups in the experiment. (A weighted average like that used for the pooled variance of the two-group *t*-test, must be used when the groups are not all the same size.)

4. The numerator of the *F* ratio, usually called MS_{bet}, is the variance of the group means multiplied by the sample size. (The formula gets a bit more complicated when the groups are not all the same size.)

5. The *F* distribution is a family of distributions that tend to be positively skewed and to have a mean that is near 1.0. The exact shape of the *F* distribution is determined by both the df for the numerator ($df_{bet} = k - 1$, where $k =$ the number of groups) and the df for the denominator ($df_w = N_T - k$, where N_T is the total number of subjects in all groups combined).

6. Knowing df_{bet} and df_w, as well as alpha, you can look up the critical *F* in Table A.7, A.8, or A.9. If the calculated *F* ratio is larger than the critical *F*, the null hypothesis can be rejected.

7. The *F*-test for ANOVA is always one-tailed, in that only *large* calculated *F* ratios (i.e., those in the positive, or righthand, tail of the *F* distribution) lead to statistical significance. *F* ratios less than 1, even if close to zero, only indicate that the sample means are unusually close together, which is not inconsistent with the null hypothesis that all the population means are equal.

8. The denominator of the *F* ratio contains only error variance (it is often called the *error term*), but the numerator contains both error variance and the treatment effect. When the *F* ratio is large enough to be significant, we reject the hypothesis that the treatment effect is zero.

9. A significant *F* tells us that there is a treatment effect (i.e., the population means are not all equal) but does not tell us which pairs of population means differ significantly. The procedures of Chapter 15 are used to test pairs of means.

The Important Points of Section B

To illustrate the calculation of a one-way independent-groups ANOVA, I will describe a hypothetical experiment involving four groups. Imagine that a social psychologist wants to know if information about a person affects the way someone judges the attractiveness of that person. This is a very general question, and it has many aspects that can be tested. For this example, I will focus on judgments made by women and create four specific conditions. In each condition a female subject looks at a photograph of a male and rates the male for attractiveness on a scale from 1 to 10. All subjects view the same photograph, which is chosen because the man pictured in it is about average in attractiveness. There are four experimental conditions, depending on the short "biography" that accompanies the photo: successful biography, failure biography, neutral biography, or no

	Control Group	Neutral Bio	Failure Bio	Success Bio
Table 14.4	4 (16)	5 (25)	4 (16)	6 (36)
	5 (25)	6 (36)	5 (25)	5 (25)
	4 (16)	5 (25)	3 (9)	7 (49)
	3 (9)	4 (16)	4 (16)	6 (36)
	6 (36)	4 (16)	2 (4)	5 (25)
	22 (102)	24 (118)	5 (25)	6 (36)
			23 (95)	35 (207)

biography at all. The effects of these four conditions can be explored by performing a one-way independent-groups ANOVA.

In this example, only six subjects will be assigned to each condition to reduce the amount of calculation. To illustrate the case of unequal sample sizes, imagine that one subject from each of two conditions had to be eliminated, because the researcher discovered later that the subject actually knew the person in the photograph. The final ns are $n_1 = 5$, $n_2 = 5$, $n_3 = 6$, and $n_4 = 6$. The attractiveness rating given by each subject, along with the squared values (in parentheses) and sums for each column, appear in Table 14.4.

I will use the raw-score formulas to calculate the F ratio; I will begin by finding SS_w and SS_{bet}, using Formulas 14.9 and 14.11, respectively. Because one of the terms is the same in both formulas, only three different terms must be calculated: ΣX^2, T_T^2/N_T, and $\Sigma(T_i^2/n_i)$.

$$\Sigma X^2 = 102 + 118 + 95 + 207 = 522$$

$$T_T = 22 + 24 + 23 + 35 = 104$$

$$T_T^2/N_T = 104^2/22 = 10{,}816/22 = 491.64$$

$$\Sigma \frac{T_i^2}{n_i} = \frac{22^2}{5} + \frac{24^2}{5} + \frac{23^2}{6} + \frac{35^2}{6} = \frac{484}{5} + \frac{576}{5} + \frac{529}{6} + \frac{1225}{6}$$

$$= 96.8 + 115.2 + 88.17 + 204.17$$

$$= 504.33$$

Now we have the values needed for Formula 14.9 to find SS_w:

$$SS_w = \Sigma X^2 - \Sigma\left(\frac{T_i^2}{n_i}\right) = 522 - 504.33 = 17.67$$

We also have the values needed to find SS_{bet} with Formula 14.11:

$$SS_{bet} = \Sigma\left(\frac{T_i^2}{n_i}\right) - \frac{T_T^2}{N_T} = 504.33 - 491.64 = 12.7$$

Table 14.5

Source	SS	df	MS	F	p
Between-groups	12.7	3	4.23	4.31	<.05
Within-groups	17.67	18	.98		
Total	30.36	21			

The next step is to find the MS components:

$$MS_{bet} = SS_{bet}/df_{bet} = 12.7/3 = 4.23$$
$$MS_w = SS_w/df_w = 17.67/18 = .98$$

Finally, we can calculate the *F* ratio:

$$F = MS_{bet}/MS_w = 4.23/.98 = 4.31$$

In order to find the appropriate critical *F* value (or the MS components), we need to know the df for both the numerator and the denominator of the *F* ratio. First we note that *k* (the number of groups) = 4. From Formula 14.4A, we find that $df_{bet} = k - 1 = 4 - 1 = 3$. Next we note that the total number of measurements is $N_T = 5 + 5 + 6 + 6 = 22$. From Formula 14.4B, we find that $df_w = N_T - k = 22 - 4 = 18$. Therefore, we look down the third column of Table A.7 until we come to the 18th row to find that the critical *F* = 3.16. Because the calculated *F* (4.31) is greater than the critical *F*, the null hypothesis can be rejected at the .05 level. Our hypothetical researcher can reject the hypothesis that all of the population means are equal, but without further testing (see Chapter 15), she cannot say which pairs of population means are different from each other. The results of the above ANOVA can be summarized as shown in Table 14.5.

The Important Points of Section C

1. A multigroup experiment can be viewed as a form of regression, in which each group mean serves as the prediction for the scores in that group. Deviations from the grand mean can be divided into deviations from the group mean (the unexplained portion) and deviations of the group mean from the grand mean (the explained portion).

2. Squaring the deviation of each score from its own group mean and adding the squared values yields the unexplained SS, which is the SS_w of ANOVA. The explained SS is the same as SS_{bet}.

3. The proportion of variance accounted for in the results of an experiment is given by SS_{bet}/SS_{total}, which is designated as η^2 (eta squared) and is identical to r_{pb}^2 in the two-group case.

4. Like r_{pb}^2, η^2 is a biased estimate of ω^2, which is the proportion of variance explained or accounted for in the entire population. The formula for η^2 can be modified to yield a much less biased estimate of ω^2 for values of F greater than 1.

5. Either η^2 or the estimate of ω^2 can serve as a very useful supplement to the calculated F ratio, because the magnitude of F provides information only about statistical significance but gives no indication about effect size.

6. Once alpha is set, the power of the F-test in ANOVA depends on the expected F value (i.e., on the *noncentrality parameter*), and the expected F depends on both ω^2 and the sample sizes.

7. As in the case of the t-test, power can be increased by increasing alpha (at the expense of increasing the Type I error rate) or by increasing the sample sizes. Power will also increase if it is possible to increase ω^2 (i.e., the effect size).

8. The F ratio can be used to test homogeneity of variance when there are two or more groups. Unfortunately, this test is sensitive to violations of the assumption that the populations are normally distributed and is therefore rarely used.

9. More robust alternative procedures have been devised for testing homogeneity of variance, but there is no universal agreement on the usefulness of these tests. In general, it is recommended that you proceed with the ANOVA if your sample sizes are equal or nearly equal and no sample variance is more than four times as large as another. If your sample sizes *and* your variances differ considerably, you should consider using an adjusted F or a data transformation.

Definitions of Key Terms

One-way analysis of variance (ANOVA) An analysis of variance in which there is only one independent variable.

Factor A factor is an independent variable and has at least two *levels*. Each group in a one-way ANOVA corresponds to a different level of one factor.

Mean-square-within (MS_w) An estimate of the population variance based on the variances within each sample of the experiment. It serves as the denominator in an independent-groups ANOVA and is often referred to as the *error term*.

Mean-square-between (MS_{bet}) When the null hypothesis is true, MS_{bet} is an estimate of the population variance based on the variance of the sample means and the sample sizes. When the null hypothesis is not true, the size of MS_{bet} depends on the magnitude of both the *treatment effect* and the *error variance*.

F ratio A ratio of two independent estimates of the same population variance that follows one of the F *distributions*. It can be used to test homogeneity of variance of an ANOVA.

***F* distribution** A mathematical distribution that is followed by the ratio of two independent estimates of the same population variance. The shape of the distribution depends on the degrees of freedom for both the numerator and the denominator.

Grand mean The mean of all the measurements in a study, regardless of group. It can be found by dividing the grand total by N_T.

Grand total The sum of all the measurements in a study, regardless of group.

Summary table A table that displays the sum of squares (SS), mean squares (MS), and degrees of freedom (df) according to source (i.e., between-groups, within-subjects, or total of both) in an ANOVA. The *F* ratio and *p* value (or significance level) are usually displayed, as well.

Eta squared (η^2) A squared correlation coefficient used to represent the proportion of variance accounted for by the independent variable in the results of a particular ANOVA. It is the same as r_{pb}^2 in the two-group case.

Noncentral *F* distribution When the null hypothesis is not true, and therefore the treatment effect in an ANOVA is not zero, the *F* ratio will not follow the *F* distribution but rather a related distribution called the noncentral *F.* The mean of the noncentral *F* distribution depends on a noncentrality parameter.

Homogeneity of variance test A comparison of sample variances used to test the assumption that two or more populations have the same variance. The simplest test involves forming a ratio of two sample variances, but more complex and more accurate tests have been devised.

Key Formulas

The *F* ratio (in terms of two independent population variance estimates):

$$F = \frac{MS_{bet}}{MS_w} \hspace{3cm} \text{Formula 14.3}$$

The degrees of freedom associated with the numerator of the *F* ratio (where k is the number of groups):

$$df_{bet} = k - 1 \hspace{3cm} \text{Formula 14.4A}$$

The degrees of freedom associated with the denominator of the *F* ratio (where N_T is the total number of measurements):

$$df_w = N_T - k \hspace{3cm} \text{Formula 14.4B}$$

The F ratio when all the groups are the same size (means and variances have already been calculated):

$$F = \frac{N \sum (\overline{X}_i - \overline{X}_G)^2/(k-1)}{\sum s^2/k}$$

Formula 14.5

Mean square within-groups estimate (variances have already been calculated):

$$MS_w = \frac{\sum (n_i - 1)s_i^2}{df_w}$$

Formula 14.6

Mean square between-groups estimate (means have already been calculated):

$$MS_{bet} = \frac{\sum n_i(\overline{X}_i - \overline{X}_G)^2}{k-1} = \frac{\sum n_i(\overline{X}_i - \overline{X}_G)^2}{df_{bet}}$$

Formula 14.7

Mean square within-groups estimate (SS has already been calculated for each group):

$$MS_w = \frac{SS_1 + SS_2 + SS_3 + \cdots + SS_k}{n_1 + n_2 + n_3 + \cdots + n_k - k} = \frac{\sum SS_i}{N_T - k} = \frac{SS_w}{df_w}$$

Formula 14.8

Sum of squares within groups, raw-score form (T_i is the total for one particular group):

$$SS_w = \sum X^2 - \sum \left(\frac{T_i^2}{n_i} \right)$$

Formula 14.9

Mean square between-groups estimate (SS_{bet} has already been calculated):

$$MS_{bet} = \frac{SS_{bet}}{df_{bet}}$$

Formula 14.10

Sum of squares between groups—raw-score form (T_T is the total of all the observations):

$$SS_{bet} = \sum \left(\frac{T_i^2}{n_i} \right) - \frac{T_T^2}{N_T}$$

Formula 14.11

The total sum of squares, in terms of its two components:

$$SS_{tot} = SS_{bet} + SS_w$$

Formula 14.12

The total sum of squares (raw-score form):

$$SS_{tot} = \Sigma X^2 - \frac{T_T^2}{N_T}$$ Formula 14.13

Eta squared (the proportion of variance accounted for in the sample data):

$$\eta^2 = \frac{SS_{bet}}{SS_{tot}}$$ Formula 14.14

Eta squared (in terms of the *F* ratio and degrees of freedom):

$$\eta^2 = \frac{df_{bet}F}{df_{bet}F + df_w}$$ Formula 14.15

Estimate of omega squared (the proportion of variance accounted for in the population):

$$\text{Estimated } \omega^2 = \frac{SS_{bet} - (k-1)MS_w}{SS_{tot} + MS_w}$$ Formula 14.16

15

CHAPTER

You will need to use the following from previous chapters:

Symbols
$\overline{X}$: Mean of a sample
k: The number of groups in a one-way ANOVA
s_p^2: The pooled variance
MS_w: Mean square within-groups
(denominator from the formula for a one-way ANOVA)

Formulas
Formula 9.5B: The pooled-variance t-test

Concepts
Homogeneity of variance
Type I and Type II errors

A

**CONCEPTUAL
FOUNDATION**

In Section A of Chapter 14, I described an experiment comparing the effects on illness of three treatments: a placebo, vitamin C, and a multivitamin supplement. In that example I rejected the null hypothesis that all three population means were equal, and I mentioned that additional tests would be required to discern which pairs of means were significantly different. A significant F in the one-way ANOVA does not tell us whether the multivitamin treatment is significantly different from vitamin C alone, or whether either vitamin treatment is significantly different from the placebo. The obvious next step would be to compare each pair of means with a t-test, performing three t-tests in all. This procedure would not be unreasonable, but it can be improved upon somewhat, as I will describe shortly. Performing all of the possible t-tests becomes more problematic as the number of conditions or groups increases. The disadvantages of performing many t-tests for one experiment, and particularly the various procedures that have been devised to modify those t-tests, comprise the main topic of this chapter.

The Number of Possible t-Tests

In order to understand the main drawback of performing multiple t-tests, consider an example of a multigroup study in which the null hypothesis is very likely to be true. Imagine a fanciful researcher who believes that the IQ of an adult depends to some extent on the day of the week on which that person was born. To test this notion, the researcher measures the mean IQ of seven different groups: one group of people who were all born on a Sunday, another group of people who were all born on a Monday, etc. As mentioned in the previous chapter, the number of possible t-tests when there are seven groups is 21. Let us see how that number can be found easily. When picking a pair of groups for a t-test, there are seven possible choices for the first member of the pair (any one of the seven days of the week). For each of those seven choices, there are six possibilities for the second member of the pair, so there are $7 \cdot 6 = 42$ pairs in all. However, half of those pairs are the same as the other half but in reverse order. For example, picking Monday first and then Thursday gives the same pair for a t-test as picking Thursday first and then Monday. Therefore there are $42/2 = 21$ different t-tests. The general formula for finding the number of possible t-tests is Formula 15.1,

$$k(k - 1)/2 \qquad\qquad \text{Formula 15.1}$$

where k is the number of groups.

Experimentwise Alpha

Now suppose that our researcher does not know about ANOVA and just goes ahead and performs all 21 t-tests, each at the .05 level. Assuming that all seven population means are indeed equal, the null hypothesis will be true for each of the 21 t-tests. Therefore, if any of the t-tests attains statistical significance (e.g., Monday turns out to be significantly different from Thursday), the researcher has made a Type I error. The researcher, not knowing that the null hypothesis is true, might try to publish the finding that, for example, people born on Mondays are smarter than those born on Thursdays—which, of course, would be a misleading "false alarm." Even if only one of the 21 t-tests leads to a Type I error, we can say that the "experiment" has produced a Type I error, and this is something researchers would like to prevent. The probability that an experiment will produce *any* Type I errors is called the **experimentwise alpha.** (Note that it is becoming increasingly popular to use the term "familywise alpha" because a family of tests can be more precisely defined. At the introductory level, however, this distinction is not important, so I will continue to use the term "experimentwise" because of its mnemonic value.) When t-tests are performed freely in a multigroup experiment, the experimentwise alpha (α_{EW}) will be larger than the alpha used for each t-test (the testwise α). Furthermore, α_{EW} will increase as the number of groups increases, because of the increasing number of opportunities to make a Type I error.

You can get an idea of how large α_{EW} can become as the number of groups increases by considering a simple case. Suppose that a researcher repeats the same totally ineffective two-group experiment (i.e., $\mu_1 = \mu_2$) 21 times. What is the chance that the results will attain significance one or more times (i.e., what is the chance that the researcher will make at least one Type I error)? The question is very tedious to answer directly—we would have to find the probability of making one Type I error, and then the probability of exactly two Type I errors, up to the probability of committing a total of 21 Type I errors! It is easier to find the probability of making *no* Type I errors; subtracting that probability from 1.0 gives us the probability of making one or more Type I errors. We begin by finding the probability of *not* making a Type I error for just one t-test. If H_0 is true, and alpha $= .05$, then the probability of not making a Type I error is just $1 - .05 = .95$. Now we have to find the probability of *not* making a Type I error 21 times in a row. If each of the 21 t-tests is independent of all the others (i.e., the experiment is repeated with new random samples each time), the probabilities are multiplied (according to the multiplication rule described in Chapter 5, Section C). Thus the probability of *not* making a Type I error on 21 independent occasions is $.95 \cdot .95 \cdot .95 \ldots$ for a total of 21 times, or .95 raised to the 21st power ($.95^{21}$), which equals about .34. So the probability of making at least one Type I error among the 21 tests is $1 - .34 = .66$, or nearly two thirds. The general formula is

$$\alpha_{EW} = 1 - (1 - \alpha)^j \qquad \qquad \text{Formula 15.2}$$

where j is the number of independent tests. Formula 15.2 does not apply perfectly to multiple t-tests within the same multigroup experiment, because the t-tests are not all mutually independent (see Section C), but it does give us an idea of how large α_{EW} can get when many t-tests are performed.

Complex Comparisons

The above calculation of α_{EW} should make it clear why it is not acceptable to perform multiple t-tests without performing some other procedure to reduce α_{EW}. The fanciful researcher in the above example would have had much more than a .05 chance of finding at least one significant difference in IQ between groups from different days of the week, even assuming that the null hypothesis is true. Before I can describe the various procedures that have been devised to keep α_{EW} at a reasonable level, I need to introduce some new terms. For instance, following an ANOVA with a t-test between two of the sample means is an example of a *comparison*. When the comparison involves only two groups it can be called a **pairwise comparison,** so this term is another way of referring to a t-test. **Complex comparisons** involve more than two means. As an example, imagine that the fanciful researcher mentioned above suggests that people born on the weekend are smarter than those born during the week. If the average of the Saturday and Sunday means is compared to the average of the remaining five days, the result is a complex comparison. This chapter will concentrate on pairwise comparisons; complex comparisons will be dealt with briefly in Section C. The α used for each test that follows an ANOVA can be called the **alpha per comparison,** or α_{pc}. (Because this term is more commonly used than the term testwise α and just as easy to remember, I will adopt it.) As you will see, adjusting α_{pc} is one way to reduce α_{EW}.

Planned Comparisons

Another important distinction is between comparisons that are planned before running a multigroup experiment and comparisons that are chosen after seeing the data. Comparisons that are planned in advance are called **a priori comparisons** and do not involve the same risk of a high α_{EW} as **a posteriori comparisons,** that is, comparisons a researcher decides on after inspecting the various sample means. Because a priori comparisons tend to be less commonly employed and may involve complex rules, they will be discussed only in Section C. The bulk of this chapter will be devoted to a posteriori comparisons, which are more often called *post hoc* (i.e., after the fact) comparisons.

Fisher's Protected *t*-Tests

In the days-of-the-week example I described a researcher who didn't know about ANOVA, in order to demonstrate what happens when multiple *t*-tests are performed freely, without having obtained a significant ANOVA. A real researcher would know that something has to be done to keep α_{EW} from becoming too high. The simplest procedure for keeping down the experimentwise alpha is not to allow multiple *t*-tests unless the *F* ratio of the one-way ANOVA is statistically significant. If this procedure were adopted, the days-of-the-week experiment would have only a .05 chance (assuming that alpha = .05) of producing a significant *F*; only about 5 out of 100 totally ineffective experiments would ever be analyzed with multiple *t*-tests. It is true that once a researcher is "lucky" enough to produce a significant *F* with a totally ineffective experiment, there is a good chance at least one of the multiple *t*-tests will also be significant, but demanding a significant *F* means that 95% (i.e., $1 - \alpha$) of the totally ineffective experiments will never be followed up with *t*-tests at all.

The procedure of following only a significant ANOVA with *t*-tests was invented by Fisher (1951), and therefore the *t*-tests are called **Fisher's protected *t*-tests.** The *t*-tests are "protected" in that they are not often performed when the null hypothesis is actually true, because a researcher must first obtain a significant *F*. Also, the *t*-tests are calculated in a way that is a little different from ordinary *t*-tests, and more powerful, as you will soon see. In order to explain the formula for calculating Fisher's protected *t*-tests, I will begin with Formula 9.5B, which appears below (without the $\mu_1 - \mu_2$ term in the numerator):

$$t = \frac{(\overline{X}_1 - \overline{X}_2)}{\sqrt{s_p^2(1/N_1 + 1/N_2)}}$$

The use of s_p^2 indicates that we are assuming that there is homogeneity of variance, so that the pooling of variances is justified. If the assumption of homogeneity of variance is valid for the entire multigroup experiment (i.e., the variances of all the populations are equal), then MS_w (also called MS_{error}) is the best estimate of the common variance and can be used in place of s_p^2 in Formula 9.5B. In particular, I will assume that for the days-of-the-week experiment, pooling all seven sample variances (i.e., MS_w) gives a better estimate of σ^2 than pooling the sample variances for only the two samples being compared in each *t*-test that follows the ANOVA. Substituting MS_w for s_p^2 yields Formula 15.3,

$$t = \frac{(\overline{X}_i - \overline{X}_j)}{\sqrt{MS_w(1/n_i + 1/n_j)}} \qquad \text{Formula 15.3}$$

where the subscripts i and j indicate that any two sample means can be compared. Note that in Formula 15.3 I have switched to lowercase n, in order to be consistent with the notation used in Chapter 14, where a lowercase n indicates the number of subjects in one sample and an uppercase N with a subscript T indicates the total number of subjects in an experiment.

If homogeneity of variance cannot be assumed, there is no justification for using MS_w. If homogeneity of variance cannot be assumed for a particular pair of conditions *and* the sample sizes are not equal, some form of separate-variances t-test must be performed for that pair (see Chapter 9, Section C). (Because matters can get rather complicated, in that case, I will deal only with analyses for which homogeneity of variance can be assumed for all pairs.) If all of the samples in the ANOVA are the same size, then both n_i and n_j in Formula 15.3 can be replaced by n without a subscript, producing Formula 15.4:

$$ t = \frac{(\overline{X}_i - \overline{X}_j)}{\sqrt{2MS_w/n}} \qquad \text{Formula 15.4} $$

Note that the denominator of Formula 15.4 is always the same, regardless of which two groups are being compared. The constancy of the denominator when all sample sizes are equal leads to a simplified procedure, called *Fisher's least significant difference (LSD) test,* which will be described more fully in Section B.

The advantage of using Formula 15.3 (or Formula 15.4) for follow-up t-tests, instead of Formula 9.5B, is that the critical t used is based on df_w, which is larger (leading to a smaller critical t) than the df for just the two groups involved in the t-test. However, the whole procedure of using protected t-tests has a severe limitation, which must be explained, if you are to understand the various alternative procedures.

Complete Versus Partial Null Hypotheses

The problem with Fisher's protected t-tests is that the "protection" that comes from finding a significant F only applies fully to totally ineffective experiments, such as the days-of-the-week example. By "totally ineffective" I mean that the null hypothesis of the ANOVA—that all of the population means are equal—is actually true. The null hypothesis that involves the equality of *all* the population means represented in the experiment is referred to as the **complete null hypothesis.** Fisher's protected t-test procedure keeps α_{EW} down to .05 (or whatever α is used for the ANOVA) only for experiments for which the complete null hypothesis is true. The "protection" does not work well if the null hypothesis is only partially true, in which case α_{EW} (and thus the rate of Type I errors) can easily become unreasonably large. To illustrate the limitations of Fisher's procedure, I will pose an extreme example below.

The Partial Null Hypothesis

Imagine that a psychologist believes that all phobics can be identified by some physiological indicator, regardless of the type of phobia they suffer from. Six types of phobics (social phobics, animal phobics, agoraphobics, claustrophobics, acrophobics, and people who fear knives) were measured on some relevant physiological variable, as was a control group of nonphobic subjects. In such a study, we could test to see if all the phobics combined differ from the control group (a complex comparison), and we could also test for differences among the different types of phobics. Depending on the variable chosen, there are many ways these seven groups could actually differ, but I will consider one simple (and extreme) pattern to make a point. Suppose that, for the physiological variable chosen, the phobic population means are different from the control population mean, but that all six phobic population means are equal to each other (i.e., H_0: $\mu_1 = \mu_2 = \mu_3 = \mu_4 = \mu_5 = \mu_6 \neq \mu_7$). In this case, the complete null hypothesis is not true, but a **partial null hypothesis** *is* true. Next, suppose that the psychologist dutifully performs a one-way ANOVA. If the control population differs only very slightly from the phobic populations, the chance of attaining a significant ANOVA may be only a little greater than alpha. However, it is quite possible that the control population differs greatly from the phobics (even though the phobics do not differ from each other), so that the ANOVA is likely to be significant.

If our psychologist finds her F ratio to be significant at the .05 level, and she adheres to the Fisher protected t-test strategy, she will feel free to conduct all the possible pairwise comparisons, each with $\alpha_{pc} = .05$. This strategy includes testing all possible pairs among the six phobic groups which (using Formula 15.1) amounts to $6(6 - 1)/2 = 30/2 = 15$ pairwise comparisons, for which the null hypothesis is true in each case. If these 15 t-tests were all mutually independent, the α_{EW} (using Formula 15.2) would become $1 - (1 - .05)^{15} = 1 - .95^{15} = 1 - .46 = .54$. Although these t-tests are not totally independent, it should be clear that there is a high chance of committing a Type I error, once the decision to perform all the t-tests has been made. (The remaining six t-tests involve comparing the control group with each of the phobic groups, and therefore cannot lead to any Type I errors in this example.) Note that without the control group, the complete null would be true, and the chance of attaining a significant ANOVA would only be alpha. There would be no drawback to using Fisher's procedure. Unfortunately, the addition of the control group can make it relatively easy to attain a significant ANOVA, thus removing the "protection" involved in Fisher's procedure and allowing α_{EW} to rise above the value that had been set for the overall ANOVA.

The Case of Three Groups

The one case for which Fisher's procedure gives adequate protection even if the complete null is *not* true is when there are only three groups. In that case the only kind of partial null you can have is one in which two population means are equal and a third is different. A significant ANOVA then leads to the testing of at most only *one* null hypothesis (i.e., the two population means that are equal), so there is no buildup of α_{EW}. However, Fisher's procedure allows a buildup of α_{EW} when there are more than three groups and the complete H_0 is not true—and the greater the number of groups, the larger the buildup in α_{EW}. For this reason, the protected *t*-test has gotten such a bad reputation that researchers are reluctant to use it even in the common three-group case for which it is appropriate. This is unfortunate, because the Fisher procedure has the most power in the three-group case, as will become clear as we analyze the various alternatives.

Tukey's HSD Test

In order to provide complete protection—that is, to keep α_{EW} at the value chosen, regardless of the number of groups or whether the null is completely or only partially true—Tukey devised an alternative procedure for testing all possible pairs of means in a multigroup experiment. His procedure is known as **Tukey's honestly significant difference (HSD) procedure,** in contrast to Fisher's least significant difference (LSD) test, mentioned above. (The term *difference,* as used in HSD and LSD, will be clarified in the next section.) The implication is that Fisher's procedure involves some "cheating" because it provides protection only when used with experiments for which the complete null hypothesis is true. To understand the protection required when all possible pairs of means are being tested, imagine that you have conducted a multigroup experiment and you are looking at the different sample means. If you are hoping to find at least one pair of means that are significantly different, your best shot is to compare the largest sample mean with the smallest. It is helpful to understand that in terms of your chance of making at least one Type I error when the complete null is true, testing the smallest against the largest mean is the same as testing all possible pairs. After all, if the two sample means that differ most do not differ significantly, then none of the other pairs will, either. If a procedure can provide protection against making a Type I error when comparing the smallest to the largest mean, then you are protected against Type I errors when testing all possible pairs. This is the strategy behind Tukey's procedure. The test that Tukey devised is based on the distribution of a statistical measure called the *studentized range statistic,* which I will explain next.

The Studentized Range Statistic

The *t* distribution is based on what happens when you draw two samples from populations with the same mean and find the difference between the two sample means (and then estimate σ^2 from the sample data). If you draw three or more samples under the same conditions and look at the difference between the smallest and largest means, this difference will tend to be greater than the difference you found when drawing only two samples. In fact, the more samples you draw, the larger the difference will tend to be between the largest and smallest means. These differences are due only to chance (since all the population means are equal). To protect ourselves from thinking that they are significant we need a critical value that accounts for the larger differences that tend to be found when drawing more and more samples. Fortunately, the distribution of the **studentized range statistic** allows us to find the critical values we need, adjusted for the number of samples in the multigroup experiment. (The statistic is "studentized" in that, like the ordinary t value—which, as you recall, is sometimes called "Student's t"—it relies on sample variances in its denominator, instead of population variances, which are usually unknown.) However, these critical values assume that all of the samples are the same size, so Formula 15.5 for performing t-tests according to Tukey's procedure looks very much like Formula 15.4 for Fisher's protected t-test with equal ns, as shown below:

$$q = \frac{(\overline{X}_i - \overline{X}_j)}{\sqrt{MS_w/n}}$$

Formula 15.5

The letter q stands for the critical value of the studentized range statistic; critical values are listed in Table A.11 in Appendix A. The use of this table will be described in Section B. You may have noticed that the number 2, which appears in the denominator of Formula 15.4, is missing in Formula 15.5. This does not represent a real difference in the structure of the test. For ease of computation Tukey decided to include the factor of 2 (actually $\sqrt{2}$, because the 2 appears under the square-root sign) in making Table A.11; thus the critical values of the studentized range statistic were each multiplied by the square root of 2 in order to produce the entries in Table A.11.

Advantages and Disadvantages of Tukey's Test

The advantage of Tukey's HSD procedure is that the alpha that is used to determine the critical value of q is the experimentwise alpha; no matter how many tests are performed, or what partial null is true, α_{EW} remains at the value set initially. If you choose to keep α_{EW} at .05, as is commonly done, then α_{pc} (the alpha for each pairwise comparison) must be reduced below .05, so that the accumulated value of α_{EW} from all possible tests does not exceed .05. The more groups, the more possible tests that can be performed, and the more α_{pc} must be reduced. The user of the HSD test does not have to determine the appropriate α_{pc}, however, because the critical value can be found directly from Table A.11 (larger critical values correspond, of course, to smaller α_{pc}). The way the critical values in Table A.11 vary as a function of the number of groups in the study will be discussed in Section B.

The disadvantage of the HSD test in the three-group case is that it results in a reduction in power, and therefore more Type II errors, as compared to the LSD test. When dealing with more than three groups, the LSD test remains more powerful than the HSD test for the same initial alpha, but this comparison is not entirely fair, in that the LSD test derives most of its extra power (when $k > 3$) from allowing α_{EW} to rise above the initially set value. By now you should be familiar with the fact that you can always increase the power of a statistical test by increasing alpha, but in doing so you are decreasing the rate of Type II errors by allowing a larger percentage of Type I errors. Because most researchers consider it unacceptable to increase power in a multigroup experiment by allowing α_{EW} to rise, the HSD test is generally preferred to the LSD test for more than three groups.

Statisticians frequently say that the Tukey test is more **conservative** than the Fisher procedure because it is better at keeping the rate of Type I errors down. All other things being equal, statistical tests that are more conservative are less powerful; as the Type I error rate is reduced, the Type II error rate (β) increases, which in turn reduces power ($1 - \beta$). Tests that are more powerful because they allow α_{EW} to build up are referred to as **liberal.** The Fisher protected *t*-test procedure is the most liberal (and for that reason the most powerful) way to conduct post hoc comparisons (other than unprotected *t*-tests). The Tukey HSD procedure is one of the more conservative.

Unlike Fisher's LSD test, the HSD test does not require that the overall ANOVA be tested for significance. Although it is unlikely, it is possible for the HSD test to find a pair of means significantly different when the overall ANOVA was not significant. Requiring a significant ANOVA before using the HSD test would reduce its power slightly—and unnecessarily, because it is already considered adequately conservative. On the other hand, it is also possible, but unlikely, to find no significant pairwise comparisons with the HSD or even the LSD test when following up a significant ANOVA. The only guarantee that follows from the significance of an ANOVA is that some comparison among the means will be significant, but the significant comparison could turn out to be a complex one. As mentioned above, complex comparisons will be discussed further in Section C.

Finally, one minor disadvantage of Tukey's HSD test is that its accuracy depends on all of the samples being the same size. The small deviations in sample sizes that most often occur accidentally in experimental studies can be dealt with by a procedure that will be described in Section C. If there are large discrepancies in sample sizes, some alternative post hoc comparison procedure must be used.

The Advantage of Planning Ahead

There is an advantage to planning particular comparisons before collecting the data, as compared to performing all possible comparisons or selecting comparisons after seeing the data (which is essentially the same as performing all possible comparisons). This advantage is similar to the advantage involved in planning a one-tailed instead of a two-tailed test: You can use a smaller critical value. As in the case of the one-tailed test, planned comparisons are less appropriate when the research is of an exploratory rather than a confirmatory nature. Moreover, the validity of planned comparisons depends on a "promise" that the researcher truly planned the stated comparisons and no others; researchers are aware of how easy it is, after seeing the data, to make a comparison that appears to have been planned. The advantages and limitations of a priori tests will be discussed further in Section C.

Exercises

*1. How many different pairwise comparisons can be tested for significance in an experiment involving a) five groups? b) Eight groups? c) Ten groups?

2. If a two-group experiment were repeated with a different independent pair of samples each time, and the null hypothesis were true in each case ($\alpha = .05$), what would be the probability of making at least one Type I error a) in five repetitions? b) In ten repetitions?

*3. In exercise 14A5, the introversion means and standard deviations for students seated in three classroom locations ($n = 80$ per group) were as follows:

	Front	Middle	Back
$\bar{X}$	28.7	34.3	37.2
s	11.2	12.0	13.5

a) Use Formula 15.4 to calculate a t value for each pair of means.
b) Which of these t values exceed the critical t based on df_w, with alpha = .05?

4. Assume that the standard deviations from exercise 3 were doubled.
a) Recalculate the t value for each pair of means.
b) Which of these t values now exceed the critical t?
c) What is the effect on the t value of doubling the standard deviations?

*5. a) Recalculate the t values of exercise 3 above for a sample size of $n = 20$.
b) What is the effect on the t value of dividing the sample size by four?

6. Compared to performing ordinary t-tests after a multigroup experiment, Fisher's protected t-tests a) require that the samples be of equal size; b) Are less powerful; c) Use a smaller α_{pc}; d) Use a smaller critical value.

*7. Compared to Fisher's LSD test, Tukey's HSD test a) is more conservative; b) Is more powerful; c) Is less likely to keep α_{EW} from building up; d) Uses a smaller critical value.

8. What would be the implication for post hoc comparisons in a multigroup experiment if there were only two possibilities concerning the null hypothesis: either the complete null hypothesis is true (all population means are equal) or all the population means are different from each other (no partial null is possible)?
a) Tukey's test would become more powerful than Fisher's.
b) Fisher's protected t-tests would be sufficiently conservative.
c) Neither Fisher's nor Tukey's test would be recommended.
d) There would be no difference between the Fisher and Tukey procedures.

In Section A of Chapter 14, I described a three-group experiment to test the effects of vitamins on sick days taken off from work. The F ratio was significant, allowing us to reject the null hypothesis that the three population means were equal. However, the researcher would probably want to ask several more specific questions, involving two population means at a time, such as, is there a significant difference between the vitamin C group and the multivitamin group? Because the F ratio was significant, it is acceptable to answer such questions with Fisher's protected t-tests.

**BASIC
STATISTICAL
PROCEDURES**

Calculating Protected *t*-Tests

Protected *t*-tests can be calculated using Formula 15.3 or Formula 15.4, but first we need some pieces of information from the ANOVA. We need to know MS_w as well as the mean and size of each group. For the vitamin example, $MS_w = 9.17$, and all the groups are the same size: $n = 10$. The means are as follows: $\overline{X}_{Plac} = 9$, $\overline{X}_{VitC} = 7$, and $\overline{X}_{MVit} = 5.5$. Inspecting the means, you can see that the largest difference is between $\overline{X}_{Plac}$ and $\overline{X}_{MVit}$, so we test that difference first. When all of the groups are the same size, the largest difference gives us our best chance of attaining statistical significance. Because the samples being tested are the same size, we can use Formula 15.4 to find the *t* value that corresponds to $\overline{X}_{MVit}$ versus $\overline{X}_{Plac}$.

$$t = \frac{(\overline{X}_i - \overline{X}_j)}{\sqrt{2MS_w/n}} = \frac{9 - 5.5}{\sqrt{2(9.17)/10}} = \frac{3.5}{\sqrt{1.834}} = \frac{3.5}{1.354} = 2.58$$

The critical *t* for any protected *t*-test is determined by the df corresponding to MS_w, that is, df_w (which is also called df_{error}). In this case, $df_w = N_T - k = (3 \cdot 10) - 3 = 30 - 3 = 27$. From Table A.2, using $\alpha = .05$, two-tailed, we find that $t_{crit} = 2.052$. Because the calculated *t* (2.58) is greater than the critical *t*, the difference between $\overline{X}_{MVit}$ and $\overline{X}_{Plac}$ is declared statistically significant; we can reject the hypothesis that the two population means represented by these groups are equal. The next largest difference is between $\overline{X}_{VitC}$ and $\overline{X}_{Plac}$, so we test it next.

$$t = \frac{9 - 7}{\sqrt{2(9.17)/10}} = \frac{2}{\sqrt{1.834}} = \frac{2}{1.354} = 1.477$$

The critical *t* is, of course, the same as in the test above (2.052), but this time the calculated *t*(1.477) is less than the critical *t*, and therefore the difference cannot be declared significant at the .05 level.

You may have noticed some redundancy in the two tests above. Not only is the critical *t* the same, the denominator of the two *t*-tests (1.354) is the same. As you can see from Formula 15.4, the denominator will be the same for all the protected *t*-tests following a particular ANOVA if all of the samples are the same size. The fact that both the critical *t* and the denominator are the same for all of these tests suggests a simplified procedure. There has to be some difference between means that when divided by the constant denominator is exactly equal to the critical *t*. This difference is called **Fisher's least significant difference (LSD).** This relationship is shown below:

$$t_{crit} = \frac{LSD}{\sqrt{2MS_w/n}}$$

Calculating Fisher's LSD

If the difference between any two means is less than LSD, it will correspond to a t value that is less than t_{crit} and therefore the difference will not be significant. Any difference of means greater than LSD *will* be significant. In order to calculate LSD, it is convenient to solve for LSD in the expression above by multiplying both sides by the denominator, to produce Formula 15.6:

$$\text{LSD} = t_{crit} \sqrt{\frac{2MS_w}{n}} \qquad \text{Formula 15.6}$$

Note that LSD can be calculated only when all the sample sizes are the same, in which case the n in Formula 15.6 is the size of any one of the samples. Let us calculate LSD for the vitamin ANOVA.

$$\text{LSD} = 2.052 \sqrt{\frac{2(9.17)}{10}} = 2.052\sqrt{1.834} = 2.052(1.354) = 2.78$$

Once LSD has been calculated for a particular ANOVA, there is no need to calculate any t-tests. All you need to do is calculate the difference between every pair of sample means and compare each difference to LSD. A simple way to display all the differences is to make a table, like Table 15.1 for the vitamin experiment. The asterisk next to the difference between $\bar{X}_{Plac}$ and $\bar{X}_{MVit}$ indicates that this difference is larger than LSD (2.78) and is therefore significant, whereas the other differences are not. It is easy to see that in an experiment with many groups, the calculation of LSD greatly streamlines the process of determining which pairs of means are significantly different. However, if the groups are not all the same size, you will need to perform each protected t-test separately, using Formula 15.3. If there are differences in sample size, and it is not reasonable to assume homogeneity of variance, you cannot use MS_w, and each protected t-test must be performed as a separate-variances t-test, as described in Chapter 9.

Table 15.1

	$\bar{X}_{VitC}$	$\bar{X}_{MVit}$
$\bar{X}_{Plac}$	2	3.5*
$\bar{X}_{VitC}$		1.5

Calculating Tukey's HSD

For the vitamin experiment, you could calculate Tukey's HSD instead of LSD. I don't recommend using HSD when there are only three groups because in that case the procedure is unnecessarily conservative. However, for comparison purposes I will calculate HSD for the example above. We begin with Formula 15.5, replacing the difference of means with HSD and the value of q with q_{crit}, as follows:

$$q_{crit} = \frac{HSD}{\sqrt{MS_w/n}}$$

Next we solve for HSD to arrive at Formula 15.7:

$$HSD = q_{crit}\sqrt{\frac{MS_w}{n}} \qquad\qquad \text{Formula 15.7}$$

To calculate HSD for the vitamin example, we must first find q_{crit} from Table A.11 (assuming $\alpha = .05$). We look down the column labeled 3, because there are three groups in the experiment, and we look down to the rows labeled 24 and 30. Because df_w (or df_{error}) = 27, which is midway between the rows for 24 and 30, we take as our value for q_{crit} a value that is midway between the entries for 24 (3.53) and 30 (3.49)—so q_{crit} is about equal to 3.51. (This value is approximate because we are performing linear interpolation, and the change in q is not linear.) The values for MS_w and n are the same, of course, as for the calculation of LSD. Plugging these values into Formula 15.7, we get

$$HSD = 3.51\sqrt{\frac{9.17}{10}} = 3.51(.958) = 3.36$$

Referring back to Table 15.1, you can see that the difference between $\bar{X}_{MVit}$ and $\bar{X}_{Plac}$ (3.5) is greater than HSD (3.36) and is therefore significant, according to the Tukey procedure. The other two differences are less than HSD and therefore not significant. In this case, our conclusions about which pairs of population means are different do not change when switching from Fisher's to Tukey's method. This will often be the case. However, note that HSD (3.36) is larger than LSD (2.78) and that the difference of 3.5 between the multivitamin and the placebo groups, which easily exceeded LSD, only barely surpassed HSD. Had there been two sample means that differed by 3, these two means would have been declared significantly different by Fisher's LSD, but not significantly different by Tukey's HSD. As discussed in the previous section, the Tukey procedure is more conservative (too conservative, in this case), and using it makes it harder for a pairwise comparison to attain significance.

Interpreting the Results of the LSD and HSD Procedures

Given that the F ratio for the vitamin ANOVA was just barely significant, it is not surprising that the largest difference of means was just barely greater than HSD. In fact, if you are using Tukey's HSD procedure, it is unusual to find a pair

of means that are significantly different when the overall ANOVA is *not* significant. Tukey's HSD procedure keeps α_{EW} down to about .05 (or whatever level is
used to determine q_{crit}), regardless of whether the ANOVA is significant at the
.05 level. On the other hand, it is not uncommon for pairs of sample means to
exceed LSD even when the overall ANOVA is not significant. That is why it is
important to have a significant F ratio before proceeding with the LSD test.
Otherwise, α_{EW} can get unacceptably large even for an experiment with only
three groups. With more than three groups, Fisher's LSD test is not recommended, for the reasons discussed in the previous section.

What are the implications of the statistical conclusions we drew from our
pairwise comparisons in the vitamin example? First, we have some confidence in
recommending multivitamins to reduce sick days. However, we cannot say with
confidence that vitamin C alone reduces sick days. A researcher might be
tempted to think that if multivitamins differ significantly from the placebo, but
vitamin C alone does not, then the multivitamins must be significantly better than
vitamin C. However, we have seen that the difference between vitamin C alone
and multivitamins is *not* statistically significant. It is not that we are asserting
that vitamin C alone is no different from the placebo, or that multivitamins are
no different from vitamin C alone. But we must be cautious; the differences just
mentioned are too small to rule out chance factors with confidence. It is only in
the comparison between multivitamins and the placebo that we have sufficient
confidence to declare that the population means are different.

Assumptions of the Fisher and Tukey Procedures

The Fisher and the Tukey tests rest on the same assumptions that underlie the
independent-groups *t*-test and one-way ANOVA, as described more fully in
Chapters 9 and 14, respectively: (1) independent, random samples; (2) normal
distribution for each population; and (3) homogeneity of variance. In addition,
the following considerations apply to each test separately.

For Fisher's Protected *t*-Tests:

1. It is assumed that the overall ANOVA is statistically significant.
2. Equal sample sizes are not required, but the procedure is simplified if LSD
   can be calculated.
3. The experimentwise alpha is kept to the level set for the overall ANOVA
   only when the complete null hypothesis is true, or there are only three
   groups.

For Tukey's HSD Procedure:

1. Equal sample sizes are assumed. If there are small differences in the sample
   sizes, the harmonic mean of the sample sizes (see Section C) can be used
   to replace n in Formula 15.7. If there are large differences in sample sizes,
   the HSD procedure is not recommended.

2. It is assumed that all possible pairwise comparisons are being performed, or that the researcher chose pairwise comparisons after seeing the data. If a relatively small set of pairwise comparisons can be planned in advance, a priori tests should be used instead (see Section C).

Other Procedures for Post Hoc Pairwise Comparisons

Scheffé Test

A particularly flexible way to conduct post hoc comparisons is to use the **Scheffé test.** The Scheffé test is versatile, because it can be used just as easily for complex comparisons as for pairwise comparisons. No special tables are required, as the F distribution can be used to test for significance. In fact, this is the only post hoc test that allows you to keep α_{EW} at a fixed level when performing not only any of the possible pairwise comparisons but any of the possible complex comparisons, as well. However, if you are interested only in pairwise comparisons, the Scheffé test is unnecessarily conservative; the Tukey HSD test is more powerful for pairwise comparisons and yet keeps α_{EW} at the level initially set (e.g., .05). This leads to a simple recommendation: If you are looking at any complex comparisons after seeing the data, use the Scheffé test for all of your post hoc tests. If you have no interest in complex comparisons, do not use the Scheffé test for any of your post hoc tests. By the way, you do not have to test the overall ANOVA for significance before using the Scheffé test. If the ANOVA is not significant, the Scheffé test will not yield any significant results, either. As mentioned above, if the ANOVA is significant, some comparison (not necessarily pairwise) will be significant, and the Scheffé test will help you find it.

The Newman-Keuls Test

The leading competitor of Tukey's HSD test is the procedure known as the **Newman-Keuls** (N-K) **test,** also called the *Student-Newman-Keuls test,* because its critical values come from the studentized range statistic. The major advantage of this test is that it is usually somewhat more powerful than HSD but more conservative than LSD. Therefore, many researchers consider the N-K test a good compromise between LSD and HSD. The chief disadvantage of the N-K test used to be that it was more complicated to apply. Rather than using the same critical value for each pairwise comparison, you place the means in order and use the *range* between any two of them (two adjacent means in the order have a range of 2, if there is one mean between them the range is 3, etc.) instead of the number of groups in the overall ANOVA to look up the critical value in Table A.11. Now that most computer packages supply the N-K test as an option after an ANOVA, the computational complexities are not important. However, a more serious drawback of the N-K test is that, unlike Tukey's HSD, it does not keep

α_{EW} at the level used to determine the critical value of the studentized range statistic. For that reason, conservative researchers do not recommend it. Nonetheless, its apparent edge over Tukey's HSD in power, together with a not too extreme inflation of α_{EW}, has made the N-K test fairly popular in the psychological literature.

Duncan's New Multiple Range Test

One test that is very similar in structure to the N-K test but far less conservative, and therefore much more powerful, is **Duncan's new multiple range test.** The major drawback of Duncan's test is that its power is derived from allowing α_{EW} to build up steadily as the number of groups increases, much as Fisher's protected *t*-tests do. Both Duncan's and Fisher's tests are considered too liberal by most researchers, especially when there are many groups, so neither test is encountered often in the literature.

Dunnett's Test

Section A described an example involving six different phobic groups and one nonphobic control group. If the experimenter wished to compare each of the phobic groups to the control group but did *not* wish to compare any of the phobic groups with each other, the best method for pairwise comparisons would be the one devised by Dunnett (1964). However, **Dunnett's test** requires the use of special tables of critical values, and the situation to which it applies rarely occurs, so I will not describe it here. Be aware, though, that Dunnett's test is performed by some computerized statistical packages, and, when it applies, it is the most powerful test available that does not allow α_{EW} to rise above its preset value.

Which Pairwise Comparison Procedure Should You Use?

Ultimately which procedure you use is a matter of judgment, which requires experience in the field. Mostly the choice is a matter of judging the relative disadvantages of Type I and Type II errors, but it is also a matter of judging which test best applies to your own particular circumstances. If your situation matches the one for which Dunnett's test was tailor-made, then that is the test you should use. If you are dealing with any complex comparisons, Scheffé's test is highly recommended. With only three groups, Fisher's protected *t*-tests are probably the best choice. When performing only pairwise comparisons with more than three groups, either the N-K test or Tukey's HSD would be the most reasonable, with Tukey's test having the edge for maintaining closer control over α_{EW}. In general, however, the most powerful *and* conservative way to conduct comparisons in a multigroup experiment is not to use any of the methods for post hoc comparisons, but rather to plan particular pairwise or complex comparisons in advance of collecting or looking at the data. Such a priori tests will be discussed in Section C.

Exercises

1. What is the critical value ($\alpha = .05$) of the studentized range statistic (q) for an experiment in which there are a) four groups of six subjects each? b) Six groups of four subjects each? c) Eight groups of 16 subjects each?

* 2. Use the results of the ANOVA you performed in exercise 14B4 to calculate Fisher's LSD and Tukey's HSD at the .05 level.

a) Which pairs of means exceed LSD?
b) Which pairs of means exceed HSD?
c) Which procedure seems to have more power in the three-group case?

3. Would it be permissible to follow the ANOVA you performed in exercise 14B5 with Fisher's protected t-tests? Explain.

* 4. In exercise 14A7, the following means and standard deviations were given as the hypothetical results of an experiment involving the effects of four different drugs ($n = 8$ subjects per group).

	Mari-juana	Ampheta-mine	Valium	Alcohol
X	7	8	5	4
s	3.25	3.95	3.16	2.07

a) Calculate Fisher's LSD ($\alpha = .05$), whether or not it is permissible.
b) Calculate Tukey's HSD ($\alpha = .05$).

5. Recalculate Fisher's LSD and Tukey's HSD for the data in exercise 4, assuming that the number of subjects per group was 16. a) What effect does increasing the number of subjects have on the size of LSD and HSD?

b) What conclusions can you draw from the LSD test?
c) Does the HSD test lead to different conclusions?
d) Which test is recommended in the four-group case and why?

* 6. a) Calculate Fisher's protected t-tests for each pair of groups in exercise 14B10, using alpha = .05.
b) What specific conclusions can you draw from the tests in part a? What do you know from these tests that you did not know just from rejecting the null hypothesis of the ANOVA?

7. In exercise 14B1, an experiment involving five different antidepressants yielded the following means and standard deviations: $\overline{X}_1 = 23$, $s_1 = 6.5$; $\overline{X}_2 = 30$, $s_2 = 7.2$; $\overline{X}_3 = 34$, $s_3 = 7$; $\overline{X}_4 = 29$, $s_4 = 5.8$; $\overline{X}_5 = 26$, $s_5 = 6$.

a) Assuming that none of the original subjects were lost (i.e., $n = 15$), calculate Tukey's HSD for this experiment.
b) Which pairs of means differ significantly?

* 8. A clinical psychologist is studying the role of repression in eight different types of phobias: four animal phobias (rats, dogs, spiders, and snakes) and four nonanimal phobias (fear of meeting strangers at a party, fear of speaking to groups, claustrophobia, and acrophobia). Sixteen subjects were sampled from each of the eight phobic groups, and their mean repression scores were as follows: $\overline{X}_{rat} = 42.8$; $\overline{X}_{dog} = 44.1$; $\overline{X}_{spider} = 41.5$; $\overline{X}_{snake} = 42.1$; $\overline{X}_{party} = 28.0$; $\overline{X}_{speak} = 29.9$; $\overline{X}_{claus} = 36.4$; $\overline{X}_{acro} = 38.2$. Assume that $MS_w = 18.7$.

a) Calculate Tukey's HSD at the .05 level.
b) Which pairs of means differ significantly?
c) What conclusions can you draw about the relationship between repression and type of phobia?

The Harmonic Mean

Although Tukey's HSD procedure is based on all samples being the same size, slight differences in sample size—as frequently result from the accidental loss of a few subjects from one group or another—can be tolerated. When the sample sizes differ slightly, the n in Formula 15.7 can be found by taking the harmonic mean of all the samples in the experiment. Chapter 10, Section C presented a simplified formula (Formula 10.9) that calculates the harmonic mean when only two groups are involved. The general formula for the harmonic mean, which applies to any number of groups, is given as Formula 15.8 below,

$$\text{Harmonic mean} = \frac{k}{\sum 1/n_i} \qquad \text{Formula 15.8}$$

where k is the number of groups. The result is the value used in place of n in Formula 15.7.

To demonstrate the use of this formula, I will use the example described in Chapter 14, Section D. In that experiment, four groups of subjects rated the attractiveness of a photograph under different conditions: the person in the photograph was described as successful; the person was described in negative terms, implying he was a failure; the person was described in neutral terms; or the person was not described at all. The ns were as follows: $n_1 = 5, n_2 = 5, n_3 = 6, n_4 = 6$. The harmonic mean of these four sample sizes is found by applying Formula 15.8:

$$\text{Harmonic mean} = \frac{4}{1/5 + 1/5 + 1/6 + 1/6} = \frac{4}{.2 + .2 + .167 + .167}$$
$$= \frac{4}{.733} = 5.45$$

Thus 5.45 is used as n in Formula 15.7, rather than the ordinary arithmetic mean (5.5). To continue with this example, recall that MS_w in the attractiveness experiment was equal to .98. Because df = 18 and there are four groups, $q_{.05} = 4.0$. Using Formula 15.7, we find that

$$\text{HSD} = 4.0 \sqrt{\frac{.98}{5.45}} = 4\sqrt{.18} = 4(.424) = 1.7$$

Only the difference between the "success" and "failure" conditions, which was 2 points, is larger than HSD (1.7), so this is the only significant difference of means, according to Tukey's procedure.

Complex Comparisons

The concept of the complex comparison was introduced in the context of the experiment described in Section A, in which a researcher wanted to compare the IQs of people born on the weekend with those of people born during the week. This could be done by taking the average IQ for Saturday and Sunday and comparing it to the average of the other days. Symbolically, a pairwise comparison can be written as $\overline{X}_i - \overline{X}_j = d$, where d is tested to see if it is significantly different from zero. Using this format, the comparison of weekend with weekdays can be written as

$$\frac{\overline{X}_1 + \overline{X}_2}{2} - \frac{\overline{X}_3 + \overline{X}_4 + \overline{X}_5 + \overline{X}_6 + \overline{X}_7}{5} = d$$

Dividing each term appropriately leads to the following expression:

$$\tfrac{1}{2}\overline{X}_1 + \tfrac{1}{2}\overline{X}_2 - \tfrac{1}{5}\overline{X}_3 - \tfrac{1}{5}\overline{X}_4 - \tfrac{1}{5}\overline{X}_5 - \tfrac{1}{5}\overline{X}_6 - \tfrac{1}{5}\overline{X}_7 = d$$

For complex comparisons, d is often replaced by another symbol such as L or ψ (the Greek letter psi). The expression above is called a linear combination of means (linear in the same sense as linear regression—the means are only being multiplied or divided by constants), and the fractions in front of each sample mean are called coefficients. Notice that the coefficients add up to zero ($\tfrac{1}{2} + \tfrac{1}{2} - \tfrac{1}{5} - \tfrac{1}{5} - \tfrac{1}{5} - \tfrac{1}{5} - \tfrac{1}{5} = 0$). When the sum of the coefficients is zero, the comparison can be called a **linear contrast.** Because linear contrasts have desirable properties, it is common to express complex comparisons in this way.

Planned (A Priori) Comparisons

If you are looking through the data of a multigroup experiment, and you realize that a particular complex comparison, which you had not planned before seeing the data, seems both meaningful and likely to be statistically significant, the generally recommended procedure is to use Scheffé's test. Post hoc complex comparisons are not performed very often, however, which is the main reason I have not taken the space to detail the mechanics of Scheffé's test in this text. A complex comparison that is only decided on after all the data from an experiment have been examined is likely to seem contrived, and results based on such comparisons usually will not be taken seriously unless they can be replicated. Fortunately, the Scheffé test is so conservative that it is not easy to attain statistical significance with a complex comparison that is conducted post hoc. On the other hand, if you can plan a complex comparison before seeing the data, the critical value you need to exceed is considerably less, as in the case of a one-tailed as compared to a two-tailed t-test. What follows is an example of a planned complex comparison.

In a four-group experiment, two groups receive different forms of psychotherapy, while a third group reads books about psychotherapy, and the fourth group receives no treatment at all. The latter two groups can both be considered control groups, and the researcher's chief interest is to compare the average improvement in the control groups with the average improvement in the two therapy groups. The researcher is entitled to test this planned comparison with the usual alpha level, regardless of whether the ANOVA for the four groups turns out to be significant. There is no danger that the researcher is taking unfair advantage of chance fluctuations in sampling. After all, only one comparison is being planned in advance; the researcher is not sifting through the data ("data-snooping," as it is often called) to find any comparison that looks good.

In fact, the designer of a four-group experiment is entitled to plan more than one comparison and to test each with the usual alpha (regardless of the significance of the ANOVA). However, there must be some limit. Certainly, if the researcher planned to test every possible comparison, there would be ample opportunity for α_{EW} to accumulate, and there would not be any difference between planning and data-snooping. One way to limit the number of planned comparisons is to require that each comparison be independent of all the others.

Orthogonal Contrasts

In the example above, suppose that the first planned comparison is $\frac{1}{2}\overline{X}_1 + \frac{1}{2}\overline{X}_2 - \frac{1}{2}\overline{X}_3 - \frac{1}{2}\overline{X}_4$ (i.e., the difference between the average score of the two control groups and the average score of the two therapy groups). It is likely that the researcher would then be interested in comparing the two therapy groups with each other. The important thing to note is that this second comparison is independent of the first; the difference between the average score of the two control groups and the average score of the two therapy groups has no effect on the difference between the scores of the two therapy groups, and vice versa (see Figure 15.1). The researcher might also have some interest in testing the difference between the two control groups. If so, this third comparison would also be independent of the first as well as the second. It would not be possible, however, to find a fourth comparison that would be independent of the other three (except for the rather trivial comparison in which the average of all four groups is compared to zero, which is the same as testing whether the grand mean differs from zero).

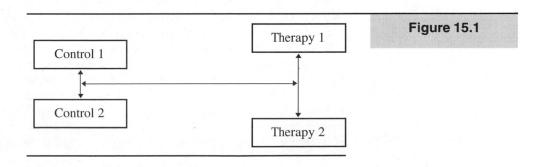

Figure 15.1

Comparisons that test mutually independent pieces of information are called **orthogonal comparisons** (or *orthogonal contrasts*). The number of orthogonal contrasts in a set is related to the number of groups in the experiment; if there are k groups, it will be possible to find $k - 1$ mutually orthogonal contrasts, but no more. For our four-group example there are only $4 - 1 = 3$ mutually independent comparisons. Bear in mind, however, that we could have found a completely different set of three mutually independent comparisons—but never a set of four. Perhaps you noticed that the number of contrasts in an orthogonal set is the same as df_{bet} for the ANOVA; this is not a coincidence. Each orthogonal contrast represents a "degree of freedom." It is considered acceptable, as an alternative to the one-way ANOVA (which tests MS_{bet}), to test each of the comparisons in an orthogonal set using the usual alpha for each (provided, of course, that the comparisons have been specified in advance). In both cases, you are using the same number of degrees of freedom.

The three orthogonal contrasts described above seem to capture the important information in the psychotherapy experiment. There are four additional pairwise comparisons that could be tested (each therapy group compared to each control group), but these comparisons are less meaningful. To test these additional pairwise comparisons without letting α_{EW} get too high, you should use Tukey's HSD. (Note that whenever you have four groups, six pairwise comparisons are possible, but all six cannot be mutually independent.) Because the Tukey procedure in effect reduces the alpha for each pairwise comparison, it would take larger differences to attain significance with the additional comparisons than with the ones that were planned—but this is only fair, because larger differences are easier to find post hoc.

Although we found a set of orthogonal contrasts that captured the most important information in our hypothetical example, this will not always be possible in other experiments. The most interesting comparisons may not be mutually independent, and there may be more than $k - 1$ truly meaningful comparisons to test. If the number of planned comparisons is more than $k - 1$ but less than the total number of possible pairwise comparisons, there is an a priori comparison procedure that may be preferable to any of the post hoc procedures.

Bonferroni's Inequality

In Section A, I presented Formula 15.2, which shows how large α_{EW} becomes when testing a number of independent comparisons. I mentioned that this equation is not accurate when the tests are not all mutually independent, such as when you are performing all possible pairwise comparisons. It is especially difficult to calculate α_{EW} precisely when performing a mixture of planned comparisons, some orthogonal to others and some not, some pairwise and some complex. However, there is an upper limit to α_{EW} that will never be exceeded. Based on the work of the mathematician Bonferroni, we can state that for a given number of comparisons (which will be symbolized as j), the experimentwise alpha will never be more than c times the alpha used for each comparison. This is one form of the *Bonferroni inequality,* and it can be symbolized by Formula 15.9:

$$\alpha_{EW} \leq j\alpha_{pc} \qquad \qquad \text{Formula 15.9}$$

The inequality expressed in the above formula can be used as the basis for a procedure for a priori comparisons.

Bonferroni *t*, or Dunn's Test

The Bonferroni inequality (Formula 15.9) suggests a very simple procedure for planned comparisons that has long been used. The logic is that if α/j is used to test each individual comparison, then α_{EW} will be no more than $j\,(\alpha/c) = \alpha$. Therefore, α is set to whatever value is desired for α_{EW} (usually .05), and then it is divided by j to find the proper alpha to use for each comparison, as shown in Formula 15.10.

$$\alpha_{pc} = \alpha_{EW}/j \qquad \text{Formula 15.10}$$

For instance, if $\alpha_{EW} = .05$ is desired, and five tests have been planned, α_{pc} would be .05/5 = .01. When five tests are performed (H_0 being true for each test), each at the .01 level, the probability of making at least one Type I error among the five tests is not more than $5 \cdot .01 = .05$. The comparisons are tested by ordinary *t*-tests (using MS_w for each s^2), except that the critical t is based on α_{pc} as found by Formula 15.10. Therefore, the procedure is often called the *Bonferroni t*.

The major difficulty involved in using the Bonferroni procedure outlined above is that the α that must be used for each comparison is often some odd value not commonly found in tables of the t distribution. For instance, if $\alpha_{EW} = .05$ is desired but four tests are planned, the α for each test will be .05/4 = .0125. How do you look up the critical t corresponding to $\alpha = .0125$? There are equations that can be used to approximate the critical t, but Dunn (1961) worked out tables to make this test easier to use, which is why the test is also called *Dunn's test*, or the **Bonferroni-Dunn test.** Of course, computers now make such tests easy to perform; most statistical programs will give exact p levels, and many offer the Bonferroni test, as well as a variety of post hoc tests, as options when running an ANOVA.

A more serious drawback of the Bonferroni test is that it is very conservative, often keeping the actual α_{EW} *below* the level that was initially set. Recall that the test is based on an inequality, and that the preset α_{EW} is an upper limit; especially when there are many tests planned, that upper limit will not be reached. This is why the Bonferroni test is not optimal when you are performing all possible pairwise comparisons. For instance, if you had run a five-group experiment and planned to perform all ten possible pairwise comparisons, the Bonferroni test would set α_{pc} to .05/10 = .005, but the Tukey test effectively sets α_{pc} to about .0063. However, if you can eliminate as few as three of the ten pairwise comparisons from consideration, the Bonferroni test becomes more powerful than Tukey's HSD.

A major advantage of the Bonferroni test is that it is very flexible, allowing you to mix pairwise and complex comparisons, whether orthogonal or not. You can even divide your α_{EW} unevenly, allocating more to some comparisons than others. In addition to its flexibility, a modification suggested by Keppel (1991) can make the Bonferroni test quite powerful. Keppel presents the following line of reasoning. As mentioned above, when there are k groups it is acceptable to perform $k - 1$ orthogonal comparisons, each at the predetermined alpha—no correction (i.e., reduction) in alpha is required. Therefore, Keppel reasons, it is reasonable to set α_{EW} to $(k - 1)\alpha$, where α is the value set for each orthogonal comparison, usually .05. Then, if more than $k - 1$ comparisons are planned, α_{pc} is set to α_{EW}/j, which equals $(k - 1)\alpha/j$, where j is the number of planned comparisons. For instance, if five tests are planned for a five-group experiment, $\alpha_{pc} = (k - 1)\alpha/j = (5 - 1) \cdot .05/5 = .2/5 = .04$. Whether you use Keppel's modified Bonferroni test or not, however, bear in mind the general principle underlying the Bonferrroni test: The fewer the number of comparisons you plan, the larger α_{pc} is for each, and the more powerful your tests are.

When to Use A Priori Tests

When you are conducting an experiment with many groups, there are quite a few opportunities for chance fluctuations to produce statistically significant results, and you have to be careful not to take advantage of those opportunities. That is why post hoc tests must be so conservative. On the other hand, if you can plan a few specific comparisons before seeing the data, the possibility of "capitalizing" on sampling fluctuation is reduced. In fact, if you can plan a set of mutually orthogonal comparisons, you need the least amount of "protection" from accidental findings—each comparison can be performed with the conventional alpha (usually .05). If the comparisons you have planned are not all mutually orthogonal (if you plan more than $k - 1$ comparisons, they cannot all be mutually orthogonal), but you do not plan to perform all possible tests, the Bonferroni test will probably give you more power than any of the post hoc tests (especially if you use Keppel's modified Bonferroni test). In any case, you are under no obligation to use the Bonferroni test if a post hoc test that you consider acceptable gives you significance more easily, even if the tests were planned in advance. On the other hand, once you have tested your small set of planned comparisons, you are free to test all other possible comparisons, as long as you use the appropriate post hoc procedure.

When beginning a new area of research, it may make sense to try many different conditions, and then use post hoc comparisons to go on a "fishing expedition" to see what turns up. However, because of the extra power associated with planned comparisons, it makes sense to plan your analyses in advance whenever theoretical predictions and past research make it possible.

Exercises

* 1. Calculate the harmonic mean for the following sets of numbers and compare it to the arithmetic mean.
a) 6, 12   b) 3, 15   c) 3, 6, 12, 15   d) 6, 6, 6, 36
e) 4, 5, 6, 7, 8

2. Calculate Tukey's HSD for the data in exercise 14B10. Which pairs of means differ significantly at the .05 level?

* 3. Calculate Tukey's HSD for the data in exercise 14B1. Which pairs of means differ significantly at the .05 level?

4. For the experiment described in exercise 14B8,
a) describe a meaningful set of orthogonal comparisons.
   b) Describe a different set of orthogonal comparisons.
   c) Express the set of comparisons in either part a or part b as a set of linear contrasts.

* 5. For the experiment described in exercise 15B8,
a) what is the largest number of comparisons that can be mutually orthogonal?
   b) Describe a meaningful set of mutually orthogonal comparisons that is as large as possible.
   c) Express these comparisons in terms of linear contrasts.
   d) If you had set α_{EW} to .05, what α would you use to test each of the comparisons in your orthogonal set?

6. What α_{pc} would you use if you had decided to perform the Bonferroni test with $\alpha_{EW} = .05$ a) to test a total of eight comparisons for one experiment?
   b) To compare all possible pairs of means in a six-group experiment?

c) To compare a third of the possible pairs of means in a seven-group experiment?

* 7. For the data in exercise 14B9, a) use the Bonferroni test to compare the sixth cultural group with each of the others; use Formula 15.3 and the appropriate α_{pc}. Which groups differ significantly from the sixth one?
   b) Use Tukey's HSD ($\alpha = .05$) to make the same comparisons. Which groups differ significantly from the sixth one with this test?
   c) Which of the two tests seems to have more power when applied in this manner?
   d) Is there another test that would have even more power when used for this set of tests? Explain.

8. According to Keppel's modified Bonferroni test, a) if you normally use $\alpha = .05$, what would be the appropriate α_{EW} for a four-group experiment? For a six-group experiment?
   b) What α_{pc} should you use if you have planned six tests for a four-group experiment?
   c) What α_{pc} should you use if you have planned ten tests for a six-group experiment?

SUMMARY

The Important Points of Section A

1. If there are k groups in an experiment, the number of different t-tests that are possible is $k(k - 1)/2$.

2. If all of the possible t-tests in an experiment are performed, the chances of making at least one Type I error (i.e., the experimentwise alpha, or α_{EW}) will be larger than the alpha used for each of the t-tests (i.e., the alpha per comparison, or α_{pc}). The α_{EW} will depend on α_{pc}, the number of tests, and the extent to which the tests are mutually independent.

3. If all possible t-tests are performed, or if t-tests are selected after seeing the results of the experiment, α_{EW} can easily become unacceptably high. To keep

α_{EW} down, a procedure for *post hoc* (or *a posteriori*) comparisons is needed. If you can plan the comparisons before seeing the data, *a priori* procedures can be used.

4. The simplest procedure for post hoc comparisons is to begin by performing the one-way ANOVA and proceed with *t*-tests only if the ANOVA is significant. The MS_w term from the ANOVA is used to replace the sample variances in these *t*-tests, which are generally referred to as *Fisher's protected t-tests*. When all the groups are the same size, *Fisher's least significant difference* (LSD) can be calculated, thus streamlining the procedure.

5. If all of the population means represented in a multigroup experiment are actually equal, the *complete null hypothesis* is true. Fisher's procedure provides "protection" against Type I errors only in this case. Protection is not adequate if a partial null is true (i.e., some, but not all, of the population means are equal); in this case Fisher's procedure allows α_{EW} to become unacceptably high when there are more than three groups.

6. Tukey devised a procedure—the *honestly significant difference* (HSD) test— that allows α_{EW} to be set before conducting any *t*-tests and assures that α_{EW} will not rise above the initially set value no matter how many groups are involved in the experiment, even if a partial null is true. This test is based on the *studentized range statistic* (*q*).

7. The HSD test is more *conservative* than the LSD test, because it does a better job of keeping Type I errors to an acceptably low level. All else being equal, conservative tests are less powerful than tests that are more liberal, because the reduction of Type I errors comes at the expense of Type II errors.

8. Comparisons that can be planned in advance will have greater power than post hoc comparisons.

The Important Points of Section B

A social psychologist is interested in knowing how the quality of life differs in the various geographical regions of the United States. In a preliminary study, he randomly samples 100 people from each of the four corners of the country: Northeast (NE), Southeast (SE), Northwest (NW), and Southwest (SW). Each of the subjects is measured on a 100-point scale of life satisfaction (LS). The means and standard deviations for each region are given in Table 15.2.

With four groups, a total of $4(3)/2 = 12/2 = 6$ pairwise comparisons can be tested. Although it would be considered acceptable to proceed with all of these comparisons using a conservative post hoc procedure, such as Tukey's HSD, it is far more common to begin by conducting a one-way ANOVA. MS_w is found by squaring each standard deviation to obtain the variance and then averaging the four variances, as follows: $(324 + 324 + 400 + 361)/4 = 1409/4 = 352.25$. MS_{bet} is found by calculating the variance of the four sample means (you should

Table 15.2

	NE	SE	NW	SW
$\bar{X}$	75	78	82	81
s	18	18	20	19

verify that this variance equals 10) and then multiplying by the sample size; thus $MS_{bet} = (10)(100) = 1000$. Finally, the F ratio is equal to $MS_{bet}/MS_w = 1000/352.25 = 2.84$. The critical F for $df_{bet} = 3$, $df_w = 396$, $\alpha = .05$ is 2.60. (df_w is so large that we look at the row corresponding to infinite degrees of freedom, symbolized by ∞.) Because the calculated F is larger than the critical F, the one-way ANOVA is statistically significant.

The significance of the ANOVA implies that Fisher's protected t-tests can be employed. Although many researchers feel that this procedure is too liberal when there are more than three groups, I will proceed with the calculations to illustrate the procedure. As all the samples are the same size ($n = 100$), LSD can be found using Formula 15.6:

$$\text{LSD} = t_{crit} \sqrt{\frac{2MS_w}{n}} = 1.96 \sqrt{\frac{2(352.25)}{100}} = 1.96(2.65) = 5.20$$

Note that for this problem the sample sizes are so large that t_{crit} is the same as z_{crit}. Now that we know the least significant difference, we can construct a table of the difference between pairs of means and determine which pairs differ significantly; see Table 15.3.

Only two of the differences exceed LSD; Fisher's procedure reveals that people in the Northeast differ significantly in life satisfaction from people in both the Northwest and the Southwest, but no other difference can be considered reliable.

Now we will test the same pairwise comparisons using Tukey's more conservative HSD test. Because all of the ns are equal, it is easy to find HSD by Formula 15.7:

$$\text{HSD} = q_{crit} \sqrt{\frac{MS_w}{n}} = 3.63 \sqrt{\frac{352.25}{100}} = 3.63(1.88) = 6.81$$

The value for q_{crit} can be found from Table A.10 by looking down the column for $k = 4$ to the bottom, where $df_{error} = \infty$.

Table 15.3

	NE	SE	NW	SW
NE		3	7*	6*
SE			4	3
NW				1

Looking at the table of differences, you can see that only the largest difference—between the NE amd the NW—is significant according to the HSD procedure. The NE-SW difference (6), which was larger than LSD (5.2), is not larger than HSD (6.81). If we choose to adopt the more conservative HSD procedure, then only the NE-NW difference can be declared statistically significant.

Tukey's HSD test is recommended when (1) there are more than three groups; (2) all the groups are the same size, or nearly so; (3) only pairwise comparisons are performed; and (4) the pairwise comparisons are chosen after seeing the results. For only three groups, Fisher's protected *t*-tests or the Newman-Keuls test are sufficiently conservative and more powerful. If you choose to make complex comparisons after seeing the results of your experiment, the Scheffé test should be used. If a limited number of comparisons can be planned, an a priori procedure, such as the Bonferroni test, should be used.

The Important Points of Section C

1. If sample sizes differ slightly, HSD can still be calculated by taking the *harmonic mean* of all the *n*s. This test is not valid, however, if the sample sizes differ considerably.

2. If a comparison can be expressed as a linear combination of means, and the coefficients add up to zero, it is called a *linear contrast.* Two linear contrasts that are statistically independent (the size of one does not affect the size of the other) are said to be *orthogonal* to each other.

3. In an experiment with *k* groups, there is at least one set of $k - 1$ contrasts that are mutually orthogonal, but not a larger set. As an alternative to the one-way ANOVA, you can plan a set of orthogonal contrasts, testing each at the conventional alpha (usually .05).

4. If you have planned a set of comparisons that are not all mutually orthogonal, it may be advantageous to use an *a priori* procedure such as the *Bonferroni-Dunn* test. In this test, the desired α_{EW} is divided by the number of comparisons that have been planned, and this smaller α is used for each comparison.

5. In general, as the number of planned tests decreases, the power of each test increases.

Definitions of Key Terms

Experimentwise alpha (α_{EW}) The probability of making at least one Type I error among all of the statistical tests that are used to analyze an experiment. (Sometimes the term *familywise alpha,* which has a more restricted definition, is preferred.)

Pairwise comparison A statistical test involving only two sample means.

Complex comparison A statistical test involving a weighted combination of any number of sample means.

Alpha per comparison (α_{pc}) The alpha level used for each particular statistical test in a series of comparisons.

A priori comparison A comparison that is planned before an experiment is run (or before the data are seen).

A posteriori comparison More often called a *post hoc* (i.e., after the fact) comparison; a comparison chosen after seeing the data.

Fisher's protected *t*-tests *t*-tests that are performed only after the one-way ANOVA has attained statistical significance. If homogeneity of variance can be assumed, MS_w is used to replace s^2 in the two-group *t*-test formula.

Complete null hypothesis An hypothesis that states that all of the populations sampled in the experiment have the same mean.

Partial null hypothesis An hypothesis that specifies that some of the population means are equal to each other but others are not.

Tukey's honestly significant difference (HSD) procedure Like LSD, this procedure calculates the smallest difference between two sample means that can be statistically significant, but this difference is adjusted (i.e., made larger) so that α_{EW} remains at the level set initially, regardless of the number of groups and whether the complete null or only a partial null hypothesis is true.

Studentized range statistic The distribution of the largest difference of means when more than two equal-sized samples from the same population are compared. The critical value from this distribution depends on the number of groups compared and the size of the groups.

Conservative A term that refers to procedures that are strict in keeping the Type I error rate down to a preset limit.

Liberal A term that refers to statistical procedures that allow relatively more Type I errors, and therefore produce fewer Type II errors.

Fisher's least significant difference (LSD) The difference between two sample means that produces a calculated *t* that is exactly equal to the critical *t*. Any larger difference of means is statistically significant. This test is applicable if the overall ANOVA is significant and all the groups are the same size.

Scheffé test A post hoc comparison test that maintains strict control over experimentwise alpha even when complex comparisons are involved. It is unnecessarily conservative when only pairwise comparisons are being performed.

Newman-Keuls test Like Tukey's HSD, this test is based on the studentized range statistic, but it uses different critical values for different pairwise comparisons depending on the "range" that separates two sample means.

Duncan's new multiple range test A more liberal version of the Newman-Keuls test. Most researchers find this test too liberal to justify its use.

Dunnett's test This test applies to the special case when one particular sample mean (usually that of a control group) is being compared to each of the other means. When this test is applicable, it is usually the most powerful.

Linear contrasts Weighted combinations of means in which the weights (or coefficients) sum to zero.

Orthogonal comparisons (or contrasts) Comparisons that are independent of each other. As many as $k - 1$ comparisons (where k is the number of groups) can be mutually independent. Testing a set of orthogonal comparisons can be used as an alternative to the one-way ANOVA.

Bonferroni-Dunn test A test that is only appropriate for planned (a priori) comparisons and is based on the *Bonferroni inequality*. The desired α_{EW} is divided by the number of planned comparisons to find the appropriate α_{pc}.

Key Formulas

The number of different t-tests that are possible among k sample means:

$$k(k - 1)/2 \qquad \text{Formula 15.1}$$

Experimentwise alpha for j independent comparisons (does not apply to the case when all possible pairwise comparisons are being performed, because they cannot all be mutually independent):

$$\alpha_{EW} = 1 - (1 - \alpha)^j \qquad \text{Formula 15.2}$$

Fisher's protected t-test (it is only "protected" if the one-way ANOVA is significant):

$$t = \frac{(\overline{X}_i - \overline{X}_j)}{\sqrt{MS_w(1/n_i + 1/n_j)}} \qquad \text{Formula 15.3}$$

Fisher's protected t-test (when both samples involved are the same size):

$$t = \frac{(\overline{X}_i - \overline{X}_j)}{\sqrt{2MS_w/n}} \qquad \text{Formula 15.4}$$

Post hoc *t*-test, according to Tukey's procedure (critical values for *q* are listed in Table A.11):

$$q = \frac{(\bar{X}_i - \bar{X}_j)}{\sqrt{MS_w/n}} \qquad \text{Formula 15.5}$$

Fisher's LSD procedure (a simplified procedure for performing Fisher's protected *t*-tests when all the samples are the same size):

$$LSD = t_{crit} \sqrt{\frac{2MS_w}{n}} \qquad \text{Formula 15.6}$$

Tukey's HSD procedure (assumes all samples are the same size; if the sample sizes differ slightly, Formula 15.8 can be used to find *n*):

$$HSD = q_{crit} \sqrt{\frac{MS_w}{n}} \qquad \text{Formula 15.7}$$

The harmonic mean for *k* groups (can be used to find *n* in Formula 15.7, when the sample sizes do not differ very much):

$$\text{Harmonic mean} = \frac{k}{\Sigma 1/n_i} \qquad \text{Formula 15.8}$$

The Bonferroni inequality, applied to a set of comparisons:

$$\alpha_{EW} \leq j\alpha_{pc} \qquad \text{Formula 15.9}$$

The Bonferroni inequality (Formula 15.9) rearranged to express the alpha per comparison for a fixed number of comparisons and a desired level of experimentwise alpha (forms the basis of the Dunn, or Bonferroni-Dunn, test):

$$\alpha_{pc} = \alpha_{EW}/j \qquad \text{Formula 15.10}$$

You will need to use the following from previous chapters:

Symbols
k: Number of groups in a one-way ANOVA
s^2: Unbiased variance of a sample
SS_{bet}: Sum of squared deviations based on group means
SS_W: Sum of squared deviations within groups
T_i: Sum of scores in the ith group
T_T: Sum of all observations in an experiment
N_T: Total number of all observations in an experiment

Formulas
Formula 14.5: F ratio for equal-sized groups
Formula 14.9: SS_W from raw scores
Formula 14.11: SS_{bet} from raw scores
Formula 14.13: SS_{total} from raw scores

Concepts
The F distribution
One-way ANOVA

16

CHAPTER

**CONCEPTUAL
FOUNDATION**

Calculating a two-way ANOVA is very similar to calculating two one-way ANOVAs, except for one extra sum of squares that almost always appears. This extra SS is the *interaction* of the two independent variables, and sometimes it is the most interesting and important part of a two-way ANOVA. Occasionally, it is a nuisance. In any case, the primary goal of this chapter is to help you understand the possible interactions of two independent variables as well as the simpler aspects of the two-way ANOVA.

Calculating a Simple One-Way ANOVA

To introduce the concept of a two-way ANOVA, I will start with a simple one-way ANOVA and show how it can be changed into a two-way ANOVA. For the one-way ANOVA, suppose we are conducting a three-group experiment to test the effects of a new hormone-related drug for relieving depression. Six depressed patients are chosen at random for each of the groups; one group receives a moderate dose of the drug, a second group receives a large dose, and a third group receives a placebo. The depression ratings for all three groups of patients after treatment appear in Table 16.1. The total of all the scores (T_T) is $198 + 195 + 167 = 560$, and the total of all squared scores ($\sum X^2$) equals 17,776. You can verify

	Placebo	Moderate Dose	Large Dose		Table 16.1
	38	33	23		
	35	32	26		
	33	26	21		
	33	34	34		
	31	36	31		
	28	34	32		
	$\Sigma = 198$	$\Sigma = 195$	$\Sigma = 167$		

the latter sum for yourself for practice. Now we can find SS_W and SS_{bet} using Formula 14.9 and Formula 14.11, respectively.

$$SS_W = 17{,}776 - \left(\frac{198^2}{6} + \frac{195^2}{6} + \frac{167^2}{6}\right)$$
$$= 17{,}776 - (6534 + 6337.5 + 4648.17)$$
$$= 17{,}776 - 17{,}519.67 = 256.33$$
$$SS_{bet} = 17{,}519.67 - \frac{560^2}{18} = 17{,}519.67 - 17{,}422.22 = 97.44$$

Next we find MS_{bet} and MS_W:

$$MS_{bet} = SS_{bet}/df_{bet} = 97.44/2 = 48.72$$
$$MS_W = SS_W/df_W = 256.33/15 = 17.089$$

The F ratio for the one-way ANOVA equals $MS_{bet}/MS_W = 48.72/17.089 = 2.85$. The critical F for $\alpha = .05$, $df_{bet} = 2$, and $df_W = 15$ is 3.68. Thus the calculated F is *not* greater than the critical F, and the null hypothesis cannot be rejected. The variability within the groups is too large compared to the differences among the group means to reach statistical significance. But as you may have guessed, this is not the end of the story.

Adding a Second Factor

What I neglected to mention in the initial description of the experiment is that within each group, half the patients are men and half are women. It will be easy to modify Table 16.1 to show this aspect of the experiment, because the first three subjects in each group are women. Table 16.2 (on page 522) shows the depression ratings grouped by gender and the new group totals.

You may have already noticed the advantage of this new grouping: The variability within these new groups is smaller. In the three-group case, differences between men and women contributed to the variability within groups, but in Table

Table 16.2		Placebo	Moderate Dose	Large Dose
Women		38	33	23
		35	32	26
		33	26	21
		$\Sigma = 106$	$\Sigma = 91$	$\Sigma = 70$
Men		33	34	34
		31	36	31
		28	34	32
		$\Sigma = 92$	$\Sigma = 104$	$\Sigma = 97$

16.2, male-female differences now contribute to differences *between* groups. This fact can be illustrated dramatically by recalculating the one-way ANOVA with the six new groups. The recalculation is made easier by the fact that the sum of all the scores, and the sum of all the squared scores, are the same.

$$SS_W = 17,776 - \left(\frac{106^2}{3} + \frac{91^2}{3} + \frac{70^2}{3} + \frac{92^2}{3} + \frac{104^2}{3} + \frac{97^2}{3} \right)$$
$$= 17,776 - 17,702 = 74$$
$$SS_{bet} = 17,702 - 17,422.22 = 279.78$$

It is important to note that the total SS does not change because of the regrouping. SS_T depends on the grand mean and each score's distance from it, and these values do not change no matter how the scores are grouped. What does change is how SS_T breaks down into SS_{bet} and SS_W. We can see a drastic reduction in SS_W and a complementary increase in SS_{bet}, because of the regrouping. Now let us see how the calculated F is affected. First we must calculate MS_{bet} and MS_W. Remember that df_{bet} is now 5 because the number of groups, k, is now 6. Thus $MS_{bet} = 279.78/5 = 55.956$. Similarly, $df_W = N_T - k = 18 - 6$, and $MS_W = 74/12 = 6.1667$. Finally, the new F ratio is $55.956/6.1667 = 9.074$, which is much higher than the new critical F (based on $\alpha = .05$, $df_{bet} = 5$, and $df_W = 12$), which is 3.11.

The large calculated F allows us to conclude that the six population means are not all the same, but how are we to interpret this information? The problem is that the six groups differ in *two ways*: on the basis of both drug treatment and gender. To sort out these differences we need to perform a *two-way* ANOVA.

New Terminology

Before I describe the mechanics of the two-way ANOVA, I need to introduce some new terms. An ANOVA is referred to as a **two-way ANOVA** if the groups differ on *two* independent variables. In the above example, the two independent

	Drug Treatment		
	Placebo	**Moderate Dose**	**Large Dose**
Women			
Men			

Table 16.3

variables are drug treatment and gender. Of course, gender is not a truly independent variable—it is not a condition created by the experimenter—but as you saw for the *t*-test and one-way ANOVA, grouping variables like gender can be treated as independent variables as long as you remember that you cannot conclude that the grouping variable is the cause of changes in the dependent variable.

An independent variable in an ANOVA is often called a **factor.** The simplest way to combine two or more factors in an experiment is to use what is called a **completely crossed factorial design.** In the case of two factors, a completely crossed design means that every level of one factor is combined with every level of the other factor and that every possible combination of levels corresponds to a group of subjects in the experiment. The example above is a completely crossed design because each combination of drug treatment and gender is represented by a group of subjects. If, for instance, there were no men in the placebo group, the design would not be completely crossed. The completely crossed factorial design is more commonly called a *factorial* design, for short. The two-way factorial design can always be represented as a matrix in which the rows represent the levels of one factor and the columns the levels of the other. Our example could be represented by the matrix shown in Table 16.3. Each box in the matrix is called a *cell* and represents a different combination of the levels of the two factors. Another way to define a completely crossed design is to say that none of the cells in the matrix is empty; there is at least one subject in each cell. If all of the cells have the same number of subjects, the design is said to be a **balanced design.** Factorial designs that are not balanced can be analyzed in more than one way; this complication will be discussed in Section C. In this section as well as in Section B, I will deal only with balanced factorial designs. A common way of describing a factorial design is in terms of the number of levels of each factor. The design represented in Table 16.3 can be referred to as a 2×3 (pronounced "2 by 3") or a 3×2 ANOVA (the order is arbitrary). If the subjects in each cell are selected independently from the subjects in all other cells, the design is a two-way independent-groups ANOVA. This is the only kind of two-way design that will be considered in this chapter.

Calculating the Two-Way ANOVA

Now we are ready to calculate a two-way ANOVA for our drug treatment $\times$ gender design. A useful way to begin is to fill in the cells of Table 16.3 with the

Table 16.4		Placebo	Moderate Dose	Large Dose	Row Means
Women		35.33	30.33	23.33	29.67
		2.517	3.786	2.517	
Men		30.67	34.67	32.33	32.56
		2.517	1.155	1.528	
Column Means		33	32.5	27.83	Grand mean = 31.11

mean and standard deviation of each group, as shown in Table 16.4. The means within the cells are called, appropriately, *cell means.* The means below each column and to the right of each row are called the **marginal means,** and they are the means of each column and row, respectively. Because all the cells have the same n (for this example, $n = 3$), the mean of each row is just the ordinary average of the cell means in that row (similarly for the columns). The number in the lower right corner of Table 16.4 is the grand mean, and it can be found by averaging all the column means or all the row means. For this example I will calculate the two-way ANOVA based only on the cell means, cell variances, and cell sizes. In Section B, I will present more convenient computational formulas. However, it is easy to lose sight of the structure of the two-way ANOVA when using these computational formulas. The following calculations, which use the definitional formulas, will help demonstrate the logic of the two-way ANOVA and its close relation to the one-way ANOVA procedures described in Chapter 14.

Calculating MS_W

I will begin by focusing on MS_W, the computation of which changes very little from the one- to the two-way ANOVA. For the one-way ANOVA calculated earlier in this section, MS_W was based on the variability within each drug treatment condition, including both the men and the women. For the two-way ANOVA, MS_W is based on the variances of each *cell.* In a balanced design all the cell sizes are equal, so MS_W is simply the average of all the cell variances. To calculate MS_W we need to square the cell standard deviations (given in Table 16.4) to obtain the cell variances, and then take their average. Therefore, $MS_W = (2.5166^2 + 3.786^2 + 2.5166^2 + 2.5166^2 + 1.155^2 + 1.528^2)/6 = (6.33 + 14.33 + 6.33 + 6.33 + 1.33 + 2.335)/6 = 37/6 = 6.166$. It should come as no surprise that this is exactly the same value we found for MS_W when we calculated the one-way ANOVA for the six separate groups, because those six groups are the six cells of our two-way ANOVA. The fact that this MS_W is so much smaller than the one calculated for the three-group ANOVA (before separating the genders) is one of the possible advantages of the two-way ANOVA, about which I will have more to say at the end of Section A.

Calculating MS_{bet} for the Drug Treatment Factor

Next we need to calculate MS_{bet} for the drug treatment factor. Because equal numbers of patients were in the three drug conditions, the numerator of Formula 14.5 can be used:

$$MS_{bet} = \frac{n\Sigma(\overline{X}_i - \overline{X}_G)^2}{k - 1}$$

In this case, $\overline{X}_i$ represents the mean of any of the three drug conditions (that is, any one of the three column means in Table 16.4) and k represents the number of levels for this factor, which is 3. Also note that n is the number of patients in each of the three drug conditions, so $n = 6$ (because the cell size is 3 and there are 2 cells in each drug condition). Plugging these values into Formula 14.5, we get

$$MS_{bet} = \frac{6[33 - 31.11)^2 + (32.5 - 31.11)^2 + (27.83 - 31.11)^2]}{2}$$

$$= 3[(1.89^2 + 1.39^2 + (-3.28)^2] = 3(16.24) = 48.72$$

It is not a coincidence that this value for MS_{bet} is the same value we found in the one-way ANOVA at the beginning of this chapter. Although we used a different formula this time, we are measuring the same thing: the variance of the three drug treatment means multiplied by their common sample size. However, the F ratio is now different because MS_W has been reduced. The new F ratio is $48.72/6.167 = 7.90$, which is much higher than the previous F of 2.85. The critical F we need to compare to is also higher, but fortunately it is only slightly higher. The reason for the change in critical F is that whereas df_{bet} has not changed, df_W has. When MS_W was based on the variability within each of the three drug groups, the appropriate df_W was $18 - 3 = 15$. However, in the two-way ANOVA, MS_W is based on the variability within *cells,* and the appropriate df_W involves losing one df for each cell (N_T − number of cells), so $df_W = 18 - 6 = 12$. The critical F for $\alpha = .05$, $df_{bet} = 2$ and $df_W = 12$ is 3.89.

The F ratio comparing the drug treatments (7.90) is now significant at the .05 level, and this time we can reject the null hypothesis that the population means corresponding to the three drug conditions are equal. The increase in the calculated F compared to the one-way ANOVA far outweighs the slight increase in the critical F. Separating the men from the women in this two-way ANOVA reduced the error term (i.e., MS_W) sufficiently to attain statistical significance.

Calculating MS_{bet} for the Gender Factor

The next step in the two-way ANOVA is to perform a one-way ANOVA on the other factor—in this case, gender. Because gender has only two levels, a t-test could be used to test for statistical significance, but to be consistent, I will use Formula 14.5 again. The denominator of this formula is the same MS_W (for cells) calculated above, so I will focus on the numerator. The $\overline{X}_i$ are the means for the women and the men; k is the number of genders, which equals 2; and n is the number of subjects in each gender, which equals 9. Plugging these values into Formula 14.5, we find

$$\text{MS}_\text{bet} = \frac{9\Sigma(\overline{X}_i - \overline{X}_G)^2}{2 - 1} = 9[(29.67 - 31.11)^2 + (32.56 - 31.11)^2]$$

$$= 9[(-1.44)^2 + 1.45^2] = 9(2.074 + 2.103) = 9(4.17) = 37.55$$

The F ratio for the gender difference is $37.55/6.167 = 6.09$. The critical F based on $\alpha = .05$, $\text{df}_\text{bet} = 1$, and $\text{df}_\text{W} = 12$ is 4.75, so this F ratio is also statistically significant. The women are, on the average, less depressed than the men in this experiment. However, if you look at the columns of Table 16.4, you'll see that women are actually *more* depressed than men in the placebo group. The row means are therefore somewhat misleading. This problem will be addressed shortly.

As you have seen, the two-way ANOVA begins with two one-way ANOVAs, and the numerator of each one-way ANOVA is found simply by ignoring the existence of the other factor. The denominator of both one-way ANOVAs is the same: the MS_W based on the variability within each cell. To the extent that the population variance within the cells is smaller than the variance within an entire row (or column) of the two-way table, these one-way ANOVAs are more powerful than ordinary one-way ANOVAs. But this is not the whole story of the two-way ANOVA, or even the most interesting part. A third MS must be calculated, and it corresponds to the amount of **interaction** between the two independent variables. Though the amount of interaction is easy to quantify, it can be hard to understand using only mathematical formulas. Before I show you how to calculate the interaction part, I want you to be able to visualize the size of an interaction.

Graphing the Cell Means

Earlier I pointed out that although the women are less depressed than the men for either drug dosage, the reverse is true for the placebo condition. It is these kinds of reversals that can lead to a significant amount of interaction. However, there are many ways that an interaction can arise; any particular way can be seen by drawing a graph of the cell means. A graph of the depression example will make it easier to explain the concept of interaction. The dependent variable (depression scores) is always placed on the Y axis. For the X axis we choose whichever of the two independent variables makes it easiest to interpret the graph. If one of the two factors has more levels, it is usually convenient to place that one on the X axis. For this example, I will place drug treatment on the X axis. Now, the cell means for each level of the second factor are plotted separately. Each level of the second factor becomes a different line on the graph when the cell means are connected; see Figure 16.1.

For the present example, the cell means for the male groups form one line and the cell means for the female groups form a second line. When the two lines are

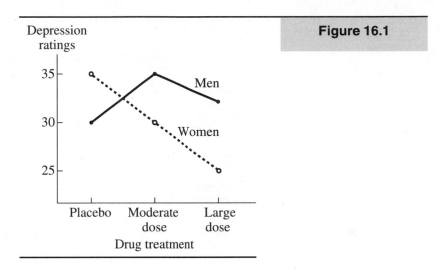

Figure 16.1

perfectly parallel, SS for interaction will always be zero, as will the *F* ratio for testing the interaction. When the lines diverge or converge strongly, or even cross each other, as in Figure 16.1, the *F* for interaction can be quite large, but whether it will be statistically significant depends on the relative size of MS_W.

The Case of Zero Interaction

To understand better what interactions involve, it will be helpful to start with the simplest case: parallel lines, zero interaction. Suppose the cell means for our example were as shown in Table 16.5 (on page 528). The graph of these cell means would consist of two parallel lines; see Figure 16.2 below. Notice that

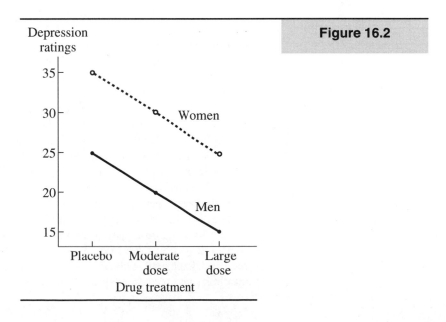

Figure 16.2

Table 16.5		Placebo	Moderate Dose	Large Dose	Row Means
	Women	35	30	25	30
	Men	25	20	15	20
	Column Means	30	25	20	25

because the lines are parallel, the male-female difference is always the same (10 points), regardless of the drug condition. Similarly, the difference between any two drug conditions is the same for men as it is for women.

The Additive Model of ANOVA

Another way to describe the lack of interaction in Table 16.5 is in terms of separate column and row effects relative to the grand mean. As you can see from the marginal means of Table 16.5, the effect of being a woman is to add 5 points to the depression score (as compared to the grand mean of 25), whereas the effect of being a man is to subtract 5 points. The effect of the placebo is to add 5 points, and the effect of the large dose is to subtract 5 points compared to the grand mean. (This doesn't mean that the placebo makes you more depressed, but rather that depression in the placebo group is higher than in the groups that get a dose of the drug.) The total lack of interaction between the factors assures that the mean for any cell will equal the sum of the row and column effects for that cell. For example, to find the cell mean for women in the placebo group, we start with 25 (the grand mean) and add 5 points for being a woman and 5 more points for taking the placebo: 25 + 5 + 5 = 35. We can write this as a general equation: Cell mean = Grand mean + Row effect + Column effect.

The above equation only applies when there is zero interaction, and it is one way of defining a total lack of interaction. The more general equation for a two-way ANOVA is Cell mean = Grand mean + Row effect + Column effect + Interaction effect. This way of describing an ANOVA is referred to as the *additive model,* for the obvious reason that the components are added to yield the final result. The more complete version of the model specifies each score in the ANOVA in terms of this equation: Score = Grand mean + Row effect + Column effect + Interaction effect + Error.

The difference between a subject's score and his or her cell mean is considered error, because it is totally unpredictable—it is not related to any manipulation in the experiment. The equation above is also referred to as the *general linear model,* and it is an extension of the equation for simple linear regression presented in Chapter 12. One way to think of any ANOVA is as a special case of multiple regression (see Chapter 12, Section C). Because it is convenient to work in terms of deviations from the grand mean, I will subtract the grand mean from both sides of the previous equation to create the following equation: Score − Grand mean = Row effect + Column effect + Interaction effect + Error. If these quantities are squared and summed for all subjects, the result is a very useful way to analyze a two-way design:

$$SS_{Total} = SS_{Row} + SS_{Column} + SS_{Interaction} + SS_{Error}$$

You are already familiar with SS_{Total} from the one-way ANOVA, and SS_{Error} is just another name for SS_W. SS_{Row} is the SS_{bet} obtained from the row means, and SS_{Column} is the SS_{bet} obtained from the column means. Together the SS_{bet} for the two factors (i.e., SS_{Row} and SS_{Column}) and $SS_{Interaction}$ will add up to the SS_{bet} you would get by treating each cell as a separate group in a one-way ANOVA (like our six-group analysis).

I will use the abbreviations SS_R, SS_C, and SS_{inter} to refer to SS_{Row}, SS_{Column}, and $SS_{Interaction}$, respectively. Some statisticians prefer to use SS_A and SS_B to refer to the two factors, because this notation easily generalizes to any number of factors (which can be referred to as C, D, etc). However, as I will never be dealing with more than two factors in the same ANOVA, I will use the row and column notation for its mnemonic value.

Calculating the Variability Due to Interaction

Now it is time to calculate the variablity due to interaction in the drug-gender example. Because we calculated the MS for each factor directly without first calculating SS, and the MS components do not add up conveniently as do the SS components, I will have to back up and get each SS by multiplying the appropriate MS by the corresponding df (recall that SS = df · MS). SS_R, which in the above example is the SS due to the difference in gender means, equals $1 \cdot (MS_{gender}) = 1 \cdot 37.55 = 37.55$. (There are only two levels of this factor, and therefore only one degree of freedom.) To find SS_C for the same example, we need to multiply MS_C (the MS_{bet} for the drug factor, which is 48.72) by the appropriate df (which is 2, because the drug factor has three levels). So $SS_C = 2 \cdot 48.72 = 97.44$. To find the amount of variability associated with the interaction, we start with the SS_{bet} that is calculated as though all the cells were separate groups in a one-way ANOVA and then subtract SS_R and SS_C, as defined above. (That is why I took the time to calculate the six-group analysis.) The remainder is SS_{inter}. For our example, SS_{inter} equals $279.78 − 37.55 − 97.44 = 144.78$. Next we need to divide SS_{inter} by df_{inter}. The df for the interaction is always equal to the df_{bet} for one factor multiplied by the df_{bet} for the other factor. (Symbolically, $df_{inter} = df_{R \times C} = df_R \times df_C$.) In this case, $df_{inter} = 2 \times 1 = 2$. Therefore, $MS_{inter} = SS_{inter}/df_{inter} = 144.78/2 = 72.39$. Finally, to find the F ratio for the interaction, we use MS_W once again as the denominator (i.e., error term), so $F = 72.39/6.167 = 11.74$. Because the df are the same, the critical F for the interaction is the same as the critical F for the drug treatment factor (3.89), so we can conclude that the interaction is statistically significant. However, that conclusion does not tell us anything about the pattern of the cell means. We would need to refer to a graph of the cell means in order to see how all of the effects in the experiment look when they are combined.

Types of Interactions

The row and column effects of the two-way ANOVA are called the *main effects*. When there is no interaction, the two-way ANOVA breaks into two one-way ANOVAs with nothing left over. All you have are the two main effects. However, in addition to the two main effects, normally you have some degree of interaction. To describe the different ways that the main effects can combine with the interaction to form a pattern of cell means, I will concentrate on the simplest type of two-way ANOVA, the 2 × 2 design. I'll begin with an example in which the interaction is more important than either of the main effects.

Suppose a psychiatrist has noticed that a certain antidepressant drug works well for some patients but hardly at all for others. She suspects that the patients for whom the drug is not working are bipolar depressives (i.e., manic-depressives) who are presently in the depressed part of their cycle. She decides to form two equal groups of patients: unipolar depressives and bipolar depressives. Half of the subjects in each group get a placebo, and the other half get the drug in question. If the psychiatrist is right, a graph of the results (in terms of depression scores) might look like the one in Figure 16.3. The divergence of the lines suggests the presence of an interaction. The main effects are usually easier to grasp in a table that includes the marginal means. Table 16.6 corresponds to Figure 16.3. As you can see from the marginal means, the two main effects are equally large and might be statistically significant. However, the psychiatrist would not be concerned with the significance of either main effect in this example. If the drug had been proven effective many times before, the main effect of drug versus placebo would not be new information. The main effect of bipolar depression versus unipolar depression would be meaningless if the psychiatrist deliberately chose subjects to make the groups comparable in initial depression

Figure 16.3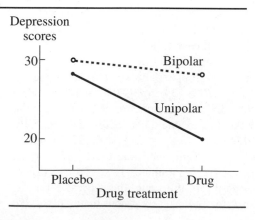

	Placebo	Drug	
Bipolar	30	28	29
Unipolar	28	20	24
	29	24	26.5

Table 16.6

scores. On the other hand, a significant interaction can imply that the psychiatrist found what she expected: The drug works for unipolar depressives but not bipolar depressives. The practical application of this result should be obvious. Furthermore, the large amount of interaction tells us to be cautious in interpreting the main effects. In this example, the main effect of drug vs. placebo is misleading, as the drug does not work equally well for both types of depressives. The main effect of unipolar depression versus bipolar depression is also misleading, as the difference only occurs under the drug condition. On the other hand, the main effects are not *entirely* misleading, because the *direction* of the drug effect is the same for both depressive groups; it is the amount of effect that differs considerably. (Similarly, the direction of the bipolar-unipolar difference is the same for both drug conditions.) This type of interaction, in which the direction (i.e., the order) of the effects is consistent is called an **ordinal interaction.**

As an illustration of an interaction that totally obliterates the main effects, consider the following hypothetical experiment. Researchers have found that amphetamine drugs, which act as strong stimulants to most people, can have a paradoxically calming effect on hyperactive children. Imagine that equal-sized groups of hyperactive and normal children are each divided in half, and the subjects are given placebos or amphetamines. A graph of the results (in terms of activity scores) could look like Figure 16.4. The nonparallel lines again indicate the presence of an interaction. The *crossing* of the lines indicates a **disordinal interaction,** that is, an interaction in which the direction of the effects reverses

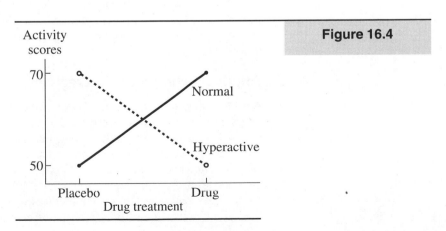

Figure 16.4

Table 16.7		Placebo	Drug	
	Hyperactive	70	50	60
	Normal	50	70	60
		60	60	60

for different subgroups. Note, however, that the lines do not have to cross to reveal a disordinal interaction; it is sufficient if one line slants downward and the other slants upward.

There are two ways to graph the cell means, depending on which of the two factors is placed on the X axis. When the interaction is disordinal, the lines will cross on at least one of the two graphs. It is also important to note that, whereas parallel lines indicate that the interaction is zero and therefore cannot be significant, lines that suggest a large amount of interaction (whether they cross or not) cannot tell you whether the interaction will be statistically significant.

The marginal means in Table 16.7 show what a disordinal interaction can do to the main effects. The SS for the numerators of the two main effects will be zero, as will the corresponding F ratios, so of course, neither of the main effects is significant. This lack of significance can be misleading if it is taken to imply either that the drug is simply ineffective or that hyperactive children do not differ from normal children on a measure of activity. A significant interaction tells you not to take the main effects at face value. In fact, even if the interaction fell just short of significance, the reversal of cell means in Table 16.7 suggests that you should use caution in interpreting the results of the main effects. Like the study in the previous example, the point of this study would be to find an interaction. A major advantage of the two-way ANOVA is the opportunity to test the size of such an interaction. The interaction of two independent variables is called a *two-way interaction*. In order to test whether three variables mutually interact, a three-way ANOVA would be required. Similarly, a four-way ANOVA is required to test a four-way interaction, and so on. ANOVAs involving more than two factors are called *higher-order designs,* and they are beyond the scope of this text. A very readable description of higher-order factorial ANOVAs can be found in Keppel (1991).

Another Definition of Interaction

A useful way to define an interaction in a two-way ANOVA is to say that an interaction is present when the effects of one of the independent variables change with different levels of the other independent variable. For instance, in one of the examples above, the effects of an antidepressant drug changed depending on whether the patients were unipolar depressives or bipolar depressives. For the unipolar level of the grouping variable, the drug treatment had a dramatic effect, but for the bipolar level the drug effect was very small. Of course, we can always look at a two-way interaction from the reverse perspective. We can say that the unipolar-bipolar effect (i.e., difference) depends on the level of the drug treatment. With the placebo, there is only a very small unipolar-bipolar difference, but with the drug this difference is large. The preferred way to view an interaction depends on which perspective makes the most sense for that experiment.

The *F* Ratio in a Two-Way ANOVA

The structure of each of the three *F* ratios in a two-way ANOVA is the same as that of the *F* ratio in a one-way ANOVA. For each *F* ratio in a two-way ANOVA, the denominator is an estimate of the population variance based on the sample variances (all the populations are assumed to have the same variance). Each numerator is also an estimate of the population variance when the appropriate null hypothesis is true, but these estimates are based on differences among row means, column means, or the cell means (after row and column mean differences have been subtracted out), multiplied by the cell size. These three numerators are statistically independent; that is, the size of any one numerator is totally unrelated to the size of the other two numerators. Moreover, the significance of the interaction in a two-way ANOVA implies nothing about whether either of the main effects is significant, and vice versa. The significance of any one of the three *F* ratios that are tested in a two-way ANOVA does not depend on the significance of any other *F* ratio. For instance, the interaction can be significant while both main effects are not, or both main effects can be significant while the interaction is not. There are eight possible combinations of the three *F* ratios ($2 \times 2 \times 2$) with respect to which are significant; these are illustrated in Figure 16.5 (on page 534).

The significance of the interaction *does* have implications for interpreting the main effects, which is why it makes sense to test the *F* for interaction first. If the interaction is *not* significant, you can test the main effects with two separate one-way ANOVAs. But if the interaction *is* significant, you must be cautious in interpreting the significance of the main effects. It is quite possible that the main effects are misleading. The way you follow up a two-way ANOVA with more specific (e.g., pairwise) comparisons differs depending on the significance of the interaction (see Section C).

Advantages of the Two-Way Design

When one factor involves an experimental manipulation (e.g., type of drug administered) and the other factor is a grouping variable (e.g., gender), one advantage of the two-way ANOVA is the likely reduction in MS_W. This advantage is lost, however, if the grouping variable is not relevant to the dependent variable being measured. If, in our drug-gender example, there were virtually no male-female differences in depression under any conditions, there would be very little reduction in MS_W. In such a case, the two-way ANOVA can actually produce a disadvantage, because df_W is reduced (one df is lost for each cell), which causes an increase in the critical *F*. However, an even more important possible advantage

Figure 16.5

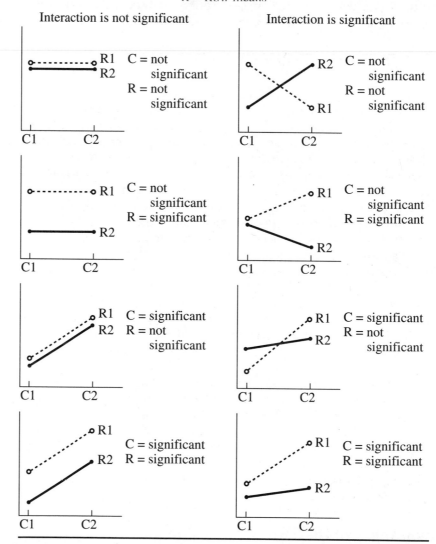

C = Column means
R = Row means

of a two-way ANOVA in such a situation is the opportunity to observe the interaction of the grouping factor with the experimental factor. A significant interaction serves to increase our understanding of how the experimental manipulation works with different groups of people; an almost complete lack of an interaction increases the generality of the experimental effect.

When both factors involve experimental manipulations, the lack of an interaction means that we have conducted two one-way ANOVAs in an efficient and economical way; I will elaborate on this point using an example in Section B. On the other hand, many two-way ANOVAs are performed with the primary goal of discovering whether there is any interaction between two experimental factors. An example of such an experiment is described in Section D.

Exercises

* 1. How many cells are there and what is the value of df_W in the following ANOVA designs? (Assume $n = 6$ subjects per cell.) a) 2×5 b) 4×4 c) 6×3

2. a) Graph the cell means in the table below and find the marginal means.
   b) Which effects (i.e., F ratios) might be significant, and which cannot be significant?

	Factor A, Level 1	Factor A, Level 2
Factor B, Level 1	75	70
Factor B, Level 2	60	65

* 3. a) Graph the cell means in the table below and find the marginal means.
   b) Which effects might be significant, and which cannot be significant?

	Factor A, Level 1	Factor A, Level 2
Factor B, Level 1	70	70
Factor B, Level 2	60	65

4. a) Graph the cell means in the table below and find the marginal means.
   b) Which effects might be significant, and which cannot be significant?

	Factor A, Level 1	Factor A, Level 2
Factor B, Level 1	75	70
Factor B, Level 2	75	70

* 5. a) Graph the cell means in the table below and find the marginal means.
   b) Which effects might be significant, and which cannot be significant?

	Factor A, Level 1	Factor A, Level 2
Factor B, Level 1	75	70
Factor B, Level 2	70	75

6. A researcher is studying the effects of both regular exercise and a vegetarian diet on resting heart rate. A 2×2 matrix was created to cross these two factors (exercisers versus nonexercisers and vegetarians versus nonvegetarians) and ten subjects were found for each cell. The mean heart rates and standard deviations for each cell are given in the table below.

	Exercisers	Non-exercisers
Vegetarians	$\bar{X} = 60$ $s = 15$	$\bar{X} = 70$ $s = 18$
Nonvegetarians	$\bar{X} = 65$ $s = 16$	$\bar{X} = 75$ $s = 19$

a) What is the value of MS_w?
b) Calculate the three F ratios. (Hint: Check to see if there is an interaction. If there is none, the calculation is simplified.) State your conclusions.
c) How large would these F ratios be if there were 40 subjects per cell? Compare these values to the ones you calculated for part b. What can you say about the effect on the F ratio of increasing the sample size? (What is the exact mathematical relationship?)

d) What conclusions can you draw based on the *F* ratios found in part c? What are the limitations on these conclusions (in terms of causation)?

* 7. The interaction for a particular two-way ANOVA is statistically significant. This implies that a) neither of the main effects is significant.

b) both of the main effects are significant.

c) the lines on a graph of cell means will intersect.

d) the lines on a graph of cell means will not be parallel.

8. When there is no interaction among the population means in a two-way design, the numerator of the *F* ratio for interaction in the data from an experiment is expected to be a) about the same as MS_W.

b) zero.

c) about the same as the sum of the numerators for the two main effects.

d) dependent on the number of cells in the design.

BASIC STATISTICAL PROCEDURES

The drug treatment–gender example in Section A represents a common situation for which the two-way ANOVA is appropriate. In this type of design, only one of the independent variables is controlled by the experimenter; in our example, subjects were randomly assigned to either one of the drug dosages or the placebo group. The other independent variable was a grouping variable; the experimenter could *select* equal numbers of males and females at random, but the experimenter had no control over whether a particular subject belonged in the male group or the female group. The main advantages to this design are that the grouping variable can reduce the within-group variability, as represented by MS_W, and the effect of the experimental variable can be observed in different groups of people. The interaction may be interesting, but if it is very small, we know that the effects of the experimental variable can perhaps be generalized to different types of subjects.

When both factors involve actual experimental manipulations, there will probably not be any reduction of MS_W, because subjects are not being separated according to preexisting differences (see the discussion of the advantages of this type of design later in this section). However, there is, potentially, a different advantage in this case. Suppose that an industrial psychologist wants to study the impact of environmental stress on work output. He would like to study the effects of both noise and temperature on a clerical task, such as adding up hours on a time sheet. Three levels of noise are chosen: a normal level (ordinary background noise), moderately loud popular music, and very loud "heavy metal" music. Three temperatures are chosen: room temperature, moderately warm, and very warm. The dependent variable is the number of time sheets completed correctly during the 1-hour experimental session. The simplest approach is to create a 3 × 3 factorial design, so that there are nine combinations of treatments (i.e., cells), and randomly assign an equal number of subjects to each cell. A potential advantage of this design is that it is economical because each subject serves in both the noise and the temperature comparisons. To perform a two-way ANOVA on the data from such an experiment, we can use the basic six-step procedure for hypothesis testing that we have used in previous chapters.

Step 1. State the Hypotheses

In a two-way ANOVA there are three null hypotheses: one corresponding to each main effect and one for the interaction. The null hypothesis for each main effect is the same as it would be for the corresponding one-way ANOVA. For noise level, H_0: $\mu_{NL} = \mu_{ML} = \mu_{VL}$; for temperature, H_0: $\mu_{RT} = \mu_{MW} = \mu_{VW}$. As for the null hypothesis concerning the interaction, there is no simple way to state it symbolically. We can say that this H_0 implies that the effects of the two factors will be additive, or that the effects of one factor do not depend on the levels of the other factor. The alternative hypotheses for the main effects follow the same format as for the one-way ANOVA. If a factor has more than two levels, H_A is simply a statement that the corresponding H_0 is not true. If a factor has only two levels, H_A is generally stated in the two-tailed form: $\mu_1 \neq \mu_2$. The H_A for the interaction is that the effects of one factor *do* depend on the levels of the other factor (i.e., the effects are *not* additive).

Step 2. Select the Statistical Test and the Significance Level

We are comparing population means along two dimensions, or factors, so the two-way ANOVA is appropriate. The same alpha level is used for testing all three *F* ratios, usually .05. We will use .05 for the present example.

Step 3. Select the Samples and Collect the Data

For this design nine independent random samples of four subjects each should be selected and then randomly assigned to the nine combinations of treatment levels. In actual practice, 36 subjects would be found and then randomly assigned to the nine cells, with the restriction that four subjects be assigned to each cell. Table 16.8 (on page 538) lists the number of time sheets completed correctly by each subject in this hypothetical 3×3 experiment and gives the sum for each cell.

Step 4. Find the Regions of Rejection

Because there are three null hypotheses, we need to find three critical values. The appropriate distribution in each case is the *F* distribution, so we need to know the appropriate df for the numerator and denominator corresponding to each *F* ratio tested. Our task is simplified by the fact that the df for the denominator is df_W for all three tests. To find df_W we need to know that the total number of subjects (N_T) is equal to the number of cells (9) times the size of each cell (4), so $N_T = 9 \times 4 = 36$. Then df_W is N_T − the number of cells = $36 - 9 = 27$.

Table 16.8	Normal Noise Level	Moderately Loud Music	Very Loud Music	Row Sums
Room Temperature	32	30	29	
	29	29	25	
	27	25	25	
	24	24	21	
	112	108	100	320
Moderately Warm	30	29	26	
	28	27	22	
	26	25	20	
	23	23	16	
	107	104	84	295
Very Warm	27	25	19	
	24	21	18	
	23	20	14	
	18	17	13	
	92	83	64	239
Column Sums	311	295	248	854

The df for the numerator for the F test of the noise factor is 1 less than the number of noise levels, so this df $= 3 - 1 = 2$. The df for the numerator corresponding to the temperature factor is also 2, because there are three temperature levels. Finally, these two df are multiplied together to give the df for the numerator for the F ratio that tests the interaction, so $df_{inter} = 2 \times 2 = 4$.

Now we can find the critical Fs to test each of the three null hypotheses. For the two main effects we need to look up $F_{.05}(2, 27)$, which equals 3.35. For the interaction we need $F_{.05}(4, 27)$, which equals 2.73.

Step 5. Calculate the Test Statistics

For a two-way ANOVA there are three different F ratios to calculate. Because they all have the same denominator (i.e., MS_W), I will calculate that term first. MS_W could be calculated by first finding the variance within each cell and then finding the average of these nine variances. However, in this section I will illustrate the use of raw-score formulas, which allow the calculation of the ANOVA components more directly from the data. The strategy is to compute each of the

SS components that comprise the total SS, before finding the appropriate MS components and F ratios. So, instead of computing MS_W directly, I will begin by finding SS_W, which is the sum of the SS within each cell, using Formula 16.1, which is Formula 14.9 modified to deal with equal-sized cells rather than the groups of a one-way ANOVA.

$$SS_W = \Sigma X^2 - \frac{1}{n}\Sigma(cs_i^2) \qquad \text{Formula 16.1}$$

The symbol cs_i (for cellsum$_i$) is the sum of scores within any cell. The symbol n does not require a subscript, because all of the cells are the same size; for this example, $n = 4$. As in the one-way ANOVA, ΣX^2 is obtained simply (but tediously) by squaring all of the scores in the experiment and adding them up. Table 16.8 does not show the squared scores, but you can verify for yourself that $\Sigma X^2 = 21,036$. We can use this value and the sums for each cell in Table 16.8 in Formula 16.1 to calculate SS_W.

$$SS_W = 21,036 - \frac{1}{4}(112^2 + 108^2 + 100^2 + 107^2 + 104^2$$
$$+ 84^2 + 92^2 + 83^2 + 64^2)$$
$$= 21,036 - \frac{1}{4}(82,978) = 21,036 - 20,744.5 = 291.5$$

Next, we calculate SS_{bet} as though this were a one-way ANOVA and all the cells were different levels (nine in all) of a single independent variable. The formula I will use (Formula 16.2) is based on Formula 14.11, also modified for use with equal-sized cells:

$$SS_{bet} = \frac{1}{n}\Sigma(cs_i^2) - \frac{T_T^2}{N_T} \qquad \text{Formula 16.2}$$

The sum of all the scores (T_T) can be found by adding all the cell sums ($112 + 108 + 100 + 107 + 104 + 84 + 92 + 83 + 64 = 854$), or even more easily by adding either the row or column totals. The term $1/n\,\Sigma cs_i^2$ was already calculated in finding SS_W. Plugging these values into Formula 16.2, we find that

$$SS_{bet} = 20,744.5 - \frac{854^2}{36} = 20,744.5 - 20,258.78 = 485.72$$

Together, SS_{bet} and SS_W equal SS_{Total}, which is defined for the two-way ANOVA exactly as it was for the one-way ANOVA (see Formula 14.13, page 467). In fact, so far all we have done is divide SS_{Total} into SS_{bet} and SS_W, as though the cells of the two-way ANOVA were the nine groups of a one-way ANOVA. To make this analysis into a two-way ANOVA, we need to divide SS_{bet} into three components: SS_{temp}, SS_{noise}, and SS_{inter}. This requires modifying Formula 16.2 to deal with row totals or column totals.

To find the SS_{bet} component that corresponds to the differences among temperature levels (i.e., SS_{temp}), we will use Formula 16.3A:

$$SS_R = \frac{1}{n_R}\sum R_i^2 - \frac{T_T^2}{N_T} \qquad \text{Formula 16.3A}$$

The term SS_R indicates that this SS component is based on differences among the rows. In Table 16.8, the variable that distinguishes the rows is temperature (each row corresponds to a different level of the temperature variable), though it makes no difference which variable is chosen for the rows and which for the columns. The term n_R refers to the number of subjects (or scores) in each row; in a balanced design, there will always be the same number of scores in each row. The value of n_R is equal to the number of *columns* times n (the number of scores in each cell), so in this example, $n_R = 3 \times 4 = 12$. The term R_i stands for the sum of any row; these sums are shown in the rightmost column of Table 16.8. The term T_T^2/N_T has already been calculated as part of the computation of Formula 16.2 above. We will now use these values to find SS_{temp}:

$$SS_{temp} = \frac{1}{12}(320^2 + 295^2 + 239^2) - 20{,}258.78$$

$$= \frac{1}{12}(246{,}546) - 20{,}258.78 = 20{,}545.5 - 20{,}258.78 = 286.72$$

To find SS_{noise}, we will use Formula 16.3B, which is identical to Formula 16.3A except that it measures variation among the columns:

$$SS_C = \frac{1}{n_C}\sum C_i^2 - \frac{T_T^2}{N_T} \qquad \text{Formula 16.3B}$$

The number of subjects in each column, n_C, equals the number of *rows* times n (the cell size). In this example, $n_C = 3 \times 4 = 12$, which is the same as n_R in this case, because the number of columns equals the number of rows in this particular example. The term C_i refers to the sum of any column; these sums form the bottom row of Table 16.8. We are now ready to use Formula 16.3B to find SS_{noise}.

$$SS_{noise} = \frac{1}{12}(311^2 + 295^2 + 248^2) - 20{,}258.78$$

$$= \frac{1}{12}(245{,}250) - 20{,}258.78 = 20{,}437.5 - 20{,}258.78 = 178.72$$

Finally, the SS for the interaction of temperature and noise is found by subtracting both SS_{temp} and SS_{noise} from SS_{bet}. Formula 16.4 for the general case is

$$SS_{inter} = SS_{bet} - SS_R - SS_C \qquad \text{Formula 16.4}$$

Applying Formula 16.4 to our example, we get

$$SS_{inter} = 485.72 - 286.72 - 178.72 = 20.28$$

Together all of these SS components add up to SS_{total}, as shown in Formula 16.5:

$$SS_{total} = SS_W + SS_R + SS_C + SS_{inter} \qquad \text{Formula 16.5}$$

The degrees of freedom add up in exactly the same way, as you can see from Formula 16.6:

$$df_{total} = df_W + df_R + df_C + df_{inter} \qquad \text{Formula 16.6}$$

If we use r to represent the number of rows, c to represent the number of columns, and n to represent the number of measurements (scores, subjects, etc.) in each cell, then the df components can be found by the following equations (collectively referred to as Formula 16.7):

$$df_{total} = N_T - 1, \quad \text{or} \quad nrc - 1$$
$$df_W = N_T - rc, \quad \text{or} \quad nrc - rc$$
$$df_r = r - 1$$
$$df_c = c - 1$$
$$df_{inter} = (r - 1)(c - 1) \qquad \text{Formula 16.7}$$

For the present example, both r and c equal 3 and n equals 4. So the df are as follows:

$$df_{total} = (4)(3)(3) - 1 = 36 - 1 = 35$$
$$df_W = (4)(3)(3) - (3)(3) = 36 - 9 = 27$$
$$df_{temp} = 3 - 1 = 2$$
$$df_{noise} = 3 - 1 = 2$$
$$df_{inter} = (2)(2) = 4$$

Notice that these df add up as shown in Formula 16.6.

Now that all of the SS and df components have been found, the MS can be calculated by dividing each SS by its corresponding df, as shown below:

$$MS_W = SS_W/df_W = 291.5/27 = 10.8$$
$$MS_{temp} = SS_{temp}/df_{temp} = 286.72/2 = 143.36$$
$$MS_{noise} = SS_{noise}/df_{noise} = 178.72/2 = 89.36$$
$$MS_{inter} = SS_{inter}/df_{inter} = 20.28/4 = 5.07$$

There is no need to calculate the overall MS_{bet}, because we are only interested in testing its separate components. The final step is to calculate the three F ratios, using MS_W as the denominator of each.

$$F_{temp} = MS_{temp}/MS_W = 143.36/10.8 = 13.3$$

$$F_{noise} = MS_{noise}/MS_W = 89.36/10.8 = 8.28$$

$$F_{inter} = MS_{inter}/MS_W = 5.07/10.8 = .47$$

Step 6. Make the Statistical Decisions

There are three decisions to be made—one for each F ratio. The F ratio for the main effect of temperature is 13.3, which is greater than the corresponding critical F (3.35), so the null hypothesis with respect to temperature can be rejected. The main effect of noise is also significant, because the calculated F ratio for noise (8.28) is again greater than the corresponding critical F (3.35). Finally, the F for interaction is *not* significant, because it is *less* than the appropriate critical F (2.73). In fact, the F for interaction (.47) is less than 1.0, so we need not look up a critical value to know that the null hypothesis cannot be rejected in this case.

Interpreting the Results

The fact that the interaction is not even close to significance means that the two-way ANOVA can be viewed simply as two separate one-way ANOVAs. For this particular example, both of the one-way ANOVAs (i.e., main effects) are significant. Because each has three levels, you would want to perform post hoc comparisons for each factor to see which pairs of levels are significantly different (see Section C). However, before any post hoc comparisons are performed, the cell means can be graphed so that you can inspect the pattern of results. As you can see from Figure 16.6, the number of time sheets completed goes down as the

Figure 16.6

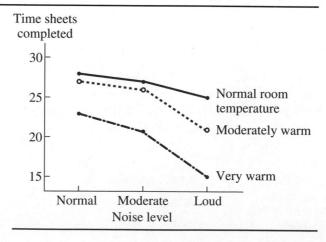

temperature rises or the noise level increases. Because the F for interaction was so small, it is not surprising to see that the lines are nearly parallel, and the effects of the two factors are, for the most part, just added together.

Assumptions of the Two-Way ANOVA for Independent Groups

The assumptions for the two-way ANOVA are the same as for the one-way ANOVA, so I will not bother to repeat them here (see Chapter 14). Bear in mind that when you are dealing with a balanced design (i.e., all cell sizes are equal) you do not have to worry about the homogeneity of variance assumption. If the design is not balanced, it may be appropriate to test for heterogeneity of variance and adjust the ANOVA according to procedures in advanced texts. However, if homogeneity of variance can be assumed for an unbalanced design, there is still more than one reasonable procedure for conducting the ANOVA, and some judgement is involved (see Section C). On the other hand, even in a balanced design, if the cell sizes are small *and* it seems that the underlying population distributions are *very* far from normal in shape, it may be appropriate to abandon the ANOVA approach entirely and use nonparametric statistics (see Part VI).

Advantages of the Two-Way ANOVA with Two Experimental Factors

Economy

I mentioned that one advantage of the experiment in the example above was economy. If we were to perform a one-way ANOVA with only the three noise levels, and we wished to have the same number of subjects at each noise level as we had in the two-way design, we would have to pick three random groups of 12 subjects each, for a total of 36. Then to perform a one-way ANOVA for the three temperature levels, we would need another 36 subjects, for a total of 72 subjects in all. The two-way design includes both one-way ANOVAs and yet uses only half of the subjects (36) that would be used in two separate one-way ANOVAs (72). The economical advantage of the two-way design comes simply from using each subject twice at the same time; each subject is assigned to both a temperature and a noise condition. This advantage is lost, however, if there is a large amount of interaction.

Exploration of Interactions

In a two-way ANOVA with two experimental factors, the interaction can be more interesting than either of the main effects. For instance, in a study of anagram solving, one factor might be the degree of time pressure (higher rewards for solving the problems in shorter amounts of time) and the other factor might be

Figure 16.7

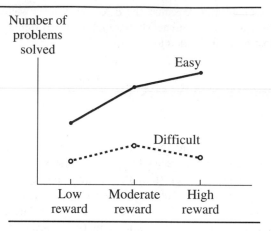

the difficulty of the anagram. Researchers may expect that subjects will solve fewer problems as difficulty increases and that they will solve more problems the greater the reward to solve quickly. What may be of interest is the interaction of those two variables. With relatively easy anagrams, time pressure should result in more problems solved, but with difficult anagrams time pressure may be counterproductive, and may even reduce the number of problems solved (see Figure 16.7).

Reduction in Error Term

In some cases, extraneous variables are not controlled in an experiment because they are thought to be totally irrelevant. If one of those variables is later found to be relevant, then adding it as another experimental factor in the ANOVA can reduce MS_W. For instance, the subjects in a social psychology experiment may be run by different research assistants, some male and some female. If the gender of the assistant affects the subjects' performance, variability will be increased. In that case, adding assistant gender as a factor can reduce MS_W. You could also reduce MS_W by using assistants of only one gender, but this could limit the generalizability of the results. Another researcher might fail to replicate your results simply because of using assistants of the opposite gender.

Advantages of the Two-Way ANOVA with One Grouping Factor

Reduction in Error Term

Even when experimental conditions are controlled as much as possible, there can be a great deal of subject-to-subject variability because of individual differences.

In some experiments, a significant amount of subject-to-subject variability can be attributed to a grouping variable, such as gender. (Remember that a grouping variable such as gender will not be relevant in all experiments; it depends on whether there are gender differences for the particular dependent variable being studied.) When a grouping variable contributes to variability, it can be added as a factor to reduce MS_W. (If instead, only one gender is used, the generalizability of the experiment is reduced.) The ultimate way to reduce subject-to-subject variability is to use the same subject in several conditions. (This is an extension of the procedure described in Chapter 13.) When that is not possible, the alternative is to match subjects on as many relevant variables as is practical. This approach is the topic of the next chapter.

Exploration of Interactions

The experimental variable (e.g., drug versus placebo) may operate differently for different levels of the grouping variable. Unless the grouping variable is included in the ANOVA, the effects of the experimental variable can be obscured in a misleading way (e.g., if the experimental variable had opposite effects on men and women, the variable might seem simply ineffective unless the genders were separated in the analysis). Analyzing the interaction can yield valuable information about the conditions under which the experimental variables are most effective. On the other hand, if the interaction is not near significance, you have increased your confidence about generalizing the experimental effect to each of the subgroups included in your study.

Advantages of the Two-Way ANOVA with Two Grouping Factors

Exploration of Interactions

When both factors in a two-way ANOVA are grouping variables, the focus is generally on the interaction. For instance, if self-esteem is being measured in a two-way design in which one factor is job level (executive versus clerical staff) and the other is gender, the primary interest would be to see if the gender difference is the same for both job levels (or, conversely, if the job difference is the same for both genders).

Qualitative Versus Quantitative Levels

As I discussed with respect to the one-way ANOVA, the levels of the independent variables can represent qualitative (i.e., categorical) or quantitative distinctions. When levels of the independent variables can be quantified precisely (e.g., drug dosage, amount of reward given, annual income), and many levels can be included, it is more helpful (and powerful) to use a regression approach or trend analysis (as suggested in Chapter 14) rather than an ordinary ANOVA (see an advanced text such as Hays, 1994).

Publishing the Results of a Two-Way ANOVA

The results of the temperature-noise experiment could be written in the following manner: "The number of time sheets completed was subjected to a 3 × 3 independent-groups ANOVA with temperature and noise as the between-subject factors. Both main effects were significant ($\underline{MSE} = 10.8$): for temperature, $\underline{F}$ (2, 27) = 13.3, $\underline{p} < .0001$, and for noise, $\underline{F}$ (2, 27) = 8.28, $\underline{p} < .005$. The interaction, however, did not approach significance, $\underline{F}$ (4,27) < 1." Because the mean square for error (MSE) will be the same for all of the F ratios in a two-way independent groups design, it need not be repeated. For an F value less than 1, there is little purpose to providing an exact F ratio or p value.

The Summary Table for a Two-Way ANOVA

The summary table for a two-way ANOVA has the same column headings as the table for a one-way ANOVA (see Chapter 14), but it is more complicated because there are more sources of variation. Each SS component represented in Formula 16.5 corresponds to a different source of variation in Table 16.9. Notice that the summary table shows the SS divided between SS_{bet} and SS_W and the SS_{bet} further subdivided into its three components. Such a presentation parallels the actual procedure for analyzing the data, and is the format used in the output of many statistical computer programs. If the summary table is published at all, it is common to leave out the first row ("Between Cells"), as that information could be easily derived by adding the SS and df for the following three rows.

Another way to view the structure of an ANOVA is to draw a **degrees of freedom (df) tree**—a diagram that shows the successive partitioning of the total degrees of freedom into all of its components. Figure 16.8 shows the df tree for the two-way factorial design as well as the df tree for the one-way design, for comparison.

Table 16.9	Source	SS	df	MS	F	p
	Between Cells	485.72	8			
	Temperature	286.72	2	143.36	13.3	< .01
	Noise	178.72	2	89.36	8.3	< .01
	Interaction	20.28	4	5.07	.47	> .05
	Within Cells	291.5	27	10.8		
	Total	777.22	35			

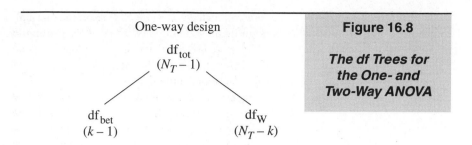

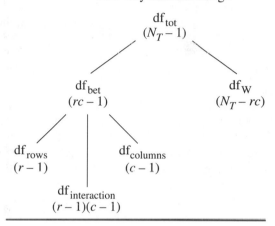

One-way design

df_{tot}
$(N_T - 1)$

df_{bet}
$(k - 1)$

df_W
$(N_T - k)$

Two-way factorial design

df_{tot}
$(N_T - 1)$

df_{bet}
$(rc - 1)$

df_W
$(N_T - rc)$

df_{rows}
$(r - 1)$

$df_{columns}$
$(c - 1)$

$df_{interaction}$
$(r - 1)(c - 1)$

Figure 16.8

The df Trees for the One- and Two-Way ANOVA

An Example from the Psychological Literature

As an example of the typical use of two-way ANOVAs in the psychological literature, I have chosen an article entitled "Evidence of codependency in women with an alcoholic parent: Helping out Mr. Wrong" (Lyon & Greenberg, 1991). One of the independent variables in this study was a grouping factor: Half the subjects "came from families in which one parent was alcohol dependent" (the codependent group), whereas the other half (the control group) did not (i.e., neither parent was alcohol dependent). The second independent variable involved an experimental manipulation. In one condition, the subject heard positive feedback about the experimenter in his absence (that the experimenter was a nurturing individual) from a "confederate" (an assistant to the experimenter, posing as another subject). In the other condition, the subject heard negative feedback (that the experimenter was an exploitive individual). Half of the subjects with alcoholic parents were randomly assigned to each condition, as were half of the control subjects. In each case, the experimenter returned to the room after the feedback and asked the subject to volunteer to help him as a research assistant. The amount of time volunteered (between 0 and 180 minutes for each subject) served as the dependent variable. The principal results were reported as follows.

We conducted a 2×2 ANOVA on the amount of time volunteered. A significant main effect was found for group membership, $F(1, 44) = 9.89$, $p < .003$, so that codependents were generally more helpful ($M = 97.9$ min) than were the members of the control group ($M = 31.3$ min). We also found a main effect for condition, $F(1, 44) = 4.99$, $p < .03$; subjects in the exploitive experimenter condition volunteered significantly more time overall ($M = 72.04$ min) than subjects in the nurturant experimenter condition ($M = 36.00$ min). The main effects were qualified, however, by a significant two-way interaction, $F(1, 44) = 43.64$, $p < .0001$, that strongly supported the primary prediction of the study; . . . codependents were significantly more helpful to the exploitive experimenter than to the nurturant experimenter, $t(23) = 6.38$, $p < .001$. In contrast, the control group was more helpful to the nurturant experimenter than to the exploitive one, $t(24) = 2.47$, $p < .05$.

The interaction is very large (F is over 40!) and disordinal, which tells us not to take the significant main effects at face value. Although one main effect suggests that codependents are "generally more helpful," this was only true when the experimenter was perceived as an exploitive individual (one who presumably resembled an alcoholic parent). The significant interaction justifies the subsequent t-tests, in which individual cells were compared. The use of t-tests to further explore significant effects from a two-way ANOVA will be discussed in the following section. The exact p levels (e.g., $p < .03$) reported by these researchers are undoubtedly a result of the availability of computer analysis.

Exercises

1. For a 3×5 ANOVA with nine subjects in each cell, a) find the value of each df component, and show that Formula 16.6 holds.

b) Find the critical value of F for each main effect and the interaction.

*2. A cognitive psychologist knows that concrete words, which easily evoke images (e.g., "sunset," "truck"), are easier to remember than abstract words (e.g., "theory," "integrity"), but she would like to know if those who frequently use visual imagery ("visualizers") differ from those with little visual imagery ("nonvisualizers") when trying to memorize these two types of words. She conducts a 2×2 experiment with five subjects in each cell and measures the number of words each subject recalls. The data appear in the table to the right.

	Visualizers	Nonvisualizers
Concrete Words	17	18
	20	19
	18	17
	21	17
	20	20
Abstract Words	14	18
	15	18
	15	17
	17	17
	16	19

a) Perform a two-way ANOVA and create a summary table.

b) Draw a graph of the cell means and describe the interaction, if any.

c) What can the cognitive psychologist conclude from her experiment?

3. A neuropsychologist is studying brain lateralization—the degree to which the control of various cognitive functions is localized more in one cerebral hemisphere than the other. This researcher is aware that for men, being right-handed seems to be associated with a greater degree of lateralization than does being left-handed, but that this relationship may not apply to women, who seem to be less lateralized in general. Various tasks are used in combination to derive a lateralization score for each subject as shown in the table below.

	Men		Women	
Left-handed	Right-handed	Left-handed	Right-handed
9	14	13	10
12	25	10	8
8	15	16	11
9	17	19	13
10	21	22	9
11	20	12	10

a) Perform a two-way ANOVA and create a summary table.

b) Graph the cell means and describe the interaction, if any.

c) Can the main effects be interpreted? Explain.

d) What can the neuropsychologist conclude?

*4. The summary table in the right column corresponds to a hypothetical experiment in which subjects must complete a mental task at five levels of difficulty and with three different levels of reward. Much of the table has deliberately been left blank, but all of the missing entries can be found by using the information in the previous sentence and the information in the table. (Assume the design is balanced).

Source	SS	df	MS	*F*	*p*
Difficulty	100				
Reward	150				
Interaction			5		
Within Cells		90			
Total	1190				

a) Complete the summary table.

b) How many subjects are in each cell of the two-way design?

5. A social studies teacher is exploring new ways to teach history to high school students, including the use of videotapes and computers. He also wants to know if these new techniques will have the same impact on average students as on gifted students. Twelve average students are randomly divided into three equal-sized groups: one instructed by the traditional method, one by videotapes, and one by an interactive computer program. In addition, twelve gifted students are divided in a similar fashion, resulting in a 3 × 2 design. At the end of the semester all students take the same final exam. The data appear in the table below.

	Traditional Method	Video-tape	Com-puter
Average Students	72	69	63
	83	66	72
	96	78	78
	79	64	59
Gifted Students	83	96	89
	95	87	93
	89	93	86
	98	86	95

a) Perform a two-way ANOVA and create a summary table.

b) Draw a graph of the cell means. Regardless of which *F* ratios were statistically significant, what does the pattern of cell means suggest about the effects of the new teaching methods in the two groups of students?

*6. A clinical psychologist is trying to find the most effective form of treatment for panic attacks. Knowing that antidepressant drugs have been surprisingly helpful in reducing the number and severity of attacks, this researcher decides to test three forms of psychological treatment, both with and without the use of antidepressants as an adjunct. The dependent variable—the total number of panic attacks during the final 30 days of treatment—is shown for each subject in the table below.

	Psycho-analysis	Group Therapy	Behavior Modifi-cation
Therapy	4	6	1
Alone	3	5	3
	2	8	0
	5	6	2
	4	8	4
	5	7	4
	5	6	2
	4	6	2
	2	7	1
	6	8	2
Therapy	1	3	0
plus	1	4	1
Drug	3	3	3
	1	4	2
	2	2	0
	1	2	0
	3	5	1
	4	3	2
	2	2	1
	3	5	0

a) Perform a two-way ANOVA. Which effects were significant at the .05 level? At the .01 level?

b) What can the clinical psychologist conclude?

*7. A college is conducting a study of its students' expectations of employment upon graduation. Students are sampled by class and major area of study and are given a score from 0 to 35 according to their responses to a questionnaire concerning their job preparedness, goal orientation, etc. The data appear in the table below.

	Human-ities	Sci-ences	Busi-ness
Freshmen	2	5	7
	4	6	8
	3	9	7
	7	10	12
Sophomores	3	10	20
	4	12	13
	6	16	16
	5	14	15
Juniors	7	14	20
	8	15	25
	7	13	22
	7	12	21
Seniors	10	16	30
	12	18	33
	9	16	34
	13	19	29

a) Perform a two-way ANOVA and create a summary table.

b) Draw a graph of the cell means. Does the interaction obscure the interpretation of the main effects?

*8. The data from exercise 14B8 for a four-group experiment on attitudes and memory are reproduced to the right. Considering the relationships among the four experimental conditions, it should be obvious that it makes sense to analyze these data with a two-way ANOVA.

a) Perform a two-way ANOVA and create a summary table of your results. (Note: You can use the summary table from exercise 14B8 as the basis for a new table.)

b) Compare your summary table to the one you produced for exercise 14B8.

c) What conclusions can you draw from the two-way ANOVA?

Incidental		Intentional	
Agree	**Disagree**	**Agree**	**Disagree**
8	2	6	7
7	3	8	9
7	2	9	8
9	4	5	5
4	4	8	7

Post Hoc Comparisons for the Two-Way Design

If the interaction between factors in an experiment is not significant and not very large, the comparisons following the ANOVA are very straightforward. The ANOVA is followed with pairwise or complex comparisons among the levels of whichever main effect is significant *and* involves more than two levels (this may be both main effects, as in the noise-temperature example). The appropriate post hoc comparison procedure should be used to keep α_{EW} from becoming too large (see Chapter 15). You may have noticed that in every two-way ANOVA, *three F* ratios are tested with no concern about the buildup of α_{EW}. This is because the three tests are generally viewed as planned comparisons. When performing post hoc comparisons, it is common to view the tests involving each factor as a separate "family" (the interaction is a third family) and to control α_{EW} within each family rather than across the entire experiment. That is why many statisticians prefer to speak about **familywise alpha** rates, rather than the experimentwise alpha. (The distinction does not arise in a one-factor design, so I did not discuss this topic in the previous chapter.) If both main effects in a 4×4 design were significant, Tukey's HSD could be used to test all six pairwise comparisons for the first factor, using $\alpha_{EW} = .05$, and then another Tukey's HSD could be used for the second factor, with α_{EW} again set to .05. Because you would be dealing with main effects, you would, of course, use the row means or column means (as appropriate) in the numerators of your *t*-tests. The MS_W used in these tests would be the same MS_W used as the error term in the two-way ANOVA.

When the interaction is significant (or quite large but just short of significance), it is usually pointless to perform comparisons on row or column means even if one or both of the main effects is significant. It is more likely that comparisons would focus on cell means, but there is more than one way to proceed in this situation. I will discuss two common approaches.

Simple Effects

In Section A, I presented an example in which the dosage-related effects of a drug differed by gender, resulting in a significant interaction. Even though the main effect of drug dose was statistically significant, it seems a bit silly to analyze it further, knowing that the men and women exhibited very different responses. It makes more sense to look at the drug effects for men and women separately; that is, to look at each row of Table 16.4 as a separate experiment. The effect of the drug dose factor at each level of the gender factor is called a **simple effect** (or sometimes the *simple main effect*). Of course, this term applies also to the columns of Table 16.4; the gender differences for each drug dosage level are also called simple effects. The main effect of drug dose is the average of the two simple effects—one for each gender. Similarly, the main effect of gender is the average of three simple effects—one at each dosage level. An interaction can now be defined as a difference among the simple effects that comprise a main effect. For instance, the simple effect of drug dose for men is different from the simple effect of drug dose for women. Because these two simple effects were significantly different, the interaction is significant. The main effect is an average of its simple effects; therefore, a main effect becomes misleading when it is averaging simple effects that are quite different.

A statistically significant interaction justifies testing each of the simple effects; each column or row of the design could be tested as a separate one-way ANOVA. Whether the simple effects analysis focuses on the rows or the columns depends on what is most meaningful for that experiment. If a simple effect that has more than two levels leads to a significant *F,* it would be appropriate to follow this test with pairwise comparisons of the levels (which would correspond to cell means) to locate the origin of the significant effect more specifically. These pairwise comparisons should be tested with an appropriate post hoc procedure, as discussed in Chapter 15.

A simpler but less logical way to follow up a statistically significant interaction is to conduct *t*-tests for every possible pair of cell means. In the drug-gender experiment there are six cell means, so $(6)(5)/2 = 30/2 = 15$ pairs can be tested. Of course, with so many pairs being tested it would be important to use a post hoc procedure, such as Tukey's HSD, to prevent α_{EW} from getting out of hand. However, some of the 15 pairwise comparisons are not very meaningful and are really not worth testing. The most meaningful comparisons are within a row or within a column. For instance, there are three drug dose comparisons that can be made just for the women (i.e., large dose versus placebo, moderate dose versus placebo, large dose versus moderate dose); the same three comparisons can be made just for the men, leading to a total of six pairwise comparisons. It is also meaningful to compare the two genders at each dosage level, which leads to an additional three comparisons. That makes a total of nine meaningful comparisons out of the 15 that are possible. The remaining six comparisons are less interpretable because they cross both row and column boundaries simultaneously. If post hoc pairwise comparisons are restricted to the meaningful comparisons, less protection is needed for α_{EW}, and Tukey's test can be modified accordingly to make it more powerful (Cicchetti, 1972).

Partial Interactions

For a 2×2 design, it is rather easy to describe the interaction (if there is one), even if you cannot explain *why* it occurred. Graphically speaking, the lines will either converge, diverge, or cross. For a larger design, such as the 2×3 drug-gender experiment, it may not be possible to describe the interaction simply. If you look again at Figure 16.1, you will see that from the placebo to the moderate dose the lines for men and women cross, but the lines do not cross from the moderate to the large dose. If the lines were parallel between the moderate and large doses, we would have an extreme case: considerable interaction in one part of the design and no interaction in another part. The significance of any interaction in a 2×3 ANOVA does not imply that there will be an interaction in each part of the design. When the interaction varies in different parts of the design, it can be of interest to test the interaction in the various 2×2 subsets of the total matrix; these 2×2 comparisons can be called **interaction contrasts** (Keppel, 1991).

The Two-Way ANOVA for Unbalanced Designs

The analysis of variance means breaking down or dividing the variance of a dependent variable into separate components. The analysis is straightforward only when those components are independent of each other (i.e., orthogonal, in the same sense that comparisons can be orthogonal; see Chapter 15, Section C). For the two-way design, Formula 16.5 implies that the total SS can be divided into four components (i.e., SS_R, SS_C, SS_{inter}, and SS_W) that do not overlap—that is, the four components are orthogonal and therefore add up to the total SS. That is why we could find SS_{inter} by subtraction. The SS components of the two-way ANOVA will only be orthogonal, however, if the cell sizes are equal (i.e., if the design is balanced). When the design is not balanced, the SS components are not mutually independent, they do not add up to SS_{total}, and SS_{inter} cannot be found by subtraction. The unbalanced design is more complicated to analyze, and there is more than one legitimate way to go about it. I will describe only the simplest way to analyze unbalanced designs.

To some extent, the procedure chosen to analyze an unbalanced design should depend on why the cell sizes are different. Probably the most common cause of unbalanced designs is the unexpected loss of a few subjects from an experiment that had been planned with equal cell sizes. There are many reasons for eliminating a subject's data: the subject may not have completed the experiment, or you may find out later that the subject misunderstood the instructions or knew too much about the purposes of the experiment. Sometimes subjects can be replaced, but there are situations in which this is not feasible. For instance, if a study is tracking the progress of subjects in psychotherapy over a 6-month period and a subject drops out of the study after 5 months, it may not be practical to replace the subject.

As long as the loss of subjects is random and unrelated to the experimental conditions, you can analyze the results and draw valid conclusions. On the other hand, if the loss of subjects is systematically related to the experimental conditions, you are dealing with a confounding variable that may make it impossible to draw valid conclusions. As a simple example of the nonrandom loss of subjects, consider an experiment comparing a psychotherapy group with a control group (involving some mock therapy) over a 6-month period. Suppose that the patients with the most severe psychological problems become frustrated with the control group and drop out, while similar patients in the real therapy group experience slight progress and stay in the experiment. At the end of 6 months, it may seem that the control group has even less pathology than the therapy group, because those with the highest pathology scores have left the control group. This kind of problem can affect any of the experimental designs we have studied.

Whenever subjects are selected for a study and then are lost, it is important to determine whether that loss is random or related to the independent variable. Although truly random samples are rare in psychological experiments, the random assignment of subjects to conditions is critical. If subjects are lost for reasons that are not random but rather related to the experimental conditions, the random assignment is compromised, and no method of statistical analysis can guarantee valid conclusions.

For unbalanced designs in which the cell sizes are not very different, and the differences are due mainly to chance, the ordinary two-way ANOVA procedure can be performed with only a slight modification. First, MS_W is calculated exactly as it would be for a one-way ANOVA with different-sized groups; a weighted average of the cell variances is calculated. (This is the same as adding the SS from all of the cells and then dividing by df_W.) Next, the SS_{bet} components can be found from the cell means, as shown in Section A, except that the actual cell sizes are not used. Instead, the harmonic mean of all the cell sizes is computed using Formula 15.8, and the analysis proceeds as though the design were balanced, with the cell size equal to the harmonic mean. If you prefer to calculate the two-way ANOVA using the cell sums, as illustrated in Section B, then you would multiply each cell mean by the harmonic mean of the cell sizes, rather than using the actual cell sums. This method for dealing with unbalanced designs is called the **analysis of unweighted means.** For a 3×4 design in which you have lost one subject in one of the 12 cells, it should be obvious that the analysis of unweighted means would be preferable to throwing out one subject at random from each of the other 11 cells in order to balance the design.

The method described in Chapter 14 for the computation of the one-way ANOVA with unequal sample sizes is called the *analysis of weighted means.* You should be aware that some researchers and statisticians strongly recommend the analysis of *unweighted* means, even for a one-way ANOVA, when the sample size discrepancies are relatively small and accidental. (You would use the procedure for equal sample sizes, substituting the harmonic mean of all the sample sizes for *n*.) However, as reasonable as that recommendation may be, the method of unweighted means for the one-way ANOVA is not included in the major statistical software packages and most introductory texts do not even mention it.

Though the weighted-means solution is commonly used in the one-factor design, it is not a reasonable option in an unbalanced factorial design. On the other hand, though the unweighted-means solution is sometimes a reasonable option for an unbalanced factorial design, it is certainly not the only option. For some experiments, the discrepancies in cell sizes can be considerable and may reflect the relative sizes of preexisting groups. Consider a 3 × 2 design in which one factor is the location of brain damage within the left hemisphere and the other factor is handedness. Right-handed patients are much more common than left-handed patients, and there may be considerable differences in how many patients will have damage in each of the three brain areas. Patients may be hard to come by and much power could be lost by keeping all groups down to the size of the smallest brain damage group. When sample sizes represent true differences in population sizes, the method of unweighted means is not appropriate; the recommended approach is the *method of least squares,* in which a multiple regression analysis is used to partition the sums of squares. Unfortunately, there is more than one way to apply the multiple regression method; choosing a regression strategy and interpreting the results requires a good deal of theoretical and statistical sophistication, so these topics will not be discussed here (see Hays, 1994 or Keren, 1993).

Exercises

* 1. In exercise 16B7, ignore the significance of the interaction (the interaction does not distort the main effects, anyway) and use Tukey's HSD to test all possible pairs of marginal means for each main effect. (Be sure to use the *n* for each row or column, as appropriate.)

2. To follow up the significant interaction in exercise 16B8, perform all of the meaningful pairwise comparisons using protected *t*-tests (Hint: This is the same as testing all of the simple effects.)

* 3. To localize the source of the significant interaction in exercise 16B6, test all of the possible 2 × 2 partial interactions (Hint: Eliminate one level at a time of the three-level factor, then perform each ANOVA, testing only the *F* ratio for interaction.)

4. A psychologist suspects that emotions affect people's eating habits, but perhaps eating habits of obese people are affected differently. Ten obese subjects are randomly divided into two equal-sized groups, one of which views a horror movie while the other watches a comedy. Ten normal-weight control subjects are similarly divided. The number of ounces of popcorn eaten by each subject during the movie is recorded. Unfortunately, one obese subject who had watched the horror movie admitted later that it was his favorite movie, and therefore that subject's data had to be eliminated (after it was too late to run additional subjects). The data for the remaining subjects appear below.

| Obese Group | | Control Group | |
Horror Movie	Comedy	Horror Movie	Comedy
5	8	15	13
2	6	19	12
3	5	17	9
7	11	16	16
6	9		15

a) Conduct a two-way ANOVA using the unweighted means solution.

b) What can you say about the effects of emotion on eating, and how does this statement change depending on whether obese or nonobese people are considered?

* 5. A cognitive psychologist has spent 3 months teaching an artificial language to 32 children, half of whom are bilingual and half of whom are monolingual. Half the children in each group are 5 years old and half are 8 years old. The table shows the number of errors made by each child on one of the comprehension tests for the new language. Unfortunately, as you can see by the cell sizes, some of the children had to drop out so late in the study that they could not be replaced.

	Five-year-olds	Eight-year-olds
Monolingual Group	18	8
	19	10
	18	12
	16	9
	11	6
	17	12
		11
Bilingual Group	8	5
	11	8
	11	6
	10	10
	13	4
	18	5
		6
		4

a) Conduct a two-way ANOVA using the unweighted means solution.

b) What conclusions can be drawn from this study?

SUMMARY

The Important Points of Section A

1. The groups in a *two-way* ANOVA design vary according to two independent variables (often referred to as *factors*), each having at least two *levels*. In a *completely crossed factorial* ANOVA (usually referred to as just a *factorial* ANOVA), each level of one factor is combined with each level of the other factor to form a *cell,* and none of the cells is empty. The two-way ANOVA is often described by the number of levels of each factor (e.g., a 3×4 design has two factors, one with three levels and the other with four).

2. In a *balanced* design all of the cells contain the same number of subjects (observations, measurements, scores, etc.). If the design is not balanced, there is more than one legitimate way to analyze it (see Section C).

3. The cells can be arranged in a matrix so that the rows are the levels of one factor and the columns are the levels of the other factor. The means of the rows and columns are called *marginal means,* and they show the effect of each factor, averaging across the other.

4. The two-way ANOVA results in three F ratios, each of which is compared to a critical F to test for statistical significance. The denominator of each ratio (called the error term) is MS_W, which is the average of all the cell variances.

5. As in the one-way ANOVA, SS_{tot} can be divided into SS_W and SS_{bet}, with the latter referring to between-cell variability in the two-way ANOVA. The SS_{bet} can be further subdivided into three components: SS_R (based on the variability of row means), SS_C (based on the variability of column means), and SS_{inter} (the variability left over after SS_R and SS_C are subtracted from SS_{bet}). Dividing by the appropriate df yields MS_R, MS_C, and MS_{inter}.

6. The F ratios based on MS_R and MS_C are like the Fs from two one-way ANOVAs, each ignoring (averaging over) the other factor, except that MS_W for the two-way ANOVA may be smaller than in either one-way ANOVA, because it is based on variability within cells rather than entire rows or columns. The two one-way ANOVAs are called the *main effects* of the two-way ANOVA.

7. The F ratio based on MS_{inter} is used to test for the significance of the *interaction* between the two factors. If the amount of interaction is zero, a graph of the cell means will produce parallel lines. If the amount of interaction is relatively small and not close to statistical significance, the two-way ANOVA can be viewed as an economical way to perform two one-way ANOVAs (i.e., the main effects can be taken at face value).

8. If the amount of interaction is relatively large and statistically significant, the main effects may be misleading. If the lines on a graph of the cell means cross or slant in opposite directions, there is a *disordinal* interaction, and the results of the main effects should probably be ignored. Even if the interaction is *ordinal,* a large amount of interaction means that the main effects should be interpreted cautiously.

The Important Points of Section B

To review the calculation of the two-way ANOVA, I have created the following hypothetical experiment. A clinical psychologist wants to know whether the number of sessions per week has an impact on the effectiveness of psychotherapy, and whether this impact depends on the type of therapy. The psychologist

Table 16.10

| | **Number of Sessions Per Week** | | | | |
	1	**2**	**3**	**4**	**Row Sums**
Classical Psychoanalysis	4 (16)	2 (4)	6 (36)	8 (64)	
	6 (36)	4 (16)	5 (25)	6 (36)	
	3 (9)	5 (25)	8 (64)	6 (36)	
	1 (1)	3 (9)	6 (36)	8 (64)	
	4 (16)	5 (25)	5 (25)	5 (25)	
	18	19	30	33	100
Client-Centered Therapy	6 (36)	4 (16)	4 (16)	5 (25)	
	3 (9)	3 (9)	7 (49)	4 (16)	
	4 (16)	6 (36)	2 (4)	4 (16)	
	4 (16)	6 (36)	4 (16)	6 (36)	
	3 (9)	4 (16)	3 (9)	4 (16)	
	20	23	20	23	86
Column Sums	38	42	50	56	186

decides that therapy effectiveness will be measured by an improvement rating provided by judges who are blind to the conditions of the experiment. Four levels are chosen for the first factor: one, two, three, or four sessions per week; two types of therapy are chosen for the second factor: classical psychoanalysis and client-centered therapy. When these two factors are completely crossed, eight cells are formed. For this example there are five randomly selected subjects in each cell, selected independently of the subjects in any other cell. The improvement rating for each subject is presented in Table 16.10, along with column and row sums (the squared values are included in parentheses).

The simplest strategy is to begin by calculating SS_{bet} and SS_W, as in the one-way ANOVA, and then divide SS_{bet} into SS_R, SS_C, and SS_{inter}. In order to use the raw-score computational formulas, we need the sums for each cell, row, and column, as shown in Table 16.10. Before we can use Formula 16.1 to calculate SS_W, we must add up all the squared scores; $\sum X^2 = 970$. Inserting this value into Formula 16.1, we obtain

$$SS_W = \sum X^2 - \frac{1}{n}\sum(cs_i^2) = 970 - \frac{1}{5}(18^2 + 19^2 + 30^2 + 33^2$$
$$+ 20^2 + 23^2 + 20^2 + 23^2)$$
$$= 970 - \frac{1}{5}(4532) = 970 - 906.4 = 63.6$$

To use Formula 16.2 to calculate SS_{bet}, we need to know the sum of all the scores (T_T), which equals 186 (see the lower right corner of Table 16.10). The first term of the equation has already been calculated as part of the computation of SS_W above.

$$SS_{bet} = \frac{1}{n}\Sigma(cs_i^2) - \frac{T_T^2}{N_T} = 906.4 - \frac{186^2}{40}$$

$$= 906.4 - \frac{34,596}{40} = 906.4 - 864.9 = 41.5$$

Next, we find the SS components that correspond to the sessions and therapy factors. In Table 16.10, type of therapy is represented by the rows, so we use Formula 16.3A to find the SS for the therapy factor. The second term in the formula has already been calculated as part of the computation of SS_{bet} above. [Note that n_R = cell size times the number of columns = (5)(4) = 20.]

$$SS_R = \frac{1}{n_R}\Sigma R_i^2 - \frac{T_T^2}{N_T} = \frac{1}{20}(100^2 + 86^2) - 864.9$$

$$= \frac{1}{20}(17,396) - 864.9 = 869.8 - 864.9 = 4.9$$

To find the SS for the sessions factor, we use Formula 16.3B [with n_C = (5)(2) = 10].

$$SS_C = \frac{1}{n_C}\Sigma C_i^2 - \frac{T_T^2}{N_T} = \frac{1}{10}(38^2 + 42^2 + 50^2 + 56^2) - 864.9$$

$$= \frac{1}{10}(1444 + 1764 + 2500 + 3136) - 864.9$$

$$= \frac{1}{10}(8844) - 864.9 = 884.4 - 864.9 = 19.5$$

At this point we can find SS_{inter} by subtracting SS_R and SS_C from SS_{bet}, as shown in Formula 16.4:

$$SS_{inter} = SS_{bet} - SS_R - SS_C = 41.5 - 4.9 - 19.5 = 41.5 - 24.4 = 17.1$$

Having broken the total SS into its components, we divide each SS by the appropriate df. The total df for this example is broken down into the following components, according to Formula 16.7:

$$df_W = N_T - rc = 40 - (2)(4) = 40 - 8 = 32$$

$$df_r = r - 1 = 2 - 1 = 1$$

$$df_c = c - 1 = 4 - 1 = 3$$

$$df_{inter} = (r - 1)(c - 1) = (1)(3) = 3$$

Notice that these df add up to df_{total}, which equals $nrc - 1 = (5)(2)(4) - 1 = 40 - 1 = 39$. Now we can find the MS by dividing each SS by its df.

$$MS_W = SS_W/df_W = 63.6/32 = 1.99$$

$$MS_R = SS_R/df_R = 4.9/1 = 4.9$$

$$MS_C = SS_C/df_C = 19.5/3 = 6.5$$

$$MS_{inter} = SS_{inter}/df_{inter} = 17.1/3 = 5.7$$

Finally, we can find the three *F* ratios as follows:

$$F_{therapy} = MS_R/MS_W = 4.9/1.99 = 2.46$$

$$F_{sessions} = MS_C/MS_W = 6.5/1.99 = 3.27$$

$$F_{inter} = MS_{inter}/MS_W = 5.7/1.99 = 2.87$$

The critical *F* ($\alpha = .05$) for testing the therapy effect is based on 1 and 32 degrees of freedom and equals 4.15 (see Table A.8). The critical *F*s for the sessions effect and the interaction are both based on 3 and 32 df, and therefore equal 2.90. The *F* for the main effect of therapy (2.46) is not greater than the critical *F*, so the null hypothesis cannot be rejected in this case. On the other hand, the *F* for the main effect of sessions (3.27) *is* greater than the critical value, so for this factor the null hypothesis *can* be rejected. The interaction falls only slightly short of statistical significance; the calculated *F* (2.87) is just under the critical *F*.

Interpreting the Results

First, we check the significance of the interaction as an indication of whether the main effects will be interpretable and of which approach should be taken to follow up the ANOVA with additional comparisons. Although the interaction is not significant at the .05 level, it is so close to significance ($p < .052$) that you should use caution in the interpretation of the main effects. Graphing the cell means is usually helpful; the graph for this example is shown in Figure 16.9. Notice that the lines in Figure 16.9 cross, indicating a disordinal interaction. Whereas client-centered therapy produces slightly more improvement than psychoanalysis when therapy is given one or two times per week, this trend reverses direction when there are three or four sessions. The significant main effect of sessions is clearly due to the psychoanalytic groups. Therefore, to state that increasing the number of sessions per week makes psychotherapy more effective is misleading, in that the results show this statement to be true for psychoanalysis but not for client-centered therapy. On the other hand, the fact that the main effect of type of therapy is *not* significant is also misleading; there *are* differences between the two types of therapy, but which type of therapy is more effective depends on the number of sessions per week. If you are surprised that the interaction falls short of significance, especially after inspecting Figure 16.9, bear in mind that the

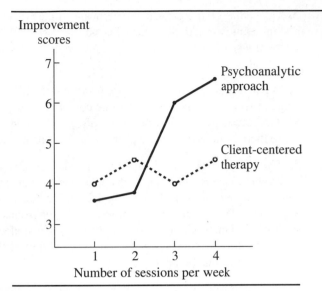

Improvement scores

Figure 16.9

Number of sessions per week

sample sizes are quite small, which provides little power unless the amount of interaction is very large.

The Advantages of the Two-Way ANOVA

1. Exploration of the interaction of two independent variables
2. Economy (if interaction is not significant)
3. Greater generalization of results (if interaction is not significant)
4. Reduction in MS_W (i.e., error term), usually due to inclusion of a grouping variable

The Important Points of Section C

1. If the interaction in an experiment is neither significant, nor disordinal, nor very large, then it is reasonable to follow up any significant main effect that has more than two levels with pairwise comparisons according to the procedures of Chapter 15.

2. Any one row or column in a two-way design is called a *simple effect*. When an interaction is significant it means that the simple effects, which are averaged together to create the main effects, differ significantly. In this case, it makes more sense to perform follow-up comparisons on significant simple effects than on the main effects. Another reasonable procedure for analyzing a significant interaction is to perform multiple comparisons on any cell means within the same column or row, adjusting α_{pc} accordingly.

3. Depending on the pattern of cell means, it may be more informative to follow a significant interaction by focusing on the various 2×2 subsets (i.e., partial interactions) of the overall design, in order to locate the interaction more specifically.

4. When a two-way ANOVA is not balanced, the SS for interaction is not independent from SS_R and SS_C; the three components do *not* have to add up to the SS for between-cell variability (i.e., SS_{bet}). This complicates the analysis, because there is more than one legitimate way to partition the total sum of squares.

5. The simplest way to analyze an unbalanced design is to use analysis of *unweighted means*. This method applies best when the cell sizes differ only slightly and by accident. If the cell sizes differ considerably and systematically (e.g., the cell sizes reflect the relative sizes of subgroups in the population), more sophisticated methods are called for. If the loss of subjects from different cells is related to the experimental conditions, the validity of the experimental conclusions becomes questionable.

6. The analysis of unweighted means proceeds as though the design were balanced with a cell size equal to the *harmonic mean* of all the cell sizes and with MS_W calculated as it is in a one-way ANOVA with groups of unequal size.

Definitions of Key Terms

Two-way (independent-groups) ANOVA A statistical procedure in which independent groups of observations differ on two independent variables, but only one dependent variable is being analyzed.

Factor An independent variable with at least two different levels; it may involve either conditions created by the experimenter or preexisting differences among subjects.

Completely crossed factorial design In the two-factor case, a design in which every level of the first factor is combined with every level of the second factor to form a matrix of *cells,* and there is at least one observation in every cell. This design, often just called a *factorial* design, can easily be generalized to accommodate any number of factors.

Balanced design A factorial design in which all of the cells contain the same number of observations.

Marginal mean The mean for any one level of one factor, averaging across all the levels of the other factor. If a two-way design is represented as a matrix, the marginal means are the means of each row and each column.

Interaction The variability among cell means not accounted for by variability among the row means or the column means. If the effects of one factor change at different levels of the other factor, some amount of interaction is present.

Ordinal interaction An interaction in which the effects of one factor change in amount, but not direction, with different levels of the other factor. In a graph of cell means, the lines would slant in the same direction but at different angles and would not cross. Despite the presence of this type of interaction, the main effects may still be interpretable.

Disordinal interaction An interaction in which the direction (or order) of the effect for one factor changes for different levels of the other factor. In a graph of cell means, the lines would either cross or slant in different directions. This type of interaction usually renders the main effects meaningless.

Degrees of freedom (df) tree A diagram that shows the number of degrees of freedom associated with each component of variation in an ANOVA design.

Familywise alpha In a two-way ANOVA, it is appropriate to follow up any of the three F ratios that are significant with post hoc comparisons. The comparisons that follow one of the F ratios can be thought of as a *family* of comparisons, so the two-way ANOVA consists of three such families. It is considered reasonable to control the overall alpha separately for each family of comparisons. The resulting probability of making at least one Type I error in an entire family of comparisons is the familywise alpha, α_{FW}. It is often preferable to control α_{FW} than to control the more general experimentwise alpha, α_{EW}.

Simple effect The effect of one factor at only one level of the other factor. In a two-way design, any single row or column represents a simple effect.

Interaction contrast A 2×2 comparison that is a subset of a larger design. When the interaction is significant for the full design, it may be appropriate to test the interaction in the various 2×2 subsets to localize the effect.

Analysis of unweighted means A procedure for analyzing a design that was intended to be balanced but in which cell sizes differ slightly as a result of incidental losses of data.

Key Formulas

SS within cells (raw-score form):

$$SS_W = \Sigma X^2 - \frac{1}{n}\Sigma(cs_i^2)$$

Formula 16.1

SS between-cells, as though the cells were groups in a one-way ANOVA (this component is eventually subdivided into SS_R, SS_C, and SS_{inter}):

$$SS_{bet} = \frac{1}{n}\Sigma(cs_i^2) - \frac{T_T^2}{N_T}$$

Formula 16.2

SS corresponding to the variability among the row means:

$$SS_R = \frac{1}{n_R}\Sigma R_i^2 - \frac{T_T^2}{N_T}$$

Formula 16.3A

SS corresponding to the variability among the column means:

$$SS_C = \frac{1}{n_C}\Sigma C_i^2 - \frac{T_T^2}{N_T}$$

Formula 16.3B

SS for interaction in terms of previously computed SS components (for a balanced design):

$$SS_{inter} = SS_{bet} - SS_R - SS_C$$

Formula 16.4

The total sum of squares in terms of its components (when the design is balanced):

$$SS_{total} = SS_W + SS_R + SS_C + SS_{inter}$$

Formula 16.5

The total df in terms of its components:

$$df_{total} = df_W + df_R + df_C + df_{inter}$$

Formula 16.6

The df components in a two-way ANOVA:

$$df_{total} = N_T - 1, \quad \text{or} \quad nrc - 1$$
$$df_W = N_T - rc, \quad \text{or} \quad nrc - rc$$
$$df_r = r - 1$$
$$df_c = c - 1$$
$$df_{inter} = (r - 1)(c - 1)$$

Formula 16.7

REPEATED MEASURES ANOVA

You will need to use the following from previous chapters:

Symbols
s^2: Unbiased variance of a sample
R_i: Sum of scores in the ith row of a two-way ANOVA
C_i: Sum of scores in the ith column of a two-way ANOVA
T_i: Sum of scores in the ith group
T_T: Sum of all observations in an experiment
N_T: Total number of observations in an experiment

Formulas
Formula 14.9: SS_w from raw scores
Formula 14.11: SS_{bet} from raw scores
Formula 16.2: SS_{total} from raw scores
Formulas 16.3A and 16.3B: SS_R and SS_C from the two-way ANOVA
Formula 16.4: SS_{inter} from the two-way ANOVA (found by subtraction)

Concepts
Matched t-test
SS components of the one-way ANOVA
SS components of the two-way ANOVA
Interaction of factors in a two-way ANOVA

17

C H A P T E R

A

**CONCEPTUAL
FOUNDATION**

In describing various experimental designs for which the matched t-test would be appropriate (see Chapter 13, Section A), I mentioned a hypothetical experiment in which the number of emotion-evoking words (e.g., "funeral," "vacation") a subject could recall was to be compared with the number of neutral words recalled. Suppose, however, that the researcher is concerned that the number of words evoking positive emotions that a subject recalls might be different from the number evoking negative emotions. The researcher would like to design the experiment with three types of words: positive, negative, and neutral. Of course, three matched t-tests could be performed to test each possible pair of word types, but, as you remember from Chapter 14, this approach does not control Type I errors, and the potential for error would get worse if more specific types of words were added (e.g., words evoking sadness or anger). What is needed is a statistical procedure that expands the matched t-test to accommodate any number of conditions, just as the one-way ANOVA expands the t-test for independent groups. There is such a procedure, and it is called the repeated measures (RM) ANOVA. This chapter will deal only with the **one-way repeated measures ANOVA** (i.e., one factor or independent variable); two-way and higher RM ANOVAs are not uncommon, but their analysis is beyond the scope of this text (except for the two-way mixed design; see Chapter 18). As with the matched

Table 17.1				
	Subject	**Neutral Words**	**Positive Words**	**Negative Words**
	1	12	16	15
	2	10	11	13
	3	14	14	18
	4	8	9	11
	5	12	13	15
	6	16	15	18
	Column Totals	72	78	90

t-test, the analysis is the same whether the measures are repeated on the same subjects or the subjects are matched across conditions.

Suppose that the above-mentioned experiment was performed with six subjects, each of whom was presented with all three types of words mixed together. Table 17.1 shows the number of words each subject correctly recalled from each category.

Calculation of an Independent-Groups ANOVA

To show the advantage of the RM ANOVA, I will first calculate the ordinary one-way ANOVA, as though the scores were not connected across categories—that is, as though three separate groups of subjects were presented with only one type of word each. First I will calculate SS_w using Formula 14.9. This requires squaring all of the scores and adding them up: $\Sigma X^2 = 3340$. I will also use the sum for each treatment (T_i), given at the bottom of each column in Table 17.1, to yield the following result:

$$SS_w = \Sigma X^2 - \Sigma\left(\frac{T_i^2}{n_i}\right) = 3340 - \left(\frac{72^2}{6} + \frac{78^2}{6} + \frac{90^2}{6}\right) = 3340 - 3228 = 112$$

Next I calculate SS_{bet} using Formula 14.11, but I will change the name of this component to SS_{RM} (short for $SS_{repeated-measures}$). (This will help to prevent confusion when I compare the one-way to a two-way design, in which SS_{bet} has a broader meaning.) To use Formula 14.11 I need to know the total of all scores (T_T). A quick way to get T_T is by finding $\Sigma T_i = 72 + 78 + 90 = 240$. The first term of the formula was already calculated as part of the computation of SS_w.

$$SS_{RM} = \Sigma\left(\frac{T_i^2}{n_i}\right) - \frac{T_T^2}{N_T} = 3228 - \frac{240^2}{18} = 3228 - \frac{57,600}{18} = 3228 - 3200 = 28$$

The MSs are as follows:

$$MS_{RM} = SS_{RM}/df_{RM} = 28/2 = 14$$

$$MS_W = SS_W/df_W = 112/15 = 7.47$$

Finally, the F ratio equals $MS_{RM}/MS_W = 14/7.47 = 1.87$.

The critical F for 2, 15 df ($\alpha = .05$) is 3.68, so the calculated F is not even near statistical significance. The null hypothesis that the population means for the three types of words are all equal cannot be rejected. This should not be surprising considering how small the groups are and how small the differences among groups are compared to the variability within each group.

The One-Way RM ANOVA as a Two-Way Independent ANOVA

When you realize that the scores in each row of Table 17.1 are all from the same subject, you can notice a kind of consistency. For instance, the score for negative words is the highest of the three conditions for all but one of the subjects. In the two-group case (Chapter 13) we were able to take advantage of such consistency by focusing on the difference scores, thus converting a two-group problem to a design involving only one group of scores. There is also a simple way to take advantage of this consistency within a one-way RM ANOVA, but in this case we take a one-factor (RM) design and treat it as though it were a two-way (independent-groups) ANOVA. The levels of the second factor are the different subjects. Table 17.2 is a modification of Table 17.1 that highlights the use of the (six) different subjects as levels of a second factor. Notice that the table looks like it corresponds to a two-way completely crossed factorial design with one measure per cell (so it is also balanced).

After studying the main example of Section A of Chapter 16, the idea of using subjects as different levels of a factor should not seem strange. It made sense in

Subject	Neutral Words	Positive Words	Negative Words	Row Totals	Table 17.2
1	12	16	15	43	
2	10	11	13	34	
3	14	14	18	46	
4	8	9	11	28	
5	12	13	15	40	
6	16	15	18	49	
Column Totals	72	78	90	240	

that example to separate men and women as two levels of a second factor, because the two genders differed consistently on the dependent variable. When each subject receives the same set of related treatments, individual differences also tend to have a type of consistency that can be exploited. A subject who scores highly at one level of the treatment tends to score highly at the other levels, and vice versa. For instance, in Table 17.1, subject number 6 scores highly at all three levels, whereas subject number 4 has consistently lower scores. It is not surprising that some subjects have better recall ability than others and that this ability applies to a variety of conditions; subjects who recall more positive words than other subjects are likely to be above average in recalling negative and neutral words, as well. Stated another way, the correlation should be high for any pair of treatment levels. You may remember that higher correlations meant higher t values in the matched t-test. The same principle applies here. With three treatment levels, however, the situation is more complicated. There are three possible pairs of levels, and therefore three correlation coefficients to consider. Moreover, if you would like to use the direct-difference method of Chapter 13, three sets of difference scores could be created. And with even more treatment levels, the complexity grows quickly. Fortunately, the two-way ANOVA approach makes it all rather simple, as described below.

To perform a one-way RM ANOVA on the data of Table 17.2, we can perform a two-way ANOVA (as described in the previous chapter) to obtain the SS components, but then we need to do something different with those SSs. In Section B, I will modify the notation of the two-way ANOVA formulas to make them usable for the one-way RM ANOVA. In the meantime I will use the appropriate computational formulas from Chapter 16.

Calculating the SS Components of the RM ANOVA

The first thing you should notice when trying to perform the two-way ANOVA with the data from Table 17.2 is that you can calculate neither SS_W nor MS_W, because there is only one measure per cell, and hence no within-cell variance. This would be a problem if we were planning to complete the two-way ANOVA procedure, but we are going to use the SS components in a different way. As it turns out, instead of being a problem, the lack of an MS_W term just simplifies the calculations. Unable to calculate SS_W using Formula 16.1, we proceed with the calculation of SS_{bet} using Formula 16.2. Bear in mind, however, that the sum of each cell (cs_i) is just the single score that is in that cell (i.e., $n = 1$). Therefore, $\Sigma(cs_i^2)$ is the same as ΣX^2, and the formula for SS_{bet} becomes exactly the same as the formula for SS_{tot}. (See Formula 14.13, page 467, which applies to any of the ANOVA designs in this text.)

$$SS_{tot} = \Sigma X^2 - \frac{T_T^2}{N_T} = 3340 - \frac{240^2}{18} = 3340 - 3200 = 140$$

Next, we can calculate the SS for rows and columns using Formulas 16.3A and 16.3B. SS_R corresponds to the variability among subject means, so in the RM ANOVA it is called $SS_{subject}$ (or SS_{sub}, for short); SS_C corresponds to the variability among treatment means, the component that I have already identified as SS_{RM}. The number of scores in each row (n_R) is equal to the number of columns (i.e., treatment levels), so $n_R = 3$. The number of scores in each column (n_C) is equal to the number of rows (i.e., the number of different subjects), so $n_C = 6$.

$$SS_R = \frac{1}{n_R}\Sigma R_i^2 - \frac{T_T^2}{N_T} = \frac{1}{3}(43^2 + 34^2 + 46^2 + 28^2 + 40^2 + 49^2) - 3200$$

$$= 3302 - 3200 = 102$$

$$SS_C = \frac{1}{n_C}\Sigma C_i^2 - \frac{T_T^2}{N_T} = \frac{1}{6}(72^2 + 78^2 + 90^2) - 3200$$

$$= 3228 - 3200 = 28$$

Now we can find SS_{inter} by subtraction, using Formula 16.4 (note that SS_{bet} is the same as SS_{tot}, as calculated above):

$$SS_{inter} = SS_{bet} - SS_{sub} - SS_{RM} = 140 - 102 - 28 = 10$$

In order to calculate the MSs we need to know the df components. The df for rows (df_R or df_{sub}) is one less than the number of rows (i.e., the number of subjects), which equals $6 - 1 = 5$; df_C (or df_{RM}) is one less than the number of columns (i.e., the number of treatment levels), which equals $3 - 1 = 2$. The df for interaction (df_{inter}) equals df_{sub} multiplied by $df_{RM} = 5 \times 2 = 10$.

The MS for subjects would be SS_{sub}/df_{sub}, but this MS plays no role in the RM ANOVA, so it is never calculated. The rationale for ignoring MS_{sub} will be discussed below. The MS for the independent (i.e., treatment) variable, MS_{RM}, equals $SS_C/df_C = 28/2 = 14$. The MS for the interaction equals $SS_{inter}/df_{inter} = 10/10 = 1$.

The F Ratio for the RM ANOVA

We are finally ready to calculate the F ratio for the RM ANOVA. The only F ratio we will calculate is the F ratio that tests differences among the treatment levels. The numerator for this F ratio, therefore, is MS_{RM}. Because there is no MS_W it may not be obvious, however, which MS should be used for the denominator. Had I not just mentioned that MS_{sub} is generally not calculated, you might have guessed that this MS would be used as the error term. As it turns out, for reasons to be explained shortly, the MS used for the error term is MS_{inter}. Thus the F ratio for the one-way RM ANOVA is MS_{RM}/MS_{inter}. For this example, $F = 14/1 = 14$.

Comparing the Independent ANOVA with the RM ANOVA

The new calculated F ratio is considerably higher than the critical F for the one-way independent ANOVA, but the degrees of freedom have changed, requiring us to find a new critical F, as well. The numerator df remains the same for both types of ANOVA (in this case, $df_{num} = 2$), but for the RM ANOVA the denominator df becomes smaller. In this example, df_{denom} (i.e., df_{inter}) is 10, whereas df_{denom} for the independent ANOVA (i.e., df_W) was 15. The new critical F is 4.10, which is *higher* than the previous critical F (3.68), because of the reduced df for the error term. However, whereas the critical F is slightly higher for the RM ANOVA, the calculated F is much higher (14) than it was for the independent ANOVA (1.87). Consequently, the RM ANOVA is significant—and would be even if an alpha much smaller than .05 were used, implying that the null hypothesis concerning the treatment levels can be rejected with much confidence. As is often the case, the increase in the calculated F that results from the matching of scores far outweighs the small increase in the critical F. The advantage of the RM design can be seen more clearly by comparing the SS components for this design with those from the independent ANOVA.

The total SS for the one-way independent ANOVA is divided into only two pieces: SS_{bet} and SS_W. The total SS for the one-way RM ANOVA is divided into three pieces: SS_{sub}, SS_{RM}, and SS_{inter}. First, the calculations above can be used to show that SS_{tot} is the same for both designs; regrouping the scores does not change the total SS, as you saw in the previous chapter. For the RM design, $SS_{tot} = 140$. For the independent design, $SS_{tot} = SS_{bet} + SS_W = 28 + 112 = 140$. It should come as no surprise that $SS_{bet} = SS_{RM}$. Therefore, the comparison becomes quite simple:

$$\text{Independent ANOVA: } SS_{tot} = SS_{bet} + SS_W$$

$$\text{RM ANOVA: } SS_{tot} = SS_{RM} + SS_{sub} + SS_{inter}$$

Because SS_{tot} is the same in both designs and $SS_{bet} = SS_{RM}$, it must be that the sum of SS_{sub} and SS_{inter} will always equal SS_W. This is true for the present example, in which $SS_{sub} = 102$ and $SS_{inter} = 10$; adding these two components we get $102 + 10 = 112$, which equals SS_W.

The Advantage of the RM ANOVA

Now it is easy to see the possible advantage of the **repeated measures (RM) design.** The SS_W that would be calculated for an RM design if independent groups were assumed (i.e., if the matching were ignored) is divided into two pieces when the scores are matched. One of these pieces, SS_{sub}, is ignored. The error term is based on SS_{inter}, which will thus be smaller than SS_W, unless SS_{sub} is zero. As you will see below, the better the matching, the smaller SS_{inter} becomes, and the larger SS_{sub} becomes. When the matching is totally ineffective (i.e., zero correlation between any pair of treatment levels), SS_W is divided equally into SS_{sub} and SS_{inter}. However, the df are also divided, so that MS_W and MS_{inter} become equal when the matching is random. To the extent that the matching gets any better than random, MS_{inter} becomes smaller than MS_W. MS_{inter} actually gets larger than MS_W in the unusual case when the matching works in reverse (i.e., the correlation is negative between pairs of treatment levels). If the correlation is positive but very low, the slight increase in the calculated F ratio may not outweigh the loss in degrees of freedom and the resulting increase in the critical value for F. As with the matched t-test, poor matching combined with fairly small sample sizes can actually mean that the results will be closer to significance with the independent-groups test. Conceptually, matching works the same way in the RM ANOVA as it does for the matched t-test, but it does get more complicated when there are more than two treatment levels. A graph can make the effect of matching more understandable.

Picturing the Subject by Treatment Interaction

If we are going to calculate the RM ANOVA as though it were a two-way ANOVA, it can be helpful to extend the analogy by graphing the cell means (which are just the individual scores, in this case). Although either factor could be placed on either axis, the graph will be more meaningful if we place the treatment levels along the X axis and draw separate lines for each subject, as in Figure 17.1 (which uses the data from Table 17.1). By noting the degree to which the lines are parallel, you can estimate whether SS_{inter} will be relatively large or small. (If the lines were all parallel, SS_{inter} would equal zero.) In Figure 17.1 you can see that the lines are fairly close to being parallel. The more the lines tend to be parallel, the smaller the SS for the interaction, and therefore the smaller the denominator of the F ratio for the RM ANOVA. Of course, a smaller denominator means a larger calculated F ratio and a greater chance of statistical significance.

Interactions are usually referred to in terms of the two factors involved, such as the subject by treatment (or subject $\times$ treatment) interaction. A relatively small subject $\times$ treatment interaction indicates a consistent pattern from subject

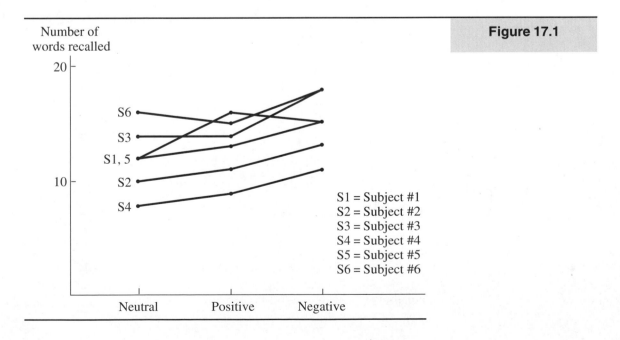

Figure 17.1

to subject. That is, different subjects are responding in the same general way to the different treatment conditions. In Figure 17.1, with a couple of exceptions, the number of words subjects recall goes up a little from neutral to positive words and then increases to a greater degree from positive to negative words. This consistency is what the RM ANOVA capitalizes on—just as the matched *t*-test was based on the similarity of the difference scores. Note that the overall level of each subject can be very different (compare subjects 4 and 6 in Figure 17.1) without affecting the interaction term. Differences in overall level between subjects (i.e., average recall over the three word types) contribute to the size of SS_{sub}, but fortunately this component does not contribute to the error term. Again, you can draw a comparison to the matched *t*-test, in which subjects' absolute scores are irrelevant, and only the consistency of the difference scores is important.

Comparing the RM ANOVA to a Matched *t*-Test

The similarity between the RM ANOVA and a matched *t*-test may be more obvious if you realize that in the case of only two treatment levels, either procedure can be applied (just as either the two-group *t*-test or the one-way ANOVA can be applied to two independent groups). As an illustration, I will calculate the RM ANOVA for the data from the matched *t*-test in Section D of Chapter 13. I begin by finding the row and column totals, as shown in Table 17.3. For these data, $\sum X^2 = 445$ and $T_T = 69$. Therefore,

$$SS_{tot} = 445 - \frac{69^2}{12} = 445 - \frac{4761}{12} = 445 - 396.75 = 48.25$$

The SS components are found using Formulas 16.3A and B, as in the RM ANOVA example above.

$$SS_{sub} = \frac{1}{2}(196 + 100 + 121 + 81 + 81 + 256) - 396.75$$

$$= 417.5 - 396.76 = 20.75$$

$$SS_{RM} = \frac{1}{6}(39^2 + 30^2) - 396.75 = \frac{1}{6}(2421) - 396.75$$

$$= 403.5 - 396.75 = 6.75$$

$$SS_{inter} = 48.25 - 20.75 - 6.75 = 20.75$$

The relevant MS values are

$$MS_{RM} = SS_{RM}/k - 1 = 6.75/1 = 6.75$$

$$MS_{inter} = SS_{inter}/(k - 1)(n_s - 1) = 20.75/5 = 4.15$$

Finally, the *F* ratio $= MS_{RM}/MS_{inter} = 6.75/4.15 = 1.63$.

Table 17.3

Subject	Rating of Same Gender	Rating of Opposite Gender	Row Totals
1	9	5	14
2	5	5	10
3	8	3	11
4	4	5	9
5	6	3	9
6	7	9	16
Column Totals	39	30	69

The matched t value calculated for the same data was 1.28, which when squared is equal (within rounding error) to the F ratio calculated above, as will always be the case. The standard deviation of the difference scores (s_D) was 2.88 (see Chapter 13), which when squared equals 8.3—exactly twice the value of MS_{inter} (this will always be true). Thus in the two-group case, the variability of the difference scores is a direct function of the amount of subject × treatment interaction.

This relationship can also be shown graphically. Figure 17.2 graphs the data from the example above, and you can see a good deal of subject by treatment interaction. This interaction helps to explain why the matched t was not significant. (The matched t will always correspond to the same p value as the F from

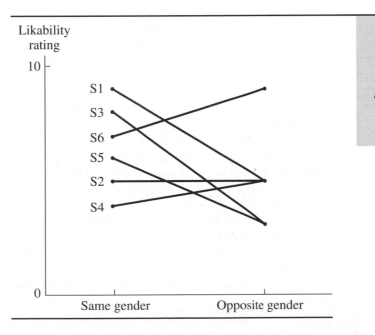

Figure 17.2

Graph of Data from Table 17.3, Depicting Subject by Condition Interaction

the RM ANOVA.) In fact, the interaction is so large that the correlation between the two sets of scores is zero (that is why $SS_{sub} = SS_{inter}$). The slope of each line represents the size of a difference score (negative slopes indicate negative difference scores); more variability in difference scores means more variability in the slopes of the lines, which in turn means a larger interaction term. When the differences are all the same, there is no variability in difference scores and the lines are all parallel, indicating no interaction. When there are more than two treatment levels, there is more than one set of difference scores. The MS for interaction is a function of the average of the variances for each set of difference scores.

The issues to consider in deciding when the repeated measures design is appropriate are the same as those discussed for the matched *t*-test, except that some matters become more complicated when there are more than two treatment levels. Perhaps the least problematic repeated-measures design is one in which the different treatment levels are presented simultaneously, or mixed randomly together, as in the word recall example above. Including more than two levels does not automatically create any problems. On the other hand, successive repeated measures designs can get complicated.

Dealing with Order Effects

In describing the successive repeated measures design with two levels (see Chapter 13), I pointed out the danger of **order effects.** For example, if the same subject is to calculate math problems during both happy and sad pieces of music, whichever type of music is played second will have an unfair advantage (practice) or disadvantage (fatigue) for the subject, depending on the details of the study. The solution in Chapter 13 was to use *counterbalancing*: Half the subjects get the happy music first and the other half get the sad music first. Counterbalancing becomes more complicated (though no less necessary) when there are more than two treatment levels.

If you wish to study the effects of classical music, jazz, and heavy metal rock music on math calculations, six different orders are possible (three choices for the first piece of music times two choices for the second times one choice for the third). If 48 subjects were in the study, 8 would be assigned to each of the six orders. However, if a fourth type of music were added (e.g., new age music), there would be 24 possible orders, and only two of the 48 subjects could be assigned to each order. For any study with five or more levels to be administered successively, complete counterbalancing, as described above, becomes impractical. Fortunately, there are clever counterbalancing schemes, such as the *Latin square design,* that do not require all possible orders to be presented, and yet do eliminate simple order effects. Because it greatly simplifies experimental procedures, the Latin square design is commonly employed even when there are only three or four treatment levels. Such designs are discussed in greater detail in Section C.

Differential Carryover Effects

Whereas counterbalancing schemes can eliminate the confounding effects of practice, fatigue, and other phenomena that result from the order of the treatment levels, such schemes cannot be counted upon to eliminate **carryover effects,** which can differ for each particular order of treatments. For instance, if the rock music is played loudly enough, it might affect the subject's ability to appreciate (or even to hear) the next piece of music played, thus attenuating the effect of that music. This effect would not be symmetrical (e.g., hearing classical music would not affect a subject's ability to hear a piece of rock music right after), and would therefore not be balanced out by counterbalancing.

With more than two treatment levels, the possibility of complex, asymmetrical carryover effects increases. In some cases, leaving more time between the presentation of treatment levels or imposing some neutral, distracting task between the experimental conditions can eliminate carryover effects. In other cases, carryover cannot be eliminated. For instance, a cognitive psychologist may want to know which of three different types of hints is the most helpful to subjects who are solving a particular problem. The psychologist cannot expect a subject to forget a previous hint and use only the present one; different subjects must be used to test different hints.

The Randomized-Blocks Design

When carryover effects make it inappropriate to use repeated measures, the best alternative is to match subjects as closely as possible. Suppose you wanted to compare three methods for teaching algebra to 14-year olds. The students could be matched into groups of three, so that all three would be as similar as possible in mathematical ability (based on previous math tests and grades). The three matched students in each group would then be randomly assigned to the three different teaching methods. (This is the same procedure I described for the matched *t*-test, in which each group consisted of a pair of subjects.) At the end of a semester all students would be given the same algebra test; the scores would be analyzed using a RM ANOVA, as though the three students in each group were really the same subject measured under three different conditions. Of course, the matching procedure would be the same for four teaching methods, or any other number of treatment levels. In this way, repeated measures can be viewed as the closest possible form of matching; the same subject measured under three conditions will usually produce scores that are more similar than three matched subjects (although a particular subject will vary in performance somewhat from time to time). Fortunately, when repeated measures are not appropriate, good matching can yield much of the same advantage.

I have been talking about the goal of good matching in terms of producing similar scores for the different conditions, but it is important to remember that we also want the scores to differ between the conditions, with some conditions associated with consistently higher scores than others. We expect the scores from the same subject (or matched subjects) to be more similar than scores from entirely different subjects under the same conditions, but it is also important that the treatment means differ, and that each subject produce the same general pattern of responses to the different treatment conditions (i.e., parallel lines on a graph of the scores but not flat, or horizontal, lines).

In the context of ANOVA, a group of subjects who are matched together on some relevant variable is called a *block*. A design in which members of a block are randomly assigned to different experimental conditions is known as a **randomized-blocks (RB) design**; the matched pairs design described in Chapter 13 is the simplest form of randomized-blocks design. The ANOVA performed on data from this type of design is called a randomized-blocks ANOVA (RB ANOVA). When the number of subjects in a block is the same as the number of treatment levels, the RB ANOVA is performed in exactly the same way as the RM ANOVA (just as the matched *t*-test is calculated exactly like a repeated-measures *t*-test). Therefore, I will use the term RM ANOVA, regardless of whether it is applied to an RM design or a simple RB design. It is possible to create more complicated RB designs, in which the number of subjects in a block is some multiple of the number of treatment levels, but such designs are not common, and I will not consider their analysis. An example of the simple type of RB design will be presented in Section D; the formulas and procedures used to analyze that RB design will be described in detail in Section B.

Exercises

1. The data below were originally presented in exercise 11B8 and were reproduced in exercise 13B4, for which you were asked to perform a matched *t*-test. Now perform an RM ANOVA on these data, and compare the calculated F ratio with the t value you found in exercise 13B4.

Segment	Male Rater	Female Rater
1	2	4
2	1	3
3	8	7
4	0	1
5	2	5
6	7	9

* 2. The data in the table below are from an experiment on short-term memory involving three types of stimuli: digits, letters, and a mixture of digits and letters.

Draw a graph of these data and describe the degree of interaction between the various pairs of levels.

Subject	Digit	Letter	Mixed
1	6	5	6
2	8	7	5
3	7	7	4
4	8	5	8
5	6	4	7
6	7	6	5

3. The progress of 20 patients is measured every month for 6 months. An RM ANOVA on these measurements produced the following sums of squares: $SS_{total} = 375$; $SS_{RM} = 40$; $SS_{subject} = 185$.

a) Complete the analysis and find the F ratio for the RM ANOVA.

b) Calculate the F ratio that would be obtained for an independent-groups ANOVA.

*4. In exercise 14A7, eight subjects were tested for problem-solving performance in each of four drug conditions, yielding the following means and standard deviations.

	Mari-juana	Amphet-amine	Valium	Alcohol
$\overline{X}$	7	8	5	4
s	3.25	3.95	3.16	2.07

If the same eight subjects were tested in all four conditions, and if SS_{inter} were equal to 93, how large would the *F* ratio for the RM ANOVA be?

5. Which of the following tends to increase the size of the *F* ratio in an RM ANOVA?
   a) Larger subject-to-subject differences (averaging across treatment levels)
   b) Better matching of the scores across treatments
   c) Larger subject by treatment interaction
   d) Smaller differences between treatment means

*6. An RM ANOVA on a certain set of data did not attain statistical significance. What advantage could be derived from performing an independent-groups ANOVA on the same data?
   a) A larger alpha level would normally be used.
   b) The numerator of the *F* ratio would probably be larger.
   c) The critical *F* would probably be smaller.
   d) The error term would probably be larger.

7. The chief advantage of the randomized blocks design (compared to the RM design) is that
   a) differential carryover effects are eliminated.
   b) the matching of scores is usually better.
   c) the critical *F* is reduced.
   d) all of the above.

*8. The chief advantage of counterbalancing is that
   a) subject by treatment interaction is reduced.
   b) simple order effects are prevented from contributing to differences among treatment means.
   c) differential carryover effects are eliminated.
   d) all of the above.

In the previous section I dealt with a repeated-measures design in which it was reasonable to mix together items from different treatment levels (i.e., the emotional content of the words). In this section I will consider a design in which such a mixture would be detrimental to performance. For quite some time, cognitive psychologists have been interested in how subjects solve transitive inference problems, such as the following. Bill is younger than Nancy. Tom is older than Nancy. Who is the youngest? There is some debate about the roles of various cognitive mechanisms (e.g., visual imagery) in solving such problems. It appears that the way the problem is presented may affect the choice of mental strategy, which in turn may affect performance. Consider a study comparing three modes of presentation: visual-simultaneous (both premises and the question appear simultaneously on a computer screen until the subject responds); visual-successive (only one of the three lines appears on the screen at a time); and auditory (the problem is read aloud to the subject—which, of course, is a successive mode of presentation). The subject is given the same amount of time to answer each problem, regardless of presentation mode. The dependent variable is the number of problems correctly answered out of the 20 presented in each mode.

Changing the presentation mode randomly for each problem would be disorienting to the subject, and probably would not be an optimal test of the research hypothesis. A reasonable approach would be to present three sets of transitive inference problems to each subject, each set being presented in a different mode.

BASIC STATISTICAL PROCEDURES

Table 17.4	Subject	First Block	Second Block	Third Block
	1	visual-sim	visual-succ	auditory
	2	visual-sim	auditory	visual-succ
	3	visual-succ	visual-sim	auditory
	4	visual-succ	auditory	visual-sim
	5	auditory	visual-sim	visual-succ
	6	auditory	visual-succ	visual-sim

However, if the three sets were presented in the same order to all the subjects, this alone could determine which condition would lead to the best performance. The solution is to counterbalance. In a completely counterbalanced design, all six of the possible orders would be used; see Table 17.4. As usual, I will minimize the calculations by creating the simplest possible design: one subject for each of the six orders.

The research hypothesis is that the mode of presentation will affect the subjects' performance on transitive inference problems. (The influence of a given presentation mode might be mediated by changes in mental strategy, but that aspect would require further exploration, perhaps in a separate study.) This research hypothesis can be tested by using the usual six-step procedure.

Step 1. State the Hypotheses

The null hypothesis for the one-way RM ANOVA is the same as for the one-way independent-groups ANOVA. With three treatment levels, H_0: $\mu_1 = \mu_2 = \mu_3$. The alternative hypothesis is also the same; H_A can only be stated indirectly by stating that H_0 is not true.

Step 2. Select the Statistical Test and the Significance Level

Because each subject is observed under all three conditions, and we expect a good degree of matching across conditions, the RM ANOVA is the appropriate procedure. The conventions with respect to alpha are no different for this design, so we will once again set $\alpha = .05$.

Step 3. Select the Samples and Collect the Data

Because this is an RM design, there is only one sample to select, which ideally should be a random sample from the population of interest. (More often, the sample is one of convenience.) Because we are considering the *same* subject's

Subject	Vis-sim	Vis-succ	Auditory	Row Sums
1	20	10	17	47
2	15	12	13	40
3	14	15	16	45
4	17	11	12	40
5	12	5	7	24
6	18	7	8	33
Column Sums	96	60	73	229

Table 17.5

performance at each treatment level, the means for the treatment levels cannot be attributed to differences in samples of subjects. The data for this hypothetical study appear in Table 17.5.

Step 4. Find the Region of Rejection

If the null hypothesis is true, the F value calculated for the RM ANOVA will represent the ratio of two independent estimates of the population variance, and will therefore follow one of the F distributions. To determine which F distribution is appropriate, we need to know the degrees of freedom for both the numerator and the denominator. The df for the numerator is the same as in the independent ANOVA; it is one less than the number of treatments, which for this example is $3 - 1 = 2$. The df for the denominator is the df for the subject $\times$ treatment interaction, which is one less than the number of subjects times one less than the number of treatments; in this case, $df_{denom} = (6 - 1)(3 - 1) = 5 \times 2 = 10$. The critical F with (2, 10) df and $\alpha = .05$ is 4.10. Thus the region of rejection is the tail of the $F(2, 10)$ distribution above the critical value of 4.10.

Step 5. Calculate the Test Statistic

We begin by calculating the various SS components. As it was for the two-way ANOVA, the total SS is divided into SS_R, SS_C, and SS_{inter}, but with no SS_W component. The formula for SS_{tot} is the same as the one used for the independent ANOVA (Formula 14.13); it is presented below as Formula 17.1

$$SS_{tot} = \Sigma X^2 - \frac{T_T^2}{N_T}$$

Formula 17.1

To calculate SS_{tot}, we need to know that $\Sigma X^2 = 3213$ and $T_T = 229$. Thus

$$SS_{tot} = 3213 - \frac{229^2}{18} = 3213 - \frac{52,441}{18} = 3213 - 2913.4 = 299.6$$

The formula for SS_{sub} is the same as Formula 16.3A with the labels changed. The term n_R is actually the number of treatment levels. In keeping with the notation of Chapter 16, I will use c (the number of columns in the data table) to refer to the number of levels of the repeated-measures factor. Thus Formula 17.2 reads as follows,

$$SS_{sub} = \frac{1}{c}\Sigma S_i^2 - \frac{T_T^2}{N_T}$$

Formula 17.2

where S_i represents any of the subject sums (i.e., the sum of all scores for any particular subject). S_i is the same as a row sum when the subjects form the rows of a matrix and the treatment levels are the columns, as in Table 17.5. We can now apply this formula to our example. Note that the second term has already been calculated above.

$$SS_{sub} = \frac{1}{3}(47^2 + 40^2 + 45^2 + 40^2 + 24^2 + 33^2) - 2913.4$$

$$= \frac{1}{3}(9099) - 2913.4 = 119.6$$

The formula for SS_{RM} is the same as Formula 16.3B, except that n_C is the number of different subjects, which we will represent as n_s. (This should be more clear than using r for the number of rows.) The total for each treatment level can be represented as T_i as it was for the one-way independent ANOVA. Thus Formula 17.3 appears as follows:

$$SS_{treat} = \frac{1}{n_s}\Sigma T_i^2 - \frac{T_T^2}{N_T}$$

Formula 17.3

Applying Formula 17.3 to the present example, we get

$$SS_{RM} = \frac{1}{6}(96^2 + 60^2 + 73^2) - 2913.4 = \frac{1}{6}(18,145) - 2913.4 = 110.8$$

The subject by treatment interaction (SS_{inter}) can now be found by subtraction, using a relabeled version of Formula 16.4, which I will call Formula 17.4:

$$SS_{inter} = SS_{tot} - SS_{sub} - SS_{RM}$$

Formula 17.4

For the present example, $SS_{inter} = 299.6 - 119.6 - 110.8 = 69.2$.

The next step is to divide the total number of degrees of freedom into components that match the SS. As usual, $df_{tot} = N_T - 1$ (where N_T is the total number of observations, or scores, which equals c times n_s). For this example, N_T equals $18 - 1 = 17$. The other components are found using the following formulas, which, collectively, are Formula 17.5:

$$df_{sub} = n_s - 1$$
$$df_{RM} = c - 1$$
$$df_{inter} = (n_s - 1)(c - 1) \qquad \text{Formula 17.5}$$

In this example, $n_s = 6$ and $c = 3$, so

$$df_{sub} = 6 - 1 = 5$$
$$df_{RM} = 3 - 1 = 2$$
$$df_{inter} = 5 \cdot 2 = 10$$

Note that the above df components add up to 17, which equals df_{tot}.

The two MSs of interest can be found by dividing the appropriate SS and df components, as follows:

$$MS_{RM} = SS_{RM}/df_{RM}$$
$$MS_{inter} = SS_{inter}/df_{inter}$$

For this example, the MS are

$$MS_{RM} = 110.8/2 = 55.4$$
$$MS_{inter} = 69.2/10 = 6.92$$

Finally, the F ratio for the RM ANOVA is found by Formula 17.6:

$$F = \frac{MS_{RM}}{MS_{inter}} \qquad \text{Formula 17.6}$$

The F ratio for the transitive inference experiment is $F = 55.4/6.92 = 8.0$.

Step 6. Make the Statistical Decision

Because the observed F ratio (8.0) is larger than the critical F ratio (4.10), our experimental result falls in the region of rejection for the null hypothesis—we can say that the result is statistically significant at the .05 level. The research hypothesis, that the population means corresponding to the different presentation modes would not be the same, has been supported. However, as with any one-way ANOVA with more than two groups, rejecting the complete null hypothesis does not tell us which pairs of population means differ significantly. Post hoc comparisons are usually conducted to locate significant differences more specifically. The procedures of Chapter 15 can be used, substituting MS_{inter} for MS_W. However, this substitution is only valid if a particular assumption (i.e., homogeneity of covariance) is satisfied. This issue will be discussed further under the topic of Assumptions later in this section and in Section C.

The Residual Component

When you are dealing with repeated measures, issues arise that you don't have to think about when using independent groups. For instance, consider the null hypothesis for the example above—that the three population means are all equal. There are two different ways this can occur in an RM design: with or without interaction. Consider just two subjects. In Figure 17.3, the two subjects differ in overall problem-solving ability, but the three treatment means are equal—and there is no interaction. Now look at Figure 17.4. The treatment means are *still* equal (they are the same as they were before), but now there is less difference between the two subjects—and there *is* an interaction. Expanding this description to populations of subjects, we can say that three population means can be equal because each subject has the same score under all three conditions, or they can be equal because subjects differ from treatment to treatment, but in such a way that these differences cancel out.

The latter possibility is considered a more realistic version of the null hypothesis (e.g., some subjects benefit when problems are presented in the auditory mode, some are hindered by it, with a net effect of zero). If we adopt this way of viewing the RM design (i.e., we do *not* assume the subject × treatment interaction to be zero), there is, fortunately, no change in the way the F ratio for the treatment effect is calculated. However, there is a difference in the way we view the structure of the F ratio.

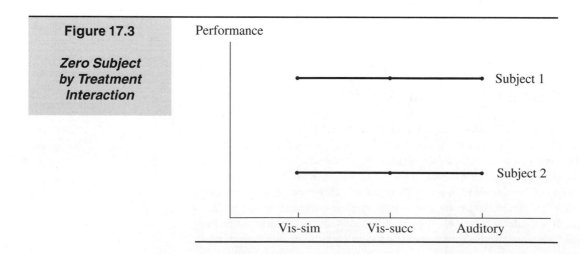

Figure 17.3

Zero Subject by Treatment Interaction

Performance

Subject 1

Subject 2

Vis-sim   Vis-succ   Auditory

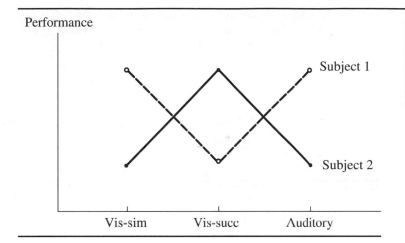

Performance

Vis-sim   Vis-succ   Auditory

Figure 17.4

Nonzero Subject by Treatment Interaction

For the independent ANOVA, I pointed out that F represents the following ratio:

$$F = \frac{\text{treatment effect} + \text{error}}{\text{error}}$$

In the case of the RM ANOVA, the denominator of the F ratio is based on MS_{inter}. If MS_{inter} is assumed to be zero in the population, then the size of the denominator could be attributed entirely to error. However, if MS_{inter} is *not* assumed to be zero, the denominator would be the sum of the actual MS_{inter} in the population plus error. The amount of MS_{inter} would also show up in the numerator along with the amount of error, because both components can affect the variability of the treatment means in a particular experiment. So, for the RM ANOVA, the F ratio is usually viewed in the following manner:

$$F = \frac{\text{treatment effect} + \text{interaction} + \text{error}}{\text{interaction} + \text{error}}$$

Notice that according to the null hypothesis, the treatment effect is expected to be zero, and therefore the F ratio is expected to be around 1.0.

Because the denominator of the F ratio is generally considered to be a sum of both the true amount of subject $\times$ treatment interaction and the error, the denominator is commonly referred to as $MS_{residual}$ (or MS_{res}, for short) rather than MS_{inter}. The term *residual* makes sense when you consider that SS_{inter} was found by subtracting both SS_{sub} and SS_{RM} from SS_{total}; what is left over is usually thought to contain a mixture of interaction and error components, and is therefore referred to in a more neutral way as $SS_{residual}$. Because there is no separate estimate of error (i.e., no SS_w), we cannot separate the effects of interaction and error on $SS_{residual}$, so this term is often preferred to the term SS_{inter} when dealing with the RM design. However, it is still common to refer to $SS_{residual}$ as the error term, or SS_{error}, so you can use either one when referring to SS_{inter} in the RM ANOVA.

Assumptions of the RM ANOVA

The assumptions underlying the hypothesis test conducted above are essentially the same as those for the independent ANOVA, except for an important additional assumption.

Independent Random Sampling

In the RM design, only one sample is drawn and ideally it should be selected at random from the population of interest, with each subject being selected independently of all others. In the RB design, subjects are matched together in small groups (usually the size of a block is equal to the number of treatment levels), and then the subjects within each block are *randomly assigned* to the different treatment conditions. Although ideally the subjects should be chosen at random from the general population, it is the random assignments of subjects to treatment conditions that is vital if the results of the RB design are to be considered valid.

Normal Distributions

As usual, it is assumed that the dependent variable follows a normal distribution in the population for each treatment level. In addition, the joint distribution that includes all levels of the independent variable is assumed to be a multivariate normal distribution. I described the bivariate normal distribution in Chapter 11. This is the distribution that is assumed in the two-group case of RM ANOVA (or the matched *t*-test). *Multivariate normal distribution* is the generic term for a joint normal distribution that can contain any number of variables. Unfortunately, when there are more than two variables (or more than two conditions in an RM ANOVA), the multivariate normal distribution that applies is impossible to depict in a two-dimensional drawing. Fortunately, the RM ANOVA is not very sensitive to departures from the multivariate normal distribution, so this assumption is rarely a cause for concern. If, however, this assumption is severely violated and the samples are fairly small, you may prefer to use nonparametric tests (such as the Friedman test, described in Chapter 21, Section C).

Homogeneity of Variance

In the RM or RB design, there is always the same number of observations at each treatment level, so you generally need have little concern about violating the assumption of homogeneity of variance.

Homogeneity of Covariance

The last assumption, **homogeneity of covariance,** does not apply when the groups are independent, or when an RM or RB design has only two treatment levels. However, when a matched (or repeated) design has more than two levels, the covariance (as defined in Chapter 11) can be calculated for each pair of levels. Homogeneity of covariance exists in the population only when all pairs of treatment levels have the same amount of covariance.

The implications of this last assumption can be difficult to understand, and there is some debate about what to do if it is violated. First, I should point out that if the third and fourth assumptions above are both true, then the population displays a condition called **compound symmetry.** When compound symmetry exists, the population correlation (ρ) between any pair of treatment levels is the same as between any other pair. Compound symmetry is certainly a desirable condition; when it is true for an RM ANOVA (along with the first two assumptions) you can find critical Fs as described earlier in this section. However, compound symmetry is a stricter assumption than is necessary. The third and fourth assumptions can be relaxed, as long as the variances and covariances follow a pattern that is called *sphericity,* or *circularity.*

Sphericity is usually defined mathematically in terms of the matrix of variances and covariances that apply to the various treatment levels and pairs of levels, but an easy way to understand this assumption is in terms of the amount of interaction (in the population) between any two levels of the independent variable: Sphericity implies that all of these interactions will be equally large. (This is the same as requiring that the variance of the difference scores will be the same no matter which pair of treatment levels you look at.) By graphing the results of the transitive inference experiment for each subject, we can see that the amount of interaction in the samples is not very large and seems to be about the same for any pair of levels (see Figure 17.5). By contrast, look at the hypothetical results depicted in Figure 17.6 (on page 586). Notice that there is very

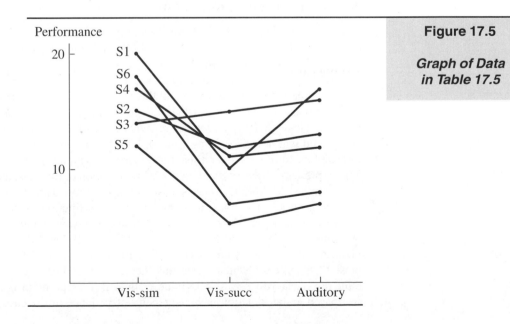

Figure 17.5

Graph of Data in Table 17.5

Figure 17.6

Graph of Hypothetical Data Illustrating a Lack of Sphericity

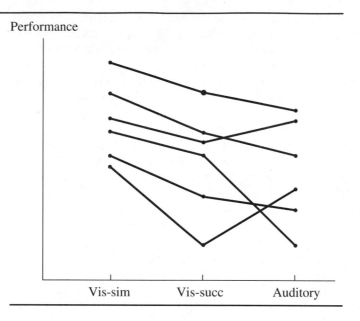

Performance

Vis-sim   Vis-succ   Auditory

little interaction between vis-sim and vis-succ in Figure 17.6, but a great deal between vis-succ and auditory (and therefore between vis-sim and auditory, as well). The total amount of subject × treatment interaction for the experiment would be moderately large—an average of the three pairwise amounts of interaction. But that average would be misleading.

Violations of the Assumptions

Recall that in the independent ANOVA the error term is MS_W, which is an average of the sample variances. The use of this average was justified by assuming homogeneity of variance. Similarly, the use of MS_{inter} as the error term in the RM ANOVA is justified by assuming that the pairwise interactions (or the variability of the difference scores) are all the same size (in the population)—that is, by assuming that sphericity exists in the population. However, it is important to note that although a population is somewhat more likely to show sphericity than compound symmetry, for many types of RM designs, especially those involving repeated measurements over time, the sphericity assumption will rarely be true. This represents a serious problem, for a combination of two reasons.

First, the RM ANOVA is *not* robust with respect to violations of sphericity, as it is with respect to violations of the normal distribution or homogeneity of variance assumptions; when sphericity does not apply, your null hypothesis will not conform to the F distribution that you think it will. Second, the effect of this violation is to inflate your calculated F; the F ratio is said to be "positively biased" in this case, which means that the actual Type I error rate will be larger than the alpha you use to determine critical F. This problem is more serious compared to the violation of homogeneity of variance in the case of two independent samples, because it is more likely to occur and more likely to produce a large increase in alpha.

Two very different solutions to this problem have become popular. One involves performing the RM ANOVA procedure described in this chapter but adjusting the df before looking up the critical value of *F.* The other involves a completely different statistical approach called the **multivariate analysis of variance (MANOVA).** The latter approach usually has more power, especially when the sample size is large, but it was not popular until recently because it was so tedious to calculate. Now that statistical software makes it easy to compute a MANOVA, this procedure is on its way to replacing the univariate RM ANOVA (as the procedure in this chapter is known) for most repeated-measures designs. Both solutions to the problem of lack of sphericity are discussed further in Section C, and MANOVA is further described in Chapter 18, Section C.

When to Use a Repeated-Measures or Randomized-Blocks Design

Basically, the RM design is almost always more powerful than an independent-groups design and should probably be employed whenever repeated measures on the same subject do not produce complications. The major complication is the problem of differential carryover effects. The effects of one particular treatment level may linger long enough to interfere with the effect of the next treatment level. Sometimes leaving a long enough interval between treatments or interposing a neutral task can eliminate the carryover, but sometimes a treatment level can permanently change the way a subject views the type of task being presented or even the entire experimental situation.

When the RM design is problematic, the RB design can supply much of the advantage of repeated-measures without the possibility of carryover effects. The RB design is helpful only when there is a good basis on which to match the subjects, however, and this is not always available. (For example, in a study of altruism under several different conditions, it is not likely that subjects can be matched on tendencies toward altruism before the study is conducted.) When there is insufficient justification for an RB design, independent groups must be used.

As with the matched *t*-test, the two basic types of RM design are *simultaneous* and *successive.* Unfortunately, the successive design can be complicated when there are more than two treatment levels. With more than three or four treatment levels, complete counterbalancing is not practical; alternatives are discussed in Section C.

The two basic types of RB design depend on whether the blocks are matched by the experimenter or occur naturally. An example of the latter case would be a study of families containing three children that explored the effect of birth order on some personality variable. The disadvantage of using natural groups is that you cannot infer that the change in the dependent variable is *caused* by whatever distinguishes the groups. Because of the difficulty involved in matching blocks of subjects, the experimental RB design is not very common, especially when the independent variable consists of more than three levels. However, because of the added power of RB designs and their avoidance of carryover effects, these designs should probably be used more often.

Finally, the levels of the independent variable in an RM ANOVA, as with an independent ANOVA, can be qualitative (e.g., different modes of presenting words: visual, auditory, tactile), or quantitative (e.g., different lengths of word lists to memorize). However, in the latter case, other methods such as regression or trend analysis may be more appropriate and powerful ways to analyze the data than the simple RM ANOVA.

Publishing the Results of an RM ANOVA

The results of a one-way RM ANOVA are reported in the same way as those for a one-way independent ANOVA. For the example above, the results could be presented in the following manner: "Simultaneous visual presentation produced the most accurate performance (M = 16), then auditory (M = 12.17), followed by visual-successive (M = 10). A repeated-measures ANOVA determined that these means were significantly different, F (2, 10) = 8.0, p <.05."

The Summary Table for RM ANOVA

The components of a repeated-measures analysis are sometimes displayed in a summary table that follows the general format described for the independent ANOVA, but with a different way of categorizing the sources of variation. In the independent ANOVA, the total sum of squares is initially divided into between-groups and within-groups components; for the two-way ANOVA, the between-groups SS is then divided further. However, in the RM ANOVA there is only one group of subjects, so the SS is initially divided into a *between-subjects* and a *within-subjects* component. For the one-way RM ANOVA, the between-subjects SS consists only of the component I have been referring to as SS_{sub}, and it is not of interest in this design. On the other hand, the within-subjects SS can be divided into two SS components. Within each subject there are several scores (one for each treatment level), and these will usually vary. There are two reasons for this variation. One reason is the fact that each treatment level may have a different mean (as the researcher usually hopes), and the other reason is interaction (each subject may have his or her own reactions to the treatment levels that do not conform exactly with the differences in means). Thus the within-subjects SS is divided into SS_{RM} and SS_{inter} (or SS_{res}) as shown in Table 17.6.

The successive division of the variation in a one-way RM design is also evident in the df tree that corresponds to this design (see Figure 17.7). The one-way RM ANOVA is fairly common in the psychological literature (though not as common as more complicated designs). The following excerpt describes the results of a typical example of the use of the one-way RM ANOVA in psychological research.

Table 17.6

Source	SS	df	MS	F	p
Between-Subjects	119.6	5			
Within-Subjects					
Between-Treatments	110.8	2	55.4	8.0	< .01
Interaction (Residual)	69.2	10	6.92		
Total	299.6	17			

Harte and Eifert (1995) measured urinary hormone concentrations for each of ten marathon runners under three different running conditions (outdoors; on an indoor treadmill while listening to environmental sounds; on an indoor treadmill while listening to amplified sounds of their own breathing), as well as in baseline (resting) and control (reading sports magazines) conditions. The main findings were as follows: "One-way repeated measures ANOVA . . . confirmed significant treatment effects for adrenaline ($F[4, 36] = 21.11$, $p < .0001$, $\epsilon = 0.6628$), noradrenaline ($F[4, 36] = 32.06$, $p < .0001$, $\epsilon = 0.3983$), and cortisol ($F[4, 36] = 41.08$, $p < .0001$, $\epsilon = .5309$). Newman-Keuls tests show that levels of adrenaline, noradrenaline, and cortisol were significantly higher after all three running conditions as compared with baseline levels or the control activity. However, Newman-Keuls tests indicate that there were no differences between hormonal concentrations at baseline and following the control session. There were also no hormonal differences among any of the three running conditions except that levels of noradrenaline and cortisol were significantly higher after the indoor run with internal attention focus [listening to breathing] than after the outdoor run" (pp. 52–53).

Note that the epsilon (ϵ) values following the p levels are based on estimates of the degree to which the sphericity assumption has been violated (see Section C). The extremely large F ratios are due to the dramatic differences in the levels of hormones secreted while running and while at rest. This experimental design is one in which a partial null hypothesis seems likely—that is, the three running

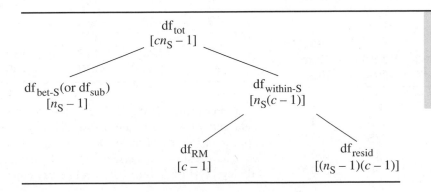

Figure 17.7

df Tree for the One-Way RM Design

conditions are equal; the baseline and control conditions are equal; but these two subsets of conditions are not equal to each other.

Exercises

* 1. In Section B of Chapter 16, the main example involved the effects of two factors on the performance of a clerical task. This exercise uses data from a hypothetical repeated-measures experiment based on just one of those factors, the level of noise: ordinary background noise, moderately loud popular music; and very loud (heavy metal) music. The number of tasks completed by each of five subjects under all three noise conditions is shown below.

Subject	Background Noise	Popular Music	Heavy Metal Music
1	10	12	8
2	7	9	4
3	13	15	9
4	18	12	6
5	6	8	3

a) Perform an RM ANOVA on the data. Is the F ratio significant at the .05 level? At the .01 level?
b) Present the results in a summary table.

2. To illustrate the effect of increasing subject-to-subject variance on the F ratio of the RM ANOVA, the data from exercise 1 have been modified to produce the table below. Ten points were added to each of the scores for subject 1, 20 points for subject 3, and 30 points for subject 4.

Subject	Background Noise	Popular Music	Heavy Metal Music
1	20	22	18
2	7	9	4
3	33	35	29
4	48	42	36
5	6	8	3

a) Perform an RM ANOVA on the data.
b) Present the results in a summary table.
c) Compare the summary table for this exercise with the one you made for exercise 1. Which of the SS components reflects the change in subject-to-subject variability? What effect does an increase in subject-to-subject variability have on the F ratio?

* 3. A psychophysiologist wishes to explore the effects of public speaking on the systolic blood pressure of young adults. Three conditions are tested. The subject must vividly imagine delivering a speech to one person; to a small class of twenty persons; or to a large audience consisting of hundreds of fellow students. Each subject has his or her blood pressure measured (mm Hg) under all three conditions. Two subjects are randomly assigned to each of the six possible treatment orders. The data appear in the table below.

Subject	One Person	Twenty Persons	Large Audience
1	131	130	135
2	109	124	126
3	115	110	108
4	110	108	122
5	107	115	111
6	111	117	121
7	100	102	107
8	115	120	132
9	130	119	128
10	118	122	130
11	125	118	133
12	135	130	135

a) Perform an RM ANOVA on the blood pressure data and write the results in words, as they would appear in a journal article. Does the size of the audience have a significant effect on blood pressure at the .05 level? (Hint: Subtract 100 from every entry in the above table before computing any of the SS. This will make your work easier without changing any of the SS components or *F* ratios.)

b) What would you do to minimize the possibility of carryover effects?

* 4. A statistics professor wants to know if it really matters which textbook she uses to teach her course. She selects four textbooks that differ in approach, and then matches her 36 students into blocks of four based on their similarity in math background and aptitude. Each student in each block is randomly assigned to a different text. At some point in the course, the professor gives a surprise 20-question quiz. The number of questions each student answers correctly appears in the table below.

Block	Text A	Text B	Text C	Text D
1	17	15	20	18
2	8	6	11	7
3	6	5	10	6
4	12	10	14	13
5	19	20	20	18
6	14	13	15	15
7	10	7	14	10
8	7	7	11	6
9	12	11	15	13

a) Perform an RM ANOVA on the data. Does it make a difference which textbook the professor uses?

b) Present the results of your ANOVA in a summary table.

c) Considering your answer to part a, what type of error could you be making (Type I or Type II)?

5. a) Perform a one-way independent-groups ANOVA on the data from exercise 4 above. (Note: You do not have to calculate any SS components from scratch; you can use the SSs from your summary table for exercise 4.)

b) Does choice of text make a significant difference when the groups of subjects are considered to be independent (i.e., the matching is ignored)?

c) Comparing your solution to this exercise with your solution to exercise 4 above, which part of the *F* ratio remains unchanged? What can you say about the advantages of matching in this case?

* 6. A neuropsychologist is exploring short-term memory deficits in people who have suffered damage to the left cerebral hemisphere. He suspects that memory for some types of material will be more affected than memory for other types. To test this hypothesis he presented six brain-damaged subjects with stimuli consisting of strings of digits, strings of letters, and strings of digits and letters mixed. The longest string that each subject in each stimulus condition could repeat correctly is presented in the table below. (One subject was run in each of the six possible orders.)

Subject	Digit	Letter	Mixed
1	6	5	6
2	8	7	5
3	7	7	4
4	8	5	8
5	6	4	7
6	7	6	5

a) Perform an RM ANOVA. Is your calculated *F* value significant at the .05 level?

b) Assuming that the three types of stimuli are recalled equally well by subjects without brain damage, what can you conclude about the effects on memory of damage to the left cerebral hemisphere?

c) Based on the graph you drew of these data for exercise 17A2, would you say that the RM ANOVA is appropriate for these data? Explain.

* 7. A school psychologist is interested in determining the effectiveness of an antidrug film on the attitudes of eighth-grade students. Each student's antidrug attitude is measured on a scale from 0 to 20 (20 representing the strongest opposition to drug use) at four times: the day before the film; as students are sitting in class waiting for the film to start; immediately after the film; and the next day in class. The data for eight subjects appear in the table below.

Subject	Day Before	Prior to Film	After Film	Day After
1	14	15	18	17
2	10	13	19	17
3	15	15	18	18
4	13	16	20	18
5	7	9	15	16
6	10	9	14	11
7	16	17	19	19
8	8	10	16	15

a) Perform an RM ANOVA on these data to determine if there is a difference in attitude over time at the .01 level.

b) Present the results of your ANOVA in a summary table.

8. A clinical psychologist wants to test the effects of exercise and meditation on moderate depression. She matched 30 of her patients into blocks of three based on the severity of their depression and their demographic characteristics. Patients were randomly assigned from each block to one or another of the following three treatments: aerobic exercise; medita-

tion; and reading inspirational books (control). After two daily 30-minute sessions of the assigned treatment over the course of a month, each subject's level of depression was measured with a standard questionnaire. The depression scores for each condition appear in the table below.

Block	Aerobic Exercise	Meditation	Control
1	22	25	28
2	13	13	12
3	24	31	34
4	11	14	13
5	18	19	26
6	16	19	23
7	15	15	18
8	10	13	14
9	15	14	18
10	9	10	16

a) Perform an RM ANOVA on these data. Is your obtained F significant at the .05 level?

b) What can you conclude about the two treatments for depression?

OPTIONAL MATERIAL

Dealing with a Lack of Sphericity

When all of the assumptions of the RM ANOVA are true as stated in Section B, the RM ANOVA is usually the most powerful procedure for testing the equality of population means while keeping the Type I error rate to the alpha used for finding the critical value of F. However, when the sphericity assumption does not apply, the possible rise in the Type I error rate for the ordinary RM ANOVA can be too large to ignore. The problem is further complicated by the fact that it can be difficult to decide whether it is reasonable to assume sphericity in the population. The variances and covariances in your sample can be used to make an inference about sphericity in the population, but the available tests for this purpose have little power for small samples, which is when they are needed most. Because it is likely that the most common RM designs will not exhibit sphericity, you should not use the ordinary RM ANOVA without some precautions. One possibility, which will be discussed shortly, is to abandon the RM ANOVA approach completely in favor of MANOVA, whose validity is not affected by heterogeneity of variances and covariances. However, if you definitely want to use an RM ANOVA, the safest way to proceed is to begin by assuming the worst case: total lack of sphericity.

The Conservative Geisser-Greenhouse *F* Test

Geisser and Greenhouse (1958) determined that when there is a total lack of sphericity, the proper critical *F* for testing the RM ANOVA is found by assuming that $df_{num} = 1$ (regardless of the number of treatment levels) and $df_{denom} = n_s - 1$ (where n_s is the number of different subjects or blocks). (Their approach is known as the **Geisser-Greenhouse *F* test.**) With this reduction in degrees of freedom the critical *F* will get larger. If the observed *F* exceeds even this larger critical *F*, then you simply do not have to worry about the sphericity assumption. Regardless of the degree of sphericity, you can safely reject the null hypothesis without the Type I error rate rising above the value used for determining critical *F*.

Let us consider again the example in Section B. Because the *F* ratio was significant, there is the possibility that we are making a Type I error. Of course, this is always a possibility, but it is also possible that the population did not exhibit sphericity and that Type I errors are more likely than we think. Assuming maximum violation of sphericity, the appropriate critical *F* would be based on 1, 5 df (this is called the conservative Geisser-Greenhouse correction). For $\alpha = .05$, the "corrected" critical $F = 6.61$. The calculated *F* of 8.0 is still larger than the conservatively corrected critical *F*, so we can be assured that the *F* ratio for our experiment will be validly significant regardless of the degree of sphericity.

On the other hand, if your observed *F* is not significant by the ordinary test (for example, if it is less than the critical *F* value of 4.10 found in the Section B example), there is nothing else to do. It does not matter if your *F* is inflated by a lack of sphericity, because you are not rejecting the null hypothesis anyway. However, if your observed *F* were about 5, you would have a problem. Whether the *F* ratio could be declared significant would depend on the assumption regarding sphericity. Assuming a high degree of sphericity, you could reject the null hypothesis; assuming very little sphericity, you should accept (or at least not reject) the null. It is for these intermediate cases—when the calculated *F* ratio falls between the usual critical value and the maximally adjusted value—that we need to estimate the actual degree to which sphericity holds.

Estimating Sphericity in the Population

The various pairwise interactions in the experimental data can be used to calculate a factor called epsilon (ϵ), which is determined by estimating the degree of sphericity in the population. Epsilon ranges from 1.0 when sphericity holds down to $1/(k - 1)$ when the samples suggest a total lack of sphericity. The ordinary df components from the RM ANOVA are multiplied by ϵ, and then the adjusted df components are used to find the critical value of *F*. Notice that when ϵ is at its lowest, the adjustment to the df leads to the same critical *F* as the conservative Geisser-Greenhouse adjustment. Calculating ϵ is not trivial, and there are at least two reasonable methods that lead to different answers—one developed by Greenhouse and Geisser (1959), and a somewhat less conservative adjustment devised by Huynh and Feldt (1976), which is a bit more powerful when ϵ is not very far from 1.0. Moreover, the adjusted df components are not likely to appear in any standard table of the F distribution. Fortunately, most computer statistical packages now calculate ϵ whenever a repeated factor has more than two levels and give an exact p level for the observed F, regardless of how the degrees of freedom come out. In fact, some packages calculate ϵ by both methods mentioned above and give p levels for both. If you do not have easy access to one of these statistical packages, the approach described below is recommended.

The Modified Univariate Approach

One way of dealing with the sphericity assumption is the **modified univariate approach,** which can be summarized in terms of the following three-step procedure.

Step 1 Compare the calculated F to the usual critical F (see Section B). If F is not statistically significant, no further steps are required—you will not be able to reject the null hypothesis, regardless of what you assume about sphericity. If F exceeds the critical value, proceed to Step 2.

Step 2 Compare the calculated F to the conservative adjustment devised by Geisser and Greenhouse: Critical F is based on $(1, n_s - 1)$ df. If F is statistically significant, no further steps are required—you *will* be able to reject the null hypothesis regardless of how completely sphericity is violated. If F is not significant by this stricter test, proceed to Step 3.

Step 3 If your observed F exceeds the usual critical value but not the conservatively adjusted value, a more precise correction is needed to determine statistical significance (you can compute ϵ by hand if you do not have the appropriate statistical software package to do the calculations; see Howell, 1992).

The Multivariate Approach to Repeated Measures

The multivariate approach views the measurements at each level of the independent variable as constituting a different dependent variable. For instance, if three types of words are used to test each subject's recall (positive, negative, and neutral), then there are three dependent variables—positive recall, negative recall, and neutral recall—and no independent variables. We assume multivariate normality, but make no assumption about sphericity. MANOVA can be used to test whether all three variables have the same mean in the population, but the details of that analysis are quite different from the tests described in this text, so I will not attempt to outline the procedure here. A general description of the structure of MANOVA and some of the ways MANOVA can be used to test statistical hypotheses will be offered in Section C of Chapter 18. At this point, I will mention only that MANOVA, which is very tedious to compute by hand, is relatively easy to use when running any of the major statistical software packages. Because the results of MANOVA are unaffected by an absence of sphericity, you need not worry about the inflated Type I error rates that can occur with the univariate RM ANOVA. However, the modified univariate approach described above can also control Type I error rates adequately, so how can you decide which procedure to use? If you are using a statistical package, computational effort is not a factor; therefore the decision should be based on which procedure can be expected to yield the most power. Unfortunately, neither procedure has greater power under all conditions, so further discussion is required.

The Relative Power of Univariate and Multivariate Approaches

As mentioned above, when all of the assumptions of the RM ANOVA are met, it is usually more powerful than the MANOVA for detecting differences in population means; the smaller the sample size, the greater the advantage of using RM ANOVA. Unfortunately, it is not uncommon for the sphericity assumption to be violated. A design that is likely to lack sphericity is one in which the same variable is measured at several intervals in time. Measurements taken at shorter time intervals are likely to have a higher covariance than measurements separated by greater intervals of time. Another design likely to lack sphericity is one in which there are two or more sets of conditions, and the conditions within a set are more similar than conditions from different sets. Bear in mind that whereas switching from an RM to an RB design may solve carryover problems, the differences among covariances produced in distinct sets of conditions may well be evident when different subjects are used. Because you can rarely be sure that sphericity is a reasonable assumption to make, you should probably use the modified univariate approach whenever you perform the RM ANOVA, which will unfortunately reduce your power relative to MANOVA. When sphericity is violated, either method can have more power depending on which means and which covariances are different (Davidson, 1972). However, for large sample sizes it is rare for MANOVA to have much less power than the modified univariate approach, but not rare for the MANOVA to have considerably more power. The general recommendation is that when your sample size is at least 20 more than the number of conditions ($n_s \geq c + 20$), your best bet is MANOVA. On the other hand, if your sample size is less than the number of conditions (e.g., four subjects are run under each of five conditions), you cannot even compute the MANOVA, so you must use the modified univariate approach. Decisions involving intermediate sample sizes require a closer look at which means and covariances can be expected to be different.

Post Hoc Comparisons

The optimal method to use for post hoc comparisons depends on whether it is reasonable to assume sphericity. If it is reasonable to make this assumption (e.g., your sample size is large and a test for sphericity did not reject the null hypothesis), you can proceed with multiple comparisons by any of the methods described in Chapter 15, substituting MS_{inter} (i.e., MS_{resid}) for MS_w. With three conditions, Fisher's protected *t*-tests are recommended if the *F* for the overall RM ANOVA is significant. With more than three conditions, Tukey's HSD would be acceptable, regardless of the significance of the overall *F*. On the other hand, if it is not safe to assume sphericity, the use of the pooled error term, MS_{resid}, is not justified for each pairwise test; the experimentwise Type I error rate can easily become unacceptably high.

The simplest and safest procedure for pairwise comparisons when sphericity cannot be assumed is to perform the ordinary matched *t*-test for each pair of conditions using the Bonferroni test (Formula 15.10) to control experimentwise alpha. The error term for each test depends only on the two conditions involved in that test, Type I error rates are adequately controlled, and the amount of power compares favorably to that of other procedures (Lewis, 1993). It is not even necessary to perform an overall RM ANOVA or MANOVA, especially if particular pairwise comparisons have been planned. If you perform a MANOVA and it is statistically significant, the appropriate way to perform pairwise comparisons is by performing the MANOVA on two conditions at a time. However, the MANOVA for two conditions is identical to the matched *t*-test, so the recommended follow-up procedure is the same no matter which way you begin to analyze the overall RM design.

Power of the RM ANOVA

If a population exhibits compound symmetry, and therefore each pair of treatment levels has the same correlation in the population (ρ), then the power of the RM ANOVA depends, in part, on the value of ρ, just as in the matched *t*-test. As ρ gets higher, the amount of subject by treatment interaction decreases (i.e., the error term of the *F* ratio decreases), making the expected *F* value larger and the observed *F* more likely to be significant. If a population does not show compound symmetry, it will probably not exhibit sphericity either, and the power of the RM test becomes difficult to predict. If some pairs of means differ more than others, and some interactions are larger than others, power will be influenced by whether the larger mean differences are associated with the larger interactions or the smaller ones.

Of course, the basic factors affecting the power of the one-way independent-groups ANOVA apply to the RM ANOVA, as well. Power increases with a larger sample size, larger differences between population means, and a larger alpha. The only factor involved in the independent-groups ANOVA that does not affect the RM ANOVA is the subject-to-subject variability within groups. It is important to bear in mind that when each subject exhibits approximately the same pattern (or profile) across the treatment levels, then even small differences in treatment levels can lead to large *F* values, despite considerable differences in overall performance from subject to subject. That is why a conservatively adjusted *F* test for RM ANOVA can still easily have more power than an independent-groups ANOVA that has even larger sample sizes. Reducing the amount of subject by treatment interaction has the same effect on power as increasing the sample size. As with the matched *t*-test, or independent designs with very large samples, you must be careful to look at the actual differences in treatment means produced by an RM design and not be overly impressed with the *p* level associated with the overall *F.*

Counterbalancing

As I mentioned in Section A, if a successive RM design is associated only with simple order effects, counterbalancing can preserve the validity of the experiment. If there are only two or three treatment levels, complete counterbalancing is a reasonable approach. To observe the nature of any order effects, the ordinal position of a treatment level can be used as a second factor in a two-way mixed-design ANOVA (see Chapter 18).

For an example, take another look at Table 17.4. Notice that any particular condition (e.g., vis-sim) occurs twice in first place, twice as the second treatment, and twice as the last treatment. To plot position and treatment on the same graph, we begin by marking off the three positions along the *X* axis. Then we plot a line for each treatment level. To find the value for vis-sim in position 1, we average together the scores for the two subjects who received the vis-sim condition first. We do the same for the other positions and the other treatments. The order in which each subject received the three conditions was not shown in the original example, but one possible pattern of results is shown in Figure 17.8.

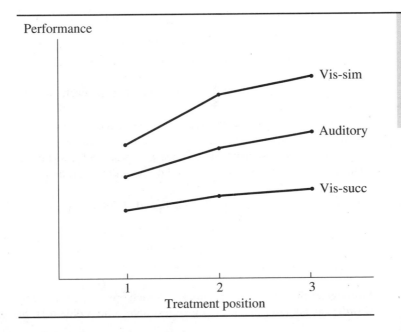

Figure 17.8

Graph of Hypothetical Data Illustrating Simple Order Effects

Note the almost total lack of interaction. This indicates that only simple order effects are present. You can also see that all treatments increase at position 2, and increase a bit more at position 3. This shows a practice effect. There is a certain advantage to being in position 2 (and more advantage for position 3), and the amount of the advantage does not depend on the treatment level. Counterbalancing literally balances out order effects (such as practice effects) by ensuring that each treatment level appears at each temporal position an equal number of times. On the other hand, if the position by treatment graph indicates a considerable amount of interaction, differential carryover effects are likely to be present, and the validity of the RM design becomes questionable.

Whereas counterbalancing succeeds in averaging out simple order effects, so that no treatment level attains an unfair advantage by virtue of its ordinal position, the order effects do lead to an increase in the error variance. Different subjects receive a particular treatment level in different positions in the order, so some of the variability in scores at a given treatment level is due to the different orders. There are statistical procedures that can account for this added variance and remove it from the error term, thus further increasing power (see Keppel, 1991).

Latin Square Designs

Complete counterbalancing becomes a nuisance when there are four levels, and becomes impractical when there are five or more levels. Fortunately, complete counterbalancing is not necessary for averaging out simple order effects. One of various schemes for partial counterbalancing, such as the **Latin square design,** will suffice. For the transitive inference experiment, only three orders are needed:

vis-sim, vis-succ, auditory

vis-succ, auditory, vis-sim

auditory, vis-sim, vis-succ

Notice that each treatment level appears once in each of the three ordinal positions.

With four treatment levels, four orders are needed (in the Latin square design, the number of orders always equals the number of treatment levels), but there are a variety of schemes that will balance out order effects. For example, if the treatment levels are labeled a, b, c, and d, the following four orders can be used:

a, b, c, d

b, c, d, a

c, d, a, b

d, a, b, c

Although each treatment level appears in each ordinal position only once, this set of orders is not considered the most desirable. The problem is that a particular level is always preceded by the same level (e.g., b is always preceded by a, except, of course, when b comes first). A more desirable set of orders is the following:

a, b, c, d

c, a, d, b

b, d, a, c

d, c, b, a

Now each treatment level appears in each ordinal position only once, *and* each level is preceded by each other level once (e.g., b is preceded by a in the first order, then by d in the second, and finally by c in the fourth). This design is said to be **digram balanced.** When there is an even number of treatments (c), a digram-balanced Latin square with c orders can be created. When c is odd, however, two Latin squares of c orders each must be constructed to achieve digram balancing. In the case of three conditions, the only way to have digram balancing is to use all six possible orders. As there is little advantage to using three orders instead of six, complete counterbalancing is generally recommended for $c = 3$. For five conditions, a total of ten orders is required for digram balancing, but this is still an enormous reduction from the 120 orders that are possible.

In some experimental designs, the number of possible orders is extremely large. For instance, there are over three million possible orders in which you can present a list of ten words for memorization. When balancing the orders seems impractical, but you want the generalizability created by giving different orders to different subjects, you can select one order randomly (out of the many available) for each subject. Whereas the random selection of orders is not likely to represent any kind of perfect balancing of the orders, it is also not likely to represent any particular kind of bias.

Exercises

* 1. For each of the specified exercises, use the three-step procedure in Section C to decide whether the results are not significant, significant regardless of the sphericity assumption, or in need of further testing. a) Exercise 17B1 b) Exercise 17B4 c) Exercise 17B6 d) Exercise 17B7

2. a) For exercise 17B3, test all the pairs of means with protected *t*-tests using the error term from the RM ANOVA. Which pairs differ significantly at the .05 level?

   b) Repeat part a for exercise 17B4, using Tukey's HSD.

* 3. a) For exercise 17B6, test all the pairs of means with separate matched *t*-tests (or two-group RM ANOVAs) at the .05 level.

   b) For exercise 17B7, test all the pairs of means that are adjacent in time with separate matched *t*-tests (or two-group RM ANOVAs) at the .05 level. According to the Bonferroni test, what alpha should be used for each pairwise comparison?

4. In exercise 17B3, if you wanted to use a Latin square design for counterbalancing, instead of all possible orders, a) how many different orders would you have to use?

b) Write out all of the treatment orders you would use.

c) Given the sample size for that exercise, how many subjects would be assigned to each order?

*5. In exercise 17A4, each subject solved problems under four different drug conditions: marijuana, amphetamine, valium, and alcohol. Write out all the orders of the drug conditions that would be used in a Latin square design that balanced for preceding condition as well as ordinal position.

SUMMARY

The Important Points of Section A

1. The one-way RM ANOVA expands the matched *t*-test to accommodate any number of levels of the independent (i.e., treatment) variable simultaneously.

2. The one-way RM ANOVA can be viewed as a two-way ANOVA in which the subjects are the different levels of the second factor and there is only one subject per cell. Although you cannot calculate SS_w, you can find the other SS components of a two-way ANOVA.

3. Using the two-way ANOVA approach, there is no MS_W to act as the denominator of the F ratio. Instead, MS_{RM} (which corresponds to MS_{bet} in a one-way ANOVA) is divided by $MS_{interaction}$. The latter term is often called $MS_{residual}$, because it is usually thought to comprise both the true amount of subject by treatment interaction in the population and an error component.

4. Compared to the corresponding one-way independent-groups ANOVA, the advantage of the RM ANOVA is that the error term is smaller (i.e., MS_{inter} will be smaller than MS_W, unless the matching is no better than random). The better the matching, the smaller the error term gets, and the larger the F ratio becomes for the RM ANOVA.

5. One disadvantage of the RM ANOVA is the reduced df in the error term, which leads to a larger critical F than the independent ANOVA. However, the increase in the calculated F for RM ANOVA that results from matching usually outweighs the increase in the critical F.

6. When there are only two treatment levels, either a matched *t*-test or RM ANOVA can be performed. The variance of the difference scores in the matched *t*-test is always twice as large as the MS_{inter} for the RM ANOVA on the same data. The *t* value is equal to the square root of the F ratio from the RM ANOVA.

7. If the treatment levels in an RM design are presented in succession and always in the same order, the effects of order (practice, fatigue, etc.) may contribute to the treatment effects. Counterbalancing provides a system for averaging out the order effects.

8. If the effects of particular treatment levels interfere (i.e., carry over) with subsequent levels, it is unlikely that these carryover effects will be compensated for by counterbalancing. In this case, matching subjects into blocks will eliminate the problem. The randomized-blocks (RB) design is analyzed in exactly the same way as the RM design, as long as the number of subjects in each block equals the number of treatment levels. If the matching is effective, the RB design will have much of the power associated with the RM design. (The RM design usually represents the best possible matching, because the same subject at different times is usually more similar than well-matched but different subjects.)

The Important Points of Section B

The RM ANOVA procedures described in Section B can be used to analyze the results of an RB design as well as an RM design as the example below will illustrate. Imagine that a consumer organization wants to answer the question, are all weight-loss programs equally effective? Four well-known programs are chosen for comparison. Subjects are matched into blocks of four based on having the same initial weight, a similar build, and a similar dieting history. Within each block, the four subjects are randomly assigned to the different weight-loss programs. There are five blocks of four subjects each. The number of pounds each subject has lost after 3 months of dieting is recorded in Table 17.7.

The first step is to calculate SS_{total}. To do this we need to square all the values in Table 17.7 and take the sum. In this case, $\Sigma X^2 = 1002$. The sum of all the values (T_T) is 130. The total number of values (N_T) = 20. Now we can use Formula 17.1.

$$SS_{tot} = \Sigma X^2 - \frac{T_T^2}{N_T} = 1002 - \frac{130^2}{20} = 1002 - \frac{16,900}{20} = 1002 - 845 = 157$$

Table 17.7

Block	Program I	Program II	Program III	Program IV	Row Sums
	Number of Pounds Lost				
1	2	4	5	1	12
2	6	3	7	5	21
3	7	9	6	7	29
4	5	8	8	6	27
5	10	10	13	8	41
Column Sums	30	34	39	27	130

Next, we use Formula 17.2 to find SS_{sub}:

$$SS_{sub} = \frac{1}{c}\sum S_i^2 - \frac{T_T^2}{N_T} = \frac{1}{4}(12^2 + 21^2 + 29^2 + 27^2 + 41^2) - 845$$

$$= \frac{1}{4}(3836) - 845 = 959 - 845 = 114$$

We find SS_{RM} using Formula 17.3:

$$SS_{RM} = \frac{1}{n_s}\sum T_i^2 - \frac{T_T^2}{N_T} = \frac{1}{5}(30^2 + 34^2 + 39^2 + 27^2) - 845$$

$$= \frac{1}{5}(4306) - 845 = 861.2 - 845 = 16.2$$

The last SS component is found by subtraction using Formula 17.4:

$$SS_{inter} = SS_{tot} - SS_{sub} - SS_{RM} = 157 - 114 - 16.2 = 26.8$$

The df components are calculated according to Formula 17.5 (in this case, n_s equals the number of blocks):

$$df_{RM} = c - 1 = 4 - 1 = 3$$
$$df_{inter} = (c - 1)(n_s - 1) = (4 - 1)(5 - 1) = 3 \cdot 4 = 12$$

The MSs that we need for the F ratio are as follows:

$$MS_{RM} = \frac{SS_{RM}}{df_{RM}} = \frac{16.2}{4 - 1} = \frac{16.2}{3} = 5.4$$

$$MS_{resid} = MS_{inter} = \frac{SS_{inter}}{df_{inter}} = \frac{26.8}{12} = 2.23$$

Finally, the F ratio (Formula 17.6) equals $MS_{RM}/MS_{resid} = 5.4/2.23 = 2.42$.

The critical F for (3, 12) df and $\alpha = .05$ is 3.49. Because our calculated F (2.42) is less than the critical F, we cannot reject the null hypothesis. We cannot conclude that there are any differences in effectiveness among the four weight-loss programs tested. Even with reasonably good matching, our sample was too small to give us much power, unless the weight-loss programs actually differ a good deal. In a real experiment, a much larger sample would be used; otherwise the experimenters would have to be very cautious in drawing any conclusions from a lack of significant results.

Summary Table

The components of the analysis performed above are often presented in the form of a summary table, as shown in Table 17.8:

Table 17.8

Source	SS	df	MS	F	p
Between-Subjects	114	4			
Within-Subjects					
Between-Treatments	16.2	3	5.4	2.42	> .05
Interaction (Residual)	26.8	12	2.23		
Total	157	19			

Assumptions of the RM ANOVA

1. *Independent random sampling.* In an RM design, subjects should all be selected independently. In an RB design, subjects in different blocks should be independent, but subjects *within* a block should be matched.
2. *Multivariate normal population distribution*
3. *Homogeneity of variance*
4. *Homogeneity of covariance.* When this assumption and assumption 3 are true simultaneously, the population exhibits compound symmetry, which is desirable, but not necessary for the RM ANOVA to be valid. Assumptions 3 and 4 can be relaxed as long as a condition called sphericity (or circularity) can be assumed. Sphericity is said to exist when the variability of the difference scores between any two levels of the independent variable in the population is the same as the variability of the difference scores for any other pair of levels.

When to Use the RM or RB Design

Repeated measures. If it is feasible, the RM design is the most desirable, as it generally has more power than independent groups or randomized blocks. For obvious reasons, it can only be used with an independent variable that involves experimental manipulations, and not a grouping variable. The two types of repeated measures design are simultaneous and successive.

In a simultaneous design the different conditions are presented as part of the same stimulus (e.g., different types of pictures in a collage), or as randomly mixed trials (e.g., trials of different difficulty levels mixed together). A successive design usually requires counterbalancing to avoid simple order effects and is not valid if there are differential carryover effects.

Randomized blocks. The RB design has much of the power of the RM design and avoids the possible carryover effects of a successive RM design. The two types of randomized-block design are experimental and naturally occurring.

In the simplest experimental RB design, the number of subjects per block is the same as the number of experimental conditions; the subjects in each block are assigned to the conditions at random. The drawback to the naturally occurring design is that you cannot conclude that your independent variable *caused* the differences in your dependent variable.

The Important Points of Section C

1. The assumption of sphericity implies that the amount of interaction (in the population) between any two levels of the independent variable is the same as for any other two levels. This is the same as assuming that the variance of the difference scores for any pair of levels is the same as for any other pair.

2. When the null hypothesis is true but the population does *not* exhibit sphericity, the *F* ratios tend to be larger than normally expected (there is a positive bias to the test), and therefore the Type I error rate tends to be larger than the alpha used to determine the critical *F.* Unfortunately, sphericity is likely to be violated in an RM design, and tests of the sphericity assumption have little power to reveal this when sample sizes are small.

3. You need not worry about the sphericity assumption if the *F* ratio is less than the usual critical *F* or greater than a very conservatively adjusted *F* (the Geisser-Greenhouse *F* test). If the *F* ratio lands between these extremes, an adjustment of the degrees of freedom based on a coefficient called epsilon is called for and can be obtained easily using statistical software.

4. The multivariate approach to repeated measures is unaffected by violations of the sphericity assumption and compares favorably to the RM ANOVA in terms of power, except when the sample size is not much larger than the number of treatments.

5. When sphericity can be assumed, post hoc comparisons are based on $MS_{residual}$ in place of MS_w. When this assumption is not reasonable, the safe approach is to conduct separate matched *t*-tests for each pair of conditions, adjusting alpha according to the Bonferroni test.

6. As with the matched *t*-test, the power of the RM ANOVA increases as the matching improves—that is, as the amount of subject by treatment interaction decreases. It is important to be aware that good matching has a similar effect as increasing the sample size considerably: The *F* ratio may become very large even when there is little difference between the means of the treatment levels.

7. If only simple order effects are present in a successive RM design (i.e., there is no interaction between order and treatment and no differential carryover effects have occurred), counterbalancing will average them out, so that no treatment level has an unfair advantage as a result of its ordinal position. Advanced statistical techniques can remove the extra amount of error due to variations in the ordinal positions of the treatment levels, providing added power to the RM ANOVA.

8. When there are only two or three treatment levels, complete counterbalancing is a practical solution. When there are four or more levels, a partial counterbalancing design such as the *Latin square* is recommended. The number of orders required by the Latin square design is equal to the number of treatment levels. If digram balancing is desired and there is an odd number of treatments, a combination of two Latin squares is required.

Definitions of Key Terms

One-way repeated-measures ANOVA ANOVA with one independent variable, all levels of which are presented to each subject (or to a block of matched subjects).

Repeated measures (RM) design An experiment in which each subject receives all levels of the independent variable.

Order effects In the simplest case, subjects in certain treatment levels have an advantage or disadvantage because of the ordinal position of the treatment (regardless of which treatment level or levels were presented earlier). Such order effects can be caused by practice or fatigue and can be averaged out by counterbalancing.

Carryover effects The influence of the effect of a treatment level on the treatment levels that follow it. When the carryover effects are not symmetric (e.g., the effect of level B is different when it is preceded by level A, whereas the effect of level A is the same whether or not it is preceded by level B), these effects are often called differential carryover effects. Differential carryover effects cannot be averaged out by counterbalancing.

Randomized-blocks (RB) design An experiment in which subjects are matched in groups (called blocks) and then randomly assigned to the different levels of the independent variable. If the number of subjects per block equals the number of levels of the independent variable, then the RB design is analyzed in exactly the same way as an RM design.

$SS_{residual}$ The sum of squares left over in an RM ANOVA after SS_{sub} and SS_{RM} have been subtracted from SS_{total}. This SS component forms the basis of the error term for the F ratio in a one-way RM ANOVA and is usually thought to contain an estimate of the subject by treatment interaction in the population, as well as a component due to sampling error.

Homogeneity of covariance A condition that exists in a population when the amount of covariance between any two treatment levels in a study is the same as for any other pair of levels.

Compound symmetry A condition that exists in a population when both homogeneity of variance and homogeneity of covariance are true. The assumption of this condition is sufficient to justify the use of the usual RM ANOVA procedure, but it is more stringent than necessary.

Multivariate analysis of variance (MANOVA) An analysis of variance procedure in which more than one dependent variable is used simultaneously to test for a difference in population means. One application of this procedure is the analysis of data from a one-way RM design.

Geisser-Greenhouse F test A very conservative approach to the RM ANOVA, which assumes a total lack of sphericity in the population and adjusts the degrees of freedom accordingly before determining the critical F.

Modified univariate approach If the observed F ratio from the RM ANOVA falls between the usual critical F and the maximally adjusted critical F, the modified univariate approach can provide a more precise estimate of the degree of sphericity to calculate a coefficient (epsilon), which is then used to reduce the df before finding the critical F.

Latin square design A system for counterbalancing in which the number of different sequences of treatment levels equals the number of treatment levels. Each treatment level appears once in each ordinal position.

Digram-balanced design A Latin square design with an even number of treatment levels in which each treatment level is preceded by each other level only once. A combination of two Latin squares is required in order to achieve digram-balancing when the number of treatment levels is odd.

Key Formulas

The total sum of squares, raw-score form (can be used with any analysis of variance design):

$$SS_{tot} = \sum X^2 - \frac{T_T^2}{N_T} \qquad \text{Formula 17.1}$$

The sum of squares due to subject-to-subject differences, raw-score form (this component is generally subtracted out and not used):

$$SS_{sub} = \frac{1}{c}\sum S_i^2 - \frac{T_T^2}{N_T} \qquad \text{Formula 17.2}$$

The sum of squares due to differences in treatment means, raw-score form (same as SS_{bet} in a one-way independent ANOVA):

$$SS_{RM} = \frac{1}{n_s}\sum T_i^2 - \frac{T_T^2}{N_T} \qquad \text{Formula 17.3}$$

The SS for the subject-by-treatment interaction (found by subtraction):

$$SS_{inter} = SS_{tot} - SS_{sub} - SS_{RM}$$ Formula 17.4

The degrees of freedom for the components of the RM ANOVA:

$$df_{sub} = n_s - 1$$
$$df_{RM} = c - 1$$
$$df_{inter} = (n_s - 1)(c - 1)$$ Formula 17.5

The *F* ratio for a one-way RM ANOVA:

$$F = \frac{MS_{RM}}{MS_{inter}}$$ Formula 17.6

Two-Way Mixed Design ANOVA

You will need to use the following from previous chapters:

Symbols
k: Number of independent groups in a one-way ANOVA
$\overline{X}_i$: Mean of the ith group
$\overline{X}_G$: Grand mean (the mean of all observations in an experiment)
S_i: Sum of scores for the ith subject
T_i: Sum of scores in the ith group
cs_i: Sum of scores in the ith cell of a two-way ANOVA
T_T: Sum of all observations in an experiment
N_T: Total number of observations in an experiment

Formulas
All of the computational formulas for the RM ANOVA (Chapter 17)
Formula 14.7: MS_{bet}, definitional form (from one-way ANOVA)
Formula 16.2: SS_{bet}, raw score form (from two-way ANOVA)
Formulas 16.3A and 16.3B: SS_R and SS_C from the two-way ANOVA
Formula 16.4: SS_{inter} (by subtraction) from the two-way ANOVA

Concepts
Advantages and disadvantages of the RM ANOVA
SS components of the one-way RM ANOVA
SS components of the two-way ANOVA
Interaction of factors in a two-way ANOVA

CHAPTER

CONCEPTUAL FOUNDATION

If you want the economy of a two-way factorial design, or its ability to detect the interaction of two independent variables, and at the same time you want the added power of repeated measures, you can use a two-way repeated measures design. The analysis of this design, using a two-way RM ANOVA, parallels the analysis of a three-way factorial design, just as the one-way RM ANOVA resembles the two-way ANOVA for independent groups. A description of three-way designs goes beyond the scope of this text; similarly, I will not discuss the analysis of a two-way RM ANOVA. However, at least as common as the two-way RM design is a two-way factorial design in which one of the factors involves repeated measures (or matched subjects) and the other factor involves independent groups of subjects. For obvious reasons, this design is often called a **mixed design,** though this designation is not universal. Mixed designs are sometimes called *split-plot* designs, a description that arises from their early use in agricultural research. (A potential source of confusion is that the term *mixed design ANOVA* sounds similar to the *mixed model ANOVA,* which involves a mixing of random and fixed effects in the same design; see Chapter 14. However, the details of the design that are specified will quickly make it obvious which type of mixing is being discussed, so there is usually little confusion.)

Because mixed designs can have any number of **within-subjects factors** and **between-subjects factors,** it is usually necessary to specify the number of each type of factor. For instance, a three-way mixed design can have either two RM factors and one between-groups factor or one RM factor and two between-groups factors. In this chapter you will encounter only the simplest mixed design: the two-way mixed design. So long as the appropriate assumptions can be made, the data from a mixed design experiment can be analyzed with the mixed design ANOVA described in this chapter.

The One-Way RM ANOVA Revisited

In order to demonstrate how the mixed design ANOVA works, I will return to the experiment described at the beginning of Chapter 17, in which each of six subjects must try to recall three types of words: positive, negative, and neutral. I will change the data, however, to suit the purposes of this chapter; see Table 18.1.

First I will conduct a one-way repeated measures ANOVA for these data. We begin as usual by finding SS_{total}. Given that $\sum X^2 = 3452$ and $T_T = 232$,

$$SS_{tot} = \sum X^2 - \frac{T_T^2}{N_T} = 3452 - \frac{232^2}{18} = 3452 - 2990.2 = 461.8$$

Next we find SS_{RM}, which involves the totals for each treatment (the column totals in Table 18.1):

$$SS_{RM} = \frac{1}{n_s}\sum T_i^2 - \frac{T_T^2}{N_T} = \frac{1}{6}(86^2 + 73^2 + 73^2) - 2990.2$$
$$= 3009 - 2990.2 = 18.8$$

Now we need to calculate SS_{sub} so it can be subtracted from the total. We will use the totals for each subject (the row totals in Table 18.1):

Subject	Neutral	Positive	Negative	Row Totals	
					Table 18.1
1	20	21	17	58	
2	16	18	11	45	
3	17	13	18	19	
4	15	10	13	38	
5	10	4	10	24	
6	8	7	4	48	
Column Totals	86	73	73	232	

	Source	SS	df	MS	*F*
Table 18.2	Between-Subjects	367.8	5		
	Within-Subjects	94	12		
	Treatment	18.8	2	9.4	1.25
	Residual	75.2	10	7.52	
	Total	461.8	17		

$$SS_{sub} = \frac{1}{c}\Sigma S_i^2 - \frac{T_T^2}{N_T} = \frac{1}{3}(58^2 + 45^2 + 19^2 + 38^2 + 24^2 + 48^2) - 2990.2$$

$$= 3358 - 2990.2 = 367.8$$

Finally, we can find $SS_{residual}$ (i.e., SS_{inter}) by subtraction:

$$SS_{resid} = 461.8 - 18.8 - 367.8 = 75.2$$

To find the *F* ratio, we first find that $MS_{RM} = SS_{RM}/df_{RM} = 18.8/2 = 9.4$, and $MS_{resid} = SS_{resid}/df_{resid} = 75.2/10 = 7.52$. Therefore, $F = 9.4/7.52 = 1.25$. The results of the RM ANOVA are summarized in Table 18.2. Because the critical $F_{.05}$ (2, 10) equals 4.1, we cannot reject the null hypothesis in this case. As you may have guessed, this is not the end of the story for this experiment. What I didn't tell you is that three of the subjects were selected because of high scores on a depression inventory, whereas the remaining three showed no signs of depression. The depressed subjects are graphed as dashed lines in Figure 18.1; solid lines represent the nondepressed subjects.

Converting the One-Way RM ANOVA to a Mixed Design ANOVA

The one-way RM ANOVA performed above can be transformed into a two-way mixed ANOVA by adding a between-groups factor with two levels: depressed and nondepressed. However, the advantage of this design may not be obvious from looking at Figure 18.1. On average, the two groups do not differ much in overall recall ability, nor does the variability within each group seem less than the total variability. The difference between the groups becomes apparent, though, when you focus on the subject × treatment interaction.

In Figure 18.2, the two groups are graphed in separate panels to make it obvious that the subject × treatment (S × T) interaction within each group is much smaller than the total amount of interaction when all subjects are considered together. The calculation of the mixed ANOVA can take advantage of this smaller S × T interaction by analyzing the SS components of the RM ANOVA further.

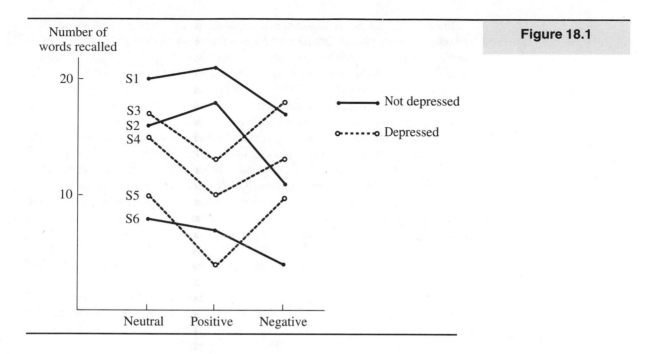

Figure 18.1

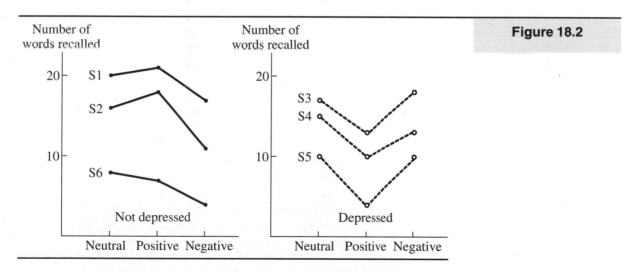

Figure 18.2

Analyzing the Between-Subjects Variability

Even though we would normally know about the depressed and nondepressed subgroups before analyzing our data, it would be reasonable to begin the mixed ANOVA by ignoring this distinction and calculating the RM ANOVA as above. The next step is to analyze further the between-subjects SS component shown in

Table 18.2. This is the component we happily throw away in the one-way RM design, because we do not care about subject-to-subject variability. However, in the mixed design some of the subject-to-subject variability is due to the difference in the means of the two (or more) groups. This SS component is like the SS_{total} for a one-way independent ANOVA; it can be divided into SS_W (variability within each group) and SS_{bet} (variability of the group means). To avoid confusion I will refer to SS_{bet} as SS_{groups} in discussing the mixed design.

The easy way to obtain SS_{groups} is to find the means for the two subgroups and use the numerator of Formula 14.7. First we note that the grand mean, $\overline{X}_G$, equals $T_T/N_T = 232/18 = 12.89$. The mean for the depressed subjects ($\overline{X}_D$) is 12.22, and the mean for the nondepressed subjects ($\overline{X}_N$) is 13.55. Next, it is important to remember that for n in Formula 14.7 we must use the number of *observations* in each group (9), rather than the actual number of subjects (3). The number of groups (k) is, of course, 2. Using the numerator of Formula 14.7, we get

$$SS_{groups} = \sum n_i (\overline{X}_i - \overline{X}_G)^2 = 9(12.22 - 12.89)^2 + 9(13.55 - 12.89)^2 = 4.0 + 4.0$$

Thus $SS_{groups} = 8.0$. To find SS_W, we need only subtract SS_{groups} from the total SS for this part of the analysis, which in this case is 367.8. (The total SS for this part of the analysis is the between-subjects SS from the original RM analysis, which is also SS_{sub}). So, $SS_W = 367.8 - 8.0 = 359.8$.

These SS components can now be converted to variance estimates so they can be put into an F ratio. Note that N_T in the MS_W formula refers to the total number of *subjects* in this case, which is 6.

$$MS_{groups} = \frac{SS_{groups}}{k - 1} = \frac{8.0}{1} = 8.0$$

$$MS_W = \frac{SS_W}{N_T - k} = \frac{359.8}{4} = 89.9$$

$$F = MS_{groups}/MS_W = 8.0/89.9 = .09$$

This F value (.09) allows us to test whether the means for the two groups differ significantly. Because we are conducting a factorial ANOVA, we can say that this F ratio tests the *main effect* of depression. Clearly, we do not have to look up a critical F to know that the null hypothesis cannot be rejected. The fact that our observed F is much less than 1.0 (indeed, unusually so) is of no interest to us. This just tells us that the means for the depressed and nondepressed subjects are surprisingly close together, given all of the variability within each group.

Analyzing the Within-Subjects Variability

Although we already calculated the F for the main effect of word type when we performed the one-way RM ANOVA, this F ratio must be recalculated for the mixed design, to take into account the separation of subjects into subgroups. As you might have guessed, the numerator of the F ratio for word type won't change; this numerator depends on the separation of the means for the three word types, and it does not change just because we have regrouped the subjects within conditions. On the other hand, the denominator (the error term) does change, for reasons that can be seen by comparing Figure 18.2 to Figure 18.1. Notice that the subject $\times$ treatment interaction is fairly small within each group (Figure 18.2), but it looks rather large when all subjects are considered together (Figure 18.1). We can say that most of the S $\times$ T interaction is really due to a group by word type interaction, which should be removed from the total S $\times$ T interaction. This is the same as taking the two much smaller interactions of the subgroups and averaging them. (The mean of these two smaller interactions can be considerably less than the interaction involving all subjects, as is the case in this example.) Mathematically, what we need to do is further analyze the $SS_{residual}$ from the RM ANOVA into smaller components.

In order to subtract the *group* by word type interaction from the *subject* by word type interaction, we must calculate the former. This is done as it is for a two-way ANOVA. First, we calculate the $SS_{between}$ for the two-way ANOVA by using Formula 16.2 or its equivalent. For this step we need to know the sum for each combination of group and word type. Table 18.3 shows the data from Table 18.1 rearranged into cells and the cell sums. These cell sums can be inserted into Formula 16.2, as follows:

$$SS_{bet} = \frac{1}{n}cs_i^2 - \frac{T_T^2}{N_T} = \frac{1}{3}(44^2 + 46^2 + 32^2 + 42^2 + 27^2 + 41^2) - 2990.2$$

$$= \frac{1}{3}(9250) - 2990.2 = 3083.33 - 2990.2 = 93.13$$

We have already calculated the SS for word type (i.e., SS_{RM}) and the SS for groups, so we are ready to find the SS for the group by treatment interaction ($SS_{g \times RM}$) by subtraction.

$$SS_{g \times RM} = 93.13 - 18.8 - 8.0 = 66.33$$

Group	Neutral	Positive	Negative	Table 18.3
Nondepressed	20	21	17	
	16	18	11	
	8	7	4	
Cell Sums	44	46	32	
Depressed	17	13	18	
	15	10	13	
	10	4	10	
Cell Sums	42	27	41	

Now we can go back to the original SS_{resid} from the one-way RM ANOVA and subtract $SS_{g \times RM}$ to get the new smaller SS_{resid} for the mixed design (the sum of the subject by treatment interactions *within each group*). This new error component can be referred to simply as SS_{resid}, or, more specifically, as the SS for the interaction of the subject factor with the repeated-measures factor, $SS_{s \times RM}$.

$$SS_{s \times RM} = 75.22 - 66.33 = 8.89$$

Now we have calculated all the SS components needed to complete the mixed design analysis. To recap: The one-way RM ANOVA gave us three components, SS_{RM}, SS_{sub}, and SS_{resid}. SS_{sub} was then further divided to yield SS_{groups} and SS_W. SS_{resid} was also divided into two components: $SS_{g \times RM}$ and $SS_{s \times RM}$. SS_{RM} was left alone. We have already tested the main effect of groups by forming an F ratio from MS_{groups} and MS_W. Now we can recalculate the F ratio for the repeated factor. The numerator MS is the same as in the one-way analysis: 9.4. But the new error term is based on $SS_{s \times RM}$ (8.89) divided by $df_{s \times RM}$ (8). So $MS_{s \times RM}$ = 8.89/8 = 1.11. Thus the new F for testing the main effect of word type is

$$F = \frac{MS_{RM}}{MS_{resid}} = \frac{9.4}{1.11} = 8.45$$

The critical F is based on 2 and 8 degrees of freedom; $F_{.05} = 4.46$. Because of the smaller error term in the mixed design, the main effect of word type is now significant at the .05 level.

Two-Way Interaction in the Mixed Design ANOVA

Like any other two-way ANOVA, the mixed design ANOVA can give us one more F ratio: a test of the interaction of the two factors. We find $MS_{g \times RM}$ by dividing $SS_{g \times RM}$ by $df_{g \times RM}$. We have already found that $SS_{g \times RM} = 66.33$, and you will have to take on faith for the moment that $df_{g \times RM} = 2$. So $MS_{g \times RM}$ = 66.33/2 = 33.17. (Formulas for the mixed design ANOVA, including those for the df, will be given in Section B.) To complete the F ratio, however, we need to know which MS to use as the error term. So far both MS_W and $MS_{s \times RM}$ have been used as error terms. Can we use one of these error terms, or is there some third error term to use? The answer is that $MS_{s \times RM}$ is the appropriate error term, for reasons that I will make clear shortly. To test whether depression and word type interact significantly, we form the following F ratio:

$$F = \frac{MS_{g \times RM}}{MS_{s \times RM}} = \frac{33.17}{1.11} = 29.85$$

You should not have to look up a critical F to know that such a large observed F ratio must be significant at the .05 level (except in the ridiculous case when some of your groups contain only one subject).

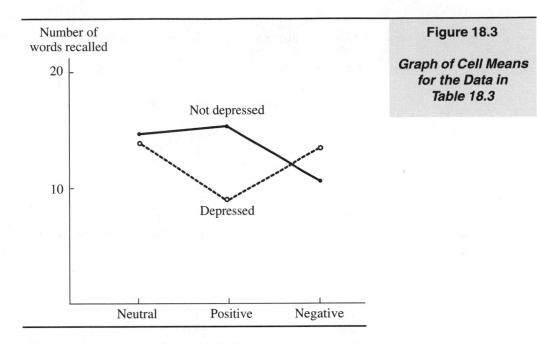

Figure 18.3

*Graph of Cell Means
for the Data in
Table 18.3*

To understand why $MS_{s \times RM}$ is the appropriate error term for the interaction of the two factors, it may help to compare a graph of the cell means (see Figure 18.3) to a graph of the data from individual subjects (see Figure 18.2). Notice that each line in Figure 18.3 represents the average number of words recalled by the subjects in that group. To the extent that the individual subjects in a group differ from the average, and from each other, $MS_{s \times RM}$ increases. As $MS_{s \times RM}$ increases, so does the chance of finding an accidental interaction between the groups (just as increased variation within groups leads to a greater chance of finding a large difference between two group means in an independent-samples *t*-test). Thus $MS_{g \times RM}$ contains error arising from $MS_{s \times RM}$ in addition to any real interaction between the factors. When the null hypothesis is true, there is no real interaction contributing to $MS_{g \times RM}$, so the *F* ratio—$MS_{g \times RM}/MS_{s \times RM}$—should equal about 1.0.

Summarizing the Mixed Design ANOVA

The results of our mixed ANOVA are summarized in Table 18.4 (on page 616). The structure of the table tells us a great deal about the structure of our analysis. First, the table is broken into two distinct sections. The upper section, "Between-Subjects," deals with the variation of scores within each *column* of Table 18.1; if all the scores were the same for a particular word type, the SS components for this section would all be zero. The differences that do exist between subjects can be divided into the difference between the subgroups (SS_{groups}) and the differences between subjects within each group (SS_W); the latter forms the basis for the error term for this part of the table. The number of degrees of

Table 18.4

Source	SS	df	MS	F	p
Between-Subjects	367.8	5			
Groups	8.0	1	8.0	.09	> .05
Within-groups	359.8	4	89.9		
Within-Subjects	94	12			
Repeated measures (RM)	18.8	2	9.4	8.45	< .05
Group × RM	66.3	2	33.2	29.8	< .01
Residual (S × RM)	8.9	8	1.1		
Total	461.8	17			

freedom for between-subjects variation is one less than the total number of different subjects, which equals $6 - 1 = 5$. These 5 df break down into 1 df for group differences ($k - 1 = 2 - 1 = 1$) and 4 df for differences within groups (number of subjects − number of groups = $6 - 2 = 4$).

The lower section of Table 18.4, "Within-Subjects," deals with the variation of scores within each *row* of Table 18.1; if each subject produced the same recall score for all word types, the SS components in this section of the table would all be zero (even though subjects might differ from each other). The differences that do exist between levels of the repeated factor (in this case, word type) can be divided into average differences among the word types (SS_{RM}) as well as the interaction of groups with word type ($SS_{g \times RM}$) and the interaction of individual subjects with word type ($SS_{s \times RM}$). For instance, if a particular subject recalls more positive than negative words, it may be because subjects tend to do this in general, or because subjects in a particular subgroup tend to do this, or because this particular subject has an individual tendency to do this. The last of these sums of squares ($SS_{s \times RM}$) forms the basis of the error term for the lower section of Table 18.4. The number of degrees of freedom for within-subjects variation equals the total number of subjects (6) times one less than the number of repeated measures ($c - 1 = 3 - 1 = 2$), which equals 12. These 12 df break down into 2 df for treatment differences ($c - 1$), 2 df for the group by treatment interaction [$(k - 1)(c - 1)$], and the remainder, 8 df, for the residual [$k(c - 1)$ (number of subjects in each subgroup − 1)], in this case, $2 \times 2 \times 2$.

Interpreting the Results

The between-groups factor in our example was not found to be significant, whereas the within-subjects factor was. This pattern is probably more common than the reverse pattern, because the test of the repeated factor is likely to have greater power, for reasons discussed in Chapter 17. However, we should interpret the results of the main effects cautiously, because of the significance of the interaction. The effect of the repeated factor is different for the two subgroups, and further analysis would be appropriate to localize these differences (see Section C). As with other two-way ANOVAs, a graph of the cell means can help you understand how the main effects combine with the interaction to produce the pattern of results. You can see from Figure 18.3 that the depressed subjects have relatively poor recall for positive words, whereas it is the recall of negative words that is weak for nondepressed subjects.

The reduction in MS_{resid} caused by separating the depressed and nondepressed subjects is reminiscent of the reduction in MS_W that followed the separation of men and women in the example of a depression drug in Chapter 16. Although separating subjects into groups is useful when the between-subjects factor is a grouping variable based on preexisting individual differences, that is not generally the reason for using a two-way design. Often it is the interaction of the two factors that is most interesting. In the example above, it is interesting that depression affects the types of words that subjects remember most easily.

Types of Mixed Designs

In the example above, the RM factor involves an experimental manipulation (i.e., type of word), whereas the between-groups factor is based on preexisting individual differences (i.e., depression). This is a common form of the mixed design. Often a researcher expects the between-groups factor to interact with the RM factor, so there is little interest in the main effect of the between-groups factor. The chief purpose of such a design is to determine whether the effects of the RM factor are the same for various subgroups in the population.

Another type of mixed design arises when the between-groups factor involves an experimental manipulation that does not lend itself to repeated measures. For instance, consider an experiment in which each subject completes a series of tasks of varying difficulty. In one condition, the subjects are told that the tasks come from a test of intelligence and that their performance will give an indication of their IQ. In another condition, subjects are given monetary rewards for good performance, and in a third condition, subjects are simply asked to work as hard as they can. It should be clear that once the researcher has run a subject in the IQ condition, running the same subject again in a different condition (using similar tasks) would yield misleading results. Nor could you always run the IQ condition last without confounding your results with an order effect. The usual solution is the mixed design, with the three motivational conditions as the levels of a between-groups factor and task difficulty as a within-subjects factor.

One purpose of the experiment described above could be to explore the interaction between the motivational condition and task difficulty. (For instance, are performance differences between difficulty levels the same for different types of motivation?) However, if there were no interaction, it might still be interesting to examine the main effect of motivational condition. (The main effect of difficulty would merely confirm that the difficulty manipulation worked.) In general, when there is no interaction between the factors of a mixed design, the analysis reduces to two one-way ANOVAs: an independent-groups ANOVA and an RM ANOVA.

The levels of the RM factor in a mixed design can be administered in several ways, as discussed for the one-way RM design. The levels can be interspersed for a presentation that is virtually simultaneous, or the levels can be presented successively. Successive presentations usually produce order effects that can be removed by counterbalancing. If successive treatments produce differential carry-over effects, the problem must be corrected. If the carryover effects cannot be removed, it may be possible to match subjects into blocks and attain much of the benefit of repeated measures by using randomized blocks instead. Finally, if carryover cannot be avoided, and no basis for matching can be found, both factors would have to consist of independent samples, and the two-way ANOVA described in Chapter 16 would be appropriate.

When the between-groups factor involves an experimental manipulation, the repeated-measures factor can be as simple as the passage of time (e.g., measurements can be taken before and after some treatment or at several points during a treatment). An example of this type of design will be used in Section B to illustrate the procedures of a two-way mixed design.

Exercises

1. a) Devise a mixed design experiment in which the between-subject variable is quasi-independent.

b) Devise a mixed design experiment in which the between-subjects variable is manipulated by the experimenter.

c) Devise a mixed design experiment in which the within-subjects variable involves matched subjects rather than repeated measures.

*2. A researcher tested two groups of subjects—six alcohol abusers and six moderate social drinkers—on a reaction time task. Each subject was measured twice: before and after drinking 4 ounces of vodka. A mixed-design ANOVA produced the following SS components: $SS_{groups} = 88$; $SS_W = 1380$; $SS_{RM} = 550$; $SS_{g \times RM} = 2.0$; $SS_{s \times RM} = 134$. Complete the analysis and present the results in a summary table.

3. Exercise 17B4 described a randomized-blocks experiment involving four different textbooks and nine blocks of subjects. The RM ANOVA produced the following SS components: SS_{treat} (SS_{RM}) = 76.75; $SS_{subject} = 612.5$; and $SS_{resid} = 27.5$. Now suppose that the nine blocks of subjects can be separated into three subgroups on the basis of overall ability, and that the mixed design ANOVA yields $SS_{groups} = 450$ and $SS_{g \times RM} = 8.5$. Complete the analysis and present the results in a summary table.

*4. The table below shows the number of ounces of popcorn consumed by each subject while viewing two emotion-evoking films: one evoking happiness and one evoking fear. Half the subjects eat a meal just before the film (preload condition), whereas the others do not (no load condition). Graph the data for all of the subjects on one graph.

	Happiness	Fear
Preload	10	12
	13	16
	8	11
	16	17
No Load	26	20
	19	14
	27	20
	20	15

a) Does there appear to be about the same amount of subject by treatment interaction in each group?

b) Does there appear to be a considerable amount of group by repeated-measure interaction?

5. If you calculate an RM ANOVA and then assign the subjects to subgroups to create a mixed design, the observed F ratio for the RM factor may get considerably larger. Under which of the following conditions is this likely?

a) The degrees of freedom associated with the error term are reduced considerably.

b) There is a good deal of subject by RM treatment interaction.

c) There is a good deal of (sub)group by RM treatment interaction.

d) There is a good deal of subject to subject variability.

In Chapter 13, I pointed out the weakness of the simple before-after design. Even if the before-after difference turns out to be statistically significant for some treatment, without a control group it is difficult to specify the cause of the difference. Was the treatment really necessary to produce the difference, or would just the act of participating in an experiment be sufficient? When you add a control group and continue to measure your variable twice in both groups, you have created a mixed design. For example, suppose that you have devised a new treatment for people afraid of public speaking. To show that the effects of your new treatment are greater than a placebo effect, half the subjects (all of whom have this phobia) are randomly assigned to a control group; their treatment consists of hearing inspirational talks about the joys of public speaking. Suppose further that you wish to demonstrate that the beneficial effects of your treatment last beyond the end of the treatment period. Consequently, you measure the degree of each subject's phobia not only before and after the treatment period, but also 6 months after the end of treatment (follow-up). We will apply our usual six-step hypothesis testing procedure to this experiment.

BASIC STATISTICAL PROCEDURES

Step 1. State the Hypotheses

The design in this example consists of two factors that have been completely crossed (i.e., a two-way factorial design). As such, the design involves three independent null hypotheses that can be tested. The H_0 for the main effect of treatment (the between-subjects factor) is $\mu_{exp} = \mu_{con}$. The H_0 for the main effect of time (the within-subjects, or repeated, factor) is $\mu_{bef} = \mu_{aft} = \mu_{fol}$. The H_0 for the interaction of the two factors is a statement that the experimental-control difference will be the same at each point in time (before, after, and at follow-up), or more simply, that the effects of one factor are independent of the effects of the other factor. The alternative hypothesis for the main effect of treatment can be stated either as one-tailed (e.g., $\mu_{exp} > \mu_{con}$) or as two-tailed ($\mu_{exp} \neq \mu_{con}$). I will take the more conservative approach and use the two-tailed H_A. For the main effect of time there are three levels, so the only simple way to state H_A is to state that H_0 is not true. Also, for simplicity the H_A for the interaction is a statement that the H_0 is not true.

Step 2. Select the Statistical Test and the Significance Level

The time factor involves three measures on each subject, but the treatment factor involves different subjects. Because our purpose is to detect differences in population means along these two different dimensions, a mixed design ANOVA is appropriate. The conventional approach is to use .05 as alpha for all three null hypothesis tests.

Step 3. Select the Samples and Collect the Data

To minimize the calculations I will assume that only eight phobic subjects are available for the experiment and that four are selected at random for each treatment group. The dependent variable will be a 10-point rating scale of phobic intensity with respect to public speaking (from 0 = relaxed when speaking in front of a large audience to 10 = incapable of making a speech in front of more than one person). Because each subject is measured three times, there will be a total of 8 × 3 = 24 ratings or observations, as shown in Table 18.5.

Table 18.5

Group	Subject	Before	After	Follow-Up	Row Totals
Phobia Treatment	1	8	4	6	18
	2	9	6	5	20
	3	6	3	5	14
	4	7	5	4	16
Cell Sums		30	18	20	68
Control Group	5	9	8	7	24
	6	7	7	8	22
	7	7	6	7	20
	8	6	4	7	17
Cell Sums		29	25	29	83
Column Totals		59	43	49	151

Step 4. Find the Regions of Rejection

Given that certain assumptions have been met (these will be discussed shortly), it is appropriate to use the F distribution to find a critical value for each null hypothesis. However, we need to know the degrees of freedom that apply in each case. The breakdown of the df can get complicated for a mixed design, so a df tree can be especially helpful when dealing with this type of design (see Figure 18.4). In Figure 18.4, I use k to represent the number of different groups (i.e., the number of levels for the between-groups factor), as I did for the one-way independent ANOVA, and c to represent the number of treatment conditions presented to each subject (i.e., the number of levels for the within-subjects factor) as I did for the one-way RM ANOVA. Thus $k \times c$ (or just kc) is the number of

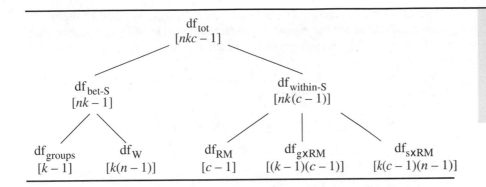

Figure 18.4

*Degrees of
Freedom Tree for
Two-Way Mixed
Design*

cells in the two-way mixed design. I will deal only with the case in which each group has the same number of subjects, so n can be used to represent the number of subjects assigned to each level of the between-groups factor (i.e., the number of subjects per group); thus kn is the total number of different subjects in the experiment. However, there are c measurements, or scores, for each subject, so nkc is the total number of observations in the experiment (i.e., N_T). Now we can express the total df as

$$df_{tot} = nkc - 1 \text{ (or } N_T - 1)$$

The total df are divided into the df associated with variation between subjects ($df_{between-S}$) and the df associated with variation within subjects ($df_{within-S}$). The $df_{between-S}$ are simply the total number of different subjects minus 1; the $df_{within-S}$ are equal to the total number of *observations* minus the total number of different subjects. These relationships are expressed in Formula 18.1A):

$$df_{between-S} = nk - 1$$
$$df_{within-S} = nk(c - 1) \text{ or } nkc - nk \qquad \text{Formula 18.1A}$$

As you can see from Figure 18.4, the df are further subdivided in each branch. The $df_{between-S}$ are divided into the following two components (Formula 18.1B),

$$df_{groups} = k - 1$$
$$df_W = k(n - 1) \text{ or } nk - k \qquad \text{Formula 18.1B}$$

where df_W stands for the df associated with subject-to-subject variation within each group. The $df_{within-S}$ are divided into three components (Formula 18.1C):

$$df_{RM} = c - 1$$
$$df_{g \times RM} = (k - 1)(c - 1)$$
$$df_{s \times RM} = k(c - 1)(n - 1) \qquad \text{Formula 18.1C}$$

The df for the interaction of the two factors, $df_{g \times RM}$, is equal to the df for groups times the df for the repeated measures factor. The $df_{s \times RM}$ component is the df corresponding to the sum of the df for the subject by repeated measure interactions for each group.

For the present example, $k = 2$, $c = 3$, and $n = 4$, so $df_T = 4 \cdot 2 \cdot 3 - 1 = 24 - 1 = 23$, which can be broken down as follows:

$$df_{\text{between-S}} = (4 \cdot 2) - 1 = 8 - 1 = 7$$
$$df_{\text{within-S}} = 4 \cdot 2(3 - 1) = 8 \cdot 2 = 16$$

$df_{\text{bet-S}}$ can be then divided as follows:

$$df_{\text{groups}} = 2 - 1 = 1$$
$$df_W = 2(4 - 1) = 6$$

Similarly, $df_{\text{within-S}}$ can be divided:

$$df_{RM} = 3 - 1 = 2$$
$$df_{g \times RM} = 1 \cdot 2 = 2$$
$$df_{s \times RM} = 2(3 - 1)(4 - 1) = 2 \cdot 2 \cdot 3 = 12$$

Now we can find the critical F value for each of our three null hypothesis tests. For the main effect of phobia treatment, the df are df_{groups} (1) and df_W (6); $F_{.05}(1, 6) = 5.99$. For the main effect of time, the df are df_{RM} (2) and $df_{s \times RM}$ (12); $F_{.05}(2, 12) = 3.89$. And for the interaction of the two factors, the df are $df_{g \times RM}$ (2) and $df_{s \times RM}$ (12); therefore, the critical F for this test is also 3.89.

Step 5. Calculate the Test Statistics

For each of the df components delineated above, there is a corresponding SS component. The total SS is calculated first, using the now familiar formula, identified here as Formula 18.2:

$$SS_{\text{tot}} = \Sigma X^2 - \frac{T_T^2}{N_T} \qquad \text{Formula 18.2}$$

For the data in Table 18.5, $\Sigma X^2 = 1003$ and $T_T = 151$, so

$$SS_{\text{tot}} = 1009 - \frac{151^2}{24} = 1009 - \frac{22,801}{24} = 1009 - 950.04 = 58.96$$

Next we calculate $SS_{\text{between-S}}$, which depends on the totals for each subject (S_i); (the row totals in Table 18.5). The formula is

$$SS_{\text{between-S}} = \frac{1}{c}\Sigma S_i^2 - \frac{T_T^2}{N_T} \qquad \text{Formula 18.3}$$

Note that Formula 18.3 is essentially the same as Formula 17.2 (SS_{sub}). For this example we have

$$SS_{\text{between-S}} = \frac{1}{3}(18^2 + 20^2 + 14^2 + 16^2 + 24^2 + 22^2 + 20^2 + 17^2) - 950.04$$
$$= 24.96$$

Now we find the SS for the grouping factor, which depends on the totals for each group (G_i), according to Formula 18.4:

$$SS_{\text{groups}} = \frac{1}{cn}\Sigma G_i^2 - \frac{T_T^2}{N_T} \qquad \text{Formula 18.4}$$

For this example,

$$SS_{\text{groups}} = \frac{1}{12}(68^2 + 83^2) - 950.04 = 959.42 - 950.04 = 9.38$$

By subtracting SS_{groups} from $SS_{\text{between-S}}$, we obtain SS_W (Formula 18.5):

$$SS_W = SS_{\text{between-S}} - SS_{\text{groups}} \qquad \text{Formula 18.5}$$

So,

$$SS_W = 24.96 - 9.38 = 15.58$$

One branch of the total SS has now been analyzed into its components. To find the value of the other branch we subtract $SS_{\text{between-S}}$ from SS_{tot} to find $SS_{\text{within-S}}$:

$$SS_{\text{within-S}} = SS_{\text{tot}} - SS_{\text{between-S}} \qquad \text{Formula 18.6}$$

Thus,

$$SS_{\text{within-S}} = 58.96 - 24.96 = 34$$

We now turn our attention to analyzing $SS_{\text{within-S}}$ into its components, starting with the SS for the repeated measures factor (SS_{RM}), which is based on the totals of the columns (RM_i) in Table 18.5. Formula 18.7 is as follows:

$$SS_{RM} = \frac{1}{kn}\Sigma RM_i^2 - \frac{T_T^2}{N_T} \qquad \text{Formula 18.7}$$

For this example,

$$SS_{RM} = \frac{1}{8}(59^2 + 43^2 + 49^2) - 950.04 = 966.37 - 950.04 = 16.33$$

The SS for the interaction of the two factors ($SS_{g \times RM}$) is a bit trickier to calculate. We need to use the method detailed in Chapter 16, in which we found the overall SS_{bet} from the cell sums and then subtracted the SS for the two main effects. I will use Formula 18.8, which is identical to Formula 16.2; however, I will call this component $SS_{between-cells}$ to avoid confusion.

$$SS_{between-cells} = \frac{1}{n}\Sigma cs_i^2 - \frac{T_T^2}{N_T} \qquad \text{Formula 18.8}$$

Based on the cell sums shown in Table 18.5, we have

$$SS_{bet-cells} = \frac{1}{4}(30^2 + 18^2 + 20^2 + 29^2 + 25^2 + 29^2) - 950.04$$
$$= 982.75 - 950.04 = 32.71$$

Now we can use Formula 18.9 to obtain $SS_{g \times RM}$:

$$SS_{g \times RM} = SS_{bet-cells} - SS_{groups} - SS_{RM} \qquad \text{Formula 18.9}$$

Thus,

$$SS_{g \times RM} = 32.71 - 9.38 - 16.33 = 7.0$$

The third component of $SS_{within-S}$ is found by subtraction, using Formula 18.10:

$$SS_{s \times RM} = SS_{within-S} - SS_{RM} - SS_{g \times RM} \qquad \text{Formula 18.10}$$

Therefore,

$$SS_{s \times RM} = 34 - 16.33 - 7.0 = 10.67$$

To obtain the variance estimates (i.e., MS) that we will need to form our three *F* ratios, we must divide each of the five final SS components we found above by the corresponding df found in the previous step of our procedure.

$$MS_{groups} = \frac{SS_{groups}}{df_{groups}} = \frac{9.38}{1} = 9.38$$

$$MS_W = \frac{SS_W}{df_W} = \frac{15.58}{6} = 2.6$$

$$MS_{RM} = \frac{SS_{RM}}{df_{RM}} = \frac{16.33}{2} = 8.17$$

$$MS_{g \times RM} = \frac{SS_{g \times RM}}{df_{g \times RM}} = \frac{7}{2} = 3.5$$

$$MS_{s \times RM} = \frac{SS_{s \times RM}}{df_{s \times RM}} = \frac{10.67}{12} = .89$$

Finally, we can form the F ratios to test each of the three null hypotheses stated in Step 1. To test the main effect of groups, we create the following ratio:

$$F_{groups} = \frac{MS_{groups}}{MS_W} = \frac{9.38}{2.6} = 3.61$$

To test the main effect of the within-subject factor, we use $MS_{s \times RM}$ as our error term:

$$F_{RM} = \frac{MS_{RM}}{MS_{s \times RM}} = \frac{8.17}{.89} = 9.19$$

Finally, the same error term is used in the F ratio to test the two-way interaction:

$$F_{g \times RM} = \frac{MS_{g \times RM}}{MS_{s \times RM}} = \frac{3.5}{.89} = 3.93$$

Step 6. Make the Statistical Decisions

The observed F for the main effect of the phobia treatment is 3.61, which is less than the critical F (5.99), so we cannot reject the null hypothesis for this factor. On the other hand, the F ratio for the time factor (before versus after versus follow-up) is 9.19, which is well above the critical F (3.89), so the null hypothesis for this effect can be rejected. For the interaction, the observed F ratio (3.93) is only slightly above the critical F (3.89), but that is all that is required to reject this null hypothesis, as well.

Interpreting the Results

At first, the lack of statistical significance for the main effect of treatment group may seem discouraging; it seems to imply that the phobia treatment didn't work, or that at best it was no more effective than the control procedure. However, the significant interaction should remind you to graph the cell means before trying to interpret the results of the main effects. You can see from Figure 18.5 that the two groups are very similar in phobic intensity before the treatment (which is to be expected with random assignment) but diverge considerably after treatment. Despite the similarity of the groups before treatment, the two later measurements might have caused the main effect of group to be significant had not the samples been so small. The *F* ratio for groups is sensitive to the variability from subject to subject, and with small samples power is likely to be low.

On the other hand, subject to subject variability does not affect the *F* ratio for the time factor. As long as the subjects exhibit similar patterns over time within each group, $MS_{s \times RM}$ will tend to be small and F_{RM} will tend to be large, as is the case in this example. However, the significance of the time factor must also be interpreted cautiously, given that the interaction is significant. The significance of $F_{g \times RM}$ tells you that the effect of time is different for the two groups. From Figure 18.5 you can see that the before-after reduction is much larger for the experimental group, and the increase from after to follow-up is somewhat greater for the control group.

It is likely that after obtaining these results a researcher would think of some more specific hypotheses to test—for instance, are the two groups significantly different just after the treatment? Or, is the before-after phobia reduction significant for the control group alone? Unless these specific questions have been planned in advance (see Chapter 15, Section C), the researcher should use procedures for post hoc comparisons. The choice of procedure depends on whether

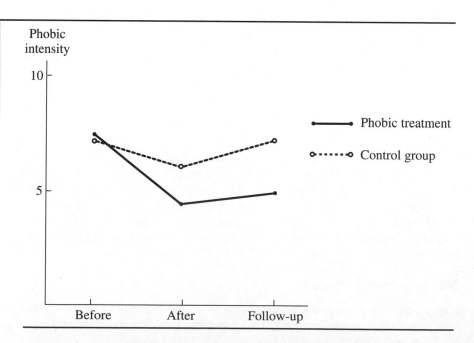

Figure 18.5

Graph of Cell Means for the Data in Table 18.5

comparisons are being made among a series of repeated measures or among different groups and on whether the interaction is significant (see Section C).

Assumptions of the Mixed Design ANOVA

Because the mixed design involves performing an independent samples ANOVA *and* a repeated measures ANOVA, the assumptions underlying each of these statistical procedures are required for the appropriate factor. Also, the procedures in this chapter are based on the assumption that, as with a two-way ANOVA, the mixed design is balanced and that, as with the one-way RM design, the sample sizes are equal across the levels of the within-subject factor. If data on a subject are missing for one condition but not another, the simplest and most common solutions are either to replace the subject entirely or to use some technique to estimate the most likely value for the missing data point. In this text I will assume that one of these solutions has already been used. On the other hand, it is not unusual to analyze data for an experiment in which the groups are of different sizes. If the differences in sample sizes across the levels of the between-groups factor are small and have occurred in a random fashion, I suggest using analysis of unweighted means, as described in Chapter 16, Section C. Otherwise, more complex methods involving multiple regression are recommended (see Hays, 1994).

Independent Random Sampling

Each group should consist of an independent random sample chosen from the population of interest, and the subjects in one group should be chosen independently of the subjects in any other group. If randomized blocks are used in place of repeated measures, then the subjects from each block must be randomly assigned to the levels of the RM factor.

Normal Distributions

The population distribution is assumed to be normal for each cell of the mixed design, though the *F* tests for each null hypothesis are known to be robust with respect to violations of this assumption.

Homogeneity of Variance

The population variance is assumed to be the same for all cells in the mixed design. With a balanced design, this assumption is usually ignored.

Homogeneity of Covariance

If the within-subject factor has more than two levels, then the covariance of any pair of levels is assumed to be the same as the covariance for any other pair. However, homogeneity of covariance applies in an additional way to mixed designs—a way that is applicable even if the within-subjects factor has only two levels: The amount of covariance between any two levels of the within-subjects factor in a particular group should be the same as that in any other group.

As I described in detail in Chapter 17, requiring that the third and fourth assumptions be satisfied simultaneously (i.e., compound symmetry) is stricter than is necessary to ensure the validity of the F ratio testing the RM factor (or the interaction of the two factors); the critical assumption is sphericity (or circularity). However, for the mixed design the assumption of sphericity must be extended to comparisons between any pair of groups. For instance, the amount of before-after interaction in the treatment group of the above example should be the same as the amount of before-after interaction in the control group (and similarly for the after–follow-up and before–follow-up interactions).

When to Use the Mixed Design

Whenever it is desirable to combine two independent variables in a factorial design, you should consider repeating measures on one or both of these variables, because of the large increase in power that is likely to occur. If it is feasible to repeat measures on both variables, a two-way RM design will probably lead to the most power. However, in some cases it may make sense to repeat measures on one factor but not the other. The example described at the end of Section A (involving tasks of varying difficulty and different levels of motivation) fits this description. Another situation involving two experimental factors that calls for the mixed design is the one illustrated in this section; wherein you are comparing two or more treatment conditions over time.

One of the most common mixed designs involves one factor that cannot be repeated because it is a grouping variable. The main effect of the grouping variable is not likely to be interesting (it is often obvious); the focus is usually on the interaction: Do both (or all) of the groups display the same pattern of reactions to the different levels of the repeated measures variable?

Finally, I should point out that the mixed design also applies when the repeated measures are not actually repeated but instead presented to a set of matched subjects. For instance, a researcher may want to know which of three weight-training procedures is most effective and whether the relative effectiveness of the procedures differ for men and women. Rather than trying to have each subject trained in all three procedures, one after the other (imagine the carryover problem!), the men could be matched in groups of three (based on height, build, current strength, etc.), as could the women. The members of each trio would be randomly assigned to the three weight-training procedures, and maximum strength after the training would be measured as the dependent variable. The mixed design ANOVA would proceed exactly as described in this section; the three matched subjects in each trio would be treated as a single subject measured three times.

A Special Case: The Before-After Mixed Design

The simplest possible mixed design is one in which an experimental group and a control group are measured before and after some treatment. Some statisticians have argued that a two-way ANOVA is unnecessary in this case and may even prove misleading (Huck & McLean, 1975). The main effect of group is misleading because the group difference *before* the treatment (which is expected to be very close to zero) is being averaged with the group difference *after* the treatment. The main effect of time is equally misleading, because the before-after difference for the experimental group is being averaged with the before-after difference for the control group. The only effect worth testing is the interaction, which tells us whether the before-after differences for the experimental group are different from the before-after differences for the control group. We do not need ANOVA to test this interaction—we need only find the before-after difference for each subject and then conduct a *t*-test of two independent samples (experimental versus control group) on these difference scores. Squaring this *t* value will give the *F* that would be calculated for the interaction in the mixed design ANOVA. If before and after measurements have been taken on more than two groups, then a one-way independent-groups ANOVA on the difference scores yields the same *F* ratio as a test of the interaction in the mixed design. Of course, you would probably want to follow a significant *F* with pairwise tests on the difference scores to determine which groups differ significantly.

Although testing the interaction of the mixed design, or testing the difference scores, as described above, seem to be the most common ways of evaluating group differences in a before-after design, they are not the most powerful. A more sensitive test of group differences is based on a procedure known as the *analysis of covariance (ANCOVA)*. Instead of simply subtracting the before score from the after score, you can use linear regression to predict the after score from the before score. You can then perform a *t*-test or one-way ANOVA on the residual scores (after score minus predicted after score) rather than on the difference scores. The residual scores always have less variance (unless the regression slope happens to be 0 or 1.0) and therefore tend to yield a higher *t* or *F.* The logic of ANCOVA will be explained further in Section C.

Publishing the Results of a Mixed ANOVA

The results of the phobia treatment experiment analyzed above could be reported in a journal article in this manner: "The phobia intensity ratings were submitted to a 2×3 mixed design ANOVA, in which treatment group (experimental versus placebo control) served as the between-subjects variable, and time (before versus after versus follow-up) served as the within-subjects variable. The main effect of treatment group did not attain significance, $\underline{F}(1, 6) = 3.61$, MSE = 2.6, $\underline{p} > .05$, but the main effect of time did reach significance, $\underline{F}(2, 12) = 9.19$, MSE = .89, $\underline{p} < .05$. The results of the main effects are qualified, however, by a significant

Source	SS	df	MS	F	p
Between-Subjects	24.96	7			
Groups	9.38	1	9.38	3.61	> .05
Within-groups	15.58	6	2.6		
Within-Subjects	34	16			
Time	16.33	2	8.2	9.19	< .05
Group × Time	7.0	2	3.5	3.94	< .05
Residual (S × Time)	10.67	12	.89		
Total	58.96	23			

Table 18.6

group by time interaction, $\underline{F}(2, 12) = 3.94$, MSE $= .89$, $\underline{p} < .05$. The cell means reveal that the before-after decrease in phobic intensity was greater, as predicted, for the phobia treatment group, and that this group difference was maintained at follow-up. In fact, at follow-up, the control group's phobic intensity had nearly returned to its level at the beginning of the experiment."

The above paragraph would very likely be accompanied by a table or graph of the cell means and followed by a report of more specific comparisons, such as testing for a significant group difference just at the follow-up point (see Section C). Although it is not likely that an ANOVA summary table would be included in the report of your results, such tables are produced by most statistical software packages, and as they are instructive, I include a summary table (Table 18.6) for the phobia treatment example.

Example from the Literature

The following example of a mixed design ANOVA in the psychological literature comes from an article entitled "Affective valence and memory in depression: Dissociation of recall and fragment completion" (Denny & Hunt, 1992). This study contains several ANOVAs, but the one I have chosen resembles the example in Section A, except that it does not include neutral words. Thus word valence refers to whether a word tended to evoke positive or negative affect. The results were reported as follows.

> Recall data were subjected to an analysis of variance (ANOVA) with group as a between-subjects variable and word valence as a within-subjects variable. The results revealed a significant main effect for group: Recall level was higher for the nondepressed group than for the depressed group, $F(1, 30) = 30.21$, $MS_e = 2.05$, $p < .0001$. This effect was qualified, however, by a highly significant Group × Word Valence interaction, $F(1, 30) = 30.29$, $MS_e = 1.45$, $p < .0001$. The results of t-tests indicated that, as predicted, the depressed group recalled more negative than positive words, $t(15) = 4.45$, $SE_{difference} = .393$, $p < .001$. Within the nondepressed group, the opposite pattern was observed. Recall of positive words was significantly higher than that of negative words, $t(15) = 3.42$,

$SE_{difference} = .456, p < .01$. Finally, comparisons revealed a significant between-groups difference in recall of positive words, $t(30) = 6.99$, $SE_M = .517, p < .001$, but not negative words, $t(30) = .75, SE_M = .411$, $p < .20$.

Note that the article gives the error term (MS_e) for each F ratio after stating that ratio; this is the recommended practice so that a reader is equipped to perform his or her own follow-up analyses. Similarly, the denominator is given for the two-sample independent t-test (SE_M) and for the matched t-test ($SE_{difference}$). The authors refer to the interaction as being "highly significant." Be aware that statistical purists abhor this expression; they argue that a result is either significant (i.e., p is less than the chosen alpha) or it is not. What the authors meant in this case is that the p level for the interaction was very small; the interaction would have been significant even if a very small alpha (e.g., .0001) had been set.

Figure 18.6 is a graph of the cell means reported by Denny and Hunt. The figure illustrates the nature of the significant interaction and should make it obvious why the main effect of group was significant but the main effect of word valence was not.

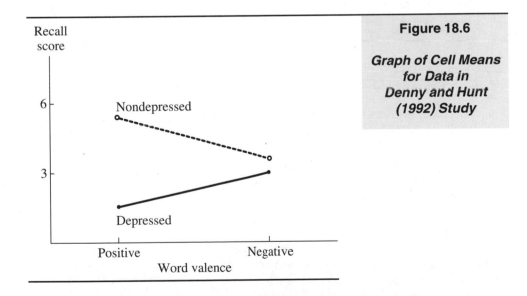

Figure 18.6

Graph of Cell Means for Data in Denny and Hunt (1992) Study

Exercises

*1. Imagine a study conducted to compare the effects of three types of training on the acquisition of a motor skill. Thirty-six subjects are divided equally into three groups, each receiving a different type of training. The performance of each subject is measured at five points during the training process.

a) Construct a df tree (like the one in Figure 18.4) for the mixed design ANOVA that would be used to analyze this experiment.

b) Find the critical values of F for each of the F ratios that would be calculated.

2. In exercise 16B2, a group of visualizers and a group of nonvisualizers were each divided in half, and all subjects were presented with either concrete or abstract words to recall. A more powerful way to conduct that experiment would be to give each subject the same list containing a mixture of concrete and abstract words. The data from exercise 16B2 are reproduced below, rearranged to make it easier to see the number of concrete and abstract words each subject recalled.

	Concrete	Abstract
Visualizers	17	14
	20	15
	18	15
	21	17
	20	16
Nonvisualizers	18	18
	19	18
	17	17
	17	17
	20	19

a) Perform a two-way mixed design ANOVA on the data above (Note: You can save time by using some of the SS that you calculated for exercise 16B2.)

b) Present your results in a summary table and compare it to the summary table you created for exercise 16B2.

*3. A psychologist is studying the relationship between emotion and eating. All of his subjects view the same two film segments. One segment evokes happiness and one segment evokes fear; the order in which subjects view the film segments is counterbalanced. Half of the subjects are randomly assigned to a condition that requires them to eat a full meal just before viewing the film segments (preload condition); the remaining half are not permitted to eat during the four hours preceding the experiment (no load condition). The subjects are offered an unlimited amount of popcorn while viewing the film segments. The amount of popcorn (in ounces) consumed by each subject in each condition appears in the table below.

	Happiness	Fear
Preload	10	12
	13	16
	8	11
	16	17
No Load	26	20
	19	14
	27	20
	20	15

a) Perform a mixed design ANOVA.

b) Draw a graph of the cell means. (Note that you already graphed the data for the individual subjects in exercise 18A4.) Describe the nature of each significant effect.

c) Calculate the happiness-fear difference score for each subject, and then perform a two-group independent *t*-test on these difference scores. Which of the *F* ratios calculated in part a is related to the *t* value you just found? What is the relationship?

*4. In exercise 17B1, subjects performed a clerical task under three noise conditions. Now suppose a new group of subjects is added in order to study the effects of the same three conditions on the performance of a simpler, more mechanical task. The data from exercise 17B1 are reproduced below, along with the data for the mechanical task. Perform a mixed design ANOVA, and display the results in a summary table.

	Background Noise	Popular Music	Heavy Metal Music
Clerical Task	10	12	8
	7	9	4
	13	15	9
	18	12	6
	6	8	3
Mechanical Task	15	18	20
	19	22	23
	8	12	15
	10	10	14
	16	19	19

*5. Dr. Jones is investigating various conditions that affect mental effort—which, in this experiment, involves solving anagrams. Subjects were randomly assigned to one of three experimental conditions. Subjects in the first group were told that they would not be getting feedback on their performance. Subjects in the second and third groups were told they *would* get feedback, but only subjects in the third group were told (erroneously) that anagram solving was highly correlated with intelligence and creativity (Dr. Jones hoped this information would produce ego involvement.) The list of anagrams given to each subject contained a random mix of problems at four levels of difficulty determined by the number of letters presented (five, six, seven, or eight). The number of anagrams correctly solved by each subject in each condition and at each level of difficulty is given in the table below.

	5 Letters	6 Letters	7 Letters	8 Letters
No Feedback	2	4	6	9
	3	4	7	10
	5	7	9	12
Feedback	12	15	16	19
	11	11	15	19
	14	17	20	22
Feedback + Ego	21	22	25	30
	23	27	30	31
	24	28	32	34

a) Perform a mixed analysis of variance, and display the results in a summary table.

b) Draw a degrees of freedom tree for this experiment.

*6. A psychiatrist is comparing various treatments for chronic headache: biofeedback; an experimental drug; self-hypnosis; and a control condition in which subjects are asked to relax as much as possible. Subjects are matched in blocks of four on headache frequency and severity and then randomly assigned to the treatment conditions. To check for any gender differences among the treatments, four of the blocks contain only women and the other four blocks contain only men. The data appear in the table below.

Blocks	Control	Biofeedback	Drug	Self-Hypnosis
Men	14	5	2	11
	11	8	0	4
	20	9	7	13
	15	12	3	10
Women	15	7	1	10
	10	8	1	4
	14	8	8	12
	17	14	4	11

a) Before conducting the ANOVA, graph the cell means and guess which of the three *F* ratios is (are) likely to be significant and which is (are) not.

b) Test the significance of each of the three *F* ratios.

*7. A market researcher is comparing three types of commercials to determine which will have the largest positive effect on the typical consumer. One type of commercial is purely informative, one features a celebrity endorsement, and the third emphasizes the glamour and style of the product. Six different subjects are randomly assigned to watch each of the three types of commercial. Subjects rate their likelihood of buying the product on a scale from 0 (very unlikely to buy) to 10 (very likely to buy) both before and after viewing the assigned commercials. The data appear in the table below.

Informative		Celebrity Endorsement		Glamour/ Style	
Before	**After**	**Before**	**After**	**Before**	**After**
3	5	6	8	5	8
6	6	6	9	7	8
5	7	4	4	5	5
7	8	5	6	5	7
4	6	7	8	6	7
6	5	2	4	3	6

a) Perform a mixed design ANOVA and test the three F ratios at the .01 level.

b) Calculate the before-after difference score for each subject, and then perform a one-way independent-groups ANOVA on these difference scores. Which of the F ratios calculated in part a is the same as the F ratio you just found? Explain the connection.

8. Exercise 17B6 described a neuropsychologist studying subjects with brain damage to the left cerebral hemisphere. Such a study would probably include a group of subjects with damage to the right hemisphere and a group of control subjects without brain damage. The data from exercise 17B6 (the number of digit or letter strings each subject recalled) follow, along with data for the two comparison groups just mentioned.

	Digits	Letters	Mixed
Left Brain Damage	6	5	6
	8	7	5
	7	7	4
	8	5	8
	6	4	7
	7	6	5

	Digits	Letters	Mixed
Right Brain Damage	9	8	6
	8	8	7
	9	7	8
	7	8	8
	7	6	7
	9	8	9
Control Group	8	8	7
	10	9	9
	9	10	8
	9	7	9
	8	8	8
	10	10	9

a) Perform a mixed design ANOVA and test the three F ratios at the .05 level.

b) What can you conclude about the effects of brain damage on short-term recall for these types of stimuli?

c) Draw a graph of these data, subject by subject. Do the assumptions of the mixed design ANOVA seem reasonable in this case? Explain.

Dealing with a Lack of Sphericity in Mixed Designs

OPTIONAL MATERIAL

In Chapter 17, I made the point that a lack of sphericity increases the likelihood of a significant F ratio, even if all of the population means for the repeated treatments are equal. A three-step procedure was described to keep the Type I error rate at the level set initially. This procedure can and should be applied to the RM factor of a mixed design, as well. If F_{RM} does not meet the usual criterion for significance, you need do nothing further. For the conservatively adjusted critical F, $df_{num} = 1$ and $df_{denom} = df_W = k(n - 1)$. If the F ratio for the RM factor (F_{RM}) surpasses this larger critical F, it can be considered statistically significant without assuming sphericity. However, if F_{RM} falls between the usual critical F and the conservative F, you can compute the epsilon (ϵ) coefficient described in Chapter 17 to estimate the degree of sphericity in the population and adjust the degrees of freedom accordingly.

The F ratio for the interaction of the two factors ($F_{g \times RM}$) uses the same error term as F_{RM} and can therefore be similarly biased when sphericity is violated. The three-step procedure should also be applied to the test of $F_{g \times RM}$, except that the degrees of freedom for the numerator of the adjusted critical F are $df_{num} = df_{groups} = k - 1$ (df_{denom} still equals df_W). This modified univariate approach can be replaced, however, by a multivariate analysis of both the RM factor and the interaction of factors.

A lack of sphericity does not affect the test of the between-groups factor, and if the groups are equal in size, a lack of homogeneity of variance can be ignored, as well. It is only when the groups are not very large and differ considerably in size that you cannot ignore large differences in within-group variances. In such a case, a separate-variances t-test or a modified one-way ANOVA may be required, if a data transformation does not seem appropriate or helpful.

Post Hoc Comparisons

When the Two Factors Do Not Interact

If the interaction of the two factors is not large or statistically significant, the significant main effects can be explored in a straightforward manner. If the RM factor is significant and has more than two levels, a pairwise comparison method (e.g., Tukey's HSD) can be used to test each pair of levels for significance (collapsing across the between-groups factor—see part a of Figure 18.7, on page 636). Assuming sphericity in the population, $MS_{s \times RM}$ can be used as the error term for testing each pair. If the sphericity assumption does not seem reasonable, an error term based on difference scores for only the two levels being compared should be used. (This is just a matched t-test between the two levels, ignoring the fact that the subjects come from different groups.) Similarly, a significant between-groups factor with more than two levels would be followed with pairwise comparisons using the between-groups error term, MS_W—unless it seemed unreasonable to assume homogeneity of variance, in which case separate-variance error terms could be used. The alpha level is usually set at .05 for each "family" of comparisons; in this case, each main effect represents a separate family.

When the Two Factors Interact

If $F_{g \times RM}$ is statistically significant, or nearly so, a different approach to post hoc comparisons should be taken. The most common approach is an analysis of simple main effects, as described in Chapter 16. You can test the effects of the RM factor separately for each group, or test the effects of the between-groups factor separately at each level of the RM factor, or conduct tests in both ways (see part b of Figure 18.7). Unfortunately, the interaction of the two factors complicates the choice of an error term for each of the simple effects, as I will explain.

The simplest and safest way to test simple effects involving the RM factor is to conduct a separate one-way RM ANOVA for each group, as though the other groups do not exist. If there is homogeneity among the groups in terms of subject by treatment interaction, more power can be gained by using $MS_{s \times RM}$ from the

Figure 18.7

*Pairwise
Comparisons in a
Mixed Design*

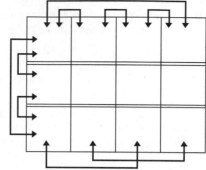

*Pairwise comparisons
between groups
(collapsing across levels
of the RM factor)*

*Pairwise comparisons between levels of the
RM factor (collapsing across groups)*

a. Pairwise comparisons following a mixed design with significant main
   effects, but no significant interaction between the factors

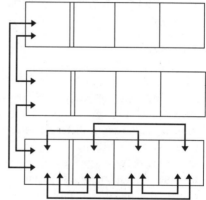

*Pairwise comparisons
for a significant simple
effect of the RM factor*

*Pairwise comparisons for a significant
simple effect of the between-groups factor*

b. Pairwise comparisons following a mixed design with a significant
   interaction

overall analysis, rather than the $MS_{s \times RM}$ for just the group being tested. However, the more conservative approach, in light of a significant interaction between the factors, is to sacrifice the small amount of extra power from the pooled error term, in favor of not increasing the risk of a Type I error, in case there actually is no homogeneity among the groups in the population.

To follow up a significant simple effect, you would probably want to conduct pairwise comparisons among the different RM levels (e.g., before versus after for the control group; after versus follow-up for the experimental group). If you had a strong reason to assume sphericity within the significant simple effect, you could use the $MS_{s \times RM}$ for that group as the error term for each pairwise test.

However, as the sphericity assumption is generally considered quite risky, especially for pairwise comparisons, it is strongly recommended that you base your error term only on the two levels being tested. (This is equivalent to performing a simple matched *t*-test between a pair of RM levels for one of the groups.) The Type I error rate can be controlled by using the Bonferroni test, setting alpha at .05 for the entire family of possible RM comparisons.

Simple effects can also be tested by comparing the different groups separately for each level of the RM factor. However, the proper error term for each of these effects, given that the interaction of the two factors is significant, is not obvious. Assuming homogeneity of variance exists, you might expect to use MS_W as the error term for between-group comparisons, but the interaction of the two factors contributes to between-groups differences and that contribution is related to the $MS_{s \times RM}$ error term. It turns out that the required error term for the between-groups simple effects is a combination of both MS_W and $MS_{s \times RM}$, which is referred to as $MS_{within-cell}$. Actually, $MS_{within-cell}$ is the error term you would get if you analyzed the mixed design as though it were a two-way independent-groups design, using the formulas of Chapter 16. To complete the analysis of the mixed design, you would have to divide the $SS_{within-cell}$ (which was called simply SS_W in Chapter 16) into SS_W and $SS_{s \times RM}$, as defined in this chapter, to create the two error terms.

Unfortunately, there is a strong likelihood in a mixed design that $MS_{within-cell}$ will pool unequal variance components and therefore lead to a biased *F* ratio. For large sample sizes this bias will be negligible, but for samples smaller that 30 an adjustment of the degrees of freedom (similar to the adjustment in the separate-variances *t*-test) is recommended before you determine the critical *F*. The problem can be avoided, with some loss of power, by pooling error terms for only the cells involved in the analysis. For instance, if you are comparing only "after" measurements for two experimental groups and a control group, the error term can be based on pooling the MS_W for just those cells, rather than being based on $MS_{within-cell}$ from the entire mixed design. This localized error term can then be used to test pairs of groups within a particular level of the RM factor, whenever the simple effect of groups is significant at that level of the RM factor. (Of course, if there are only two groups in the design, there are no follow-up tests to be done on that factor.)

Analysis of Covariance (ANCOVA)

I introduced the two-way ANOVA in Chapter 16 by taking a one-way ANOVA and then dividing the subjects along a second factor (gender). Because the second factor was very relevant to the dependent variable, the division by gender caused a dramatic reduction in the error term (MS_W). And because gender is a categorical variable, it was easy to make it a factor in an ANOVA. But what if there were another equally relevant variable that was continuous, such as the number of years each patient had been depressed at the time of the study? If we were to break the number of years into a few broad categories to create a second factor (e.g., fewer than 5 years; between 5 and 10 years; more than 10 years), we could lose a good deal of information (i.e., there could be important differences *within* a category). On the other hand, if we rounded off to the nearest year and included a level for each number of years, there would be a large number of levels for this factor, and some of the cells would be empty! There is an easier way to incorporate a continuous variable (in addition to the dependent variable) in an ANOVA; the method is called **analysis of covariance (ANCOVA).**

As a simple example, imagine that a researcher suspects that married men weigh more than bachelors, and to test this hypothesis he collects a sample of each group and weighs each man. The variability within each group is likely to be large compared to the difference between group means—that is, the effect size is likely to be small in the experiment and the population, as a whole—so unless the samples are very large there does not seem to be a good chance of rejecting the null hypothesis. But the actual weights of the men are not what the researcher is likely to be interested in. Part of the variation in weight within each group would be due to height and body frame. These parts of the variation won't interest the researcher if he is concerned with the degree to which each man is under- or overweight. A more accurate variable, in this case, would be each man's deviation from his ideal weight (based on height and body frame).

If each married man were about 2 pounds overweight and each bachelor were close to his ideal weight, a *t*-test on the actual weights would have little power to detect this difference, because of the subject-to-subject variability. However, if we assign each subject an "overweight score," the married men are all close to $+2$ and the bachelors are all close to 0. The within-group variability may be reduced enough to obtain a significant *t* for the 2-pound group difference, even with small sample sizes. This effect is similar to the benefit of the matched t-test (matching each bachelor with a married man as similar as possible in height and frame), but the method is closely related to the analysis of covariance.

To perform an ANCOVA on the example above, we would view weight as the dependent variable and height as the covariate (I will ignore body frame for a moment). Using the methods of Chapter 12, we would find the regression slope for predicting weight from height in each of the two groups and then pool the results to come up with an average regression slope. The regression slope based on the pooled results would be used to predict the weight of each man in the study based on his height. The predicted weight would then be subtracted from each man's actual weight to obtain the residual (a measure of under- or overweight, in this case), and then a *t*-test would be performed on the residuals. To the extent that height is correlated with weight, the residuals will have less variance than the original weights. [If the correlation between height and weight within each group is r, then the variance of the residuals will be $(1 - r^2)\text{MS}_\text{W}$, where MS_W is the average within-group variance of the original weights.] Assuming that the heights of married men and bachelors do not differ, the *t* value for the residuals will be considerably higher than the *t* for the original weights.

In this example we used the weight predicted from height as the ideal weight. However, if you have a second covariate that does not overlap too much with the first—such as body frame, in this case—you can use both covariates simultaneously (using the methods of multiple regression) to make an even better prediction. Any number of covariates can be used, but you lose a degree of freedom in your error term for each covariate, so it can hurt more than help to add a covariate that is highly correlated with one or a combination of the covariates already being used.

The usual assumptions of linear regression apply to ANCOVA (e.g., linear relationships, homoscedasticity), together with an important additional assumption: **homogeneity of regression.** This assumption states that the regression slope for predicting the dependent variable from the covariate(s) is the same in each population that is sampled (i.e., at each level of the between-groups factor); this assumption is the justification for pooling those regression slopes. This homogeneity assumption becomes increasingly complex and unlikely to be satisfied as the number of covariates increases, which is another reason to be cautious about adding covariates. It is also assumed that the covariate has been measured accurately. Using a covariate based on a questionnaire whose reliability and validity are not very high can increase the error variance of the experiment and even bias the results, especially if the questionnaire is not measuring exactly what it was designed to measure.

ANCOVA is commonly used to control "nuisance" variables, such as individual differences that affect the dependent variable but are extraneous to the phenomenon being studied. For instance, you might be comparing the persuasiveness of two types of written arguments in two different groups of subjects, only to find that reading comprehension accounts for some of the variability in persuasion in each group. (To the extent that a subject can't understand what she has read, she will not be persuaded.) If you can measure reading comprehension accurately and you know that it is linearly related to persuasion, you can remove its effects from the dependent variable and reduce the error term. This way of controlling for differences in reading comprehension is called *statistical control.* The type of control of extraneous variables that is generally preferred is *experimental control.*

Experimental control can take several forms. The simplest is to make sure that all subjects are alike on the nuisance variable. For instance, for the persuasion experiment, a reading comprehension test can be used to screen subjects. This approach seems reasonable for the persuasion example, but when the nuisance variable is something like age or IQ, this kind of control is not desirable. If all the subjects are the same age or IQ, it can be difficult to generalize the results of the experiment to other groups. (For instance, would older subjects or brighter subjects be affected the same way by the independent variable?) Another form of experimental control is to add a factor to an ANOVA design. The limitations of this approach when dealing with continuous variables were discussed above. Perhaps the best form of experimental control is matching subjects into blocks (or using the same subject in all conditions, if possible). For the persuasion example, subjects could be matched in pairs for reading comprehension and then randomly assigned to one of the two persuasion conditions.

Unfortunately, experimental control is sometimes not possible or practical. You may discover a nuisance variable after the experiment has been completed.

If the nuisance variable had been measured on the subjects before the experiment, it could be used as a covariate. It is even acceptable to measure the nuisance variable after the experiment, provided that it represents some stable trait that could not have been changed by the experimental treatments. Sometimes practical considerations make it expensive and/or time-consuming to match subjects, and sometimes it is virtually impossible (e.g., it may be necessary to randomly assign patients as they arrive at a clinic). In such cases, ANCOVA can produce, retroactively, some of the beneficial effects of experimental control.

I have discussed the effect of ANCOVA in reducing error variance, but there is another potential benefit of ANCOVA: the adjustment of group means to remove the confounding effects of the covariate. For instance, suppose that married men are not only heavier but also taller than bachelors. The ANCOVA procedure would adjust the mean weights of both groups according to how much of the mean weight is predictable from the mean height. In this case, ANCOVA would reduce not only the error term but the numerator of the F ratio, as well. (Once the effects of height are accounted for, there may be little mean weight difference between married men and bachelors.) Depending on exactly how the scores come out, the entire F ratio may be increased or reduced. On the other hand, ANCOVA can increase the separation of the group means. If there is little difference in the group means for weight, but the bachelors are found to be considerably taller, the adjusted means will show a greater weight difference for the married men, as they would be more overweight (about the same weight as the single men, but shorter).

When the independent variable involves experimental manipulations on randomly assigned subjects, the adjustment of group means by ANCOVA can be helpful, but it is not likely to be large. The usually small, accidental differences in group means that are due to covariate differences will be removed by ANCOVA. The random assignment of subjects ensures that large differences in the covariate (which should be unaffected by the independent variable) will be very rare. If the subjects are not randomly assigned to groups, but rather preexisting groups (e.g., different sections of an introductory psychology class) are used for the different levels of the independent variable, ANCOVA can be used to compensate for mean differences on the covariate. However, there may be other important differences between the groups that are not captured by any of the covariates that have been measured. Caution in drawing conclusion from such studies is advised.

Moreover, the use of ANCOVA with quasi-independent variables can be controversial. For instance, suppose researchers follow a group of moderate social drinkers and a group of nondrinkers for a number of years to determine longevity. If the social drinkers are found to live longer, various other differences between the two groups may be contributing to their longevity. (Unless the groups have been carefully matched, there may be differences in diet, exercise, stress, etc.) ANCOVA can be used to adjust the two groups for differences on the covariates chosen. (When the necessary assumptions are met, ANCOVA answers the question, how would these groups differ on the dependent variable, if they were equal on all the covariates?) However, an ANCOVA demonstrating that the groups still differ on longevity after being equated on several relevant covariates would still not be conclusive. It is possible that the groups differ in some way other than in alcohol consumption that is not captured by any of the covariates. Perhaps the social drinkers are more sociable and this helps them deal with stress (and, incidentally, leads them to drink moderately in social situations). Only a true experiment, with subjects randomly assigned to moderate and zero drinking conditions, can settle the issue decisively. On the other hand, using ANCOVA can be an improvement over merely comparing preexisting groups; ANCOVA helps to rule out some of the possible explanations for the group differences on the dependent variable.

There is one application for which ANCOVA is clearly superior to the statistical method currently most popular. This is the experimental design in which subjects are measured both before and after either an experimental treatment or a control treatment (any number of experimental or control treatments can be used, with an independent sample of subjects for each). As I mentioned at the end of Section B, it is common to perform either a mixed design ANOVA (focusing on the interaction) or just a one-way ANOVA (or *t*-test) on the difference scores. However, except in some unlikely circumstances, ANCOVA provides a more powerful test to compare the different treatments. Instead of subtracting each subject's "before" score from his or her "after" score to form a difference score, ANCOVA uses the correlation between the before and after scores to predict an after score. Unless the regression slope is 1.0 or zero, the predicted after scores will be closer to the actual after scores than are the before scores; therefore, the residuals from the regression will have less variance than the before-after difference scores, and the ANCOVA *F* will tend to be higher than the *F* for comparing the difference scores (the latter is the same as the *F* for the interaction in the mixed design ANOVA). The use of before-after differences can be seen as a crude form of ANCOVA, in which the regression slope is assumed to be 1.0, without calculating its value in each group. Now that statistical software makes it easy to apply ANCOVA, there is no longer any excuse for using the cruder, traditional analysis.

Multivariate Analysis of Variance (MANOVA)

ANCOVA is a rather specialized procedure that is used when you have one dependent variable that you expect to be affected by the independent variable and one or more concomitant variables (i.e., covariates) that contribute to the dependent variable but are not expected to be directly affected by the independent variable. However, sometimes you can have more than one dependent variable that you expect will be changed by your experimental treatments. If these dependent variables are really just different ways to measure the same underlying variable, there is little to be gained by using them together in the same analysis. On the other hand, if these dependent variables are not highly related to each other, testing them together can give you more power to distinguish your experimental treatments than any one dependent variable, by itself. The general method for testing two or more dependent variables simultaneously in an ANOVA design is called multivariate analysis of variance (MANOVA). The details of calculating a MANOVA involve matrix algebra and are well beyond the scope of this text, but the concepts of MANOVA can be explained in terms of procedures already described in this text.

The simplest form of MANOVA involves two independent samples and two dependent variables. For an example, let us return to the question of vitamin C's effect on the common cold. It has been claimed that people who take vitamin C experience not only shorter colds but less severe symptoms. To test this claim, we will imagine that subjects in one random sample receive vitamin C for a year, while subjects in another take a placebo. Each cold that a subject experiences is measured for both length and severity; at the end of the year each subject gets a total duration score and a total severity score. Suppose that the vitamin C group has a lower mean for both duration and severity and that each of these dependent variables is tested with an ordinary *t*-test. It is possible that each *t*-test could fall just short of significance. In that case, it is likely that a combination of the two variables would reach significance.

The multivariate approach is to take a weighted combination of the two variables to create one composite variable, and then to calculate the *t* value for the composite. The complicated part is to figure out how much weight to give to each variable. The goal, however, is simple. The weights that are chosen are the ones that produce the largest *t* value possible. When the resulting *t* value is squared, it is known as Hotelling's T^2, after the statistician who devised it. The squared *t* value follows one of the *F* distributions, but because trying all possible combinations of the dependent variables increases the chance of Type I error, the *F* value must be reduced accordingly before being tested for significance. Adding dependent variables does not change the basic analysis, but the *F* used to test T^2 must be reduced further for each variable added, so adding dependent variables that are redundant with those already being tested could hurt more than help.

The weighted combination of dependent variables that produces the largest *t* value is called the discriminant function. If the discriminant function sounds like a multiple regression equation, it's not surprising—that's what it is. In fact, if each group (vitamin C and placebo) is considered the predicted variable and coded (using 0 and 1 or any two numbers), and duration and severity are viewed as the predictors, the multiple regression equation will give the same relative weights to the two predictors as the discriminant function does. (Of course, you would have no need to "predict" which group a subject had been in, but the discriminant function shows you just how much subjects from the two groups can differ when you choose the optimal weights for your dependent variables.) Moreover, the *F* that would be used to test the significance of R^2 in the multiple regression case is the same as the *F* that would be used to test Hotelling's T^2. Recall that in the univariate case, the point-biserial *r* can be calculated and tested for significance, in place of the ordinary *t*-test. In the multivariate case, R^2 can be calculated as an alternative to T^2 [$R^2 = T^2/(T^2 + \mathrm{df})$].

MANOVA is a general method that can be used in conjunction with any ANOVA design that has two or more dependent variables. The simple example discussed above is equivalent to a one-way MANOVA; just as a one-way ANOVA with two groups leads to the same conclusion as the corresponding *t*-test, the one-way MANOVA with two groups reduces to Hotelling's T^2. With three or more groups the one-way MANOVA becomes considerably more complicated. Instead of just one discriminant function, $k - 1$ such functions can be found, all of which are mutually orthogonal. The primary discriminant function (or greatest characteristic root) is the one whose composite variable leads to the largest F ratio possible for the given set of data. Although it is legitimate to test only the greatest characteristic root for statistical significance, there are alternative significance tests that incorporate all of the $k - 1$ discriminant functions.

It is beyond the scope of this text to delve any more deeply into the complexities of MANOVA, but it is important to emphasize a few more points. First, MANOVA helps to control the experimentwise error rate, because a single test replaces separate univariate ANOVAs for each dependent variable. If the MANOVA is not significant, none of the univariate tests should be conducted. On the other hand, a significant MANOVA is usually followed by univariate tests on each dependent variable, but the alpha for each test should probably be adjusted for multiple comparisons. Second, the MANOVA is usually more powerful than any univariate test, because it can find significance for a combination of dependent variables, even when no individual dependent variable would lead to a significant F. (This statement implies that a significant MANOVA does *not* guarantee that any of the univariate tests that follow it will reach significance.)

Finally, it is important to reemphasize the usefulness of MANOVA as a replacement for the RM ANOVA. Now that I have explained a little bit about the structure of MANOVA, I can explain this particular use of MANOVA further. In a one-way RM ANOVA there is only one group of subjects; the several repeated measures can be considered different dependent variables. If the RM factor has j levels, the simplest way to proceed is to form $j - 1$ difference scores for each subject (e.g., for $j = 3$, you could find differences between levels 1 and 2 and between levels 2 and 3). These difference scores become the set of dependent variables for the MANOVA test. As there is only one group of subjects, the MANOVA is a one-sample test; in this case you would be comparing the set of difference scores to zero. Note that if the RM factor has only two levels, the MANOVA test reduces to one set of difference scores tested against zero—in other words, an ordinary matched t-test. If there are more than two levels the MANOVA finds the combination of difference scores that produces the highest F ratio in a comparison with zero. As mentioned in Chapter 17, this MANOVA test does not require the sphericity assumption, and for large sample sizes its power compares favorably with that of the RM ANOVA.

Exercises

*1. a) If the interaction in exercise 18B3 were not significant, would any post hoc tests be called for?

b) Given that the interaction was significant, which post hoc comparisons would be justified? Perform those tests.

2. Given the results of the mixed design ANOVA in exercise 18B5, a) list the post hoc tests that would be justified.

b) Describe the type of error term that would be used for each of the tests listed in part a.

* 3. Given the results of the mixed design ANOVA in exercise 18B6, perform the appropriate pairwise comparisons using the conservative error term.

4. a) Describe how you would follow up the results of the mixed design ANOVA in exercise 18B7 with post hoc tests.

b) Describe the type of analysis that should have been performed for the data in that exercise, and explain why it would have been more powerful.

* 5. Given the results of the mixed design ANOVA in exercise 18B4, perform the appropriate post hoc comparisons using the conservative error term.

SUMMARY

The Important Points of Section A

1. The two-way mixed design (also called a *split-plot* design) includes one between-subjects factor and one within-subjects factor; the latter involves either repeated measures or matched blocks of subjects. Only balanced designs are considered in this chapter.

2. The total variability in a two-way mixed design can be initially divided into between-subjects and within-subjects variation.

3. The between-subjects variation can be further subdivided into a portion that depends on the separation of the group means and a portion that depends on subject-to-subject variation within each group. The within-group variability is used as the error term for testing the main effect of the between-subjects factor.

4. The within-subjects variation can be divided into three components: one that reflects variation among the means of the repeated conditions; one that reflects the interaction of the two factors; and one that reflects the interaction of subjects with the RM factor within each group. The last of these components is used as the error term for testing both the main effect of the within-subjects factor and the interaction of the two factors.

5. One possible advantage of the mixed design is that it can reduce the error term of a one-way RM ANOVA if the subject by treatment interaction is smaller within each subgroup than across all groups (this means that there is some group by RM factor interaction). In addition, the group by RM factor interaction is likely to be the most interesting effect in the analysis.

6. The within-subjects factor in a mixed design has the same advantages and disadvantages as the within-subjects factor in a one-way RM design. This factor is likely to have greater power because subject-to-subject variability is ignored; however, degrees of freedom are lost, and homogeneity of covariance must be dealt with if this factor has more than two levels. Moreover, order and carryover effects must be considered if the conditions are repeated sequentially; matching subjects in blocks, however, can be used as an alternative.

7. The between-subjects (or grouping) factor usually has less power than the within-subjects factor in a mixed design, and its main effect is therefore less likely to be significant. However, the grouping factor is often added to the design to test for an interaction with the repeated factor. The grouping factor may be based on preexisting individual differences or on some experimental condition that is not easily repeated.

The Important Points of Section B

In this section I will present an example of a mixed design in which both variables involve experimental manipulations. The example involves a hypothetical school that is comparing three methods for teaching the sixth grade: the traditional method; a method that uses computers in the classroom; and a method using computers at home. To simplify the analysis, three samples of three students each are drawn and randomly assigned to the three teaching methods. At the end of the school year, each pupil is given final exams in four subject areas: math, English, science, and social studies. The final exams are scored from zero to 100; because each pupil takes four final exams, there are a total of $9 \times 4 = 36$ scores. To make the data more manageable, 70 points were subtracted from each score. (70 was chosen because it was the largest even number below which there were no scores.) The data are shown in Table 18.7.

We will begin by finding the number of degrees of freedom for each component in the analysis, using Formula 18.1 (A, B, and C). For this example,

Table 18.7

Method	Math	English	Science	Social Studies	Row Totals
Traditional	15	26	20	17	78
	10	23	16	12	61
	5	18	4	1	28
Cell Sums	30	67	40	30	167
Classroom Computers	27	25	26	28	106
	18	20	20	23	81
	16	17	12	10	55
Cell Sums	61	62	58	61	242
Home Computers	25	20	23	27	95
	22	15	19	25	81
	17	9	10	14	50
Cell Sums	64	44	52	66	226
Column Totals	155	173	150	157	635

k (number of groups) $= 3$, c (number of repeated measures) $= 4$, and n (number of subjects per group) $= 3$. Therefore,

$$df_{tot} = nkc - 1 = 3 \cdot 3 \cdot 4 - 1 = 36 - 1 = 35$$

$$df_{between\text{-}S} = nk - 1 = 3 \cdot 3 - 1 = 9 - 1 = 8$$

$$df_{within\text{-}S} = nk(c - 1) = 3 \cdot 3(3) = 27$$

$$df_{groups} = k - 1 = 3 - 1 = 2$$

$$df_W = k(n - 1) = 3(2) = 6$$

$$df_{RM} = c - 1 = 4 - 1 = 3$$

$$df_{g \times RM} = (k - 1)(c - 1) = (2)(3) = 6$$

$$df_{s \times RM} = k(c - 1)(n - 1) = 3(3)(2) = 18$$

To save space I will not show the calculation of the SS components (although you might find this a useful exercise). Assume they have already been calculated and that you are now ready to obtain the variance estimates (i.e., MS), as follows:

$$MS_{groups} = \frac{SS_{groups}}{df_{groups}} = \frac{260}{2} = 130$$

$$MS_W = \frac{SS_W}{df_W} = \frac{913.6}{6} = 152.3$$

$$MS_{RM} = \frac{SS_{RM}}{df_{RM}} = \frac{33}{3} = 11$$

$$MS_{g \times RM} = \frac{SS_{g \times RM}}{df_{g \times RM}} = \frac{383.3}{6} = 63.9$$

$$MS_{s \times RM} = \frac{SS_{s \times RM}}{df_{s \times RM}} = \frac{94.4}{18} = 5.24$$

Finally, we can form the F ratios to test each of our three null hypotheses. To test the main effect of groups, we create the following ratio:

$$F_{groups} = \frac{MS_{groups}}{MS_W} = \frac{130}{152.3} = .95$$

To test the main effect of the within-subject factor, we use $MS_{s \times RM}$ as our error term:

$$F_{RM} = \frac{MS_{RM}}{MS_{s \times RM}} = \frac{11}{5.24} = 2.1$$

Finally, the same error term is used in the F ratio to test the two-way interaction:

$$F_{g \times RM} = \frac{MS_{g \times RM}}{MS_{s \times RM}} = \frac{63.9}{5.24} = 12.2$$

To test the main effect of method, we are looking for a critical F based on df_{groups}, $df_W = F_{.05}(2, 6) = 5.14$. For the main effect of subject area we need $F(df_{RM}, df_{s \times RM}) = F_{.05}(3, 18) = 3.16$. For the interaction, we are looking for $F(df_{g \times RM}, df_{s \times RM}) = F_{.05}(6, 18) = 2.66$.

Because the observed F is so low, the main effect of teaching method is not significant at the .05 level. Similarly, the F for the main effect of the repeated measures factor (2.1) also falls short of its critical value, so this null hypothesis cannot be rejected, either. The observed F for the interaction of the two factors (12.2), however, is greater than its critical value, so this effect is statistically significant.

Interpreting the Results

The significant interaction suggests that the different teaching methods do make some difference in how much students learn, but that this difference is not uniform among the different subject areas. That is, students given a particular teaching method benefit in some subject areas more than others, and which subject area is affected more depends on which teaching method the pupils receive. For instance, from Figure 18.8 you can see that students studying English benefit the most from the traditional method and least from learning on a home computer. A significant main effect for the grouping variable would help us decide on a single teaching method, but a significant main effect for subject area would be meaningless. (If, for example, grades are higher overall in science than in the other three subjects, does that mean students are better at science than the other subjects, or

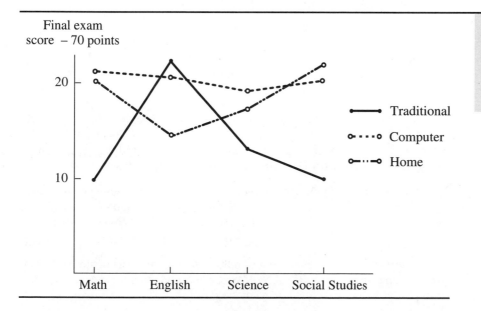

Final exam score – 70 points

Figure 18.8

Graph of Cell Means for Data in Table 18.7

	Source	SS	df	MS	*F*	*p*
Table 18.8	Between-Subjects	1173.6	8			
	Methods	260	2	130	.85	> .05
	Within-Groups	913.6	6	152.3		
	Within-Subjects	510.7	27			
	Subject Area	33	3	11	2.09	> .05
	Method × Area	383.3	6	63.9	12.2	< .05
	Residual (S × Area)	94.4	18	5.24		
	Total	1684.3	35			

merely that the science final exam was relatively easy?) The results of the mixed design ANOVA are summarized in Table 18.8.

Assumptions of the Mixed Design ANOVA

Because the mixed design involves performing an independent samples ANOVA *and* a repeated measures ANOVA, the assumptions underlying both of these statistical procedures are required for the appropriate factors.

1. *Independent random sampling*
2. *Normal distributions*
3. *Sphericity within the RM factor.* This assumption is the same as the sphericity assumption for the one-way RM ANOVA.
4. *Homogeneity between groups.* The amount of subject by treatment interaction for any pair of levels in one group is the same as the amount of interaction for that pair of levels in any other group.

The Important Points of Section C

1. A violation of the sphericity assumption affects tests of both the main effect of the RM factor and the interaction of the two factors. To find the critical *F* for the Geisser-Greenhouse conservative test, use df_W for df_{denom} and either 1 (RM main effect) or df_{groups} (interaction effect) for df_{num}.

2. If the two factors in a mixed design do not significantly interact, significant main effects comprising more than two levels can be subjected to pairwise tests. Usually, it is safe to use MS_W for all between-group pairs, but pairs of RM levels should use an error term based on the two levels involved rather than $MS_{s \times RM}$.

3. A significant interaction is usually followed by an analysis of simple (main) effects. Separate one-way RM ANOVAs can be conducted for each group, followed by matched *t*-tests for pairs of levels, if the simple effect is significant. One-way ANOVAs can also be conducted between the groups for each level of the RM factor. The most cautious approach is to pool the variances only for the cells involved. This limited pooled error term can then be used to test pairs of groups, assuming that the corresponding simple effect reached significance.

4. In the analysis of covariance (ANCOVA), a covariate is used to predict the dependent variable by means of linear regression; the predicted part is discarded and the ANOVA is performed on the residuals. The residuals usually have less variance than the original scores, so the analysis is more likely to attain significance.

5. Any number of covariates can be used, but covariates that are highly correlated with other covariates add little, and may hurt the analysis. Also, several assumptions must be satisfied for ANCOVA to be valid, including homogeneity of regression (the slope of the regression line should be the same in each group of the design).

6. Although direct experimental control is preferable, ANCOVA can provide statistical control of nuisance variables and can adjust group means that have been confounded by a concomitant variable. Whereas the use of ANCOVA with a truly independent variable is straightforward, ANCOVA can lead to misleading results when applied to designs that use preexisting groups or populations.

7. An important application of ANCOVA is the analysis of a before-after design with two or more independent groups. Instead of conducting the analysis on the before-after difference scores, ANCOVA uses residuals based on predictions from the before scores. The residuals nearly always lead to less error variance, and therefore a more powerful test, than the before-after difference scores.

8. Multivariate analysis of variance (MANOVA) is a general method for analyzing any ANOVA design using two or more dependent variables simultaneously. The analysis uses a linear combination of the dependent variables (called a discriminant function) that maximizes the F ratio.

9. A one-way MANOVA with only two groups is equivalent to Hotelling's T^2 test, in which the dependent variables are combined so as to produce the largest possible t value. The discriminant function is closely related to the multiple regression equation for predicting group membership from a combination of the dependent variables (R^2 is easily obtained from T^2, and vice versa).

10. MANOVA helps to prevent the build-up of experimentwise error that would occur from repeating univariate ANOVAs for each dependent variable. Also, MANOVA can detect group differences involving combinations of dependent variables, even when none of the dependent variables would result in a significant F by itself. Finally, MANOVA provides a safe and powerful alternative to RM ANOVA.

Definitions of Key Terms

Mixed (or split-plot) design An experimental design that contains one or more between-subjects factors along with one or more within-subjects factors.

Within-subjects factor An independent variable for which each subject participates at every level, or subjects are matched across levels (also called a repeated measures factor or variable).

Between-subjects factor An independent variable for which each subject participates at only one level (also called a between-groups factor or variable).

Analysis of covariance (ANCOVA) A form of ANOVA in which a concomitant variable (i.e., covariate) that is linearly related to the dependent variable but not the independent variable is used to remove unwanted variance from the dependent variable and the group means.

Homogeneity of regression The assumption in ANCOVA that the regression slope for predicting the dependent variable from the covariate is the same at each level of the between-groups variable (or in each cell of a factorial design).

Key Formulas

$$\text{df}_{\text{between-S}} = nk - 1$$

$$\text{df}_{\text{within-S}} = nk(c - 1) \text{ or } nkc - nk \qquad \text{Formula 18.1A}$$

$$\text{df}_{\text{groups}} = k - 1$$

$$\text{df}_{\text{W}} = k(n - 1) \text{ or } nk - k \qquad \text{Formula 18.1B}$$

$$\text{df}_{\text{RM}} = c - 1$$

$$\text{df}_{\text{g} \times \text{RM}} = (k - 1)(c - 1)$$

$$\text{df}_{\text{s} \times \text{RM}} = k(c - 1)(n - 1) \qquad \text{Formula 18.1C}$$

The total sum of squares, raw-score form (can be used with any analysis of variance design):

$$\text{SS}_{\text{tot}} = \Sigma X^2 - \frac{T_T^2}{N_T} \qquad \text{Formula 18.2}$$

The sum of squares due to subject-to-subject differences, raw-score form (this component is further divided into subcomponents corresponding to the differences between groups and the differences among subjects within groups):

$$SS_{\text{between-S}} = \frac{1}{c}\Sigma S_i^2 - \frac{T_T^2}{N_T} \qquad \text{Formula 18.3}$$

The sum of squares due to differences in group means, raw-score form:

$$SS_{\text{groups}} = \frac{1}{cn}\Sigma G_i^2 - \frac{T_T^2}{N_T} \qquad \text{Formula 18.4}$$

The sum of squares due to subject-to-subject variability within groups, found by subtraction:

$$SS_W = SS_{\text{between-S}} - SS_{\text{groups}} \qquad \text{Formula 18.5}$$

The sum of squares due to variation among the several measurements within each subject (this component is further divided into subcomponents corresponding to mean differences between levels of the RM factor, the interaction of the two factors, and the interaction of the subjects with the RM factor within each group):

$$SS_{\text{within-S}} = SS_{\text{tot}} - SS_{\text{between-S}} \qquad \text{Formula 18.6}$$

The sum of squares due to differences among the levels of the repeated factor, raw-score form:

$$SS_{\text{RM}} = \frac{1}{kn}\Sigma RM_i^2 - \frac{T_T^2}{N_T} \qquad \text{Formula 18.7}$$

The sum of squares due to differences in cell means (this component can be divided into subcomponents corresponding to differences from each main effect and the interaction of the two factors):

$$SS_{\text{between-cells}} = \frac{1}{n}\Sigma cs_i^2 - \frac{T_T^2}{N_T} \qquad \text{Formula 18.8}$$

The sum of squares due to the interaction of the two factors, found by subtraction:

$$SS_{g \times \text{RM}} = SS_{\text{bet-cells}} - SS_{\text{groups}} - SS_{\text{RM}} \qquad \text{Formula 18.9}$$

The sum of squares due to the interaction of subjects with the repeated factor within each group, found by subtraction:

$$SS_{s \times \text{RM}} = SS_{\text{within-S}} - SS_{\text{RM}} - SS_{g \times \text{RM}} \qquad \text{Formula 18.10}$$

Nonparametric Statistics

$\mathbf{P}$art VI presents several of the most commonly used nonparametric statistical procedures. In order to explain what nonparametric statistics are, I will begin by reminding you that the critical part of a statistical test is finding the null hypothesis distribution (NHD)—the relative probabilities of different experimental outcomes when the null hypothesis is true. For the parametric tests described earlier in this text, finding the NHD was made relatively easy by first assuming that the dependent variable followed a normal distribution in the population and then using simple statistical laws that apply to choosing random samples from a normal distribution. The tests were directed at drawing inferences about the parameters of the hypothetical normal distributions. Nonparametric tests are based on null hypothesis distributions that are found by applying probability rules to random events, but these NHDs are *not* derived by assuming that the dependent variable is normally distributed, and the statistical tests are not directed at drawing inferences about the parameters of the distribution of the dependent variable. For this reason nonparametric tests are often called *distribution-free tests*.

There are two main situations when it is not reasonable to base your statistical tests on normal distributions in the population. The first is when your measurement scale does not have the interval property—that is, when you are using either an ordinal or a nominal scale. The second case is when you are using an interval or ratio scale, but the distribution of your variable is very unlike the normal distribution and your sample size is small. In both of these situations a nonparametric test is preferable. Although there is no controversy about using nonparametric tests with a nominal or ordinal scale (though there is some debate about which psychological scales have the interval property and which are merely ordinal), there is some disagreement concerning when population distributions sufficiently resemble the normal distribution to justify parametric tests. Computer simulations have shown that some parametric tests (e.g., the *t*-test) are robust with respect to violations of the normality assumption (that is, they are relatively unaffected by a lack of normality in the population); others (e.g., the F test for homogeneity of variance) are not robust in this way.

In general, the larger the sample size, the less you have to worry about the shape of the population distribution. Some statisticians would argue that with intermediate sample sizes (around 20 to 40 subjects), using nonparametric tests is the more cautious approach, as you are rarely certain about the shape of the population distribution in such cases. Other statisticians point out that there is some loss of power in using nonparametric tests if the population distribution does happen to be normal; nonparametric tests throw away some of the quantitative information in the data. The difficulty of finding appropriate tables for the NHDs of various nonparametric tests used to be a factor in deciding which test to use, but the advent of readily available statistical software has virtually eliminated that factor. I do not expect the beginning student to be able to judge which type of test is best in an intermediate situation, but it is important to understand the basis of the controversy.

Before reading the chapters of Part VI it will be useful to review the material on measurement scales in Chapter 1, Section A, and the rules of probability discussed in Chapter 5, Section C.

Chapter 19 introduces the topic of nonparametric statistics as applied to dichotomous events (i.e., only two outcomes are possible). Section A shows how the binomial distribution is constructed and how it can be used to draw inferences about the probabilities of dichotomous events. Section B describes an important application of the binomial distribution: the sign test, which is a nonparametric alternative to the matched t-test. In Section C the rules of probability are reviewed and applied to the case of discrete variables. Counting techniques—combinations and permutations—are explained, as these are used to construct the binomial distribution in the general case.

Chapter 20 deals with the case in which all of your variables have been measured on a nominal scale, so you must perform some type of chi-square test. Section A develops the concepts of the chi-square test in terms of the one-variable, or goodness-of-fit, test. These concepts are extended to the more common and more interesting two-variable case in Section B, which discusses the test for the independence of two variables. Section C is devoted primarily to ways of assessing the strength of association between two variables measured on a nominal scale. Two special topics are also briefly discussed: the Fisher Exact test for when your samples are very small, and the log-linear model, for when there are more than two categorical variables in your analysis.

Chapter 21 presents several statistical tests that are appropriate when one or more of your variables have been measured on an ordinal scale. Section A introduces concepts relevant to rank-ordering observations for any ordinal test and for the Mann-Whitney test for independent groups, in particular. Section B includes two important statistical tests dealing with ordinal data. It outlines the step-by-step hypothesis testing procedure, as applied to the Mann-Whitney test and introduces the Spearman correlation coefficient (which is Pearson's r applied to ranked data), as applied to hypothesis testing. Section C describes two ordinal tests that are required when an experiment includes more than two levels of an independent variable: the Kruskal-Wallis test for independent samples and the Friedman test for matched samples.

19

CHAPTER

A

**CONCEPTUAL
FOUNDATION**

You will need to use the following from previous chapters:

Symbols
μ: Mean of a population
$\overline{X}$: Mean of a sample
σ: Standard deviation of a population

Formulas
Formula 5.1: The z score

Concepts
Null hypothesis distribution
Normal distribution
One- and two-tailed hypothesis tests
Addition rule for mutually exclusive events
Multiplication rule for independent events

I will begin the discussion of nonparametric statistics by describing a situation in which you could not use any of the statistical procedures already presented in this text. For this example I need to bring back our psychic friend from Chapter 7. This time, instead of predicting math aptitude, the psychic claims that he can predict the gender of a child soon after its conception. To test his claim we find four women who very recently became pregnant, and ask the psychic to make a prediction about each child. (Perhaps he places his hand on each woman's abdomen and feels "vibrations".) Then we wait, of course, to find out if the psychic is correct in each instance. Suppose that the psychic is correct in all four instances. Should we believe that the psychic has some special ability? (By the way, I am not using this example because I am certain that psychic powers don't exist, but rather because it is easy to believe that there are at least some people claiming psychic powers who have no special abilities at all.)

By now you must know what Dr. Null would say about this claim. Dr. Null would say that the psychic was just lucky this time, and that he has no psychic ability. Before we challenge Dr. Null's assessment, we would like to know the probability of Dr. Null's making us look foolish. Dr. Null cannot beat the psychic in this case, but if Dr. Null also makes four correct predictions in a row, our psychic will no longer seem so impressive, because Dr. Null makes his predictions at random—perhaps by flipping a coin: heads, it's a boy, tails, it's a girl. In this simple case it is not difficult to calculate the probability of making four correct predictions in a row by chance. First, note that the probability of being correct about any one gender prediction by chance is .5. (Although the number of boys that are born is not exactly equal to the number of girls, I will assume they are equal for this example.) To find the probability of being correct four times in a row, we need to use the multiplication rule for independent events (see

Chapter 5, Section C). We multiply .5 by itself four times: $p = .5 \cdot .5 \cdot .5 \cdot .5 = .0625$.

Now we know that Dr. Null has a .0625 chance of predicting just as well as the psychic. This is not a very large probability (1 out of 16), but if we had set alpha at .05 we could not reject Dr. Null's hypothesis that the psychic has no ability. Fortunately, it was very easy in this example to calculate Dr. Null's probability of duplicating the psychic's performance, but you will not always be quite this fortunate. Suppose that the psychic makes ten predictions and is right in eight of the cases. This level of performance sounds pretty good, but we need to find Dr. Null's chance of being right eight, nine, or all ten times, and this involves more than a simple application of the multiplication rule.

The Origin of the Binomial Distribution

The approach to null hypothesis testing that I have taken in previous chapters is to find the null hypothesis distribution (NHD) and then to locate particular experimental results on that distribution. Constructing the NHD when you are dealing with a total of only four dichotomous predictions is not difficult, but it will require the application of some simple probability rules. Constructing the NHD when there are ten predictions to be made is considerably more tedious, but it uses the same principles. In both cases, the NHD is a form of the **binomial distribution**; there is a different binomial distribution for each number of predictions to be made. Of course, in real life you would not have to construct the binomial distribution yourself. You would either find a table of the distribution, use an approximation of the distribution if appropriate, or analyze your data by computer. However, in order to understand this distribution and its use, it will help to see how the distribution arises.

A binomial distribution may arise whenever events or observations can be classified into one (and only one) of two categories, each with some probability of occurrence (e.g., male or female, right or wrong); such events are called **dichotomous events.** The probabilities corresponding to the two categories are usually symbolized as P and Q. Because P and Q must add up to 1.0, knowing P automatically tells you what Q is; in fact, the two probabilities are often referred to as P and $1 - P$. (Note that I am using an uppercase P for the probability of a single event, whereas I will continue to use a lowercase p to represent a p level for testing a null hypothesis.) The simplest case is when $P = Q = .5$, so this is the case I will consider first. Fortunately, this case frequently represents the null hypothesis, such as when you are flipping a coin to see if it is fair (i.e., the probability of heads equals the probability of tails).

To get a binomial (or Bernouilli) distribution you need to have a sequence of *Bernouilli trials* (after Jacques Bernouilli, 1654–1705, a Swiss mathematician). Bernouilli trials are dichotomous events that are independent of each other, and for which P and Q do not change as more and more trials occur. The total number of trials is usually symbolized as N; the binomial distribution is a function of both P and N. The reason you have a distribution at all is that whenever there are N

Figure 19.1

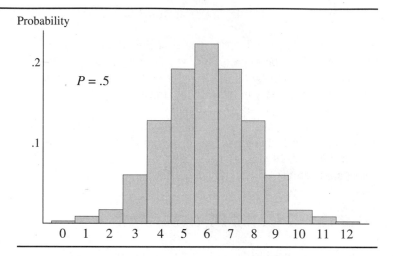

trials, some of them will fall into one category and some will fall into the other, and this division into categories can change for each new set of N trials. The number of trials that fall into the first category (the one with probability P) is usually called X; this is the variable that is distributed. For instance, P can stand for the proportion of women eligible for jury duty, and Q for the proportion of men. Assuming that each jury selected consists of twelve people ($N = 12$), X (the number of women on a particular jury) can vary from 0 to 12. If you look at thousands of juries, you will see that some values of X are more common than others; you are not likely to get a uniform distribution. If juries were selected at random with respect to gender, then an infinite number of juries would give you a perfect binomial distribution, as shown in Figure 19.1. To show you how this distribution is constructed, and why the middle values of X turn out to be the most frequent, I will begin with a case that has a small N.

The Binomial Distribution with $N = 4$

A very simple "experiment" consists of flipping the same coin four times to see if it is fair. With only four flips it is not difficult to write out every possible sequence of heads and tails that can occur. Because there are two possible outcomes for each toss (H or T) and four tosses, there are $2 \cdot 2 \cdot 2 \cdot 2 = 16$ different sequences that can occur. They are shown in Table 19.1.

If our focus is on the number of heads (X) in each sequence, then some of the sequences shown in Table 19.1 can be lumped together. For instance, the following four sequences are alike, because each contains just one head: HTTT, THTT, TTHT, TTTH. For each possible number of heads (zero to four) we can count the number of sequences, as follows: 0H: one sequence; 1H: four sequences; 2H: six sequences; 3H: four sequences; 4H: one sequence. If the probability of heads is P and the probability of tails is Q, the probability of any particular

HHHH	HTHH	THHH	TTHH
HHHT	HTHT	THHT	TTHT
HHTH	HTTH	THTH	TTTH
HHTT	HTTT	THTT	TTTT

Table 19.1

sequence is found by multiplying the appropriate Ps and Qs. For example, the probability of flipping HTTH would be $P \cdot Q \cdot Q \cdot P$. Matters are greatly simplified if you assume that the coin is fair (i.e., H_0 is true). In this case, $P = Q = .5$, so every sequence has the same probability, $.5 \cdot .5 \cdot .5 \cdot .5 = .0625$. Note that we could have arrived at the same answer by observing that there are 16 possible sequences, and if each is equally likely (as is the case when $P = Q$), then the probability of any one sequence occurring is 1 out of 16, which equals $1/16 = .0625$.

Now, if we want to know the probability of flipping just one head in four flips we have to add the probabilities of the four sequences that contain only one head each (using the addition rule of probability; see Chapter 5, Section C). So, the probability of one head $= .0625 + .0625 + .0625 + .0625 = .25$. Again, we could have arrived at the same conclusion by noting that 4 out of the 16 equally likely sequences shown in Table 19.1 contain one head, and $4/16 = .25$.

By finding the number of sequences with zero, one, two, three, and four heads we obtain the frequency distribution shown in part a of Figure 19.2. Dividing each number of sequences by the total number of sequences (16), we can convert the frequency distribution into a probability distribution: the binomial distribution with $P = .5$, $N = 4$ (see part b of Figure 19.2). Looking at the figures, the first thing you might notice about these distributions is that they are symmetrical; this is because $P = Q$. If X were the number of blond children two brunette parents might have, then P would be less than Q, and three blondes in a family of four siblings would be much less likely than one blond in four. The distribution

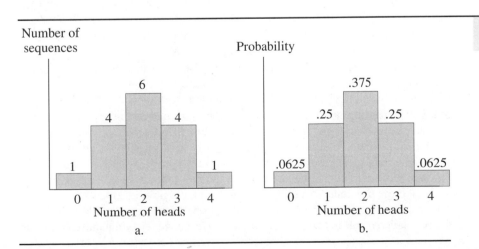

Figure 19.2

a.

b.

would be positively skewed. The next thing you might notice about Figure 19.2 is that the symmetrical binomial distribution looks a little like the normal distribution. This resemblance increases as N gets larger; for example, take another look at Figure 19.1, in which $P = .5$ and $N = 12$.

The Binomial Distribution with $N = 12$

The binomial distribution with $N = 12$ can be constructed using the same method we employed for $N = 4$. However, with $N = 12$ there are a total of 4096 sequences to write out (2 multiplied by itself 12 times, or 2^{12}). Then we would have to count how many contained one head, two heads, etc. There is a mathematical shortcut, but it is based on combinations and permutations, which I will not discuss until Section C. Table A.12 shows the probabilities for all the binomial distributions from $N = 1$ to $N = 15$ (assuming $P = .5$). To show how we can use these probabilities to test statistical hypotheses, I will return to the example involving juries ($N = 12$), in which X represents the number of women.

For the sake of the example, assume that each jury is selected at random from the list of registered voters and that this list contains an equal number of women and men. Suppose that the docket includes a case involving a divorced couple, and the husband is claiming that the jury—nine women and three men—has been unfairly stacked against him. If the null hypothesis is that the jury is just a random selection from the voter list (i.e., $P = Q$), we can use Table A.12 to find the p level that corresponds to a nine-woman-three-man jury. Remember that the p level can be viewed as Dr. Null's chance of beating your experimental result, and a tie goes to Dr. Null. In this example Dr. Null has to randomly select a jury that is just as unbalanced or even more unbalanced than the one in question. First, you see from the table that when $N = 12$ and $X = 9$, $p = .0537$. That is, Dr. Null has a .0537 chance of selecting such a jury at random. But we must also include the chance that Dr. Null will select a jury of ten women, eleven women, or twelve women—because he "beats" us in each of those cases. Summing the probabilities for $X = 9, 10, 11,$ and 12, we have $.0537 + .0161 + .0029 + .0002 = .0729$. This is not a large probability, but if we use an alpha of .05 to make our decision about fairness, then $p > .05$ and we cannot reject the null hypothesis that this jury was selected at random (from a pool of an equal number of women and men).

In the above example, I performed a one-tailed test. I looked only at the chance of the jury's having an excess of women. If the jury had had nine men and three women the husband might not have complained, but from the standpoint of testing for random selection this jury would be just as unbalanced as the one tested above. In most cases a two-tailed test is the more reasonable approach. To perform a two-tailed test, we need to find the two-tailed p value. All we have to do is take the one-tailed p value found above and double it: $.0729 \cdot 2 = .1458$. This makes the p value even larger, so if the one-tailed p was not less than .05, the two-tailed p will not be significant either. Could we reject the null hypothesis with a jury of ten women and two men (or vice versa)? In that case, $p = .0161 + .0029 + .0002 = .0192$, and the two-tailed $p = .0192 \cdot 2 = .0384$. Even with a two-tailed test, p is less than .05, so we can reject the null hypothesis. Of course, this doesn't mean that a jury of ten women and two men (or vice versa) could *not* have been selected at random—but assuming a human element of bias could have entered the jury selection process, finding the p level can help us decide how seriously to consider the possibility that the jury was not randomly selected.

The binomial distribution can also be applied to the evaluation of experiments, such as the gender predictions of the psychic. If the psychic makes 12 independent gender predictions (for 12 different, unrelated pregnant women) and is wrong only twice, we know from the above calculations that p (two-tailed) $< .05$, so we can reject the null hypothesis that the psychic has no special ability. On the other hand, if the psychic were wrong 3 out of 12 times, then $p > .05$ (even for a one-tailed test) so we could not reject the null hypothesis.

When the Binomial Distribution Is Not Symmetrical

So far I have been dealing only with the symmetrical binomial distribution, but there are plenty of circumstances in which P is less than or more than .5. For instance, in the jury problem, imagine a city whose citizens belong to either of two racial or ethnic groups, which I will refer to as X and Y. Forty percent of the people belong to group X (i.e., $P = .4$) and 60% belong to group Y (i.e., $Q = .6$). In this case, selecting a jury consisting of four Xs and eight Ys would be fairly likely and much more likely than selecting eight Xs and four Ys. Unfortunately, Table A.12 could not help us find p levels for different jury combinations. We would need a version of Table A.12 constructed for $P = .4$.

Complete tables for various values of P (other than .5) are not common, but there are tables that will give you the critical value for X as a function of both N and P for a particular alpha level. For instance, for the example above ($N = 12$, $P = .4$, $\alpha = .05$), the critical value of X would be 9, which means that randomly selecting a jury with nine Xs has a probability of less than .05 (so would a jury with even more Xs, of course). Bear in mind that this kind of table is set up for a one-tailed test. When the distribution is not symmetrical the two-tailed p is not simply twice the size of the one-tailed p. I have not included such a table in this text, because these tables are rarely used. Alternative procedures for finding p levels in such cases will be discussed below and in the next chapter.

I pointed out earlier that the symmetrical binomial distribution bears some resemblance to the normal distribution, and that this resemblance increases as N gets larger. In fact, the binomial distribution becomes virtually indistinguishable from the normal distribution when N is very large, and the two distributions become truly identical when N is infinitely large. For any particular value of P, even when P is not equal to .5, the binomial distribution becomes more symmetrical and more like the normal distribution as N gets larger. However, as P gets further from .5 it takes a larger N before the distribution begins to look symmetrical. When N is large enough, the binomial distribution resembles a normal distribution that has a mean of NP and a standard deviation of $\sqrt{NPQ}$. This resemblance can be used to simplify null hypothesis testing in situations that would otherwise call for the binomial distribution. An example of the normal approximation to the binomial distribution is given below.

The Normal Approximation to the Binomial Distribution

Consider a grand jury, which can contain as many as 48 individuals. Assume again that equal numbers of women and men are eligible to serve as jurors. With $N = 48$ and $P = .5$, the binomial distribution looks a lot like the normal distribution; see Figure 19.3. Indeed, this binomial distribution can be approximated, without much error, by a normal distribution with a mean of 24 ($\mu = NP = 48 \cdot .5$) and a standard deviation of 3.46 ($\sigma = \sqrt{48 \cdot .5 \cdot .5} = \sqrt{12}$). Assuming that we are now working with a normal distribution and we want to find the p level associated with a particular value of X, we need to convert X to a z score. According to Formula 5.1,

$$z = \frac{X - \mu}{\sigma}$$

If we substitute expressions for μ and σ in terms of N, P, and Q, we create the following new formula, Formula 19.1, for finding z scores when the normal distribution is being used to approximate a binomial distribution.

$$z = \frac{X - NP}{\sqrt{NPQ}} \qquad \text{Formula 19.1}$$

The above formula can be used to determine how likely it is to select a jury with a number of women equal to X or more. For example, a particular grand

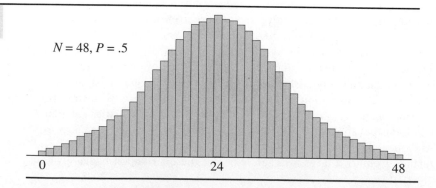

Figure 19.3

$N = 48, P = .5$

0   24   48

jury contains 30 women. What is the likelihood of randomly selecting a jury with this many women or more? First, we find the z score according to Formula 19.1:

$$z = \frac{X - 24}{3.46} = \frac{30 - 24}{3.46} = \frac{6}{3.46} = 1.73$$

Looking up this z score in Table A.1, we see that the p level (i.e., the area beyond z) is equal to .0418. With a one-tailed test we could reject the null hypothesis at the .05 level, but not with a two-tailed test.

Another Example of the Normal Approximation

The z score formula above can be used in the same way no matter what the value of P is, as long as N is sufficiently large. Returning to the example of two ethnic or racial groups in a city, assume that group X comprises 30% and group Y comprises 70% of the population (i.e., $P = .3$, $Q = .7$). A grand jury is found to have only 10 Xs and 38 Ys. What is the chance of randomly selecting only ten Xs (or fewer) for a grand jury? We begin by using Formula 19.1:

$$z = \frac{X - NP}{\sqrt{NPQ}} = \frac{10 - 48(.3)}{\sqrt{48(.3)(.7)}} = \frac{10 - 14.4}{\sqrt{10.08}} = \frac{-4.4}{3.17} = -1.38$$

The p level corresponding to this z score is .0838, which does not permit us to reject the null hypothesis at the .05 level. Selecting only 10 Xs for a jury of 48 is not extremely unusual given that the X group is only 30% of the population. ($NP = 14.4$ is the number of Xs to be expected on the average grand jury, in this example.)

I have not said how large N must be to justify using the normal distribution as an approximation of the binomial distribution. There is no exact answer, of course, but most researchers agree that when $P = .5$, N does not have to be more than 20 or 25 before the error of approximation becomes negligible. As P gets closer to 0 or 1, a larger N is needed to maintain a good approximation. For example, when P is only .1, an N of at least 100 is preferred. As a general rule, when P is not near .5, NPQ should be at least 9.

The z Test for Proportions

The results of large-scale studies, such as national surveys or polls, are usually presented in terms of percentages or proportions, rather than as actual frequencies. For instance, a newspaper might report that 58% of those sampled favor candidate A, whereas 42% favor candidate B. Knowing N, we could determine the actual number of people favoring one of the candidates and use Formula

19.1 to test the hypothesis that the two candidates are equally favored in the population. On the other hand, we can modify Formula 19.1 to get a formula for testing proportions directly. All we need to do is divide both the numerator and the denominator of Formula 19.1 by N, as shown below.

$$z = \frac{\frac{1}{N}(X - NP)}{\frac{1}{N}\sqrt{NPQ}} = \frac{X/N - \frac{NP}{N}}{\frac{\sqrt{NPQ}}{N}}$$

This leads to Formula 19.2,

$$z = \frac{p - P}{\sqrt{\frac{PQ}{N}}} \qquad\qquad \text{Formula 19.2}$$

where p (not to be confused with a p level) is the proportion in the X category, and P, as before, is the population proportion, according to the null hypothesis. To use this formula, you must convert data given as percentages to proportions, but that only entails dividing the percentage by 100. For example, if 58% of those polled favor candidate A, then $p = 58/100 = .58$. If the null hypothesis is that the two candidates are equally favored, then $P = .5$. Assuming the sample consists of 200 people ($N = 200$), Formula 19.2 yields the following result:

$$z = \frac{.58 - .5}{\sqrt{(.5)(.5)/200}} = \frac{.08}{\sqrt{.00125}} = \frac{.08}{.0353} = 2.26$$

This z score is large enough to allow us to reject the null hypothesis and conclude that there is a preference for candidate A in the population. Note that we would have obtained the same z score by finding that $X = 116$ (58% of 200), $NP = 100$, $NPQ = 50$, and inserting these values in Formula 19.1. Formula 19.2, however, gives us the convenience of dealing directly with proportions.

If there are three or more candidates in an election poll, and you want to know if they are all equally favored, then you cannot use the binomial distribution at all; you are dealing with a *multinomial* situation and need to use the methods described in Chapter 20. In the next section of this chapter, I will focus on a particular application of the binomial distribution for null hypothesis testing: an alternative to the matched t-test.

Exercises

1. a) Write out all the possible gender sequences for five children born into one family.

b) Assuming that $P = Q$ for each birth, construct the binomial distribution for the genders of the five children.

* 2. A particular woman has given birth to 11 children: nine boys and two girls. Assume $P = .5$ and use Table A.12 to answer the following questions.

a) What is the probability of having nine or more boys?

b) What is the probability of having nine or more children of the same gender?

c) Would you reject the null hypothesis (i.e., $P = .5$) at the .05 level, with a one-tailed test? With a two-tailed test?

3. Fourteen infants are shown a picture of a human face and a colorful ball of the same size, simultaneously (relative positions of the pictures are varied).

a) If ten of the infants spend more time looking at the face than the ball, can the null hypothesis (no preference between the face and the ball) be rejected at the .05 level (two-tailed test)?

b) In a two-tailed test, how many infants must spend more time looking at the face than the ball to allow the researcher to reject the null hypothesis at the .05 level? At the .01 level?

* 4. He didn't get a chance to study, so Johnny guessed on all 100 questions of his true-or-false history test.

a) If Johnny scored 58 correct, can we conclude ($\alpha = .05$, two-tailed) that he actually knew some of the answers and wasn't guessing (i.e., can we reject $P = .5$)?

b) How many questions would Johnny have to get correct for us to conclude that he was not just guessing randomly?

5. Jane's history test consisted of 50 multiple-choice questions (four choices for each question).

a) If she gets 20 correct, can we conclude that she wasn't merely guessing?

b) How many questions would Jane have to get correct for us to conclude that she was not just guessing?

* 6. Suppose that 85% of the population is right-handed (Q) and 15% are left-handed (P).

a) If out of 120 randomly selected civil engineers, 27 are found to be left-handed, what is the z score for testing the null hypothesis? Can we reject the null hypothesis that $P = .15$ for this profession?

b) If 480 civil engineers are sampled and 108 are found to be left-handed, what is the z score for testing the null hypothesis?

c) How does the z score in part a compare to the z score in part b? Can you determine the general rule that is being illustrated?

7. Fifty women involved in abusive marriages filled out a questionnaire. The results indicated that thirty of these women had been abused as children. If you know that 20% of all women were abused as children, test the null hypothesis that the women in the study are a random selection from the general population (use $\alpha = .05$, two-tailed). What can you conclude about the likelihood that women abused as children will end up in abusive marriages?

* 8. In the town of Springfield, 70% of the voters are registered as Republicans and 30% as Democrats. If 37% of 80 voters polled at random say that they plan to vote for Bert Jones, a Democrat, for mayor, can we conclude ($\alpha = .05$, two-tailed) that the people of Springfield are not going to vote strictly along party lines?

One important application of the binomial distribution arises when two stimuli are being compared but the comparison cannot be easily quantified. For instance, a music teacher can listen to two students play the same piece of music and be quite sure which student is superior without being able to quantify the difference—especially if the difference is slight. This kind of comparison can form the basis of an experiment. Suppose that music students are closely matched in pairs so that the two students in each pair are virtually indistinguishable in their ability to play a particular piece of music. Then one member of each pair is chosen at random to participate in a mental practice session, in which the subject is guided

BASIC STATISTICAL PROCEDURES

in using visual, auditory, and kinesthetic imagery to rehearse the piece of music selected for the study. The other member of each pair is exposed to a control procedure (perhaps some relaxation exercises) for the same period of time. Then each pair of subjects comes before a panel of well-trained judges, who do not know which subject received imagery training and which did not. One at a time, the two subjects play the same piece of music, and the judges must decide which subject had mastered the piece more successfully. (The order in which the subjects play is determined randomly for each pair.) In each pair, either the experimental subject or the control subject is rated superior (ties are avoided by having an odd number of judges on the panel). If the difference between the members of each pair could be reliably quantified (e.g., if each member of the pair could be given a rating, and then one rating could be subtracted from the other), then a matched *t*-test could be performed on the difference scores. If it is only possible to judge which member of each pair is better, an alternative to the matched *t*-test, called the **sign test,** can be performed. The six-step procedure I have been using to test null hypotheses based on parametric statistics can also be applied to nonparametric statistics, as I will now demonstrate.

Step 1. State the Hypotheses

The null hypothesis is that subjects given imagery training will play no better (or worse) than the control subjects. If *P* represents the probability of the experimental subject of the pair being rated superior, the null hypothesis can be stated symbolically as H_0: $P = .5$. The two-tailed alternative hypothesis would be stated as H_A: $P \neq .5$, whereas a one-tailed H_A could be stated as $P > .5$ or $P < .5$.

Step 2. Select the Statistical Test and the Significance Level

Because the difference in each pair of subjects will not be measured precisely but only categorized in terms of direction, the appropriate test is the sign test. The same considerations concerning the selection of alpha for parametric statistics apply to nonparametric procedures as well. Therefore, we will stay with the convention of setting alpha at .05.

Step 3. Select the Samples and Collect the Data

It may not be feasible to obtain subjects through random selection from the entire population of interest. However, as usual it is critical that the assignment of subjects within each pair be random. In order to use the normal distribution as an approximation of the null hypothesis distribution, you need to have at least about 20 pairs. Imagine that you have 20 pairs of subjects who are to be tested. Each pair is given a plus ($+$) if the experimental subject is judged superior and a minus ($-$) if the control subject is judged superior. The total number of $+$ signs is referred to as X. It is because we are only considering the direction, or sign, of the difference in each pair that the test is called the sign test. For this example, we will assume that there are 15 pluses and 5 minuses (and no ties), so $X = 15$.

Step 4. Find the Region of Rejection

Because we are using the normal distribution and have set alpha equal to .05 (two-tailed), the region of rejection is the portion of the distribution above $z = +1.96$ or below $z = -1.96$.

Step 5. Calculate the Test Statistic

The appropriate statistic, if we are using the normal approximation, is z as calculated by Formula 19.1. However, using the normal curve, which is smooth and continuous, to approximate the binomial distribution, which is discrete and step-like, becomes quite crude when N is fairly small. The biggest problem is that for the binomial distribution the probability of some value, say 3, is equivalent to the area of the rectangular bar that extends from 2 to 3 (the first bar goes from 0 to 1, the second from 1 to 2, etc.), whereas for the normal distribution the probability of 3 corresponds to the area enclosed by a range from 2.5 to 3.5. This discrepancy is reduced by subtracting half of a unit in the numerator of Formula 19.1. This is called the *correction for continuity*. Because sometimes the numerator will be negative (when $X < NP$), we need to take the absolute value of the numerator before subtracting .5; otherwise we would actually be making the numerator larger whenever it was negative. Including the correction factor in Formula 19.1 yields Formula 19.3:

$$z = \frac{|X - NP| - .5}{\sqrt{NPQ}} \qquad \text{Formula 19.3}$$

As N increases, the "steps" of the binomial distribution get smaller, and the discrepancy between the normal and binomial distributions becomes smaller, as well. Therefore, for large N the continuity correction makes very little difference and is usually ignored. As usual, there is some disagreement about how large N should be before the continuity correction can be ignored. It is safe to say that when N is greater than 100, the continuity correction makes too little difference to worry about. For the example in this section, N is small enough that the continuity correction makes a noticeable difference, so we will use Formula 19.3 to find our *z* score, as shown below.

$$z = \frac{|X - NP| - .5}{\sqrt{NPQ}} = \frac{|15 - 20(.5)| - .5}{\sqrt{20(.5)(.5)}} = \frac{|15 - 10| - .5}{\sqrt{5}}$$

$$= \frac{5 - .5}{2.24} = \frac{4.5}{2.24} = 2.01$$

Step 6. Make the Statistical Decision

Because the calculated *z* falls in the region of rejection (2.01 > 1.96), we can reject the null hypothesis that *P* = .5. Having rejected the null hypothesis we can conclude that the imagery training has had more of an effect than the control procedure. Too many experimental subjects have performed better than their control subject counterparts for us to conclude that this is merely coincidence.

Interpreting the Results

You might argue that the subjects in this hypothetical study are not representative of the entire population of interest, and that this limits our conclusion to saying that the imagery training only works on subjects with some musical background, or with other characteristics that resemble the subjects in our study. However, the random assignment of subjects to type of training, the random order of playing within each pair, and the blindness of the judges to the subjects' assignment all ensure that the results are internally valid (that is, within the sample of subjects that was tested).

Assumptions of the Sign Test

Dichotomous Events

It is assumed that the outcome of each simple event or trial must fall into one of two possible categories, and that these two categories are mutually exclusive and exhaustive. That is, the event cannot fall into both categories simultaneously, and there is no third category. Therefore, if the probabilities for the two categories are *P* and *Q,* then *P* + *Q* will always equal 1.0. In terms of the example above, we are assuming that one member of each pair must be superior; a tie is not possible. In reality, the sign test is sometimes performed by discarding any trials that result in a tie, thus reducing the sample size (e.g., if there were two ties out of the 20 pairs, *N* would equal 18) and therefore the power of the test. Bear in mind, however, that if more than a few ties occur, the validity of the sign test is weakened.

Independent Events

It is assumed that the outcome of one trial in no way influences the outcome of any other trial. In the example above, each decision about a pair of subjects constitutes a trial. The assumption holds for our example, as there is no connection between the decision for one pair of subjects and the decision for any other pair.

Stationary Process

It is also assumed that the probability of each category remains the same throughout the entire experiment. Whatever the value is for P on one trial, it is assumed to be the same for all trials. There is no reason for P to change over the course of trials in the experiment used for the example in this section. It is possible, however, to imagine a situation in which P changes over time, even if the trials are independent. For instance, if the trials are mayoral elections in a small city, and P is the probability of the mayor's belonging to a particular ethnic group, P can change between elections because of shifts in the relative percentages of various ethnic groups in that city.

Normal Approximation

To use the normal distribution to represent the NHD, as we did in our example, we need to assume that there is a negligible amount of error involved in this approximation. If $P = .5$ (as stated by H_0) and N is at least 25, the normal distribution is a good approximation.

Distribution-Free Tests

Note that we do *not* have to assume that the dependent variable is normally distributed. (In fact, the dependent variable in the example above is dichotomous, being just a $+$ or a $-$ for each pair of musicians, and so is not normally distributed.) That is why the sign test is sometimes called a **distribution-free test.** On the other hand, to perform a matched t-test you need to assume that the difference scores from all the pairs of subjects are normally distributed. For the sign test, we do need to know that the binomial distribution will represent the null hypothesis distribution (NHD), but this is guaranteed by the first three assumptions above.

The Gambler's Fallacy

The assumption of independent events will be true in a wide range of situations, including the mechanics underlying games of chance. For instance, successive flips of the same coin can be assumed to be independent, except in rather strange circumstances, such as the following. A sticky coin that is flipped can accumulate some dirt on the side that lands face down, and the coin can therefore develop an increasing bias; the outcome of a trial would then depend to some extent on the outcome of the trial before it (e.g., a string of heads leads to a buildup of dirt on the tail side, making heads even more likely). On the other hand, even when random trials are completely independent, it can appear to some people as though they are not independent. This is called the *gambler's fallacy* (as mentioned in Chapter 5, Section C). In one version of the fallacy, it seems that after a run of trials that fall in one category, the probability of future trials falling in the other category increases, so that there will tend to be equal numbers of trials in each category. It "feels" as though there is some force of nature trying to even out the two categories (if $P = .5$)—so after a string of heads, for example, you expect that the chance of a tail has increased. However, if the trials are truly independent (as they usually are in such situations), this will not happen; after a "lucky" streak of ten heads in a row, the probability of a tail is still equal to .5. Although there tends to be an equal number of heads and tails after a coin has been flipped many times, there is no process that compensates for an unusual run of trials in the same category—the coin has no memory of its past flips.

When to Use the Binomial Distribution for Null Hypothesis Testing

The Sign Test

The sign test is an appropriate alternative to the matched t-test when only the direction, but not the magnitude, of the difference between two subjects in a pair (or two measurements of the same subject) can be determined. The sign test can also be applied in cases where a matched t-test had been originally planned but the difference scores obtained are so far from resembling a normal distribution that this assumption of the matched t-test appears to be violated. This violation of the normality assumption can threaten the accuracy of a t-test, especially if the number of paired scores is less than about 20. If you are worried that the assumptions of the matched t-test will not be met, you can ignore the magnitudes of the difference scores and add up only their signs.

For example, in an experiment involving 15 trials, if 12 difference scores are positive and about equally large, and the remaining difference scores are just as large but negative, a normal distribution in the population does not seem likely. Looking only at the direction of the differences, there would be 12 pluses and 3 minuses. The binomial distribution ($N = 15$, $P = .5$) could be used to make a decision. From Table A.12 you can see that even using a two-tailed test, $X = 12$ leads to rejection of the null hypothesis. On the other hand, the sign test will usually have considerably less power than the matched t-test in a situation where both tests could apply. Therefore, if each of the assumptions of the matched t-test is true (or nearly true), the matched t-test is to be preferred over the sign test. A third alternative, which is usually intermediate in power between the two tests just mentioned, is the Wilcoxon signed-rank test. The Wilcoxon test can only be applied, however, when the difference scores can be rank ordered; this approach will be discussed further in Chapter 21.

Correlational Research

There are many grouping variables that consist of two, and only two, mutually exclusive categories (e.g., male or female, married or not married). If the proportion of each category in the population is fairly well known, it is possible to test a subpopulation to see if it differs from the general population with respect to those categories. For instance, imagine that you test 20 leading architects for handedness and find that eight are left-handed. If the instance of left-handedness is only 10% in the population (i.e., $P = .1$), the binomial distribution with $N = 20$, $P = .1$ can be used to represent the null hypothesis for the population of architects. You would need to use a table for this particular binomial distribution (or you could construct one with the methods described in Section C); the normal approximation would not be sufficiently accurate in this case. Whichever method you used, you would find that H_0 could be rejected, so you could conclude that the proportion of left-handers is higher among architects than in the general population. However, you could *not* determine whether left-handedness contributes to becoming an architect based on this finding. You have only observed a correlation between handedness and being an architect.

Experimental Research

In some cases, a subject's response to an experimental task or condition can fall into only one or the other of two categories (e.g., the subject solved the problem or not; used a particular strategy for solution or not). Consider a study of seating preferences in a theater. One subject at a time enters a small theater (ostensibly to see a short film); the theater contains only two seats. Both seats are at the same distance from the screen, but one is a bit to the left of center, whereas the other is equally to the right of center. The variable of interest is which seat is chosen, and this variable is dichotomous, having only two values (i.e., left and right). The null hypothesis is that the left and right seats have an equal probability of being chosen. If one side is chosen by more subjects than the other, the binomial distribution can be used to decide whether the null hypothesis can be rejected. (Assume that only right-handers are included in the study.) If the experiment involved two different groups of subjects—such as left-handers and right-handers—and you observed seating preferences for both groups, you would have a two-factor experiment (handedness vs. seating preference), in which both factors are dichotomous. To analyze the results of such an experiment, you would need the methods discussed in Chapter 20.

Exercises

*1. Perform the sign test on the data from exercise 13B5, using the same alpha level and number of tails. Did you reach the same conclusion with the sign test as with the matched *t*-test? If not, explain the discrepancy.

2. Perform the sign test on the data from exercise 13B6, using the same alpha level and number of tails. Did you reach the same conclusion with the sign test as with the matched *t*-test? If not, explain the discrepancy.

* 3. Six students create two paintings each. Each student creates one painting while listening to music and the other while listening to white noise in the background. If for five of the six students, the painting produced with music is judged to be more creative than the other, can you reject the null hypothesis ($\alpha = .05$, two-tailed) that music makes no difference? (Use Table A.12.)

4. Imagine that the experiment described in the previous exercise involved 60, instead of 6, subjects, and that for 50 of the students the painting produced with music was rated more highly.

   a) Use the normal approximation to test the null hypothesis.

   b) How does the experiment in this exercise compare with the one in exercise 3, in terms of the proportion of "music" paintings judged more highly? Why are the conclusions different?

* 5. Does the mental condition of a chronic schizophrenic tend to deteriorate over time spent in a mental institution? To answer this question, nine patients were assessed clinically after 2 years on a ward, and again 1 year later. These clinical ratings appear in the table below.

Patient	1	2	3	4	5	6	7	8	9
Time 1	5	7	4	2	5	3	5	6	4
Time 2	3	6	5	2	4	4	6	5	3

Assume that these clinical ratings are so crude that it would be misleading to calculate the difference score for each subject and perform a matched *t*-test. However, the direction (i.e., sign) of each difference is considered meaningful, so the sign test can be performed. Test the null hypothesis ($\alpha = .05$, two-tailed) that there is no difference over time for such patients.

6. One hundred and fifty schizophrenics have been taking an experimental drug for the last 4 months. Eighty of the patients exhibited some improvement; the rest did not. Assuming that half the patients would show some improvement over 4 months without the drug, use the sign test ($\alpha = .01$) to determine whether you can conclude that the new drug has some effectiveness.

* 7. In a simulated personnel selection study, each male subject interviews two candidates for the same job: an average-looking man and a male model. Two equally strong sets of credentials are devised, and the candidates are randomly paired with the credentials for each interview. Thirty-two subjects choose the male model for the job, and 18 choose the average-looking man. Can you reject the null hypothesis (at the .05 level) that attractiveness does not influence hiring decisions with a one-tailed test? With a two-tailed test?

8. Suppose that in recent years 30 experiments have been performed to determine whether violent cartoons increase or decrease aggressive behavior in children. Twenty-two studies demonstrated increased aggression and 8 produced results in the opposite direction. Based on this collection of studies, can you reject the null hypothesis (at the .05 level) that violent cartoons have no effect on aggressive behavior?

OPTIONAL MATERIAL

The Classical Approach to Probability

Chapter 5 discussed probability in terms of the normal distribution. Events were defined as some range of values (e.g., over 6 feet tall; IQ between 90 and 110); the probability of any very precise value (e.g., exactly 6 feet tall) was assumed to be virtually zero. The probability of an event was defined in terms of the relative amount of area under the distribution that is enclosed by the event (e.g., if 20% of the distribution represents people above 6 feet tall, then the probability of randomly selecting someone that tall is .2). When you are dealing with discrete events (e.g., getting either heads or tails in a coin toss), probability can be defined in a different way. Specific events can be counted and compared to the total number of possible events. However, the counting can get quite complicated, as you will soon see.

You have already seen how simple events (trials), such as individual coin tosses, can pile up in different ways to form the binomial distribution. Another example that was mentioned is the birth of a child. If a family has one child, only two gender outcomes are possible: boy or girl. However, if the family has four children, quite a few outcomes are possible. Each possible sequence of genders represents a different outcome; having a boy first followed by three girls (BGGG) is a different outcome from having a boy *after* three girls (GGGB). As I pointed out in Section A (in the context of tossing a coin four times), the total number of different outcomes would be $2^4 = 2 \cdot 2 \cdot 2 \cdot 2 = 16$.

Complex events can be defined in terms of the number of different outcomes they include. For instance, if an event is defined as a family with four children, only one of which is a boy, four different outcomes are included (BGGG, GBGG, GGBG, GGGB). If all of the specific outcomes are equally likely, then the probability of an event can be found by counting the outcomes included in the event and counting the total number of outcomes (this is known as the **classical approach to probability**). The probability of an event A [i.e., $p(A)$] is defined as a proportion, or ratio, as follows:

$$p(A) = \frac{\text{number of outcomes included in A}}{\text{total number of possible outcomes}} \qquad \text{Formula 19.4}$$

Using Formula 19.4, the probability that a family with four children will contain exactly one boy is $4/16 = .25$. The probability of having two boys and two girls is $6/16 = .375$. (You can use Table 19.1 to count the number of sequences that contain two of each category.) This counting method is sufficient as long as we can assume that all of the particular outcomes or sequences are equally likely, and this assumption will be true if $P = Q$. In reality, the birth rates for boys and girls are not exactly equal, but for our purposes the difference is too tiny to bother with. On the other hand, there are interesting cases in which P and Q are clearly different, and these cases require more complex calculations (to be described later in this section).

The Rules of Probability Applied to Discrete Variables

Let's review the addition and multiplication rules of probability (first presented in Chapter 5, Section C), as they apply to discrete events. One of the easiest examples for illustrating these rules involves selections from an ordinary deck of 52 playing cards, 13 cards in each of four suits (hearts, diamonds, clubs, spades). Ten of the 13 cards in each suit are numbered (assuming that the ace counts as 1), and the other three are picture cards (jack, queen, and king). The probability of selecting a heart on the first draw is, according to the classical approach to probability, 13 (the number of simple events that are classified as a heart) divided by 52 (the total number of simple events), which equals .25.

The Addition Rule

We can use the addition rule to find the answer to the more complex question, What is the probability of picking either an ace or a picture card on the first draw? The probability of picking an ace is 4/52 = .077, and the probability of choosing a picture card is 12/52 (three picture cards in each of four suits) = .231. Because these two events are mutually exclusive (one card cannot be both an ace and a picture card), we can use Formula 5.5:

$$p(A \text{ or } B) = p(A) + p(B) = .077 + .231 = .308$$

On the other hand, if we want to know the probability that the first selection will be either a club ($p = .25$) or a picture card ($p = .231$), the above addition rule is not valid. These two events are not mutually exclusive: a card can be a club *and* a picture card at the same time (there are three such cards). The addition rule modified for overlapping events (Formula 5.6) must be used instead:

$$p(A \text{ or } B) = p(A) + p(B) - p(A \text{ and } B) = .25 + .231 - .058 = .423$$

The Multiplication Rule

If our question concerns more than one selection from the deck of cards, we will need to use some form of the multiplication rule. If we want to know the probability of drawing two picture cards in a row, and we *replace* the first card before drawing the second, we can use the multiplication rule for independent events (Formula 5.7):

$$p(A \text{ and } B) = p(A)\, p(B) = .231 \cdot .231 = .0533$$

If we draw two cards in succession *without* replacing the first card, the probability of the second card will be altered according to which card is selected first. If the first card drawn is a picture card ($p = .231$), and it is not replaced, there will be only 11 picture cards and a total of only 51 cards left in the deck for the second draw. The probability of drawing a picture card on the second pick is 11/51 = .216. The probability of drawing two picture cards in a row without replacement is given by the multiplication rule for dependent events (Formula 5.8), expressed in terms of conditional probability:

$$p(A \text{ and } B) = p(A)p(B \mid A) = .231 \cdot .216 = .050$$

Permutations and Combinations

In Section A I mentioned that the problem with counting is that when you get to as many as twelve flips of a coin there are $2^{12} = 4096$ possible outcomes to write out. To understand how to do the counting mathematically (without writing out all the sequences), you need to know something about permutations and combinations. A **permutation** is just a particular ordering of items. If you have four different items, for instance, they can be placed in a variety of orders, or permutations. Suppose you are organizing a symposium with four speakers, and you have to decide in what order they should speak. How many orders do you have to choose from? You have four choices for your opening speaker, but once you have made that choice you have only three choices left for the second presenter. Then there are only two choices left for the third slot; after that choice is made the remaining speaker is automatically placed in the fourth slot. Because *each* of the four choices for opening speaker can be paired with any one of the three remaining speakers for the second slot, we multiply 4 by 3. Then we have to multiply by 2 for the third slot choice, and finally by 1 for the fourth slot (of course, this last step does not change the product). The number of orders (permutations) is $4 \cdot 3 \cdot 2 \cdot 1 = 24$. The general rule is that you take the number of items, multiply by that number minus 1, subtract 1 again and multiply, and continue this process until you are down to the number 1. This multiplication sequence arises often enough in mathematics that its product is given a name: **factorial.** The symbol for a factorial is the exclamation point. For example, 5! (pronounced "five factorial") stands for $5 \cdot 4 \cdot 3 \cdot 2 \cdot 1$, which is equal to 120. If you had five different speakers, 120 orders are possible.

You may wonder how permutations can apply to 12 flips of a coin; in that case you do not have 12 different items. In fact, you have only two different categories into which the 12 flips fall. This is where **combinations** enter the picture. I'll begin with an easier example consisting of six flips of a coin. The kind of question we are usually interested in (especially if we are constructing a binomial distribution) is, How many of the different possible sequences created by the six flips contain exactly X heads? If we used permutations to decide how many different orderings there were, the answer would be $6! = 720$. But we know this is too many. If $X = 2$, then two of the flips will be in the same category: heads. It doesn't matter if you reverse the two heads, so 6! is twice as large as it should be. But if there are two heads there are four tails and the order of the four tails is also irrelevant, so the 6! contains 4! that should also be removed. The solution is to divide the 6! by both 2! and 4!, which gives us

$$\frac{6!}{2!4!} = \frac{720}{(2)(24)} = \frac{720}{48} = 15$$

Fifteen of the possible sequences created when you flip a coin six times will have exactly two heads. Because the total number of possible outcomes is $2^6 = 64$, the probability of getting exactly two heads is $15/64 = .23$.

As a general rule, when you have N items and X of them fall into one category and the remainder $(N - X)$ fall into a second category, the number of different sequences is given by Formula 19.5:

$$\binom{N}{X} = \frac{N!}{X!(N - X)!}$$

Formula 19.5

The term on the left is often referred to as a **binomial coefficient,** which can be symbolized as either $\binom{N}{X}$, or $_NC_X$, and is expressed as "N taken X at a time." Note that $_NC_X$ will always yield the same value as $_NC_{N-X}$. In the above example, the number of sequences with exactly two tails (and four heads) is the same as the number of sequences with exactly two heads (and four tails).

Combinations have many uses. I used combinations, without mentioning the term, in Chapter 15, when I showed how many different t-tests could be performed with a particular number of group means. Because X is always equal to two when dealing with t-tests, the combination formula can be simplified to

$$\frac{N!}{2!(N - 2)!}$$

Now note that $N! = N \cdot (N - 1)! = N \cdot (N - 1) \cdot (N - 2)!$ and so on (e.g., $5! = 5 \cdot 4! = 5 \cdot 4 \cdot 3!$). So the above formula can be rewritten as

$$\frac{N \cdot (N - 1) \cdot (N - 2)!}{2! \cdot (N - 2)!}$$

Canceling out the $(N - 2)!$ and noting that $2! = 2$, we have reduced the formula to $N(N - 1)/2$, which is equivalent to Formula 15.1.

Constructing the Binomial Distribution

Combinations can be used to construct a binomial distribution. If $N = 6$, we find $_NC_0$, $_NC_1$, $_NC_2$, $_NC_3$, $_NC_4$, $_NC_5$, and $_NC_6$, and divide each by the total number of outcomes (which, in the case of six tosses of a coin, is 64). ($_NC_4$, $_NC_5$, and $_NC_6$ will be the same as $_NC_2$, $_NC_1$ and $_NC_0$, respectively.) To find $_NC_0$ you must know that $0!$ is defined as equal to 1, so $_NC_0$ always equals 1. Counting the outcomes for a particular X and dividing by the total number of outcomes is appropriate only when all the outcomes are equally likely, and this is only true when $P = Q = .5$. When P and Q are not equal, a bit more work is required to find the binomial distribution.

Suppose we are interested in the eye color, rather than the gender, of the children in a family. Assuming that each parent has one gene for brown eyes and one gene for blue eyes, elementary genetics tells us that the chance of having a blue-eyed child is .25. If the parents have four children, the probability that exactly one child will have blue eyes is *not* the same as the probability that exactly one child will *not* have blue eyes, even though there are four possible outcomes in each case. Using the multiplication rule for independent events, the probability that the first child will have blue eyes and the next three will not is $.25 \cdot .75 \cdot .75 \cdot .75 = .105$. All four sequences that contain one blue-eyed child will have the same probability, so the probability of having exactly one blue-eyed child is $.105 \cdot 4 = .42$. On the other hand, the probability of having three blue-eyed children in a row followed by one who does not have blue eyes is $.25 \cdot .25 \cdot .25 \cdot .75 = .012$. Again, there are four (equally likely) sequences with three blue-eyed children, so the probability of having exactly one child who is not blue-eyed is $.012 \cdot 4 = .048$. Note the asymmetry. Having one blue-eyed child is much more likely ($p = .42$) than having only one child who does not have blue eyes ($p = .048$). When $P \neq Q$, all outcomes are not equally likely; the probability of an outcome depends on the value of X. Only outcomes that share the same value of X will have the same probability. Formula 19.6 is the general formula for constructing the binomial distribution:

$$p(x) = \binom{N}{X} P^X Q^{N-X} \qquad \text{Formula 19.6}$$

We can use this formula to find the probability corresponding to any value of X, when P is not equal to Q. (Note that when $P = Q$, the second part of the formula, $P^X Q^{N-X}$, always comes out to $.5^N$, or $1/2^N$.) For instance, if we want to find the probability of obtaining two blue-eyed children out of six ($P = .25$, as in the example above), we can use Formula 19.6:

$$\binom{6}{2} P^2 Q^4 = \frac{6!}{2!4!}(.25^2)(.75^4) = \frac{6 \cdot 5 \cdot 4!}{2!4!}(.0625)(.32) = \frac{6 \cdot 5}{2 \cdot 1}.02 = 15(.02) = .3$$

If we continue this procedure for all possible values of X (in this case, from 0 to 6), we can use the resulting values to graph the binomial distribution for $N = 6$, $P = .25$, as shown in Figure 19.4.

The Empirical Approach to Probability

The classical approach to probability is particularly appropriate when dealing with games of chance based on mechanical devices (e.g., a roulette wheel) or a countable set of similar objects (e.g., a deck of playing cards). On the other hand, when I mentioned at the end of Section B that the probability of selecting a left-hander from the population might be .1, I was alluding to a case in which it is very unlikely that the classical approach to probability would be employed. To use this approach you would have to know the exact number of people in the population and the exact number of left-handers. Considering that it would be difficult to obtain such exact information for real populations, it is more likely

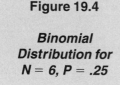

Figure 19.4

Binomial Distribution for N = 6, P = .25

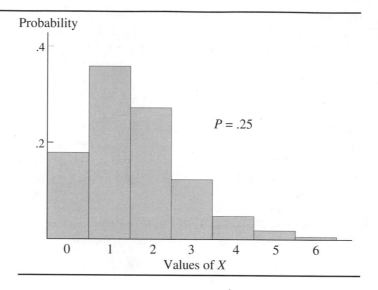

Probability

$P = .25$

Values of X

that you would use the **empirical approach to probability.** The empirical approach is based on sampling the population in order to estimate the proportion that falls into some category, such as left-handers. This estimated proportion is then used as the probability that a random selection from the population will fall in the category of interest. Some sort of estimate would most likely be used, as well, to determine the probability corresponding to each ethnic group in the grand jury selection examples described in Section A.

Exercises

* 1. On your first draw from a deck of 52 playing cards, a) what is the probability of selecting a numbered card higher than 5?

 b) What is the probability of selecting either a red card (a heart or a diamond) or a spade?

 c) What is the probability of selecting either a red card or a numbered card higher than 5?

2. If schizophrenics make up 40% of the psychiatric population, a) what is the probability that the next three patients selected at random from the psychiatric population will all be schizophrenic?

 b) What is the probability that none of the next four patients selected at random from the psychiatric population will be schizophrenic?

* 3. On your first two draws from a deck of 52 playing

cards (without replacement), a) what is the probability of selecting two hearts?

 b) What is the probability of selecting a heart and a spade?

 c) What is the probability of selecting two numbered cards higher than 5?

4. If a disc jockey has eight songs to play in the next half hour, in how many different orders can he play the songs?

* 5. In a seminar class of 15 students, three must be chosen to make a joint presentation in the next class.

 a) How many different three-person groups can be formed?

 b) If five students are to present jointly, how many different groups can be formed?

6. If a family has sixteen children, a) what is the probability that all of them will be of the same gender (assuming that boys and girls are equally likely to be born)?

b) What is the probability that there will be an equal number of each gender?

* 7. If a quiz consists of ten multiple-choice questions with five possible answers for each, and a student guesses on all ten questions, a) what is the probability that the student will obtain a score of 8?

b) What is the probability that the student will obtain a score of 8 or more?

8. If it is estimated that 15% of all fifth-grade boys have some form of dyslexia, a) what is the probability that there will be exactly three dyslexics in a random group of eight fifth-grade boys?

b) What is the probability that there will be no dyslexics in a random group of eight fifth-grade boys?

SUMMARY

The Important Points of Section A

1. The binomial distribution arises from a series of independent, dichotomous events. The two possible outcomes of each event have probabilities P and Q, which sum to 1.0 (i.e., $Q = 1 - P$).

2. Each simple event is commonly called a trial, and the total number of trials is symbolized by N. The number of trials falling in the category with probability P is labeled X. The distribution of X, which follows the binomial distribution, is a function of both N and P.

3. When $P = Q = .5$, the binomial distribution will be symmetrical. As N increases, this distribution more closely resembles the normal distribution. Even when $P \neq Q$, the resemblance to the normal distribution increases as N increases; as N becomes infinitely large the binomial distribution becomes identical to the normal distribution.

4. The binomial distributions for $P = Q = .5$ and N ranging from 1 to 15 are given in Table A.12. This table can be used to test the null hypothesis that $P = .5$. For example, if three women are members of a jury of 12, look at the binomial distribution for $N = 12$ and find the probability corresponding to $X = 3$. Then also find the probabilities for more extreme values of X (in this case, $X = 2, 1,$ and 0). The sum of these probabilities is the one-tailed p level for testing whether $P = .5$.

5. To perform a two-tailed test, double the p level you find using the above procedures. If $P \neq Q$, the binomial distribution will not be symmetrical, and it will not be easy to perform a two-tailed test.

6. When N is sufficiently large, the binomial distribution closely resembles a normal distribution with a mean of NP and a standard deviation of $\sqrt{NPQ}$. The probability that X or more trials will land in the P category can be approximated by calculating the appropriate z score and looking up the area beyond that z score.

7. If $P = .5$, the normal distribution becomes a good approximation when N reaches about 25. If P is considerably more or less than .5, a larger N is needed for a good approximation. A good rule of thumb is that when P is not near .5, NPQ should be at least 9 if you wish to use the normal approximation.

The Important Points of Section B

The Sign Test

The sign test can be used in place of a matched t-test when the amount of difference between the members of a matched pair cannot be determined, but the direction of that difference can be. Because the direction of each difference must fall into one of only two categories (such as $+$ or $-$), the binomial distribution can be used to determine whether a given imbalance between those categories is likely to occur by chance. (Generally, an equal number of pluses and minuses are expected under the null hypothesis.)

Imagine an example in which there are 14 pairs of matched subjects, and 11 of the differences between subjects are in the negative direction, whereas only three differences are positive. Table A.12, under $N = 14$, shows that the probability for $X = 11$ is .0222. The probabilities for $X = 12$, 13, and 14 are .0056, .0009, and .0001, respectively. Summing these probabilities, we obtain .0222 + .0056 + .0009 + .0001 = .0288. Doubling this sum, we get .0288 · 2 = .0576; this is the two-tailed p level. In this case, we could not reject the null hypothesis at the .05 level, two-tailed.

The Correction for Continuity

Although you would not ordinarily calculate the normal approximation with such a small N, I will do so below to review the procedure. I will use Formula 19.3 because it includes a correction for continuity.

$$z = \frac{|X - NP| - .5}{\sqrt{NPQ}} = \frac{|11 - 7| - .5}{\sqrt{3.5}} = \frac{4 - .5}{1.87} = \frac{3.5}{1.87} = 1.87$$

The above z score leads to the same conclusion we reached with the sign test based directly on the binomial distribution; the null hypothesis cannot be rejected because the calculated z (1.87) is less than the critical z (1.96). If you use Table A.1 to look up the p level (the area beyond z) that corresponds to the calculated z score, you will see that $p = .0307$, which is quite close to the one-tailed p level (.0288) found from the binomial distribution. Even with a sample size as small as 14, the correction for continuity provides a fairly good normal approximation.

Assumptions of the Sign Test

Dichotomous events. Each simple event or trial can fall into only one or the other of two categories—not both categories simultaneously or some third category. The probabilities of the two categories, *P* and *Q*, must sum to 1.0.

   Independent events. The outcome of one trial does not influence the outcome of any other trial.

   Stationary process. The probabilities of each category (i.e., *P* and *Q*) remain the same for all trials in the experiment.

When to Use the Binomial Distribution for Null Hypothesis Testing

The sign test. The binomial distribution can be used as an alternative to the matched or repeated measures *t*-test, when it is possible to determine the direction of the difference between paired observations but not the amount of that difference. The sign test can be planned (as when you make no attempt to measure the amount of difference but assess only its direction) or unplanned (as when you plan a matched *t*-test but the sample size is fairly small and the difference scores are very far from following a normal distribution).

   Correlational research. The binomial distribution applies when this kind of research involves counting the number of individuals in each of two categories within a specified group (e.g., counting the number of smokers and nonsmokers in an intensive cardiac care unit). The values of *P* and *Q* are based on estimates of the proportion of each category in the general population (e.g., if it is estimated that 30% of the population smoke, then $P = .3$ and $Q = .7$).

   Experimental research. The binomial distribution is appropriate when the dependent variable is not quantifiable but can be categorized as one of two alternatives. (For example, given a choice between two rooms in which to take a test—identical except that one is painted red and the other blue—do equal numbers of subjects choose each one?)

The Important Points of Section C

1. Given a series of *N* trials with only two possible outcomes for each, a total of 2^N different sequences can occur (e.g., for $N = 6$ the number of sequences is $2^6 = 64$).

2. A complex event can be defined in such a way that several different sequences are included. For instance, if a coin is flipped six times and the event is defined as obtaining exactly two heads, there are 15 different sequences included in that event.

3. If all the sequences of the two outcomes are equally likely (i.e., $P = .5$), the *classical approach* to probability can be used to determine the probability of a particular event. The probability of the event is a proportion: the number of outcomes (or sequences) included in that event divided by the total number of possible outcomes. Using this approach the probability of obtaining exactly two heads out of six flips of a fair coin is 15/64 = .23.

4. A *permutation* is an ordering of distinct items. The number of permutations that are possible for N items is $N!$ (read as "N factorial"), which equals $N \cdot (N - 1) \cdot (N - 2) \cdot (N - 3)$, and so on, down to the number 1 (e.g., $6! = 6 \cdot 5 \cdot 4 \cdot 3 \cdot 2 \cdot 1 = 720$). Thus six people in a line can be arranged into 720 different orders.

5. If out of N items, X fall into one category and the remainder (i.e., $N - X$) fall into a second category, *combinations* (symbolized as $\binom{N}{X}$ or $_NC_X$) can be used to determine the number of different possible sequences. The number of possible sequences is given by $N!$ divided by the product of $X!$ and $[(N - X)!]$. Thus $_NC_X$ will always yield the same value as $_NC_{N-X}$. (Note that $0! = 1$ by definition.)

6. When $P = .5$, combinations can be used to find the binomial distribution. First, find the total number of possible sequences. Then find $_NC_X$ for each value of X from 0 to N. Finally, divide each $_NC_X$ by the total number of sequences to find the probability corresponding to each value of X.

7. When $P \neq .5$, all sequences are not equally likely. The probability of a sequence depends on the value of X; only sequences that share the same value for X will have the same probability. To find the probability corresponding to a value of X, use the binomial coefficient to determine the number of sequences, then multiply P and Q the appropriate number of times each to calculate the probability of each sequence (see Formula 19.6).

8. The *empirical approach* to probability uses sampling to estimate proportions corresponding to different categories in a population, rather than exhaustive counting, as in the classical approach.

Definitions of Key Terms

Binomial distribution For N independent, dichotomous simple events, this distribution gives the probability for each value of X, where X is the number of simple events falling into one of the two categories (X ranges from 0 to N).

Dichotomous events Consisting of two distinct, mutually exclusive categories (e.g., heads and tails; on and off).

Sign test An alternative to the matched t-test in which only the sign of each difference score is used. The probability of a true null hypothesis producing the observed results is derived from the binomial distribution.

Distribution-free test An hypothesis test in which no assumption about the distribution of the observed variables is necessary. In fact, the dependent variable can be dichotomous, as in the sign test. Nonparametric tests generally fall in this category.

Classical approach to probability An approach in which the probability of a complex event is determined by counting the number of outcomes included in that event and dividing by the total number of possible outcomes (assuming all outcomes are equally likely). This approach is especially appropriate when dealing with games of chance.

Permutation A particular order in which some number of distinct items can be placed.

Factorial A mathematical expression, symbolized by an exclamation point following an integer (i.e., $N!$), that expresses the product of a descending string of integers starting with N and ending with 1 (i.e., $N \cdot (N - 1) \cdot (N - 2) \cdot \ldots \cdot 1$).

Combination An ordering of N items that fall into two distinct categories (items in the same category are considered identical), X belonging to one category, and $N - X$ to the other.

Binomial coefficient A mathematical expression that gives the number of combinations for a given value of N and X.

Empirical approach to probability An approach in which probabilities are estimated by using sampling, rather than exhaustive counting.

Key Formulas

z score when the normal distribution is used to approximate the binomial distribution. Because this formula does not contain a correction for continuity, it is most appropriate when N is very large:

$$z = \frac{X - NP}{\sqrt{NPQ}}$$

Formula 19.1

z score for testing a proportion (this is the same as Formula 19.1, except that all terms have been divided by N so that the formula is expressed in terms of proportions instead of frequencies):

$$z = \frac{p - P}{\sqrt{\dfrac{PQ}{N}}}$$

Formula 19.2

z score for the binomial test (this is the same as Formula 19.1, except that it includes a correction for continuity):

$$z = \frac{|X - NP| - .5}{\sqrt{NPQ}}$$

Formula 19.3

The classical approach to determining the probability of an event:

$$p(A) = \frac{\text{number of outcomes included in A}}{\text{total number of possible outcomes}}$$

Formula 19.4

The number of combinations when X out of N items fall into one category and the remainder fall into a second category:

$$\binom{N}{X} = \frac{N!}{X!(N-X)!}$$

Formula 19.5

The probability of X for a binomial distribution for any values of P and N:

$$p(x) = \binom{N}{X} P^X Q^{N-X}$$

Formula 19.6

You will need to use the following from previous chapters:

Symbols:
N: Sample size
X: Frequency of first category
P: Probability corresponding to first category
Q: Probability corresponding to second category

Formulas:
Formula 19.1: Normal approximation to the binomial distribution

Concepts:
Binomial distribution
Chi-square distribution (Chapter 8, Section C)
Correction for continuity

CHAPTER

In the previous chapter I showed how the binomial distribution can be used to determine the probability of randomly selecting a jury that misrepresents the population. This probability estimate could help you decide whether a particular jury is so lopsided that you should doubt that the selection was truly random. I pointed out that if there were more than two subpopulations to be considered in the same problem, you could not use the binomial distribution, and the problem would become more complicated. In this chapter I will show how to consider any number of subpopulations with just a small increase in the complexity of the statistical procedure compared to the two-group case.

A

**CONCEPTUAL
FOUNDATION**

The Multinomial Distribution

Imagine that the population of a city is made up of three ethnic groups, which I will label A, B, and C. The proportion of the total population belonging to each group is designated $P, Q,$ and $R,$ respectively ($P + Q + R = 1.0$, so $R = 1 - P - Q$). A grand jury ($N = 48$) is randomly selected from the population; the jury consists of X members from group A, Y members from group B, and Z members from group C ($X + Y + Z = N$, so $Z = N - X - Y$). Suppose that $P = .5, Q = .33,$ and $R = .17,$ and that $X = 28, Y = 18,$ and $Z = 2$. A spokesperson for group C might suggest that the overrepresentation of group A ($X = 28$) and group B ($Y = 18$) at the expense of group C ($Z = 2$) is the result of deliberate bias. Certainly, X and Y are larger than expected based on the proportions P and Q (expected $X = PN = .5 \cdot 48 = 24$; expected $Y = QN = .33 \cdot 48 = 16$), and Z is less than expected ($RN = .17 \cdot 48 = 8$), but is this jury composition very unlikely to occur by chance?

To calculate the probability of selecting a jury with the above composition, you could use an extension of the procedures described in Section C of Chapter 19. You could test the null hypothesis that $P = .5$ and $Q = .33$, but you would not be dealing with a binomial distribution. Because there are more than two categories in which the jurors can fall, you would be calculating values of a **multinomial distribution.** Just as the binomial distribution depends on the values of N and P, the multinomial distribution in the case of three categories depends on N, P, and Q. As the number of categories increases, so does the complexity of the multinomial distribution. In fact, even with only three categories, the calculations associated with the multinomial distribution are so tedious that they are universally avoided. The easy way to avoid these calculations is to use an approximation—just as we used the normal distribution to approximate the binomial distribution in Chapter 19.

The Chi-Square Distribution

To get from a binomial approximation to a multinomial approximation I will employ the same trick that I used in Chapter 14 to get from the *t*-test to ANOVA: I will take the formula for the two-group case and square it. Squaring Formula 19.1 (leaving out the correction for continuity at this point for the sake of simplicity) gives the following expression:

$$z^2 = \frac{(X - NP)^2}{NPQ}$$

Assuming that the z scores follow a normal distribution before squaring, the squared z scores follow a different mathematical distribution, called the **chi-square distribution** (introduced in Chapter 8, Section C). The symbol for chi-square is the Greek letter chi (pronounced "kie" to rhyme with "pie") squared (χ^2). As with the *t* distribution, there is a whole family of chi-square distributions; the shape of each chi-square distribution depends on the number of degrees of freedom. However, chi-square distributions tend to be positively skewed, because the value of χ^2 cannot fall below zero. The distribution of squared z scores referred to above is a chi-square distribution with one degree of freedom [$\chi^2(1)$]. A more typical chi-square distribution (df = 4) is shown in Figure 20.1.

Figure 20.1

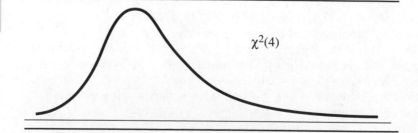

$\chi^2(4)$

If the above formula for z^2 is used instead of Formula 19.1, then instead of finding the critical value from the normal table (Table A.1), we need to look in Table A.13 under df = 1. For instance, for a .05 two-tailed test, we would use ± 1.96 as our critical z, but for $\chi^2(1)$ the corresponding critical value would be 3.84. Note that $3.84 = 1.96^2$. Squaring the test statistic means squaring the corresponding critical value, as well.

Expected and Observed Frequencies

When there are only two categories, it doesn't matter whether you use Formula 19.1 or its squared version. (Note that if the correction for continuity is appropriate, you would make the correction before squaring.) The advantage of squaring is that it readily leads to an expanded formula that can accommodate any number of categories. However, to accommodate more categories, it is convenient to adopt some new symbols. Consider the numerator of Formula 19.1: $X - NP$. The term NP is the number of trials we *expected* to fall in the category with probability P, and X is the number of trials that actually did. In the terminology of the chi-square test, NP is the expected frequency (f_e) and X is the frequency that was actually obtained, or observed (f_o). I will apply this terminology to the example I posed earlier of three ethnic groups being represented on a grand jury. I mentioned that the expected frequencies for groups A, B, and C were 24, 16, and 8, respectively. The obtained frequencies were 28, 18, and 2. It is customary to put both the obtained and expected frequencies in a table, as shown in Table 20.1.

	Group A	Group B	Group C
f_o	28	18	2
f_e	24	16	8

Table 20.1

The Chi-Square Statistic

Now we need a formula (similar to Formula 19.1 squared, but with new symbols) to measure the discrepancy between the obtained and expected frequencies. The appropriate formula is Formula 20.1

$$\chi^2 = \sum \frac{(f_o - f_e)^2}{f_e}$$
Formula 20.1

I left out the indexes on the summation sign, but the sum goes from 1 to k, where k is the number of groups, or cells ($k = 3$ in this example). Thus the formula is applied k times, once for each pair of observed and expected frequencies. The test statistic produced by the formula is called the chi-square statistic, because it follows the chi-square distribution (approximately) when the null hypothesis is true (and certain assumptions, to be discussed later, are met). It is also referred to as **Pearson's chi-square statistic,** after Karl Pearson, who devised the formula. (Pearson is the same mathematician whose name is associated with the correlation coefficient described in Chapter 11.) Applying Formula 20.1 to our example, we get

$$\chi^2 = \frac{(28 - 24)^2}{24} + \frac{(18 - 16)^2}{16} + \frac{(2 - 8)^2}{8} = \frac{16}{24} + \frac{4}{16} + \frac{36}{8} = 5.42$$

Critical Values of Chi-Square

Before we can make any decision about the null hypothesis, we need to look up the appropriate critical value of the chi-square distribution. To do this we need to know the number of degrees of freedom. In general, if there are k categories, $df = k - 1$, because once $k - 1$ of the f_e have been fixed, the last one has to be whatever number is needed to make all of the f_e add up to N. (The expected frequencies must add up to the same number as the obtained frequencies.) In this example, $df = 3 - 1 = 2$. Looking in Table A.13 (or Table 20.2, which is a portion of Table A.13) we find that χ^2_{crit} for $df = 2$ and alpha $= .05$ is 5.99. As you can see from Figure 20.2, the calculated χ^2 (5.42), being less than the critical χ^2, does not fall in the region of rejection. Although it may seem that group C is underrepresented on the jury, we cannot reject the null hypothesis that this jury is a random selection from the population (given that $P = .5$, $Q = .33$, and $R = .17$, as stated in the example).

If you look closely at Table 20.2, you will see that the critical values increase as alpha is reduced, just as they do for the other statistics in this text. However, unlike t or F, the critical value of χ^2 gets larger as the degrees of freedom increase. (The df play a different role for χ^2, being based only on the number of categories rather than on the number of subjects.) You may also notice that the χ^2 table does not give you a choice between one- and two-tailed tests. This is the same situation we encountered in using the tables of the F distributions and I will return to this issue later in this section.

Table 20.2					
	Area in the Upper Tail				
df	.10	.05	.025	.01	.005
1	2.71	3.84	5.02	6.63	7.88
2	4.61	5.99	7.38	9.21	10.60
3	6.25	7.81	9.35	11.34	12.84
4	7.78	9.49	11.14	13.28	14.86
5	9.24	11.07	12.83	15.09	16.75

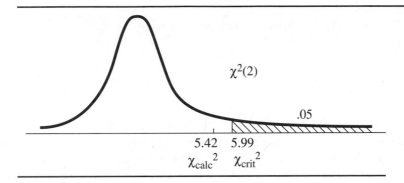

Figure 20.2

Tails of the Chi-Square Distribution

Note that the rejection zone appears in only one tail of the distribution, the positive tail. Recall that the positive tail of the *F* distribution represents ANOVAs in which the group means are unusually spread out, and the negative tail represents cases in which the group means are unusually close together. Similarly, unusually large discrepancies between expected and obtained frequencies lead to chi-square values in the positive tail, whereas studies in which the obtained frequencies are unusually close to their expected values lead to values in the negative tail (near zero). The rejection zone is placed entirely in the positive tail of the chi-square distribution, because only large discrepancies between expected and obtained values are inconsistent with the null hypothesis; extremely small discrepancies, no matter how unusual, serve only to support the null hypothesis. Thus in one sense, all chi-square tests are one-tailed.

In another sense chi-square tests are multitailed, because there are several ways to get large discrepancies. Imagine Dr. Null trying to duplicate our somewhat lopsided grand jury. He would take a truly random selection from the population (with $P = .5$, $Q = .33$, and $R = .17$), and then calculate the χ^2 corresponding to his jury. If his χ^2 exceeded ours, Dr. Null would "win." But it doesn't matter what kind of imbalance led to Dr. Null's large χ^2 —he can "win" by selecting a jury that *over*represents group C, or even one in which group C's representation is exactly as expected but there is an unexpected imbalance between groups A and B.

Expected Frequencies Based on No Preference

In the example above, the expected frequencies were based on the known proportions of the different subgroups in a population. This is not a very common situation. More often, especially in an experimental context, the expected frequencies are what you want to test. For instance, imagine that a developmental psychologist is studying color preference in toddlers. Each child is told that he or she can take one toy out of four that are offered. All four toys are identical except for color: red, blue, yellow, or green. Forty children are run in the experiment,

	Red	Blue	Yellow	Green
Table 20.3				
f_o	13	9	15	3
f_e	10	10	10	10

and their color preferences are as follows: red, 13; blue, 9; yellow, 15; and green, 3. These are the obtained frequencies. The expected frequencies depend on the null hypothesis. If the null hypothesis is that toddlers in general have no preference for color, then we would expect choices of colors to be equally divided among the entire population of toddlers. Hence, the expected frequencies would be 10 for each color (the f_e must add up to N). The f_o and f_e are shown in Table 20.3. Applying Formula 20.1 to the data in the table, we get

$$\chi^2 = \frac{(13 - 10)^2}{10} + \frac{(9 - 10)^2}{10} + \frac{(15 - 10)^2}{10} + \frac{(3 - 10)^2}{10}$$

$$= \frac{9}{10} + \frac{1}{10} + \frac{25}{10} + \frac{49}{10} = 8.4$$

The number of degrees of freedom is one less than the number of categories, so df = 3 for this example. Therefore, the critical value for a .05 test (from Table A.13) is 7.82. Because our calculated χ^2 (8.4) exceeds the critical value, we can reject the null hypothesis that the population of toddlers is equally divided with respect to color preference. We can conclude that toddlers, in general, have preferences among the four primary colors used in the study. (A toy company might be interested in this conclusion.)

Goodness-of-Fit Tests

The chi-square tests described thus far can be considered one-way tests, in the same sense that an ANOVA can be one-way: All of the categories are considered levels of the same independent (or quasi-independent) variable. For instance, a one-way chi-square test can involve any number of different religions or any number of different political parties. However, if the design of the study included several different religions *and* several different political parties in a completely crossed factorial design (for example, you wish to determine for each religion the number of subjects in each political party), you would be dealing with a "two-way" chi-square test, which is the subject of Section B.

The one-way chi-square test is often called a **goodness-of-fit test,** because what is being measured is the fit between observed and expected frequencies. There are several types of goodness-of-fit tests, as described below.

Population Proportions are Known

It is rare that proportions of the population are known exactly, but for many categorical variables we have excellent estimates of the population proportions. For instance, from voter registration data we could find out the proportion of citizens belonging to each political party within a particular locale. Then, by sampling the readers of a particular newspaper, for instance, and performing a chi-square test, we could decide if the politics of these readers represented a random selection of the electorate. The example above concerning three ethnic groups and their representation on a grand jury is of this type.

A related use of the goodness-of-fit test is to compare one population with another. For instance, knowing the proportions of certain mental illnesses in industrial nations, you could compare samples of the population of a less-developed nation to see if these countries exhibit significantly different proportions of the same mental illnesses.

Expected Frequencies are Hypothesized To Be Equal

In conducting tests involving games of chance or experimental research, researchers commonly hypothesize equal expected frequencies. If you were throwing a die 90 times to see if it were "loaded" (i.e., biased), you would expect each side to turn up 15 times (90/6 sides = 15 per side) if the die were fair (i.e., if the null hypothesis were true). In a social psychology experiment, descriptions of hypothetical "job applicants" could be paired with names representing different ethnic groups. If subjects are selecting the best applicant from each set without prejudice, then the various names should be selected equally often (assuming that the name-description pairings are properly counterbalanced).

Of course, games of chance do not always involve equal frequencies. For example, if a pair of dice is thrown, the number 7 is much more likely to come up than the numbers 2 or 12. The expected frequencies can be determined by the kind of counting techniques described in Chapter 19. Similarly, the choices in an experimental design might lead to predicted frequencies that are based on a theoretical model and are not equal.

The Shape of a Distribution Is Being Tested

One of the original applications of the χ^2 goodness-of-fit test was to test the shape of a population distribution for some continuous variable, based on the distribution of that variable within a sample. The expected frequencies depend on the distribution shape you wish to test. For instance, if you want to see if body weight has a normal distribution in the population, you could generate expected frequencies based on areas from the normal table. Out of a random sample of 100 subjects from a normal distribution, you would expect about 34 people to fall between $z = 0$ and $z = +1$ (and 34 people to fall between $z = 0$ and $z = -1$), about 13.5 (or 14) people to fall between $z = +1$ and $z = +2$ (and 14 between

−1 and −2), and about 2.5 (or 3) people to fall beyond $z = +2$ (or −2). These are your expected frequencies. Then you would convert the 100 body weights to z scores and sort them into the same categories to find the obtained frequencies. For a more sensitive test you would break the distribution into somewhat more than six categories.

The chi-square statistic can be used to measure how well the frequencies in your sample fit the normal distribution. However, for an interval/ratio variable like body weight, you would probably use a more appropriate procedure, such as the Kolmogorov-Smirnov test (see Hays, 1994). Rejecting the null hypothesis implies that your sample distribution is inconsistent with a normally distributed population. This could be important information if you had planned to use a parametric statistic that assumes your variable *is* normally distributed in the population. Knowing that the assumption does not hold, you can make a more informed decision about how to proceed.

Exercises

*1. A goodness-of-fit-test involves eight different categories.

a) How many degrees of freedom are associated with the test?

b) What is the critical value for χ^2 at $\alpha = .05$? At $\alpha = .01$?

2. Rework part a of exercise 19A6 using the chi-square statistic instead of the normal approximation to the binomial distribution. How does the value for χ^2 compare with the z score you calculated for that exercise?

*3. A soft drink manufacturer is conducting a blind taste test to compare its best-selling product (X) with two leading competitors (Y and Z). Each subject tastes all three and selects the one that tastes best to him or her.

a) What is the appropriate null hypothesis for this study?

b) If 27 subjects prefer product X, 15 prefer product Y, and 24 prefer product Z, can you reject the null hypothesis at the .05 level?

4. Suppose that the taste test in exercise 3 was conducted with twice as many subjects, but the proportion choosing each brand did not change (i.e., the number of subjects selecting each brand also doubled).

a) Recalculate the chi-square statistic with twice the number of subjects.

b) Compare the value of chi-square in part a with the value you found for exercise 3.

c) What general rule is being illustrated here?

*5. A gambler suspects that a pair of dice he has been playing with are loaded. He rolls one of the dice 120 times and observes the following frequencies: one, 30; two, 17; three, 27; four, 14; five, 13; six, 19. Can he reject the null hypothesis that the die is fair at the .05 level? At the .01 level?

*6. A famous logic problem has four possible answers (A, B, C, and D). Extensive study has demonstrated that 40% of the population choose the correct answer, A, 26% choose B, 20% choose C, and 14% choose D. A new study has been conducted with 50 subjects to determine whether presenting the problem in concrete terms changes the way subjects solve the problem. In the new study, 24 subjects choose A, 8 subjects choose B, 16 subjects choose C, and only 2 subjects choose D. Can you reject the null hypothesis that the concrete presentation does not alter the way subjects respond to the problem?

*7. It has been suggested that admissions to psychiatric hospitals may vary by season. One hypothetical hospital admitted 100 patients last year: 30 in the spring; 40 in the summer; 20 in the fall; and 10 in the winter. Use the chi-square test to evaluate the hypothesis that mental illness emergencies are evenly distributed throughout the year.

8. Of the 100 psychiatric patients referred to in the previous exercise, 60 were diagnosed as schizo-phrenic, 30 were severely depressed, and 10 were manic-depressives. Assuming that the national percentages for psychiatric admissions are 55% schizophrenic, 39% depressive, and 6% manic-depressive, use the chi-square test to evaluate the null hypothesis that this particular hospital is receiving a random selection of psychiatric patients from the national population.

Two-Variable Contingency Tables

B

BASIC STATISTICAL PROCEDURES

Most of the interesting questions in psychological research involve the relationship between (at least) two variables, rather than the distribution of only one variable. That is why, in psychological research, the one-way chi-square test is not used nearly as often as the two-way chi-square. When one of two variables is categorical (e.g., psychiatric diagnoses, different sets of experimental instructions) and the other is continuous, the parametric tests already discussed in this text may be appropriate. It is when both of the variables can only be measured categorically that the two-variable chi-square test is needed.

For instance, suppose a researcher believes that adults whose parents were divorced are more likely to get divorced themselves compared to adults whose parents never divorced. We cannot quantify the degree of divorce; a couple either gets divorced or they do not. To simplify the problem we will focus only on the female members of each couple. Suppose that the researcher has interviewed 30 women who have been married: ten whose parents were divorced before the subject was 18 years old and 20 whose parents were married until the subject was at least 18. As it turns out, half of the 30 women in this hypothetical study have gone through their own divorce; the other half are still married for the first time. To know whether the divorce of a person's parents makes the person more likely to divorce, we need to see the breakdown in each category—that is, how many currently divorced women come from "broken" homes and how many do not, and similarly for those still married. These frequency data are generally presented in a **contingency** (or cross-classification) **table,** in which each combination of levels from the two variables (e.g., parental divorce and subject's own divorce) is represented as a cell in the table. The example above involves two levels for each variable and therefore can be represented by a 2×2 contingency table, such as the one in Table 20.4 (on page 694). The data in such a table are often referred to as *cross-classified categorical data*.

Pearson's Chi-Square Test of Association

You can see at once from Table 20.4 that among the subjects whose parents were divorced, about twice as many have been divorced as not; the reverse trend is

Table 20.4

	Parents Divorced	Parents Married	Row Sums
Self Divorced	7	8	15
Self Married	3	12	15
Column Sums	10	20	30

evident for those whose parents were not divorced. There seems to be some association between parental divorce and one's own divorce. Of course, Dr. Null would claim that this association is accidental, and that the results are just as likely to come out in the opposite direction for the next 30 subjects. **Pearson's chi-square test of association** would allow us to decide whether to reject the null hypothesis in this case (i.e., that parental divorce is *not* associated with the likelihood of one's own divorce), but first we have to determine the expected frequency for each cell. The naive approach would be to use the same logic as the one-way chi-square and divide the total N by 4 to get the same expected frequency in each cell (i.e., $30/4 = 7.5$). However, the marginal sums in Table 20.4 remind us that twice as many subjects did *not* have divorced parents as did, and so this relationship should hold within each row of the table if the two variables are not associated. Given the marginal sums in Table 20.4, Dr. Null expects the cell frequencies to be as shown in Table 20.5. Note that whereas Dr. Null expects the same marginal sums as were actually obtained, he expects that the ratios of cell frequencies within any row or column will reflect the ratios of the corresponding marginal sums, as shown in Table 20.5. The next example will illustrate a simple procedure for finding the appropriate expected frequencies.

Table 20.5

	Parents Divorced	Parents Married	Row Sums
Self Divorced	5	10	15
Self Married	5	10	15
Column Sums	10	20	30

An Example of Hypothesis Testing with Categorical Data

A psychologist wants to know if corporal punishment (e.g., spanking) leads to increased aggressiveness in young boys. Children are observed in playground sessions by raters who do not know which of the children experience corporal punishment at home. The raters count the number of fights initiated by each boy.

The researcher may have planned to perform a *t*-test between the mean numbers of fights initiated by the two groups of boys, but the data may not lend themselves to such an analysis. If the majority of the boys have a score of zero (no fights initiated), most of the rest have scores of 1, and a few have higher scores, the distribution of scores may be so far from a normal distribution that a *t*-test is not appropriate. One simple alternative is to categorize each boy as either an "initiator" (started at least one fight during the observation periods) or a "noninitiator" (did not start any fights under observation). The boys are also categorized as having been corporally punished or not. To test whether corporal punishment is associated with the initiation of fights in the playground, the now familiar six-step hypothesis testing procedure can be used.

Step 1. State the Hypotheses

In the case of the two-variable chi-square test, the null hypothesis is that there is no association or correlation between the two variables—that is, the way that one of the variables is distributed into categories does not change at different levels of the second variable. Stated yet another way, the null hypothesis asserts that the two variables are independent of each other. Hence this test is often called the *chi-square test for independence.* For this example, the null hypothesis is H_0: corporal punishment and initiation of fights are independent. As usual, the alternative hypothesis is the negation of H_0; H_A: corporal punishment and initiation of fights are *not* independent.

Step 2. Select the Statistical Test and the Significance Level

The data consist of the frequencies in categories arranged along two dimensions (corporal punishment at home and fighting at school), so the two-way chi-square test is appropriate. As usual we will set alpha = .05.

Step 3. Select the Samples and Collect the Data

Rather than placing randomly selected boys together in a playground, a researcher would probably select a preexisting playground group. (It could be argued that the preexisting group would be not only more convenient but also a more natural situation for the subjects.) Information would be gathered on the types of punishment each boy receives at home, but this information would not be made available to the playground observers, who would simply note whether each boy initiated any fights. The data, in terms of frequencies, would be placed in a contingency table. The hypothetical data for this example are shown in Table 20.6 (on page 696).

Step 4. Find the Region of Rejection

We will be using a chi-square distribution to represent the null hypothesis distribution. But to know which chi-square distribution is appropriate, we need to know the number of degrees of freedom we are dealing with. As in the case of the one-variable chi-square test, the df depend on the number of categories rather

Table 20.6		Corporal Punishment	No Corporal Punishment	Row Sums
	Initiator	9	6	15
	Noninitiator	11	34	45
	Column Sums	20	40	60

than on the number of subjects. However, when there are two variables, the number of categories for each must be considered. The two-variable case can always be represented by an R × C contingency table, where R stands for the number of rows and C stands for the number of columns. Formula 20.2 for df can then be stated as follows:

$$df = (R - 1)(C - 1)$$

Formula 20.2

In the case of a 2 × 2 table, df = (2 − 1)(2 − 1) = (1)(1) = 1. Looking in Table A.12 for df = 1 and alpha = .05, we find that the critical value for χ^2 is 3.84. The region of rejection is that area of the $\chi^2(1)$ distribution above 3.84. As described in Section A, chi-square tests are essentially one-tailed, in that large χ^2 values can lead to statistical significance but there is no value of χ^2 so small that it would lead to the rejection of the null hypothesis.

Step 5. Calculate the Test Statistic

For chi-square tests, this step begins with finding the expected frequencies. In the one-variable case this usually involves either dividing the total sample size (N) by the number of categories, or multiplying N by the population proportion corresponding to each category. In the test for independence of two variables, however, finding the f_e is a bit more complicated. Fortunately, there is a mathematical trick that is easy to apply. The expected frequency for a particular cell is found by multiplying the two marginal sums to which it contributes and then dividing by the total N. For instance, to find the f_e for the upper left cell in Table 20.6, multiply its row sum (15) by its column sum (20) and divide by N (60): $f_e = (15 \cdot 20)/60 = 300/60 = 5$. This trick can be expressed as Formula 20.3.

$$f_e = \frac{(\text{Row sum})(\text{Column sum})}{N}$$

Formula 20.3

Applying Formula 20.3 to the lower right cell of Table 20.6, we get

$$f_e = (45 \cdot 40)/60 = 1800/60 = 30$$

Table 20.7 shows the data from Table 20.6 along with the f_e for each cell in parentheses. Note that you can check your calculation of the f_e by seeing whether

Table 20.7

	Corporal Punishment	No Corporal Punishment	Row Sums
Initiator	9 (5)	6 (10)	15
Noninitiator	11 (15)	34 (30)	45
Column Sums	20	40	60

the f_e add up to the same marginal sums as the f_o. Also, because we are considering a simple case, it is easy to see that within each row or column of Table 20.7 the f_e follow the same ratio as the marginal sums—that is, within each column the lower row is three times larger than the upper row, and within each row the right column is twice as large as the left column.

We can now apply Formula 20.1, wherein the summation sign tells us to find the value for every cell and then add up all of those values. However, because $df = 1$ for our example, it is appropriate to use the same correction for continuity that we used with the normal approximation to the binomial distribution (see Formula 19.3). When this correction, known as **Yates's correction for continuity,** is added to Formula 20.1, the result is Formula 20.4, as shown below.

$$\chi^2 = \sum \frac{(|f_o - f_e| - .5)^2}{f_e} \qquad \text{Formula 20.4}$$

Note that the vertical bars surrounding $f_o - f_e$ indicate that the absolute value (i.e., the size of the difference, ignoring the sign) is to be taken *before* subtracting .5. The corrected difference is then squared and divided by f_e. Applying Formula 20.4 to the data in Table 20.7, we obtain the following:

$$\chi^2 = \frac{(|9 - 5| - .5)^2}{5} + \frac{(|6 - 10| - .5)^2}{10} + \frac{(|11 - 15| - .5)^2}{15} + \frac{(|34 - 30| - .5)^2}{30}$$

$$= \frac{3.5^2}{5} + \frac{3.5^2}{10} + \frac{3.5^2}{15} + \frac{3.5^2}{30} = 2.45 + 1.22 + .82 + .41 = 4.90$$

The effect of Yates's correction is to make χ^2 smaller, and therefore it reduces the power of the test. It is important to note that some statisticians think that the correction is too conservative, especially in research applications such as the example above; I have included the correction here for completeness, and because it is still commonly used in the 2×2 case. For large samples the continuity correction can safely be ignored.

Step 6. Make the Statistical Decision

The calculated value of χ^2 (4.9) is larger than the critical value (3.84), and therefore it lands in the region of rejection. We can reject the null hypothesis that the two variables (corporal punishment and playground aggression) are independent. From the cell frequencies in Table 20.7 we can see that although the majority of boys do not initiate fights (even within the corporally punished group), the proportion of boys starting fights is larger among those who were corporally punished than among those who were not. However, just because the two variables are significantly related, we cannot conclude that the relation is a strong one. A tiny association between the two variables can lead to a statistically significant result, if the sample size is sufficient. Methods for assessing the degree of association between the two variables are described in Section C.

Shortcut Formulas for 2 × 2 Tables

You may have noticed some redundancy in the calculation of the χ^2 statistic above: the numerator was the same value (i.e., 3.5) for each cell. This is not a coincidence; this redundancy occurs whenever there is only one degree of freedom. In the case of a 2 × 2 table, the redundancy leads to a simplified formula for calculating χ^2, based on the cell frequencies. If the cell frequencies are labeled with letters, as in Table 20.8, the χ^2 formula reduces to Formula 20.5.

$$\chi^2 = \frac{N(|ad - bc| - N/2)^2}{(a + b)(c + d)(a + c)(b + d)} \qquad \text{Formula 20.5}$$

The term $N/2$ is subtracted in the numerator to correct for (lack of) continuity. Applying Formula 20.5 to the data in Table 20.6, we obtain

$$\chi^2 = \frac{60(|306 - 66| - 60/2)^2}{(15)(45)(20)(40)} = \frac{60(240 - 30)^2}{540,000} = \frac{2,646,000}{540,000} = 4.9$$

The value for χ^2 obtained from Formula 20.5 is the same, of course, as the value that we obtained from Formula 20.4. The reduction in computational effort may not seem like much, but if you needed to do a whole series of 2 × 2 tests by hand, the shortcut provided by Formula 20.5 could be helpful. (Notice that Formula 20.5 eliminates the need to calculate the expected frequencies.)

Table 20.8		
a	b	
c	d	

Assumptions of the Chi-Square Test

The chi-square statistic will not follow the chi-square distribution exactly unless the sample size is infinitely large. The chi-square distribution can be a reasonably good approximation, however, provided that the following assumptions have been met. These assumptions apply to both goodness-of-fit tests and tests of association.

Mutually Exclusive and Exhaustive Categories

The categories in the chi-square test must be exhaustive, so that every observation falls into one category or another (this may require an "other" category in which to place subjects not easily classified). The categories must also be mutually exclusive, so that no observation can simultaneously fall in more than one category.

Independence of Observations

We assume that observations are independent, just as we did with respect to random samples for parametric tests. This assumption is usually violated when the same subject is categorized more than once. For instance, suppose that men and women are rating romantic movies as "thumbs up" or "thumbs down." If five movies are being judged by a total of ten men and ten women, and each subject judges all five movies, then there will be a total of 100 observations (20 subjects $\times$ 5 movies). However, it is not likely that all 100 observations will be mutually independent; a subject who hates romantic movies may tend to judge all five movies harshly. Although this will not necessarily be the case, it is safest to ensure that each subject contributes only one observation (i.e., the total number of observations equals the number of different subjects).

Size of Expected Frequencies

In order for the chi-square distribution to be a reasonably accurate approximation of the chi-square statistic, the expected frequency for each cell should be at least 5. By this criterion the corporal punishment example (see Table 20.7) is an acceptable candidate for the chi-square test. Bear in mind, however, that some conservative statisticians think that when there is only one degree of freedom, an expected frequency of 10 for each cell is required for the chi-square test to be sufficiently accurate. By this stricter criterion the chi-square test should not be used for the data in Table 20.7. On the other hand, for studies with many categories, and therefore many degrees of freedom, it is acceptable for just a few categories to have a small f_e, but in no case should an f_e be less than 1. A common rule of thumb is to make sure that no f_e is less than 1, and not more than 20% of the f_e are less than 5. In the special case of a 2×2 table with small f_e, the chi-square approximation can be avoided by using an exact multinomial test. This option is discussed further in Section C.

When to Use the Chi-Square Test for Independence

The chi-square test for independence can be planned when you want to see the relationship between two variables, both of which are being measured on a categorical scale (e.g., psychiatric diagnosis and season of birth). The two-variable chi-square test can also be used for a study originally designed to be analyzed by a *t*-test or a one-way ANOVA, if the distribution of the dependent variable is very far from normal. For instance, if a floor or ceiling effect causes most of the scores to have the same low or high value (as with the number of fights initiated in the corporal punishment example), it may be reasonable to convert the original multivalued scale to a few broad categories. Bear in mind that this usually results in a considerable loss of power, so you should only change to a categorical scale when the distribution of values is too far from the normal distribution to permit the use of parametric statistics.

Correlational Research

In correlational research, one or both of the categorical variables can be grouping variables. In the corporal punishment example the punishment variable was based on preexisting differences—the researcher made no attempt to influence punishment practices. (Thus the researcher could draw no conclusions about corporal punishment *causing* aggression.) The other variable in that example involved behavior observed in the playground. For an example involving two grouping variables, imagine testing the hypothesis that agoraphobia (fear of leaving home, of being in a crowd, etc.) is associated with the early loss of a loved one. A group of agoraphobics and a comparison group of persons with some other form of neurosis (preferably from the same mental health clinic) are obtained. Each patient is then categorized as having lost a loved one or not, thus forming a 2×2 contingency table. In this example, the marginal frequencies for one of the variables would be determined beforehand; the experimenter would decide how many agoraphobics and how many comparison subjects to select. In some studies, all the marginal frequencies are free to vary. For example, suppose some gene is found to be associated with schizophrenia. A sample of the general population can be selected and then the individuals can be categorized along two dimensions: whether they possess the gene in question or not and whether they have a schizophrenic family member or not.

Experimental Research

In experimental research, one of the variables can be a truly independent variable (i.e., manipulated by the experimenter); the other variable is a dependent variable consisting of distinct categories. For instance, a social psychologist may be exploring the factors that influence "helping behavior"—specifically, the factors that make it more likely that a person finding a wallet on the street will return it to the owner. In one experiment, the factor may be the presumed income of the wallet's owner. The same wallet is used each time (containing the same modest sum of money in each case); the difference is that the card inside the wallet that shows the address of the owner indicates either a low-income, a middle-income, or a high-income section of town. For each of the three addresses the psychologist could record the number of wallets returned with the money, without the money, or not at all. The results would form a 3×3 contingency table.

Publishing the Results of a Chi-Square Test

The results of the corporal punishment experiment could be reported in the following manner: "The 2×2 contingency table revealed a statistically significant association between the use of corporal punishment at home and the tendency of boys to initiate fights in the playground, $\chi^2 (1, \underline{N} = 60) = 4.9, \underline{p} < .05$." (Notice that unlike t or F, χ^2 is not underlined; according to APA style all statistical symbols are to be underlined except for Greek letters.)

The first number in the parentheses following χ^2 is the number of degrees of freedom associated with the chi-square statistic (for this example, df = 1). The second number is the sample size. The number of degrees of freedom helps to determine the critical value needed for statistical significance; the sample size allows the reader to compute a measure of the strength of association between the two variables. The latter measure provides important information that is separate from your decision about the null hypothesis. Measures of strength of association in a contingency table are discussed in Section C.

The two-way chi-square test appears very frequently in the psychological literature; the example that follows shows how an experimental design that might have led to a t-test produced data for which a chi-square test was more appropriate. Schwartz, Slater, and Birchler (1994) randomly assigned half of a group of 34 chronic back pain patients to a stressful interview condition and the other half to a neutral conversation condition. After the stressful or neutral condition, all subjects were asked to pedal an exercise bicycle at a steady rate for 20 minutes but were instructed to stop if they experienced considerable back pain. Subjects who had been in the stressful condition were expected to stop after fewer minutes than subjects who had been in the control group. If only a few subjects had pedaled the full 20 minutes, a t-test could have been performed on the number of minutes pedaled by subjects in the two groups. However, the results did not lend themselves to such an analysis, as described in the following excerpt.

> Examination of the bicycling data revealed a skewed distribution that was due to a ceiling effect (i.e., 53% of the subjects persisted in the bicycle task for the maximum amount of time). The persistence data were dichotomized, and a contingency table analysis was performed to determine whether there were significant differences in the proportion of patients in each interview group who persisted in the physical activity task. A significantly greater proportion of patients in the stress interview condition terminated the bicycling task prematurely ($n = 11$), compared with patients in the neutral talking condition ($n = 5$), $\chi^2 (1, N = 34) = 4.25, p < .05$.

Exercises

1. Is one's personality related to one's choice of college major? To address one aspect of this question, 25 physics majors, 35 literature majors, and 45 psychology majors were tested for introversion/extraversion. The results were as follows for each major: physics, 15 introverts and 10 extroverts; literature, 17

introverts and 18 extroverts; psychology, 19 introverts and 26 extroverts. Display these results in a 3 × 2 contingency table and perform the appropriate chi-square test.

* 2. Does a diet high in sugar cause hyperactivity in young children? Ten fourth-graders whose parents admit to providing a high-sugar diet and 20 fourth-graders on a more normal diet are rated for hyperactivity in the classroom. The data are shown in the contingency table below.

	Normal Diet	High-Sugar Diet
Hyperactive	2	4
Not Hyperactive	18	6

a) Perform the chi-square test (use Formula 20.4) and decide whether the null hypothesis (i.e., hyperactivity is not associated with dietary sugar) can be rejected at the .05 level.
b) Recalculate the value of χ^2 using Formula 20.5, and check that the value is the same.
c) If the null hypothesis were rejected, could you conclude that a large amount of sugar can *cause* hyperactivity? Why or why not?

* 3. In spite of what may seem obvious, medical research has shown that being cold is not related to catching a cold. Imagine that in one such study, 40 subjects are exposed to cold viruses. Then half of the subjects are randomly assigned to remain in a cold room for several hours, and the other half remain in a room kept at ordinary room temperature. At the end of the experiment, the numbers of subjects who catch a cold are as shown in the following table.

	Cold Room	Normal Room
Caught Cold	14	12
No Cold	6	8

a) Test the null hypothesis ($\alpha = .05$) that being cold is unrelated to catching a cold.

b) Given your statistical decision in part a, which type of error (Type I or Type II) could you be making?

* 4. A social psychologist is studying whether people are more likely to help a poor person or a rich person who has fallen down. The three conditions all involve an elderly woman who falls down in a shopping mall (when only one person at a time is nearby). The independent variable concerns the apparent wealth of the woman; she is dressed to appear either poor, wealthy, or middle class. The reaction of each bystander is classified in one of three ways: ignoring her; asking if she is all right; and helping her to her feet. The data appear in the contingency table below. Test the null hypothesis at the .01 level. Is there evidence for an association between the apparent wealth of the victim and the amount of help provided by a bystander?

	Poor	Middle Class	Wealthy
Ignore Her	16	10	7
Talk to Her	8	6	5
Help Her Up	6	14	18

* 5. Is there a connection between a person's position on the death penalty and his or her opinion about gun control? Fifty people are polled about their views on both issues. The data appear below. Calculate χ^2 for these data. Can you conclude that opinions on the two issues are not independent?

	Favors No Gun Control	Favors Some Gun Control	Favors Strict Gun Control
In Favor of the Death Penalty	8	12	7
Opposed to the Death Penalty	4	10	9

6. In exercise 20A6, the responses of 50 subjects to a logic problem were compared to expected frequencies from extensive prior research. Suppose instead that the experiment were conducted with two groups of 50 subjects each and that one group is presented with the problem in the traditional (abstract) way, whereas the other group receives the more concrete presentation. Suppose that the data for the concrete subjects are the same as given in exercise 20A6, and that the abstract subjects perform exactly according to the expectations stated in that exercise, as shown below:

	A	B	C	D
Abstract	20	13	10	7
Concrete	24	8	16	2

a) Calculate χ^2 for the data above. Can you reject the null hypothesis that type of presentation and type of response to the problem are independent?

b) Compare the value for χ^2 in this problem with the value you calculated for exercise 20A6. Can you see the advantage of having expectations based on a population, instead of another group of the same size?

*7. The use of the polygraph for lie detection remains controversial. In a typical laboratory experiment to test the technique, half the subjects are told to commit a mock crime (to remove a camera from a cabinet); the other half of the subjects remain "innocent." The polygrapher, blind to each subject's condition, must make a judgment of "guilt" or "innocence" based on a subject's physiological responses to a series of questions. The hypothetical data for one such study are shown below. Test the null hypothesis ($\alpha = .05$) that the polygrapher's judgments are unrelated to the subject's "guilt" (i.e., that this form of lie detection is totally ineffective).

	"Innocent"	"Guilty"
Judged Innocent	10	3
Judged Guilty	5	12

8. In exercise 14B4, the dependent variable was the amount of time a subject listened to tape-recorded cries for help from the next room before getting up to do something. If some subjects never respond within the time allotted for the experiment, the validity of using parametric statistical techniques could be questioned. As an alternative, subjects could be classified as fast responders and slow responders (and possibly, nonresponders). The data from exercise 14B4 were used to classify subjects as fast responders (less than 12 seconds to respond) or slow responders (12 seconds or more). The resulting contingency table is shown below.

	Child's Voice	Adult Female's Voice	Adult Male's Voice
Fast Responder	5	3	1
Slow Responder	2	4	6

a) Test the null hypothesis ($\alpha = .05$) that speed of response is independent of type of voice heard.

b) How does your conclusion in part a compare with the conclusion you drew in exercise 14B4? Categorizing the dependent variable throws away information; how do you think that loss of information affects power?

*9. The director of a mental health clinic, which trains all of its therapists in a standardized way, is interested in knowing whether some of its therapists are more effective than others. Twenty patients (carefully matched for severity of problem, age, and other relevant factors) are assigned to each of five therapists for a 6-month period, at the end of which each patient is categorized as "improved," "no change," or "worse." The data appear below. Test the null hypothesis ($\alpha = .01$) that the assignment of a particular therapist is unrelated to the patient's progress.

	Dr. A	Dr. B	Dr. C	Dr. D	Dr. E
Improved	15	11	16	13	10
No Change	5	3	0	4	6
Worse	0	6	4	3	4

10. A magazine publisher is testing four different covers for her anniversary issue and is interested to

know if each one appeals equally to people of both genders. Fifty subjects are brought in, and each selects the cover that would most draw him or her to buy the magazine. The number of people of each gender who chose each cover are shown in the table below.

Gender of Subject	Cover I	Cover II	Cover III	Cover IV
Female	12	10	5	3
Male	5	9	1	5

a) Test the null hypothesis ($\alpha = .05$) that cover preference is independent of gender.

b) If you rejected the null hypothesis in part a, test the preferences of each gender separately (i.e., perform two one-way chi-square tests) to see if the preferences are equal within each gender. If you did not reject the null hypothesis in part a, ignore gender (i.e., sum the frequencies for each cover for the two genders) and perform a one-way chi-square test to see if the four covers are equally preferred.

OPTIONAL MATERIAL

Measuring Strength of Association

When you perform a chi-square test you want the value for χ^2 to be high—at least high enough to be statistically significant. But what does it mean if χ^2 is extremely high? A very high χ^2 value tells you that your p level is very small, which implies that your results would be significant even if you chose an unusually small alpha level. However, as with the t and F statistics, the size of the chi-square statistic from a two-way test tells you nothing about the strength of the association between the two variables. A very weak dependency between two variables can lead to a large χ^2 value if the sample size (N) is sufficiently large. In fact, if you multiply each cell frequency in a contingency table by the same constant, the chi-square statistic will be multiplied by that constant. This principle also applies to the one-way test. For instance, if flipping a fair coin ten times produces seven heads and three tails, χ^2 will equal a nonsignificant 1.6 (even less with Yates's correction). However, if you obtain the same proportions of heads and tails in 100 flips (i.e., 70H, 30T), the χ^2 value will be 16 ($p < .001$). It would be helpful to supplement the χ^2 value with a measure of strength of association (or an estimate of effect size) that is not affected by sample size. A number of such measures have been devised; the most widely used of these are described below.

Phi Coefficient

In the case of a 2×2 contingency table, there is a very natural choice for a measure of association: Pearson's r. Recall that in Chapter 12, Section C I described how to calculate a Pearson correlation coefficient when you have two distinct groups of scores (i.e., when one variable consists of only two categories and the other variable is continuous). The two levels of the grouping variable can be assigned any two values (0 and 1 are the most convenient), and the ordinary correlation formula can then be applied. The result is often called the point-biserial correlation (r_{pb}) as a reminder that one of the variables has only two values. A similar procedure can be applied to a 2×2 contingency table. In this case, both variables are dichotomous, so both can be assigned the arbitrary values of 0 and 1. Then any one of the Pearson correlation formulas can be applied. The result is called the **phi (φ) coefficient,** or sometimes, the **fourfold point correlation.**

If you have already found the χ^2 value, you do not have to calculate φ from scratch. The phi coefficient is a simple function of both the chi-square statistic and the total sample size, as given by Formula 20.6.

$$\phi = \sqrt{\frac{\chi^2}{N}}$$ Formula 20.6

The formula above should remind you of Formula 12.13 for finding r_{pb} from a two-group t value. Note that multiplying every cell frequency by the same constant results in N being multiplied by that constant, as well as χ^2. In Formula 20.6 both the numerator and the denominator would be multiplied by the same constant, leaving the overall value unchanged. Thus increasing the sample size without changing the relative proportions does not affect the phi coefficient. Also note that in a 2×2 table, χ^2 can never exceed N (otherwise φ would exceed 1.0); χ^2 will equal N only if all of the frequencies fall in two diagonal cells and the other two cells are empty.

Applying Formula 20.6 to the corporal punishment example in Section B, we find

$$\phi = \sqrt{\frac{4.9}{60}} = \sqrt{.082} = .286$$

The general practice is to take the positive square root in Formula 20.6, so that phi ranges between 0 and +1.0. However, it is possible in some cases to assign 0 and 1 to each variable in a meaningful way (e.g., absence and presence of alcoholism in a subject and the subject's parents) so that when calculated directly from the raw data, the sign of the phi coefficient is meaningful (e.g., a positive phi indicates that alcoholic parents are associated with children who become alcoholics, whereas a negative phi would indicate that children of alcoholics are less likely to become alcoholics). The sign of the phi coefficient becomes irrelevant, though, when strength of association is measured by squaring phi. You may recall that squaring a correlation coefficient is useful because it tells you the proportion of variance accounted for. Squaring phi gives you an analogous measure; squaring both sides of Formula 20.6 results in Formula 20.7 for phi squared, as follows:

$$\phi^2 = \frac{\chi^2}{N}$$ Formula 20.7

The shortcut formula for finding χ^2 from a 2 × 2 contingency table (Formula 20.5) can be inserted in Formula 20.7 above to create a shortcut formula for finding phi squared:

$$\phi^2 = \frac{(|ad - bc| - N/2)^2}{(a + b)(c + d)(a + c)(b + d)} \qquad \text{Formula 20.8}$$

The letters in Formula 20.8 correspond to the cell frequencies of the 2 × 2 table, as shown in Table 20.8. Applying Formula 20.8 to the corporal punishment example, we obtain

$$\phi^2 = \frac{(|306 - 66| - 60/2)^2}{(15)(45)(20)(40)} = \frac{(240 - 30)^2}{540,000} = \frac{44,100}{540,000} = .082$$

If we take the square root of phi squared, as found above, we get .286 for the value of phi, which agrees with the result we obtained using Formula 20.6.

Cramér's Phi and the Contingency Coefficient

The value of ϕ^2 provides useful information to accompany χ^2; the latter allows you to determine statistical significance, whereas the former tells you the degree of association between the two variables in your present sample. It would be nice to have this supplemental information for chi-square problems that do not involve a 2 × 2 contingency table. Fortunately, only a slight modification of Formula 20.6 is needed to create a phi coefficient that is appropriate for any contingency table. Actually, no modification of that formula is required if one of the variables has only two levels, but if both variables have more than two levels, Cramér (1946) suggested using the following statistic,

$$\phi_c = \sqrt{\frac{\chi^2}{N(L - 1)}} \qquad \text{Formula 20.9}$$

where L is either the number of rows or the number of columns, whichever is smaller. This statistic is often called **Cramér's phi,** and like ordinary phi it ranges from 0 to 1.0. A related statistic, called the **contingency coefficient,** is given by Formula 20.10 below:

$$C = \sqrt{\frac{\chi^2}{\chi^2 + N}} \qquad \text{Formula 20.10}$$

However, because C does not generally range from 0 to 1, statisticians find that Cramér's phi is preferable for descriptive purposes.

The Cross-Product Ratio (or Odds Ratio)

A strength of association measure for the 2 × 2 table that is commonly employed in biological and medical research is the *cross-product ratio,* also called the *odds ratio.* This is a very simple measure that consists of the ratio between the products of the diagonally opposite cells in the 2×2 table. In terms of the letter designations in Table 20.8, Formula 20.11 for the cross-product ratio is

$$\text{Cross-product ratio} = \frac{ad}{bc} \qquad \text{Formula 20.11}$$

For the corporal punishment example in Section B, the cross-product ratio is

$$\frac{ad}{bc} = \frac{(9)(34)}{(6)(11)} = \frac{306}{66} = 4.636$$

The ratio can vary from zero to an infinitely high number. Numbers much larger or smaller than 1.0 represent a strong degree of association; a ratio of 1.0 indicates a total lack of association between the variables. As applied to the corporal punishment example, the cross-product ratio compares the odds of initiating fights given corporal punishment (9/11) with the odds of initiating fights given no corporal punishment (6/34)—hence the alternative term for this measure, the odds ratio. In medical applications, you might want to know the odds of getting positive test results if you have the disease being tested for, compared to the odds of getting false positive results. Unless the odds ratio is very high, it might not be worthwhile to take the test.

Power as a Function of Strength of Association

As in the *t*-test or ANOVA, the power of a two-way chi-square test depends on both a measure of effect size (or the strength of the association between the variables) and the size of the sample. For a given strength of association, the sample size required to achieve a particular level of power can be estimated using tables prepared for that purpose (Cohen, 1988).

Fisher's Exact Test

It is important to remember that in normal practice, the chi-square statistic does not follow the chi-square distribution exactly, and the use of the chi-square distribution as an approximation is not recommended when one or more of the f_e is too small. A similar problem occurs with using the normal approximation to the binomial distribution. For small sample sizes (especially less than 20), it is recommended that you use the exact binomial distribution (Table A.12). Calculating exact probabilities is also recommended for 2×2 contingency tables when one or more of the f_e is less than 5. These calculations involve permutations (see Chapter 19, Section C) and are quite tedious. Moreover, as with the binomial distribution, you must find the probability corresponding to your data *and* for all (2×2) tables that exhibit a greater degree of association in the same direction. This procedure is known as **Fisher's exact test.** Fortunately, this test can be performed by computer, so I will spare you the complex mathematical details. There is some debate as to whether this test is overly conservative, but for very small sample sizes it has the distinct advantage of making no assumption that the chi-square distribution can be used as a reasonable approximation.

For tables with more than two rows or columns, in which f_e is below 5 for one or more cells, an exact test might be desirable, but it is unfortunately not feasible. The standard recommendation is to collapse categories to form a smaller table with larger f_e. However, the danger is that this restructuring will be performed in an arbitrary way that capitalizes on chance and leads to more Type I errors than your alpha level would suggest. Therefore, when collapsing categories, you should avoid the temptation to "shop around" for the scheme that gives you the highest χ^2. (Any further discussion of when and how to collapse categories is beyond the scope of this text.)

Contingency Tables Involving More Than Two Variables

If you can only look at two variables at a time, the research questions you can ask are quite limited. You saw in Chapter 16, for instance, how two independent variables can interact in their influence on a third, dependent variable. A similar situation can arise with three categorical variables. For example, the dependent variable of interest might be whether a subject returns for a second session of an experiment; this variable is categorical and has only two levels: returns and does not return. If we wanted to know the effect of the experimenter's behavior on the subject's return, we could have the experimenter treat half the subjects rudely and half the subjects politely. (Subjects would be run individually, of course.) The results would fit in a 2 × 2 contingency table.

A more interesting question, however, would explore the role of self-esteem in the relationship between experimenter behavior and subject return. Research in social psychology has shown that subjects with low self-esteem are more helpful to a *rude* experimenter, whereas the reverse is true for subjects with high self-esteem (e.g., Lyon & Greenberg, 1991). Repeating the 2 × 2 contingency table once for a group of low self-esteem subjects and once again for a group of high self-esteem subjects produces a three-variable contingency table. Rather than trying to represent a cube on a flat page, it is easier to display the 2 × 2 × 2 contingency table as two 2 × 2 tables side by side, as depicted in Table 20.9.

Table 20.9

	Low Self-Esteem		High Self-Esteem	
	Polite Experimenter	Rude Experimenter	Polite Experimenter	Rude Experimenter
Return				
Don't Return				

The major problem in applying the chi-square test to a contingency table with three or more variables is finding the appropriate expected frequencies. The trick we used in Section B is not valid when there are more than two variables. The solution that has become very popular in recent years is to use a **log-linear model.** A discussion of the log-linear model is beyond the scope of this text; I will just point out that once the expected frequencies have been estimated with such a model, the ordinary chi-square test (or the increasingly popular *likelihood ratio test*) can then be applied. Although log-linear analyses are now performed by most of the major computerized statistical packages, an advanced understanding of statistics is required to use these techniques properly.

Exercises

*1. a) Calculate the phi coefficient for the data in exercise 20B2, using the χ^2 value that you found for that exercise.

   b) Calculate ϕ again using the shortcut formula (i.e., take the square root of Formula 20.8).

   c) What can you say about the strength of the relationship between dietary sugar and hyperactivity in the data of exercise 20B2?

2. a) Calculate the phi coefficient for the data in exercise 20B3, using the χ^2 value that you found for that exercise.

   b) Calculate the cross-product ratio for the same data.

   c) What can you say about the strength of the relationship between being cold and catching cold in the data of exercise 20B3?

*3. a) Calculate the phi coefficient for the data in exercise 20B7.

   b) Calculate the cross-product ratio for the same data.

   c) What do these measures tell you about the accuracy of the lie detector test in exercise 20B7?

4. a) Calculate both Cramér's phi and the contingency coefficient for the data in exercise 20B4.

   b) What can you say about the degree of relationship between the apparent wealth of the person in need and the amount of help given in that hypothetical experiment?

*5. a) Calculate both Cramér's phi and the contingency coefficient for the data in exercise 20B9.

   b) What can you say about the degree of relationship between the therapist assigned and the amount of improvement in the patient?

6. Devise an experiment whose results could be represented by a three-way contingency table.

SUMMARY

The Important Points of Section A

1. When dealing with more than two categories, the probability that a sample of a certain size will break down into those categories in a particular way depends on the appropriate *multinomial distribution*.

2. Because it is tedious to find exact multinomial probabilities, the *chi-square* (χ^2) *distribution* is used as an approximation. The number of degrees of freedom for a one-variable chi-square test is one less than the number of categories. Knowing df and alpha, you find the critical value of χ^2 in Table A.13.

3. *Pearson's chi-square statistic* is based on finding the difference between the expected and observed frequencies for each category, squaring the difference, and dividing by the expected frequency. The sum over all categories is the χ^2 statistic, which follows the chi-square distribution (approximately) when the null hypothesis is true.

4. Large discrepancies between expected and observed frequencies produce large values of χ^2, which fall in the positive (right) tail of the distribution, and if large enough, can lead to rejection of the null hypothesis. Unusually small discrepancies can produce χ^2 values near zero, which fall in the negative (left) tail, but regardless of how small, never lead to rejection of the null hypothesis.

5. When you are testing to see whether subjects have any preference among the categories available, the null hypothesis is that the categories will be selected equally often (i.e., subjects will have no preference). The expected frequencies are found in this case by dividing the sample size by the number of categories.

6. The chi-square test involving categories of only one variable is often called a *goodness-of-fit* test; the frequencies of values in a sample may be tested to see how well they fit a hypothesized population distribution.

The Important Points of Section B

The example discussed in Section B was correlational; the experimenter noted which families practiced corporal punishment but made no attempt to control punishment practices. In this summary section I will discuss a design in which one of the variables is actually manipulated by the experimenter. Also, I will extend the chi-square test for independence to a table larger than 2×2.

Imagine that a psychiatrist has been frustrated in her attempts to help chronic schizophrenics. She designs an experiment to test four therapeutic approaches to see if any treatment is better than the others for improving the lives of her patients. The four treatments are: intense individual psychodynamic therapy; constant unconditional positive regard and Rogerian therapy; extensive group therapy and social skills training; a token economy system. The dependent variable is the patient's improvement over a 6-month period, measured in terms of three categories: became less schizophrenic (improved); became more schizophrenic (got worse); or showed no change. Eighty schizophrenics meeting certain criteria (not responsive to previous treatment, more than a certain number of years on the ward, etc.) are selected and then assigned at random to the four treatments, with the constraint that 20 are assigned to each group. After 6 months

	Psycho-dynamic Therapy	Rogerian Therapy	Group Therapy	Token Economy	Row Sums
Improved	6	4	8	12	30
Did Not Change	6	14	3	5	28
Got Worse	8	2	9	3	22
Column Sums	20	20	20	20	80

Table 20.10

Observed Frequencies

of treatment, each patient is rated as having improved, having gotten worse, or having remained the same. The data are displayed in Table 20.10, a 3 × 4 contingency table.

First we must find the expected frequencies for each of the 12 cells in the contingency table. Each f_e can be calculated using Formula 20.3, but fortunately, not all 12 cells need to be calculated in this way. Finding the number of degrees of freedom associated with Table 20.10 tells us how many f_e must be calculated.

$$df = (R - 1)(C - 1) = (3 - 1)(4 - 1) = (2)(3) = 6$$

Because df = 6, we know that only six of the f_e are free to vary; the remaining cells can be found by subtraction (within each row and column, the f_e must add up to the same number as the f_o). However, if you want to save yourself some calculation effort in this way, you will have to choose the right six cells to calculate; these are shown in Table 20.11.

To illustrate how the f_e in Table 20.11 were calculated, I will use Formula 20.3 to find the f_e for the "Rogerian Therapy, Did Not Change" cell.

$$f_e = \frac{(\text{Row sum})(\text{Column sum})}{N} = \frac{(28)(20)}{80} = \frac{560}{80} = 7$$

Now we can find the remaining f_e in Table 20.11 by subtraction. First, find the f_e for the "Got Worse" cells in the first three columns (subtract the other two f_e

	Psycho-dynamic Therapy	Rogerian Therapy	Group Therapy	Token Economy	Row Sums
Improved	7.5	7.5	7.5		30
Did Not Change	7	7	7		28
Got Worse					22
Column Sums	20	20	20	20	80

Table 20.11

Expected Frequencies

Table 20.12					
	Psycho-dynamic Therapy	Rogerian Therapy	Group Therapy	Token Economy	Row Sums
Improved	6 (7.5)	4 (7.5)	8 (7.5)	12 (7.5)	30
Did Not Change	6 (7)	14 (7)	3 (7)	5 (7)	28
Got Worse	8 (5.5)	2 (5.5)	9 (5.5)	3 (5.5)	22
Column Sums	20	20	20	20	80

from 20). Then, find each f_e in the "Token Economy" column by subtracting the other three treatments from each row sum. The 3×4 contingency table with the observed frequencies and all of the expected frequencies (in parentheses) is shown as Table 20.12.

We are now ready to apply Formula 20.1 to the data in Table 20.12 to find χ^2.

$$\chi^2 = \sum \frac{(f_o - f_e)^2}{f_e} = \frac{(6 - 7.5)^2}{7.5} + \frac{(4 - 7.5)^2}{7.5} + \frac{(8 - 7.5)^2}{7.5} + \frac{(12 - 7.5)^2}{7.5}$$

$$+ \frac{(6 - 7)^2}{7} + \frac{(14 - 7)^2}{7} + \frac{(3 - 7)^2}{7} + \frac{(5 - 7)^2}{7}$$

$$+ \frac{(8 - 5.5)^2}{5.5} + \frac{(2 - 5.5)^2}{5.5} + \frac{(9 - 5.5)^2}{5.5} + \frac{(3 - 5.5)^2}{5.5}$$

$$= .3 + 1.63 + .03 + 2.7 + .143 + 7 + 2.29 + .57$$

$$+ 1.14 + 2.23 + 2.23 + 1.14 = 21.4$$

From Table A.13, we see that the critical value of χ^2 for df = 6 and α = .05 is 12.59. Because the calculated χ^2 (21.4) is greater than the critical χ^2, we can reject the null hypothesis and conclude that the tendency toward improvement is not independent of the type of treatment; that is, the various treatments differ in the proportion of the population that experience improvement, show no change, or become worse.

Assumptions of the Chi-Square Test

Mutually exclusive and exhaustive categories. Each observation falls into one, and only one, category.

Independence of observations. Usually this assumption is satisfied by having each frequency count represent a different subject (i.e., each subject contributes only one observation).

Size of expected frequencies. The rule of thumb is that no f_e should be less than 5. This rule is relaxed somewhat if there are many categories (as long as f_e is never less than 1 and no more than 20% of the f_e are less than 5). It becomes more stringent if df = 1 (when df = 1, f_e should be at least 10).

When to Use the Chi-Square Test for Independence

Two-variable chi-square tests are generally used with one of three types of experimental designs:

1. *Two grouping variables* (e.g., proportions of left-handed and right-handed persons in various professions).
2. *One grouping variable* (e.g., proportions of babies from different cultures expressing one of several possible emotional reactions to an experimental condition, such as removing a toy).
3. *One experimentally manipulated variable* (e.g., exposing children to violent, neutral, or peaceful cartoons and categorizing their subsequent behavior as aggressive, neutral, or cooperative).

The Important Points of Section C

1. The size of the chi-square statistic in a two-variable design does not tell you the degree to which the two variables are related. A very weak association can produce a large value for χ^2 if the sample size is very large.

2. For a 2×2 contingency table, the strength of association can be measured by assigning arbitrary numbers (usually 0 and 1) to the levels of each variable and then applying the Pearson correlation formula. The result is called the *phi coefficient* or *the fourfold point correlation.*

3. Like other correlation coefficients, phi ranges from -1 to 0 to $+1$; however, the sign of phi is not always meaningful. If you are not concerned about the sign, you can find phi easily by taking the positive square root of χ^2 divided by N (the sample size). An alternative measure of association for 2×2 tables, often used in medical research, is the *cross-product ratio* or *odds ratio.* It is the product of the frequencies in one set of diagonally opposite cells divided by the product of the other diagonal cells.

4. For contingency tables larger than 2×2, a modified phi coefficient, known as *Cramér's phi,* is recommended as a measure of association. Cramér's phi is preferred to a related statistic, the *contingency coefficient,* because the former always ranges from 0 to 1, whereas the latter need not.

5. For 2×2 tables in which the expected frequencies are too low to justify the chi-square approximation, the probability of the null hypothesis producing the observed results (or more deviant results) can be calculated exactly using *Fisher's exact test.* For tables larger than 2×2 with small f_e, collapsing categories can solve the problem, but in so doing you must be cautious not to capitalize on chance.

6. If more than two categorical variables are included in a contingency table, determining the expected frequencies is not straightforward. The f_e can be estimated, however, using the *log-linear model*. The analysis is completed by using the ordinary chi-square test or the *likelihood ratio test*.

Definitions of Key Terms

Multinomial distribution A discrete mathematical distribution that gives the probability that a sample of N observations will break down into the available categories with a certain frequency for each. The multinomial distribution is a function of N, the number of categories, and the relative probabilities of each category.

Chi-square distribution A continuous mathematical distribution that is positively skewed, but whose shape depends on the number of degrees of freedom. (The chi-square distribution becomes identical to the normal distribution when the df are infinite.)

Pearson's chi-square statistic A measure of the discrepancy between expected and obtained frequencies in a sample. It follows one of the chi-square distributions approximately when the null hypothesis is true and certain assumptions are met.

Goodness-of-fit test A one-variable chi-square test; it measures the fit between the frequencies of different categories in a sample and the frequencies of those categories in a real or hypothetical population.

Contingency table A table that displays the frequencies of joint occurrence for the categories of two or more variables; in the two-variable case, each category of one variable is divided into all the categories of the other variable, and the table shows the joint frequency for each cell of the two-way matrix.

Pearson's chi-square test of association *(also test for independence)* A test in which the expected frequencies for a two-way contingency table are found based on the marginal sums and the assumption that the two variables are independent. Then Pearson's chi-square statistic is applied to measure the discrepancy between the expected and observed frequencies.

Yates's correction for continuity A procedure in which a half unit (.5) is subtracted from the absolute value of the difference between an expected and an observed frequency to reduce the discrepancy between the (discrete) distribution of the chi-square statistic and the corresponding (continuous) chi-square distribution when there is only one degree of freedom.

Phi (φ) coefficient (or fourfold point correlation) A Pearson correlation coefficient that is calculated for two variables that consist of only two values each. It measures the strength of association of the two variables.

Cramér's phi A measure of strength of association that ranges from 0 to 1 and can be applied to contingency tables larger than 2×2.

Contingency coefficient A traditional measure of strength of association that can be applied to contingency tables larger than 2×2; it is not preferred by statisticians because its maximum value is generally less than 1.0.

Fisher's exact test A procedure for determining the exact probability of obtaining particular frequencies in a 2×2 table (no chi-square approximation is used). It is used only when one or more of the expected frequencies is so small that the chi-square approximation is not appropriate.

Log-linear model A model used to estimate the expected frequencies when more than two categorical variables have been combined into a single multi-way contingency table.

Key Formulas

Pearson chi-square statistic for df > 1:

$$\chi^2 = \sum \frac{(f_o - f_e)^2}{f_e}$$

Formula 20.1

Degrees of freedom for a two-variable contingency table:

$$df = (R - 1)(C - 1)$$

Formula 20.2

Expected frequencies in a two-variable contingency table:

$$f_e = \frac{(\text{Row sum})(\text{Column sum})}{N}$$

Formula 20.3

Pearson chi-square statistic for df = 1:

$$\chi^2 = \sum \frac{(|f_o - f_e| - .5)^2}{f_e}$$

Formula 20.4

Shortcut formula for the chi-square statistic for a 2×2 table:

$$\chi^2 = \frac{N(|ad - bc| - N/2)^2}{(a + b)(c + d)(a + c)(b + d)}$$

Formula 20.5

Phi coefficient for a 2 × 2 table as a function of the chi-square statistic:

$$\phi = \sqrt{\frac{\chi^2}{N}}$$

Formula 20.6

Phi coefficient squared:

$$\phi^2 = \frac{\chi^2}{N}$$

Formula 20.7

Shortcut formula for the phi coefficient squared:

$$\phi^2 = \frac{(|ad - bc| - N/2)^2}{(a + b)(c + d)(a + c)(b + d)}$$

Formula 20.8

Cramér's phi coefficient for contingency tables larger than 2 × 2 (varies between 0 and 1):

$$\phi_c = \sqrt{\frac{\chi^2}{N(L - 1)}}$$

Formula 20.9

Contingency coefficient for tables larger than 2 × 2 (maximum value is usually less than 1):

$$C = \sqrt{\frac{\chi^2}{\chi^2 + N}}$$

Formula 20.10

Cross-product ratio:

$$\frac{ad}{bc}$$

Formula 20.11

STATISTICAL TESTS FOR ORDINAL DATA

You will need to use the following from previous chapters:

Symbols
Σ: Summation sign

Formulas
Formula 11.1: Pearson correlation coefficient

Concepts
Ordinal scales
Median
Correlation

21

C H A P T E R

A

**CONCEPTUAL
FOUNDATION**

Remember the Martian from Chapter 10 who wanted to test whether men and women differ in height? Imagine that the Martian has rounded up 10 men and 12 women but does not have any kind of ruler to measure their heights. Can he, she, or it still perform a statistical test to decide whether the populations of men and women are identical with respect to height? The answer is yes. Perhaps you're thinking that a crude ruler could be made on the spot to serve the Martian's purpose, but I have a different solution in mind—one that will introduce the concepts of this chapter.

Ranking Data

One simple solution to the Martian's dilemma is to line up all 22 of the humans in order of height. This is easy to do without any kind of ruler. Of course, it will occasionally be difficult to determine which of two people is taller, but assuming such ties are rare, the statistical procedures for ordinal data are still valid. I will discuss what to do about tied scores later in this section; for now I will assume that all 22 people have been placed in height order.

The next step is to assign *ranks* to the people standing in line. It is arbitrary whether you assign the rank of 1 to the tallest or the shortest person in line, but it is common to assign the lowest rank to the one with the "most" of that variable—so I will give the rank of 1 to the tallest person, 2 to the next tallest, and so on, assigning the rank of 22 to the shortest. If this group of adult humans is typical, we would expect the women to be clustered toward one end of the line and the men to be concentrated near the other end. Nonetheless, we expect a fair amount of mixing of the genders near the middle of the line. In terms of ranks, we expect the men to have lower ranks than the women, for the most part.

Comparing the Ranks from Two Separate Groups

Suppose that the Martian finds the sum of ranks for men to be considerably smaller than that for women. Can it conclude that the population of men differs from the population of women in height? As you know, Dr. Null would claim that this difference in the sums of ranks is the result of accidentally drawing unusual samples of men and women—that typical samples would not exhibit such a difference. What we need is a test statistic based on the sum of ranks that follows a known mathematical distribution when the null hypothesis is true. The statistic most commonly used is called *U,* and the statistical test based on *U* is called the *Mann-Whitney U-test,* after the two statisticians who devised this approach (Mann & Whitney, 1947). Wilcoxon (1949) simplified the test somewhat by working out tables of probabilities that are expressed directly in terms of sums of ranks (eliminating the necessity to calculate the *U* statistic); therefore his version of the test is sometimes referred to as the *Wilcoxon rank-sum test.* Because *U* is still commonly used, I will discuss that approach briefly. But because Wilcoxon's version is simpler, it is the one that I will follow in this chapter. Regardless of whose version is used to find the critical value or *p* level, the test is sometimes called the *Wilcoxon-Mann-Whitney test.* However, I will use the more common designation: the **Mann-Whitney test.** This name is less confusing, because another common ordinal test, based on matched samples (mentioned later in this chapter) is most often called the *Wilcoxon test.* If the *U* statistic is being used, rather than Wilcoxon's approach, I will refer to the test as the Mann-Whitney *U*-test.

The Sum of Ranks

To make it easier to illustrate the mathematics of dealing with ranks, I will use an example with very small samples: four men and five women (i.e., $n_m = 4$ and $n_w = 5$). Again, we line up all of the men and women and assign them ranks from 1 to 9. One quantity that will prove useful is the sum of the ranks (S_R) for all people in the line (i.e., $\sum R_i$). If the total number of people ranked is *N,* the sum of the ranks will be

$$S_R = \frac{N(N + 1)}{2}$$ Formula 21.1

For $N = 9$, $S_R = 9(10)/2 = 90/2 = 45$. If there is no overlap between the men and women in the line (i.e., the men are ranked 1 to 4 and the women 5 to 9), then S_R for the men can be found by using Formula 21.1:

$$S_R = \frac{n_m(n_m + 1)}{2} = \frac{4(5)}{2} = \frac{20}{2} = 10$$

The sum of ranks for the women can be found by subtracting the sum for the men from the total sum: $45 - 10 = 35$. No matter how the men and women are arranged in line, adding the sum of ranks for men to the sum of ranks for women will always give you the sum of ranks for the whole group (i.e., 1 to N), and this sum can be found by inserting N into Formula 21.1.

The U Statistic

Another way to measure the discrepancy in height order between the men and the women standing in our hypothetical line is to note how many women are shorter than each man (or conversely, how many men are shorter than each woman). I will illustrate this measure by taking the simple case above, in which men are ranked 1 to 4 and women are ranked 5 to 9. The man ranked 1 is taller than all five women, so he gets 5 "points." In fact, in this example, all four men receive 5 points each, so the men receive a total of 20 points. Because none of the women are taller than any of the men, the women are assigned 0 points. Both of these values, 0 and 20, are possible values for the U statistic mentioned above. The convention is to define U as the smaller of the two possible values, so for the example above, $U = 0$. As you might guess, $U = 0$ is the lowest possible value, and it occurs whenever there is no overlap in ranks for the two subgroups.

Determining the U value directly can be tedious when the two ns are fairly large and there is a good deal of mixing between the two subgroups. Fortunately, there is a simple formula that allows you to calculate U in terms of the sum of ranks for two groups labeled A and B, respectively.

$$U_A = n_A n_B + \frac{n_A(n_A + 1)}{2} - \Sigma R_A \qquad \text{Formula 21.2}$$

where ΣR_A represents the sum of ranks for members of group A.

You may notice that the middle term in Formula 21.2 is the sum of ranks from 1 to n_A. This happens to be the lowest value that ΣR_A can take on. If the A group has all the low ranks (i.e., there is no overlap between groups), then ΣR_A will equal the middle term, and U_A will equal $n_A n_B$. (This means $U_B = 0$, because the sum of U_A and U_B will always equal $n_A n_B$.) Generally, ΣR_A will be higher than the middle term, so that some amount will be subtracted from U_A. However, ΣR_A (or ΣR_B, as appropriate) can be used directly to complete the test, as I will show in Section B.

Dealing with Tied Scores

The statistical procedures presented in this chapter rest on the assumption that the variables involved are actually continuous, even though they are being measured on an ordinal scale. For instance, just because we have no precise way to measure beauty does not mean beauty exists at only particular levels with no gradations in between. Beauty varies continuously. Theoretically, with any two paintings that have a similar degree of beauty, we should be able to find (or create) a third painting whose beauty falls somewhere between the first two. The implication of this principle is that when we rank objects according to beauty (or some other continuous variable that is hard to measure), we should not encounter two objects that have exactly the same amount of beauty—that is, there should not be any ties.

In practice, our ability to discern differences is limited, so some ties are inevitable—but ties should be fairly rare. If ties are common in our data, then either the variable is not continuous or our ability to discern differences is quite crude. In either case the validity of the tests in this chapter would be questionable. Bear in mind, however, that if there are many ties (e.g., most boys initiate either zero or one fight, and a few initiate more than one) it may be necessary to divide your scale into a few broad categories and use statistics for qualitative data (e.g., chi-square tests). If you decide to assign ranks in spite of having several ties, you should follow the method below before applying any of the ordinal statistical procedures described in this chapter.

To illustrate ranking with tied scores, I will rank order a set of ten subjects with the following IQ scores: 109, 112, 120, 100, 95, 112, 104, 100, 112, 115. The first step is to place these scores in numerical order and assign a unique rank to each score, even though the rank seems arbitrary in the case of tied scores; see Table 21.1.

The next step is to calculate the average rank for each group of tied scores in Table 21.1. For instance, the score 112 has been given the ranks 3, 4, and 5. The average of these three ranks $(3 + 4 + 5)/3$ is 4; therefore, each score of 112 is given the same rank: 4. The score 100 has ranks 8 and 9. The average of 8 and 9 is 8.5, so each score of 100 receives the rank of 8.5. Making these changes gives the final ranks, as shown in Table 21.2.

It is important to note that the sum of the ranks is the same in Tables 21.1 and 21.2. For Table 21.1, S_R is given by Formula 21.1:

$$S_R = \frac{N(N + 1)}{2} = \frac{10(10 + 1)}{2} = \frac{110}{2} = 55$$

You can check for yourself that the ranks in Table 21.2 also add up to 55. This is why the first step—ranking all the scores and ignoring ties—is important. Otherwise there is a strong tendency to lose track of the total number of ranks and assign ranks that result in a different sum. (For example, you might tend to give the rank of 3 to all three of the 112s, and then the rank of 4 to 109, or a rank of 4 to all of the 112s, but a rank of only 5 to 109.)

Table 21.1										
Score	120	115	112	112	112	109	104	100	100	95
Rank	1	2	3	4	5	6	7	8	9	10

Table 21.2										
Score	120	115	112	112	112	109	104	100	100	95
Rank	1	2	4	4	4	6	7	8.5	8.5	10

In some cases, the subjects in a study will not produce any scores, but rather will have to be ranked directly (e.g., students in a music class who are ranked based on their ability to sing). The same two-step procedure is applied. First, you assign a unique rank to each subject (even if two students with indistinguishable singing ability must be ordered arbitrarily for the moment), then calculate and assign average ranks for each set of tied subjects.

When to Use the Mann-Whitney Test

The Mann-Whitney test is generally used to answer the same types of research questions that are analyzed by a *t*-test of two independent groups. You may wish to compare the means of two preexisting populations (e.g., smokers, and nonsmokers) on some continuous dependent variable, or to assign subjects randomly to two different treatment groups and then compare the means on some continuous dependent variable that you expect to be altered differentially by the treatments. However, there are two very different types of data situations for which you would want to substitute the Mann-Whitney test for the *t*-test described in Chapter 9. One of these situations has already been introduced: The variable of interest cannot be measured precisely, but subjects can be placed in a meaningful order. This situation can occur when a precise scale exists but is not immediately available (as when the Martian cannot find a ruler). More commonly, a precise scale for the variable has not yet been devised. The other situation occurs when the interval/ratio data collected do not meet the distributional assumptions of the *t*-test. Both of these situations will be discussed in greater detail below. When the assumptions of the *t*-test have been met, it will have more power than the Mann-Whitney test, but it is important to note that in most common situations the power of the Mann-Whitney test is nearly as high as that of a *t*-test on the same data.

Variables That Are Not Measured Precisely

Many variables do not lend themselves to precise measurement; beauty has already been mentioned as an example. For another example, we can return to the corporal punishment study described in Chapter 20. In that example, playground aggressiveness was measured in terms of number of fights initiated. The chief problem with this scale is the "floor effect"—too many children score zero. The scale is not sensitive enough to distinguish between two boys (neither of whom starts fights directly): one who teases others and one who is very shy and passive. Although it may be difficult to devise a precise aggressiveness scale for observed playground behavior, it is quite possible that observers could rank order the boys from most to least aggressive (just as ethologists rank order animals in a colony based on position in the dominance hierarchy). Then the ranks could be summed separately for those boys who receive corporal punishment and those who do not, in order to perform the Mann-Whitney test. The rank-order method would be more sensitive and have greater power to detect population differences than merely categorizing the boys as initiators or noninitiators of fights.

Some variables are measured by psychologists using self-report questionnaires. For example, anxiety is often measured by asking the subject a series of questions about different anxiety symptoms. The subject assigns a rating to each symptom (e.g., indicating how often or how intensely she or he experiences that symptom), and then all of these ratings are added to yield a single anxiety score. These scores are usually treated as though they were derived from an interval scale. (That is, the psychologist assumes that two subjects scoring 20 and 30, respectively, are as different in anxiety level as two subjects scoring 30 and 40.) Means and standard deviations are calculated and parametric statistics are performed. However, it can be argued that subjective ratings represent an ordinal scale at best. (The subject scoring 30 may be more anxious than the subject scoring 20, and the subject scoring 40 may be more anxious than the subject scoring 30, but we cannot compare the differences.) From this point of view, the calculation of means, standard deviations, and parametric statistics is not justified. This is a controversial issue in psychological research. I am not going to attempt to solve the controversy in this text. But I can tell you that if you are worried that the scale you are using does not have the interval property or does not even come close, you can rank order your scores and use the methods for ordinal data, such as the Mann-Whitney test.

Variables That Do Not Meet Parametric Assumptions

The other common situation in which the Mann-Whitney test may be appropriate occurs when a *t*-test for two independent samples has been planned but the data do not fit the assumptions of a parametric test. I will assume that an interval or ratio scale of measurement is being used, as required for the *t*-test, but that the sample data indicate a distribution that is not consistent with the assumption of a normal distribution in the population. For instance, two groups of subjects try to solve a problem that requires insight (e.g., they must see that some ordinary object can be used as a tool); subjects in one group get a hint and subjects in the other do not. The amount of time it takes to find a solution is recorded for each subject.

In this situation you may find that most subjects in the "hint" group solve the problem quickly, with a few subjects taking quite a long time, and perhaps one or two not solving the problem at all. In the "no hint" group there would probably be more subjects taking a long time or not solving the problem at all. Given such data, the assumption that solution times are normally distributed in the population seems unreasonable. However, the Central Limit Theorem implies that with sufficiently large sample sizes the sampling distribution of the mean will be approximately normal, in spite of the population distribution. But what if your samples are fairly small? You can combine the data from the two groups, rank order all the subjects, sum the ranks separately for each group, and perform the Mann-Whitney test. Even subjects who never solved the problem could be included; they would be tied at the lowest rank. Just remember that too many ties threaten the validity of the test.

Repeated Measures or Matched Samples

If each subject has been measured on an interval/ratio scale twice (or subjects have been matched in pairs), but the difference scores present a distribution that is inconsistent with a normal distribution in the population (and the sample is relatively small), the sign test described in Chapter 19 can be performed as a substitute for the matched *t*-test. Unfortunately, the sign test throws away all of the quantitative information, retaining only the direction for each pair, and therefore has considerably less power than the matched *t*-test for a given sample size. A more powerful alternative to the sign test, and one that is also distribution-free, is the **Wilcoxon matched-pairs signed-rank test.** This test operates by rank ordering the difference scores (ignoring sign) and then summing the ranks separately for each sign. The symbol *T* is used to represent the sum of ranks for one particular sign; hence the test is sometimes referred to as the *Wilcoxon T test*. This test can also be used when an interval/ratio scale has not been used but it is possible to rank the magnitude of the differences directly. (This application of the test is not very common.) The details of the Wilcoxon test will not be discussed further in this text, but descriptions of the test can be found in many introductory and advanced texts in statistics (e.g., Howell, 1992).

Exercises

1. Describe two variables that would be difficult to measure on an interval/ratio scale but on which subjects could be rank ordered.

* 2. If the top 16 tennis players in the world are ranked from 1 to 16, what is the sum of the ranks?

3. If the top seven tennis players in the previous exercise are from the United States, a) what is the sum of ranks for the U.S. players?

   b) What is the sum of ranks for the rest of the players?

* 4. Suppose that ten women and nine men are rank ordered on some variable, and the result is that the two genders alternate, with women occupying the highest and lowest ranks. If men are labeled A and women B, a) what are the values of ΣR_A and ΣR_B?

   b) What are the values of U_A, U_B, and U?

5. Rank order the following sets of numbers, assigning the average rank to each set of ties. For each set, sum the ranks to check that you have assigned the ranks correctly.

   a) 102, 99, 101, 98, 104, 99, 100, 99, 101, 100, 102, 101, 99, 103

   b) 72, 68, 72, 75, 69, 68, 73, 70, 72, 70, 76, 68, 71, 74, 72

Testing for a Difference in Ranks Between Two Independent Groups

As an example in which two groups of subjects are measured for some variable on an ordinal scale, imagine testing one aspect of Freud's theory of personality. Freud suggested that toddlers subjected to a strict form of toilet training will later exhibit an anal retentive personality, including stubbornness, stinginess, and

BASIC STATISTICAL PROCEDURES

neatness. An English teacher decides that she will test the prediction concerning neatness by rank ordering her 21 students according to the neatness and organization of their assignments. She then consults the parents of her students to find out which students had been strictly toilet trained and which had not. The difference in neatness between the "strict" group and the "lenient" group can be tested with the usual six-step method.

Step 1. State the Hypotheses

The hypotheses for inference with ordinal data usually involve statements about the entire population distributions rather than just measures of central tendency. For the present example the appropriate null hypothesis would be that the distribution of the strictly trained population (with respect to neatness) is *identical* to the distribution of the leniently trained population. The alternative hypothesis is that the two population distributions differ. However, there are many ways in which the distributions can differ. If the null hypothesis is rejected, it could mean that the distributions differ only in central tendency, but it could also mean that the distributions have the same central tendency but differ in variability and/or shape. Only if there is some basis for assuming that the two population distributions are similar in shape and variability can the rejection of the null hypothesis be taken to imply a difference in central tendency. For ordinal data a difference in central tendency would be expressed as a difference in the *medians* of the two distributions, since it is not proper to calculate means for ordinal data.

Step 2. Select the Statistical Test and the Significance Level

Because the dependent variable (neatness) has been measured on an ordinal scale and the independent variable (strict versus lenient toilet training) is categorical, the appropriate procedure for testing the difference between two groups is the Mann-Whitney test. I will use Wilcoxon's approach, as you will see in steps 4 and 5. As usual, alpha will be set to .05.

Step 3. Select the Samples and Collect the Data

In this example, the sample is obviously one of convenience. A truly random sample from the general population would increase the generalizability of the results. However, in testing a controversial hypothesis it is not uncommon to use a convenient sample for an exploratory study before launching a more comprehensive investigation.

Assume that in ranking the 21 students the teacher found that the two neatest were indistinguishable and therefore tied for the lowest rank. Also assume that the three sloppiest students were tied for last place. Students who were then found to have been strictly trained were assigned to group A; these students represent the smaller subgroup. The remainder (those in the larger subgroup) were assigned to group B, as shown in Table 21.3.

Table 21.3

Subject	Rank	Group	Subject	Rank	Group
1	1.5	A	12	12	A
2	1.5	A	13	13	B
3	3	B	14	14	B
4	4	A	15	15	B
5	5	B	16	16	A
6	6	A	17	17	B
7	7	A	18	18	B
8	8	A	19	20	B
9	9	B	20	20	B
10	10	B	21	20	B
11	11	A			

Step 4. Find the Region of Rejection

The critical values for the Mann-Whitney test can be found in Table A.14. First, we need to count the subjects in group A; we label this total n_A. Similarly, the number of subjects in group B equals n_B. In this example, $n_A = 9$ and $n_B = 12$. (If n_A were equal to n_B, then of course it would not matter which subgroup was marked A and which was marked B.) In Table A.14 we look at the panel labeled $n_A = 9$, and the row labeled $n_B = 12$. We look for our values under the column .025, because we are performing a two-tailed test with alpha = .05. The corresponding entry in the table is 71–127. This means that if our observed value is anywhere between 71 and 127, the null hypothesis *cannot* be rejected. Thus there are two regions of rejection: less than or equal to 71 and greater than or equal to 127. (Large samples cannot be found in Table A.14; for large samples you would have to use the normal approximation described below.)

Step 5. Calculate the Test Statistic

According to Wilcoxon's approach, all we have to do is sum the ranks for the smaller of the two groups (i.e., ΣR_A; if $n_A = n_B$, either sum will do). Summing the ranks marked A in Table 21.3, we get $\Sigma R_A = 1.5 + 1.5 + 4 + 6 + 7 + 8 + 11 + 12 + 16 = 67$. (Note: If we were using a table of critical values for the U statistic we could now calculate U with Formula 21.2:

$$U_A = n_A n_B + \frac{n_A(n_A + 1)}{2} - \Sigma R_A = 9 \cdot 12 + \frac{9(10)}{2} - 67$$
$$= 108 + 45 - 67 = 86$$
$$U_B = n_A n_B - U_A = 108 - 86 = 22$$

Because U_B is less than U_A, U_B (22) would be taken as the value of U.)

To check that you have assigned tied ranks correctly, find ΣR_B and add this sum to ΣR_A. The total of all ranks should equal S_R, as given by Formula 21.1,

$$S_R = \frac{N(N+1)}{2} = \frac{21(22)}{2} = \frac{462}{2} = 231$$

where $N = n_A + n_B$. Using the ranks for group B in Table 21.3, we find that $\Sigma R_B = 164$. Then we see that $\Sigma R_A + \Sigma R_B = 67 + 164 = 231$, which, fortunately, equals the S_R found above.

Step 6. Make the Statistical Decision

The value for ΣR_A (67) is less than 71, so it is in the region of rejection. The null hypothesis—that the two population distributions are identical—can be rejected. To the extent that we can assume that the two distributions are similar in shape and variability, we can conclude that those who were strictly toilet trained have neater work habits. In any case, we are entitled to say that the two populations differ in some way and that further investigation may be warranted to explore the basis for this difference. Note that we *cannot* conclude that strict toilet training *causes* neatness in later life, because the teacher had no control over the way the parents trained their children. It is possible, for instance, that neatness is hereditary, and that neat people tend to be strict about toilet training. Only by randomly assigning toddlers to toilet training conditions can we test whether strict toilet training leads to neatness later in life.

The Normal Approximation to the Mann-Whitney Test

I pointed out that Table A.14 cannot be used when n_A or n_B is larger than 15. With fairly large samples, a normal approximation can be used to perform the Mann-Whitney test. Fortunately, when N (i.e., $n_A + n_B$) is greater than about 20, the distribution of ΣR_A begins to resemble the normal distribution with a mean of $.5(n_A)(N+1)$ and a standard deviation of

$$\sigma = \sqrt{\frac{n_A n_B (N+1)}{12}}$$

Therefore, when the samples are too large to use Table A.14, the Mann-Whitney test can be performed by calculating the following test statistic:

$$z = \frac{\Sigma R_A - .5(n_A)(N+1)}{\sqrt{n_A n_B (N+1)/12}} \qquad \text{Formula 21.3}$$

The z score found by this formula should follow the standard normal distribution (approximately) when the null hypothesis is true, so you can find the region of rejection from Table A.1 (e.g., for $\alpha = .05$, two-tailed, critical $z = \pm 1.96$). For the neatness example, the z score would be

$$z = \frac{67 - .5(9)(22)}{\sqrt{(9)(12)(22)/12}} = \frac{67 - 99}{\sqrt{198}} = \frac{-32}{14.1} = -2.274$$

This z score would be significant at the .05 level (two-tailed), in agreement with our previous conclusion. The minus sign on the z score reminds us that group A has the smaller ranks (more neatness). Although Table A.14 is more accurate for the sample sizes in this example, the normal approximation becomes quite reasonable when n_B is larger than 10.

Assumptions of the Mann-Whitney Test

Independent random sampling. The observations within each set of scores should be independent, and the two sets of scores should be independent of each other.

The dependent variable is continuous. This assumption implies that ties will be rare.

Publishing the Results of the Mann-Whitney Test

The Mann-Whitney test is most likely to be used in research areas that involve fairly small samples that are likely to differ in size and variability—for instance, neuropsychology. The study described below illustrates the use of the Mann-Whitney test in neuropsychological research and the way the results of this test would be reported in a journal.

Tranel and Damasio (1994) conducted a study of skin conductance responses (SCRs) in brain-damaged subjects. Psychological SCRs were elicited by presenting emotion-evoking slides. Subjects were divided into groups based on the location of their brain damage; for example, all the subjects in Group 2 had damage in the inferior parietal area. Each group of brain-damaged subjects was compared with a group of 20 control subjects who were free of neurological and psychiatric disease. Qualitative observations of Group 2 subjects suggested that damage to the right cerebral hemisphere had a much greater effect on SCRs than damage on the left side. This led to the following statistical test: "To check the reliability of our conclusion regarding Group 2, we conducted three statistical comparisons (Mann-Whitney U tests based on psychological SCRs). The comparison of the four right parietal subjects with controls was significant ($U = 0, p < .001$); by contrast, the comparison of the six left parietal subjects with controls was nonsignificant ($U = 44, p > .05$). A comparison of the four right parietal subjects with he six left parietal subjects was significant ($U = 4, p < .05$)" (pp. 432–433).

The authors state that their use of nonparametric techniques was motivated by their small sample sizes and a lack of homogeneity of variance. Note that the U

of zero for one of the tests indicates that all four of the right parietal subjects had smaller SCRs than any of the control subjects.

Correlation with Ordinal Data

Before concluding this section, I want to introduce another statistical procedure that is used with ordinal data, because it is easy to explain and rather commonly used. For an example of this statistic, I will return to the notion of rank ordering boys in the playground according to the amount of aggressiveness displayed by each one. Perhaps you wondered, Whose judgment can we trust in ranking the boys? Even experts can disagree dramatically in making such judgments. We can train an observer to use specific criteria in ranking the boys, but how can we know that the criteria are being applied in a reliable fashion? The simplest solution is to have two raters, trained the same way, rank the boys independently (i.e., each rater is unaware of the other's rankings). Ideally, the two raters will rank each boy in the same position. Usually, however, there will be some discrepancy between the two sets of rankings. We need to measure the amount of discrepancy, or conversely, the amount of agreement, between the raters, so we can decide whether any observer's set of rankings can be trusted. Also, if the agreement between the two raters is high, we can feel justified in averaging the two sets of rankings.

The Spearman Correlation Coefficient

The amount of agreement between two sets of rankings can be determined by calculating the Pearson correlation coefficient, just as we would for any two variables measured on an interval or ratio scale. When the two variables consist of ordinal data, the correlation coefficient that results from applying Pearson's formula is often called the **Spearman correlation (r_s),** or the *rank correlation coefficient*. As with the point-biserial r (r_{pb}) and the phi coefficient (ϕ), the special symbol, r_s, reminds us that although this is a Pearson correlation coefficient, it is calculated for data that were not measured on an interval or ratio scale.

It is important to note that there are cases in which both variables are measured on interval/ratio scales and yet it is preferable to rank order both variables and find r_s, rather than calculate r directly. For instance, a few extreme outliers can have a devastating effect on r, but their impact can be reduced by converting the data to ranks. When you do this, the highest score on one variable can be very far from the next highest, but in terms of ranking it is just one rank higher. Also, recall that Pearson's r measures only the degree of *linear* relationship. If the relationship between two variables follows a curve (as in Figure 21.1), Pearson's r can be deceptively low. Transforming the data or calculating a curvilinear correlation can help, but ranking represents a simpler solution. The Spearman correlation will give the relationship credit for being *monotonic* (whenever X goes up, Y goes up, and vice versa), even if it's far from linear. In fact, if the

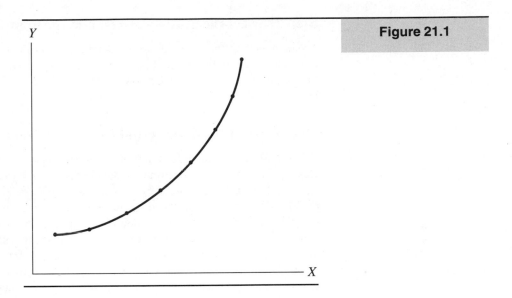

Figure 21.1

relationship is perfectly monotonic (as in Figure 21.1), r_s will equal 1.0. There-fore, if it is the degree of monotonicity, rather than the linearity of the relationship that you wish to measure, r_s will serve you better than Pearson's r.

Sometimes one of the two variables you are interested in can be measured on an interval/ratio scale but the other can only be measured ordinally. In such a case you must sacrifice the precision of one of the variables and rank order both of them. For an example, consider a study to answer the question, Do the prices of abstract paintings bear any relation to their beauty? Ten abstract paintings, all similar in size and medium (e.g., oil paint as opposed to water colors), but repre-senting a range of prices, are selected from art galleries. Students in an interme-diate art class are asked to rank the ten paintings in order based on how much they would like to have each painting hanging in their home. We will assume that somehow the class comes to an agreement about the ranks, except that two paint-ings are tied for last place. Whether there is any relationship between the stu-dents' preference and the actual retail prices of abstract paintings can be assessed with the six-step procedure for hypothesis testing.

Step 1. State the Hypotheses

For a Pearson correlation the null hypothesis is that $\rho = 0$. For normally distrib-uted variables, $\rho = 0$ implies that the two variables are independent. This impli-cation does not hold, however, if we make no assumptions about the shape of the distributions of the variables, as is the case with the Spearman correlation. So rather than $H_0: \rho_s = 0$, the null hypothesis that we are really interested in is H_0: the two variables are independent in the population. The two-tailed alternative hypothesis is H_A: the two variables are *not* independent in the population.

Step 2. Select the Statistical Test and the Significance Level

If one or both of the variables are measured on an ordinal scale, the appropriate measure of correlation is the Spearman correlation coefficient, r_s. Because the preference measure consists of ranks, r_s will be calculated. The alpha for the two-tailed test will be .05.

Step 3. Select the Sample and Collect the Data

The sample in this example is the set of ten paintings, not the students in the art class ranking the paintings. From the data in the study we want to make a general statement about the relationship between price and preference for *all* abstract paintings (comparable to those in our sample in size, medium, etc.), so our sample should be selected at random from the types of paintings to which we want to generalize. Of course, this general statement will only be true when preference is judged by people similar to those in our art class. To increase the generalizability of our conclusions, we would have to repeat the study with different types of people as judges. Table 21.4 lists the paintings, labeled from A to J, and displays the retail price along with the students' preference rank.

Step 4. Find the Region of Rejection

With ten objects to be ranked (assuming no ties) there are exactly 3,628,800 orders in which they can be placed. (This is equivalent to 10! or 10 factorial; see the discussion of permutations in Chapter 19, Section C.) Squaring this number tells us how many ways these ten objects can be ranked on *two* variables. Assuming that all of these pairs of rankings are equally likely, exact probabilities can be worked out for values of r_s. However, Table A.15 includes only the critical values for r_s for common alpha levels. Note that Table A.15 is organized by sample size (N = the number of *pairs*), rather than by df. For the present example, $N = 10$

Table 21.4	Painting	Preference	Price (in $)
	A	3	250
	B	5	1275
	C	6	375
	D	9.5	1000
	E	2	750
	F	1	575
	G	9.5	1500
	H	7	690
	I	4	500
	J	8	2000

and $\alpha = .05$ (two-tailed), so the critical value for r_s is .648. Because we are going to perform a two-tailed test, the region of rejection is greater than $+.648$ or less than $-.648$. For large N (greater than about 50) the distribution of r_s approaches the distribution for Pearson's r, so r_s can be tested with a t value (see Formula 11.6) or compared to critical values of Pearson's r (see Table A.5).

Step 5. Calculate the Test Statistic

Before we can calculate r_s, *both* variables must be ranked. The preference variable is already in the form of ranks, so we need only rank the prices; the rankings are shown in Table 21.5.

Now we must go back to the original data (Table 21.4) and replace the price of each painting in dollars with the rank for that price, as shown in Table 21.6.

Table 21.5

Painting	Price ($)	Rank
A	250	10
C	375	9
I	500	8
F	575	7
H	690	6
E	750	5
D	1000	4
B	1275	3
G	1500	2
J	2000	1

Table 21.6

Painting	Preference	Price (rank)	D	D^2
A	3	10	-7	49
B	5	3	$+2$	4
C	6	9	-3	9
D	9.5	4	$+5.5$	30.25
E	2	5	-3	9
F	1	7	-6	36
G	9.5	2	$+7.5$	56.25
H	7	6	$+1$	1
I	4	8	-4	16
J	8	1	$+7$	49
				$\Sigma D^2 = 259.5$

At this point we can apply any of the Pearson correlation formulas from Chapter 11 to the pairs of ranks in Table 21.6 to find r_s. However, as with r_{pb} and ϕ, there is a shortcut formula that eases the calculation of r_s. This shortcut formula requires the computing and squaring of the difference score for each pair of ranks, which is why Table 21.6 includes columns for D and D^2. The shortcut formula is

$$r_s = 1 - \frac{6\Sigma D^2}{N(N^2 - 1)}$$ Formula 21.4

where N is the number of pairs of ranks. Inserting ΣD^2 from Table 21.6 into Formula 21.4, we obtain

$$r_s = 1 - \frac{6(259.5)}{10(100 - 1)} = 1 - \frac{1557}{990} = 1 - 1.573 = -.573$$

Note that when the correlation is perfect, all of the Ds are zero, so the formula reduces to $1 - 0 = +1$. When there is perfect negative correlation the term subtracted from 1 reaches its maximum, 2, so $r_s = 1 - 2 = -1$.

Step 6. Make the Statistical Decision

The observed $r_s(-.573)$ does not fall in the region of rejection, so the null hypothesis cannot be rejected in this case. We must retain the hypothesis that price and preference are independent (i.e., unrelated) over the population of abstract paintings represented in the study. Note that the negative correlation indicates a tendency for lower-priced paintings to be given higher preference, when it might have been logical to expect the opposite. Although r_s was not large enough to attain statistical significance, a large enough negative correlation could have led to the rejection of the null hypothesis, because we had planned a two-tailed test. A significant negative correlation would have justified the conclusion that the raters in our study prefer less expensive abstract paintings.

Comparison to the Pearson Correlation

Although r_s can be found by applying the ordinary Pearson correlation formula to ranked data, the distribution of r_s is not the same as the distribution of r that is calculated for interval/ratio data, unless the sample size is extremely large (theoretically, infinite). In general, it is easier to obtain a high correlation with ranks than it is with interval/ratio data, so the critical values for r_s are higher than the corresponding critical values for r. (Compare Table A.5 and A.15, noting that the former lists critical values in terms of df, whereas the latter lists critical values in terms of N). Perfect Spearman correlation requires only that ordinal positions match across two variables, but perfect Pearson correlation requires, in addition, that the variables have a perfect *linear* relationship. For only four pairs, perfect Spearman correlation is common enough that the null hypothesis cannot be rejected at the .05 level (two-tailed) even if $r_s = 1$. In contrast, perfect Pearson correlation for four pairs of interval/ratio scores is rare enough to reject the null hypothesis with $\alpha = .05$ (two-tailed); in fact, any r over .95 is sufficient.

Assumptions of the Spearman Correlation

Unlike the Pearson correlation, the Spearman correlation does not require a bivariate normal distribution or even that each variable follow its own normal distribution. This is why the Spearman correlation is considered another distribution-free statistic. The relevant assumptions are those required for other tests of ordinal data.

Independent random sampling. Each pair of observations should be independent of all other pairs.

Continuous variables. Both variables are assumed to be continuous, so tied ranks are expected to be rare. When tied ranks occur, the tied items are assigned an average of their ranks, as we did for the two least-preferred paintings in the example above. If there are only a few tied ranks, the effect on r_s is negligible, but if there are many ties a correction factor should be applied to Formula 21.4 (see Siegel & Castellan, 1988).

When to Use the Spearman Correlation

The situations calling for the use of r_s can be divided into three cases, according to the measurement scale applied to each variable.

Both variables have been measured on an ordinal scale. The situation in which you wish to measure interrater reliability is a common example—for instance, when two judges rank order a set of photos of human faces for attractiveness. Another situation involves a set of subjects rank ordered for two different attributes (e.g., obedience to authority and creativity).

One variable has been measured on an ordinal scale and the other on an interval or ratio scale. The abstract painting example falls into this category. If you were to correlate the two sets of numbers without ranking the interval/ratio data, you would be "comparing apples to oranges," and you would not know what distribution could be used to represent the null hypothesis, or where to obtain critical values.

Both variables have been measured on an interval/ratio scale. In this case, you would only convert the ranks in order to calculate r_s instead of r if the distributional assumptions of r were severely violated and the sample was relatively small. Another reason to calculate r_s instead of r with interval/ratio data is if the relationship is far from linear and you just want to assess the degree to which the relationship is monotonic (i.e., whenever X goes up, Y goes up, and vice versa).

Exercises

* 1. Imagine that judges for the last Mr. Universe body-building contest discover that the second, third, fifth, sixth, and eighth ranked finalists out of the top 20 broke the rules by using steroid drugs. Use these data to test the null hypothesis that steroid drugs offer no advantage in a body-building contest at a) the .05 level; b) the .01 level.

2. Suppose that sixth-grade boys are ranked for their activity level in a classroom. Boys labeled A regularly eat a large amount of sweets, whereas the boys labeled B are allowed only a limited amount of sweets. Activity decreases from left to right.

A A B A A B A A B B A B B A B B A B B B B

Test the null hypothesis ($\alpha = .05$) that activity levels are unrelated to the amount of sweets eaten.

* 3. Table 9.1 from Chapter 9 is reproduced below. In that chapter, the data were used to illustrate the calculation of the two-group *t*-test comparing patients with left- versus right-hemisphere brain damage on the number of trials required to learn a maze. Using the same data, perform the Mann-Whitney test and compare your conclusion with the one reached after the *t*-test.

Left Damage	Right Damage
5	9
3	13
8	8
6	7
	11
	6

4. In exercise 9B9, the results of an industrial psychology experiment were given in terms of the number of clerical tasks successfully completed by the subjects in two experimental groups. The data are reproduced below.

Individual Motivation: 11, 17, 14, 10, 11, 15, 10, 8, 12, 15.

Group Motivation: 10, 15, 14, 8, 9, 14, 6, 7, 11, 13.

a) Perform the Mann-Whitney test on these data ($\alpha = .05$, one-tailed).

b) Using the normal approximation formula, find the *z* score corresponding to the Mann-Whitney test.

c) Compare the *z* score you found in part b with the *t* value that you calculated for exercise 9B9. Are the two tests comparable in this case?

* 5. A behavioral psychologist is studying the disruptive effects of a drug on maze-learning in rats. Five rats are injected with the drug before being placed in the maze, and five rats receive a control (i.e., saline) injection. The amount of time in minutes each rat takes to complete the maze is given below.

Drug: 18, 47, 15, 10, 16

Control: 8, 3, 5, 12, 9

a) Test the null hypothesis ($\alpha = .05$) that the drug has no effect (i.e., it has the same effect as saline) by performing the *t*-test for two independent groups.

b) Test the same hypothesis using the Mann-Whitney test.

c) Are your conclusions in parts a and b different? What problem in the data for this exercise is eliminated by ranking?

6. Does partying the night before a very difficult math test improve performance? Listed below are the exam scores for four students who partied all night before the test and five students who spent the evening studying.

Students Who Partied: 20, 50, 10, 60

Students Who Studied: 80, 55, 90, 45, 85

a) Test the null hypothesis ($\alpha = .05$) that partying and studying are equally helpful for math performance by conducting the *t*-test for two independent groups.

b) Test the same hypothesis using the Mann-Whitney test.

c) Are your conclusions in parts a and b different? Can you see why the Mann-Whitney test might have less power in this case?

* 7. Imagine that both Siskel and Ebert have ranked the ten top-grossing films of last year (labeled A to J) in order of preference (see table below). Calculate the Spearman correlation to assess the amount of agreement between these two movie critics.

	Siskel	Ebert
A	3	2
B	1	3
C	7	5
D	6	4
E	10	10
F	5	8
G	9	7
H	8	9
I	4	6
J	2	1

* 8. For part a of exercise 11B4, you were to calculate the Pearson correlation coefficient between number of years worked at a hypothetical company and annual salary earned. The data are reproduced below.

Years (X)	Annual Salary (Y)
5	24
8	40
3	20
6	30
4	50
9	40
7	35
10	50
2	22

a) Rank order the data for each variable and apply any of the Pearson correlation formulas to the ranks to find the Spearman correlation.

b) Recalculate the Spearman correlation using the shortcut formula.

c) Test the correlation you found in part a (or part b) for significance at the .05 level.

d) Compare the correlation coefficient you found in part a (or part b) to the correlation coefficient you found in part a of exercise 11B4.

Testing for Differences in Ranks Among Several Groups: The Kruskal-Wallis Test

OPTIONAL MATERIAL

Children raised in different cultures may differ in the intensity with which they express emotion during an upsetting experience. To compare two cultures, the emotional expressions of children from the two groups can be rank ordered together and the Mann-Whitney test can be performed. But what if you wish to compare children from three or more cultures in the same study? Fortunately, it is easy to extend the Mann-Whitney test so that it can accommodate any number of groups.

The extended test is called the *Kruskal-Wallis one-way analysis of variance by ranks,* or more commonly, the **Kruskal-Wallis test.** This test begins exactly like the Mann-Whitney test: You combine all of the subjects from all of the subgroups (e.g., cultures) into one large group, and rank order them on the dependent variable (e.g., intensity of emotional expression); ties are given average ranks, as usual. Then, as you might expect, you sum the ranks separately for each subgroup in the study. The only new thing to learn about the Kruskal-Wallis test is how to derive a convenient test statistic from the several sums of ranks.

The test statistic devised by Kruskal and Wallis (1952) is referred to as H; hence, the test is sometimes called the *Kruskal-Wallis H test.* To simplify the notation in the formula for H, I will use the symbol T_i to represent the sum of the ranks in one of the subgroups (i.e., the ith group). (You may recall that this notation parallels the use of T_i to represent the sum of scores in each group in the one-way ANOVA, as in Formula 14.11.) I will use n_i to represent the number of subjects in one of the subgroups, k to represent the number of subgroups, and N to represent the total number of subjects in the study (i.e., $\sum_{i=1}^{k} n_i$). Using this notation, Formula 21.5 for H is

$$H = \frac{12}{N(N+1)} \sum_{i=1}^{k} \left(\frac{T_i^2}{n_i} \right) - 3(N+1) \qquad \text{Formula 21.5}$$

For example, if the sum of ranks for four people from culture 1 were 40, the sum for five people from culture 2 were 50, and the sum for six people from culture 3 were 30, H would be:

$$H = \frac{12}{15(16)} \left(\frac{40^2}{4} + \frac{50^2}{5} + \frac{30^2}{6} \right) - 3(16) = \frac{1}{20}(400 + 500 + 150) - 48$$
$$= 52.5 - 48 = 4.5$$

H falls to its minimum value (near zero) when the ranks are equally divided among the groups. Usually, the researcher wants the sums of ranks to be as disparate as possible, leading to a large and (hopefully), significant H. But we need to know the distribution of H when the null hypothesis is true in order to decide when H is large enough to be statistically significant. Conveniently, when all of the populations involved are identical and the samples are not too tiny, H follows a distribution that resembles the chi-square distribution with df = $k - 1$. Therefore, we can use Table A.13 to find the critical value for H. For the example above, df = $3 - 1 = 2$, so critical H for $\alpha = .05$ is 5.99. Note that we are using only the upper tail of the chi-square distribution, as we did with the F distribution for analysis of variance (and for similar reasons).

Our observed H (4.5), being less than the critical H, does not allow us to reject the null hypothesis. Incidentally, if none of our subgroups were larger than 5, the chi-square distribution would not be considered a good enough approximation, and we would be obliged to use a special table to find the critical value for H (see Siegel & Castellan, 1988). Had our observed H for this example been greater than 5.99, we could have concluded that the cultural populations sampled are not all identical. If, in addition, we had made the common assumption that all the population distributions are similar in shape and variability, we could have concluded that the median intensity of emotional expression was not the same for all of the populations. We would still need to conduct post hoc comparisons, however, to determine which pairs of populations have different medians. Pairwise comparisons associated with the Kruskal-Wallis test are described in more advanced or specialized texts (e.g., Siegel & Castellan, 1988).

As you would guess, the Kruskal-Wallis test requires the same assumptions as the Mann-Whitney test and is used in the same circumstances. Moreover, ties are handled in a similar fashion, and there is a correction factor for H that should be used if more than about 25% of the observations result in ties (Siegel & Castellan, 1988).

Testing for Differences in Ranks among Matched Subjects: The Friedman Test

The power of the Kruskal-Wallis test compares reasonably well with ANOVA in most common situations, but suppose you want the added power of matching subjects (or repeated measures), even though you have more than two treatment conditions and your data are measured on an ordinal scale. You need a substitute for the repeated-measures ANOVA, such as the **Friedman test.**

The Friedman test is particularly easy to apply, because rather than having to rank order all of the subjects (or difference scores) with respect to each other, you only order the subjects (or repeated observations) within each matched set. For instance, if there are only three different treatment conditions (e.g., three methods for teaching children to play piano), then you would only rank order three subjects (or repeated observations) at a time. Suppose that the subject taught by method I is always the best piano player of the three matched subjects, and that method II always produces the second best, and method III, the third. This ideal situation is depicted in Table 21.7.

The data shown in Table 21.7 produce the largest possible separation in the sums of ranks for each method and the best chance of rejecting the null hypothesis (that the three methods produce identical populations). On the other hand, when the three teaching methods produce inconsistent results from subject to subject, the sums of ranks are more similar. The worst case—in which all methods produce the same sum of ranks—is shown in Table 21.8 (on page 738).

Subject	Method I	Method II	Method III
A	1	2	3
B	1	2	3
C	1	2	3
D	1	2	3
E	1	2	3
F	1	2	3
ΣR	6	12	18

Table 21.7

Subject	Method I	Method II	Method III
A	3	1	2
B	2	3	1
C	3	2	1
D	1	2	3
E	2	1	3
F	1	3	2
ΣR	12	12	12

Table 21.8

What we need is a test statistic that can measure the degree to which these sums of ranks differ. The statistic devised by Friedman (I will call it F_r) is found by Formula 21.6,

$$F_r = \frac{12}{Nc(c+1)} \sum_{i=1}^{c} T_i^2 - 3N(c+1) \qquad \text{Formula 21.6}$$

where c is the number of (repeated) treatments, N is the number of matched sets of observations (e.g., the number of rows in Table 21.7), and T_i is the sum of ranks for the ith treatment.

The resemblance between the Friedman test and the Kruskal-Wallis H statistic should be obvious. Like H, F_r follows a distribution that approaches the chi-square distribution with df $= c - 1$ when the null hypothesis is true and the total number of observations is not too small. (Tables are available to find critical values of F_r for small samples; see Siegel & Castellan, 1988.)

Applying Formula 21.6 to the data in Table 21.7, we get

$$F_r = \frac{12}{(6)(3)(4)}(6^2 + 12^2 + 18^2) - (3)(6)(4)$$
$$= .167(504) - 72 = 84 - 72 = 12$$

This is the highest F_r can get when $N = 6$ and $c = 3$. The critical value for F_r can be found in Table A.13; with df $= 3 - 1 = 2$, and $\alpha = .05$, the critical $F_r = 5.99$. Therefore, we can reject the null hypothesis for the data in Table 21.7 and conclude that the three teaching methods do *not* produce identical populations. In order to decide which pairs of methods differ significantly, you would have to make post hoc comparisons, as described by Siegel and Castellan (1988).

The Friedman test is sometimes described as an extension of the Wilcoxon matched-pairs test, but in some ways it is more an extension of the sign test. If the Friedman test is performed with just $c = 2$ conditions, then each score in a pair of scores is ranked 1 or 2. This really is no different from assigning a plus or minus to each pair according to which member of the pair rates more highly on the dependent variable. The Friedman test does not compare the sizes of the differences from one matched set to another.

On the other hand, the Friedman test resembles the RM ANOVA, in that the size of the test statistic reflects subject-to-subject consistency with respect to conditions. To the extent that subjects tend to "agree" about the different conditions (i.e., all subjects tend to respond best to the same treatment, second best to some other treatment, and so on), the data will resemble Table 21.7, producing sums of ranks that differ considerably, which, in turn, leads to a large value for F_r.

Like the RM ANOVA, the Friedman test can be used with repeated measures from the same subject or measures from sets of subjects that have been matched as closely as possible (the size of each set must equal c). Also, the Friedman test can be based either on ranking each set of observations directly (e.g., rank ordering the piano playing of each set of three matched subjects taught by different methods) or on interval/ratio data that has been converted to ranks because parametric assumptions have been severely violated. Like other procedures for ordinal data, the Friedman test is affected by the presence of tied ranks, so a correction factor should be applied to Formula 21.6 when there are more than just a few ties (see Siegel & Castellan, 1988).

Exercises

* 1. Suppose that you are comparing three architects, three musicians, and three lawyers on a spatial ability test, and you find that the three architects occupy the three highest ranks and the three lawyers occupy the three lowest ranks.

    a) Find the value of H.

    b) Even though the sample sizes are too small to justify it, use the chi-square approximation to test the significance of H at the .05 level.

2. a) Perform the Kruskal-Wallis test for the data in exercise 14B4.

    b) Does your statistical conclusion in part a differ from your conclusion for exercise 14B4? How does the power of the Kruskal-Wallis test seem to compare with the power of the one-way ANOVA?

* 3. Perform the Kruskal-Wallis test for the data in exercise 14B3.

4. a) Perform the Friedman test for the data in exercise 17B3, with alpha = .01.

    b) What would your statistical conclusion be for exercise 17B3, using alpha = .01?

    c) Does your statistical conclusion in part a differ from your conclusion in part b? How does the power of the Friedman test seem to compare with the power of the one-way RM ANOVA?

* 5. Perform the Friedman test for the data in exercise 17B7, with alpha = .05.

SUMMARY

The Important Points of Section A

1. To test whether two populations differ on some continuous variable without measuring the variable precisely (i.e., on an interval or ratio scale), you can take a mixed sample of both populations and put them in order. If you assign ranks to the entire group and then sum the ranks separately for members of each population, you can determine whether there is a large discrepancy between the two sums.

2. The discrepancy in the sums of ranks for any two groups ranked together on some variable can be tested by calculating the *Mann-Whitney U statistic,* or by using Wilcoxon's table (the Wilcoxon rank-sum test) that gives critical values for sums of ranks. In either case, the procedure is often called the *Mann-Whitney test.*

3. The sum of ranks (S_R) for any N items ranked 1 to N is equal to $N(N + 1)/2$.

4. Most ordinal tests rest on the assumption that the variable being measured is actually continuous, so that true ties are virtually impossible and apparent ties are due to the crudeness of the measurement. In assigning ranks, tied scores are initially given ranks as though they were different, and then these ranks are replaced by the average of the ranks for the tied scores.

5. The Mann-Whitney test is usually used when a variable cannot be measured precisely, but it is possible to determine which of two subjects has more of that variable (i.e., subjects can be ranked without too many ties), or when a variable has been measured precisely, but its distribution does not meet the assumptions for a two-group *t*-test.

The Important Points of Section B

The Mann-Whitney Test

To review the two statistical procedures described in Section B, consider a dependent variable that is measured on a ratio scale (time in seconds), but which has a distribution that is not consistent with the use of parametric statistics. In our hypothetical experiment, subjects are asked to find a hidden figure (a drawing of a dog) embedded in a very complex visual stimulus. The amount of time it takes each subject to trace the hidden figure is recorded. One group of subjects is exposed to a quick subliminal flash of the hidden figure before beginning the task; the other group sees an equally quick subliminal flash of an irrelevant drawing. The amount of time taken by subjects in the former ("primed") group can be compared to the time taken by subjects in the latter ("unprimed") group with the six-step procedure for null hypothesis testing. For this example, seven subjects are drawn at random for each group. The amount of time (in seconds) spent finding the figure is shown for each subject in Table 21.9.

First, we rank the data for both groups combined, giving average ranks for tied measurements, as shown in Table 21.10. Each rank is labeled with an A or a B, according to which group is associated with that rank. Because the two samples are the same size ($n_A = n_B = 7$), it is arbitrary which group is labeled

Primed Subjects	Unprimed Subjects	Table 21.9
12	10	
42	38	
8	20	
160	189	
220	225	
105	189	
22	45	

A. (I have labeled the primed group as A.) In this example, I will illustrate the use of the one-tailed test.

Now we sum the ranks for the primed group (A). $\sum R_A = 1 + 3 + 5 + 7 + 9 + 10 + 13 = 48$. As a check, we find that $\sum R_B = 57$ and note that $\sum R_A + \sum R_B = 49 + 56 = 105$, which is the same as the sum of ranks 1 to 14, as found by Formula 21.1:

$$S_R = \frac{N(N + 1)}{2} = \frac{14(15)}{2} = \frac{210}{2} = 105$$

For a one-tailed test, we look at the column labeled .05 in Table A.14 (we would use the .025 column for a two-tailed test) and find that the critical values are 39–66. As we expect the primed group (A) to have *lower* ranks, we will use only the smaller critical value, 39, for our one-tailed test. The region of rejection, therefore, is $\sum R_A$ less than or equal to 39. The $\sum R_A$ (48) is *not* less than or equal to 39, so the null hypothesis *cannot* be rejected. Had we rejected H_0 and made the usual assumption that the population distributions are similar in form, we could have asserted that the median time it takes for the primed population to find the hidden figure is less than the median time for the unprimed population. As the results stand, we must retain the possibility that the subliminal stimulation does not help subjects to find the hidden figure any faster.

Time	Rank	Group	Time	Rank	Group	Table 21.10
8	1	A	45	8	B	
10	2	B	105	9	A	
12	3	A	160	10	A	
20	4	B	189	11.5	B	
22	5	A	189	11.5	B	
38	6	B	220	13	A	
42	7	A	225	14	B	

The Normal Approximation to the Mann-Whitney Test

The $\sum R_A$ can be converted to a z score by means of Formula 21.3. The z score for the present example is

$$z = \frac{\sum R_A - .5(n_A)(N + 1)}{\sqrt{n_A n_B (N + 1)/12}} = \frac{48 - .5(7)(15)}{\sqrt{(7)(7)(15)/12}} = \frac{48 - 52.5}{\sqrt{61.25}} = \frac{-4.5}{7.83} = -.57$$

Although our samples are too small to justify using the normal approximation, it is not surprising that the z score is very far from statistical significance; this is consistent with the conclusion we drew using Table A.14.

Spearman Correlation for Ranked Data

To illustrate the Spearman correlation, I will assume that each primed subject was carefully matched with an unprimed subject and that the rows of Table 21.11 represent these matched pairs. Although Pearson's r can be calculated directly for the data in Table 21.11, assume that the distributions of the variables are not compatible with the assumptions underlying the use of Pearson's r. To circumvent these assumptions, we assign ranks separately to the data for each variable and then apply Pearson's formula to these ranks. That is, we rank the "primed" scores from 1 to 7 (giving average ranks to ties) and then rank the "unprimed" scores from 1 to 7. The original data are then replaced by these ranks. Table 21.11 includes the original data from Table 21.9, along with the corresponding ranks.

Any of the formulas for Pearson's r can be applied directly to the ranks in Table 21.11, and the result would be r_s. However, if you are calculating by hand, it is easier to compute the difference score for each pair of ranks, square each difference, find the sum of the squared differences ($\sum D^2$), and insert this sum into the shortcut formula for the Spearman correlation. These differences (D) and squared differences (D^2) have been included in Table 21.11. To find r_s we take $\sum D^2$ from Table 21.11, and plug it into Formula 21.4.

$$r_s = 1 - \frac{6 \sum D^2}{N(N^2 - 1)} = 1 - \frac{6(4.5)}{7(48)} = 1 - \frac{27}{336} = 1 - .080 = .920$$

Table 21.11	Primed Subjects	Rank	Unprimed Subjects	Rank	D	D²
	12	2	10	1	1	1
	42	4	38	3	1	1
	8	1	20	2	−1	1
	160	6	189	5.5	.5	.25
	220	7	225	7	0	0
	105	5	189	5.5	−.5	.25
	22	3	45	4	−1	1
					$\sum D^2 = 4.5$	

As you can see, r_s is very high, which indicates that the pairs of scores were very well matched. Checking the critical value in Table A.15, we find that the correlation is significant at the .01 as well as the .05 level. If the scores were actually matched in this way, the Wilcoxon matched-pairs test would be recommended in place of the Mann-Whitney test, because it could take advantage of the close matching between the two sets of scores.

Assumptions of Tests on Ordinal Data

All of the tests described in this chapter are distribution-free, in that none makes any assumption about the shape of the distribution of the dependent variable. Only the following two assumptions are required.

Independent random sampling. This is the same assumption that is made for parametric tests.

The dependent variable is continuous. This implies that tied ranks will be rare. If there are more than a few ties, you may need to use correction factors.

When to Use Tests on Ordinal Data

There are two major situations that call for the use of ordinal tests:

1. *The dependent variable has been measured on an ordinal scale.* In this case, it is not feasible to measure the dependent variable precisely, but it is possible to place items (e.g., subjects) in order of magnitude on the dependent variable and assign ranks.

2. *The dependent variable has been measured on an interval or ratio scale, but the distribution of the dependent variable does not fit the assumptions for a parametric test.* In this case the interval/ratio measurements are assigned ranks before applying ordinal tests.

The Important Points of Section C

1. *The Kruskal-Wallis test* is a straightforward extension of the Mann-Whitney test that is used when there are more than two independent groups. A statistic called H is calculated from the sums of ranks (and sample sizes) for the subgroups; H approximately follows the chi-square distribution with df $= k - 1$ (where k is the number of groups) when the null hypothesis is true. For very small samples a special table of critical values must be used. When H is statistically significant, post hoc pairwise comparisons are usually performed to determine which pairs of groups differ significantly.

2. *The Friedman test* is a good substitute for the repeated-measures ANOVA when the dependent variable is measured on an ordinal scale or the data are not consistent with the assumptions of parametric statistics. The observations are ranked separately for each matched set (there are c observations per set, where c is the number of different treatment conditions), and then the ranks are summed separately for each treatment. Consistent rankings from subject to subject lead to large values of a test statistic, F_r, which varies like the chi-square distribution (with df $= c - 1$) when the null hypothesis is true. Special tables are available for critical values of F_r when the sample is very small. As with H, significant values of F_r need to be followed by pairwise comparisons to localize the source of the effect.

Definitions of Key Terms

Mann-Whitney test Compares two independent groups when the dependent variable has been measured on an ordinal scale. Also called the *Mann-Whitney U test* (if the U statistic is computed) or the *Wilcoxon rank-sum test*.

Wilcoxon matched-pairs signed-rank test A substitute for the matched *t*-test when differences scores are not considered precise but can be rank ordered. Also called the *Wilcoxon T test,* because the test statistic is referred to as T.

Spearman correlation, r_s. The coefficient that results when you apply the Pearson correlation formula to two variables when both are in the form of ranks.

Kruskal-Wallis test A direct extension of the Mann-Whitney test that can accommodate any number of independent groups. Because it replaces the one-way ANOVA when the data are in the form of ranks, it is sometimes called the *Kruskal-Wallis one-way analysis of variance by ranks.* It is also called the *Kruskal-Wallis H test,* because the test statistic is referred to as H.

Friedman test A test that replaces the one-way repeated-measures or randomized blocks ANOVA when the data are in the form of ranks.

Key Formulas

The sum of the ranks from 1 to N:

$$S_R = \frac{N(N + 1)}{2}$$

Formula 21.1

The Mann-Whitney U statistic:

$$U_A = n_A n_B + \frac{n_A(n_A + 1)}{2} - \Sigma R_A$$

Formula 21.2

Normal approximation to the Mann-Whitney test:

$$z = \frac{\Sigma R_A - .5(n_A)(N + 1)}{\sqrt{n_A n_B (N + 1)/12}}$$

Formula 21.3

Shortcut formula for Spearman correlation:

$$r_s = 1 - \frac{6\Sigma D^2}{N(N^2 - 1)}$$

Formula 21.4

Kruskal-Wallis H statistic:

$$H = \frac{12}{N(N + 1)} \sum_{i=1}^{k} \left(\frac{T_i^2}{n_i} \right) - 3(N + 1)$$

Formula 21.5

F_r statistic for Friedman test:

$$F_r = \frac{12}{Nc(c + 1)} \sum_{i=1}^{c} T_i^2 - 3N(c + 1)$$

Formula 21.6

Statistical Tables

Table A.1

Areas under the Standard Normal Distribution

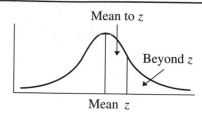

Mean to *z*

Beyond *z*

Mean *z*

z	Mean to z	Beyond z	z	Mean to z	Beyond z
.00	.0000	.5000	.34	.1331	.3669
.01	.0040	.4960	.35	.1368	.3632
.02	.0080	.4920	.36	.1406	.3594
.03	.0120	.4880	.37	.1443	.3557
.04	.0160	.4840	.38	.1480	.3520
.05	.0199	.4801	.39	.1517	.3483
.06	.0239	.4761	.40	.1554	.3446
.07	.0279	.4721	.41	.1591	.3409
.08	.0319	.4681	.42	.1628	.3372
.09	.0359	.4641	.43	.1664	.3336
.10	.0398	.4602	.44	.1700	.3300
.11	.0438	.4562	.45	.1736	.3264
.12	.0478	.4522	.46	.1772	.3228
.13	.0517	.4483	.47	.1808	.3192
.14	.0557	.4443	.48	.1844	.3156
.15	.0596	.4404	.49	.1879	.3121
.16	.0636	.4364	.50	.1915	.3085
.17	.0675	.4325	.51	.1950	.3050
.18	.0714	.4286	.52	.1985	.3015
.19	.0753	.4247	.53	.2019	.2981
.20	.0793	.4207	.54	.2054	.2946
.21	.0832	.4168	.55	.2088	.2912
.22	.0871	.4129	.56	.2123	.2877
.23	.0910	.4090	.57	.2157	.2843
.24	.0948	.4052	.58	.2190	.2810
.25	.0987	.4013	.59	.2224	.2776
.26	.1026	.3974	.60	.2257	.2743
.27	.1064	.3936	.61	.2291	.2709
.28	.1103	.3897	.62	.2324	.2676
.29	.1141	.3859	.63	.2357	.2643
.30	.1179	.3821	.64	.2389	.2611
.31	.1217	.3783	.65	.2422	.2578
.32	.1255	.3745	.66	.2454	.2546
.33	.1293	.3707	.67	.2486	.2514

(table continues)

z	Mean to z	Beyond z	z	Mean to z	Beyond z
.68	.2517	.2483	1.09	.3621	.1379
.69	.2549	.2451	1.10	.3643	.1357
.70	.2580	.2420	1.11	.3665	.1335
.71	.2611	.2389	1.12	.3686	.1314
.72	.2642	.2358	1.13	.3708	.1292
.73	.2673	.2327	1.14	.3729	.1271
.74	.2704	.2296	1.15	.3749	.1251
.75	.2734	.2266	1.16	.3770	.1230
.76	.2764	.2236	1.17	.3790	.1210
.77	.2794	.2206	1.18	.3810	.1190
.78	.2823	.2177	1.19	.3830	.1170
.79	.2852	.2148	1.20	.3849	.1151
.80	.2881	.2119	1.21	.3869	.1131
.81	.2910	.2090	1.22	.3888	.1112
.82	.2939	.2061	1.23	.3907	.1093
.83	.2967	.2033	1.24	.3925	.1075
.84	.2995	.2005	1.25	.3944	.1056
.85	.3023	.1977	1.26	.3962	.1038
.86	.3051	.1949	1.27	.3980	.1020
.87	.3078	.1922	1.28	.3997	.1003
.88	.3106	.1894	1.29	.4015	.0985
.89	.3133	.1867	1.30	.4032	.0968
.90	.3159	.1841	1.31	.4049	.0951
.91	.3186	.1814	1.32	.4066	.0934
.92	.3212	.1788	1.33	.4082	.0918
.93	.3238	.1762	1.34	.4099	.0901
.94	.3264	.1736	1.35	.4115	.0885
.95	.3289	.1711	1.36	.4131	.0869
.96	.3315	.1685	1.37	.4147	.0853
.97	.3340	.1660	1.38	.4162	.0838
.98	.3365	.1635	1.39	.4177	.0823
.99	.3389	.1611	1.40	.4192	.0808
1.00	.3413	.1587	1.41	.4207	.0793
1.01	.3438	.1562	1.42	.4222	.0778
1.02	.3461	.1539	1.43	.4236	.0764
1.03	.3485	.1515	1.44	.4251	.0749
1.04	.3508	.1492	1.45	.4265	.0735
1.05	.3531	.1469	1.46	.4279	.0721
1.06	.3554	.1446	1.47	.4292	.0708
1.07	.3577	.1423	1.48	.4306	.0694
1.08	.3599	.1401	1.49	.4319	.0681

Table A.1
(continued)

Areas under the Standard Normal Distribution

(table continues)

Table A.1 *(continued)*	z	Mean to z	Beyond z	z	Mean to z	Beyond z
Areas under the Standard Normal Distribution	1.50	.4332	.0668	1.91	.4719	.0281
	1.51	.4345	.0655	1.92	.4726	.0274
	1.52	.4357	.0643	1.93	.4732	.0268
	1.53	.4370	.0630	1.94	.4738	.0262
	1.54	.4382	.0618	1.95	.4744	.0256
	1.55	.4394	.0606	1.96	.4750	.0250
	1.56	.4406	.0594	1.97	.4756	.0244
	1.57	.4418	.0582	1.98	.4761	.0239
	1.58	.4429	.0571	1.99	.4767	.0233
	1.59	.4441	.0559	2.00	.4772	.0228
	1.60	.4452	.0548	2.01	.4778	.0222
	1.61	.4463	.0537	2.02	.4783	.0217
	1.62	.4474	.0526	2.03	.4788	.0212
	1.63	.4484	.0516	2.04	.4793	.0207
	1.64	.4495	.0505	2.05	.4798	.0202
	1.65	.4505	.0495	2.06	.4803	.0197
	1.66	.4515	.0485	2.07	.4808	.0192
	1.67	.4525	.0475	2.08	.4812	.0188
	1.68	.4535	.0465	2.09	.4817	.0183
	1.69	.4545	.0455	2.10	.4821	.0179
	1.70	.4554	.0446	2.11	.4826	.0174
	1.71	.4564	.0436	2.12	.4830	.0170
	1.72	.4573	.0427	2.13	.4834	.0166
	1.73	.4582	.0418	2.14	.4838	.0162
	1.74	.4591	.0409	2.15	.4842	.0158
	1.75	.4599	.0401	2.16	.4846	.0154
	1.76	.4608	.0392	2.17	.4850	.0150
	1.77	.4616	.0384	2.18	.4854	.0146
	1.78	.4625	.0375	2.19	.4857	.0143
	1.79	.4633	.0367	2.20	.4861	.0139
	1.80	.4641	.0359	2.21	.4864	.0136
	1.81	.4649	.0351	2.22	.4868	.0132
	1.82	.4656	.0344	2.23	.4871	.0129
	1.83	.4664	.0336	2.24	.4875	.0125
	1.84	.4671	.0329	2.25	.4878	.0122
	1.85	.4678	.0322	2.26	.4881	.0119
	1.86	.4686	.0314	2.27	.4884	.0116
	1.87	.4693	.0307	2.28	.4887	.0113
	1.88	.4699	.0301	2.29	.4890	.0110
	1.89	.4706	.0294	2.30	.4893	.0107
	1.90	.4713	.0287	2.31	.4896	.0104

(table continues)

z	Mean to z	Beyond z	z	Mean to z	Beyond z
2.32	.4898	.0102	2.71	.4966	.0034
2.33	.4901	.0099	2.72	.4967	.0033
2.34	.4904	.0096	2.73	.4968	.0032
2.35	.4906	.0094	2.74	.4969	.0031
2.36	.4909	.0091	2.75	.4970	.0030
2.37	.4911	.0089	2.76	.4971	.0029
2.38	.4913	.0087	2.77	.4972	.0028
2.39	.4916	.0084	2.78	.4973	.0027
2.40	.4918	.0082	2.79	.4974	.0026
2.41	.4920	.0080	2.80	.4974	.0026
2.42	.4922	.0078	2.81	.4975	.0025
2.43	.4925	.0075	2.82	.4976	.0024
2.44	.4927	.0073	2.83	.4977	.0023
2.45	.4929	.0071	2.84	.4977	.0023
2.46	.4931	.0069	2.85	.4978	.0022
2.47	.4932	.0068	2.86	.4979	.0021
2.48	.4934	.0066	2.87	.4979	.0021
2.49	.4936	.0064	2.88	.4980	.0020
2.50	.4938	.0062	2.89	.4981	.0019
2.51	.4940	.0060	2.90	.4981	.0019
2.52	.4941	.0059	2.91	.4982	.0018
2.53	.4943	.0057	2.92	.4982	.0018
2.54	.4945	.0055	2.93	.4983	.0017
2.55	.4946	.0054	2.94	.4984	.0016
2.56	.4948	.0052	2.95	.4984	.0016
2.57	.4949	.0051	2.96	.4985	.0015
2.58	.4951	.0049	2.97	.4985	.0015
2.59	.4952	.0048	2.98	.4986	.0014
2.60	.4953	.0047	2.99	.4986	.0014
2.61	.4955	.0045	3.00	.4987	.0013
2.62	.4956	.0044	· · ·	· · ·	· · ·
2.63	.4957	.0043	3.25	.4994	.0006
2.64	.4959	.0041	· · ·	· · ·	· · ·
2.65	.4960	.0040	3.50	.4998	.0002
2.66	.4961	.0039	· · ·	· · ·	· · ·
2.67	.4962	.0038	3.75	.4999	.0001
2.68	.4963	.0037	· · ·	· · ·	· · ·
2.69	.4964	.0036	4.00	.5000	.0000
2.70	.4965	.0035			

Table A.1 (continued)

Areas under the Standard Normal Distribution

Source: Adapted from *Statistical Methods for Psychology,* 3rd ed., by D. C. Howell. Copyright © 1992 PWS-Kent. Used with permission.

Table A.2

*Critical Values of
the t Distribution*

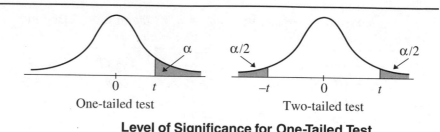

One-tailed test   Two-tailed test

	Level of Significance for One-Tailed Test					
	.10	.05	.025	.01	.005	.0005
	Level of Significance for Two-Tailed Test					
df	.20	.10	.05	.02	.01	.001
1	3.078	6.314	12.706	31.821	63.657	636.620
2	1.886	2.920	4.303	6.965	9.925	31.599
3	1.638	2.353	3.182	4.541	5.841	12.924
4	1.533	2.132	2.776	3.747	4.604	8.610
5	1.476	2.015	2.571	3.365	4.032	6.869
6	1.440	1.943	2.447	3.143	3.707	5.959
7	1.415	1.895	2.365	2.998	3.499	5.408
8	1.397	1.860	2.306	2.896	3.355	5.041
9	1.383	1.833	2.262	2.821	3.250	4.781
10	1.372	1.812	2.228	2.764	3.169	4.587
11	1.363	1.796	2.201	2.718	3.106	4.437
12	1.356	1.782	2.179	2.681	3.055	4.318
13	1.350	1.771	2.160	2.650	3.012	4.221
14	1.345	1.761	2.145	2.624	2.977	4.140
15	1.341	1.753	2.131	2.602	2.947	4.073
16	1.337	1.746	2.120	2.583	2.921	4.015
17	1.333	1.740	2.110	2.567	2.898	3.965
18	1.330	1.734	2.101	2.552	2.878	3.922
19	1.328	1.729	2.093	2.539	2.861	3.883
20	1.325	1.725	2.086	2.528	2.845	3.850
21	1.323	1.721	2.080	2.518	2.831	3.819
22	1.321	1.717	2.074	2.508	2.819	3.792
23	1.319	1.714	2.069	2.500	2.807	3.768
24	1.318	1.711	2.064	2.492	2.797	3.745
25	1.316	1.708	2.060	2.485	2.787	3.725
26	1.315	1.706	2.056	2.479	2.779	3.707
27	1.314	1.703	2.052	2.473	2.771	3.690
28	1.313	1.701	2.048	2.467	2.763	3.674
29	1.311	1.699	2.045	2.462	2.756	3.659
30	1.310	1.697	2.042	2.457	2.750	3.646
40	1.303	1.684	2.021	2.423	2.704	3.551
50	1.299	1.676	2.009	2.403	2.678	3.496
100	1.290	1.660	1.984	2.364	2.626	3.390
∞	1.282	1.645	1.960	2.326	2.576	3.291

Source: Adapted from *Statistical Methods for Psychology,* 3rd ed., by D. C. Howell, Copyright © 1992 PWS-Kent. Used with permission.

Table A.3 • Power as a Function of δ and Significance Criterion (α)

δ	One-Tailed Test (α)				Table A.3
	.05	.025	.01	.005	
	Two-Tailed Test (α)				Power as a Function of δ and Significance Criterion (α)
	.10	.05	.02	.01	
0.0	.10*	.05*	.02	.01	
0.1	.10*	.05*	.02	.01	
0.2	.11*	.05	.02	.01	
0.3	.12*	.06	.03	.01	
0.4	.13*	.07	.03	.01	
0.5	.14	.08	.03	.02	
0.6	.16	.09	.04	.02	
0.7	.18	.11	.05	.03	
0.8	.21	.13	.06	.04	
0.9	.23	.15	.08	.05	
1.0	.26	.17	.09	.06	
1.1	.30	.20	.11	.07	
1.2	.33	.22	.13	.08	
1.3	.37	.26	.15	.10	
1.4	.40	.29	.18	.12	
1.5	.44	.32	.20	.14	
1.6	.48	.36	.23	.16	
1.7	.52	.40	.27	.19	
1.8	.56	.44	.30	.22	
1.9	.60	.48	.33	.25	
2.0	.64	.52	.37	.28	
2.1	.68	.56	.41	.32	
2.2	.71	.59	.45	.35	
2.3	.74	.63	.49	.39	
2.4	.77	.67	.53	.43	
2.5	.80	.71	.57	.47	
2.6	.83	.74	.61	.51	
2.7	.85	.77	.65	.55	
2.8	.88	.80	.68	.59	
2.9	.90	.83	.72	.63	
3.0	.91	.85	.75	.66	
3.1	.93	.87	.78	.70	
3.2	.94	.89	.81	.73	
3.3	.96	.91	.83	.77	
3.4	.96	.93	.86	.80	
3.5	.97	.94	.88	.82	
3.6	.97	.95	.90	.85	

(table continues)

	One-Tailed Test (α)			
	.05	.025	.01	.005
	Two-Tailed Test (α)			
δ	.10	.05	.02	.01
3.7	.98	.96	.92	.87
3.8	.98	.97	.93	.89
3.9	.99	.97	.94	.91
4.0	.99	.98	.95	.92
4.1	.99	.98	.96	.94
4.2	.99	.99	.97	.95
4.3	**	.99	.98	.96
4.4		.99	.98	.97
4.5		.99	.99	.97
4.6		**	.99	.98
4.7			.99	.98
4.8			.99	.99
4.9			.99	.99
5.0			**	.99
5.1				.99
5.2				**

Table A.3 *(continued)*

Power as a Function of δ and Significance Criterion (α)

*Values inaccurate for *one-tailed* test by more than .01.
**The power at and below this point is greater than .995.
Source: From *Introductory Statistics for the Behavioral Sciences,* 2nd ed., by J. Welkowitz, R. B. Ewen, & J. Cohen. Copyright © 1976 Academic Press. Reprinted with permission.

	One-Tailed Test (α)			
	.05	.025	.01	.005
	Two-Tailed Test (α)			
Power	.10	.05	.02	.01
.25	0.97	1.29	1.65	1.90
.50	1.64	1.96	2.33	2.58
.60	1.90	2.21	2.58	2.83
.67	2.08	2.39	2.76	3.01
.70	2.17	2.48	2.85	3.10
.75	2.32	2.63	3.00	3.25
.80	2.49	2.80	3.17	3.42
.85	2.68	3.00	3.36	3.61
.90	2.93	3.24	3.61	3.86
.95	3.29	3.60	3.97	4.22
.99	3.97	4.29	4.65	4.90
.999	4.37	5.05	5.42	5.67

Table A.4

δ as a Function of Significance Criterion (α) and Power

Source: From *Introductory Statistics for the Behavioral Sciences,* 2nd ed., by J. Welkowitz, R. B. Ewen, & J. Cohen. Copyright © 1976 Academic Press. Reprinted with permission.

Table A.5	Levels of Significance for a One-Tailed Test			
	.05	.025	.01	.005
Critical Values of Pearson's r (*df* = *N* − 2)	Levels of Significance for a Two-Tailed Test			
df	.10	.05	.02	.01
1	.988	.997	.9995	.9999
2	.900	.950	.980	.990
3	.805	.878	.934	.959
4	.729	.811	.882	.917
5	.669	.755	.833	.875
6	.622	.707	.789	.834
7	.582	.666	.750	.798
8	.549	.632	.716	.765
9	.521	.602	.685	.735
10	.497	.576	.658	.708
11	.476	.553	.634	.684
12	.458	.532	.612	.661
13	.441	.514	.592	.641
14	.426	.497	.574	.623
15	.412	.482	.558	.606
16	.400	.468	.542	.590
17	.389	.456	.529	.575
18	.378	.444	.516	.561
19	.369	.433	.503	.549
20	.360	.423	.492	.537
22	.344	.404	.472	.515
24	.330	.388	.453	.496
26	.317	.374	.437	.479
28	.306	.361	.423	.463
30	.296	.349	.409	.449
35	.275	.325	.381	.418
40	.257	.304	.358	.393
45	.243	.288	.338	.372
50	.231	.273	.322	.354
55	.220	.261	.307	.339
60	.211	.250	.295	.325
70	.195	.232	.274	.302
80	.183	.217	.256	.283
90	.173	.205	.242	.267
100	.164	.195	.230	.254
120	.150	.178	.210	.232
150	.134	.159	.189	.208

(table continues)

	Levels of Significance for a One-Tailed Test			
	.05	**.025**	**.01**	**.005**
	Levels of Significance for a Two-Tailed Test			
df	**.10**	**.05**	**.02**	**.01**
200	.116	.138	.164	.181
300	.095	.113	.134	.148
400	.082	.098	.116	.128
500	.073	.088	.104	.115
1000	.052	.062	.073	.081

Table A.5
(continued)

***Critical Values of
Pearson's r
(df = N − 2)***

Source: From Table VI of Fisher & Yates, *Statistical Tables for Biological,
Agricultural and Medical Research* (6th ed.), published by Longman Group UK
Ltd., copyright © 1974. Used with permission. Supplementary values were
calculated at San Jose State University by K. Fernandes.

Table A.6		r	Z	r	Z	r	Z	r	Z
Table of Fisher's Transformation of r to Z		0.000	0.000	0.210	0.213	0.420	0.448	0.630	0.741
		0.005	0.005	0.215	0.218	0.425	0.454	0.635	0.750
		0.010	0.010	0.220	0.224	0.430	0.460	0.640	0.758
		0.015	0.015	0.225	0.229	0.435	0.466	0.645	0.767
		0.020	0.020	0.230	0.234	0.440	0.472	0.650	0.775
		0.025	0.025	0.235	0.239	0.445	0.478	0.655	0.784
		0.030	0.030	0.240	0.245	0.450	0.485	0.660	0.793
		0.035	0.035	0.245	0.250	0.455	0.491	0.665	0.802
		0.040	0.040	0.250	0.255	0.460	0.497	0.670	0.811
		0.045	0.045	0.255	0.261	0.465	0.504	0.675	0.820
		0.050	0.050	0.260	0.266	0.470	0.510	0.680	0.829
		0.055	0.055	0.265	0.271	0.475	0.517	0.685	0.838
		0.060	0.060	0.270	0.277	0.480	0.523	0.690	0.848
		0.065	0.065	0.275	0.282	0.485	0.530	0.695	0.858
		0.070	0.070	0.280	0.288	0.490	0.536	0.700	0.867
		0.075	0.075	0.285	0.293	0.495	0.543	0.705	0.877
		0.080	0.080	0.290	0.299	0.500	0.549	0.710	0.887
		0.085	0.085	0.295	0.304	0.505	0.556	0.715	0.897
		0.090	0.090	0.300	0.310	0.510	0.563	0.720	0.908
		0.095	0.095	0.305	0.315	0.515	0.570	0.725	0.918
		0.100	0.100	0.310	0.321	0.520	0.576	0.730	0.929
		0.105	0.105	0.315	0.326	0.525	0.583	0.735	0.940
		0.110	0.110	0.320	0.332	0.530	0.590	0.740	0.950
		0.115	0.116	0.325	0.337	0.535	0.597	0.745	0.962
		0.120	0.121	0.330	0.343	0.540	0.604	0.750	0.973
		0.125	0.126	0.335	0.348	0.545	0.611	0.755	0.984
		0.130	0.131	0.340	0.354	0.550	0.618	0.760	0.996
		0.135	0.136	0.345	0.360	0.555	0.626	0.765	1.008
		0.140	0.141	0.350	0.365	0.560	0.633	0.770	1.020
		0.145	0.146	0.355	0.371	0.565	0.640	0.775	1.033
		0.150	0.151	0.360	0.377	0.570	0.648	0.780	1.045
		0.155	0.156	0.365	0.383	0.575	0.655	0.785	1.058
		0.160	0.161	0.370	0.388	0.580	0.662	0.790	1.071
		0.165	0.167	0.375	0.394	0.585	0.670	0.795	1.085
		0.170	0.172	0.380	0.400	0.590	0.678	0.800	1.099
		0.175	0.177	0.385	0.406	0.595	0.685	0.805	1.113
		0.180	0.182	0.390	0.412	0.600	0.693	0.810	1.127
		0.185	0.187	0.395	0.418	0.605	0.701	0.815	1.142
		0.190	0.192	0.400	0.424	0.610	0.709	0.820	1.157
		0.195	0.198	0.405	0.430	0.615	0.717	0.825	1.172
		0.200	0.203	0.410	0.436	0.620	0.725	0.830	1.188
		0.205	0.208	0.415	0.442	0.625	0.733	0.835	1.204

(table continues)

r	Z	r	Z	r	Z	r	Z
0.840	1.221	0.880	1.376	0.920	1.589	0.960	1.946
0.845	1.238	0.885	1.398	0.925	1.623	0.965	2.014
0.850	1.256	0.890	1.422	0.930	1.658	0.970	2.092
0.855	1.274	0.895	1.447	0.935	1.697	0.975	2.185
0.860	1.293	0.900	1.472	0.940	1.738	0.980	2.298
0.865	1.313	0.905	1.499	0.945	1.783	0.985	2.443
0.870	1.333	0.910	1.528	0.950	1.832	0.990	2.647
0.875	1.354	0.915	1.557	0.955	1.886	0.995	2.994

**Table A.6
(continued)**

***Table of Fisher's
Transformation of
r to Z***

Source: From *Statistical Methods for Psychology,* 3rd ed., by D. C. Howell.
Copyright © 1992 PWS-Kent. Reprinted with permission.

Table A.7

Critical Values of the F Distribution for $\alpha = .05$

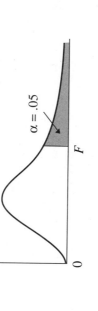

$\alpha = .05$

df Denominator	\multicolumn{19}{c}{df Numerator}
	1	2	3	4	5	6	7	8	9	10	12	15	20	24	30	40	60	120	∞
1	161.4	199.5	215.7	224.6	230.2	234.0	236.8	238.9	240.5	241.9	243.9	245.9	248.0	249.1	250.1	251.1	252.2	253.3	254.3
2	18.51	19.00	19.16	19.25	19.30	19.33	19.35	19.37	19.38	19.40	19.41	19.43	19.45	19.45	19.46	19.47	19.48	19.49	19.50
3	10.13	9.55	9.28	9.12	9.01	8.94	8.89	8.85	8.81	8.79	8.74	8.70	8.66	8.64	8.62	8.59	8.57	8.55	8.53
4	7.71	6.94	6.59	6.39	6.26	6.16	6.09	6.04	6.00	5.96	5.91	5.86	5.80	5.77	5.75	5.72	5.69	5.66	5.63
5	6.61	5.79	5.41	5.19	5.05	4.95	4.88	4.82	4.77	4.74	4.68	4.62	4.56	4.53	4.50	4.46	4.43	4.40	4.36
6	5.99	5.14	4.76	4.53	4.39	4.28	4.21	4.15	4.10	4.06	4.00	3.94	3.87	3.84	3.81	3.77	3.74	3.70	3.67
7	5.59	4.74	4.35	4.12	3.97	3.87	3.79	3.73	3.68	3.64	3.57	3.51	3.44	3.41	3.38	3.34	3.30	3.27	3.23
8	5.32	4.46	4.07	3.84	3.69	3.58	3.50	3.44	3.39	3.35	3.28	3.22	3.15	3.12	3.08	3.04	3.01	2.97	2.93
9	5.12	4.26	3.86	3.63	3.48	3.37	3.29	3.23	3.18	3.14	3.07	3.01	2.94	2.90	2.86	2.83	2.79	2.75	2.71
10	4.96	4.10	3.71	3.48	3.33	3.22	3.14	3.07	3.02	2.98	2.91	2.85	2.77	2.74	2.70	2.66	2.62	2.58	2.54
11	4.84	3.98	3.59	3.36	3.20	3.09	3.01	2.95	2.90	2.85	2.79	2.72	2.65	2.61	2.57	2.53	2.49	2.45	2.40
12	4.75	3.89	3.49	3.26	3.11	3.00	2.91	2.85	2.80	2.75	2.69	2.62	2.54	2.51	2.47	2.43	2.38	2.34	2.30
13	4.67	3.81	3.41	3.18	3.03	2.92	2.83	2.77	2.71	2.67	2.60	2.53	2.46	2.42	2.38	2.34	2.30	2.25	2.21
14	4.60	3.74	3.34	3.11	2.96	2.85	2.76	2.70	2.65	2.60	2.53	2.46	2.39	2.35	2.31	2.27	2.22	2.18	2.13
15	4.54	3.68	3.29	3.06	2.90	2.79	2.71	2.64	2.59	2.54	2.48	2.40	2.33	2.29	2.25	2.20	2.16	2.11	2.07
16	4.49	3.63	3.24	3.01	2.85	2.74	2.66	2.59	2.54	2.49	2.42	2.35	2.28	2.24	2.19	2.15	2.11	2.06	2.01
17	4.45	3.59	3.20	2.96	2.81	2.70	2.61	2.55	2.49	2.45	2.38	2.31	2.23	2.19	2.15	2.10	2.06	2.01	1.96
18	4.41	3.55	3.16	2.93	2.77	2.66	2.58	2.51	2.46	2.41	2.34	2.27	2.19	2.15	2.11	2.06	2.02	1.97	1.92
19	4.38	3.52	3.13	2.90	2.74	2.63	2.54	2.48	2.42	2.38	2.31	2.23	2.16	2.11	2.07	2.03	1.98	1.93	1.88
20	4.35	3.49	3.10	2.87	2.71	2.60	2.51	2.45	2.39	2.35	2.28	2.20	2.12	2.08	2.04	1.99	1.95	1.90	1.84
21	4.32	3.47	3.07	2.84	2.68	2.57	2.49	2.42	2.37	2.32	2.25	2.18	2.10	2.05	2.01	1.96	1.92	1.87	1.81
22	4.30	3.44	3.05	2.82	2.66	2.55	2.46	2.40	2.34	2.30	2.23	2.15	2.07	2.03	1.98	1.94	1.89	1.84	1.78
23	4.28	3.42	3.03	2.80	2.64	2.53	2.44	2.37	2.32	2.27	2.20	2.13	2.05	2.01	1.96	1.91	1.86	1.81	1.76
24	4.26	3.40	3.01	2.78	2.62	2.51	2.42	2.36	2.30	2.25	2.18	2.11	2.03	1.98	1.94	1.89	1.84	1.79	1.73
25	4.24	3.39	2.99	2.76	2.60	2.49	2.40	2.34	2.28	2.24	2.16	2.09	2.01	1.96	1.92	1.87	1.82	1.77	1.71
26	4.23	3.37	2.98	2.74	2.59	2.47	2.39	2.32	2.27	2.22	2.15	2.07	1.99	1.95	1.90	1.85	1.80	1.75	1.69
27	4.21	3.35	2.96	2.73	2.57	2.46	2.37	2.31	2.25	2.20	2.13	2.06	1.97	1.93	1.88	1.84	1.79	1.73	1.67
28	4.20	3.34	2.95	2.71	2.56	2.45	2.36	2.29	2.24	2.19	2.12	2.04	1.96	1.91	1.87	1.82	1.77	1.71	1.65
29	4.18	3.33	2.93	2.70	2.55	2.43	2.35	2.28	2.22	2.18	2.10	2.03	1.94	1.90	1.85	1.81	1.75	1.70	1.64
30	4.17	3.32	2.92	2.69	2.53	2.42	2.33	2.27	2.21	2.16	2.09	2.01	1.93	1.89	1.84	1.79	1.74	1.68	1.62
40	4.08	3.23	2.84	2.61	2.45	2.34	2.25	2.18	2.12	2.08	2.00	1.92	1.84	1.79	1.74	1.69	1.64	1.58	1.51
60	4.00	3.15	2.76	2.53	2.37	2.25	2.17	2.10	2.04	1.99	1.92	1.84	1.75	1.70	1.65	1.59	1.53	1.47	1.39
120	3.92	3.07	2.68	2.45	2.29	2.17	2.09	2.02	1.96	1.91	1.83	1.75	1.66	1.61	1.55	1.50	1.43	1.35	1.25
∞	3.84	3.00	2.60	2.37	2.21	2.10	2.01	1.94	1.88	1.83	1.75	1.67	1.57	1.52	1.46	1.39	1.32	1.22	1.00

Source: Adapted from *Biometrika Tables for Statisticians, Vol. 1*, 3rd ed, by E. Pearson & H. Hartley, Table 18. Copyright © 1966 University Press. Used with the permission of the Biometrika Trustees.

Table A.8

Critical Values of the F Distribution for $\alpha = .025$

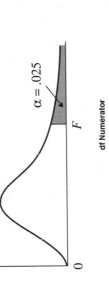

$\alpha = .025$

df Numerator

df Denominator	1	2	3	4	5	6	7	8	9	10	12	15	20	24	30	40	60	120	∞
1	647.8	799.5	864.2	899.6	921.8	937.1	948.2	956.7	963.3	968.6	976.7	984.9	993.1	997.2	1001	1006	1010	1014	1018
2	38.51	39.00	39.17	39.25	39.30	39.33	39.36	39.37	39.39	39.40	39.41	39.43	39.45	39.46	39.46	39.47	39.48	39.49	39.50
3	17.44	16.04	15.44	15.10	14.88	14.73	14.62	14.54	14.47	14.42	14.34	14.25	14.17	14.12	14.08	14.04	13.99	13.95	13.90
4	12.22	10.65	9.98	9.60	9.36	9.20	9.07	8.98	8.90	8.84	8.75	8.66	8.56	8.51	8.46	8.41	8.36	8.31	8.26
5	10.01	8.43	7.76	7.39	7.15	6.98	6.85	6.76	6.68	6.62	6.52	6.43	6.33	6.28	6.23	6.18	6.12	6.07	6.02
6	8.81	7.26	6.60	6.23	5.99	5.82	5.70	5.60	5.52	5.46	5.37	5.27	5.17	5.12	5.07	5.01	4.96	4.90	4.85
7	8.07	6.54	5.89	5.52	5.29	5.12	4.99	4.90	4.82	4.76	4.67	4.57	4.47	4.42	4.36	4.31	4.25	4.20	4.14
8	7.57	6.06	5.42	5.05	4.82	4.65	4.53	4.43	4.36	4.30	4.20	4.10	4.00	3.95	3.89	3.84	3.78	3.73	3.67
9	7.21	5.71	5.08	4.72	4.48	4.32	4.20	4.10	4.03	3.96	3.87	3.77	3.67	3.61	3.56	3.51	3.45	3.39	3.33
10	6.94	5.46	4.83	4.47	4.24	4.07	3.95	3.85	3.78	3.72	3.62	3.52	3.42	3.37	3.31	3.26	3.20	3.14	3.08
11	6.72	5.26	4.63	4.28	4.04	3.88	3.76	3.66	3.59	3.53	3.43	3.33	3.23	3.17	3.12	3.06	3.00	2.94	2.88
12	6.55	5.10	4.47	4.12	3.89	3.73	3.61	3.51	3.44	3.37	3.28	3.18	3.07	3.02	2.96	2.91	2.85	2.79	2.72
13	6.41	4.97	4.35	4.00	3.77	3.60	3.48	3.39	3.31	3.25	3.15	3.05	2.95	2.89	2.84	2.78	2.72	2.66	2.60
14	6.30	4.86	4.24	3.89	3.66	3.50	3.38	3.29	3.21	3.15	3.05	2.95	2.84	2.79	2.73	2.67	2.61	2.55	2.49
15	6.20	4.77	4.15	3.80	3.58	3.41	3.29	3.20	3.12	3.06	2.96	2.86	2.76	2.70	2.64	2.59	2.52	2.46	2.40
16	6.12	4.69	4.08	3.73	3.50	3.34	3.22	3.12	3.05	2.99	2.89	2.79	2.68	2.63	2.57	2.51	2.45	2.38	2.32
17	6.04	4.62	4.01	3.66	3.44	3.28	3.16	3.06	2.98	2.92	2.82	2.72	2.62	2.56	2.50	2.44	2.38	2.32	2.25
18	5.98	4.56	3.95	3.61	3.38	3.22	3.10	3.01	2.93	2.87	2.77	2.67	2.56	2.50	2.44	2.38	2.32	2.26	2.19
19	5.92	4.51	3.90	3.56	3.33	3.17	3.05	2.96	2.88	2.82	2.72	2.62	2.51	2.45	2.39	2.33	2.27	2.20	2.13
20	5.87	4.46	3.86	3.51	3.29	3.13	3.01	2.91	2.84	2.77	2.68	2.57	2.46	2.41	2.35	2.29	2.22	2.16	2.09
21	5.83	4.42	3.82	3.48	3.25	3.09	2.97	2.87	2.80	2.73	2.64	2.53	2.42	2.37	2.31	2.25	2.18	2.11	2.04
22	5.79	4.38	3.78	3.44	3.22	3.05	2.93	2.84	2.76	2.70	2.60	2.50	2.39	2.33	2.27	2.21	2.14	2.08	2.00
23	5.75	4.35	3.75	3.41	3.18	3.02	2.90	2.81	2.73	2.67	2.57	2.47	2.36	2.30	2.24	2.18	2.11	2.04	1.97
24	5.72	4.32	3.72	3.38	3.15	2.99	2.87	2.78	2.70	2.64	2.54	2.44	2.33	2.27	2.21	2.15	2.08	2.01	1.94
25	5.69	4.29	3.69	3.35	3.13	2.97	2.85	2.75	2.68	2.61	2.51	2.41	2.30	2.24	2.18	2.12	2.05	1.98	1.91
26	5.66	4.27	3.67	3.33	3.10	2.94	2.82	2.73	2.65	2.59	2.49	2.39	2.28	2.22	2.16	2.09	2.03	1.95	1.88
27	5.63	4.24	3.65	3.31	3.08	2.92	2.80	2.71	2.63	2.57	2.47	2.36	2.25	2.19	2.13	2.07	2.00	1.93	1.85
28	5.61	4.22	3.63	3.29	3.06	2.90	2.78	2.69	2.61	2.55	2.45	2.34	2.23	2.17	2.11	2.05	1.98	1.91	1.83
29	5.59	4.20	3.61	3.27	3.04	2.88	2.76	2.67	2.59	2.53	2.43	2.32	2.21	2.15	2.09	2.03	1.96	1.89	1.81
30	5.57	4.18	3.59	3.25	3.03	2.87	2.75	2.65	2.57	2.51	2.41	2.31	2.20	2.14	2.07	2.01	1.94	1.87	1.79
40	5.42	4.05	3.46	3.13	2.90	2.74	2.62	2.53	2.45	2.39	2.29	2.18	2.07	2.01	1.94	1.88	1.80	1.72	1.64
60	5.29	3.93	3.34	3.01	2.79	2.63	2.51	2.41	2.33	2.27	2.17	2.06	1.94	1.88	1.82	1.74	1.67	1.58	1.48
120	5.15	3.80	3.23	2.89	2.67	2.52	2.39	2.30	2.22	2.16	2.05	1.94	1.82	1.76	1.69	1.61	1.53	1.43	1.31
∞	5.02	3.69	3.12	2.79	2.57	2.41	2.29	2.19	2.11	2.05	1.94	1.83	1.71	1.64	1.57	1.48	1.39	1.27	1.00

Source: Adapted from Biometrika Tables for Statisticians, Vol. 1, 3rd ed., by E. Pearson & H. Hartley, Table 18. Copyright © 1966 University Press. Used with the permission of the Biometrika Trustees.

Table A.9

Critical Values of the F Distribution for $\alpha = .01$

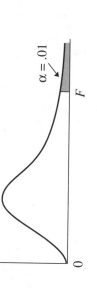

$\alpha = .01$

df Denominator	\multicolumn{19}{c}{df Numerator}
	1	2	3	4	5	6	7	8	9	10	12	15	20	24	30	40	60	120	∞
1	4052	4999.5	5403	5625	5764	5859	5928	5981	6022	6056	6106	6157	6209	6235	6261	6287	6313	6339	6366
2	98.50	99.00	99.17	99.25	99.30	99.33	99.36	99.37	99.39	99.40	99.42	99.43	99.45	99.46	99.47	99.47	99.48	99.49	99.50
3	34.12	30.82	29.46	28.71	28.24	27.91	27.67	27.49	27.35	27.23	27.05	26.87	26.69	26.60	26.50	26.41	26.32	26.22	26.13
4	21.20	18.00	16.69	15.98	15.52	15.21	14.98	14.80	14.66	14.55	14.37	14.20	14.02	13.93	13.84	13.75	13.65	13.56	13.46
5	16.26	13.27	12.06	11.39	10.97	10.67	10.46	10.29	10.16	10.05	9.89	9.72	9.55	9.47	9.38	9.29	9.20	9.11	9.02
6	13.75	10.92	9.78	9.15	8.75	8.47	8.26	8.10	7.98	7.87	7.72	7.56	7.40	7.31	7.23	7.14	7.06	6.97	6.88
7	12.25	9.55	8.45	7.85	7.46	7.19	6.99	6.84	6.72	6.62	6.47	6.31	6.16	6.07	5.99	5.91	5.82	5.74	5.65
8	11.26	8.65	7.59	7.01	6.63	6.37	6.18	6.03	5.91	5.81	5.67	5.52	5.36	5.28	5.20	5.12	5.03	4.95	4.86
9	10.56	8.02	6.99	6.42	6.06	5.80	5.61	5.47	5.35	5.26	5.11	4.96	4.81	4.73	4.65	4.57	4.48	4.40	4.31
10	10.04	7.56	6.55	5.99	5.64	5.39	5.20	5.06	4.94	4.85	4.71	4.56	4.41	4.33	4.25	4.17	4.08	4.00	3.91
11	9.65	7.21	6.22	5.67	5.32	5.07	4.89	4.74	4.63	4.54	4.40	4.25	4.10	4.02	3.94	3.86	3.78	3.69	3.60
12	9.33	6.93	5.95	5.41	5.06	4.82	4.64	4.50	4.39	4.30	4.16	4.01	3.86	3.78	3.70	3.62	3.54	3.45	3.36
13	9.07	6.70	5.74	5.21	4.86	4.62	4.44	4.30	4.19	4.10	3.96	3.82	3.66	3.59	3.51	3.43	3.34	3.25	3.17
14	8.86	6.51	5.56	5.04	4.69	4.46	4.28	4.14	4.03	3.94	3.80	3.66	3.51	3.43	3.35	3.27	3.18	3.09	3.00
15	8.68	6.36	5.42	4.89	4.56	4.32	4.14	4.00	3.89	3.80	3.67	3.52	3.37	3.29	3.21	3.13	3.05	2.96	2.87
16	8.53	6.23	5.29	4.77	4.44	4.20	4.03	3.89	3.78	3.69	3.55	3.41	3.26	3.18	3.10	3.02	2.93	2.84	2.75
17	8.40	6.11	5.18	4.67	4.34	4.10	3.93	3.79	3.68	3.59	3.46	3.31	3.16	3.08	3.00	2.92	2.83	2.75	2.65
18	8.29	6.01	5.09	4.58	4.25	4.01	3.84	3.71	3.60	3.51	3.37	3.23	3.08	3.00	2.92	2.84	2.75	2.66	2.57
19	8.18	5.93	5.01	4.50	4.17	3.94	3.77	3.63	3.52	3.43	3.30	3.15	3.00	2.92	2.84	2.76	2.67	2.58	2.49
20	8.10	5.85	4.94	4.43	4.10	3.87	3.70	3.56	3.46	3.37	3.23	3.09	2.94	2.86	2.78	2.69	2.61	2.52	2.42
21	8.02	5.78	4.87	4.37	4.04	3.81	3.64	3.51	3.40	3.31	3.17	3.03	2.88	2.80	2.72	2.64	2.55	2.46	2.36
22	7.95	5.72	4.82	4.31	3.99	3.76	3.59	3.45	3.35	3.26	3.12	2.98	2.83	2.75	2.67	2.58	2.50	2.40	2.31
23	7.88	5.66	4.76	4.26	3.94	3.71	3.54	3.41	3.30	3.21	3.07	2.93	2.78	2.70	2.62	2.54	2.45	2.35	2.26
24	7.82	5.61	4.72	4.22	3.90	3.67	3.50	3.36	3.26	3.17	3.03	2.89	2.74	2.66	2.58	2.49	2.40	2.31	2.21
25	7.77	5.57	4.68	4.18	3.85	3.63	3.46	3.32	3.22	3.13	2.99	2.85	2.70	2.62	2.54	2.45	2.36	2.27	2.17
26	7.72	5.53	4.64	4.14	3.82	3.59	3.42	3.29	3.18	3.09	2.96	2.81	2.66	2.58	2.50	2.42	2.33	2.23	2.13
27	7.68	5.49	4.60	4.11	3.78	3.56	3.39	3.26	3.15	3.06	2.93	2.78	2.63	2.55	2.47	2.38	2.29	2.20	2.10
28	7.64	5.45	4.57	4.07	3.75	3.53	3.36	3.23	3.12	3.03	2.90	2.75	2.60	2.52	2.44	2.35	2.26	2.17	2.06
29	7.60	5.42	4.54	4.04	3.73	3.50	3.33	3.20	3.09	3.00	2.87	2.73	2.57	2.49	2.41	2.33	2.23	2.14	2.03
30	7.56	5.39	4.51	4.02	3.70	3.47	3.30	3.17	3.07	2.98	2.84	2.70	2.55	2.47	2.39	2.30	2.21	2.11	2.01
40	7.31	5.18	4.31	3.83	3.51	3.29	3.12	2.99	2.89	2.80	2.66	2.52	2.37	2.29	2.20	2.11	2.02	1.92	1.80
60	7.08	4.98	4.13	3.65	3.34	3.12	2.95	2.82	2.72	2.63	2.50	2.35	2.20	2.12	2.03	1.94	1.84	1.73	1.60
120	6.85	4.79	3.95	3.48	3.17	2.96	2.79	2.66	2.56	2.47	2.34	2.19	2.03	1.95	1.86	1.76	1.66	1.53	1.38
∞	6.63	4.61	3.78	3.32	3.02	2.80	2.64	2.51	2.41	2.32	2.18	2.04	1.88	1.79	1.70	1.59	1.47	1.32	1.00

Source: Adapted from *Biometrika Tables for Statisticians, Vol. 1*, 3rd ed, by E. Pearson & H. Hartley, Table 18. Copyright © 1966 University Press. Used with the permission of the Biometrika Trustees.

Table A.10

Power of ANOVA as a Function of Effect Size and Sample Size (k = 3 groups)

		ω^2								
n	α	.1	.2	.3	.4	.5	.6	.7	.8	.9
2	.05	.07	.10	.14	.18	.25	.35	.49	.70	.95
	.01	.02	.02	.03	.04	.06	.09	.15	.26	.55
3	.05	.10	.16	.25	.38	.53	.72	.89	.99	.99
	.01	.02	.04	.08	.13	.21	.35	.56	.84	.99
4	.05	.13	.23	.38	.56	.75	.91	.99	.99	.99
	.01	.03	.07	.14	.25	.42	.64	.87	.99	.99
5	.05	.15	.31	.50	.70	.88	.98	.99	.99	.99
	.01	.04	.11	.22	.39	.61	.84	.97	.99	.99
6	.05	.18	.38	.60	.81	.95	.99	.99	.99	.99
	.01	.06	.15	.31	.53	.77	.94	.99	.99	.99
7	.05	.22	.45	.70	.89	.98	.99	.99	.99	.99
	.01	.07	.20	.40	.66	.87	.98	.99	.99	.99
8	.05	.25	.51	.77	.94	.99	.99	.99	.99	.99
	.01	.09	.25	.49	.76	.94	.99	.99	.99	.99
9	.05	.28	.58	.83	.97	.99	.99	.99	.99	.99
	.01	.10	.30	.58	.84	.97	.99	.99	.99	.99
10	.05	.31	.63	.88	.98	.99	.99	.99	.99	.99
	.01	.12	.36	.66	.89	.99	.99	.99	.99	.99
11	.05	.34	.69	.92	.99	.99	.99	.99	.99	.99
	.01	.14	.41	.73	.93	.99	.99	.99	.99	.99
12	.05	.37	.73	94	.99	.99	.99	.99	.99	.99
	.01	.16	.46	.78	.96	.99	.99	.99	.99	.99
13	.05	.40	.77	.96	.99	.99	.99	.99	.99	.99
	.01	.18	.51	.83	.98	.99	.99	.99	.99	.99
14	.05	.44	.81	.97	.99	.99	.99	.99	.99	.99
	.01	.20	.57	.87	.99	.99	.99	.99	.99	.99
15	.05	.47	.85	.98	.99	.99	.99	.99	.99	.99
	.01	.22	.61	.91	.99	.99	.99	.99	.99	.99
20	.05	.60	.95	.99	.99	.99	.99	.99	.99	.99
	.01	.34	.80	.98	.99	.99	.99	.99	.99	.99
30	.05	.81	.99	.99	.99	.99	.99	.99	.99	.99
	.01	.56	.96	.99	.99	.99	.99	.99	.99	.99

(table continues)

Table A.10
(continued)

Power of ANOVA as a Function of Effect Size and Sample Size (k = 4 groups)

n	α	ω^2								
		.1	.2	.3	.4	.5	.6	.7	.8	.9
2	.05	.07	.11	.15	.21	.30	.42	.60	.82	.99
	.01	.02	.02	.04	.05	.08	.13	.22	.40	.78
3	.05	.10	.18	.29	.44	.61	.80	.95	.99	.99
	.01	.02	.05	.09	.17	.28	.47	.72	.94	.99
4	.05	.13	.26	.43	.64	.83	.96	.99	.99	.99
	.01	.04	.09	.18	.33	.53	.78	.95	.99	.99
5	.05	.17	.35	.57	.78	.94	.99	.99	.99	.99
	.01	.05	.13	.28	.50	.74	.93	.99	.99	.99
6	.05	.20	.43	.68	.88	.98	.99	.99	.99	.99
	.01	.07	.19	.39	.65	.88	.98	.99	.99	.99
7	.05	.24	.51	.77	.94	.99	.99	.99	.99	.99
	.01	.08	.25	.50	.77	.95	.99	.99	.99	.99
8	.05	.27	.58	.85	.97	.99	.99	.99	.99	.99
	.01	.10	.31	.61	.86	.98	.99	.99	.99	.99
9	.05	.31	.65	.90	.99	.99	.99	.99	.99	.99
	.01	.12	.37	.69	.92	.99	.99	.99	.99	.99
10	.05	.35	.71	.93	.99	.99	.99	.99	.99	.99
	.01	.15	.45	.78	.96	.99	.99	.99	.99	.99
11	.05	.39	.76	.96	.99	.99	.99	.99	.99	.99
	.01	.17	.51	.83	.98	.99	.99	.99	.99	.99
12	.05	.43	.81	.98	.99	.99	.99	.99	.99	.99
	.01	.19	.56	.88	.99	.99	.99	.99	.99	.99
13	.05	.46	.85	.99	.99	.99	.99	.99	.99	.99
	.01	.22	.62	.92	.99	.99	.99	.99	.99	.99
14	.05	.49	.88	.99	.99	.99	.99	.99	.99	.99
	.01	.24	.67	.94	.99	.99	.99	.99	.99	.99
15	.05	.53	.91	.99	.99	.99	.99	.99	.99	.99
	.01	.27	.72	.96	.99	.99	.99	.99	.99	.99
20	.05	.68	.98	.99	.99	.99	.99	.99	.99	.99
	.01	.42	.89	.99	.99	.99	.99	.99	.99	.99
30	.05	.88	.99	.99	.99	.99	.99	.99	.99	.99
	.01	.68	.99	.99	.99	.99	.99	.99	.99	.99

(table continues)

n	α	.1	.2	.3	.4	.5	.6	.7	.8	.9
2	.05	.08	.11	.17	.24	.34	.49	.69	.90	.99
	.01	.02	.03	.04	.06	.10	.17	.30	.54	.92
3	.05	.11	.20	.33	.49	.69	.87	.98	.99	.99
	.01	.03	.06	.11	.21	.36	.58	.83	.98	.99
4	.05	.14	.29	.49	.71	.89	.98	.99	.99	.99
	.01	.04	.10	.22	.40	.64	.87	.99	.99	.99
5	.05	.18	.39	.63	.85	.97	.99	.99	.99	.99
	.01	.06	.16	.35	.59	.84	.97	.99	.99	.99
6	.05	.22	.48	.75	.93	.99	.99	.99	.99	.99
	.01	.07	.23	.47	.75	.94	.99	.99	.99	.99
7	.05	.26	.57	.84	.97	.99	.99	.99	.99	.99
	.01	.10	.30	.60	.86	.98	.99	.99	.99	.99
8	.05	.30	.65	.90	.99	.99	.99	.99	.99	.99
	.01	.12	.37	.70	.93	.99	.99	.99	.99	.99
9	.05	.35	.72	.94	.99	.99	.99	.99	.99	.99
	.01	.15	.45	.79	.97	.99	.99	.99	.99	.99
10	.05	.39	.78	.97	.99	.99	.99	.99	.99	.99
	.01	.17	.52	.85	.98	.99	.99	.99	.99	.99
11	.05	.43	.82	.98	.99	.99	.99	.99	.99	.99
	.01	.20	.59	.90	.99	.99	.99	.99	.99	.99
12	.05	.47	.87	.99	.99	.99	.99	.99	.99	.99
	.01	.23	.66	.94	.99	.99	.99	.99	.99	.99
13	.05	.51	.90	.99	.99	.99	.99	.99	.99	.99
	.01	.27	.72	.96	.99	.99	.99	.99	.99	.99
14	.05	.55	.93	.99	.99	.99	.99	.99	.99	.99
	.01	.29	.77	.98	.99	.99	.99	.99	.99	.99
15	.05	.59	.95	.99	.99	.99	.99	.99	.99	.99
	.01	.33	.81	.99	.99	.99	.99	.99	.99	.99
20	.05	.74	.99	.99	.99	.99	.99	.99	.99	.99
	.01	.49	.95	.99	.99	.99	.99	.99	.99	.99
30	.05	.92	.99	.99	.99	.99	.99	.99	.99	.99
	.01	.77	.99	.99	.99	.99	.99	.99	.99	.99

Note: Each entry of .99 represents power that is .99 or greater.
Source: Adapted from *Statistics,* 4th ed., by W. L. Hays. Holt, Rinehart, and Winston, 1988.

Table A.10 (continued)

Power of ANOVA as a Function of Effect Size and Sample Size ($k = 5$ groups)

Table A.11

Critical Values of the Studentized Range Statistic (q) for $\alpha = .05$

Number of Groups (or Number of Steps Between Ordered Means)

df for Error Term	2	3	4	5	6	7	8	9	10	11	12	13	14	15	16	17	18	19	20
1	17·97	26·98	32·82	37·08	40·41	43·12	45·40	47·36	49·07	50·59	51·96	53·20	54·33	55·36	56·32	57·22	58·04	58·83	59·56
2	6·08	8·33	9·80	10·88	11·74	12·44	13·03	13·54	13·99	14·39	14·75	15·08	15·38	15·65	15·91	16·14	16·37	16·57	16·77
3	4·50	5·91	6·82	7·50	8·04	8·48	8·85	9·18	9·46	9·72	9·95	10·15	10·35	10·52	10·69	10·84	10·98	11·11	11·24
4	3·93	5·04	5·76	6·29	6·71	7·05	7·35	7·60	7·83	8·03	8·21	8·37	8·52	8·66	8·79	8·91	9·03	9·13	9·23
5	3·64	4·60	5·22	5·67	6·03	6·33	6·58	6·80	6·99	7·17	7·32	7·47	7·60	7·72	7·83	7·93	8·03	8·12	8·21
6	3·46	4·34	4·90	5·30	5·63	5·90	6·12	6·32	6·49	6·65	6·79	6·92	7·03	7·14	7·24	7·34	7·43	7·51	7·59
7	3·34	4·16	4·68	5·06	5·36	5·61	5·82	6·00	6·16	6·30	6·43	6·55	6·66	6·76	6·85	6·94	7·02	7·10	7·17
8	3·26	4·04	4·53	4·89	5·17	5·40	5·60	5·77	5·92	6·05	6·18	6·29	6·39	6·48	6·57	6·65	6·73	6·80	6·87
9	3·20	3·95	4·41	4·76	5·02	5·24	5·43	5·59	5·74	5·87	5·98	6·09	6·19	6·28	6·36	6·44	6·51	6·58	6·64
10	3·15	3·88	4·33	4·65	4·91	5·12	5·30	5·46	5·60	5·72	5·83	5·93	6·03	6·11	6·19	6·27	6·34	6·40	6·47
11	3·11	3·82	4·26	4·57	4·82	5·03	5·20	5·35	5·49	5·61	5·71	5·81	5·90	5·98	6·06	6·13	6·20	6·27	6·33
12	3·08	3·77	4·20	4·51	4·75	4·95	5·12	5·27	5·39	5·51	5·61	5·71	5·80	5·88	5·95	6·02	6·09	6·15	6·21
13	3·06	3·73	4·15	4·45	4·69	4·88	5·05	5·19	5·32	5·43	5·53	5·63	5·71	5·79	5·86	5·93	5·99	6·05	6·11
14	3·03	3·70	4·11	4·41	4·64	4·83	4·99	5·13	5·25	5·36	5·46	5·55	5·64	5·71	5·79	5·85	5·91	5·97	6·03
15	3·01	3·67	4·08	4·37	4·59	4·78	4·94	5·08	5·20	5·31	5·40	5·49	5·57	5·65	5·72	5·78	5·85	5·90	5·96
16	3·00	3·65	4·05	4·33	4·56	4·74	4·90	5·03	5·15	5·26	5·35	5·44	5·52	5·59	5·66	5·72	5·79	5·84	5·90
17	2·98	3·63	4·02	4·30	4·52	4·70	4·86	4·99	5·11	5·21	5·31	5·39	5·47	5·54	5·61	5·67	5·73	5·79	5·84
18	2·97	3·61	4·00	4·28	4·49	4·67	4·82	4·96	5·07	5·17	5·27	5·35	5·43	5·50	5·57	5·63	5·69	5·74	5·79
19	2·96	3·59	3·98	4·25	4·47	4·65	4·79	4·92	5·04	5·14	5·23	5·31	5·39	5·46	5·53	5·59	5·65	5·70	5·75
20	2·95	3·58	3·96	4·23	4·45	4·62	4·77	4·90	5·01	5·11	5·20	5·28	5·36	5·43	5·49	5·55	5·61	5·66	5·71
24	2·92	3·53	3·90	4·17	4·37	4·54	4·68	4·81	4·92	5·01	5·10	5·18	5·25	5·32	5·38	5·44	5·49	5·55	5·59
30	2·89	3·49	3·85	4·10	4·30	4·46	4·60	4·72	4·82	4·92	5·00	5·08	5·15	5·21	5·27	5·33	5·38	5·43	5·47
40	2·86	3·44	3·79	4·04	4·23	4·39	4·52	4·63	4·73	4·82	4·90	4·98	5·04	5·11	5·16	5·22	5·27	5·31	5·36
60	2·83	3·40	3·74	3·98	4·16	4·31	4·44	4·55	4·65	4·73	4·81	4·88	4·94	5·00	5·06	5·11	5·15	5·20	5·24
120	2·80	3·36	3·68	3·92	4·10	4·24	4·36	4·47	4·56	4·64	4·71	4·78	4·84	4·90	4·95	5·00	5·04	5·09	5·13
∞	2·77	3·31	3·63	3·86	4·03	4·17	4·29	4·39	4·47	4·55	4·62	4·68	4·74	4·80	4·85	4·89	4·93	4·97	5·01

Source: Adapted from *Biometrika Tables for Statisticians, Vol 1*, 3rd ed., by E. Pearson & H. Hartley, Table 29. Copyright © 1966 University Press. Used with the permission of the Biometrika Trustees.

n	X	p	n	X	p	n	X	p
1	0	.5000		1	.0176	13	0	.0001
	1	.5000		2	.0703		1	.0016
2	0	.2500		3	.1641		2	.0095
	1	.5000		4	.2461		3	.0349
	2	.2500		5	.2461		4	.0873
3	0	.1250		6	.1641		5	.1571
	1	.3750		7	.0703		6	.2095
	2	.3750		8	.0176		7	.2095
	3	.1250		9	.0020		8	.1571
4	0	.0625	10	0	.0010		9	.0873
	1	.2500		1	.0098		10	.0349
	2	.3750		2	.0439		11	.0095
	3	.2500		3	.1172		12	.0016
	4	.0625		4	.2051		13	.0001
5	0	.0312		5	.2461	14	0	.0001
	1	.1562		6	.2051		1	.0009
	2	.3125		7	.1172		2	.0056
	3	.3125		8	.0439		3	.0222
	4	.1562		9	.0098		4	.0611
	5	.0312		10	.0010		5	.1222
6	0	.0156	11	0	.0005		6	.1833
	1	.0938		1	.0054		7	.2095
	2	.2344		2	.0269		8	.1833
	3	.3125		3	.0806		9	.1222
	4	.2344		4	.1611		10	.0611
	5	.0938		5	.2256		11	.0222
	6	.0156		6	.2256		12	.0056
7	0	.0078		7	.1611		13	.0009
	1	.0547		8	.0806		14	.0001
	2	.1641		9	.0269	15	0	.0000
	3	.2734		10	.0054		1	.0005
	4	.2734		11	.0005		2	.0032
	5	.1641	12	0	.0002		3	.0139
	6	.0547		1	.0029		4	.0417
	7	.0078		2	.0161		5	.0916
8	0	.0039		3	.0537		6	.1527
	1	.0312		4	.1208		7	.1964
	2	.1094		5	.1934		8	.1964
	3	.2188		6	.2256		9	.1527
	4	.2734		7	.1934		10	.0916
	5	.2188		8	.1208		11	.0417
	6	.1094		9	.0537		12	.0139
	7	.0312		10	.0161		13	.0032
	8	.0039		11	.0029		14	.0005
9	0	.0020		12	.0002		15	.0000

Table A.12

Probabilities of the Binomial Distribution

Table A.13

Critical Values of the χ^2 Distribution

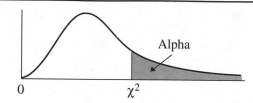

Alpha

0   χ^2

Alpha (area in the upper tail)

df	.10	.05	.025	.01	.005
1	2.71	3.84	5.02	6.63	7.88
2	4.61	5.99	7.38	9.21	10.60
3	6.25	7.81	9.35	11.34	12.84
4	7.78	9.49	11.14	13.28	14.86
5	9.24	11.07	12.83	15.09	16.75
6	10.64	12.59	14.45	16.81	18.55
7	12.02	14.07	16.01	18.48	20.28
8	13.36	15.51	17.53	20.09	21.96
9	14.68	16.92	19.02	21.67	23.59
10	15.99	18.31	20.48	23.21	25.19
11	17.28	19.68	21.92	24.72	26.76
12	18.55	21.03	23.34	26.22	28.30
13	19.81	22.36	24.74	27.69	29.82
14	21.06	23.68	26.12	29.14	31.32
15	22.31	25.00	27.49	30.58	32.80
16	23.54	26.30	28.85	32.00	34.27
17	24.77	27.59	30.19	33.41	35.72
18	25.99	28.87	31.53	34.81	37.16
19	27.20	30.14	32.85	36.19	38.58
20	28.41	31.41	34.17	37.57	40.00
21	29.62	32.67	35.48	38.93	41.40
22	30.81	33.92	36.78	40.29	42.80
23	32.01	35.17	38.08	41.64	44.18
24	33.20	36.42	39.36	42.98	45.56
25	34.38	37.65	40.65	44.31	46.93
26	35.56	38.89	41.92	45.64	48.29
27	36.74	40.11	43.19	46.96	49.64
28	37.92	41.34	44.46	48.28	50.99
29	39.09	42.56	45.72	49.59	52.34
30	40.26	43.77	46.98	50.89	53.67
40	51.81	55.76	59.34	63.69	66.77
50	63.17	67.50	71.42	76.15	79.49
60	74.40	79.08	83.30	88.38	91.95
70	85.53	90.53	95.02	100.42	104.22
80	96.58	101.88	106.63	112.33	116.32
90	107.56	113.14	118.14	124.12	128.30
100	118.50	124.34	129.56	135.81	140.17

Table A-14

Critical Values of ΣR_A for the Wilcoxon Rank-Sum Test (the Mann-Whitney Test)

n_B	$n_A = 3$ .05	.025	.01	.005	n_B	$n_A = 4$ .05	.025	.01	.005
3	6–15				4	11–25	10–26		
4	6–18				5	12–28	11–29	10–30	
5	7–20	6–21			6	13–31	12–32	11–33	10–34
6	8–22	7–23			7	14–34	13–35	11–37	10–38
7	8–25	7–26	6–27		8	15–37	14–48	12–40	11–41
8	9–27	8–28	6–30		9	16–40	14–42	13–43	11–45
9	10–29	8–31	7–32	6–33	10	17–43	15–45	13–47	12–48
10	10–32	9–33	7–35	6–36	11	18–46	16–48	14–50	12–52
11	11–34	9–36	7–38	6–39	12	19–49	17–51	15–53	13–55
12	11–37	10–38	8–40	7–41	13	20–52	18–54	15–57	13–59
13	12–39	10–41	8–43	7–44	14	21–55	19–57	16–60	14–62
14	13–41	11–43	8–46	7–47	15	22–58	20–60	17–63	15–65
15	13–44	11–46	9–48	8–49					

n_B	$n_A = 5$ .05	.025	.01	.005	n_B	$n_A = 6$ .05	.025	.01	.005
5	19–36	17–38	16–39	15–40	6	28–50	26–52	24–54	23–55
6	20–40	18–42	17–43	16–44	7	29–55	27–57	25–59	24–60
7	21–44	20–45	18–47	16–49	8	31–59	29–61	27–63	25–65
8	23–47	21–49	19–51	17–53	9	33–63	31–65	28–68	26–70
9	24–51	22–53	20–55	18–57	10	35–67	32–70	29–73	27–75
10	26–54	23–57	21–59	19–61	11	37–71	34–74	30–78	28–80
11	27–58	24–61	22–63	20–65	12	38–76	35–79	32–82	30–84
12	28–62	26–64	23–67	21–69	13	40–80	37–83	33–87	31–89
13	30–65	27–68	24–71	22–73	14	42–84	38–88	34–92	32–94
14	31–69	28–72	25–75	22–78	15	44–88	40–92	36–96	33–99
15	33–72	29–76	26–79	23–82					

(table continues)

Table A-14
(continued)

Critical Values of ΣR_A for the Wilcoxon Rank-Sum Test (the Mann-Whitney Test)

$n_A = 7$

n_B	.05	.025	.01	.005
7	39–66	36–69	34–71	32–73
8	41–71	38–74	35–77	34–78
9	43–76	40–79	37–82	35–84
10	45–81	42–84	39–87	37–89
11	47–86	44–89	40–93	38–95
12	49–91	46–94	42–98	40–100
13	52–95	48–99	44–103	41–106
14	54–100	50–104	45–109	43–111
15	56–105	52–109	47–114	44–117

$n_A = 8$

n_B	.05	.025	.01	.005
8	51–85	49–87	45–91	43–93
9	54–90	51–93	47–97	45–99
10	56–96	53–99	49–103	47–105
11	59–101	55–105	51–109	49–111
12	62–106	58–110	53–115	51–117
13	64–112	60–116	56–120	53–123
14	67–117	62–122	58–126	54–130
15	69–123	65–127	60–132	56–136

$n_A = 9$

n_B	.05	.025	.01	.005
9	66–105	62–109	59–112	56–115
10	69–111	65–115	61–119	58–122
11	72–117	68–121	63–126	61–128
12	75–123	71–127	66–132	63–135
13	78–129	73–134	68–139	65–142
14	81–135	76–140	71–145	67–149
15	84–141	79–146	73–152	69–156

$n_A = 10$

n_B	.05	.025	.01	.005
10	82–128	78–132	74–136	71–139
11	86–134	81–139	77–143	73–147
12	89–141	84–146	79–151	76–154
13	92–148	88–152	82–158	79–161
14	96–154	91–159	85–165	81–169
15	99–161	94–166	88–172	84–176

$n_A = 11$

n_B	.05	.025	.01	.005
11	100–153	96–157	91–162	87–166
12	104–160	99–165	94–170	90–174
13	108–167	103–172	97–178	93–182
14	112–174	106–180	100–186	96–190
15	116–181	110–187	103–194	99–198

$n_A = 12$

n_B	.05	.025	.01	.005
12	120–180	115–185	109–191	105–195
13	125–187	119–193	113–199	109–203
14	129–195	123–201	116–208	112–212
15	133–203	127–209	120–216	115–221

(table continues)

Table A-14
(continued)

Critical Values of ΣR_A for the Wilcoxon Rank-Sum Test (the Mann-Whitney Test)

$n_A = 13$					$n_A = 14$			
n_B	.05	.025	.01	.005	n_B	.05	.025	.01	.005

n_B	.05	.025	.01	.005
13	142–209	136–215	130–221	125–226
14	147–217	141–223	134–230	129–235
15	152–225	145–232	138–239	133–244

n_B	.05	.025	.01	.005
14	166–240	160–246	152–254	147–259
15	171–249	164–256	156–264	151–269

$n_A = 15$				
n_B	.05	.025	.01	.005

n_B	.05	.025	.01	.005
15	192–273	184–281	176–289	171–294

Source: Adapted from Table 22 of Pearson, E., & Hartley, H. (1966). *Biometrika Tables for Statisticians* (Vol. 1, 3rd ed.). Cambridge: Cambridge University Press.

Table A.15	Level of Significance for a One-Tailed Test			
	.05	.025	.01	.005
Critical Values of Spearman's Rank Correlation Coefficient	Level of Significance for a Two-Tailed Test			
n	0.10	0.05	0.02	0.01
5	0.900	—	—	—
6	0.829	0.886	0.943	—
7	0.714	0.786	0.893	0.929
8	0.643	0.738	0.833	0.881
9	0.600	0.700	0.783	0.833
10	0.564	0.648	0.745	0.794
11	0.536	0.618	0.709	0.818
12	0.497	0.591	0.703	0.780
13	0.475	0.566	0.673	0.745
14	0.457	0.545	0.646	0.716
15	0.441	0.525	0.623	0.689
16	0.425	0.507	0.601	0.666
17	0.412	0.490	0.582	0.645
18	0.399	0.476	0.564	0.625
19	0.388	0.462	0.549	0.608
20	0.377	0.450	0.534	0.591
21	0.368	0.438	0.521	0.576
22	0.359	0.428	0.508	0.562
23	0.351	0.418	0.496	0.549
24	0.343	0.409	0.485	0.537
25	0.336	0.400	0.475	0.526
26	0.329	0.392	0.465	0.515
27	0.323	0.385	0.456	0.505
28	0.317	0.377	0.448	0.496
29	0.311	0.370	0.440	0.487
30	0.305	0.364	0.432	0.478

Source: Adapted from *CRC Handbook of Tables for Probability and Statistics,* 2nd ed., by W. H. Beyer (Ed.), Chemical Rubber Company, 1968.

Review of Basic Math

I will begin by assuming that you will be using an electronic calculator with a square root key to solve the exercises in this text. You may have a statistical or scientific calculator that can compute the variance or standard deviation of a string of numbers automatically. You should not use that function, however, when the object of the exercise is to calculate the variance or standard deviation by one of the formulas in this text. Regardless of how sophisticated your calculator is, you will need to recall some of the basics of high school mathematics to solve some of the exercises and to follow some of the more involved mathematical discussions and derivations in the text. This text stresses understanding the structure of statistical formulas and the relationship between seemingly very different formulas. Given this emphasis, you may need to brush up on some basic mathematical concepts to appreciate fully what the text is trying to convey. The math review that follows covers only the concepts and procedures that you will need for this text.

To help you determine whether you to need to review basic math, I devised a two-part diagnostic quiz. The two parts of the quiz correspond to the two parts of the math review that follows. (The answers to the diagnostic quiz are at the end of the math review.) If you get more than two answers wrong in one part of the quiz, I strongly recommend that you study the math review corresponding to that part. After studying the math review and working out the practice exercises, you should take the post-quiz for the part or parts you studied. (The answers to the practice exercises and the post-quizzes are also at the end of the math review.) If you still get more than two answers wrong in either part of the post-quiz, you should consider a more extensive review of the math concepts with which you are having some difficulty. There are many excellent texts that will give you practice with the kind of mathematics that is needed for an introductory or intermediate statistics course. Following are two suggestions:

Baggaley, A. R. (1969). *Mathematics for introductory statistics.* New York: John Wiley & Sons.

Falstein, L. D. (1986). *Basic mathematics* (2nd ed.). Reading, MA: Addison-Wesley.

If your difficulty with basic math is caused by or is causing you anxiety, I recommend Tobias, S. (1994). *Overcoming math anxiety.* New York: W. W. Norton.

Diagnostic Quiz

Part I

1. $11 - 3 \cdot 2 - 8 \div 2 = ?$

2. $(11 - 3) \cdot 2 - 8 \div 2 = ?$

3. $14 + (-2) + (-5) + 4 - (-1) = ?$

4. $(-9) \cdot (-2) \div (-6) = ?$

5. Convert 6/50 to a decimal.

6. Convert .003 to a fraction.

7. What fraction corresponds to 7%?

8. What percent of 86 is 16?

9. What is the value of $4.1 \cdot 10^3$?

10. Express 5.5% as a decimal.

Part II

1. $\dfrac{F}{GX} - \dfrac{LS}{GX} = ?$

2. $\dfrac{T}{S} + \dfrac{L}{K} = ?$

3. $\left(\dfrac{R}{QP}\right) \cdot \left(\dfrac{MO}{S}\right) = ?$

4. $\left(\dfrac{B}{C}\right) \div \left(\dfrac{X}{Y}\right) = ?$

5. Simplify this fraction: $\dfrac{RX + PR}{QR}$.

6. What is the value of Y in the following equation?
$\frac{1}{2}Y - 6 = 1$

7. Solve the following equation for X: $2Y = CX + B$.

If $A = 3$ and $B = 5$, what is the value for each of the following expressions?

8. $(2A + B)^2$

9. $\sqrt{(2A + 2B)}$

10. $2BA^2$

Part I. Arithmetic Operations

Symbols you will need:

$+$	Addition	$\neq$	Not equal to
$-$	Subtraction	$>$	Greater than
$\cdot$	Multiplication	$<$	Less than

÷ or / Division $\geq$ Greater than or equal to
   ‖   Absolute value $\leq$ Less than or equal to

A. Order of Arithmetic Operations

If you have a string of numbers separated by various arithmetic operation signs, the convention is to perform the operations from left to right, except that multiplication and division take precedence over addition and subtraction. Consider the following sequence: $12 - 8 \div 2 + 9 - 4 \cdot 3$. If you performed the operations from left to right without regard for precedence, the sequence would be evaluated as follows: $\underline{12 - 8} \div 2 + 9 - 4 \cdot 3 = \underline{4 \div 2} + 9 - 4 \cdot 3 = \underline{2 + 9} - 4 \cdot 3 = \underline{11 - 4} \cdot 3 = 7 \cdot 3 = 21$. Recognizing the precedence of multiplication and division over addition and subtraction, you would evaluate the sequence differently: $12 - \underline{8 \div 2} + 9 - 4 \cdot 3 = 12 - 4 + 9 - \underline{4 \cdot 3} = \underline{12 - 4} + 9 - 12 = \underline{8 + 9} - 12 = 17 - 12 = 5$. Parentheses are used to change the conventional order of operations, or to make the order more obvious. The rule is that the operation enclosed in parentheses is performed before those surrounding it. For example, the value of the sequence above can be changed by adding parentheses in the following manner: $(12 - 8) \div 2 + (9 - 4) \cdot 3 = 4 \div 2 + (9 - 4) \cdot 3 = \underline{4 \div 2} + 5 \cdot 3 = 2 + \underline{5 \cdot 3} = 2 + 15 = 17$.

Practice

1. $(4 + 3) \cdot (8 - 6) = ?$
2. $7 \cdot (5 - 2) + 3 = ?$
3. $12 \div 6 + 15 \cdot 2 = ?$
4. $10 + 3 + 2 \cdot 8 - 5 = ?$
5. $20 \cdot 2 + 6 \cdot (9 - 5) = ?$

B. Signed Numbers

Numbers that are written without plus signs in front of them are assumed to be positive (i.e., greater than zero). A minus sign in front of a number indicates that it is negative (i.e., less than zero). When positive and negative numbers are mixed together in the same sequence, it is common to use plus signs in front of the positive numbers to reduce the possibility of confusion. The conventional way to think about numbers is to imagine a horizontal number line that extends infinitely to the left and to the right. Zero is in the middle, and the numbers become more positive (i.e., larger) as you move to the right of zero, and more negative (i.e., smaller) as you move to the left of zero. Thus larger negative numbers are said to be less than smaller negative numbers (e.g., $-7 < -5$), because they are more negative (i.e., further to the left on the number line). It can be confusing to think of -7 as being *smaller* than -5. It may help to think of a bank account; negative numbers mean you owe the bank. The larger the negative number, the more you owe, so in that sense your bank account is smaller.

On the other hand, in some cases we are interested only in the magnitude of the number (i.e., its distance from zero) and not its sign. In those cases, we are interested in the *absolute value* of the number. The mathematical symbol for absolute value is a pair of vertical lines surrounding the number. The absolute value of a positive number is the same as the number (e.g., $|+4| = +4$), but the absolute value of a negative number is the same number with the sign changed from minus to plus (e.g., $|-7| = +7$). In terms of absolute values, $|-7| > |-5|$.

To add a negative number and a positive number, find the difference between the two absolute values (subtract the smaller from the larger absolute value) and then add the sign of the number that had the larger absolute value. For instance, to add -7 and $+5$, ignore the signs, subtract the smaller from the larger number ($7 - 5 = 2$), and then add a minus sign to the result (-2), because the number with the larger absolute value (-7) has a minus sign. If you need to add a whole string of positive and negative numbers, the easiest way is to add all the positive numbers together into one positive number, and then all of the negative numbers into one negative number. (You can add a string of negative numbers by ignoring the minus signs, adding all the numbers, and then attaching the minus sign to the sum.) Once you have a single positive number and a single negative number, you can add them together according to the rule just given above. To subtract a negative number, you just change it to a positive number and add it. For instance, $5 - (-7) = 5 + 7 = 12$. The two minus signs in a row are said to cancel each other out.

There is one simple rule that covers both the multiplication and the division of signed numbers. If the two numbers have the same sign, the result will be positive; if the two numbers have different signs, the result will be negative. Ignore the signs when performing the multiplication or division, and just attach the appropriate sign to the result. (Note: The result of a multiplication is called a *product;* the result of a division is called a *quotient.*) For example, $(-5)(-7) = +35$. (Note that the multiplication dot was not needed in this problem; when two sets of parentheses are placed next to each other, multiplication is assumed.) Some additional examples should help to make the rule clear:

$$(-4)(+6) = -24$$
$$(+8) \div (-2) = -4$$
$$(-27) \div (-9) = +3$$

Practice

1. $-18 - (-5) = ?$
2. $-20 + 14 = ?$
3. $-90 \div +5 = ?$
4. $(-14)(-7) = ?$
5. $(-3) + (+11) + (-8) + (-10) + (+5) = ?$

C. Converting Fractions to Decimals and Decimals to Fractions

Finding the decimal that corresponds to a fraction is very easy to do with a calculator. The top part of the fraction is called the *numerator,* and the bottom part is called the *denominator.* The horizontal line in a fraction means "divided by." To find the decimal corresponding to a fraction, divide the numerator by the denominator. For instance, if the fraction is 2/3, find the answer to "2 divided by 3" on your calculator; you will see that 2/3 = .6666 . . . , which can be rounded off to .67 (see Chapter 1).

However, if the denominator is 10 or some power of 10 (e.g., 100, 1000, etc.), you should not need to use a calculator to convert a fraction to a decimal—you just insert the decimal point in the appropriate place in the numerator (assuming the numerator is a whole number). If the denominator is 10, insert a decimal point in the numerator so that there is only one digit to the right of the decimal point (e.g., 23/10 = 2.3) and then drop the denominator. In general, the number of digits to the right of the decimal point is determined by the number of zeroes following 1 in the denominator. For instance, when dividing by 1000 there will be three digits to the right of the decimal point, so 23/1000 becomes .023. (You need to add as many zeroes to the left of the numerator as it takes to have the required number of digits to the right of the decimal point.)

If the denominator is not a power of 10, the fraction can still be easily converted to a decimal without a calculator, if the denominator can be changed to a power of 10 by multiplying or dividing by a whole number. The trick is that whatever you do to the denominator to change it into a power of 10 you must also do to the numerator. *Multiplying or dividing both the numerator and the denominator by the same number will not change the value of the fraction.* For instance, 23/50 = 46/100 = .46. Similarly, 74/200 = 37/100 = .37. In this example the denominator was divided by 2 to become a power of 10, so the numerator had to be divided by 2, as well. If dividing the numerator does not give you a whole number, do not use division at all. You can always use multiplication instead. For example, both the numerator and denominator of 74/200 can be multiplied by 5 to yield 370/1000. If this operation gives a numerator ending in zero, you can drop the zero at the end of the numerator if you also drop the last zero in the denominator (370/1000 becomes 37/100).

It is always easy to convert a decimal into a fraction that involves a power of 10 (e.g., tenths, hundredths, etc.). The number of digits to the right of the decimal point tells you how many zeroes should follow 1 in the denominator (this is just the reverse of the method described above). For instance, .703 has three digits to the right of the decimal point, so the denominator will be 1000. The original number goes in the numerator, with the decimal point removed. Therefore, .703 = 703/1000 (which is read as "703 thousandths"). Sometimes a decimal corresponds to a simple fraction. For example, by the method above, .75 would be converted to 75/100, but this fraction is also equal to the much simpler 3/4. Reducing a fraction to its simplest form involves dividing both the numerator and the denominator by the same whole number, so that the numerator and denominator both remain whole numbers. Reducing fractions to simple terms can be a useful thing to do, but because it is not a necessary step to finding a correct answer, I will not devote space to it here.

Practice Convert each of the following to a fraction.

1. .0202
2. .881

Convert each of the following to a decimal.

3. 3/500
4. 437/10,000
5. 17/20

D. Converting to and from Percentages

A percentage is just a shorthand way of expressing a fraction in terms of hundredths (*percent* can be read as "out of or over one hundred"). In fact, the percent symbol, %, is just an alternative way of writing /100. You can always convert a percentage to a fraction by removing the percent sign and putting the percentage in the numerator of a fraction and 100 in the denominator. For instance, 83% means 83 over 100, or 83/100. If the percentage contains a decimal point (e.g., 83.5%), the fraction will not come out in a simple form (e.g., it would be 83.5/ 100). In that case, it is better to convert the percentage to a decimal first, and then convert the decimal to a fraction. To explain how to convert percentages to decimals I need to discuss multiplying and dividing decimals by a power of 10.

When you multiply a decimal by a power of 10, you move the decimal point to the right; when you divide it by a power of 10, you move the decimal point to the left. How many places (i.e., digits) you move is determined by the number of zeroes following 1. For instance, if you want to multiply .703 by 100, you move the decimal point two places to the right, so the result is 70.3. If you multiply 83.5 by 100, the result is 8350. Notice that a zero was added at the end of 83.5 so the decimal point could be moved two places to the right. (When there are no digits to the right of the decimal point, as in 8350, it is customary to not show the decimal point.) To divide 83.5 by 1000 you move the decimal point three places to the left, which gives .0835. (Notice again that a zero was added, this time on the left, so that we could move three places to the left.)

To convert a percentage to a decimal, just drop the percent sign and divide the number by 100. For example, 83.5% becomes .835, 2% becomes .02, and 0.4% becomes .004. To convert a decimal to a percentage, multiply by 100 and add the percent sign. For example, .66 becomes 66%, .066 becomes 6.6%, .0066 becomes .66%, and .00066 becomes .066%. To convert a fraction to a percentage, first convert to a decimal, and then use the method just described. For example, 3/8 = .375 = 37.5%, 3/20 = .15 = 15%, and 3/200 = .015 = 1.5%.

Practice Convert each of the following to a percentage.

1. .062
2. 7/25
3. .0076

Convert each of the following to a fraction.

4. 99.5%
5. .095%

E. Scientific Notation

If the result of a calculation has more digits to the right or left of the decimal point than can be displayed on your calculator, the number will be displayed in a form that is often called scientific notation. Most calculators will display a number between 1 and 10 together with an *exponent* that tells you which power of 10 (e.g., how many zeroes follow the 1) you must multiply that number by. For instance, the number 200 million is too large to be shown in standard form on a calculator with an eight-digit display, so it would be expressed as 2 with an exponent of 8. In scientific notation this would be written as 2×10^8. I will explain exponents further in the second part of the math review, but for now you can think of 10^8 as a shorthand way of expressing a number that consists of 1 followed by eight zeroes (i.e., 100 million). To save space, most calculators use an even briefer form by leaving out the 10 (which is always part of scientific notation) and only showing its exponent.

Converting scientific notation to a standard decimal involves multiplying by a power of 10 as explained in the previous section. For example, 3×10^9 means multiply 3 by a number consisting of a 1 followed by nine zeroes. In order to move the decimal nine places to the right of the 3, you must add nine zeroes to the right of the 3; the result is 3,000,000,000. For another example, 8.19×10^5 = 819,000. A negative exponent means that you need to divide instead of multiply. For example, 8.19×10^{-3} means that 8.19 is divided by 1000. Moving the decimal three places to the left, the result is .00819.

If a number is not between 1 and 10, it can be converted to scientific notation by moving the decimal point until the number is between 1 and 10. The number of places you had to move the decimal point to achieve this is the exponent (e.g., the power of 10); the exponent is positive if you moved the decimal point to the left and negative if you moved it to the right. For example, to convert 86,100 into a number between 1 and 10 (i.e., 8.61), you must move the decimal point four places to the left, so $86,100 = 8.61 \times 10^4$. On the other hand, to convert .00319 into a number between 1 and 10 (i.e., 3.19) the decimal point must be moved three places to the *right,* so $.00319 = 3.19 \times 10^{-3}$.

Practice Express each of the following in scientific notation.

1. 3,877,300
2. .00301
3. .0000029

Express each of the following in standard notation:

4. $5.16 \cdot 10^7$
5. $4.20 \cdot 10^{-5}$

Summary of Part I

A. Order of Arithmetic Operations

1. Arithmetic operations proceed from left to right, except that multiplication and division take precedence over addition and subtraction.

2. Operations enclosed in parentheses are performed before adjacent operations.

B. Signed Numbers

1. Numbers without signs are assumed to be positive.

2. A large number with a minus sign is considered less than a small number with a minus sign.

3. The absolute value of a positive number is the same as the original number, but the absolute value of a negative number is the same number with the sign changed from minus to plus (e.g., $|-5| = +5$).

4. To add a positive and a negative number, find the difference of the two numbers ignoring their signs, and then attach the sign of the number with the larger absolute value.

5. To add a series of numbers, some positive and some negative, add all the negative numbers separately and all the positive numbers separately, so that you have just one negative number and one positive number. Then proceed as described above.

6. To subtract a negative number, change the sign of the number to a plus and then add the number (e.g., $8 - (-4) = 8 + 4 = 12$).

7. When multiplying or dividing two signed numbers, the final result will be positive if the two numbers have the same sign and negative if the two numbers have opposite signs.

C. Converting Fractions to Decimals and Decimals to Fractions

1. To convert a fraction to a decimal on your calculator, just divide the numerator of the fraction by the denominator and round off according to your needs.

2. If the denominator of the fraction is a power of 10 (1 followed by some number of zeroes), insert a decimal point in the numerator so that the number of digits to the right of the decimal point is the same as the number of zeroes following 1 in the denominator; then drop the denominator.

3. Multiplying or dividing both the numerator and the denominator by the same number will not change the value of the fraction.

4. To convert a decimal into a fraction, drop the decimal point and create a denominator that consists of 1 followed by a number of zeroes equal to the number of digits to the right of the decimal point in the original number (e.g., $.3456 = 3456/10,000$).

D. Converting to and from Percentages

1. If the percentage is a whole number, it can easily be converted to a fraction by removing the percent sign and putting 100 in the denominator.

2. To multiply a decimal by a power of 10, move the decimal point to the right a number of places that equals the number of zeroes following 1 in the multiplier, adding zeroes on the right, as necessary.

3. To divide by a power of 10, move the decimal point to the left, adding zeroes as necessary.

4. To convert a percentage to a decimal, drop the percent sign and divide the number by 100 (i.e., move the decimal point two places to the left).

5. To convert a decimal to a percentage, multiply by 100 (i.e., move the decimal point two places to the right) and add the percent sign.

E. Scientific Notation

1. Any number can be expressed in scientific notation; the number must be converted to a part that is between 1 and 10, which is multiplied by a part that consists of 10 raised to some power.

2. The number of places the decimal point had to be moved to convert the original number to a number between 1 and 10 is the exponent (i.e., power) that is attached to 10. If the decimal point had to be moved to the left, the exponent is positive; if the decimal point had to be moved to the right, the exponent is negative (e.g., $.00319 = 3.19 \cdot 10^{-3}$; $973{,}600 = 9.736 \cdot 10^{5}$).

Part II. Algebraic Operations

A. Adding and Subtracting Fractions

In order to add or subtract fractions, you must make sure they have the same denominator. There are rules you can use to make the denominators the same, but if the fractions involve actual numbers (e.g., 3/4, 3/20) you can more easily convert them to decimals and add or subtract them on your calculator. If the fractions involve symbols (e.g., $2X/Y$), the same rules can be used to change the way the fraction looks. Because statistical formulas frequently contain fractions, and changing the form of the fraction can make the formula look very different, it is worth exploring the various algebraic rules affecting fractions. The first rule is that when two fractions have the same denominator you can add or subtract the numerators, and put the sum or difference over the common denominator. For instance,

$$\frac{X}{Z} + \frac{Y}{Z} = \frac{X + Y}{Z}$$

This rule can also be reversed; if the numerator contains addition or subtraction, the fraction can be broken apart. For example,

$$\frac{2A - 3B}{4C} = \frac{2A}{4C} - \frac{3B}{4C}$$

However, it is important to note that this rule does not work for addition or subtraction in the denominator:

$$\frac{A}{B + C}$$

cannot be broken apart.

  If the denominators do not match they must be changed so that they do match before the fractions can be added or subtracted. Of course, if you change the denominator of a fraction, you must also change the numerator in a compensating way, or you will change the value of the fraction. This leads to the next simple rule involving fractions: You can multiply or divide both the numerator and the denominator of a fraction by the same constant or variable without changing the value of the fraction. For instance,

$$\frac{A}{B} = \frac{2A}{2B}$$

$$\frac{X}{Y} = \frac{AX}{AY}$$

This rule can be used to make the denominators of two fractions match. The trick is to multiply the numerator and denominator of each fraction by the denominator of the other fraction. For instance, suppose you want to add $\frac{A}{B} + \frac{X}{Y}$. First, you multiply the top and bottom of A/B by Y to get AY/BY. Next you multiply the top and bottom of X/Y by B to get BX/BY. Now that the denominators match you can add the fractions. (Note: YB is considered the same as BY.)

$$\frac{AY}{BY} + \frac{BX}{BY} = \frac{AY + BX}{BY}$$

Practice

1. $X/4 - 3Y/4$
2. $XY/RC + 5/RC$
3. $5P/Q + 3P/T$
4. $A/5 - B/2$
5. $4/X + 8/Y$

B. Multiplying and Dividing Fractions

The rule for multiplying fractions could not be simpler. The denominators do not even have to match. Just multiply the two numerators and multiply the two denominators. For example,

$$\frac{X}{Z} \cdot \frac{B}{C} = \frac{XB}{ZC}$$

When you multiply both the numerator and the denominator of a fraction by the same constant or variable, say C, what you are really doing is multiplying the fraction by C/C. For instance, multiplying X/Y by C/C yields XC/YC. However, a fraction in which the numerator equals the denominator (e.g., C/C) is always equal to 1. Therefore, another way to understand how the form of a fraction can be changed without changing its value is to realize that multiplying a fraction by any other fraction that is equal to 1 will not change the first fraction. The new form is said to be *algebraically equivalent* to the first form.

One more rule concerning multiplication is worth mentioning. When a fraction is multiplied by a whole number, it is only the numerator that gets multiplied. For instance, if you want to multiply $\dfrac{A}{B}$ by C, you can write

$$C\frac{A}{B} \quad \text{or} \quad \frac{AC}{B}$$

The two forms above are equivalent. The rule can be reversed; whenever there is a multiplication in the numerator, one of the terms can be moved in front (or in back of) the fraction. For example, $2X/Y$ can be written as

$$2\frac{X}{Y} \quad \text{or} \quad X\frac{2}{Y} \quad \text{or} \quad \frac{X}{Y}2 \quad \text{or} \quad \frac{2}{Y}X$$

The rule for dividing fractions involves only a slight modification of the rule for multiplication. The second fraction, called the *divisor,* must be inverted, and then the two fractions are multiplied. For instance,

$$\frac{2X}{3Y} \div \frac{4D}{5E} = \frac{2X}{3Y} \cdot \frac{5E}{4D} = \frac{(2X)(5E)}{(3Y)(4D)} = \frac{10XE}{12YD}$$

One way to express the division of fractions is to put a fraction in the numerator and/or denominator of another fraction. For instance,

$$\frac{\dfrac{X}{Y}}{Z}$$

is the same as $X/Y \div Z$, which, according to the division rule, means

$$\frac{X}{Y} \cdot \frac{1}{Z} = \frac{X}{YZ}$$

When a fraction appears in the numerator of another fraction, as in the example above, the simple rule is to put the two denominators together (i.e., move the denominator in the numerator down so that it multiplies the denominator of the entire fraction).

Note that any whole number or variable (e.g., Z) can be expressed as a fraction by placing it over 1 as a denominator (e.g., $Z = Z/1$). Inverting a fraction means that the numerator and denominator trade places. In the case of a whole number, this means that 1 appears in the numerator instead of the denominator. Inverting the number means finding its *reciprocal* (e.g., the reciprocal of Z is $1/Z$).

When a fraction appears in the denominator of another fraction, as below,

$$\frac{X}{\dfrac{Y}{Z}}$$

you can again express this as a division problem. For example,

$$\frac{X}{\dfrac{Y}{Z}} = X \div \frac{Y}{Z} = X \cdot \frac{Z}{Y} = \frac{X}{1} \cdot \frac{Z}{Y} = \frac{XZ}{Y}$$

The simple rule is that the denominator of the denominator is moved up to the numerator. Finally, when there are fractions in both the numerator and the denominator,

$$\frac{\dfrac{A}{B}}{\dfrac{C}{D}}$$

it is just another way to show that two fractions are to be divided using the division rule described above. For example,

$$\frac{\dfrac{A}{B}}{\dfrac{C}{D}} = \frac{A}{B} \div \frac{C}{D} = \frac{A}{B} \cdot \frac{D}{C} = \frac{AD}{BC}$$

Practice

1. $X(A/B) = ?$

2. $(5/X)(Y/6) = ?$

3. $2D/E \div 3F/G = ?$

4. $\dfrac{\frac{3X}{7}}{Y} = ?$

5. $\dfrac{AB}{\frac{X}{3}} = ?$

C. Factoring

Consider the following expression: $A(X + Y)$. You may recall that the parentheses indicate that X and Y are to be added before being multiplied by A. However, it is also permissible to multiply both X and Y by A before adding— $A(X + Y) = AX + AY$. This rule is often used in reverse to *factor out* a common term in a complex expression. For instance, $2BX + 3BY + 4AB = B(2X + 3Y + 4A)$. Another useful rule is that terms that appear in both the numerator and denominator of a fraction can be canceled out. For example, AB/BC can be reduced to A/C. The numerator and denominator are being divided by the same factor, which doesn't change the value of the fraction. This is just the reverse of multiplying the numerator and denominator by the same factor, which also doesn't change the value of the fraction.

The two rules described above can be used together to simplify a complex fraction, such as the following:

$$(AC + BC)/DC$$

First factor C out of the numerator to yield

$$C(A + B)/DC$$

Then, cancel C out of both the numerator and the denominator to leave

$$(A + B)/D$$

Practice

1. $3 \cdot (2A + B + 4C) = ?$
2. $5 \cdot (B + .2Y) = ?$

Factor out the common term in each of the following.

3. $2XY + 2AX + 2X = ?$
4. $QPR - PR/4 + 3PRS = ?$
5. $(3CD - 3DY)/3AD = ?$

D. Exponents and Square Roots

You probably recall that in the expression X^2 the 2 is called the *exponent* (X, in this case, is called the *base*), and it means that X is to be multiplied by itself ($X \cdot X$). An exponent of 3 would indicate that a string of three Xs are to be multiplied ($X \cdot X \cdot X$), and so on. When the exponent is 2 we say that the number is being *squared*. Exponents greater than 2 are only used in one section of this text (Chapter 4, Section C), so I will not devote space to them here.

The exponent takes precedence over the arithmetic operations that we have been dealing with, so if you see XY^2, it is only the Y that is being squared (i.e., the squaring occurs before the multiplication of X and Y). If you want the product of X and Y to be squared, you can use parentheses to indicate that the multiplication is to be done first: $(XY)^2$. Once you are squaring both the X and the Y it doesn't matter whether you multiply before squaring or square before multiplying—that is, $(XY)^2 = X^2Y^2$. However, the order does make a difference when addition or subtraction is involved. For instance, $(A + B)^2$ does *not* equal $A^2 + B^2$. (You can check this with some simple numbers, such as $A = 2$ and $B = 3$.) The expression $(A + B)^2$ is the same as $(A + B) \cdot (A + B)$, which requires the multiplication of two binomials (a binomial is just two terms being added or subtracted). The product of $(A + B)$ times $(A + B)$ is $A^2 + B^2 + 2AB$, which is always larger than $A^2 + B^2$ when A and B are both positive numbers.

The square root of a number is the number that has to be squared to get the original number. In other words, $(\sqrt{X})^2 = X$. It is also true that $\sqrt{X^2} = X$. If you are trying to remember the square root of 144, for example, and you think it might be 12, you can always check by squaring 12 (i.e., $12 \cdot 12$) to see if it equals 144. The rules for mixing multiplication with square roots are similar to the rules for exponents. First, the square root applies only to the terms under the square root sign. For example, $\sqrt{X}\, Y$ means that the square root of X is multiplied by Y. On the other hand, $\sqrt{XY}$ means that X and Y are multiplied first and the square root of the product is taken. However, the order of operations in that case doesn't matter: $\sqrt{XY} = \sqrt{X}\sqrt{Y}$. As with exponents, the order of operations does make a difference when addition or subtraction is involved. For instance, $\sqrt{X - Y}$ does not equal $\sqrt{X} - \sqrt{Y}$.

The rules that concern squaring or taking the square root of a fraction are very straightforward. Squaring a fraction is equivalent to squaring both the numerator and the denominator separately. For example, $(X/Y)^2 = X^2/Y^2$. Similarly, taking the square root of a fraction is equivalent to taking the square roots of both the numerator and the denominator separately. For example, $\sqrt{(X/Y)} = \sqrt{X}/\sqrt{Y}$. Sometimes this rule is used in reverse. For instance, $A/\sqrt{B}$ can be rewritten as $\sqrt{A^2}/\sqrt{B}$ (recall that $\sqrt{A^2} = A$), and then the previous rule can be used in reverse to yield $\sqrt{A^2/B}$.

Practice: Rewrite the following expressions without parentheses.

1. $(3FG)^2$
2. $(2X/5Y)^2$
3. $(\sqrt{4BC})^2$

Rewrite the following expressions with a single square root sign and simplify.

4. $(\sqrt{5J})(\sqrt{3K})$

5. $\dfrac{\sqrt{XY}}{\sqrt{AX}}$

E. Solving Simple Equations

An equation is a statement that two mathematical expressions have the same numerical value. Equations are usually easy to solve when the mathematical expressions involved use only the four basic arithmetic operations (addition, subtraction, multiplication, and division). An equation that includes a square root or an exponent, for example, is considerably more difficult to solve, so I will confine my discussion to simple equations. If you are dealing with a simple equation with only one variable—for example, $5X - 2 = 13$—you can use a few algebraic rules to find the value of that variable. On the other hand, if a simple equation contains several variables you won't be able to find the value of each variable, but you can use the same algebraic rules to change the appearance of the equation, which is often desirable. The most fundamental rule when working with equations is that whatever manipulation is performed on one side of the equals sign must be performed on the other side. Otherwise, the two sides of the equation will no longer be equal.

When the equation has only one variable, the goal is to isolate the variable on one side of the equation, so that the other side of the equation can be reduced to a particular numerical value. For example, if the equation you are dealing with is $5X - 2 = 13$, you want to rearrange the equation into the form $X = N$, where N is some particular number. Two terms must be moved so that X can be alone: the term that is multiplying X (i.e., 5) and the term that is being subtracted from X (i.e., 2). It is usually easier to deal with addition and subtraction, so we'll consider the latter term first. To get rid of -2 we add $+2$ to both sides of the equation:

$$
\begin{aligned}
5X - 2 &= 13 \\
+\, 2 &= +2 \\
\hline
5X &= 15
\end{aligned}
$$

Then to get rid of the 5 we divide both sides of the equation by 5:

$$\frac{5X}{5} = \frac{15}{5}$$

$$X = 3$$

We can check that $X = 3$ by substituting 3 for X in the original equation: $5(3) - 2 = 13$; $15 - 2 = 13$; $13 = 13$.

These same rules can be used to change the form of an equation, even if it contains no numbers at all. Consider the following equation:

$$\frac{AX + B}{C} = DY$$

If you are interested in the value of X when values of the other variables have been filled in, it may be convenient to solve the equation for X, which means isolating X on one side of the equation. We begin by multiplying both sides of the equation by C, in order to get C out of the side of the equation that contains X:

$$\frac{C(AX + B)}{C} = CDY$$

Canceling out C in the left side of the equation, we are left with $AX + B = CDY$. Next we subtract B from both sides:

$$
\begin{array}{rcl}
AX + B & = & CDY \\
-\,B & = & -B \\
\hline
AX = CDY & - & B
\end{array}
$$

Finally, we divide both sides by A:

$$\frac{AX}{A} = \frac{CDY - B}{A}$$

$$X = \frac{CDY - B}{A}$$

Practice Find the value of Y in the following equations.

1. $\dfrac{Y}{3} + 15 = 21$
2. $.7Y - 7 = 7$

Solve for Z in the following equations.

3. $\dfrac{2Z + 8}{6} = X$
4. $\dfrac{3Z}{5} = \dfrac{A}{B}$
5. $BZ - 5X = 2AC$

Summary of Part II

A. Adding and Subtracting Fractions

1. If two fractions have the same denominator you can add (or subtract) them by adding (or subtracting) the numerators and placing the sum (or the difference) over the common denominator.

2. If the numerator and denominator of a fraction are both multiplied or divided by the same constant or variable, the value of the fraction will not be changed.

3. The denominators of two fractions can be made to match by multiplying both the numerator and the denominator of each fraction by the denominator of the other fraction.

B. Multiplying and Dividing Fractions

1. To multiply two fractions, multiply the two numerators together and the two denominators together:

$$(A/B) \cdot (C/D) = AC/BD.$$

2. When multiplying a fraction by a whole number, multiply only the numerator: $C(A/B) = CA/B$.

3. To divide two fractions, invert the second one (the divisor) and then multiply the two fractions.

4. When a fraction appears in the numerator of another fraction, the denominator of the fraction in the numerator can be moved so that it multiplies the denominator of the whole fraction:

$$\frac{\dfrac{X}{Y}}{Z} = \frac{X}{YZ}$$

5. When a fraction appears in the denominator of another fraction, the denominator of the denominator can be moved so that it multiplies the numerator:

$$\frac{X}{\dfrac{Y}{Z}} = \frac{XZ}{Y}$$

C. Factoring

1. If all the terms in a string of terms being added and/or subtracted contain a common term, that term can be *factored out* of each member of the string and placed in front of what's left of the original expression, which is enclosed in parentheses: $AX + XY - BX = X(A + Y - B)$.

2. If the numerator and denominator of a fraction contain a term in common, that term can be *canceled out*—that is, the common term can be removed from both the numerator and the denominator: $BX/AB = X/A$.

D. Exponents and Square Roots

1. The calculation of exponents and square roots takes precedence over other arithmetic operations.

2. Taking the square root and squaring are opposite operations—one can undo the other; e.g., $\sqrt{X^2} = X$ and $(\sqrt{X})^2 = X$.

3. Parentheses can be used to indicate that more than one variable is to be squared—e.g., $(XY)^2 = X^2Y^2$. However, the order in which squaring and addition/subtraction are done makes a difference: $(X + Y)^2$ does *not* equal $X^2 + Y^2$; it equals $X^2 + Y^2 + 2XY$.

4. The previous point applies to taking the square root; e.g., $\sqrt{AB} = \sqrt{A}\sqrt{B}$, but $\sqrt{A + B}$ does not equal $\sqrt{A} + \sqrt{B}$.

5. Squaring a fraction is equivalent to squaring both the numerator and the denominator: $(A/B)^2 = A^2/B^2$. Taking the square root of a fraction is equivalent to taking the square root of both the numerator and denominator: $\sqrt{A/B} = \sqrt{A}/\sqrt{B}$.

E. Solving Simple Equations

1. The most fundamental rule for solving equations is that whatever operation is performed on one side of the equation must be performed on the other side, to preserve the equality.

2. If the equation consists of numbers and only one (unknown) variable, the equation can be solved by isolating the variable on one side of the equation.

3. If the equation has several variables, you may want to solve for one of the variables, which means using the rules of algebra to isolate that variable on one side of the equation.

Post-Quiz

Part I

1. $3 \cdot 4 - 2 + 28 \div 7 - 5 = ?$

2. $3 \cdot (4 - 2) + 28 \div (7 - 5) = ?$

3. $-17 - (-9) + (+11) + (-7) - (+4) = ?$

4. $(+5) \cdot (-4) \div (-10) = ?$

5. Convert 46/200 to a decimal.

6. Convert .077 to a fraction.

7. What fraction corresponds to .66%?

8. What percent of 140 is 3?

9. Express .42% as a decimal.

10. What is the value of $3.3 \cdot 10^{-3}$?

Part II

1. $\dfrac{4X}{17} + \dfrac{2Y}{17} = ?$

2. $\dfrac{2A}{F} + \dfrac{3B}{G} = ?$

3. $\dfrac{7K}{M} \cdot \dfrac{2J}{Q} = ?$

4. $\dfrac{3X}{Y} \div \dfrac{4A}{C}$

5. Simplify this fraction. $\dfrac{2RT + 3ST}{TW}$.

6. What is the value of B in the following equation?
$\dfrac{3B}{2} + 7 = 19$

7. Solve the following equation for S: $\dfrac{5S + 7R}{2T} = 4X$

If $X = 2$ and $Y = 6$, what is the value of each of the following expressions?

8. $(2Y - 3X)^2$

9. $\left(\dfrac{3Y}{X}\right)^2$

10. $\sqrt{2XY^2}$

Answers to Diagnostic Quiz

Part I

1. +1

2. +12

3. +12

4. −3

5. .12

6. 3/1000

7. 7/100

8. 18.6%

9. 4100

10. .055

Part II

1. $\dfrac{F - LS}{GX}$

2. $\dfrac{KT + LS}{KS}$

3. $\dfrac{RMO}{QFS}$

4. $\dfrac{BY}{CX}$

5. $\dfrac{X + P}{Q}$

6. 14

7. $X = \dfrac{24 - B}{C}$

9. 4

8. 121

10. 90

Answers to Practice Exercises

Part I

A. 1. 14   2. 24   3. 32   4. 24   5. 64
B. 1. -13 2. -6 3. -18 4. 98   5. -5
C. 1. 202/10,000   2. 881/1000   3. .006   4. .0437   5. .85
D. 1. 6.2%   2. 28%   3. .76%   4. 995/1000   5. 95/100,000
E. 1. $3.8773 \cdot 10^6$ 2. $3.01 \cdot 10^{-3}$ 3. $2.9 \cdot 10^{-6}$ 4. 51,600,000   5. .000042

Part II

A. 1. $\dfrac{X - 3Y}{4}$ 2. $\dfrac{XY + 5}{RC}$ 3. $\dfrac{5PT + 3PQ}{QT}$ 4. $\dfrac{2A - 5B}{10}$ 5. $\dfrac{4Y + 8X}{XY}$

B. 1. $\dfrac{AX}{B}$ 2. $\dfrac{5Y}{6X}$ 3. $\dfrac{2DG}{3EF}$ 4. $\dfrac{3X}{7Y}$ 5. $\dfrac{3AB}{X}$

C. 1. $6A + 3B + 12C$ 2. $5B + Y$ 3. $2X(Y + A + 1)$ 4. $PR(Q - 1/4 + 3S)$

   5. $\dfrac{3D(C - Y)}{3AD} = \dfrac{C - Y}{A}$

D. 1. $9F^2G^2$ 2. $\dfrac{4X^2}{25Y^2}$ 3. $4BC$ 4. $\sqrt{15JK}$ 5. $\sqrt{Y/A}$

E. 1. 18   2. 20   3. $Z = 3X - 4$ 4. $Z = \dfrac{5A}{3B}$ 5. $Z = \dfrac{2AC + 5X}{B}$

Answers to Post-Quiz

Part I

1. 9

6. 77/1000

2. 20

7. 66/10,000

3. -8

8. 2.14%

4. $+2$

9. 0.0042

5. 0.23

10. 0.0033

Part II

1. $\dfrac{4X + 2Y}{17}$

2. $\dfrac{2AG + 3BF}{FG}$

3. $\dfrac{14KJ}{MQ}$

4. $\dfrac{3CX}{4AY}$

5. $\dfrac{2R + 3S}{W}$

6. 8

7. $\dfrac{8TX - 7R}{5}$

8. 36

9. 81

10. 12

Decision Trees

As most students will tell you, the easiest part of statistics is plugging numbers into a formula once you know which formula to use. One of the hardest parts is reading a description of an experiment, looking at some data, and then deciding which statistical test is the most appropriate. Teachers usually do not expect a student who has taken only an introductory statistics course to have this kind of understanding, although the student should be able to make an excellent guess in most common situations. Knowing which test to use comes from taking courses in experimental design or research methods, as well as from the experience you gain as you design experiments and collect data. The three charts in this appendix, however, represent a useful way to organize what you have learned from this text and can help you decide which statistical test to use with a particular experimental design.

The chart in Figure C.1 is for descriptive statistics; it can be used whether or not you plan to proceed with hypothesis testing. If you have collected data on more than one variable, you will have to consult the chart separately for each variable of interest. The next two charts (Figures C.2 and C.3) can help you to decide which type of null hypothesis test to perform. However, in order to use these charts, you must be sure you know which of your variables are independent variables and which are dependent variables. It should be fairly easy to pick out the independent variables, because they involve values that are either assigned to the subjects or used in selecting subjects. Anything else you come to know (or measure) about your subjects could be considered values of a dependent variable. Deciding which dependent variable(s) to include in an experimental design or a statistical test is often a difficult process. However, to use the charts in this appendix, you need only know how many dependent variables you want to include in the analysis.

As you look at the charts, bear in mind that as complex as they look, I have simplified them somewhat to make them readable. The notes below refer to the

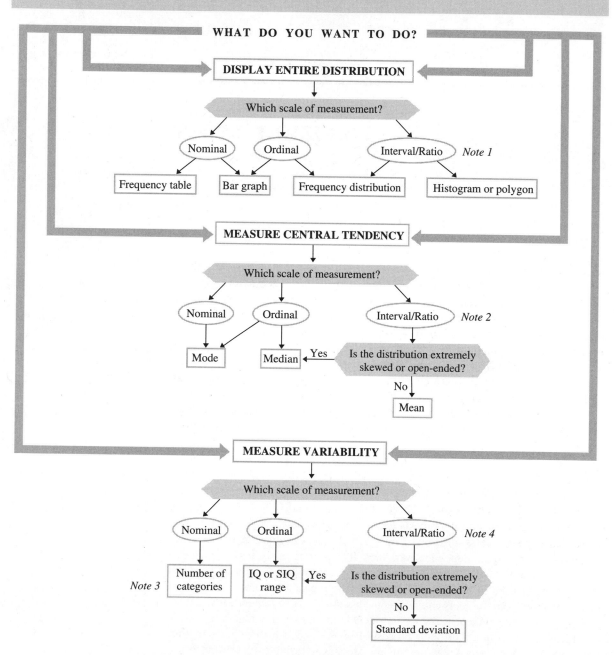

Figure C.1   Descriptive Statistics

Figure C.2   Hypothesis Testing with One Dependent Variable

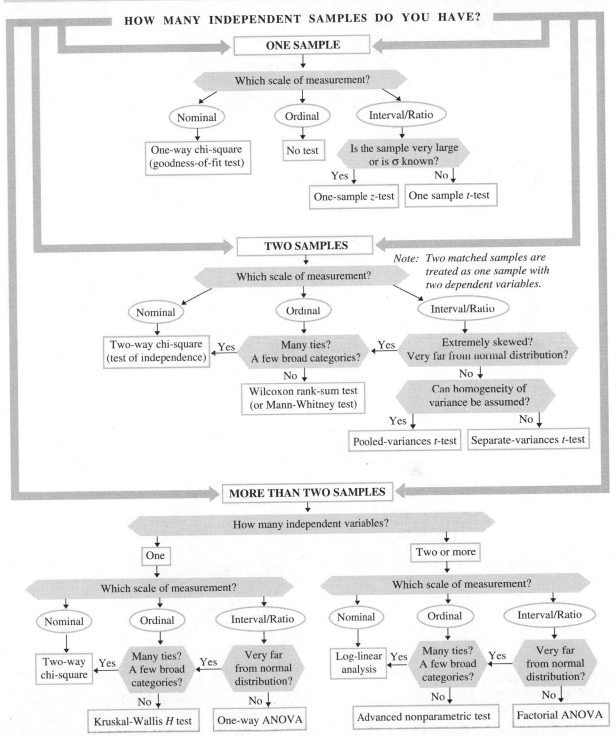

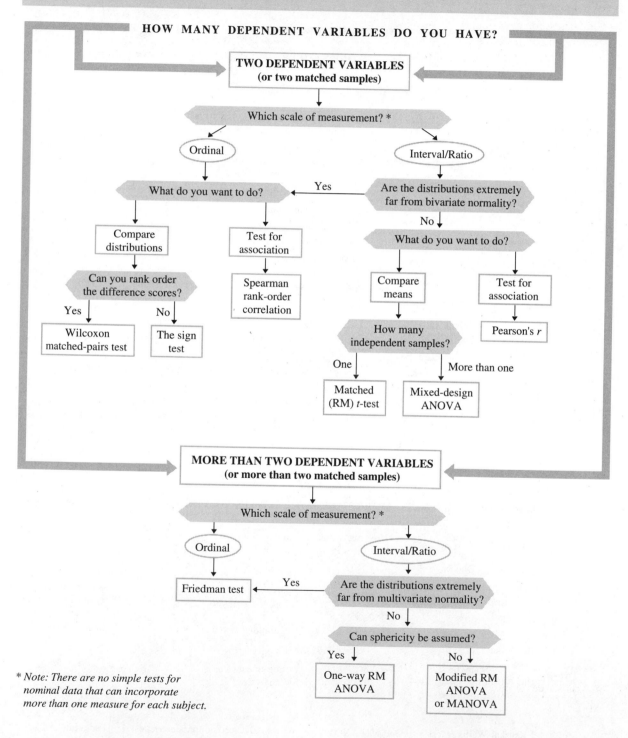

Figure C.3   Hypothesis Testing with More Than One Dependent Variable

HOW MANY DEPENDENT VARIABLES DO YOU HAVE?

TWO DEPENDENT VARIABLES
(or two matched samples)

Which scale of measurement? *

Ordinal   Interval/Ratio

Are the distributions extremely far from bivariate normality? — Yes →

What do you want to do?   No ↓

Compare distributions   Test for association   What do you want to do?

Can you rank order the difference scores?

Spearman rank-order correlation

Compare means   Test for association

Yes ↓   No ↓

Wilcoxon matched-pairs test   The sign test

How many independent samples?   Pearson's *r*

One   More than one

Matched (RM) *t*-test   Mixed-design ANOVA

MORE THAN TWO DEPENDENT VARIABLES
(or more than two matched samples)

Which scale of measurement? *

Ordinal   Interval/Ratio

Friedman test ← Yes — Are the distributions extremely far from multivariate normality?

No ↓

Can sphericity be assumed?

Yes ↓   No ↓

One-way RM ANOVA   Modified RM ANOVA or MANOVA

** Note: There are no simple tests for
nominal data that can incorporate
more than one measure for each subject.*

numbers in Figure C.1 and explain some of these simplifications. This appendix also includes some exercises that will give you practice using the charts. Answers are given at the end, so you can test your ability to choose the appropriate statistical test for a particular experimental situation.

Note 1. In addition to the traditional frequency distribution table, histogram, or polygon, J. W. Tukey (1977) has devised several innovative techniques for exploring interval/ratio data. Two of these, the stem-and-leaf plot and the box-and-whisker plot, are described in Section C of Chapter 2 and Chapter 3, respectively.

Note 2. Although it isn't shown in the chart, the mode can be used to describe the central tendency of interval/ratio data. However, because the median and the mean are more useful, as well as more reliable, the mode is rarely used for this purpose. The chief thing you will want to know about the mode of interval/ratio data is whether a single mode or two or more modes characterize the distribution.

Note 3. Although it is not a standard measure of variability, the number of categories can give an indication of variability in some situations. For instance, if citizens are being categorized by country of origin, a larger number of categories suggests greater diversity for the population being categorized.

Note 4. Although it is not indicated in the chart, the simple range (highest score minus lowest score) can be used as a measure of variability for interval/ratio data, and sometimes for ordinal data, as well. The range is especially useful when you are concerned with the width of the entire distribution and not just the bulk of the scores.

Exercises

Each of the following describes a research study. Use the charts in this appendix to determine the most appropriate statistical test for each one.

1. A classroom teacher is observing children from three types of families: families with a single parent who works; families with two parents, both of whom work; and families with two parents, only one of whom works. The teacher records the number of times each child speaks without being called on during one particular school day, to determine whether the home situation has an effect on classroom behavior. The teacher finds that most children score zero for the day, with a large number of children scoring 1 or 2, and a few having high scores.

2. A sample of private airplane pilots is tested for their ability to solve a three-dimensional maze, in order to determine whether total amount of flight time is related to maze solution time.

3. A large sample of joggers is tested to see if they consume more salt in their diets (measured in number of grams) as compared to the general population.

4. A sample of right-handers, a sample of left-handers, and a sample of ambidextrous people are measured on a tactile recognition task. Each subject is tested separately with the right hand and with the left hand so that right/left differences can be observed. The observer records the number of objects correctly identified with each hand.

5. A small group of chronic schizophrenics is compared to a larger group of general psychiatric patients to determine whether the chronic schizophrenics score lower on a test of orientation. The standard deviations for the two groups turn out to be very different.

6. A sample of men and a sample of women are obtained, and then each sample is randomly divided,

with half of the subjects assigned to a competitive condition and half assigned to a cooperative condition. The number of tasks successfully completed in the allotted time by each subject is recorded.

7. Each subject in a random sample is shown two sets of faces to remember. One set of faces matches the subject's cultural background, and the other set does not. Facial recognition is measured for each set.

8. Twenty families, each having exactly three children, are studied. In each family the three children are rank ordered for shyness, to determine whether birth order has an effect on shyness.

Answers to Exercises

1. There is one dependent variable, number of times each child speaks, so you use the chart in Figure C.2. There are three samples of children, chosen according to their parental situation, so you begin at the box for More Than Two Samples. There is only one independent variable (parental work arrangement), and the scale for the dependent variable is an interval/ratio scale. However, the distribution is very far from normal—and in fact there are many ties (many zeroes)—so the appropriate test is the two-way chi-square test.

2. There are two dependent variables (flight time and maze solution time) so you use the chart in Figure C.3. The two dependent variables are measured on ratio scales, and you want to test their association, so the appropriate statistic is Pearson's r. If you were dealing with very strange distributions, the Spearman correlation coefficient would be more appropriate.

3. There is only one dependent variable, dietary salt, so you use the chart in Figure C.2. There is only one (large) sample, and the dependent variable is measured on a ratio scale, so the appropriate test is the one-sample z-test.

4. There are actually two dependent variables, because subjects are tested with each hand, so you use the chart in Figure C.3. The scale is a ratio scale, and you want to compare means to look at differences between the two hands. There are a total of three independent samples (according to handedness), so the appropriate test is the mixed-design ANOVA.

5. There is one dependent variable, orientation, so you use the chart in Figure C.2. There are two samples. Assume for the moment that orientation is measured on an interval scale and that the distributions are not too far from normal. We come to the question,

Can homogeneity of variance be assumed? Considering that the samples are different sizes, that one of them is small, and that the sample standard deviations (and therefore the sample variances) are very different, we must answer no, and choose the separate-variances t-test. If orientation were measured on an ordinal scale, or if the distributions were quite strange, the appropriate test would have been the Wilcoxon rank-sum test (also known as the Mann-Whitney test).

6. There is only one dependent variable (number of tasks) so you use the chart in Figure C.2. There are more than two samples (there are four, because both the men and the women are divided into two sub-groups), so you look at the bottom part of the chart. There are two independent variables (gender and condition), and the dependent variable has been measured on a ratio scale, so the appropriate test is the factorial ANOVA. If the distributions were very strange and the samples not very large, you would need to use a factorial version of the Kruskal-Wallis test.

7. There are two dependent variables, because facial recognition is measured twice, so you use the chart in Figure C.3. The data are measured on a ratio scale, and you want to know if facial recognition differs between the two sets, so you want to compare means. There is only one sample of subjects, so the appropriate test is the matched (RM) t-test.

8. This is a tricky one. There are three samples of children: those born first, those born second, and those born third. But the samples are matched, because each set of three children comes from the same family. We use the bottom of the chart in Figure C.3 (the box labeled More Than Two Dependent Variables). Because the children are rank ordered, the scale is ordinal, so the appropriate test is the Friedman test.

Answers to Selected Exercises

Chapter 1

Section A

2. a) ratio c) nominal e) ordinal (but often treated as interval) g) nominal i) ratio
3. a) discrete c) discrete e) continuous
5. a) size of reward b) number of words recalled c) ratio
7. a) correlational b) correlational c) experimental d) experimental

Section B

1. a) $4 + 6 + 8 + 10 = 28$ c) $10 + 20 + 30 + 40 + 50 = 150$ e) $4 + 16 + 36 + 64 + 100 = 220$ g) $3^2 + 5^2 + 7^2 + 9^2 + 11^2 = 240$

2. a) $5 + 9 + 13 + 17 + 21 = 65$ c) $(30)(35) = 1050$ e) $(-1) + (-1) + (-1) + (-1) + (-1) = -5$ g) $9 + 11 + 13 + 15 + 17 = 65$
4. a) $9N$ c) $3\Sigma D$ e) $\Sigma Z^2 + 4N$
6. a) 144.01 c) 99.71 e) 7.35 g) 6.00
8. a) 55.6 c) 99.0 e) 1.4

Section C

2. $\displaystyle\sum_{j=1}^{12}\sum_{i=1}^{9} X_{ij}$

4. $\displaystyle\sum_{j=1}^{N}\sum_{i=1}^{6} X_{ij}$

Chapter 2

Section A

1. a)

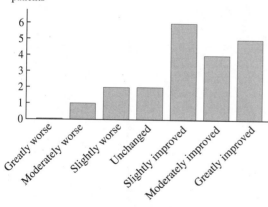

c) five patients; 25% e) moderately improved; unchanged

Progress	f	cf	rf	crf	cpf
Greatly improved	5	20	.25	1.00	100
Moderately improved	4	15	.20	.75	75
Slightly improved	6	11	.30	.55	55
Unchanged	2	5	.10	.25	25
Slightly worse	2	3	.10	.15	15
Moderately worse	1	1	.05	.05	5
Greatly worse	0	0	0	0	0

2.

Number of Words	f	cf	rf	crf	cpf
29	1	25	.04	1.00	100
28	2	24	.08	.96	96
27	0	22	0	.88	88
26	3	22	.12	.88	88
25	5	19	.20	.76	76
24	7	14	.28	.56	56
23	4	7	.16	.28	28
22	0	3	0	.12	12
21	1	3	.04	.12	12
20	0	2	0	.08	8
19	2	2	.08	.08	8

a) 28% c) 76; 88

5.

Score	f	cf	rf	crf	cpf
10	2	20	.10	1.00	100
9	2	18	.10	.90	90
8	4	16	.20	.80	80
7	3	12	.15	.60	60
6	2	9	.10	.45	45
5	2	7	.10	.35	35
4	1	5	.05	.25	25
3	2	4	.10	.20	20
2	1	2	.05	.10	10
1	1	1	.05	.05	5

a) two; 10% c) 35; 90

Section B

1.

Interval	f	cf	rf	crf	cpf
145–149	1	50	.02	1.00	100
140–144	0	49	0	.98	98
135–139	3	49	.06	.98	98
130–134	3	46	.06	.92	92
125–129	4	43	.08	.86	86
120–124	7	39	.14	.78	78
115–119	9	32	.18	.64	64
110–114	10	23	.20	.46	46
105–109	6	13	.12	.26	26
100–104	4	7	.08	.14	14
95–99	3	3	.06	.06	6

a)
Frequency

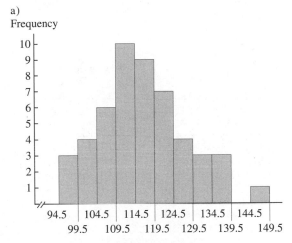

c) 40th percentile ≈ 113; 60th percentile ≈ 118

2.

Test Scores	f	cf	rf	crf	cpf
95–99	4	60	.066	1.00	100
90–94	8	56	.133	.933	93
85–89	12	48	.20	.800	80
80–84	13	36	.216	.600	60
75–79	10	23	.166	.383	38.3
70–74	5	13	.083	.216	21.6
65–69	3	8	.05	.133	13.3
60–64	3	5	.05	.083	8.3
55–59	2	2	.033	.033	3.3

b)
Number of
applicants

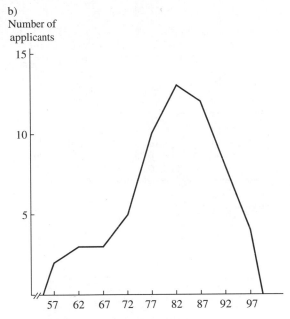

c) about 90; e) about the 75th percentile

3.

Number of Calls	f	cf	rf	crf	cpf
55–59	1	40	.025	1.00	100
50–54	2	39	.05	.975	97.5
45–49	2	37	.05	.925	92.5
40–44	2	35	.05	.875	87.5
35–39	2	33	.05	.825	82.5
30–34	3	31	.075	.775	77.5
25–29	2	28	.05	.70	70.0
20–24	3	26	.075	.65	65.0
15–19	3	23	.075	.575	57.5
10–14	5	20	.125	.50	50.0
5–9	8	15	.20	.375	37.5
0–4	7	7	.175	.175	17.5

a) 14.5; c) about 93

4.

Speed	f	cf	rf	crf	cpf
85–89	1	25	.04	1.00	100
80–84	1	24	.04	.96	96
75–79	3	23	.12	.92	92
70–74	2	20	.08	.80	80
65–69	3	18	.12	.72	72
60–64	3	15	.12	.60	60
55–59	5	12	.20	.48	48
50–54	3	7	.12	.28	28
45–49	4	4	.16	.16	16

a) about 70% c) about 61 e) about 54

5.

Number of Dreams	f	cf	rf	crf	cpf
35–39	4	40	.10	1.00	100
30–34	3	36	.075	.90	90
25–29	6	33	.15	.825	82.5
20–24	8	27	.20	.675	67.5
15–19	4	19	.10	.475	47.5
10–14	7	15	.175	.375	37.5
5–9	2	8	.05	.20	20
0–4	6	6	.15	.15	15

a)
Subjects

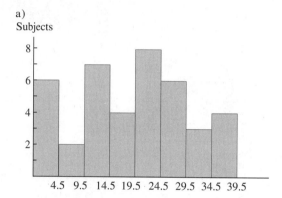

c) about 21

Section C

1.

Stem	Leaf
60*	4
60	8
70*	2
70	5 5 9
80*	1 1 1 1 3 4
80	5 5 6 6 6 8 9 9
90*	2 2 3
90	6 8

3.

Stem	Leaf
50*	
50	5 6
60*	1 3 4
60	5 8 9
70*	0 1 2 3 4
70	5 5 6 6 6 7 7 8 9 9
80*	0 0 0 1 1 2 2 2 3 3 4 4 4
80	5 5 5 5 6 6 6 7 7 7 9 9
90*	0 0 1 2 2 2 4 4
90	5 6 7 9

5. (2B1b) first quartile = 104.5 + 4.58 = 109.08; third quartile = 119.5 + 3.928 = 123.43; (2B1c) 40th percentile = 109.5 + 3.5 = 113; 60th percentile = 114.5 + 3.88 = 118.38; (2B1d) PR for 125 = 78 + .8 = 78.8; PR for 117 = 46 + 9 = 55; PR for 108 = 14 + 8.4 = 22.4

6. (2B2c) 80th percentile = 84.5 + 5 = 89.5; (2B2d) 75th percentile = 84.5 + 3.75 = 88.25; 60th percentile = 79.5 + 5.02 = 84.52; (2B2e) PR for 88 = 60 + 14 = 74; (2B2f) PR for 81 = 38.3 + 6.48 = 44.78

Chapter 3

Section A

1. b) median
3. b) ceiling effect
4. a) mean c) median e) median
6. slightly improved; slightly improved
8. negatively skewed

Section B

1. mean = 94/8 = 11.75; median = (11 + 13)/2 = 12; mode = 14
3. $\mu = 3750/35 = 107.14$
5. mean = 4854/60 = 80.9; median = 79.5 + 2.69 = 82.19; mode = 82

7. missing score = 6
9. GPA = 188/64 = 2.94

Section C

1. a) median = 114.5 + 1.11 = 115.61 b) median = 9.5 + 5 = 14.5

2.

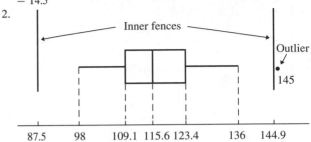

Chapter 4

Section A

1. a) SIQ range b) standard deviation c) range d) standard deviation e) SIQ range
3. d) variance
4. a) SS = 79.5; σ^2 = 79.5/8 = 9.9375 b) MD = 2.75; $\sigma = \sqrt{9.9375} = 3.152$
6. a) 15 b) df = 4 c) MD = 4.8; s = 6.67
8. a) range = 114 − 100 = 14; σ = 4.67
9. range = 36 − 17 = 19; SIQ range = (Q3 − Q1)/2 = (31 − 24)/2 = 3.5; MD = 67.6625/16 = 4.23; $\sigma = \sqrt{(457.75)/16} = 5.35$

Section B

1. a) range = 145 − 98 = 47 b) IQ range = 123.43 − 109.08 = 14.35; SIQ range = 14.35/2 = 7.175
3. a) IQ range = 8.25 − 4.5 = 3.75; SIQ range = 3.75/2 = 1.875 b) MD = 43.4/20 = 2.17; $\sigma = \sqrt{132.2/20} = 2.57$

5. $s = \sqrt{1/7(277.9)} = 6.30$
7. a) $s = \sqrt{1/7(30.87)} = 2.10$ b) The standard deviation in part a is one-third the size of the standard deviation in exercise 5. If you divide every number in a set by the same constant, the standard deviation will be divided by that constant.
9. a) 5.031 b) 4.617 c) 4.523

Section C

1. σ = 9.615; population skewness = 11,856/(9 · 888.9) = 1.48
3. a) σ − 2.40; population skewness = −12/(8 · 13.82) = −.11; b) One extreme score can have a very large effect on dispersion and skewness.
5. a) population kurtosis = .024; b) sample kurtosis = 1.857; underestimated

Chapter 5

Section A

1. b only
2. a) −1.667 c) +1.667; +3.333; −.333 e) 74 inches; 71.99 inches; 70.1 inches; 67.7 inches
4. a) 480 c) 190
5. a) +0.2 c) −2.5
6. a) 400 c) 566.7
7. a) 10 c) 50

8. a) (70 − 100)/16 = −30/16 = −1.875 c) (105 − 100)/16 = 5/16 = +.3125
9. a) .0714 c) .3106 e) .4826
10. a) .4641 c) .1469 e) .0051

Section B

1. a) 95.15% c) 4.85%
2. a) .1915 + .3413 = .5328 c) .9332 − .7734 = .1598

3. a) .1056; .5987 c) .1587 − .0516 = .1071; .4332 − .0987
= .3345; .4834 + .1480 = .6314
4. a) $X = z\sigma + \mu = (.67)(8) + 72 = 77.36$ c) between 62.8
and 81.2
5. .4406 − .2324 = .2082

2. a) (.2266)(.2266) = .0513 c) 2 · (.5987)(.6915) = .8280
4. a) (.25)(.25) = .0625 c) P(H) · P(S|H) = (.25)(13/51)
= .0637
5. a) (8/20)(7/19) = .147 c) (12/20)(8/19) + (8/20)(12/19)
= .505

Section C

1. a) .0401 + .1056 = .1457 c) .5091 + .4013 − .1354
= .775

Chapter 6

Section A

1. d
3. a) 7.83 c) 1.565
5. $\sigma_{\bar{X}}$ for variable A will be three times larger than $\sigma_{\bar{X}}$ for
variable B.
6. $\sigma_{\bar{X}} = 100/\sqrt{20} = 22.36$
8. $\sigma_{\bar{X}} = 11.34$
9. $N = 49$

4. $\bar{X} = 1190/7 = 170$; $z = (170 − 150)/(30/\sqrt{7}) = 1.76$;
$p = .0392$
6. a) $z = −4.04$ b) $−8.08$ c) If sample size is multiplied by
C, z is multiplied by $\sqrt{C}$.
8. a) $z = 3.16$ b) $z = 3.16$ c) z scores are unaffected by
(linear) changes in scale; to evaluate a z score you don't
have to know what measurement scale was used.

Section C

1. d
3. a

Section B

1. a) $z = 3.2$ b) .07%
2. a) $z = −2.81$ b) .0025

Chapter 7

Section A

1. a) .0885; .177 c) .0139; .0278
2. a) about 1.405; ±1.75 c) 2.455; ±2.70
4. a) It gets larger. b) It gets smaller.
5. $z = 2.4$, two-tailed $p = .0164$
8. d

Section B

2. no; $z_{calc} = 1.6 < z_{crit} = 1.96$
4. a) reject null; $z_{calc} = 2.05 > z_{crit} = 1.96$ b) accept null;
$z_{calc} < z_{crit} = 2.575$; as alpha becomes smaller, the chance
of attaining statistical significance decreases
6. no; $z_{calc} = (470 − 500)/(100/\sqrt{9}) = −.9, |−.9| < z_{crit} = 1.96$

8. a) accept null; $z_{calc} = 1.83 < z_{crit} = 1.96$ b) reject null;
$z_{calc} > z_{crit} = 1.645$ c) The drug would be investigated
further, even though it is worthless. d) Others might
ignore the drug even though it has some value. e) If over-
looking a valuable drug is very much worse than raising
expectations about what is actually a worthless drug, it
could justify making alpha larger.
10. $N \geq 198$

Section C

1. b
3. anywhere between 0 and 20 (there is no way to specify the
number of Type I errors in this case); none of the 20 significant experiments can be a Type II error.

Chapter 8

Section A

1. a) .524 b) .175 (one-third as large as the answer in part a)
3. a) df = 9 b) ±2.262; ±3.250 c) 1.833; 2.821
5. a) $t_{calc} = 6.20$; $t_{crit} = 2.08$ b) $t_{calc} = 4.39$; t_{calc} in part a

divided by $\sqrt{2}$ equals t_{calc} in part b (because there are half
as many subjects in part b)
7. The p level will be smaller for the experiment with the
larger sample size, because as N increases the tails of the

t distribution become thinner, and there is less area beyond a given *t* value.

Section B

1. a) accept null; $t_{calc} = 2.044 < t_{crit} = 2.132$ b) accept null; could be making a Type II error
3. a) $\mu_{lower} = 4.78$, $\mu_{upper} = 5.62$ b) $\mu_{lower} = 4.99$, $\mu_{upper} = 5.41$ c) width of CI in part a = .84, width of CI in part b = .42; If you multiply the sample size by C, you divide the width by $\sqrt{C}$ (This relationship does not hold for small samples, because changing the sample size also changes the critical values of the *t* distribution.)

4. $\mu_{lower} = 2.32$, $\mu_{upper} = 2.88$ b) Because $\mu = 3$ does not fall within the 95% CI for the American couples, we can say the American couples differ significantly from the Europeans at the .05 level.
7. a) $t_{calc} = 2.87$; Reject null at .05 level but not at .01 level b) $\mu_{lower} = 8.41$, $\mu_{upper} = 12.99$
9. a

Section C

1. b
2. d

Chapter 9

Section A

1. a) $z = (47.5 - 45.8)/\sqrt{5.5^2/150 + 4^2/100} = 2.83$ b) .0023
3. $s_p^2 = [100(12) + 50(8)]/150 = 10.67$
5. $s_p^2 = (120 + 180)/2 = 150$
7. a) $t = (27.2 - 34.4)/\sqrt{4^2/15 + 14^2/15} = -1.92$ b) $t = (27.2 - 34.4)/\sqrt{106\,(1/15 + 1/15)} = -1.92$ c) The answers to parts a and b should be the same. When the two samples are the same size the pooled-variances *t* equals the separate-variances *t*.

Section B

1. accept null; $t_{calc} = (52 - 44)/\sqrt{130.2\,(1/7 + 1/10)} = 1.42 < t_{crit} = 2.132$
2. a) accept null; $t_{calc} = (12 - 8)/\sqrt{(16 + 25)/15} = 2.42 < t_{crit} = 2.763$ b) The experimenter has no control over the vividness of a subject's imagery; also, we have to rely on each subject's own judgment of his or her imagery vividness.
3. reject null; $t_{calc} = (87.2 - 82.9)/\sqrt{(28.09 + 19.36)/12} = 2.16 > t_{crit} = 2.074$ (false expectations *can* affect student performance)
4. a) $4.4 \pm (1.65)(1.98)$; $\mu_{lower} = 1.13$, $\mu_{upper} = 7.67$ b) $4.4 \pm (1.65)(2.617)$; $\mu_{lower} = .07$, $\mu_{upper} = 8.73$

c) Because zero is not contained in either the 95% CI or the 99% CI, the null hypothesis can be rejected at both the .05 and the .01 levels.
6. reject null; $t_{calc} = 4.65 > t_{crit} = 2.042$
7. a) accept null; $t_{calc} = (21.11 - 17.14)/\sqrt{67.55(1/9 + 1/7)} = .96$ b) $\mu_{lower} = -4.91$, $\mu_{upper} = 12.85$
8. a) reject null; $t_{calc} = (21.11 - 13.83)/\sqrt{37.36(1/9 + 1/6)} = 2.26 > t_{crit} = 2.16$; the value of *t* changed from .96 to 2.26 with the removal of a single outlier, so the *t*-test seems very susceptible to outliers. b) $\bar{X}_1 - \bar{X}_2 = 7.28$ minutes; $\mu_{lower} = .325$, $\mu_{upper} = 14.24$ c) Yes, the separate-variances test would be recommended because the sample sizes are small and different, and the variances are quite different.
9. a) accept null; $t_{calc} = 1.18 < t_{crit} = 2.101$ b) could be making a Type II error

Section C

1. $t_{sep} = 1.40$; Because the separate-variances *t* value is not significant with df = $N_1 + N_2 - 2$, no further testing is appropriate—we must retain H_0.
3. a) $t_{pool} = 1.36$ b) $t_{sep} = 1.90$; the separate-variances *t* value is larger.

Chapter 10

Section A

2. $\gamma = .3$
4. $\delta = .3\sqrt{28/2} = 1.12$; δ is less than the critical value required for significance, so the results are expected to be significant less than half of the time.
6. $\gamma = 1.5$; $\delta = 4.74$
8. a) $g = .70$ b) $g = 1.70$

10. d; b and c reduce Type II errors without affecting Type I errors, and a reduces Type II errors by allowing Type I errors to increase.

Section B

1. $\beta = .68$, power = .32; $\beta = .29$, power = .71
3. 1.7; 3.25

5. a) $2 \cdot (3.1/.7)^2 = 39.2$; 40 subjects are required in each group b) 54 subjects per group
7. a) $N = 138$ b) $N = 224$
9. a) 16 subjects per group b) $\gamma = 1.32$

Section C

2. harmonic mean = 13.33; power = .54 (approximately); the experiment is probably not worth doing.
4. harmonic mean = 8.24; $\delta = 2.23$; power = .60 (approximately)

Chapter 11

Section A

3. Annual salary

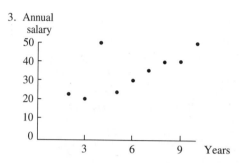

a) The general trend is toward a positive correlation.
b) misleading because of the outlier at (4, 50)

4. Number of details recalled

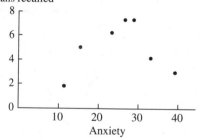

a) Low; the points do not come close to fitting on one straight line. b) Misleading; the degree of linear correlation is low, but there appears to be a meaningful (curvilinear) relationship between anxiety and number of details recalled.

5. Score

a) Negative; points slope down as you move to the right in the scatterplot. b) Moderate; the trend is noticeable, but there is considerable scatter with respect to any straight line.
7. a
9. a) -1.0 b) $-.45$

Section B

4. a) accept null; $r_{calc} = 1/8(2029 - 9 \cdot 6 \cdot 34.56)/(2.74 \cdot 11.39) = .653 < r_{crit} = .666$ b) reject null; $r_{calc} = .957 > r_{crit} = .707$ c) You find out that the person with four years at the company is related to the president of the company.
5. a) $r_{calc} = 1/9(1267 - 10 \cdot 6.7 \cdot 20.4)/(3.23 \cdot 5.13) = -.669$ b) reject null; $|r_{calc}| > r_{crit} = .632$
6. a) $r_{calc} = .845$ b) reject null; $r_{calc} > r_{crit} = .805$; no, $r_{calc} < r_{crit} = .878$
8. a) $r_{calc} = 1/5(140 - 6 \cdot 3.33 \cdot 4.83)/(3.33 \cdot 2.86) = .91$ b) reject null; $r_{calc} > r_{crit} = .882$
9. a) $r_{calc} = -.718$ b) reject null; $|r_{calc}| > r_{crit} = .707$ c) Because subjects were randomly assigned to exercise durations, we can conclude that an increase in exercise duration (within the limits of the study) *causes* a lowering of serum cholesterol level.
10. $r_{calc} = 1/11(46,451.8 - 45,495)/(1.17 \cdot 93.53) = .795$

Section C

1. a) .05 c) .31 e) .867 g) 1.832 i) .245 k) .74 m) .922 o) .987
3. a) $\delta = .5\sqrt{9} = 1.5$; power = .32 b) $\delta = .669\sqrt{9} = 2.0$; power = .52 c) $N = (3.24/.669)^2 + 1 = 24.5$; 25 schizophrenics
5. accept null; $z_{calc} = (.81 - .203)/\sqrt{1/7 + 1/12} = 1.28 < z_{crit} = 1.96$
7. a) no, $z_{calc} = (.549 - .424)/\sqrt{1/77} = 1.10 < z_{crit} = 1.96$ b) no, $z_{calc} = (.549 - .31)/\sqrt{1/77 + 1/117} = 1.63 < z_{crit} = 1.96$

Chapter 12

Section A

2. a) $z_{Y'} = .4 \cdot 1.5 = .6$ b) $z_{Y'} = .4(-.9) = -.36$
4. $Y' = 2/25X + 70 = .08X + 70$
5. a) $b_{YX} = .6(4/3) = .8$ $a_{YX} = 32 - .8(69) = -23.2$
   c) No, it is the predicted waist size of someone whose height is zero. d) $Y' = .8X - 23.2$
6. a) 34.4 c) 71.5 inches
8. $r = \sqrt{.5} = .707$
10. b

Section B

1. a) $Y' = 3.56X + 10.4$ b) $Y' = 24.6$ c) about 14 years
3. a) $r^2 = -.669^2 = .45$ b) $Y' = -1.06X + 27.51$ c) 27.51; it is the expected orientation score of someone who just entered the hospital. d) about 16.5 years
5. a) $Y' = -4.02X + 225.4$ d) 169.12 e) 169.12 $\pm (3.707)(15.27) \sqrt{1 + .125 + (14 - 6)^2/[7(13.2)]};$ $Y'_{lower} = 92.7$, $Y'_{upper} = 245.53$

7. a) $Y' = .425X + 2.95$ b) $Y' = .763X - 2.75$
8. a) $r = .87$ b) $r = .239$

Section C

1. a) $r_{pb} = 1/9\,[54 - 10(.6)(7.6)]/[(2.914)(.5164)] = .620$
   b) $t = .62\sqrt{8}/\sqrt{1 - .62^2} = 2.24$ c) The two t values are the same.
3. a) $r_{pb}^2 = 10^2/(10^2 + 38) = .725$ b) $r_{pb}^2 = .20$ c) estimated $\omega^2 = (10^2 - 1)/(10^2 + 38 + 1) = .712$; estimated $\omega^2 = .198$
5. a) $\omega^2 = .8^2/(.8^2 + 4) = .138$ b) $\omega^2 = .059$ c) $\gamma = 2$
7. a) $z_{Y'} = .28z_{X_1} - .43z_{X_2} + .61z_{X_3}$ b) $R^2 = .6354$
   c) $R = .797$
9. a) $z_{Y'} = -.176z_{X_1} - .176z_{X_2}$ b) $r_{Y(1,2)}^2 = (.176)^2 (1 - .7^2)$ $= .016$ c) $Y' = 69.6 - .059X_1 - .353X_2$ d) $Y' = 64.89$ bpm e) 62 bpm; because the club member is running the mean distance for the mean amount of time, the prediction would be the mean heart rate of the group.

Chapter 13

Section A

2. a) $t = (35 - 39)/\sqrt{(7^2 + 5^2)/10} = -1.47$ b) No ($t_{crit} = 2.101$)
3. a) $t = (35 - 39)/\sqrt{(7^2 + 5^2)/10 - 2(.1)(7)(5)/10} = -1.55$
   c) -1.74
5. a) $\overline{X} = 0$; $s = 4.62$ b) $\overline{X} = 1.44$; $s = 6.15$
8. d

Section B

2. a) reject null; $t_{calc} = 1.6/(1.075/\sqrt{10}) = 4.71 > t_{crit} = 2.262$ b) The matched t is much larger than the independent groups t (1.18) from exercise 9B9, because of the close matching.
4. a) $t = 8.5/(10.46/\sqrt{8}) = 2.30$; no, $t_{calc} < t_{crit} = 3.499$
   b) $\mu_D = 8.5 \pm 3.499 \cdot 3.70$; $\mu_{lower} = -4.44$, $\mu_{upper} = 21.44$

5. $t_{calc} = \dfrac{4.83 - 3.33}{\sqrt{(11.07 + 8.17)/6 - 2(.91)(3.33)(2.858)/6}} = 2.645$

6. accept null; $t = \dfrac{.6}{7.575/\sqrt{10}} = .25$

8. reject null; $t = \dfrac{1.67}{2.345/\sqrt{9}} = 2.13 > t_{crit} = 1.860$

Section C

1. a) $\delta = (.4)\sqrt{1/(1 - .5)}\sqrt{25/2} = 2$; power = .52
   b) $\delta = 2.58$; power = about .74 c) power = about .51
3. a) $\delta = 2.63 = .3\sqrt{1/(1 - .6)}\sqrt{N/2}$; $N = 62$ b) $\delta = 3.25$; $N = 94$
5. d

Chapter 14

Section A

1. $MS_w = (100 + 225 + 144 + 121 + 100)/5 = 138$
3. Five groups with 17 subjects in each group
5. $F = 80 \cdot 18.67/150.6 = 9.92$
6. $F = 80 \cdot 18.67/602.3 = 2.48$; doubling the standard deviations causes the F ratio to be divided by 4. (In general, if s is multiplied by C, then F is divided by C^2.)

7. a) $F_{calc} = 8 \cdot 3.33/10.11 = 2.64$ b) $F_{crit} = 2.95$ c) The null hypothesis cannot be rejected.
9. d

Section B

1. $F_{calc} = 229.24/42.88 = 5.346$; $F_{.05}(4, 63) = 2.52$ (approximately); reject the null hypothesis.

3. a) $F = 0/365 = 0$ c) Because all three sample means are the same, you would have known that MS_{bet} must be zero, and that therefore F must be zero.
4. a) $F_{calc} = 99.2/26.5 = 3.75$ b) $F_{crit} = 3.55$ c) reject the null hypothesis
5. a) $F_{calc} = 5.06/6.37 = .79$ b) $F_{.01} = 6.36$ c) accept null

d)

Source	SS	df	MS	F	p
Between Groups	10.11	2	5.06	.794	$> .05$
Within Groups	95.5	15	6.37		

8. c)

Source	SS	df	MS	F	p
Between Groups	64.2	3	21.4	9.11	$< .01$
Within Groups	37.6	16	2.35		

9. a) $F_{calc} = 35.51/9.74 = 3.64$ b) $F_{crit} = 2.45$ c) reject null
10. a) $F_{calc} = 9.827/1.176 = 8.36$

Section C

1. a) $\eta^2 = 2(5)/(2(5) + 27) = .27; .15; .10$
3. a) $\eta^2 = 10.11/(10.11 + 95.5) = .096$; you cannot estimate ω^2 using Formula 14.16, because F is less than 1.
5. a) $k = 5, n = 6, \omega^2 = .3, \alpha = .05$, therefore power $= .75$; for $\alpha = .01$, power $= .47$.
7. a) .2
9. a) reject homogeneity of variance assumption; $F_{calc} = .49/.16 = 3.063 > F_{.025}(11,19) = 2.77$ (interpolated)

Chapter 15

Section A

1. a) $5(5 - 1)/2 = 10$ c) 45
3. a) front vs. middle: $t = (34.3 - 28.7)/\sqrt{2 \cdot 150.6/80} = 2.886$; middle vs. back: $t = 1.49$; front vs. back: $t = 4.38$
   b) $t_{crit} = 1.96$; front vs. middle and front vs. back exceed the critical t
5. a) front vs. middle: $t = 1.443$; middle vs. back: $t = .747$; front vs. back: $t = 2.19$
   b) The t value is divided by 2. (In general, if n is divided by C, t will be divided by $\sqrt{C}$.)
7. a) is more conservative

Section B

2. LSD $= 5.78$; HSD $= 7.02$; a) only child vs. adult male b) same as part a c) LSD; e.g., female vs. male comes much closer to LSD than to HSD
4. a) LSD $= 3.26$ b) HSD $= 3.87\sqrt{10.11/8} = 4.35$
6. a) <u>athletes vs. controls</u>: reject null, $t_{calc} = (14 - 11.57)/\sqrt{1.176 (1/7 + 1/6)} = 4.03 > t_{crit} = 2.201$; athletes vs. musicians: reject null, $t_{calc} = 2.5 > t_{crit} = 2.306$; musicians vs. controls: accept null, $t_{calc} = 1.0$ b) Instead of merely concluding that all three populations do not have the same mean, you can conclude that the athletes differ from both the musicians and controls, but that the latter two groups do not differ (significantly) from each other.
8. a) HSD $= 4.714$ b) All but 11 pairs differ significantly. None of the six comparisons involving two animal phobias is significant, and acrophobia does not differ significantly from spider, snake, rat phobias or claustrophobia. Also, the two social phobias do not differ significantly.
   c) Amount of repression does not differ among subjects who suffer from the animal phobias, but does differ between the social phobias on one hand and the animal phobias and claustrophobia and acrophobia, on the other.

Section C

1. a) $2/(1/6 + 1/12) = 8$ c) 6.15 e) 5.65
3. HSD $= 3.98\sqrt{42.88/13.46} = 7.10$; $\overline{X}_3$ differs significantly from both $\overline{X}_1$ and $\overline{X}_5$
5. a) $8 - 1 = 7$ b) animal vs. nonanimal; mammal vs. non-mammal; rat vs. dog; spider vs. snake; social vs. environmental; party vs. speak; claustrophobia vs. acrophobia.
   c) $1/4\overline{X}_{rat} + 1/4\overline{X}_{dog} + 1/4\overline{X}_{spider} + 1/4\overline{X}_{snake} - 1/4\overline{X}_{party} - 1/4\overline{X}_{speak} - 1/4\overline{X}_{claus} - 1/4\overline{X}_{acro}; 1/2 \overline{X}_{rat} + 1/2 \overline{X}_{dog} - 1/2 \overline{X}_{spider} - 1/2 \overline{X}_{snake}; \overline{X}_{rat} - \overline{X}_{dog}; \overline{X}_{spider} - \overline{X}_{snake}; 1/2 \overline{X}_{party} + 1/2 \overline{X}_{speak} - 1/2 \overline{X}_{claus} - 1/2 \overline{X}_{acro}; \overline{X}_{party} - \overline{X}_{speak}; \overline{X}_{claus} - \overline{X}_{acro}$ d) .05
7. a) $\alpha_{pc} = .05/5 = .01$; $\overline{X}_3$ vs. $\overline{X}_6$: $t = (17.8 - 12)/\sqrt{9.74(1/8 + 1/7)} = 3.59 > t_{.01}(13) = 3.012$; only $\overline{X}_3$ differs significantly from $\overline{X}_6$ b) HSD $= 4.83$; only $\overline{X}_3$ c) The Dunnett test is designed specifically for testing all group means against the same comparison group mean.

Chapter 16

Section A

1. a) 10; 50 c) 18; 90
3. a)

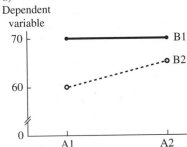

Dependent variable

The marginal means are $A_1 = 65$, $A_2 = 67.5$, $B_1 = 70$, $B_2 = 62.5$ b) All three effects could be significant.

5. Dependent variable

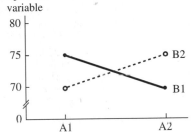

The marginal means are $A_1 = A_2 = B_1 = B_2 = 72.5$
b) The interaction might be significant, but neither of the main effects can be significant.
7. d) The lines on a graph of cell means will not be parallel.

Section B

2. a)

Source	SS	df	MS	F	p
Imagery group	2.45	1	2.45	1.53	> .05
Word type	22.05	1	22.05	13.8	< .01
Interaction	14.45	1	14.45	9.03	< .01
Within-cells	25.6	16	1.6		

b) Number of words recalled

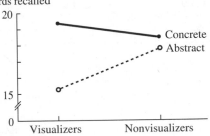

Whereas the recall of concrete words is slightly higher for visualizers than for nonvisualizers, the recall of abstract words is considerably higher for nonvisualizers than for visualizers. c) She can conclude that recall is better for concrete words in general, but that the concrete-abstract difference is significantly larger for visualizers than for nonvisualizers. However, because the experimenter did not create the imagery groups, she must be cautious in concluding that the use of imagery "caused" any differences in recall.

4. a)

Source	SS	df	MS	F	p
Difficulty	100	4	25	2.5	< .05
Reward	150	2	75	7.5	< .01
Interaction	40	8	5	.5	> .05
Within-cells	900	90	10		
Total	1190	104			

b) 7

6. a)

Source	SS	df	MS	F	p
Type of therapy	119.7	2	59.85	43.7	< .01
Presence of drug	68.27	1	68.27	49.8	< .01
Interaction	13.63	2	6.82	4.97	< .05
Within-cells	74.0	54	1.37		

b) The drug reduces the number of panic attacks significantly; this trend holds for each type of therapy. Type of therapy also makes a significant difference, with behavior modification being best and group therapy being worst. The interaction is due to the fact that type of therapy makes more of a difference without the drug.

7. a)

Source	SS	df	MS	F	p
Year	1114.5	3	371.5	87.7	< .01
Major	1314.1	2	657.1	155.1	< .01
Interaction	336.9	6	56.15	13.3	< .01
Within-cells	152.5	36	4.24		

b) No, the interaction does not obscure the interpretation of the main effects.

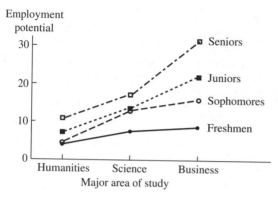

8. a)

Source	SS	df	MS	F	p
Agreement	20.0	1	20.0	8.51	< .05
Intent	24.2	1	24.2	10.3	< .01
Interaction	20.0	1	20.0	8.51	< .05
Within-cells	37.6	16	2.35		

Section C

1. main effect of year: $HSD = 3.82\sqrt{4.24/12} = 2.27$; all six pairs of marginal means differ significantly at the .05 level; main effect of major: $HSD = 3.46\sqrt{4.24/16} = 1.78$; all three pairs of marginal means differ significantly at the .05 level

3. psychoanalysis/group $\times$ drug/no drug: interaction is not significant, $F_{calc} = 4.12 < F_{crit} = 4.125$; group/behavior $\times$ drug/no drug: interaction is significant, $F_{calc} = 10.11 > F_{crit} = 4.125$; psychoanalysis/behavior $\times$ drug/no drug: interaction is not significant, $F_{calc} < 1$

5. a)

Source	SS	df	MS	F	p
Age	264.9	1	264.9	38.1	< .01
Languages	116.8	1	116.8	16.8	< .01
Interaction	1.51	1	1.51	.22	> .05
Within-cells	159.8	23	6.95		

b) The older children make significantly fewer errors than the younger children, and bilinguals make significantly fewer errors than monolinguals. There is very little interaction between these factors (not close to significance). However, one should be cautious about concluding that bilingualism is the "cause" of fewer errors in the task, because this factor was not under the control of the experimenter.

Chapter 17

Section A

2. Longest string recalled

There is a fairly small amount of interaction between digit and letter, and a considerably larger amount of interaction between letter and mixed.

4. $MS_{inter} = 93/21 = 4.43$; $F = 8 \cdot 3.33/4.43 = 6.02$
6. c
8. b

Section B

1. b)

Source	SS	df	MS	F	p
Type of Music	83.73	2	41.87	8.32	< .05
Residual	40.27	8	5.03		

3. a) The size of the imagined audience had a significant effect on systolic blood pressure, as shown by a one-way repeated measures ANOVA, $F(2,22) = 7.07, p < .05$.
b) allow time for blood pressure to return to baseline level before presenting the next condition.

4. a) Yes

b)

Source	SS	df	MS	F	p
Text	76.75	3	25.58	22.3	< .01
Residual	27.50	24	1.15		

c) Type I error

6. a) accept null; $F_{calc} = 3.17/1.77 = 1.79 < F_{crit} = 4.10$
   b) Nothing can be concluded from the analysis in part a.
   c) No; it looks like sphericity is unlikely to exist in the population.

7. b)

Source	SS	df	MS	F	p
Time	178.1	3	59.36	30.7	< .01
Residual	40.66	21	1.94		

Section C

1. a) F_{calc} is significant based on the standard critical F, so it is compared to the adjusted critical F. $F_{calc} = 8.32 >$ adjusted F_{crit} (1,4) = 7.71, so this result is significant regardless of whether sphericity can be assumed. c) F is not significant based on the standard critical F, so it will also not be significant by the adjusted critical F.

3. a) $t_{crit} = 2.571$; digits vs. letters: $t = 3.16$, reject null; digits vs. mixed: $t = 1.66$, accept null; letters vs. mixed: $t = .16$, accept null b) $t_{crit} = 2.365$; before vs. start: $t = 2.76$, reject null; start vs. end: $t = 7.74$, reject null; end vs. after: $t = 2.16$, accept null; Bonferroni $\alpha_{pc} = .05/3 = .0167$ (before vs. start would not be significant by this criterion)

5. One set of orders that meets the requirements is as follows: mar, amph, val, alc; alc, val, amph, mar; val, mar, alc, amph; amph, alc, mar, val.

Chapter 18

Section A

2.

Source	SS	df	MS	F	p
Groups	88	1	88	.64	> .05
Within-groups	1380	10	138		
Time	550	1	550	41.04	< .01
Group × Time	2	1	2	.15	> .05
Subject × Time	134	10	13.4		

4. Ounces of popcorn eaten

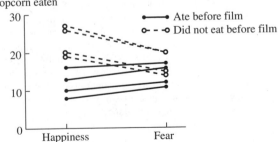

a) yes b) yes

Section B

1. a)

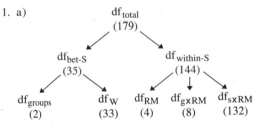

b) main effect of groups: $F_{.05}$ (2, 33) = 3.29 (approximately); main effect of time: $F_{.05}$ (4,132) = 2.45 (approximately); interaction of group and time: $F_{.05}$ (8,132) = 2.02 (approximately)

3. a) $F_{load} = 210.25/23.46 = 8.96$, $p < .05$; $F_{emotion} = 12.25/.46 = 26.73$, $p < .01$; $F_{inter} = 64.0/.46 = 139.64$, $p < .01$
   b) Ounces of popcorn eaten

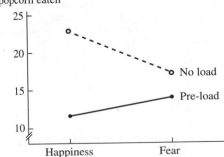

The main effect of groups is due to the larger amounts eaten by the no load group during both films. The main effect of emotion is due to a larger amount eaten during the happy film compared to the fear-evoking film. However, this main effect is qualified by a very large disordinal interaction. Happiness leads to more consumption only for the no load group; the effect reverses for the preload group.
c) $t = [(5.75 - (-2.25)]/\sqrt{(.917 + .917)/4} = 11.817$; $11.817^2 = 139.64$, which is the F ratio for the interaction of the two factors.

4.

Source	SS	df	MS	F	p
Type of task	333.33	1	333.33	7.87	<.05
Within-groups	338.67	8	42.33		
Type of music	16.07	2	8.03	2.58	>.05
Task × Music	120.87	2	60.43	19.44	<.01
Subject × Music	49.73	16	3.11		

5.

Source	SS	df	MS	F	p
Condition	2590.7	2	1295.4	71.85	<.01
Within-groups	108.2	6	18.03		
Difficulty level	315.8	3	105.3	124.9	<.01
Condition × Difficulty	5.06	6	.84	1.0	>.05
Subject × Difficulty	15.17	18	.84		

6. b) $F_{gender} = 0/28.5 = 0, p > .05$; $F_{treat} = 169.25/5.31 = 31.9, p < .01$; $F_{inter} = 1.25/5.31 = .24, p > .05$

7. a) $F_{type} = .36/4.16 = .09, p > .01$; $F_{time} = 17.36/.69 = 25.0, p < .01$; $F_{inter} = .36/.69 = .52, p > .01$

Section C

1. a) No. In the absence of an interaction, significant main effects can lead to follow-up tests, but not if a main effect has only two levels. b) Given a significant interaction, it makes sense to test the simple effects, which in this case involves four t-tests: happiness vs. fear for preload subjects, $t = 4.7, p < .05$; happiness vs. fear for no load subjects, $t = 12.0, p < .01$; preload vs. no load in happi-

ness condition, $t = 4.18, p < .01$; preload vs. no load in fear, $t = 1.49, p > .05$.

3. The interaction is not significant, so it is appropriate to follow up any main effect that is significant. In this case, the significant main effect (type of treatment) has four levels, so six pairwise comparisons are possible. This amounts to six matched t-tests (both the men and women are used in these tests; the lack of an interaction allows us to ignore gender in this case). The conservative approach is to base the error term of each matched t-test on the variability of difference scores only for the two levels involved in the test. To control Type I error, the Bonferroni test is employed, so $\alpha_{pc} = .05/6 = .0083$. The t-tests are as follows: control vs. biofeedback, $t = 4.71$; control vs. drug, $t = 12.22$; control vs. self-hypnosis, $t = 8.02$; biofeedback vs. drug, $t = 4.44$; biofeedback vs. self-hypnosis, $t = .34$; drug vs. self-hypnosis, $t = 7.55$. All of the pairwise comparisons are significant using the Bonferroni criterion, except for biofeedback vs. self-hypnosis.

5. Because the interaction is significant, the simple effects are tested for each factor. First, a one-way RM ANOVA is performed for each group of subjects. For the clerical task, $F = 8.32, p < .05$ (this is the result of exercise 17B1); for the mechanical task, $F = 26.6/1.18 = 22.5, p < .01$. Each significant simple effect can be followed by three matched t-tests to localize the effect. So as to be conservative and not assume sphericity, each error term is based only on the pair of levels involved. For the clerical task: background vs. popular, $t = .25$; background vs. heavy metal, $t = 2.63$; popular vs. heavy metal, $t = 13.9$. For the mechanical task: background vs. popular, $t = 3.83$; background vs. heavy metal, $t = 6.78$; popular vs. heavy metal, $t = 2.83$. If the Bonferroni test is used for these six matched t-tests, $\alpha_{pc} = .0083$, and the only significant differences are background vs. heavy metal for the mechanical task, and popular vs. heavy metal for the clerical task. Finally, the simple effects of the between-groups factor are tested; the clerical and mechanical groups are compared for each type of music (using a pooled error term just for the levels involved in the test). Background: $t = .94$; popular: $t = 1.95$; heavy metal: $t = 6.07$. Using the Bonferroni correction, $\alpha_{pc} = .0167$. The between-groups difference is significant only for heavy metal music. As there are only two levels for the between-groups factor, there is no further testing to be done.

Chapter 19

Section A

2. a) $p = .0269 + .0054 + .0005 = .0328$ b) $.0328 \cdot 2 = .0656$ c) yes; no

4. a) no, $z = (58 - 50)/\sqrt{100 \cdot .5 \cdot .5} = 1.6$ b) $(X - 50)/$

$\sqrt{100 \cdot .5 \cdot .5} = 1.96$; $X = 59.8$, so Johnny would have to get sixty questions right.

6. a) $z = [(27 - (120 \cdot .15)]/\sqrt{120 \cdot .15 \cdot .85} = 2.30$; yes, at the .05 level (two-tailed) b) $z = [(108 - (480 \cdot .15)]/\sqrt{480 \cdot .15 \cdot .85} = 4.60$ c) It is half as large. If N is multi-

plied by C, but the proportion obtained in the P category remains the same, z is <u>multiplied</u> by $\sqrt{C}$.

8. No, $z = (.37 - .30)/\sqrt{(.3 \cdot .7)/80} = 1.37$

Section B

1. From Table A.12 ($N = 6$), $p = (.0938 + .0156) \cdot 2 = .1094 \cdot 2 = .2188$; $p > .05$, so accept null; no, even if you used a .05, one-tailed test, you would have rejected the null hypothesis for exercise 13B5, but not for the sign test. The sign test throws away much of the information in the data, and therefore has considerably less power than the matched t-test.

3. no, $p = .2188 > .05$ (see calculation for exercise 19B1)

5. From Table A.12 [$N = 9 - 1$ (tie) $= 8$], $p = (.2188 + .1094 + .0312 + .0039) \cdot 2 = .3633 \cdot 2 = .727$; accept null

7. a) yes, $z = (32 - 25)/\sqrt{50 \cdot .5 \cdot .5} = 1.98 > z_{crit} = 1.645$
   b) yes, just barely

Section C

1. a) $20/52 = .385$ b) $26/52 + 13/52 = .75$ c) $26/52 + 20/52 - 10/52 = 36/52 = .69$

3. a) $(13/52)(12/51) = .0588$ b) $(13/52)(13/51) \cdot 2 = .127$
   c) $(20/52)(19/51) = .143$

5. a) $_{15}C_3 = 15!/(3!12!) = 455$ b) $_{15}C_5 = 15!/(5!10!) = 3{,}003$

7. a) $p(8) = {}_{10}C_8 \cdot 2^8 \cdot 8^2 = .0000737$ b) $p(8) + p(9) + p(10) = .0000737 + .000004 + .0000001 = .0000778$

Chapter 20

Section A

1. a) df $= 7$ b) 14.07, 18.48

3. a) Each of the three products will be selected one-third of the time. b) no, $\chi_{calc}^2 = (27 - 22)^2/22 + (15 - 22)^2/22 + (24 - 22)^2/22 = 3.55 < \chi_{crit}^2 = 5.99$

5. $\chi_{calc}^2 = (30 - 20)^2/20 + (17 - 20)^2/20 + (27 - 20)^2/20 + (14 - 20)^2/20 + (13 - 20)^2/20 + (19 - 20)^2/20 = 12.2 > \chi_{.05}^2 = 11.07$ (but not significant at the .01 level)

6. yes, $\chi_{calc}^2 = 9.89 > \chi_{.05}^2 = 7.81$

7. reject null; $\chi_{calc}^2 = 20 > \chi_{.05}^2 = 7.81$

Section B

2. a) accept null, $\chi_{calc}^2 = (|2 - 4| - .5)^2/4 + (|4 - 2| - .5)^2/2 + (|18 - 16| - .5)^2/16 + (|6 - 8| - .5)^2/8 = 2.11 < \chi_{.05}^2 = 3.84$ c) No, because the experimenter had no control over the diets of the children.

3. a) accept null; $\chi_{calc}^2 = .11 < \chi_{.05}^2 = 3.84$ b) possible Type II error

4. accept null; $\chi_{calc}^2 = 10.45 < \chi_{.01}^2(4) = 13.28$

5. $\chi_{calc}^2 = (8 - 6.48)^2/6.48 + (12 - 11.88)^2/11.88 + (7 - 8.64)^2/8.64 + (4 - 5.52)^2/5.52 + (10 - 10.12)^2/10.12 + (9 - 7.36)^2/7.36 = 1.45$; no, the null hypothesis cannot be rejected.

7. reject null; $\chi_{calc}^2 = 4.89 > \chi_{.05}^2(1) = 3.84$

9. accept null; $\chi_{calc}^2 = 13.54 < \chi_{.01}^2(8) = 20.09$

Section C

1. a) $\phi = \sqrt{(2.11/30)} = .265$ c) The relationship is not very strong.

3. a) $\phi = \sqrt{(4.89/30)} = .40$ b) cross-product ratio $= (10 \cdot 12)/(3 \cdot 5) = 8$ c) It is not extremely accurate, but a moderate <u>amount of accuracy</u> has been demonstrated.

5. a) $\phi_c = \sqrt{13.54/[100 \cdot (3 - 1)]} = .26$; $C = \sqrt{13.54/[(13.54 + 100)]} = .345$ b) The relationship is not very strong.

Chapter 21

Section A

2. $S_R = 16(17)/2 = 136$

4. a) $\Sigma R_A = 90$; $\Sigma R_B = 100$ b) $U_A = 9(10) + 9(10)/2 - 90 = 45 = U_B = U$

Section B

1. If steroid users are labeled A, then $n_A = 5$, $n_B = 15$, and $\Sigma R_A = 24$. a) For a one-tailed test at $\alpha = .05$, the critical value from Table A.14 is 33, so the null hypothesis can be rejected at this level. b) For $\alpha = .01$ (one-tailed), the critical value is 26, so the null hypothesis can be rejected at this level (the results would not be significant, however, with a .01, two-tailed test).

3. If left damage is labeled A, then $n_A = 4$, $n_B = 6$, and $\Sigma R_A = 13$ (assigning a rank of 1 to the smallest number). The critical values are 12 and 32, so the null hypothesis cannot be rejected. This agrees with the decision in the example in Chapter 9, Section D. Note that in both cases the test statistic approaches statistical significance. In most cases, the Mann-Whitney test is only a little less powerful than the corresponding t-test.

5. a) $t = \dfrac{21.2 - 7.4}{\sqrt{\dfrac{14.7^2 + 3.51^2}{5}}} = \dfrac{13.8}{\sqrt{45.67}} = 2.04 < t_{.05} = 2.306;$

therefore the null hypothesis cannot be rejected.
b) $n_A = n_B = 5$. If you assign lower ranks to smaller amounts of time and arbitrarily label the drug group as A, $\Sigma R_A = 5 + 7 + 8 + 9 + 10 = 39$. From Table A.14, the rejection zones (.05, two-tailed) are ≤ 17 or ≥ 38. Because $SR_A > 38$, the null hypothesis can be rejected.
c) Yes. The drug group contains an extreme outlier, which adversely affects the t-test by inflating the variance of one of the groups. Ranking minimizes the impact of the outlier, resulting in a significant Mann-Whitney test.
7. $r_S = .806$
8. a) .6808 b) .683 (the slight difference from the answer to part a is due to two pairs of ties in annual salary) c) The critical value of r_s from Table A.15 is .648; reject the null hypothesis, because the correlation in part a (or part b) is greater than the critical value. d) The Spearman correlation coefficient is slightly larger than Pearson's r in this case.

Section C

1. a) $H = 12/9(10)[36/3 + 225/3 + 576/3] - 3(10) = 7.2$
   b) $7.2 > \chi_{.05}^2(2) = 5.99$, so reject null hypothesis
3. $H = 12/18(19)[47.5^2/5 + 56^2/6 + 67.5^2/7] - 3(19) = .01$; accept null
5. $F_r = 12/8(4)(4 + 1)[9.5^2 + 14.5^2 + 30^2 + 26^2] - 3(8)$
   $(4 + 1) = 20.74 > \chi_{.05}^2(3) = 7.81$; reject null hypothesis

References

Brown, G. E., Wheeler, K. J., & Cash, M. (1980). The effects of a laughing versus a nonlaughing model on humor responses in preschool children. *Journal of Experimental Child Psychology, 29,* 334–339.

Cicchetti, D. V. (1972). Extension of multiple range tests to interaction tables in the analysis of variance: A rapid approximate solution. *Psychological Bulletin, 77,* 405–408.

Cohen, J. (1988). *Statistical power analysis for the behavioral sciences* (2nd ed.). Hillsdale, NJ: Lawrence Erlbaum.

Cohen, J. (1994). The earth is round ($p < .05$). *American Psychologist, 49,* 997–1003.

Cowles, M. (1989). *Statistics in psychology: An historical perspective.* Hillsdale, NJ: Lawrence Erlbaum.

Cramér, H. (1946). *Mathematical methods of statistics.* Princeton, NJ: Princeton University Press.

Davidson, E. S., & Schenk, S. (1994). Variability in subjective responses to marijuana: Initial experiences of college students. *Addictive Behaviors, 19,* 531–538.

Davidson, M. L. (1972). Univariate versus multivariate tests in repeated measures experiments. *Psychological Bulletin, 77,* 446–452.

Denny, E. B., & Hunt, R. R. (1992). Affective valence and memory in depression: Dissociation of recall and fragment completion. *Journal of Abnormal Psychology, 101,* 575–580.

Driskell, J. E., & Salas, E. (1991). Group decision-making under stress. *Journal of Applied Psychology, 76,* 473–478.

Dunn, O. J. (1961). Multiple comparisons among means. *Journal of the American Statistical Association, 56,* 52–64.

Dunnett, C. W. (1964). New tables for multiple comparisons with a control. *Biometrics, 20,* 482–491.

Ekman, P. (Ed.). (1982). *Emotion in the human face* (2nd ed.). London: Cambridge University Press.

Fisher, R. A. (1951). *The design of experiments* (6th ed.). Edinburgh: Oliver and Boyd.

Fisher, R. A. (1970). *Statistical methods for research workers* (14th ed.). Edinburgh: Oliver and Boyd.

Geisser, S., & Greenhouse, S. W. (1958). An extension of Box's results on the use of the F distribution in multivariate analysis. *The Annals of Mathematical Statistics, 29,* 885–891.

Greenhouse, S. W., & Geisser, S. (1959). On methods in the analysis of profile data. *Psychometrika, 24,* 95–112.

Griggs, R. A., & Cox, J. R. (1982). The elusive thematic-materials effect in Wason's selection task. *British Journal of Psychology, 73,* 407–420.

Harte, J. L., & Eifert, G. H. (1995). The effects of running, environment, and attentional focus on athletes' catecholamine and cortisol levels and mood. *Psychophysiology, 32,* 49–54.

Hartley, H. O. (1950). The maximum F-ratio as a short-cut test for heterogeneity of variance. *Biometrika, 37,* 308–312.

Hays, W. L. (1994). *Statistics* (5th ed.). New York: Harcourt Brace College Publishing.

Hedges, L. V. (1981). Distribution theory for Glass's estimator of effect size and related estimators. *Journal of Educational Statistics, 6,* 107–128.

Hedges, L. V. (1982). Estimation of effect size from a series of independent experiments. *Psychological Bulletin, 92,* 490–499.

Hogg, R. V., & Craig, A. T. (1970). *Introduction to mathematical statistics* (3rd ed.). New York: Macmillan.

Howell, D. C. (1992). *Statistical methods for psychology* (3rd ed.). Boston: PWS-Kent.

Huck, S. W., & McLean, R. A. (1975). Using a repeated measures ANOVA to analyze the data from a pretest-posttest design: A potentially confusing task. *Psychological Bulletin, 82,* 511–518.

Huynh, H., & Feldt, L. S. (1976). Estimation of the Box correction for degrees of freedom from sample data in randomized block and split-plot designs. *Journal of Educational Statistics, 1,* 69–82.

Johnson-Laird, P. N., Legrenzi, P., & Legrenzi, M. S. (1972). Reasoning and a sense of reality. *British Journal of Psychology, 63,* 395–400.

Kaye, K. L., & Bower, T. G. R. (1994). Learning and intermodal transfer of information in newborns. *Psychological Science, 5,* 286–288.

Keppel, G. (1991). *Design and analysis: A researcher's handbook* (3rd ed.). Englewood Cliffs, NJ: Prentice-Hall.

Keren, G. (1993). A balanced approach to unbalanced designs. In G. Keren & C. Lewis (Eds.), *A handbook for data analysis in the behavioral sciences: Statistical issues* (pp. 95–127). Hillsdale, NJ: Lawrence Erlbaum.

Kruskal, W. H., & Wallis, W. A. (1952). Use of ranks in one-criterion variance analysis. *Journal of the American Statistical Association, 47,* 583–621.

Levene, H. (1960). Robust tests for the equality of variances. In I. Olkin (Ed.), *Contributions to probability and statistics.* Palo Alto, CA: Stanford University Press.

Lewis, C. (1993). Analyzing means from repeated measures data. In G. Keren & C. Lewis (Eds.), *A handbook for data analysis in the behavioral sciences: Statistical issues* (pp. 73–94). Hillsdale, NJ: Lawrence Erlbaum.

Lykken, D. E. (1968). Statistical significance in psychological research. *Psychological Bulletin, 70,* 151–159.

Lyon, D., & Greenberg, J. (1991). Evidence of codependency in women with an alcoholic parent: Helping out Mr. Wrong. *Journal of Personality and Social Psychology, 61,* 435–439.

Mann, H. B., & Whitney, D. R. (1947). On a test of whether one of two random variables is stochastically larger than the other. *Annals of Mathematical Statistics, 18,* 50–60.

O'Brien, R. G. (1981). A simple test for variance effects in experimental designs. *Psychological Bulletin, 89,* 570–574.

Pollard, P., & Richardson, J. T. E. (1987). On the probability of making type I errors. *Psychological Bulletin, 102,* 159–163.

Rosenthal, R. (1979). The "file drawer problem" and tolerance for null results. *Psychological Bulletin, 86,* 638–641.

Rosenthal, R. (1993). Cumulating evidence. In G. Keren & C. Lewis (Eds.), *A handbook for data analysis in the behavioral sciences. Methodological issues* (pp. 519–559). Hillsdale, NJ: Lawrence Erlbaum.

Rozeboom, W. W. (1960). The fallacy of the null hypothesis significance test. *Psychological Bulletin, 57,* 416–428.

Schwartz, L., Slater, M. A., & Birchler, G. R. (1994). Interpersonal stress and pain behaviors in patients with chronic pain. *Journal of Consulting and Clinical Psychology, 62,* 861–864.

Siegel, S., & Castellan, N. J., Jr. (1988). *Nonparametric statistics for the behavioral sciences* (2nd ed.). New York: McGraw-Hill.

Sternberg, S. (1966). High-speed scanning in human memory. *Science, 153,* 652–654.

Tranel, D., & Damasio, H. (1994). Neuroanatomical correlates of electrodermal skin conductance responses. *Psychophysiology, 31,* 427–438.

Tukey, J. W. (1977). *Exploratory data analysis.* Reading, MA: Addison-Wesley.

van Lawick-Goodall, J. (1971). *In the shadow of man.* Boston: Houghton Mifflin.

Wilcoxon, F. (1949). *Some rapid approximate statistical procedures.* Stamford, CT: American Cyanamid Company, Stamford Research Laboratories.

Yuen, K. K., & Dixon, W. J. (1973). The approximate behavior and performance of the two-sample trimmed *t. Biometrika, 60,* 369–374.

INDEX

TO THE OWNER OF THIS BOOK:

I hope that you have enjoyed *Explaining Psychological Statistics* as much as I've enjoyed writing it. I'd like to know as much about your experiences with the book as you care to offer. Only through your comments and comments of others can I learn how to make a better book for the future.

School and address: _____

Department: _____

Instructor's name: _____

1. What I like most about this book is: _____

_____

_____

2. What I like least about this book is: _____

_____

_____

3. My general reaction to this book is: _____

_____

4. The name of the course in which I used this book is: _____

_____

5. Were all of the chapters of the book assigned for you to read? _____

   If not, which ones weren't? _____

6. In the space below, or on a separate sheet of paper, please write specific suggestions for improving this book and anything else you'd care to share about your experience in using the

_____

_____

_____

_____

Optional:

Your name: _____ Date: _____

May Brooks/Cole quote you, either in promotion for *Explaining Psychological Statistics* or in future publishing ventures?

Yes: _____ No: _____

Sincerely,

Barry Cohen

- FOLD HERE

BUSINESS REPLY MAIL

FIRST CLASS   PERMIT NO. 358   PACIFIC GROVE, CA

POSTAGE WILL BE PAID BY ADDRESSEE

ATT: *Barry Cohen* _____

Brooks/Cole Publishing Company
511 Forest Lodge Road
Pacific Grove, California 93950-9968

FOLD HERE